Technology and Management for Sustainable Buildings and Infrastructures

Technology and Management for Sustainable Buildings and Infrastructures

Editor

Sunkuk Kim

MDPI • Basel • Beijing • Wuhan • Barcelona • Belgrade • Manchester • Tokyo • Cluj • Tianjin

Editor
Sunkuk Kim
Kyung Hee University
Korea

Editorial Office
MDPI
St. Alban-Anlage 66
4052 Basel, Switzerland

This is a reprint of articles from the Special Issue published online in the open access journal *Sustainability* (ISSN 2071-1050) (available at: https://www.mdpi.com/journal/sustainability/special_issues/Technology_sensors).

For citation purposes, cite each article independently as indicated on the article page online and as indicated below:

LastName, A.A.; LastName, B.B.; LastName, C.C. Article Title. *Journal Name* **Year**, *Volume Number*, Page Range.

ISBN 978-3-0365-2069-8 (Hbk)
ISBN 978-3-0365-2070-4 (PDF)

Contents

About the Editor

Sunkuk Kim studied Construction Engineering and Management at the department of architectural engineering, Seoul National University. He had joined three Korean construction firms, Dealim Industrial Co., Ltd., Deadong Coporation Co., Ltd. and Seoktop Construction Co., Ltd., for 12 years. As a visiting scholar, he researched about the construction management & organization at the department of civil engineering, Stanford University from 1994 to 1995. Since September, 1995, he has served at Kyung Hee University as a professor.

Kim served as a dean of the Graduate School of Technology Management from 2015 to 2018. In addition, he was an organization chair in three international conferences including ICCEPM (2009) and MOC (2015). He was also a vice president of Korea Institute of Ecological Architecture and Environment and Korean Council on Tall Buildings and Urban Habitat (K-CTBUH). He also served as a guest editor for two special issues of Modular & Offsite Construction (2017) and Global Convergence in Construction (2011), Automation in Construction and special issue of Global Convergence in Construction (2010), International Journal of Strategic Property Management. Currently he is the guest editor for special issue of Technology and Management for Sustainable Buildings and Infrastructures, Sustainability.

Kim has concentrated on the research such as health performance evaluation of buildings, development of sustainable construction technology and management, simulation, optimization and risk management, construction information technology. Especially, for about a decade, he has participated in the development of SMART frame, a sustainable structural system, and production technology of free-form concrete panels.

 *sustainability*

Editorial

Technology and Management for Sustainable Buildings and Infrastructures

Sunkuk Kim

Department of Architectural Engineering, Kyung Hee University, Yongin-si 17104, Gyeonggi-do, Korea; kimskuk@khu.ac.kr; Tel.: +82-31-201-2922

Citation: Kim, S. Technology and Management for Sustainable Buildings and Infrastructures. *Sustainability* **2021**, *13*, 9380. https://doi.org/10.3390/su13169380

Received: 13 August 2021
Accepted: 19 August 2021
Published: 20 August 2021

Publisher's Note: MDPI stays neutral with regard to jurisdictional claims in published maps and institutional affiliations.

According to a report published in 2019 by the United Nations Environment Program (UNEP), the building sector accounts for 38% of all energy-related CO_2 emissions when adding building construction industry emissions [1]. Yudelson (2008) argued that the building sector accounted for 45% to 65% of landfill waste [2].

Given this fact, the building sector must be one of the major causes of global warming and the resulting climate catastrophe. Therefore, research on the technology and management of the entire process including design, construction, O&M, and decommissioning is urgently needed for sustainable buildings and infrastructure that minimize energy use throughout their life cycle. At this point, it is judged that it was timely to hold a Special Issue under the topic of "Technology and Management for Sustainable Buildings and Infrastructures".

At the time that the world is struggling with the COVID-19 pandemic, this special issue has been published in 27 research papers [3–29], 1 review paper [30], and 2 technical notes [31,32], and with the help of many research colleagues and reviewers. A total of 30 papers were published. A total of 104 authors from 9 countries including Korea [3,5,6,8,12–15,17,19,20,22,23,25–28,30–32], Spain [11,18,21], Taiwan [4,24], USA [16,17,25], Finland [10], China [29], Slovenia [9], the Netherlands [7], and Germany [21] participated in writing and submitting very excellent papers that were finally published after the review process had been conducted according to very strict standards.

Among the published papers, 13 papers directly addressed words such as sustainable, life cycle assessment (LCA) and CO_2 [5–7,11,12,14,16,19,20,22,25,27,28], and 17 papers indirectly dealt with energy and CO_2 reduction effects [3,4,8–10,13,15,17,18,21,23,24,26,29–32]. Sustainability research related to CO_2 and the resulting climate change started in the construction field more than 20 years ago. Although life cycle cost analysis (LCCA) has dealt with the energy use of buildings for more than 40 years, it focuses on cost rather than CO_2 reduction. In the 21st century, research on net zero or near zero energy use of buildings has been conducted, but research on embodied CO_2 resulting from the design and construction stage has been excluded because it is limited to the operation and maintenance stage. Until recently, many design and construction studies focused on maximizing economic benefits, and rarely focused on carbon neutrality or CO_2 emission minimization. As a result, there are not yet many papers directly dealing with energy and CO_2 reduction throughout the construction project life cycle.

Among the published papers, there are 6 papers [4,6,9,18,29,32] dealing with construction technology, but a majority, 24 papers [3,5,7,8,10–17,19–28,30,31] deal with management techniques. The reason is that construction management can be approached more easily than construction technology when considering research cost, time, and effort. Among all the papers, 15 studies focused on buildings [7,8,10,12,14–16,18,19,22,25–27,31,32], 9 studies on infrastructures [6,9,11,13,17,20,24,28,29], and 6 papers could apply to both [3–5,21,23,30]. With the development of science and technology, there is a tendency for buildings to become taller, larger, and more luxurious, and the energy use tends to increase rapidly. In particular, this trend is conspicuous in the developed countries where most of the papers

1

have been submitted. In addition, in the developed countries, infrastructures such as roads, bridges, seaports, airports, and power plants are sufficiently established. Therefore, it is presumed that there are more papers on sustainable buildings than on sustainable infrastructures.

The authors of the published papers used various analysis techniques to obtain the suggested solutions for each topic. Listed by key techniques, various techniques such as Analytic Hierarchy Process (AHP) [3,12], the Taguchi method [4], machine learning including Artificial Neural Networks (ANNs) [5,28], Life Cycle Assessment (LCA) [6,7], regression analysis [13,17,19,25,28], Strength–Weakness–Opportunity–Threat (SWOT) [11], system dynamics [16,26], simulation and modeling [10,19,22–24,29,31,32], Building Information Model (BIM) with schedule [21,24,27], and graph and data analysis after experiments and observations [8,9,14,15,18,20,27,29–32] are identified.

As mentioned above, although the construction sector is a key influencer that harms the global environment, many studies have been focused on cost, time, quality, and safety. However, future research should be conducted on the basis of carbon neutrality or CO_2 emission reduction. For example, previous cost minimization studies should be conducted as cost optimization studies based on CO_2 emission reduction or minimization. As such, if all research is conducted in the direction of pursuing sustainable buildings and infrastructures, the global environment will be gradually improved.

Finally, I would like to thank Maggie Sun of MDPI and others for their active cooperation in making this Special Issue successful, research colleagues who submitted excellent papers, and reviewers who have been active in the review process.

Author Contributions: Conceptualization, S.K.; methodology, S.K.; validation, S.K.; formal analysis, S.K.; writing—original draft preparation, S.K.; writing—review and editing, S.K.; supervision S.K. The author has read and agreed to the published version of the manuscript.

Funding: This research received no external funding.

Conflicts of Interest: The author declares no conflict of interest.

References

1. UNEP. Building Sector Emissions Hit Record High, but Low-Carbon Pandemic Recovery Can Help Transform Sector. Press Release 16 December 2020. Available online: https://www.unep:news-and-stories/press-release/building-sector-emissions-hit-record-high-low-carbon-pandemic (accessed on 12 August 2021).
2. Yudelson, J. *The Green Building Revolution*, 1st ed.; Island Press: Washington, DC, USA, 2008.
3. Kim, D.; Oh, W.; Yun, J.; Youn, J.; Do, S.; Lee, D. Development of Key Performance Indicators for Measuring the Management Performance of Small Construction Firms in Korea. *Sustainability* **2021**, *13*, 6166. [CrossRef]
4. Chen, H.-J.; Lin, H.-C.; Tang, C.-W. Application of the Taguchi Method for Optimizing the Process Parameters of Producing Controlled Low-Strength Materials by Using Dimension Stone Sludge and Lightweight Aggregates. *Sustainability* **2021**, *13*, 5576. [CrossRef]
5. Jang, Y.; Son, J.; Yi, J.S. Classifying the Level of Bid Price Volatility Based on Machine Learning with Parameters from Bid Documents as Risk Factors. *Sustainability* **2021**, *13*, 3886. [CrossRef]
6. Seol, Y.; Lee, S.; Lee, J.-Y.; Han, J.-G.; Hong, G. Excavation Method Determination of Earth-Retaining Wall for Sustainable Environment and Economy: Life Cycle Assessment Based on Construction Cases in Korea. *Sustainability* **2021**, *13*, 2974. [CrossRef]
7. Pujadas-Gispert, E.; Vogtländer, J.G.; Moonen, S.P.G. Environmental and Economic Optimization of a Conventional Concrete Building Foundation: Selecting the Best of 28 Alternatives by Applying the Pareto Front. *Sustainability* **2021**, *13*, 1496. [CrossRef]
8. Yang, K.; Kim, K.; Go, S. Towards Effective Safety Cost Budgeting for Apartment Construction: A Case Study of Occupational Safety and Health Expenses in South Korea. *Sustainability* **2021**, *13*, 1335. [CrossRef]
9. Bizjak, K.F.; Likar, B.; Lenart, S. Using Recycled Material from the Paper Industry as a Backfill Material for Retaining Walls near Railway Lines. *Sustainability* **2021**, *13*, 979. [CrossRef]
10. Kurvinen, A.; Saari, A.; Heljo, J.; Nippala, E. Modeling Building Stock Development. *Sustainability* **2021**, *13*, 723. [CrossRef]
11. Fernández-Pérez, V.; Peña-García, A. The Contribution of Peripheral Large Scientific Infrastructures to Sustainable Development from a Global and Territorial Perspective: The Case of IFMIF-DONES. *Sustainability* **2021**, *13*, 454. [CrossRef]
12. Kim, J.Y.; Lee, D.S.; Kim, J.D.; Kim, G.H. Priority of Accident Cause Based on Tower Crane Type for the Realization of Sustainable Management at Korean Construction Sites. *Sustainability* **2021**, *13*, 242. [CrossRef]
13. Kim, Y.; Oh, J.; Kim, S. The Transition from Traditional Infrastructure to Living SOC and Its Effectiveness for Community Sustainability: The Case of South Korea. *Sustainability* **2020**, *12*, 10227. [CrossRef]

14. Lee, D.; Kim, S. Energy and CO_2 Reduction of Aluminum Powder Molds for Producing Free-Form Concrete Panels. *Sustainability* **2020**, *12*, 9613. [CrossRef]
15. Oh, Y.; Kang, M.; Lee, K.; Kim, S. Construction Management Solutions to Mitigate Elevator Noise and Vibration of High-Rise Residential Buildings. *Sustainability* **2020**, *12*, 8924. [CrossRef]
16. Lim, J.; Kim, J.J. Dynamic Optimization Model for Estimating In-Situ Production Quantity of PC Members to Minimize Environmental Loads. *Sustainability* **2020**, *12*, 8202. [CrossRef]
17. Yum, S.-G.; Son, K.; Son, S.; Kim, J.-M. Identifying Risk Indicators for Natural Hazard-Related Power Outages as a Component of Risk Assessment: An Analysis Using Power Outage Data from Hurricane Irma. *Sustainability* **2020**, *12*, 7702. [CrossRef]
18. Carretero-Ayuso, M.J.; Rodríguez-Jiménez, C.E.; Bienvenido-Huertas, D.; Moyano, J. Cataloguing of the Defects Existing in Aluminium Window Frames and Their Recurrence According to Pluvio-Climatic Zones. *Sustainability* **2020**, *12*, 7398. [CrossRef]
19. Kim, J.-M.; Son, S.; Lee, S.; Son, K. Cost of Climate Change: Risk of Building Loss from Typhoon in South Korea. *Sustainability* **2020**, *12*, 7107. [CrossRef]
20. Kim, K.; Park, Y. Development of Design Considerations as a Sustainability Approach for Military Protective Structures: A Case Study of Artillery Fighting Position in South Korea. *Sustainability* **2020**, *12*, 6479. [CrossRef]
21. Leicht, D.; Castro-Fresno, D.; Dìaz, J.; Baier, C. Multidimensional Construction Planning and Agile Organized Project Execution— The 5D-PROMPT Method. *Sustainability* **2020**, *12*, 6340. [CrossRef]
22. Lee, D.; Son, S.; Kim, D.; Kim, S. Special-Length-Priority Algorithm to Minimize Reinforcing Bar-Cutting Waste for Sustainable Construction. *Sustainability* **2020**, *12*, 5950. [CrossRef]
23. Palikhe, S.; Yirong, M.; Choi, B.Y.; Lee, D.-E. Analysis of Musculoskeletal Disorders and Muscle Stresses on Construction Workers' Awkward Postures Using Simulation. *Sustainability* **2020**, *12*, 5693. [CrossRef]
24. Wu, I.; Lin, Y.C. Evaluation of Space Service Quality for Facilitating Efficient Operations in a Mass Rapid Transit Station. *Sustainability* **2020**, *12*, 5295. [CrossRef]
25. Yum, S.-G.; Kim, J.-M.; Son, K. Natural Hazard Influence Model of Maintenance and Repair Cost for Sustainable Accommodation Facilities. *Sustainability* **2020**, *12*, 4994. [CrossRef]
26. Kang, S.; Kim, S.; Kim, S.; Lee, D. System Dynamics Model for the Improvement Planning of School Building Conditions. *Sustainability* **2020**, *12*, 4235. [CrossRef]
27. Kim, S.; Kim, S.; Lee, D.E. Sustainable Application of Hybrid Point Cloud and BIM Method for Tracking Construction Progress. *Sustainability* **2020**, *12*, 4106. [CrossRef]
28. Park, S.-S.; Ogunjinmi, P.D.; Woo, S.-W.; Lee, D.-E. A Simple and Sustainable Prediction Method of Liquefaction-Induced Settlement at Pohang Using an Artificial Neural Network. *Sustainability* **2020**, *12*, 4001. [CrossRef]
29. Gao, X.; Tian, W.-P.; Zhang, Z. Analysis of Deformation Characteristics of Foundation-Pit Excavation and Circular Wall. *Sustainability* **2020**, *12*, 3164. [CrossRef]
30. Kwon, K.; Kim, D.; Kim, S. Cutting Waste Minimization of Rebar for Sustainable Structural Work: A Systematic Literature Review. *Sustainability* **2021**, *13*, 5929. [CrossRef]
31. Kang, S.; Kim, S.; Lee, D.; Kim, S. Inter-Floor Noise Monitoring System for Multi-Dwelling Houses Using Smartphones. *Sustainability* **2020**, *12*, 5065. [CrossRef]
32. Kim, S.; Lee, D.-E.; Kim, Y.; Kim, S. Development and Application of Precast Concrete Double Wall System to Improve Productivity of Retaining Wall Construction. *Sustainability* **2020**, *12*, 3454. [CrossRef]

 *sustainability*

Article

Development of Key Performance Indicators for Measuring the Management Performance of Small Construction Firms in Korea

Doyeong Kim [1], Wongyun Oh [2], Jiyeong Yun [2], Jongyoung Youn [2], Sunglok Do [2] and Donghoon Lee [2,*

[1] Department of Architectural Engineering, Kyung Hee University, Suwon 17104, Korea; dream1968@khu.ac.kr
[2] Department of Architectural Engineering, Hanbat National University, Daejeon 34158, Korea; ok0786@hanmail.net (W.O.); 9736jy@naver.com (J.Y.); 97colin@naver.com (J.Y.); sunglokdo@hanbat.ac.kr (S.D.)
* Correspondence: donghoon@hanbat.ac.kr; Tel.: +82-42-821-1123

Abstract: Large construction firms execute management control in preparation for a fast-paced business environment, but small ones are unable to do so. This is because there is no management control model tailored to them. The current study derived Management Performance Evaluation Indicators (MAPEIs) for small construction firms for measuring the management performance of construction firms with 10 or fewer employees, considering the characteristics of small construction firms. MAPEIs consist of BSC (Balanced Scorecard), performance, and the hierarchy and weighted value of KPIs (Key Performance Indicators). After an interview with an expert, based on the management performance indicators of large construction firms, a final hierarchy of small construction firms was constructed through modification and supplementation. The KPIs of the hierarchy were analyzed through a survey using the AHP (Analytic Hierarchy Process) method to finalize MAPEIs for small construction firms in Korea. The final MAPEIs underwent a feasibility evaluation to apply them to real life. It is expected that they can be used as fundamental resources for system development for small construction firm management performance and control. In addition, further studies to resolve the limitations would improve the competitiveness of small construction firms.

Keywords: management performance evaluation indicators (MAPEIs) for small construction firms; AHP; key performance indicators (KPIs); corporation management; small construction firms

Citation: Kim, D.; Oh, W.; Yun, J.; Youn, J.; Do, S.; Lee, D. Development of Key Performance Indicators for Measuring the Management Performance of Small Construction Firms in Korea. *Sustainability* **2021**, *13*, 6166. https://doi.org/10.3390/su13116166

Academic Editor: Ali Bahadori-Jahromi

Received: 12 March 2021
Accepted: 24 May 2021
Published: 30 May 2021

Publisher's Note: MDPI stays neutral with regard to jurisdictional claims in published maps and institutional affiliations.

1. Introduction

Recently, the Construction and Economy Research Institute of Korea concluded that the construction industry of Korea has officially been in a depression since the second half of 2018 and anticipated that it would likely continue until the early to mid-2020s. They mentioned that it would be necessary to develop management strategies suitable for the period. The downturn's impact on the construction industry is greater for smaller firms compared to larger firms. Despite radical changes, the number of construction firms registered in Korea increased by about 120% from 10,921 in 2013 to 13,168 in 2020 [1].

Statistics Korea classified the construction firms in Korea into scales based on the number of full-time employees. The number of firms with fewer than 50 employees was 97,314 out of 100,654. This means that small construction firms account for 96.7% of Korea's construction market according to the criteria of the Construction Association of Korea, and most of the construction firms being added to the list are small ones [1,2].

These days, there is not much call for construction work, and the number of projects to bid for is very limited. An increase in the number of small construction firms increases competition and makes the probability of winning a bid very slim. The Construction and Economy Research Institute of Korea released a BSI Report in 2019. The average BSI (Business Survey Index) of small construction firms was 79.5 and did not exceed 100 in 2019, reflecting the overall worsening of the industry and greater burdens on the management of firms [2].

The management difficulties experienced by small construction firms are caused by both external and internal factors. First, there is not a sufficient management control system. Large construction firms analyze management characteristics, along with the external environment and internal capabilities, using an adequate management control system to establish management strategies and plans according to the management characteristics. Management performance is measured to verify whether the objectives have been met. However, small construction firms are incapable of identifying the causes of difficulties due to a lack of management control systems and difficulties in measuring the management performance. Second, it is difficult to respond to and prepare for changes in the external environment. Due to the difficulties in management control, it is difficult to respond to the fast-paced environment of the construction industry, which involves a high rate of unpredictability, and impossible to prepare for the changes that they may encounter. Third, the unorganized structure of firms is a challenge. Unlike large construction firm, small construction firms find it hard to organize because there are only a few members and they lack management expertise. As mentioned above, it is impossible for small construction firms to execute management control due to difficulties in management. Therefore, the first step to take would be to understand the current circumstances of small construction firms. The position and status of each firm shall be identified by analyzing the problems and measuring the management performance. Then, sustainable management control shall be executed. For that purpose, this study was conducted to derive the MAPEIs (Management Performance Evaluation Indicators for Small Construction Firms) to analyze the management characteristics of small construction firms in Korea and measure their management performance.

In order to obtain the performance indicators of small construction firms, experts were interviewed based on the management performance indicators of large construction firms derived from preceding studies. The first expert survey was conducted with the top managers of five small construction firms, and the second survey was conducted with 33 top managers and engineers. The indicators were supplemented and modified to create a final hierarchy suitable for the scope of this study. KPIs (Key Performance Indicators) of the hierarchy were analyzed through a survey using the AHP method to finally derive the MAPEIs of management performance of small construction firms. Finally, the MAPEIs were tested in a real-life environment.

2. Preliminary Study

The balanced scorecard (BSC) of Kaplan and Norton is a strategic management system developed to measure the management performance of companies. BSC measures and controls performance in four balanced perspectives of finances, customers, internal processes, and learning/growth. Various studies have been conducted by companies, organizations within companies, and other areas that require individual competence and strategic systemization since the development of BSC (Balanced Scorecard) [3,4]. Small construction firms also need to manage intangible assets as well as tangible assets. This study applies BSC to measure the management performance of small construction companies.

Kim (2010) interviewed the management officers of companies to develop a model for analyzing the management performance of large Korean construction firms. He adopted a program to feed back on goals, management strategies, management plans, and management performance evaluation to suggest the importance of effective management strategies and efficient management control. Additionally, he identified errors and misses when applied to real life, unlike the preceding studies, and analyzed the findings to suggest the general process of management performance for construction firms and the errors and solutions to consider when measuring performance [5]. Jung (2005) comparatively analyzed the weighted value according to the scale of companies to measure the management performance of construction firms. An AHP analysis was applied to calculate the weighted value of performance indicators and analyze their importance for small/medium versus large companies. The study of Jung is different from other studies in terms of the subject of analysis, survey method, and findings [6]. Yu (2004) analyzed past cases of other countries to suggest that it is necessary to develop key performance indicators for the construction firms in Korea along with a PMS framework to control them [7]. In other countries, KPIs were applied to the PMS Islam Bank based on BSC and AHP [8], and BSC was used to improve the efficiency of company operations for Luka Koper and d.d. Company [9]. Among the top 1000 companies named by *Fortune*, a U.S. magazine on economics, about 60% are assumed to have adopted the concept of BSC [10–12].

Management diagnosis refers to hiring an outside management expert to address management issues that cannot be resolved internally or to identify directions for future development. Management diagnosis models are mostly used by consulting firms or individual companies [13]. The management diagnosis model for small construction firms in Korea is still in the theoretical development stage and the only available models are modified forms of generalized models.

The Korea Small Business Institute has suggested a corporate diagnosis model to select businesses for a small business support project. The model suggested indicators of diagnosis for categories including attractiveness, competence, systems, and CEO.

The government of Korea is also developing various evaluation indicators, such as "the Small/Medium Business Healthcare System" and the "INNO-BIZ Evaluation Model," with continued efforts to enhance the management control capabilities of small construction firms. Management performance was analyzed according to the scale of construction firms and characteristics of organization, and a model for management performance evaluation has been developed. Additionally, there have been continued efforts to develop management diagnosis models to enhance the management control capacities of small construction firms to enhance their competitiveness. However, the study of performance evaluation models for small construction firm management control has not been sufficient, as there are many limitations when applying the management diagnosis models developed for small construction firms.

Research has been conducted into the management of small and medium-sized companies in Korea, with differences in the target companies and objectives from this study [14]. We conducted a study to evaluate the management performance of small construction companies with fewer than 10 employees. There are differences between management diagnosis strategies and management performance evaluation models. Management diagnosis is the process of identifying problems with a company's management, identifying the cause of the problem, and deriving improvements to these problems. It is a good tool for improving current management problems and providing future management directions in corporate management. However, this differs in purpose and process from the assessment of management performance. In addition, BSC has been used to develop a framework for small/medium businesses and a performance control system has been constructed by a small nonprofit organization using BSC. As a result, BSC made it possible to search for and correct problems, but there is a limitation in that it cannot be used in many areas [15,16]. It is necessary to continue studying various models suitable for small companies [17,18]. Therefore, the current study suggested KPIs to measure the management performance of small construction firms in order to improve their competitiveness and pursue gradual corporate growth.

Kim proposed MAPEIs to evaluate the management performance of large construction firms. The MAPEIs are composed of the hierarchy and weighted value of BSC, performance, and KPIs to derive the management performance evaluation indicators of small construction firms. Figure 1 shows the basic structure of MAPEIs [19]. In addition, MAPEIs were established as a practical evaluation management system. This study applied the concept of measuring the management performance of construction firms.

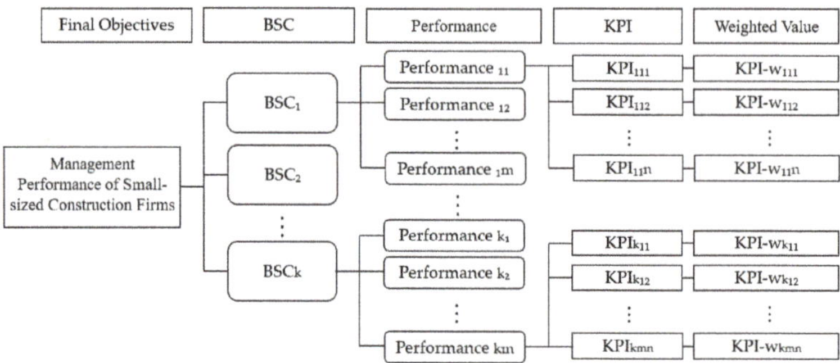

Figure 1. Basic structure of MAPEIs [19].

3. Methodology

Management Performance Evaluation Indicators (MAPEIs) for small firms consist of evaluation indicators with various hierarchies and weighted values for each KPI. The management performance evaluation models for construction firms vary according to each firm's business and scale, knowledge informatization level, brand value in Korea or abroad, and soundness of management control. Therefore, it is necessary to provide appropriate indicators. The current study's MAPEIs, as mentioned above, may be applied to small construction firms in Korea. The study was limited to small construction firms in Korea with no more than 10 full-time employees, no construction projects abroad, and businesses not including civil works and plants.

Construction firms have many factors to consider when measuring performance due to the uncertainties in the market environment. Therefore, the current study applied a Balance Scorecard (BSC). A BSC consists of four areas—finances, customers, internal processes, and learning/growth—and is applied to the management control of firms in good standing in Korea and abroad [19]. The BSC of construction firms is the same as that of other companies, but performance and KPIs differ due to corporate characteristics. Management characteristics also vary according to the size of corporations, even if they are in the same industry. Therefore, the current study considered the characteristics of small construction firms to develop the hierarchy of MAPEIs. KPIs were derived by analyzing the characteristics of small construction firms, and all items for performance evaluation had weighted values. MAPEIs serve as KPIs to measure the management performance of small construction firms.

We selected the items used to evaluate performance (Figure 2) and deleted and supplemented items through expert interviews to configure the hierarchy. The survey was performed based on the hierarchy and the weighted values were tabulated by analyzing the importance of each item to derive the MAPEIs of small construction firms.

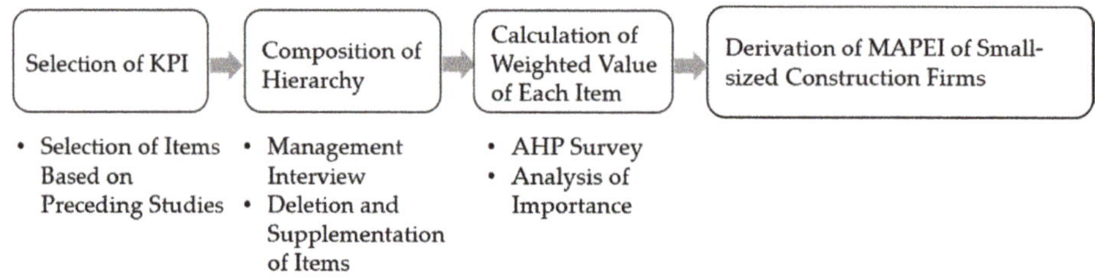

Figure 2. How to derive MAPEIs.

4. MAPEIs of Small Construction Firms

4.1. Selection of Management Performance Evaluation Indicators for Small Construction Firms

The current study was conducted to derive KPIs for the management performance evaluation of small construction firms. Preceding studies analyzed the strategies, plans, and goals of construction companies to configure the performance areas of a subject to identify KPIs. Kim (2010) conducted a survey on managers of large construction companies and derived the following evaluation items and weights by an AHP analysis. This indicated the evaluation items of large construction firms that allow for systemized management control but cannot be KPIs of small construction firms. As shown in Table 1, we used the hierarchy of Kim (2010) as the preliminary indicators to derive MAPEIs [19].

Table 1. Weighted value of factors of MAPEIs [19].

	Weighted Value	Performance Areas	Weighted Value	KPI	Weighted Value
Finance	0.28	Profitability	0.24	ROIC	0.28
				Cost of Sale Ratio	0.37
				Ordinary Profit	0.35
		Growth	0.18	Increase in Revenues in Korea	0.49
				Increase in Revenues Abroad	0.51
		Stability	0.13	Debt Ratio	0.48
				Achievement of Collection Goal	0.52
		Activity	0.13	Turnover Ratio of Total Liabilities and Net Worth	1.00
Customers	0.34	Orders	0.32	Number of New Orders	1.00
		Satisfaction of External Customers	0.38	Awards Won in Competitions	0.15
				Customer Satisfaction	0.28
				Corporate Image	0.38
				Social Contribution	0.19
		Satisfaction of Internal Customers	0.28	Transfer Rate of Employees	0.42
				Work Environment and Corporate Culture	0.58
		Market Share	0.34	Market Share of Orders in Korea	0.49
				Market Share of Orders Abroad	0.51
Internal Process	0.14	Investment in R&D	0.33	R&D Cost to Revenue	0.49
				Effect of New Technology to Cost of Development	0.51
		Technology Capacities	0.38	Application of Internal Technology Development	0.57
				Possession of Intellectual Property Rights	0.43
		Work Efficiency	0.29	Selling and Administrative Expenses to Revenue	0.25
				Compliance with Guidelines	0.21
				Accident rate	0.32
				Reuse/Recycling of Waste	0.22
Learning and Growth	0.24	Manpower training	0.38	Index of Excellent Workforce	0.30
				Cost of Training per Employee	0.33
				Satisfaction of Trainees	0.37
		Organizational Capacity	0.38	Knowledge Sharing	0.40
				Productivity of Employees	0.60
		Informatization	0.24	Informatization Capacity Index	1.00

The primary tier of the hierarchy consists of 14 performance areas and 31 KPIs. These are MAPEIs for large construction firms and cannot be indicators for the management performance evaluation of small construction firms. Therefore, there must be KPIs suitable for small construction firms. In order to configure a hierarchy for the purpose, the top management of five small Korean construction firms were interviewed and the findings are shown in Table 2. The respondents to this survey were top management who had managed construction companies for a long time.

Table 2. Details of survey subjects.

	Details of Survey
Classification of Firms	Construction Firms in Korea
Scale of Firms	Small (No More than 10 Full-time Employees)
Classification of Industries	Construction of Other Nonresidential Buildings, Office/Commercial Use, Public Organizations
Number of Subjects Surveyed	5 Companies
Position in Organization	Top Management (More than 20 years of experience)

This study removed unnecessary KPIs following interviews. The interviews surveyed the items that realistically reflect the management performance of current companies among the items in the primary hierarchy. Tables 3–6 show the performance evaluation of small construction firms. The respondents selected items necessary for management evaluation. The score is the sum of the choices. For a maximum score of 5, all the respondents of the five companies analyzed deem that KPI is relevant. In fact, it is not easy for small construction firms to analyze management performance in various areas. Therefore, this study deleted items selected by fewer than half of the companies.

Table 3. Reflection of management performance evaluation on finance.

BSC	Performance Areas	KPI	Reflection of Management Performance Evaluation (Point)
Finance	Profitability	ROIC	1
		Cost of Sale Ratio	4
		Ordinary Profit	3
	Growth	Increase in Revenues in Korea	3
		Increase in Revenues Abroad	0
	Stability	Debt Ratio	3
		Achievement of Collection Goals	4
	Activity	Turnover Ratio of Total Liabilities and Net Worth	3
	Orders	Number of New Orders	5

Table 4. Reflection of management performance evaluation on customers.

BSC	Performance Areas	KPI	Reflection of Management Performance Evaluation (Point)
Customers	Satisfaction of External Customers	Awards Won in Competitions	0
		Customer Satisfaction	4
		Corporate Image	5
		Social Contribution	1
	Satisfaction of Internal Customers	Transfer Rate of Employees	4
		Work Environment and Corporate Culture	4
	Market Share	Market Share of Orders in Korea	3
		Market Share of Orders Abroad	0

Table 5. Reflection of management performance evaluation on internal process.

BSC	Performance Areas	KPI	Reflection of Management Performance Evaluation (Point)
Internal Process	Investment in R&D	R&D Cost to Revenue	2
		Effect of New Technology to Cost of Development	0
	Technology Capacities	Application of Internal Technology Development	2
		Possession of Intellectual Property Rights	3
	Work Efficiency	Selling and Administrative Expenses to Revenue	2
		Compliance with Guidelines	3
		Accident rate	4
		Reuse/Recycling of Waste	0

Table 6. Reflection of management performance evaluation on learning and growth.

BSC	Performance Areas	KPI	Reflection of Management Performance Evaluation (Point)
Learning and Growth	Manpower training	Index of Excellent Workforce	3
		Cost of Training per Employee	2
		Satisfaction of Trainees	2
	Organizational Capacity	Knowledge Sharing	2
		Productivity of Employees	4
	Informatization	Informatization Capacity Index	3

In the KPIs of preceding studies, finance consists of five areas, including profitability, growth, stability, activity, and order, as in Table 3. Profitability areas consist of ROIC (Return on Invested Capital), cost of sale ratio, and ordinary profit. Growth consists of an increase in revenue in Korea and increase in revenue abroad. Stability includes the debt ratio and achievement of collection goals, while activity includes the turnover ratio of total liabilities and net worth. Orders consist of amounts of new orders. In each area of performance, the number of new orders was selected as a major KPI by all five companies. Cost of sale ratio and achievement of collection goals were also representative. Achievement of collection goals was widely reflected, as poor collection is likely to lead to poor performance, inactivity, or unprofitability for small companies. On the other hand, ROIC and increase in revenues abroad are rarely representative. ROIC is a return on invested capital and may be evaluated based on the cost of sale ratio or ordinary profit as it is the actual assets invested in projects. This is mostly applied to companies where responsible management is possible, so it is difficult to use with small companies that lack systemized management control. The increase in revenues abroad is unrealistic for small construction firms that receive few orders from abroad.

Customers, as shown in Table 4, account for three performance areas, including satisfaction of external customers, satisfaction of internal customers, and market share. Satisfaction of external customers consists of awards won in competitions, customer satisfaction, corporate image, and social contribution, while satisfaction of internal customers is composed of employee transfer rate, work environment, and corporate culture. In the customer area, corporate image, customer satisfaction, employee transfer rate, work environment, and corporate culture are widely reflected. However, awards won in competitions and social contributions that have an additional impact on corporate image are not reflected as much and the market share of orders abroad is also rarely reflected as small construction firms receive few orders from abroad, as seen in Table 3.

The internal process consists of three performance areas—investment in R&D, technology capacities, and work efficiency—as shown in Table 5. Investment in R&D consists of the cost of R&D to revenue, and the effect of new technology on the cost of development. Technology capacity consists of the application of internally developed technology and intellectual property rights, while work efficiency consists of selling and administrative expenses to revenue, compliance with guidelines, accident rates, and reuse/recycling of waste. The KPIs of the internal process were generally reflected less frequently than other areas were. On the other hand, the accident rate of efficiency area was widely reflected. This is because construction projects are generally large in scale and the losses related to accidents may be massive. Therefore, the accident rate is frequently applied to small construction firms.

Learning and growth, as shown in Table 6, consist of three performance areas: training, organizational capacity, and informatization. Manpower training includes index of excellent workforce, cost of training per employee, and satisfaction of trainees. Organizational capacity includes the knowledge sharing and productivity of employees, while informatization includes the informatization capacity index. In learning and growth, the productivity of employees was widely reflected. The number of employees is smaller than it is for large construction firms. Therefore, each member has a great impact on the organization, and the productivity of employees is significant. The index of excellent workforce is also frequently reflected because the competence of each individual employee is significant due to the smaller scale of firms. In a fast-paced business environment, informatization knowledge of construction is used as a strategic resource for the construction market and plays a major function. Therefore, the informatization capacity index is widely used for the evaluation of firms.

Based on the preliminary hierarchy, the top managers of firms were interviewed to survey the reflection of KPIs. In order to configure the evaluation indicators suitable for small construction firms based on the surveyed resources, the items that could not be assigned 3 points or more were deleted to configure the hierarchy. The secondary hierarchy of MAPEIs, configured based on the aforementioned standards, consisted of 13 performance areas and 18 KPIs.

However, there are many differences in management methods between large companies and small companies, and different sets of evaluation items apply for appropriate management control. In order to bridge the differences, the items that are considered most important by small construction firms for performance evaluation were assessed in addition to the evaluation indicators of large construction firms. Major MAPEIs of small construction firms included 10 indicators: net profit of construction projects, accident rate, complaint handling capacity, possibility of open bidding, construction performance rate, cost of construction, employees' task-processing capacity, revenue, gain, and accident-free rate.

The items' similarity to the pre-existing evaluation indicators was analyzed through interviews with experts. Net profit of construction projects, cost of construction, revenue, and gain refer to the profitability of companies and overlap with the cost of sale ratio and the ordinary profit of profitability area under finance heading. The possibility of open bidding and construction performance rate are items that evaluate the profitability, growth, and number of orders of companies and are similar to the detailed items of finance. The accident rate was similar to the accident rate of the internal process area, while employees' task-processing capacity was similar to the index of excellent workforce in learning and growth. However, the complaint handling capacity, although it may be considered part of corporate image, was judged to be a new item for evaluating the management performance of small construction firms based on corporate characteristics.

The corporate image of large construction firms includes quality, brand, customer service, market reputation, stock prices, corporate value, and defects, as in Figure 3. These are auxiliary factors of corporate image for management performance evaluation and do not have a significant impact on performance evaluation. However, they may have a significant impact on small construction firms. In other words, the factors of corporate image can be a significant indicator for small construction firms. Therefore, complaint-processing capacity was included in the work efficiency area of internal process as an indicator of performance evaluation.

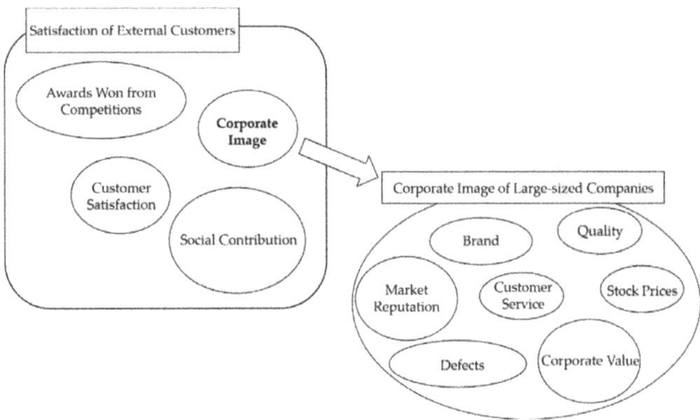

Figure 3. Corporate image of large companies.

As mentioned above, the performance evaluation indicators were analyzed for deletion, modification, and supplementation. Then, the findings were used to derive the final hierarchy. The final hierarchy became the hierarchy of MAPEIs and consisted of 13 performance areas and 19 KPIs, as in Figure 4.

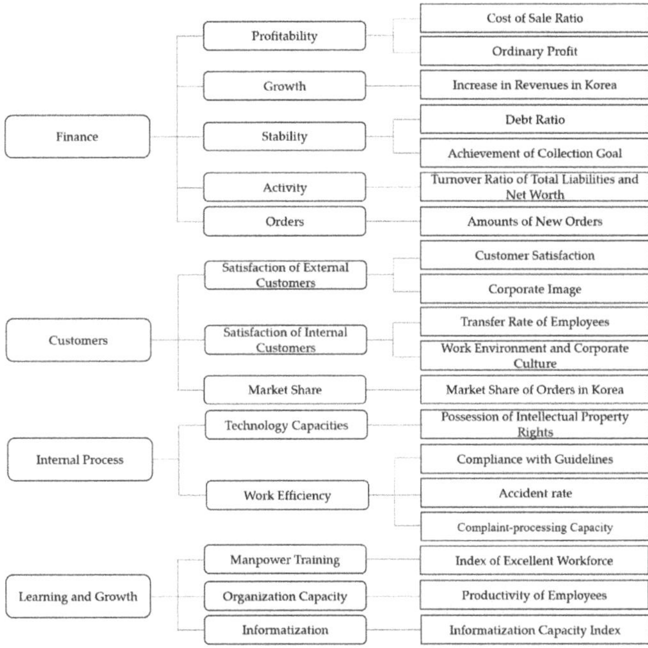

Figure 4. Hierarchy of MAPEIs.

4.2. Tabulation of Weighted Values

Each item comprising the hierarchy of MAPEIs becomes an indicator for the management performance evaluation. However, not all performance areas and evaluation indicators have equal weight. Therefore, each item shall be assigned a weighted value to evaluate management performance by considering the weight of each indicator.

An AHP survey was performed to assign a weighted value to each item. The survey was comprised of an importance analysis of each BSC area, an importance analysis of the performance of each BSC area, and an importance analysis of KPIs of each performance area, and the overview of the survey is as shown in Table 7.

Table 7. Details of subjects for tabulation of weighted values.

	Details of Survey
Subject of Survey	Importance Analysis of MAPEIs for Small Construction Firms
Survey Period	10 February–4 March 2020
Survey Members	33
Subjects of Survey	Top Management and Engineers of Small Construction Firms
Method of Application	AHP

The survey took about one month and the subjects were 33 members of top management or engineers of small construction firms. As the AHP survey was conducted, the consistency of responses was verified. The validity range of the consistency index was limited to 0.1 and the number of questions satisfying the consistency index was identified. The survey results satisfying the consistency index were analyzed for relative importance through a paired comparison analysis.

Figure 5 shows the weighted value of performance areas of BSC. The weighted value of the finance area was the highest at 0.379 and for the customer area it was 0.217. A weighted value of 0.198 was assigned to internal process and 0.206 to learning and growth. The importance of BSC of small construction firms was in the following order: finance, customers, learning and growth, and internal process. The highest weighted value of performance in finance was 0.115, assigned to profitability, followed by stability, orders, growth, and activity. Performance in the customers area assigned the highest value of 0.089 to satisfaction of external customers, followed by satisfaction of external customers, market share, and satisfaction of internal customers. Performance in the internal process area assigned 0.100 to work efficiency, which was a weighted value greater than that of technological capacity. The highest weighted value of 0.078 was assigned to the organizational capacity area in terms of performance on learning and growth, followed by manpower training and informatization. Table 8 lists the weighted values of all items tabulated through an importance analysis with AHP.

As mentioned above, the weighted value of finance was highest in BSC. In detail, profitability was assigned to the highest weighted value in finance, satisfaction of external customers in customers, work efficiency in internal process, and organizational capacity in learning and growth.

MAPEIs were compared between Tables 1 and 8. The importance of items for large construction firms is different from that for small construction firms. For large construction firms, the weighted value of customers was 0.34 and highest in BSC, followed by finance and learning and growth. However, the highest weighted value was 0.379 for finance, followed by customers, learning and growth, and internal process, for small construction firms. In the performance area of finance, the importance of orders and profitability was high for large construction firms, whereas the importance of profitability and stability was high for small construction firms. Unlike large construction firms, where orders are considered important with a weighted value of 0.32, small construction firms assigned greater importance to stability with a weighted value of 0.246 when the value of orders is

0.184. This is because stability is considered a very important indicator due to the constantly decreasing orders for small construction firms. This shows that the priority of performance evaluation indicators varies even for companies within the construction industry, according to their scale, management environment, and management characteristics.

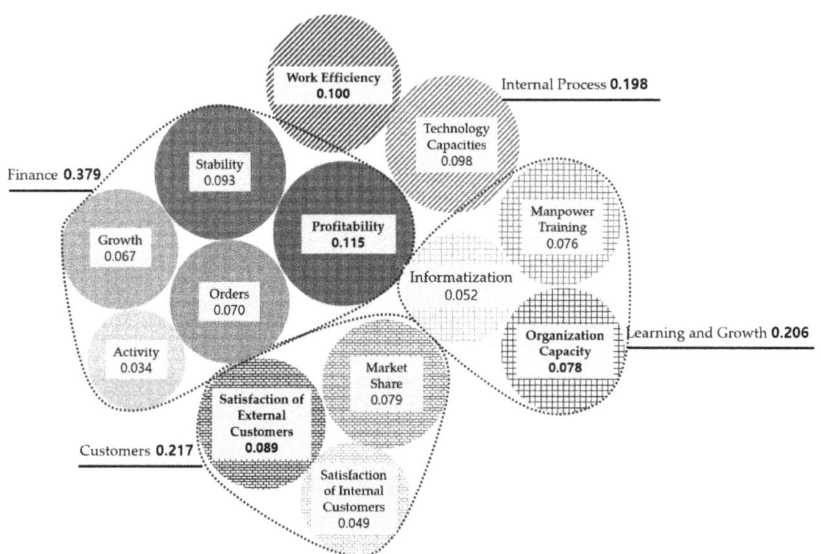

Figure 5. Performance areas.

Table 8. Weighted values of all MAPEIs.

BSC	Weighted Value	Performance Areas	Weighted Value	KPI	Weighted Value
Finance	0.379	Profitability	0.303	Cost of Sale Ratio Ordinary Profit	0.541 0.459
		Growth	0.177	Increase in Revenues in Korea	1.000
		Stability	0.246	Debt Ratio Achievement of Collection Goal	0.428 0.572
		Activity	0.090	Turnover Ratio of Total Liabilities and Net Worth	1.000
		Orders	0.184	Amounts of New Orders	1.000
Customers	0.217	Satisfaction of External Customers	0.412	Customer Satisfaction Corporate Image	0.661 0.339
		Satisfaction of Internal Customers	0.223	Transfer Rate of Employees Work Environment and Corporate Culture	0.415 0.585
		Market Share	0.365	Market Share of Orders in Korea	1.000
Internal Process	0.198	Technology Capacities	0.494	Possession of Intellectual Property Rights	1.000
		Work Efficiency	0.506	Compliance with Guidelines Accident rate Complaint-processing Capacity	0.329 0.379 0.292
Learning and Growth	0.206	Manpower training	0.370	Index of Excellent Workforce	1.000
		Organizational Capacity	0.377	Productivity of Employees	1.000
		Informatization	0.253	Informatization Capacity Index	1.000

4.3. Derivation of MAPEIs for Small Construction Firms

MAPEIs are the KPIs for management performance evaluation of small construction firms and consist of the hierarchy and weighted value of BSC, performance areas, and KPIs. There are four BSC areas, 13 performance areas, and 19 KPIs, and the weighted value of each item is as follows.

BSC-W refers to the weighted value of BSC, and the sum of BSC-W assigned to finance, customers, internal process, and learning and growth is 1. Performance-W refers to the weighted value of performance and is the product of BSC-W and the weighted value of performance, as in Equation (1). The sum of all weighted values of Performance-W is 1. KPI-W refers to the weighted values of KPI and is the product of Performance-W and the weighted value of KPIs, as in Equation (2). The sum of all weighted values of 'KPI-W' is 1.

$$\text{Performance} - W = \text{Weighted Value of Performance of BSC} - W \tag{1}$$

$$\text{KPI} - W = \text{Weighted Value of KPI of Performance} - W \tag{2}$$

MAPEIs are the most detailed items and the key indicators of management performance. Generally, BSC-W and Performance-W were highest in finance and profitability, so KPI-W would be highest for the items of finance. However, KPI-W was highest for possession of intellectual property rights in technology capacity at 0.098, as in Table 9. As KPI is applied to the hierarchy of performance, however, the items evaluating profitability were further categorized to reduce the weight of each item. The KPI of technology capacity applies to the possession of intellectual property rights only, whereas profitability was divided into two items of cost of sale ratio and ordinary profit. This implies that finance is important for evaluating the management performance of companies and the many evaluation indicators allow for accurate evaluation.

Table 9. Reflection of management performance evaluation on internal process.

BSC	BSC-W	Performance Areas	Weighted Values of Performance Areas	Performance Areas -W	KPI	Weighted Values of KPI	KPI-W
Finance	0.379	Profitability	0.303	0.115	Cost of Sale Ratio	0.541	0.062
					Ordinary Profit	0.459	0.053
		Growth	0.177	0.067	Increase in Revenues in Korea	1.000	0.067
		Stability	0.246	0.093	Debt Ratio	0.428	0.040
					Achievement of Collection Goal	0.572	0.053
		Activity	0.090	0.034	Turnover Ratio of Total Liabilities and Net Worth	1.000	0.034
		Orders	0.184	0.070	Number of New Orders	1.000	0.070
Customers	0.217	Satisfaction of External Customers	0.412	0.089	Customer Satisfaction	0.661	0.059
					Corporate Image	0.339	0.030
		Satisfaction of Internal Customers	0.223	0.049	Transfer Rate of Employees	0.415	0.020
					Work Environment and Corporate Culture	0.585	0.029
		Market Share	0.365	0.079	Market Share of Orders in Korea	1.000	0.079
Internal Process	0.198	Technology Capacities	0.494	0.098	Possession of Intellectual Property Rights	1.000	0.098
		Work Efficiency	0.506	0.100	Compliance with Guidelines	0.329	0.033
					Accident rate	0.379	0.038
					Complaint-processing Capacity	0.292	0.029
Learning and Growth	0.206	Manpower training	0.370	0.076	Index of Excellent Workforce	1.000	0.076
		Capacity	0.377	0.078	Productivity of Employees	1.000	0.078
		Informatization	0.253	0.052	Informatization Capacity Index	1.000	0.052
Total	1.000	Total		1.000	Total		1.000

4.4. Evaluation of Applicability of MAPEIs (Small Construction Firms)

The current study evaluated management performance to verify the applicability and necessity of MAPEIs. The subject applying MAPEIs was one small construction firm within the scope of study and three years' management performance was evaluated using a five-point scale. Figure 6 gives the MAPEI scores applying the weighted values and the MAPEI scores not applying the weighted values based on the evaluation results of the firm.

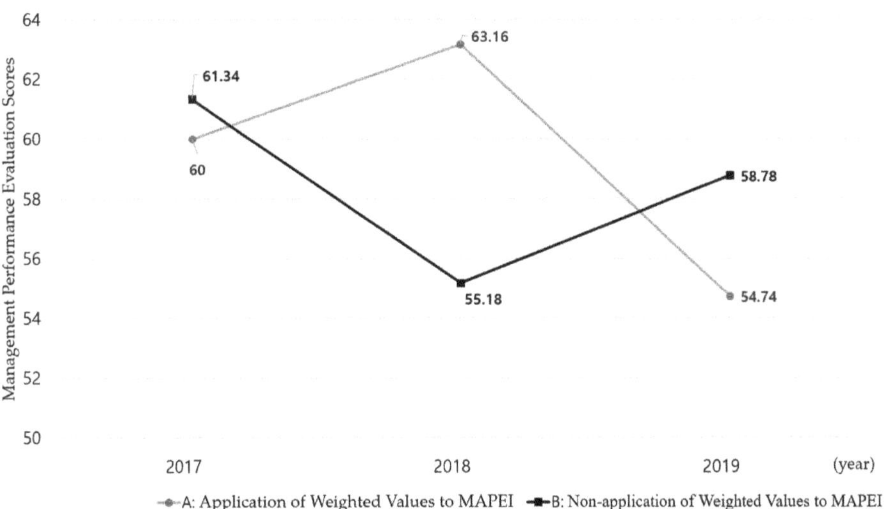

Figure 6. Results of management performance evaluation with/without weighted values of MAPEIs. (A) Results when not applying the weighted values to MAPEIs and assigning 60 points to 2017, 63.16 points to 2018, and 58.78 points to 2019; (B) results applying the weighted values to MAPEIs and assigning 61.34 points to 2017, 55.18 points to 2018, and 54.74 points to 2019. In (A), the management performance evaluation score of 2018 was 6.16%P lower than the previous year and showed a worsening of performance. The evaluation in (A) could not reflect the decline in management that was identified when analyzed by KPIs and importance (B). The management performance evaluation score of 2019 decreased by 8.42%P compared to the previous year in (A) but increased by 3.60%P with (B).

The management performance of companies varied greatly according to the application of weighted values to MAPEIs. This is because the results are distorted by applying the same weight value to all items affecting the management of firms. When the same weight is applied to all items, the performance of items with minimal impact is exaggerated and the performance of items with greater impact is lessened, which can cause errors. In other words, critical situations that may have a negative impact on management may be misinterpreted as an improvement in management. Therefore, it is important to apply weighted values to the evaluation items for the accurate evaluation of management performance.

5. Conclusions

The construction market in Korea is constantly being depressed due to the poor management of construction firms in Korea, and this has a significant impact on management performance. However, most firms in the construction market are small and the impact on management performance is tremendous. Additionally, small construction firms lack sufficient management control systems, response to and preparation for changes in the management environment, and structure of organization to improve the management. Therefore, the current study derived the MAPEIs (Management Performance Evaluation Indicators) for small construction firms for management performance evaluation.

The current study applied the management performance evaluation indicators of large construction firms from preceding studies as preliminary indicators to derive MAPEIs. Five small construction firms in Korea were selected, and we interviewed the top management about the items that are realistically adopted by companies for management performance evaluation. A secondary hierarchy was created by analyzing the items surveyed, and items besides preliminary indicators were surveyed to finalize the hierarchy through deletion and supplementation. Complaint-handling capacity was added to the final hierarchy. This was derived from the corporate characteristics of small construction firms. The final hierarchy of MAPEIs consisted of 13 performance areas and 19 KPIs.

Not all performance areas and evaluation items of the final hierarchy have equal weight values. When the same weight value is applied to all items, the management performance of companies may be distorted. Therefore, the weight values of items shall be tabulated for accurate evaluation. An AHP survey was conducted for top management and engineers and the weight values of items were tabulated through a paired comparison. The survey involved analysis of BSC and performance areas of BSC and analysis of importance of KPIs of each performance area. As a result of the importance analysis, the highest values were applied to finance of BSC, profitability of performance, and possession of intellectual property rights of KPIs. This shows that a performance evaluation based on financial factors is more important than customer-centered performance for small construction firms.

MAPEIs are KPIs for the management performance evaluation of small construction firms and consist of the hierarchy and weighted values of BSC, performance areas, and KPIs. In order to verify the feasibility of MAPEIs, one small construction firm in Korea was selected for the applicability evaluation. The evaluation results varied according to the application of weighted values to MAPEIs and the need to apply weighted values to MAPEIs was confirmed as the management performance evaluation scores were distorted when the same weighted value was applied to all indicators.

The current study analyzed the characteristics of small construction firms and selected the evaluation items through an actual corporate survey to derive a weighted value for each item. Additionally, applicability was evaluated to verify the feasibility and applicability of MAPEIs. MAPEIs are fundamental to the study of management control in small construction firms; KPIs can be applied to construction companies with no more than 10 full-time employees. However, the items can be modified and supplemented to fit the characteristics of each company, and further studies and the development of performance evaluation systems for the performance evaluation of small construction firms to resolve the limitations would improve their management evaluation and achieve competitive management control. Therefore, the results of this study can be used as basic data not only for measuring management performance, but also for developing a system for the management of small construction firms. In addition, this study has limitations because it was conducted for construction companies in Korea. This study used the MEPAI model, which is the result of existing research on the management performance of construction companies. Not applying various models can be another limitation of this study.

Author Contributions: Conceptualization, D.K. and D.L.; methodology, D.K., and D.L.; software, D.L. and W.O.; validation, D.K., W.O., J.Y. (Jiyeong Yun), J.Y. (Jongyoung Youn), and S.D.; formal analysis, W.O.; investigation, J.Y. (Jiyeong Yun); resources, D.K.; data curation, J.Y. (Jiyeong Yun); writing—original draft preparation, D.K., W.O., and D.L.; writing—review and editing, D.K., J.Y. (Jongyoung Youn), W.O., J.Y. (Jiyeong Yun), S.D., and D.L. All authors have read and agreed to the published version of the manuscript.

Funding: This work was supported by the National Research Foundation of Korea (NRF) grant funded by the Korea government (MSIT) (No. 2020R1C1C1012600).

Institutional Review Board Statement: Not applicable.

Informed Consent Statement: Not applicable.

Data Availability Statement: Not applicable.

Conflicts of Interest: The authors declare no conflict of interest.

References

1. The Construction & Economy Research Institute of Korea. Searching for Successful Management Strategies of Construction Companies during Recession. 2019. Available online: http://www.cerik.re.kr/report/issue/detail/2323 (accessed on 1 December 2019).
2. The Statistics Korea, Korean Statistical Information Service. 2016. Available online: https://kosis.kr/eng/ (accessed on 1 December 2019).
3. Kaplan, R.S.; Norton, D.P. The Balanced Scorecard-Measures that Drive Performance. *Harv. Bus. Rev.* **1992**. Available online: https://hbr.org/1992/01/the-balanced-scorecard-measures-that-drive-performance-2 (accessed on 1 December 2019).
4. Kaplan, R.S.; Norton, D.P. Using the Balanced Scorecard as a Strategic Management System. *Harv. Bus. Rev.* **1996**, *74*, 75–85. [CrossRef]
5. Lee, D.-H.; Kim, S.-K.; Kim, M.-K.; Shin, N.-H. An Analysis of Managerial and Organizational Status of Korean Construction Firms. *Korean J. Constr. Eng. Manag.* **2010**, *11*, 38–47. [CrossRef]
6. Jung, W.J.; Yu, I.H.; Kim, K.R.; Shin, D.W. Analysis of the Weights of Performance Measurement Index according to the Size of Construction Companies. *J. Archit. Inst. Korea Struct. Constr.* **2005**, *21*, 121–128. Available online: http://www.dbpia.co.kr/journal/articleDetail?nodeId=NODE00618929 (accessed on 1 December 2019).
7. Yu, I.H.; Kim, K.R.; Jung, Y.S.; Chin, S.Y.; Kim, Y.S. A Framework of the Comparable Performance Measurement in the Construction Industry. *Korea J. Constr. Eng. Manag.* **2004**, *5*, 172–182. Available online: https://www.kci.go.kr/kciportal/ci/sereArticleSearch/ciSereArtiView.kci?sereArticleSearchBean.artiId=ART001159458 (accessed on 1 December 2019).
8. Shahhoseini, M.A.; Khassehkhan, S.; Shanyani, N. Identifying Key Performance Indicators of an Iranian Islamic Bank Based on BSC and AHP. *J. Am. Sci.* **2012**, *8*, 64–73. [CrossRef]
9. Janeš, A.; Faganel, A. Instruments and methods for the integration of company's strategic goals and key performance indicators. *Kybernetes* **2013**, *42*, 928–942. [CrossRef]
10. Silk, S. Automating the balanced scorecard. *Manag. Account.* **1998**, *79*, 38–44. Available online: https://search.proquest.com/openview/7e12dd71a2486f45ee40384b6e6aa500/1?pq-origsite=gscholar&cbl=48426 (accessed on 1 December 2019).
11. Malmi, T. Balanced scorecards in Finnish companies: A research note. *Manag. Account. Res.* **2001**, *12*, 207–220. [CrossRef]
12. Rigby, D. Management tools and techniques: A survey. *Calif. Manag. Rev.* **2001**, *43*, 139–160. [CrossRef]
13. Kim, Y.D. *A Study on Development of the Management Diagnosis Model for Small and Medium Construction Companies*; Construction & Economy Research Institute of Korea, 2014; pp. 1–143. Available online: http://scholar.dkyobobook.co.kr/searchDetail.laf?barcode=4010023816045 (accessed on 1 December 2019).
14. The Construction & Economy Research Institute of Korea. A Study on the Development of a Management Team Model to Enhance the Competitiveness of Small and Medium-Sized Construction Companies. 2014. Available online: http://www.cerik.re.kr/report/research/detail/1703 (accessed on 1 December 2019).
15. Ahmed, A.M.M.; Abonamah, A.A. Innovative System for Measuring SMEs Performance. *Int. J. Cust. Relatsh. Mark. Manag.* **2013**, *4*. [CrossRef]
16. Manville, G. Implementing a balanced scorecard framework in a not for profit SME. *Int. J. Product. Perform. Manag.* **2007**, *56*, 162–169. [CrossRef]
17. Kemzūraitė, D. Balanced Scorecard for Startups and SME: Case of Laserpas. Master's Thesis, Vytautas Magnus University, Kaunas, Lithuania, 18 June 2020. Available online: https://www.vdu.lt/cris/handle/20.500.12259/106969 (accessed on 1 December 2019).
18. Chong, P.; Ong, T.; Abdullah, A.; Choo, W. Internationalisation and innovation on balanced scorecard (BSC) among Malaysian small and medium enterprises (SMEs). *Manag. Sci. Lett.* **2019**, *9*, 1617–1632. [CrossRef]
19. Kim, M.K. A Study on the Management Performance Evaluation Model of Construction Firms. Ph.D. Thesis, Kyung Hee University, Seoul, Korea, 2010. Available online: http://www.riss.kr/link?id=T12173249 (accessed on 1 December 2019).

 *sustainability*

Application of the Taguchi Method for Optimizing the Process Parameters of Producing Controlled Low-Strength Materials by Using Dimension Stone Sludge and Lightweight Aggregates

How-Ji Chen [1], Hsuan-Chung Lin [1] and Chao-Wei Tang [2,3,4,*]

[1] Department of Civil Engineering, National Chung-Hsing University, No. 250, Kuo Kuang Road, Taichung 402, Taiwan; hojichen@dragon.nchu.edu.tw (H.-J.C.); abb7028@gmail.com (H.-C.L.)

[2] Department of Civil Engineering and Geomatics, Cheng Shiu University, No. 840, Chengching Road, Niaosong District, Kaohsiung 83347, Taiwan

[3] Center for Environmental Toxin and Emerging-Contaminant Research, Cheng Shiu University, No. 840, Chengching Road, Niaosong District, Kaohsiung 83347, Taiwan

[4] Super Micro Mass Research and Technology Center, Cheng Shiu University, No. 840, Chengching Road, Niaosong District, Kaohsiung 83347, Taiwan

* Correspondence: tangcw@gcloud.csu.edu.tw; Tel.: +886-7-735-8800

Abstract: In view of the increasing concerns over non-renewable resource depletion and waste management, this paper studied the development of low-density controlled low-strength material (CLSM) by using stone sludge and lightweight aggregates. First, the investigation was performed at a laboratory scale to assess the effects of the composition on the properties of the resulting low-density CLSM. The Taguchi method with an $L_9(3^4)$ orthogonal array and four controllable three-level factors (i.e., the stone sludge dosage, water to binder ratio, accelerator dosage and lightweight aggregate dosage) was adopted. Then, to optimize the selected parameters, the analysis of variance method was used to explore the effects of the experimental factors on the performance (fresh and hardened properties) of the produced low-density CLSM. The test results show that when the percentage of stone sludge usage was increased from 30% to 60%, the initial setting time approximately doubled on average. Moreover, at the age of 28 days, the compressive strength of most specimens did not exceed the upper limit of 8.83 MPa stipulated by Taiwan's Public Construction Commission. Further, the material cost per cubic meter of the produced CLSM was about NT$ 720.9 lower than that of the ordinary CLSM, which could reduce the cost by 40.6%. These results indicate that the use of stone sludge as a raw material to produce CLSM could achieve environmental sustainability. In other words, the use of stone sludge and lightweight aggregates to produce low-density CLSM was extremely feasible.

Keywords: stone sludge; lightweight aggregates; controlled low-strength materials; Taguchi method

Citation: Chen, H.-J.; Lin, H.-C.; Tang, C.-W. Application of the Taguchi Method for Optimizing the Process Parameters of Producing Controlled Low-Strength Materials by Using Dimension Stone Sludge and Lightweight Aggregates. *Sustainability* 2021, 13, 5576. https://doi.org/10.3390/su13105576

Academic Editor: Sunkuk Kim

Received: 1 April 2021
Accepted: 15 May 2021
Published: 17 May 2021

1. Introduction

The continuous progress of science and technology has improved social productivity and material standard of living, and has achieved unprecedented prosperity and development of human society. However, the widespread application of science and technology in various fields has also caused a certain degree of environmental pollution, ecological destruction, and resource scarcity, which may cause devastating potential threats to the entire earth of human life. According to the second edition of NACE [1], the total amount of waste generated by the economic activities of households and businesses in the EU in 2018 was 2.609 billion tons, and in 2008, the amount of recyclable waste was 202 million tons. In view of this, a key element of the EU's environmental policy is to manage waste in an environmentally sound manner and make full use of the auxiliary materials contained therein. Similarly, with the rise of interest in the concept of sustainable development and the awareness of environmental protection, it has become increasingly difficult to obtain raw materials for concrete production in Taiwan. For these reasons, Taiwan promulgated (amended on 21 January 2009) the Resource Recycling Act [2] on 3 July 2002, to reduce the use of natural resources, reduce waste generation, promote the recycling and reuse of materials, reduce the environmental load and build a society in which resources can be used continuously. Currently, the Ministry of Economic Affairs of the Republic of China in Taiwan has announced 57 types of industrial waste for reuse, which can be divided into engineering, agricultural and other uses. Among them, there are 22 types of recycling for engineering purposes (such as waste casting sand, stone sludge, stone waste, coal ash, waste wood, waste glass, waste pottery, waste porcelain, waste bricks, waste tiles, etc.). These can be recycled for the production of civil engineering and construction materials.

According to the definition of the American Concrete Institute (ACI), controlled low-strength materials (CLSM) is a self-filling, self-leveling and cementitious material that is mainly used to replace traditional backfill soil and structural fillers. Generally, the compressive strength of CLSM at 28 days is 8.3 MPa or lower [3]. Although CLSM's requirements for its constituent materials are not as strict as those of ordinary concrete materials, it provides suitable engineering characteristics at a lower cost for most needs and is in line with energy-saving and ecological benefits [4]. Essentially, CLSM has excellent rheological properties and anti-segregation ability, and it can be easily filled in a narrow excavation surface without any tamping. Moreover, CLSM has a low and sufficient load-bearing strength, which can facilitate future excavation [5–7]. Therefore, the application of CLSM has become more and more common in countries around the world and in Taiwan.

The stone processing industry is an important industry in Taiwan (with an annual output value of nearly 1.42 billion U.S. dollars), but the impact of its wastewater, solid waste and dust in the environment cannot be underestimated. In the cutting process of natural stone, a large amount of stone chips and waste will be produced. In order to cool the cutting saw blade, a large amount of cooling water must be used, resulting in a considerable amount of mud-like waste (a mixture wastewater and stone chips), which is called dimension stone waste mud. The dimension stone waste mud is discharged to the wastewater treatment plant, and the waste produced after sedimentation, separation and dehydration is called stone sludge. In Europe, the amount of annual stone sludge is estimated to be 5 million tons [8]. The annual output of Taiwan's stone sludge and stone waste exceeds 1.1 million tons [9], which in general comprises industrial wastes. However, due to the relatively low cost of the burial treatment of stone sludge or the poor marketing of resource-recycling products, the amount of resource treatment is quite limited, and the benefits of its resource application cannot be brought into full play.

Because concrete has the advantages of durability, fire resistance, high compression resistance, good shape ability, economy, etc., it is currently the most widely used man-made material. However, the production of cement is the main source of greenhouse gas emissions, which has severely damaged the earth's climate and environment and threatened the sustainable survival of mankind. From the perspective of maintaining the ecological sustainability of the natural environment and the overall economic benefits of society, actively developing alternative sources of construction materials to replace some of the traditional main sources is an important issue that cannot be delayed. In terms of the sustainable development of concrete materials, we can start with the design of materials and mix proportions. For example, the use of renewable resources to replace part of cement or aggregates will not only greatly contribute to energy saving and carbon reduction, but also improve the fresh property and durability of concrete materials. Martínez-García et al. [10] evaluated the viability of incorporating fine recycled concrete aggregates (FRCA) from urban demolition and construction waste for the manufacture of cement-based mortars. The results showed that the optimal percentage of substitution of fine natural aggregates for FRCA was 25% with respect to compressive and flexural strength tests. López Boadella et al. [11] analyzed the feasibility of using waste from a granite quarry to replace the micronized quartz in ultra-high-performance concrete (UHPC). The results showed that when the substitution rate was 35%, the flexural strength and tensile strength increased, and the values obtained even for 100% substitution was acceptable. This confirmed that granite cutting waste instead of commonly used micronized quartz powder was a viable alternative to the expected more sustainable UHPC. Zamora-Castro et al. [12] reviewed the latest research on the performance of different sustainable concrete types. They recommend the use of standardized testing to ensure reliable results of the impact of sustainable materials on the physical and mechanical properties of concrete specimens. In addition, because recycled materials mixed into concrete showed high absorption capacity, it could cause workability problems of fresh concrete, thereby affecting its mechanical strength. Therefore, they recommend finding the best combination of materials from different sources to improve these properties of sustainable concrete. In addition, Chen et al. [13–15] used reservoir sediments, paper sludge and tile grinding sludge to produce lightweight aggregates, turning waste into renewable resources.

In order to achieve the purpose of waste reduction and resource reuse, renewable resources, and industrial wastes from all over the world have been used in large quantities in the production of CLSM, such as coal ash and new pozzolanic materials [16], cement kiln dust [17], stockpiled circulating fluidized bed combustion ashes [18], circulating fluidized bed combustion ash and recycled aggregates [19], bottom ash of municipal solid waste incinerator and water filter silt [20], solid wastes/byproducts from paper mills [21], circulating fluidized bed combustion ash [22,23], waste oyster shells [24], treated oil sand waste [25], alum sludge and green materials [26], waterworks sludge [27], water purification sludge [28], etc. Hung et al. [29] established a prediction model for the compressive strength and surface resistivity of controlled low-strength desulfurization slag. They suggested that expanding the use of unqualified raw materials and man-made waste as secondary raw materials could be one of the most important directions for creating a waste-free process to ensure the most reasonable use of natural resources and reduce the negative impact on environmental conditions. Park and Hong [30] analyzed the influence of the mixing conditions of wastepaper sludge ash (WPSA) on the strength and bearing capacity of controlled low-strength materials (CLSM). The results showed that CLSM and WPSA could be used as backfill materials for sewage pipes, which could ensure higher stability compared with soil backfill.

In the past, the disposal cost of stone sludge in Taiwan was relatively low, and the related reuse or volume reduction technology has not received much attention from the industry. However, due to the lack of natural resources and the increasing difficulty of finding waste disposal sites, stone sludge should be prioritized for resource utilization. On the other hand, the composition of stone sludge and stone waste is not significantly different from the parent stone processed [31], and its output is huge, representing a large resource that can be recycled and reused with great resource potential and economic value. If stone sludge replaces the fine particles in CLSM as an alternative material source, the cost of raw materials can be reduced. In addition, replacing ordinary aggregates with lightweight aggregates can produce low-density CLSM, which can reduce the load on the underground structure and the soil or filler below it.

In view of the above, the development of low-density CLSM by using stone sludge and lightweight aggregates was explored in this study. First, the investigation was performed on a laboratory scale to assess the effects of the composition on the properties of the resulting low-density CLSM. The Taguchi method with an $L_9(3^4)$ orthogonal array and four controllable three-level factors (i.e., stone sludge, water/binder ratio, accelerating agent and lightweight aggregate content) was adopted. Then, in order to optimize the selected parameters, the analysis of variance method was used to explore the effects of the experimental factors on the performances (fresh and hardened properties) of the produced low-density CLSM. This study confirmed that the use of stone sludge and lightweight aggregates to produce low-density CLSM was extremely feasible. Especially, in view of the various engineering requirements of CLSM, the Taguchi method could be used to optimize the process parameters of using size stone sludge and lightweight aggregates to produce controlled low-strength materials.

2. Materials and Methods

2.1. Materials

A large amount of stone sludge in Taiwan has created an environmental burden. Therefore, this study uses stone sludge in the production of CLSM, and its purpose is to treat stone sludge through recycling and reuse to avoid secondary pollution. The materials used in this study included cement, ground-granulated blast-furnace slag, water, ordinary fine aggregates, lightweight aggregates, stone sludge, accelerating agent and air-entraining agent. The cement used was Type I Portland cement with a specific gravity of 3.15 produced by Taiwan Cement Corporation. The ground-granulated blast-furnace slag was produced by CHC Resources Corporation and its specific gravity was 2.9. Ordinary fine aggregates were taken from the nearby sand and gravel plant, with a specific gravity of 2.6 and a water absorption rate of 1.1%. Lightweight aggregates were purchased from Ming Chun Ceramic Corporation, and their specific gravity was 1.6. Two kinds of stone sludge were taken from Stone and Resource Industry Research and Development Center in eastern Taiwan, and their chemical composition is shown in Table 1. These stone sludge is characterized by a very fine size distribution, which is mainly made up of the same compounds as the processed stones. Since the particle size of marble stone sludge was larger than that of granite stone sludge, its particle size distribution is shown in Figure 1. Therefore, this study used marble stone sludge with a specific gravity of 2.6 to replace part of the ordinary fine aggregates. The accelerating agent and air-entraining agent were purchased from Guanghui Building Materials Company, in line with the Chinese national standards or the American Society for Testing and Materials specifications, and their specific gravity was 1.05.

Table 1. The composition of the stone sludge initially examined for research purpose.

Marble Stone Sludge		Granite Stone Sludge	
Chemical Composition	**(%)**	**Chemical Composition**	**(%)**
SiO_2	5.47	SiO_2	65.31
Al_2O_3	1.22	Al_2O_3	11.86
Fe_2O_3	1.10	Fe_2O_3	4.23
Na_2O	0.18	Na_2O	3.53
CaO	46.62	CaO	5.55
K_2O	0.33	K_2O	5.32
P_2O_5	0.02	P_2O_5	0.14
Cl	ND	Cl	ND
SO_3	0.09	SO_3	0.06
CuO	0.01	CuO	0.02
NiO	0.01	NiO	0.01
MnO	0.02	MnO	0.08
MgO	4.35	MgO	1.48
SrO	0.04	ZnO	0.01
TiO_2	0.08	SrO	0.06
ZrO_2	0.01	TiO_2	0.51
Cr_2O_3	0.02	ZrO_2	0.06
-	-	Cr_2O_3	0.01
-	-	Rb_2O	0.02
Loss on Ignition	40.43	Loss on Ignition	1.74
Sum	100.00	Sum	100.00

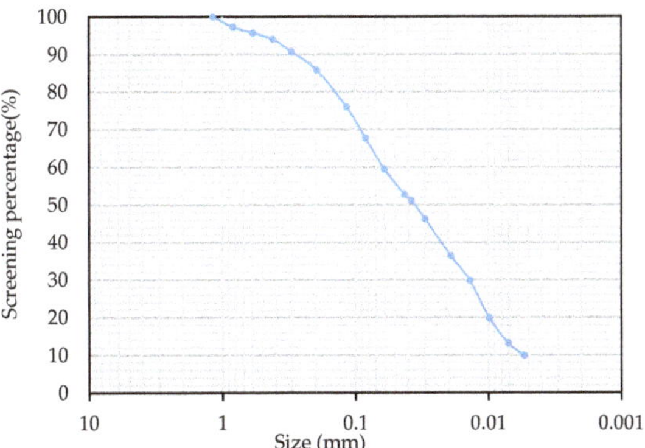

Figure 1. Grain size distribution of the used marble stone sludge.

2.2. Experimental Design

This study aimed to produce low-density CLSM using stone sludge and lightweight aggregates. The main component of stone sludge was suspended solids in the raw water—that is, fine sand particles with a specific gravity between 2.6 and 2.8. Because the particle size of stone sludge was relatively fine, this study used it as a filler to replace part of the fine aggregates.

Depending on the application field, such as backfill, utility bedding, void fill, and bridge approach, the important characteristics a CLSM must have are different. The mix proportions of the ingredients in a CLSM mixture depends on the required properties of the CLSM in two states, namely the plastic state and the hardened state. Basically, a CLSM mixture is made by mixing cementitious materials, aggregates, and water in a designed ratio. If necessary, chemical admixtures or mineral admixtures can be used to ensure that the CLSM meets the requirements of fluidity, setting time, and low strength. According to the setting time of CLSM, its mix design is divided into early-strength type and general type. The early-strength CLSM is developed for pipeline projects in urban areas and traffic hubs. It can solve urban construction troubles and keep traffic flow; in addition, it can effectively ensure project quality and construction safety. For conventional CLSM, it is developed for projects that do not require early-strength, non-emergency, and more cost considerations. It is more suitable for pipeline projects in suburbs and industrial areas. Compared with traditional backfilling methods, it can not only improve the quality and safety of general backfilling projects, but also is very economical.

The use of different types and sources of renewable resources is the most important factor that affects the water demand of CLSM. In addition, the water-cement ratio and the type of renewable resources are variables that affect the compressive strength of CLSM. Because CLSM contains a large number of materials that exceed traditional specifications, there is currently no unanimously accepted mix design method. Therefore, based on previous experience and trial and error, we screened out four test variables that needed to be investigated; namely, the percentage of the stone sludge to replace the fine aggregates, the water–binder ratio, the percentage of the accelerating agent and the dosage of the lightweight aggregates. After the experimental control factors were selected, the level of each factor was set to keep the level value within a reasonable range as far as possible. The evaluation indicators of the test included the initial setting time, final setting time, slump, slump flow, unit weight, air content and compressive strength of the CLSM produced.

The main performance characteristics of CLSM include high fluidity and a controllable low strength, and the characteristics of the constituent materials and their proportion in the mixture are the main parameters that affect the performance of CLSM. So far, there is no standard method for CLSM ratio design. The different control levels of each control factor are shown in Table 2 to explore the performance of each control factor level combination on the characteristics of CLSM and evaluate the best treatment. The literature showed that renewable resources with fine particle size could be used to replace fine aggregates, and the replacement percentage was usually between 0–90% [14,17,21]. Due to the high-water absorption characteristics of stone sludge, this study believes that the percentage of stone sludge to replace fine aggregates should not be too high. Under the condition of a fixed amount of sand (1450 kg per cubic meter), we planned a total of three stone sludge dosages, and the weight percentages of the stone sludge to replace the fine aggregates were 0%, 30% and 60%. Of these, the proportion of 0% was the control group, and the other proportions made up the experimental group. In addition, for economic cost considerations, the amount of cement was fixed at 125 kg per cubic meter, the amount of ground-granulated blast-furnace slag was fixed at 50 kg per cubic meter, the amount of accelerating agent was between 2% to 4% of the binder content and the amount of air-entraining agent was fixed at 1% of the binder content. It is worth mentioning that the use of accelerating agent was to enhance the early strength of CLSM. However, among various raw materials, the unit price of accelerating agent and air-entraining agent was relatively expensive, so their usage was less.

Table 2. Factors and design levels for test mixtures.

Experimental Control Factor	Levels of Factor			Performance Parameter
	1	2	3	
Stone sludge dosage, A (%)	0	30	60	Initial setting time, final
Water–binder ratio, B	0.9	1.0	1.1	setting time, slump, slump
Accelerator dosage, C (%)	2	3	4	flow, unit weight, air content,
Lightweight aggregate dosage, D (kg/m^3)	250	300	350	and compressive strength.

Under the condition of four factors and three levels for each factor, if the full factor experiment was carried out, the scale of the experiment would be very large (with 3^4 experimental combinations). In this study, an experimental orthogonal array $L_9(3^4)$ was selected to arrange the test plan, as shown in Table 3. Then, through the use of range analysis and analysis of variance, it was possible to quickly analyze the effect factors that had a significant impact on the experimental characteristic indexes among many factors to determine the factor combination that could obtain the best characteristic indexes.

Table 3. Orthogonal array ($L_9(3^4)$) for test mixtures.

Mix Number	Factor (Level)			
	Sludge Content (%)	Water-Binder Ratio	AA Content (%)	LWA Content (%)
M1	0 (1) *	0.9 (1)	2 (1)	250 (1)
M2	0 (1)	1.0 (2)	3 (2)	300 (2)
M3	0 (1)	1.1 (3)	4 (3)	350 (3)
M4	30 (2)	0.9 (1)	3 (2)	350 (3)
M5	30 (2)	1.0 (2)	4 (3)	250 (1)
M6	30 (2)	1.1 (3)	2 (1)	300 (2)
M7	60 (3)	0.9 (1)	4 (3)	300 (2)
M8	60 (3)	1.0 (2)	2 (1)	350 (3)
M9	60 (3)	1.1 (3)	3 (2)	250 (1)

Notes: * The numbers in parentheses indicate the level of the factor. AA = accelerating agent; LWA = lightweight aggregate.

2.3. Mix Proportions Design

According to the experimental combination of CLSM in the orthogonal design table, the amount of each constituent material was calculated as shown in Table 4. According to the design of the aforementioned various combinations, a horizontal twin-shaft mixer was used to mix CLSM. After the mixing of each group of CLSM was completed, the fresh properties (slump, slump flow, setting time, unit weight and air content) were measured and recorded; after that, ninety-one cylindrical test body (100 mm in diameter and 200 mm in height) was cast. The specimens were disassembled according to the planned age and then placed in a water bath for curing. They were not taken out until the day before the test age for the compressive strength test.

Table 4. Mix proportions of test mixtures.

Mix Number	Cement (kg/m³)	Slag (kg/m³)	Water (kg/m³)	LWA (kg/m³)	Sludge (kg/m³)	FA (kg/m³)	AA (kg/m³)	AE (kg/m³)
M1	133.90	53.60	168.70	267.80	0.00	1553.50	1.88	3.75
M2	127.10	50.80	177.90	304.90	0.00	1473.90	1.78	5.34
M3	120.90	48.30	186.10	338.40	0.00	1402.10	1.69	6.77
M4	125.30	50.10	157.90	350.90	436.10	1017.50	1.75	5.26
M5	131.00	52.40	183.40	262.00	455.90	1063.70	1.83	7.34
M6	125.00	50.00	192.60	300.10	435.20	1015.40	1.75	3.50
M7	129.10	51.70	162.70	309.90	898.80	599.20	1.81	7.23
M8	123.40	49.30	172.70	345.40	858.50	572.30	1.73	3.45
M9	128.90	51.50	198.40	257.70	896.90	597.90	1.80	5.41

Notes: LWA = lightweight aggregate; FA = fine aggregate; AA = accelerating agent; AE = air-entraining agent.

2.4. Test Methods and Data Analysis

All the mixtures were evaluated in terms of their initial setting time, final setting time, slump, slump flow, unit weight, air content and compressive strength. The slump of the mixture was measured using Chinese National Standard (CNS) 1176 [32], while the slump flow of the mixture was measured using CNS 14842 [33]. The time of setting the mixture was measured using American Society for Testing and Materials (ASTM) C403/C403M-16 [34]. The unit weight and air content of the fresh mixture were measured using ASTM D 6023 [35]. The compressive strengths of the hardened mixture were measured at curing ages of 12 h, 24 h, and 28 days, respectively. Three specimens from each mixture were tested for compressive strength. Only one specimen was taken for other test items. The preparation and testing of the mixture specimens were in accordance with CNS 1232 [36].

In the process of discussing optimization, the most important factor is to find the objective function that can best express the quality characteristics. For example, maintaining the overall average value of the product close to the set value or reducing the variation between products can be used as an objective function to improve quality. During the product life cycle, the total price paid by the entire society is called quality loss. The less the quality loss, the higher the quality. In Taguchi's quality concept, parameter design is the most important step to achieve high-quality and low-cost goals. Taguchi believes that the quality characteristics should be concentrated around the target value, and the further away from the target value, the greater the loss. Dr. Taguchi used the loss function of the quadratic curve to measure quality characteristics [37]. When the quality characteristic completely met the target value, the quality loss was zero. When the quality characteristic deviated from the target value, the quality loss increased at the speed of a quadratic curve. The quality loss function (L) is a second-order function, which is defined as: "The quality loss is equal to the square of the difference between the actual value and the target value, multiplied by a quality loss coefficient." It is a criterion for evaluating the quality of a product. Its mathematical function can be expressed as follows [37]:

$$L(y) = k(y - m)^2 \tag{1}$$

where L = quality loss function; y = quality characteristic; m = target value; k = quality loss coefficient. The total quality loss can be calculated as follows:

$$Total\ quality\ loss = \sum_{i=1}^{n} k(y_i - m)^2 \tag{2}$$

where y_i = test value; n = measurement or observation times; m = target value.

There are three standard types of calculation loss function: smaller-the-better, larger-the-better, and nominal-the-better. The signal-to-noise ratio (S/N) is an important evaluation index in Taguchi's quality engineering approach. According to different quality characteristics, the calculation of the S/N ratio (η) can be divided into three types [37,38]:

$$\text{Smaller} - \text{the} - \text{better} : \eta = -10 \times \log\left(\frac{1}{n}\sum_{i=1}^{n} y_i^2\right) \tag{3}$$

$$\text{Larger} - \text{the} - \text{better} : \eta = -10 \times \log\left(\frac{1}{n}\sum_{i=1}^{n} \frac{1}{y_i^2}\right) \tag{4}$$

$$\text{Nominal} - \text{the} - \text{better} : \eta = -10 \times \log\left(\frac{1}{n}\sum_{i=1}^{n} (y_i - y_0)^2\right) \tag{5}$$

where n is the number of repetitions or observations, y_i is the observed data, and y_0 is the nominal value desired.

In this study, the objective functions of the mixture specimens were set separately according to engineering requirements. The objective function of the initial setting time, final setting time and unit weight was a smaller-the-better type, while the objective function of the slump, slump flow, air content and compressive strength was a larger-the-better type. On the other hand, the statistical range analysis method was used to explore the relationship between CLSM properties and various control factors. The range was used to represent the measures of variation in statistical data. The difference between the maximum value and the minimum value was taken, which was the data equal to the maximum value minus the minimum value. The range analysis benefits from a simple calculation and intuitiveness and was simple and easy to understand; thus, it is the most used method for the analysis of orthogonal test results. The range analysis process for the test results included the following steps: calculating the range of each factor, determining the order of importance of the factors, drawing a trend graph of the factors and indicators, and determining the optimal level and the optimal level combination of the test factors.

Moreover, an analysis of variance (ANOVA) was used to detect the optimization of the observed values. This was accomplished by separating the total variation of the S/N ratios into contributions by each of the process parameters and the error [39]. In other words, the total variation can be decomposed into two parts: variation due to changes in various factors, and variation due to experimental errors. The different experimental values measured under exactly the same process conditions are all attributed to "experimental error", or simply "error". Analytically, the total sum of square deviation (SS_T) of the S/N ratio can be calculated as [37]:

$$SS_T = \sum_{i=1}^{n} (\eta_i - \eta_m)^2 \tag{6}$$

where n is the number of experiments in the orthogonal array; η_i is the mean S/N ratio for the ith experiment; and η_m is the grand mean of the S/N ratio. Then, the sum of squares of the measured parameter Z (SS_Z) can be calculated as [37]:

$$SS_Z = \sum_{j=1}^{r} \frac{Z_j^2}{t} - \frac{1}{n}\left(\sum_{i=1}^{n} \eta_i\right)^2 \tag{7}$$

where Z represents one of the tested parameters; j is the level number of parameter Z; r is the number of levels of parameter Z; t is the number of repetitions of each level of parameter Z; and Z_j is the sum of the S/N ratio involving parameter Z and level j. The sum of squares of the error parameter (SS_e) can be calculated as follows [37]:

$$SS_e = SS_T - SS_F \tag{8}$$

where SS_F represents the sum of squared deviations due to each parameter.

On the other hand, the application of statistical F-test can determine which process parameters have a significant impact on performance characteristics. In order to perform the F test, it is necessary to calculate the average of the squared deviation (variation) due to each process parameter and error term, as shown below [37]:

$$MS_Z = SS_Z / df_Z \tag{9}$$

$$MS_e = SS_e / df_e \tag{10}$$

Among them, MS_Z is the average value of the square deviation attributed to the parameter Z; df_Z is the degree of freedom of the parameter Z; MS_e is the average value of the square deviation due to the error term; and df_e is the degree of freedom of the error term. Then, the F value of the parameter Z (F_Z) can be calculated according to the following formula [37]:

$$F_Z = MS_Z / MS_e \tag{11}$$

In an orthogonal array experiment, when the F value of a control factor is large, it means that the control factor is influential (important). The corrected sum of squares (SS_Z^*) can be calculated as follows [37]

$$SS_Z^* = SS_Z - MS_e \times df_Z \tag{12}$$

Finally, the percentage contribution of parameter Z (P_Z) can be calculated as follows [37]:

$$P_Z = SS_Z^* / SS_T \tag{13}$$

P_Z can be used as a simple indicator to represent the influence of a factor's change, so it can be used as an indicator of the "importance" of a factor.

3. Results and Discussion

3.1. Fresh Properties of Tested Mixtures

The fresh properties (slump, slump flow, setting time, air content and unit weight) of mixture specimens are shown in Table 5. As with ordinary concrete, the fresh properties of CLSM were also affected by its composition. It can be seen from Table 5 that the fresh properties of the prepared CLSM were a function of the amount of stone sludge, the water–binder ratio, the accelerating agent, and the lightweight aggregate. The slump of tested mixtures was between 2.0 and 22.5 cm, as shown in Figure 2. Of these, the slump of the M4 mixture was the smallest, and the slump of the M9 mixture was the largest. On the other hand, the slump flow of tested mixtures was between 20 and 38 cm, as shown in Figure 2. Of these, the slump flow of the M2 and M4 mixtures was the smallest, and the slump flow of the M9 mixture was the largest. Regarding the setting time of tested mixtures, the initial setting time was between 159 and 600 min, and the final setting time was between 396 and 1855 min, as shown in Figure 3. Of these, the initial setting time of the M4 mixture was the shortest, the initial setting time of the M3 mixture was the longest, the final setting time of the M6 mixture was the shortest, and the final setting time of the M3 mixture was the longest. In terms of the air content of tested mixtures, its value was between 2.4% and 5%. The air content of the M1 and M2 mixtures was the smallest, and the air content of the M5 mixture was the largest, as shown in Figure 4. In terms of the unit weight of tested mixtures, its value was between 1961–2346 kg/m^3. The unit weight of the M3 mixture was the smallest, and the unit weight of the M4 mixture was the largest, as shown in Figure 4. From this point of view, the M1–M3 and M9 mixtures can be seen to produce low-density CLSM.

Table 5. Fresh properties and the corresponding signal-to-noise ratios of tested mixtures.

Mix No.	Experimental Results								Corresponding *S/N* Ratio (dB)			
	Setting Time (min.)		Slump (cm)	Slump Flow (cm)	Air Content (%)	Unit Weight (kg/m^3)	Setting Time		Slump	Slump Flow	Air Content	Unit Weight
	Initial Setting	Final Setting					Initial Setting	Final Setting				
M1	520	1807	6.0	21	2.4	1962.9	−54.32	−65.14	15.56	26.44	7.60	−65.86
M2	400	1256	3.0	20	2.4	2084.3	−52.04	−61.98	9.54	26.02	7.60	−66.38
M3	600	1855	6.5	21	3.0	1961.4	−55.56	−65.37	16.26	26.44	9.54	−65.85
M4	183	476	2.0	20	4.4	2345.7	−45.25	−53.55	6.02	26.02	12.87	−67.41
M5	212	456	3.5	21	5.0	2325.7	−46.53	−53.18	10.88	26.44	13.98	−67.33
M6	159	396	4.0	21	2.9	2338.6	−44.03	−51.95	12.04	26.44	9.25	−67.38
M7	326	1284	8.0	23	4.5	2272.9	−50.26	−62.17	18.06	27.23	13.06	−67.13
M8	409	1295	18.5	32	3.1	2210.0	−52.23	−62.25	25.34	30.10	9.83	−66.89
M9	347	1012	22.5	38	3.2	2198.6	−50.81	−60.10	27.04	31.60	10.10	−66.84

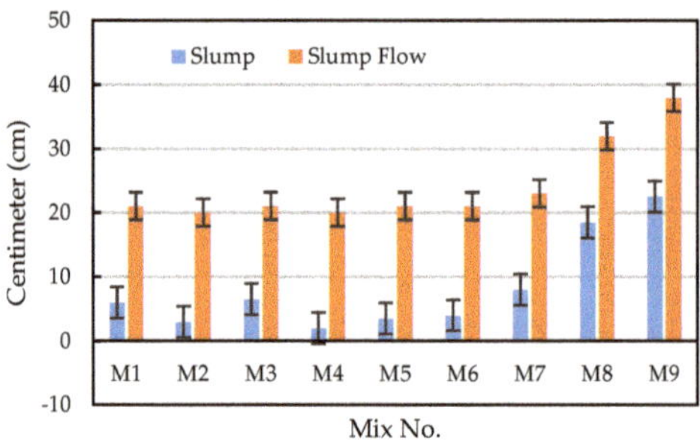

Figure 2. Slump and slump flow test results of tested mixtures.

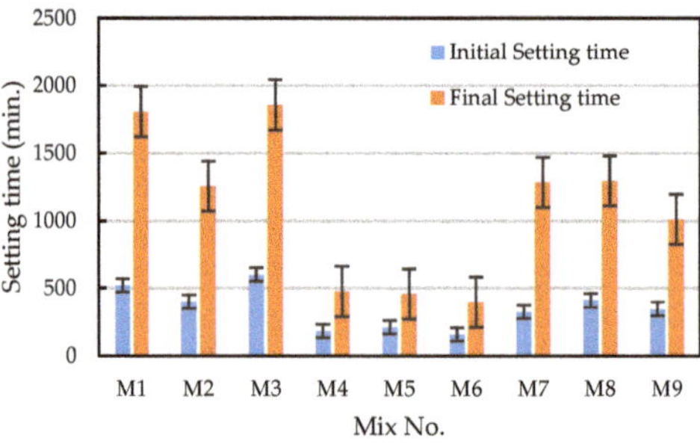

Figure 3. Setting time test results of tested mixtures.

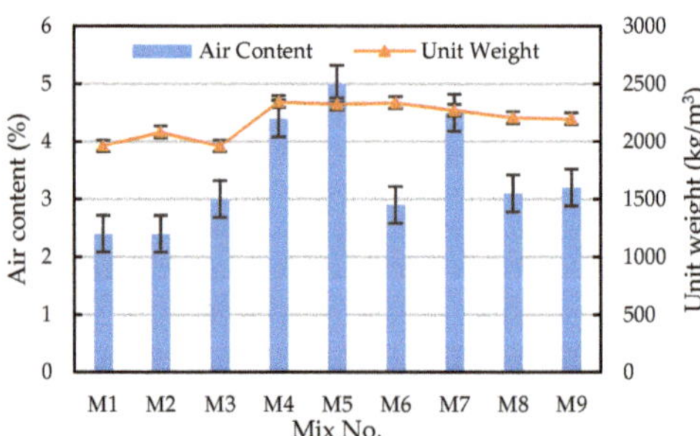

Figure 4. Air content and unit weight test results of tested mixtures.

On the other hand, Kaliyavaradhan et al. [40] pointed out that researchers who used different types of wastes with high water absorption rates as fine aggregate substitutes in CLSM often observed negative effects on the fresh properties. For example, the test results showed that when the percentage of stone sludge usage was increased from 30% to 60%, the initial setting time approximately doubled on average. In other words, the setting time increased with the increase of stone sludge dosage in the CLSM mixture. In order to further understand the appropriateness of the properties of the prepared CLSM specimens, a comparison was made with the test results of other researchers. Although different researchers used different composition materials, the test results were still available for reference, as shown in Table 6. On the whole, the results of the fresh properties of this study were included in the scope of the existing literature. Taking the slump flow as an example, the results of this study ranged from 20 to 38 cm, while the literature range was 0 to 65 cm. This means that the various properties of the produced CLSM were within a reasonable range. From this point of view, it is feasible to produce CLSM from stone sludge.

Table 6. Comparison of test results and literature.

Literature	Initial Setting	Final Setting	Slump (cm)	Slump Flow (cm)	Air Content (%)	Unit Weight (kg/m^3)	Compressive Strength		
							12 h	One Day	28 Days
This research	159–600	396–1855	2–22.5	20–38	2.4–5.0	1961.4–2345.7	0.02–0.62	0.20–1.49	1.37–9.86
Lachemi et al. [17]	-	-	-	0–53.5	1.2–2.8	1787–2028	-	-	0.7–4.0
Shon et al. [18]	300–660	-	-	21.1–25.5	-	-	-	-	0.70–0.95
Lin et al. [19]	-	-	-	41.9–54.9	-	-	-	-	4.0–8.5
Kuo & Gao [20]	-	-	26.5–26.8	56.9–65.3	-	-	-	-	5.87–6.25
Kuo et al. [24]	-	-	21.5–26.5	41–60	-	-	-	0.96–1.67	4.21–6.63
Fang et al. [27]	-	210–270	-	21–23	-	-	-	-	1.18–1.58
Hung et al. [29]	-	-	-	-	-	-	-	0.2–1.3	2.7–7.2
Park & Hong [30]	-	-	-	-	-	-	-	0.04–1.4	0.53–5.84
Trejo et al. [41]	-	-	-	13–25	0.5–28.5	1382–2291	-	-	-
Do & Kim [42]	386–586	-	-	24–30	-	-	-	-	0.5–0.85
Razak et al. [43]	360–1608	-	-	-	-	1528–1560	-	-	0.12–1.73
Türkel [44]	-	-	-	20.5–22	-	1944–2064	-	-	0.85–1.15
Taha et al. [45]	-	-	22–25	-	-	1380–2140	-	-	0.084–0.85

In order to analyze the effect of each factor, we calculated the average value of the S/N ratio of each factor at the same level, and then calculated the main effect value (delta) of the factor level. In this way, these data could be made into an auxiliary table as shown in Table 7. If the main effect value of one of the factors was larger, it meant that the influence of the factor on the whole system was greater, and the quality of improvement was also greater. It can be seen from Table 7 that the influence of A_1 (i.e., the level of A factor was under the condition of the first level) was reflected in the experimental combination of M1–M3, the influence of A_2 was reflected in the experimental combination of M4–M6, and the influence of A_3 was reflected in the experimental combination of M7–M9. In Table 7, delta represents the calculated range; that is, the difference between the maximum value and the minimum value in levels 1–3. In the same way, the influence of the water–binder ratio, the accelerator dosage and the lightweight aggregate dosage on the performance parameters can be analyzed. In principle, the greater the delta value, the greater the influence of the level change of the factor on the performance parameters; that is, the more important the factor. On the other hand, a statistical method, ANOVA, was used to further explore the test results. The results of the ANOVA for the fresh properties of tested mixtures are shown in Table 8. The total variation represents the possible total quality loss. In addition, the F value and P value obtained under a confidence level of 95% and the contribution percentage of each parameter are also listed in the table, which represents the proportion of the variation of the factor to the total quality loss.

Table 7. Range analysis for fresh properties of tested mixtures.

Performance Parameter	Experimental Control Factor	Mean S/N Ratio (η, Unit: dB)			Delta (Max. η−Min. η)	Rank
		Level 1	Level 2	Level 3		
Initial setting time	Stone sludge dosage, A (%)	−53.97	−45.27	−51.10	8.707	1
	Water–binder ratio, B	−49.94	−50.27	−50.13	0.323	4
	Accelerator dosage, C (%)	−50.19	−49.37	−50.78	1.419	3
	LWA dosage, D (kg/m^3)	−50.55	−48.78	−51.02	2.238	2
Final setting time	Stone sludge dosage, A (%)	−64.16	−52.90	−61.51	11.267	1
	Water–binder ratio, B	−60.29	−59.13	−59.14	1.153	4
	Accelerator dosage, C (%)	−59.78	−58.55	−60.24	1.694	2
	LWA dosage, D (kg/m^3)	−59.47	−58.70	−60.39	1.686	3
Slump	Stone sludge dosage, A (%)	13.79	9.65	23.48	13.835	1
	Water–binder ratio, B	13.22	15.26	18.45	5.233	2
	Accelerator dosage, C (%)	17.65	14.20	15.07	3.447	4
	LWA dosage, D (kg/m^3)	17.83	13.22	15.87	4.614	3
Slump flow	Stone sludge dosage, A (%)	26.30	26.30	29.64	3.341	1
	Water–binder ratio, B	26.57	27.52	28.16	1.595	2
	Accelerator dosage, C (%)	27.66	27.88	26.71	1.171	4
	LWA dosage, D (kg/m^3)	28.16	26.57	27.52	1.595	3
Air content	Stone sludge dosage, A (%)	8.25	12.03	11.00	3.782	1
	Water–binder ratio, B	11.18	10.47	9.63	1.548	3
	Accelerator dosage, C (%)	8.89	10.19	12.20	3.302	2
	LWA dosage, D (kg/m^3)	10.56	9.97	10.75	0.774	4
Unit weight	Stone sludge dosage, A (%)	−66.03	−67.37	−66.95	1.342	1
	Water–binder ratio, B	−66.80	−66.87	−66.69	0.175	3
	Accelerator dosage, C (%)	−66.71	−66.88	−66.77	0.168	4
	LWA dosage, D (kg/m^3)	−66.68	−66.96	−66.71	0.286	2

Note: S/N = signal-to-noise ratio; LWA = lightweight aggregate.

Table 8. Analysis of variance and F test for fresh properties of tested mixtures.

Performance Parameter	Experimental Control Factor	Sum of Square (SS_Z)	Degree of Freedom	Variation (MS_Z)	F Value (F_Z)	Percentage Contribution (P_Z)
Initial setting time	Stone sludge dosage, A (%)	118.10	3	39.37	748.00	90.95
	Water–binder ratio, B	0.16	3	0.05	1.00	0.00
	Accelerator dosage, C (%)	3.05	3	1.02	19.31	2.72
	LWA dosage, D (kg/m^3)	8.37	3	2.79	53.00	6.33
	All others/error	0.16	3	0.05	–	–
	Total	129.67	12	43.22	–	100.00
Final setting time	Stone sludge dosage, A (%)	208.15	3	69.38	78.78	93.55
	Water–binder ratio, B	2.64	3	0.88	1.00	0.00
	Accelerator dosage, C (%)	4.60	3	1.53	1.74	5.70
	LWA dosage, D (kg/m^3)	4.28	3	1.43	1.62	0.74
	All others/error	2.64	3	0.88	–	–
	Total	219.68	12	73.23	–	100.00
Slump	Stone sludge dosage, A (%)	302.55	3	100.85	15.68	71.57
	Water–binder ratio, B	41.73	3	13.91	2.16	5.67
	Accelerator dosage, C (%)	19.30	3	6.43	1.00	19.50
	LWA dosage, D (kg/m^3)	32.18	3	10.73	1.67	3.26
	All others/error	19.30	3	6.43	–	–
	Total	395.76	12	131.92	–	100.00

Table 8. *Cont.*

Performance Parameter	Experimental Control Factor	Sum of Square (SS_Z)	Degree of Freedom	Variation (MS_Z)	F Value (F_Z)	Percentage Contribution (P_Z)
Slump flow	Stone sludge dosage, A (%)	22.33	3	7.44	9.57	61.73
	Water–binder ratio, B	3.87	3	1.29	1.66	4.74
	Accelerator dosage, C (%)	2.33	3	0.78	1.00	28.80
	LWA dosage, D (kg/m^3)	3.87	3	1.29	1.66	4.74
	All others/error	2.33	3	0.78	–	–
	Total	32.39	12	10.80	–	100.00
Air content	Stone sludge dosage, A (%)	22.92	3	7.64	23.36	49.74
	Water–binder ratio, B	3.60	3	1.20	3.67	5.94
	Accelerator dosage, C (%)	16.61	3	5.54	16.92	44.32
	LWA dosage, D (kg/m^3)	0.98	3	0.33	1.00	0.00
	All others/error	0.98	3	0.33	–	–
	Total	44.11	12	14.70	–	100.00
Unit weight	Stone sludge dosage, A (%)	2.83	3	0.94	65.89	90.95
	Water–binder ratio, B	0.05	3	0.02	1.09	0.12
	Accelerator dosage, C (%)	0.04	3	0.01	1.00	5.61
	LWA dosage, D (kg/m^3)	0.14	3	0.05	3.37	3.33
	All others/error	0.04	3	0.01	–	–
	Total	3.07	12	1.02	–	100.00

Note: LWA = lightweight aggregate.

3.1.1. Setting Time

For the setting time of tested mixtures, the objective function was a smaller-the-better type. From the range analysis results in Table 7, in order to shorten the initial setting time of tested mixtures, the order of importance of the control factors was the stone sludge dosage (factor A), the lightweight aggregate dosage (factor D), the accelerator dosage (C factor) and the water–binder ratio (B factor); the corresponding delta values were 8.707, 2.238, 1.419 and 0.323, respectively. Moreover, the S/N response graph for the initial setting time of tested mixtures is shown in Figure 5. When the use percentage of stone sludge decreased from 30% to 0%, the S/N ratio decreased significantly, which reflected a significant increase in the initial setting time. It can be seen from the test results that the initial setting time was increased by approximately 2.7 times on average. In addition, when the percentage of stone sludge usage was increased from 30% to 60%, the S/N ratio decreased significantly, which reflected a significant increase in the initial setting time. It can be seen from the test results that the initial setting time approximately doubled on average. Therefore, the stone sludge dosage had the greatest impact and was the main factor. On the other hand, from the ANOVA results in Table 8, the most significant factor affecting the initial setting time was the stone sludge dosage, and its contribution percentage (P_Z) was 90.95%. The purpose of the prepared CLSM is to emphasize the need to shorten the initial setting time; that is, the shorter the initial setting time, the better. According to the experimental results, the optimal combination is $A_2B_3C_1D_2$, and the shortest initial setting time was 159 min. However, the best combination estimated by the range analysis and ANOVA was $A_2B_1C_2D_2$, i.e., stone sludge dosage at level 2, water–binder ratio at level 1, accelerator dosage at level 2, and lightweight aggregate dosage at level 2.

In terms of shortening the final setting time of the tested mixtures, the range analysis results in Table 7 show that the order of importance of the control factors was the stone sludge dosage, the accelerator dosage the accelerator dosage, the lightweight aggregate dosage, and the water–binder ratio; the corresponding delta values were 11.267, 1.694, 1.686 and 1.153, respectively. In addition, the S/N response graph for the final setting time of tested mixtures is shown in Figure 6. When the use percentage of stone sludge decreased from 30% to 0%, the S/N ratio decreased significantly, which reflected a significant increase in the final setting time. It can be seen from the test results that the final setting time was increased by approximately 2.7 times on average. In addition, when the percentage of stone sludge usage was increased from 30% to 60%, the S/N ratio decreased significantly, which reflected a significant increase in final setting time. It can be seen from the test results that the final setting time increased by approximately 2.7 times on average. From this point of view, the stone sludge dosage (factor A) had the greatest impact and was the main factor, while the accelerator dosage was the secondary factor. On the other hand, from the ANOVA results in Table 8, the most significant factor affecting the final setting time was the stone sludge dosage, and its contribution percentage (P_Z) was 93.55%. The purpose of the prepared CLSM is to emphasize the need to shorten the final setting time. That is to say, the shorter the final setting time, the better. According to the experimental results, the optimal combination was $A_2B_3C_1D_2$, and the shortest final setting time was 396 min. In contrast, the best combination estimated by range analysis and ANOVA was $A_2B_2C_2D_2$, i.e., stone sludge dosage at level 2, water–binder ratio at level 2, accelerator dosage at level 2, and lightweight aggregate dosage at level 2.

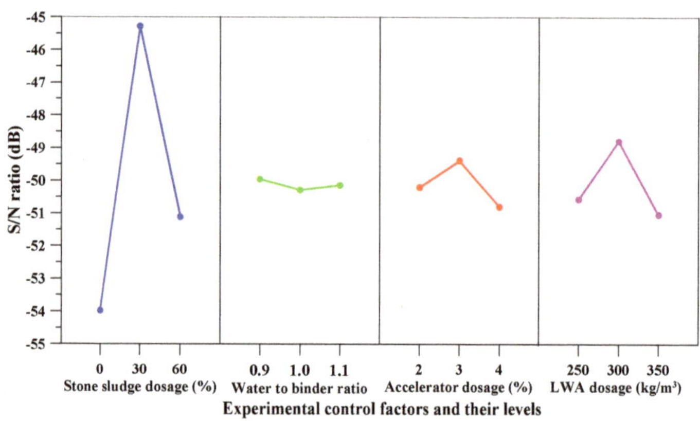

Figure 5. Signal-to-noise (S/N) response graph for the initial setting time of tested mixtures.

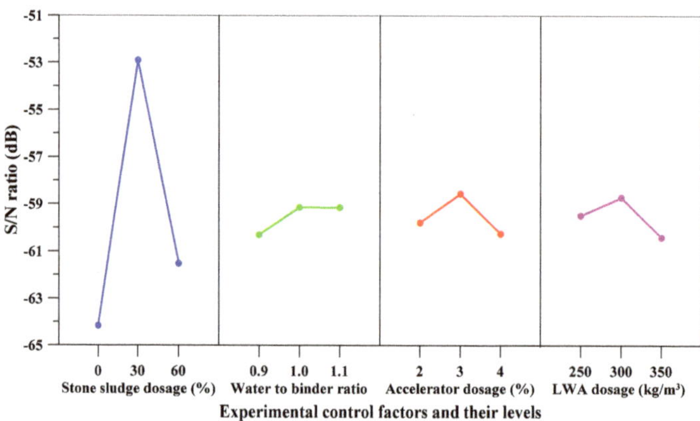

Figure 6. Signal-to-noise (S/N) response graph for final setting time of tested mixtures.

3.1.2. Slump and Slump Flow

For the slump and slump flow of tested mixtures, the objective function was a larger-the-better type. From the results of the range analysis in Table 7 when increasing the slump of CLSM, the order of importance of the control factors was the stone sludge dosage (factor A), the water–binder ratio (factor B), the lightweight aggregate dosage (factor D) and the accelerator dosage (factor C); the corresponding delta values were 13.835, 5.233, 4.614 and 3.447, respectively. In addition, the S/N response graph for the slump of tested mixtures is shown in Figure 7. When the use percentage of stone sludge decreased from 30% to 0%, the S/N ratio increased, which reflected an increase in the slump. It can be seen from the test results that the slump was increased by approximately 1.6 times on average. In addition, When the percentage of stone sludge usage increased from 30% to 60%, the S/N ratio increased significantly, which reflected a significant increase in the slump. In addition, as the water–binder ratio increased, the S/N ratio increased significantly, which reflected a significant increase in the slump. Therefore, the stone sludge dosage (factor A) had the greatest impact and was the main factor, while the water–binder ratio (factor B) was the secondary factor. Furthermore, from the ANOVA results in Table 8, the most significant factor affecting the slump was the stone sludge dosage, and its contribution percentage (P_Z) was 71.57%. The purpose of the prepared CLSM is to emphasize the need for increasing slump. The larger the slump, the better. According to the experimental results, the optimal combination was $A_3B_2C_2D_1$, and the maximum slump was 22.5 cm. However, the best combination estimated by range analysis and ANOVA was $A_3B_3C_1D_1$, i.e., stone sludge dosage at level 3, water–binder ratio at level 3, accelerator dosage at level 1, and lightweight aggregate dosage at level 1.

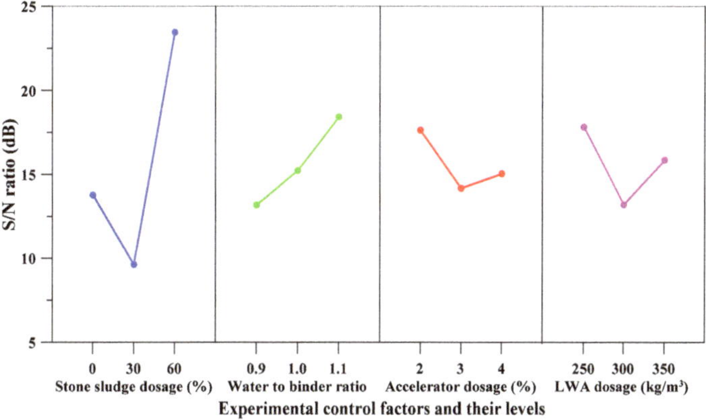

Figure 7. Signal-to-noise (S/N) response graph for slump of tested mixtures.

It can be seen from Table 7 that when increasing the slump flow of tested mixtures, the order of importance of the control factors was the stone sludge dosage (factor A), the water–binder ratio (factor B), the lightweight aggregate dosage (factor D) and the accelerator dosage (factor C). Moreover, the S/N response graph for the slump flow of tested mixtures is shown in Figure 8. When the use percentage of stone sludge was reduced from 30% to 0%, the signal-to-noise ratio did not change, which reflected that there was basically no change in the slump flow. In addition, when the percentage of stone sludge usage increased from 30% to 60%, the S/N ratio increased significantly, which reflected a significant increase in the slump flow. Moreover, as the water–binder ratio increased, the S/N ratio increased significantly, which reflected a significant increase in the slump flow. Therefore, the stone sludge dosage (factor A) had the greatest impact and was the main factor, while the water–binder ratio (factor B) was the secondary factor. Furthermore, from the ANOVA results in Table 8, the most significant factor affecting the slump was the stone sludge dosage, and its contribution percentage (P_Z) was 71.57%. The purpose of the prepared CLSM is to emphasize the need for increasing slump. The larger the slump, the better. According to the experimental results, the optimal combination was $A_3B_3C_2D_1$, and the maximum slump flow was 38 cm. Coincidentally, the best combination estimated by range analysis and ANOVA was also $A_3B_3C_2D_1$.

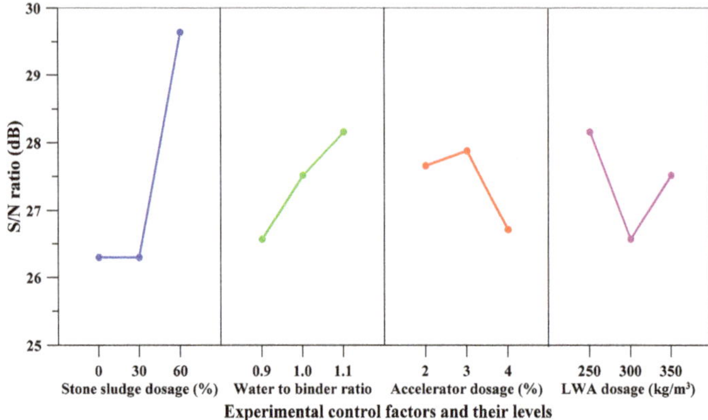

Figure 8. Signal-to-noise (S/N) response graph for slump flow of tested mixtures.

3.1.3. Air Content and Unit Weight

For the air content of tested mixtures, the objective function was a larger-the-better type. From the range analysis results in Table 7, in order to increase the air content of CLSM, the order of importance of the control factors was the stone sludge dosage (factor A), the accelerator dosage (factor C), the water to binder ratio (factor B) and the lightweight aggregate dosage (factor D); the corresponding delta values were 3.782, 3.302, 1.548 and 0.774, respectively. Moreover, the S/N response graph for the air content of tested mixtures is shown in Figure 9. When the use percentage of stone sludge decreased from 30% to 0%, the S/N ratio decreased significantly, which reflected a significant decrease in the air content. Moreover, when the percentage of stone sludge usage increased from 30% to 60%, the S/N ratio decreased significantly, which reflected a significant decrease in the air content. In addition, as the accelerator dosage increased, the S/N ratio increased significantly, which reflected a significant increase in the air content. Therefore, the stone sludge dosage had the greatest impact and was the main factor, while the accelerator dosage was the secondary factor. On the other hand, from the ANOVA results in Table 8, the most significant factor affecting the air content was the stone sludge dosage, and its contribution percentage (P_Z) was 49.74%. The purpose of the prepared CLSM is to emphasize the need to increase the air content; that is, the larger the air content, the better. According to the experimental results, the optimal combination was $A_2B_2C_3D_1$, and the largest air content was 5%. However, the best combination estimated by range analysis and ANOVA was $A_2B_1C_3D_3$, i.e., stone sludge dosage at level 2, water–binder ratio at level 1, accelerator dosage at level 3, and lightweight aggregate dosage at level 3.

For the unit weight of tested mixtures, the objective function was a smaller-the-better type. From the range analysis results in Table 7, in order to reduce the unit weight of CLSM, the order of importance of the control factors was the stone sludge dosage (factor A), the lightweight aggregate dosage (factor D), the water–binder ratio (factor B), and the accelerator dosage (factor C); the corresponding delta values were 1.342, 0.286, 0.175 and 0.168, respectively. Moreover, the S/N response graph for the unit weight of tested mixtures is shown in Figure 10. When the use percentage of stone sludge decreased from 30% to 0%, the S/N ratio increased significantly, which reflected a significant decrease in the unit weight. In addition, when the percentage of stone sludge usage increased from 30% to 60%, the S/N ratio increased significantly, which reflected a significant decrease in the unit weight. Therefore, the stone sludge dosage had the greatest impact and was the main factor, while the lightweight aggregate dosage was the secondary factor. On the other hand, from the ANOVA results in Table 8, the most significant factor affecting the unit weight was the stone sludge dosage, and its contribution percentage (P_Z) was 90.95%. The purpose of the prepared CLSM is to emphasize the need to reduce the unit weight; that is, the lighter the unit weight, the better. According to the experimental results, the optimal combination was $A_1B_3C_3D_3$, and the lightest unit weight was 1961.4 kg/m^3. However, the best combination estimated by range analysis and ANOVA was $A_1B_3C_1D_1$, i.e., stone sludge dosage at level 1, water–binder ratio at level 3, accelerator dosage at level 1, and lightweight aggregate dosage at level 1.

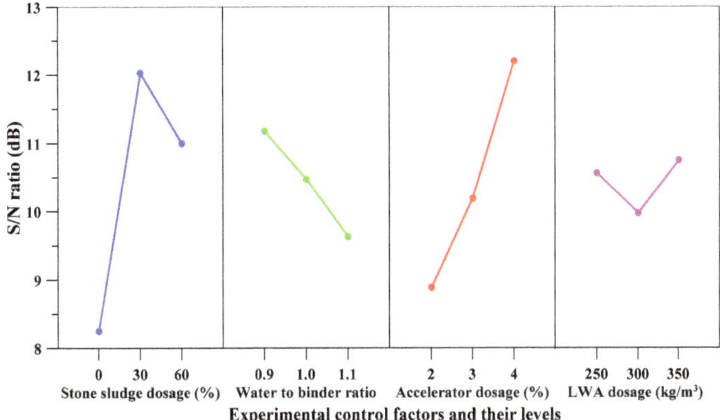

Figure 9. Signal-to-noise (S/N) response graph for air content of tested mixtures.

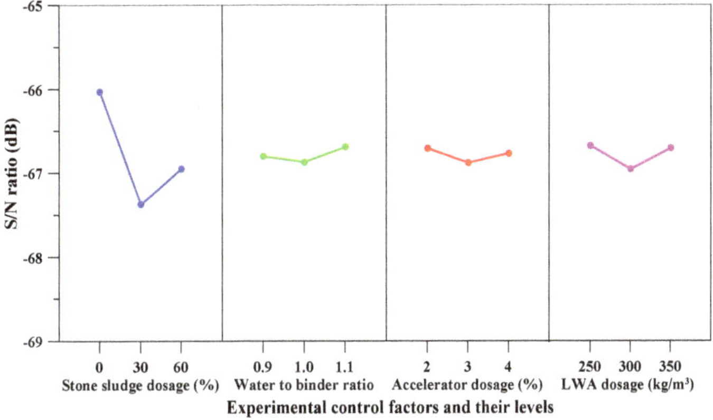

Figure 10. Signal-to-noise (S/N) response graph for unit weight of tested mixtures.

3.2. Compressive Strength of Tested Mixtures

The test results of the compressive strength of mixture specimens are shown in Table 9. It can be observed from the table that the compressive strengths of tested mixtures at 12 h, 1 day and 28 days of age were between 0.02–0.62, 0.2–1.49, and 1.37–9.86 MPa, respectively. In addition, it can be seen from Figure 11 that the compressive strength of the M1–M3 mixtures without stone sludge at each age was generally lower, and the 28-day compressive strength of the age was between 1.37–2.13 MPa. The weight percentage of the stone sludge to replace the fine aggregates in the M4–M6 mixtures was 30%. These mixtures had a higher compressive strength at all ages, and the 28-day compressive strength was between 8.56–9.86 MPa. The weight percentage of the stone sludge to replace the fine aggregates in the M7–M9 mixtures was 60%. Their compressive strength at each age was roughly between the first two groups, and the compressive strength at 28 days was between 3.88 and 6.61 MPa.

Table 9. Compressive strengths and the corresponding signal-to-noise ratio of tested mixtures.

Mix No.	Experimental Results (MPa)			Corresponding *S/N* Ratio (dB)		
	12 h	One Day	28 Days	12 h	One Day	28 Days
M1	0.24 (0.04)	0.52 (0.05)	1.78 (0.03)	−12.25	−5.68	4.98
M2	0.10 (0.01)	0.33 (0.03)	2.13 (0.07)	−20.00	−9.58	6.55
M3	0.02 (0.01)	0.20 (0.02)	1.37 (0.23)	−34.42	−13.81	2.73
M4	0.59 (0.02)	1.49 (0.05)	9.86 (0.18)	−4.63	3.46	19.88
M5	0.52 (0.05)	1.39 (0.01)	8.69 (0.43)	−5.75	2.84	18.78
M6	0.62 (0.04)	1.47 (0.09)	8.56 (0.14)	−4.15	3.35	18.65
M7	0.59 (0.05)	1.19 (0.07)	6.61 (0.32)	−4.66	1.49	16.40
M8	0.55 (0.01)	1.01 (0.02)	5.83 (0.31)	−5.19	0.09	15.31
M9	0.33 (0.02)	0.65 (0.02)	3.88 (0.12)	−9.68	−3.68	11.77

Note: Data in parentheses indicate standard deviation.

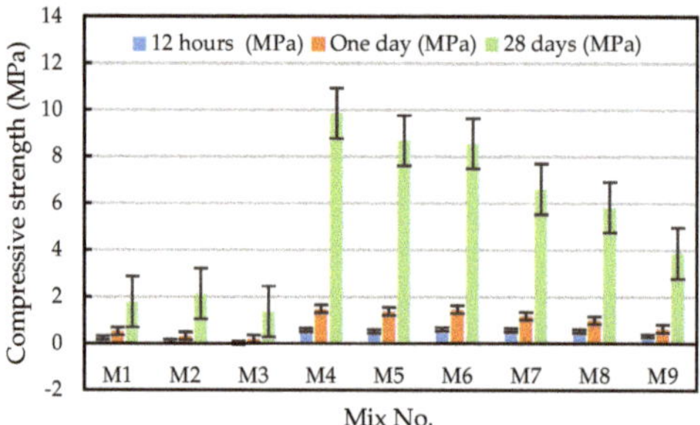

Figure 11. Compressive strength test results of tested mixtures.

On the other hand, in many cases, CLSM must be designed to have a strength equivalent to the surrounding soil after hardening, so that it can be used for future maintenance and excavation operations. Therefore, the excavability of hardened CLSM in the late age is an important consideration for many projects. In order to further understand the appropriateness of the harden properties of the prepared CLSM specimens, a comparison was made with the research results of other researchers, as shown in Table 6. It can be seen from Table 6 that the compressive strength results of this study were included in the scope of the existing literature. Taking the one-day compressive strength as an example, the results of this study ranged from 0.20 to 1.49 MPa, while the literature range was 0.04 to 1.67 MPa. In order to facilitate the subsequent construction of the paving layer, most of Taiwan's CLSM regulations require that the early strength of CLSM used for road trench backfilling be greater than 0.69 MPa. The results of the M4–M8 specimens in this study all met this requirement. As for the 28-day compressive strength, except for the M4 specimen, the 28-day compressive strength of the remaining specimens did not exceed 8.83 MPa, which was consistent with most of Taiwan's CLSM regulations. From this point of view, the compressive strength range of CLSM produced by using stone sludge could meet the needs of engineering operations.

The corresponding *S/N* ratios of the compressive strength of mixture specimens is shown in Table 9. Then, the *S/N* ratios data in Table 9 are compiled into Table 10 to analyze the impact of each level of various experimental control factors on the compressive strengths of tested mixtures. On the other hand, the results of the ANOVA of the compressive strengths of tested mixtures are shown in Table 11.

Table 10. Range analysis for compressive strengths of tested mixtures.

Performance Parameter	Experimental Control Factor	Mean *S/N* Ratio (η, Unit: dB)			Delta (Max. η−Min. η)	Rank
		Level 1	Level 2	Level 3		
12-h compressive strength	Stone sludge dosage, A (%)	−22.23	−4.84	−6.51	17.384	1
	Water–binder ratio, B	−7.18	−10.31	−16.09	8.908	2
	Accelerator dosage, C (%)	−7.20	−11.44	−14.94	7.744	3
	LWA dosage, D (kg/m^3)	−9.23	−9.60	−14.75	5.521	4
One-day compressive strength	Stone sludge dosage, A (%)	−9.69	3.22	−0.70	12.905	1
	Water–binder ratio, B	−0.24	−2.22	−4.71	4.466	2
	Accelerator dosage, C (%)	−0.75	−3.26	−3.16	2.518	3
	LWA dosage, D (kg/m^3)	−2.17	−1.58	−3.42	1.842	4
28-day compressive strength	Stone sludge dosage, A (%)	4.76	19.10	14.49	14.347	1
	Water–binder ratio, B	13.75	13.55	11.05	2.702	2
	Accelerator dosage, C (%)	12.98	12.73	12.64	0.345	4
	LWA dosage, D (kg/m^3)	11.85	13.87	12.64	2.020	3

Note: LWA = lightweight aggregate.

Table 11. Analysis of variance and *F* test for compressive strengths of tested mixtures.

Performance Parameter	Experimental Control Factor	Sum of Square (SS_Z)	Degree of Freedom	Variation (MS_Z)	F Value (F_Z)	Percentage Contribution (P_Z)
12-h compressive strength	Stone sludge dosage, A (%)	551.93	3	183.98	9.67	60.22
	Water–binder ratio, B	122.50	3	40.83	2.15	7.96
	Accelerator dosage, C (%)	90.22	3	30.07	1.58	31.82
	LWA dosage, D (kg/m^3)	57.10	3	19.03	1.00	0.00
	All others/errors	57.10	3	19.03	–	–
	Total	821.75	12	273.92	–	100.00
One-day compressive strength	Stone sludge dosage, A (%)	262.68	3	87.56	49.49	82.97
	Water–binder ratio, B	30.05	3	10.02	5.66	7.98
	Accelerator dosage, C (%)	12.17	3	4.06	2.29	9.06
	LWA dosage, D (kg/m^3)	5.31	3	1.77	1.00	0.00
	All others/errors	5.31	3	1.77	–	–
	Total	310.21	12	103.40	–	100.00
28-day compressive strength	Stone sludge dosage, A (%)	321.90	3	107.30	1693.26	94.10
	Water–binder ratio, B	13.58	3	4.53	71.41	3.92
	Accelerator dosage, C (%)	0.19	3	0.06	1.00	0.22
	LWA dosage, D (kg/m^3)	6.21	3	2.07	32.68	1.76
	All others/errors	0.19	3	0.06	–	–
	Total	341.88	12	113.96	–	100.00

Note: LWA = lightweight aggregate.

3.2.1. Twelve-Hour Compressive Strength.

For the compressive strength of tested mixtures, the objective function was a larger-the-better type. From the range analysis results in Table 10, in order to increase the 12-h compressive strength of CLSM, the order of importance of the control factors was the stone sludge dosage (factor A), the water–binder ratio (factor B), the accelerator dosage (factor C) and the lightweight aggregate dosage (factor D); the corresponding delta values were 17.384, 8.908, 7.744 and 5.521, respectively. Moreover, the *S/N* response graph for the compressive strength of tested mixtures is shown in Figure 12. When the use percentage of stone sludge decreased from 30% to 0%, the *S/N* ratio decreased significantly, which reflected a significant decrease in the 12-h compressive strength. Moreover, when the percentage of stone sludge usage increased from 30% to 60%, the *S/N* ratio decreased, which reflected a decrease in the 12-h compressive strength. In addition, as the water–binder ratio increased, the *S/N* ratio decreased significantly, which reflected a significant decrease in the 12-h compressive strength. Therefore, the stone sludge dosage had the greatest impact and was the main factor, while the water–binder ratio was the secondary factor. On the other hand, from the ANOVA results in Table 11, the most significant factor affecting the 12-h compressive strength was the stone sludge dosage, and its contribution percentage (P_Z) was 60.22%. The purpose of the prepared CLSM is to emphasize the need to increase the 12-h compressive strength; that is, the larger the compressive strength, the better. According to the experimental results, the optimal combination was $A_2B_3C_1D_2$, and the 12-h compressive strength was 0.62 MPa. However, the best combination estimated by range analysis and ANOVA was $A_2B_1C_1D_1$, i.e., the stone sludge dosage at level 2, water–binder ratio at level 1, accelerator dosage at level 1, and lightweight aggregate dosage at level 1.

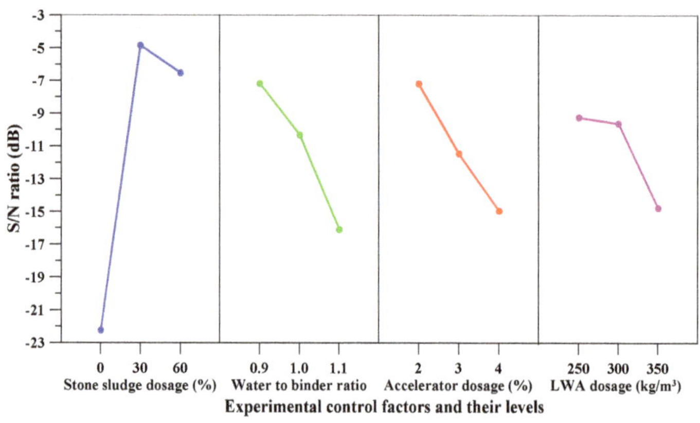

Figure 12. Signal-to-noise (S/N) response graph for the 12-h compressive strength of tested mixtures.

3.2.1. One-Day Compressive Strength

From the range analysis results in Table 10, in order to increase the one-day compressive strength of CLSM, the order of importance of the control factors was the stone sludge dosage (factor A), the water–binder ratio (factor B), the accelerator dosage (factor C) and the lightweight aggregate dosage (factor D). The corresponding delta values were 12.905, 4.466, 2.518, and 1.842, respectively. This result is consistent with the result of the 12-h compressive strength. Moreover, the S/N response graph for the one-day compressive strength of tested mixtures is shown in Figure 13. When the use percentage of stone sludge decreased from 30% to 0%, the S/N ratio decreased significantly, which reflected a significant decrease in the one-day compressive strength. Moreover, when the percentage of stone sludge usage increased from 30% to 60%, the S/N ratio decreased significantly, which reflected a significant decrease in one-day compressive strength. In addition, as the water–binder ratio increased, the S/N ratio decreased significantly, which reflected a significant decrease in one-day compressive strength. Therefore, the stone sludge dosage had the greatest impact and was the main factor, while the water–binder ratio was the secondary factor. On the other hand, from the ANOVA results in Table 11, the most significant factor affecting the one-day compressive strength was the stone sludge dosage, and its contribution percentage (P_Z) was 82.97%. The purpose of the prepared CLSM is to emphasize the need to increase the one-day compressive strength. That is to say, the larger the one-day compressive strength, the better. According to the experimental results, the optimal combination was $A_2B_1C_2D_3$, and the compressive strength was 1.49 MPa. However, the best combination estimated by range analysis and ANOVA was $A_2B_1C_1D_2$, i.e., stone sludge dosage at level 2, water–binder ratio at level 1, accelerator dosage at level 1 and lightweight aggregate dosage at level 2.

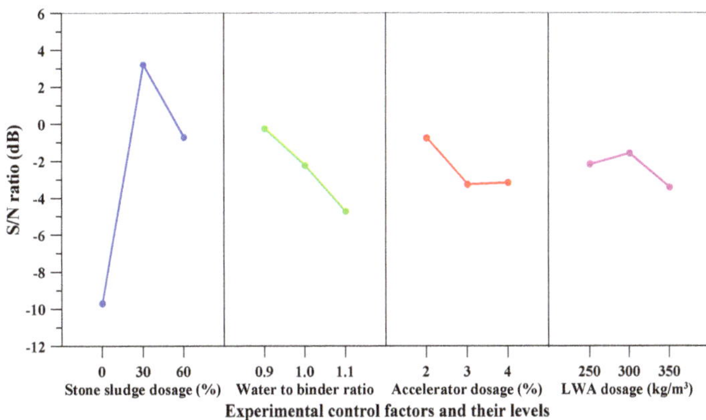

Figure 13. Signal-to-noise (S/N) response graph for one-day compressive strength of tested mixtures.

3.2.2. Twenty-Eight-Day Compressive Strength

From the range analysis results in Table 10, in order to increase the 28-day compressive strength of CLSM, the order of importance of the control factors was the stone sludge dosage (factor A), the water–binder ratio (factor B), the lightweight aggregate dosage (factor D) and the accelerator dosage (factor C); the corresponding delta values were 14.347, 2.702, 2.020 and 0.345, respectively. Moreover, the S/N response graph for the 28-day compressive strength of tested mixtures is shown in Figure 14. When the use percentage of stone sludge decreased from 30% to 0%, the S/N ratio decreased significantly, which reflected a significant decrease in the 28-day compressive strength. Moreover, when the percentage of stone sludge usage increased from 30% to 60%, the S/N ratio decreased significantly, which reflected a significant decrease in the 28-day compressive strength. In addition, as the water–binder ratio increased, the S/N ratio decreased significantly, which reflected a significant decrease in the 28-day compressive strength. Therefore, the stone sludge dosage had the greatest impact and was the main factor, while the water–binder ratio was the secondary factor. On the other hand, from the ANOVA results in Table 11, the most significant factor affecting the 28-day compressive strength was the stone sludge dosage, and its contribution percentage (P_Z) was 94.1%. The purpose of the prepared CLSM is to emphasize the need to increase the 28-day compressive strength. That is to say, the larger the 28-day compressive strength, the better. According to the experimental results, the optimal combination was $A_2B_1C_2D_3$, and the compressive strength was 9.86 MPa. However, the best combination estimated by range analysis and ANOVA was $A_2B_1C_1D_2$, i.e., stone sludge dosage at level 2, water–binder ratio at level 1, accelerator dosage at level 1, and lightweight aggregate dosage at level 2.

3.3. Confirmation Test

The results of this study show that in order to meet the requirements of most CLSM regulations in Taiwan that the early strength of CLSM used for road trench backfilling needs to be greater than 0.69 MPa, the percentage of stone sludge to replace fine aggregates can be 30%. However, in order to facilitate future maintenance and excavation, the late strength development of CLSM should not be too great. As the percentage of stone sludge substituted for fine aggregates increased from 30% to 60%, the compressive strength of CLSM at all ages decreased, and the 28-day compressive strength did not exceed 9 MPa. This indicates that stone sludge is a viable material for CLSM.

For various performance parameters, the experimental optimal combination and estimated optimal combination of CLSM prepared by this research are shown in Table 12. Of these, the experimental optimal combination of slump fluidity was consistent with the estimated optimal combination. The experimental optimal combination of other performance parameters was inconsistent with the estimated optimal combination. In order to verify that the best combination of experimental control factors can be obtained using the Taguchi method, four sets of confirmation test combinations were planned as shown in Table 12. It can be seen from Table 13 that the initial setting time of the confirmation test combination $A_2B_1C_2D_2$ was 143 min, which was shorter than the optimal combination of the experiment. The unit weight of the confirmation test combination $A_1B_3C_1D_1$ was 1950.2 kg/m^3, which was less than the optimal combination of the experiment. The 12-h compressive strength of the confirmation test combination $A_2B_1C_1D_1$ was 0.78 MPa, which was higher than the experimental best combination. The one-day compressive strength of the confirmation test combination $A_2B_1C_1D_2$ was 1.52 MPa, which was higher than the experimental best combination. The confirmation test results showed that the optimal combination of experimental control factors proposed by the Taguchi method could obtain the best results for the performance parameters.

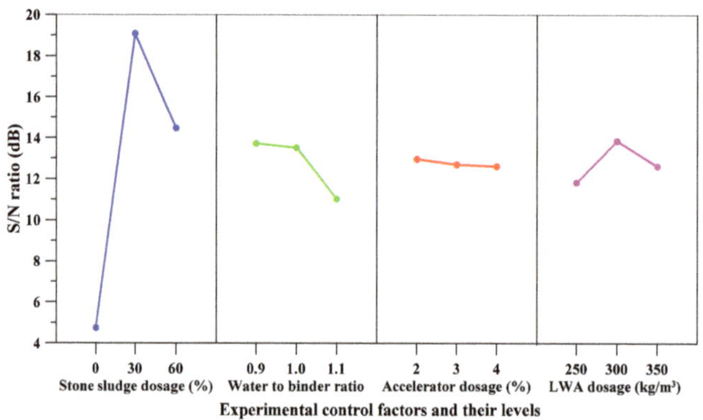

Figure 14. Signal-to-noise (S/N) response graph for 28-day compressive strength of tested mixtures.

Table 12. Combinations of confirmation test.

Performance Parameters	Experimental Optimal Combination	Estimated Optimal Combination	Combinations of Confirmation Test
Initial setting time	$A_2B_3C_1D_2$	$A_2B_1C_2D_2$	$A_2B_1C_2D_2$
Unit weight	$A_1B_3C_3D_3$	$A_1B_3C_1D_1$	$A_1B_3C_1D_1$
12-h compressive strength	$A_2B_3C_1D_2$	$A_2B_1C_1D_1$	$A_2B_1C_1D_1$
One-day compressive strength	$A_2B_1C_2D_3$	$A_2B_1C_1D_2$	$A_2B_1C_1D_2$

Table 13. Results of confirmation test.

Performance Parameters	Experimental Optimal Combination	Result of the Optimal Combination	Combinations of Confirmation Test	Results of Confirmation Test	Standard Deviation
Initial setting time	$A_2B_3C_1D_2$	159 (minute)	$A_2B_1C_2D_2$	143 (minute)	-
Unit weight	$A_1B_3C_3D_3$	1961.4 (kg/m^3)	$A_1B_3C_1D_1$	1950.2 (kg/m^3)	-
12-h Compressive Strength	$A_2B_3C_1D_2$	0.62 (MPa)	$A_2B_1C_1D_1$	0.78 (MPa)	0.025 (MPa)
One-day Compressive Strength	$A_2B_1C_2D_3$	1.49 (MPa)	$A_2B_1C_1D_2$	1.52 (MPa)	0.04 (MPa)

3.4. Cost Analysis of Production CLSM with Stone Sludge and Lightweight Aggregates

This research aimed to produce and characterize low-density CLSM using stone sludge. There were four parameters selected in this study and each had three levels, which were the amount of stone sludge (0%, 30%, and 60%), the water–binder ratio (0.9, 1.0 and

1.1), the amount of accelerator (2%, 3%, and 4%) and the amount of lightweight aggregate (250, 300 and 350 kg per cubic meter). Considering the performance of the fresh properties, compressive strength, utilization rate of renewable resources and economy, the following mixture design of the large-scale production CLSM containing stone sludge was selected: a water–binder ratio of 1.1, an amount of accelerator agent of 3%, replacement of 60% fine aggregates with stone sludge and a lightweight aggregate content of 250 kg per cubic meter. These mixture proportions had a higher amount of stone sludge, and the mechanical performance could also meet the requirements of CLSM.

Table 14 shows the unit price of various materials required for the production of CLSM. The processing fee of stone sludge is currently NT$1000–1500 per ton because it is a general industrial waste. Therefore, the unit price of stone sludge in the table is NT$-1.0/kg, which means that the cost can be reduced. The unit price of general commercial CLSM in Taiwan varies from region to region, and the price per cubic meter applied to public works ranges from NT$1230 to NT$2200. Table 14 shows the mix proportions and material cost analysis of ordinary CLSM and stone sludge CLSM. The material cost per cubic meter of ordinary CLSM is about NT$1777.3. The material cost of stone sludge CLSM is NT$697.8 per cubic meter. The material cost per cubic meter of stone sludge CLSM is about NT$1079.5 lower than that of ordinary public works CLSM, which can reduce the cost by 60.7%. In terms of economy, the production of CLSM from stone sludge is indeed quite competitive.

3.5. The Sustainability Effect of CLSM with Stone Sludge and Lightweight Aggregates

In Taiwan, it is quite common to use slices of stone such as marble or granite as decoration materials for building structures. However, the stone sludge produced by stone slicing in Taiwan exceeds one million metric tons every year. The sludge produced by cutting is extremely small and has a high-water content, which causes great trouble in the terminal disposal. At present, only a small part of it was recycled into raw cement, and most of it became earthwork. The amount of stone sludge that has not been reused and piled-up is already huge, which has caused an environmental burden in Taiwan. On the other hand, Taiwan's sludge treatment regulations are becoming stricter, and the cost of sludge disposal is gradually increasing. Therefore, stone sludge must be managed from the source, recycled and reused, and strengthened in the final treatment and management, so that the stone sludge can be reduced, stabilized, harmless, and recycled, thereby avoiding secondary pollution.

Although the definition of sustainability has not yet been consistent, it is usually defined as the process and action by which humans avoid the depletion of natural resources to maintain ecological balance without reducing the quality of life in modern society. Especially, it is composed of economy, society, and environment. Taiwan is small in area, densely populated, and limited land resources. In the face of the difficulty of finding disposal sites, reducing waste production and recycling resources utilization has become an important direction for current disposal. For this, in order to achieve the purpose of sustainable environmental management, the disposal and reuse methods of stone sludge need to be developed urgently.

Since CLSM is mostly used in backfill sites that require re-excavation, the required strength is not high for the purpose of continuous re-excavation. In view of this, the Taiwanese government encourages the use of recycled materials to manufacture CLSM. In this study, the back-end waste (i.e., stone sludge) produced by stone cutting was reused as a renewable material in CLSM. Its appearance is not much different from the general CLSM, as shown in Figure 15. From the test results and economic cost analysis, the use of stone sludge to produce CLSM could not only achieve waste recycling, but also had economic feasibility. These results also showed that waste reduction and resource disposal could be achieved, and the number of fine aggregates used could also be reduced. In other words, the production of CLSM with stone sludge as a raw material could reduce the impact of carbon emissions on the environment, thereby achieving the goal of environmental sustainability.

Table 14. Unit price of controlled low-strength material (CLSM) materials and cost analysis of CLSM.

					Unit Price					
Cement (NT\$/kg)	Slag (NT\$/kg)	Water (NT\$/kg)	AA (NT\$/kg)	LWA (NT\$/kg)	CA (NT\$/kg)	FA (NT\$/kg)	Sludge (NT\$/kg)	WRA (NT\$/kg)	AE (NT\$/kg)	
3.0	1.5	0	20	2.188	0.6	0.7	-1.0	15	15	

		Cost Analysis of CLSM										
Type of CLSM	W/B	Binder (kg/m^3)		Water (kg/m^3)	AA (kg/m^3)	LWA (kg/m^3)	CA (kg/m^3)	FA (kg/m^3)	Sludge (kg/m^3)	WRA (kg/m^3)	AE (kg/m^3)	Unit Price (NT\$/m^3)
		Cement	Slag									
Standard CLSM	1.20	100.0	150.0	300	12.50	-	350	1089	0	2.00	-	1777.3
Stone sludge CLSM	1.10	128.9	51.5	198.4	5.41	250	-	597.9	896.9	2.00	1.8	697.8

Notes: AA = accelerator agent; LWA = lightweight aggregate; CA = coarse aggregate; FA = fine aggregate; WRA = water-reducing agent; AE = air-entraining agent.

(a)   (b)

Figure 15. Appearance of CLSM samples: (**a**) CLSM produced by stone sludge and (**b**) CLSM produced by reservoir sediments.

4. Conclusions

In this study, dimension stone sludge was used to replace fine aggregates, and lightweight aggregates were used to replace ordinary coarse aggregates to explore the feasibility of producing low-density CLSM. The test results showed that the use of stone sludge and lightweight aggregates to produce low-density CLSM was extremely feasible. Based on the aforementioned test results and analysis, the following conclusions can be drawn.

The increase in the percentage of stone sludge to replace fine aggregates prolonged the setting time. Moreover, at the age of 28 days, the compressive strength of most specimens did not exceed the upper limit of 8.83 MPa stipulated by Taiwan's Public Construction Commission.

In view of the various engineering requirements of CLSM, the Taguchi method can be used for optimizing the process parameters of producing controlled low-strength materials by using dimension stone sludge and lightweight aggregates. When reducing the unit weight of CLSM, the order of importance of the control factors was the stone sludge dosage, the lightweight aggregate dosage, the water–binder ratio, and the accelerator dosage. Moreover, the ANOVA results showed that the most significant factor affecting the unit weight was the stone sludge dosage, and its contribution percentage was 90.95%.

For the improvement of the 12-h compressive strength of CLSM, the order of importance of the control factors was the stone sludge dosage, the water–binder ratio, the accelerator dosage and the lightweight aggregate dosage. In addition, the ANOVA results showed that the most significant factor affecting the compressive strength was the stone sludge dosage, and its contribution percentage was 60.22%.

Considering the performance of the fresh properties, compressive strength, utilization rate of renewable resources and economy, the following mixture design of the large-scale production CLSM containing stone sludge was selected: a water–binder ratio of 1.1, an amount of accelerator agent of 3%, replacement of 60% fine aggregates with stone sludge, and a lightweight aggregate content of 250 kg/m^3. These mixture proportions had a higher amount of stone sludge, and the mechanical performance could also meet the requirements of CLSM. The material cost per cubic meter of stone sludge CLSM is about NT$1079.5 lower than that of ordinary public works CLSM, which can reduce the cost by 60.7%. In economic terms, the production of CLSM from stone sludge is indeed quite competitive.

Author Contributions: Conceptualization, H.-J.C. and C.-W.T.; methodology, H.-J.C. and C.-W.T.; software, C.-W.T.; validation, H.-C.L. and C.-W.T.; formal analysis, C.-W.T.; investigation, H.-C.L. and C.-W.T.; resources, C.-W.T.; data curation, H.-J.C. and C.-W.T.; writing—original draft preparation, H.-J.C. and C.-W.T.; writing—review and editing, C.-W.T.; visualization, C.-W.T.; supervision, C.-W.T.; project administration, C.-W.T.; funding acquisition, C.-W.T. All authors have read and agreed to the published version of the manuscript.

Funding: This research was funded by the Ministry of Science and Technology of Taiwan grant number MOST 108-2622-E-230-003-CC3.

Institutional Review Board Statement: Not applicable.

Informed Consent Statement: Not applicable.

Data Availability Statement: The data presented in this study are available on request from the corresponding author.

Acknowledgments: The authors are grateful to the Department of Civil Engineering of National Chung-Hsing University for providing experimental equipment and technical support.

Conflicts of Interest: The authors declare no conflict of interest.

References

1. An Official Website of the European Union. Available online: https://ec.europa.eu/environment/index_en (accessed on 24 April 2021).
2. Resource Recycling Act. Available online: http://law.epa.gov.tw/zh-tw/laws/962396701.html (accessed on 24 April 2021).
3. *229R-13 Report on Controlled Low-Strength Materials*; ACI Committee 229; American Concrete Institute: Farmington Hills, MI, USA, 2013.
4. *Study on the Application of Controlled Low-Strength Materials in Civil. Engineering*; Construction and Planning Agency, Ministry of the Interior: Taiwan, 2002.
5. Adaska, W.S. Controlled low strength materials. *Concr. Int.* **1997**, *19*, 41–43.
6. Funston, J.J.; Krell, W.C.; Zimmer, F.V. Flowable fly ash, a new cement stabilized backfill. *Civ. Eng. ASCE* **1984**, *4*, 22–26.
7. *The Design and Application of Controlled Low Strength Material (Flowable Fill)*; ASTM STP 1331; Howard, A.K.; Hitch, A.K.J.L. (Eds.) ASTM: West Conshohocken, PA, USA, 1998; ASTM STP 1331.
8. Graziani, A.; Giovannelli, G. *Materiali da Costruzione i Lapidei Struttura del Settore e Tendenze Innovative*; Centro Studi Osservatorio Fillea Grandi Imprese e Lavoro: Roma, Italy, 2015.
9. Chen, P.L. Reuse of Stone Sludge in Control Low Strength Material. Master's Thesis, Department of Civil Engineering & Environmental Resource Management, Dahan Institute of Technology, Taiwan, 2014.
10. Martínez-García, R.; de Rojas, M.I.S.; Morán de Pozo, J.M.; Fraile-Fernández, F.J.; Juan-Valdés, A. Evaluation of Mechanical Characteristics of Cement Mortar with Fine Recycled Concrete Aggregates (FRCA). *Sustainability* **2021**, *13*, 414. [CrossRef]
11. López Boadella, Í.; López Gayarre, F.; Suárez González, J.; Gómez-Soberón, J.M.; López-Colina Pérez, C.; Serrano López, M.; de Brito, J. The Influence of Granite Cutting Waste on The Properties of Ultra-High Performance Concrete. *Materials* **2019**, *12*, 634. [CrossRef] [PubMed]
12. Zamora-Castro, S.A.; Salgado-Estrada, R.; Sandoval-Herazo, L.C.; Melendez-Armenta, R.A.; Manzano-Huerta, E.; Yelmi-Carrillo, E.; Herrera-May, A.L. Sustainable Development of Concrete through Aggregates and Innovative Materials: A Review. *Appl. Sci.* **2021**, *11*, 629. [CrossRef]
13. Chen, H.J.; Yang, Y.; Tang, C.W.; Wang, S.Y. Producing synthetic lightweight aggregates from reservoir sediments. *Constr. Build. Mater.* **2012**, *28*, 387–394. [CrossRef]
14. Chen, H.J.; Hsueh, Y.C.; Peng, C.F.; Tang, C.W. Paper Sludge Reuse in Lightweight Aggregates Manufacturing. *Materials* **2016**, *9*, 876. [CrossRef]
15. Chen, H.J.; Chang, S.N.; Tang, C.W. Application of the Taguchi Method for Optimizing the Process Parameters of Producing Lightweight Aggregates by Incorporating Tile Grinding Sludge with Reservoir Sediments. *Materials* **2017**, *10*, 1294. [CrossRef]
16. Naik, T.R.; Kraus, R.N.; Siddique, R. Controlled low-strength materials containing mixtures of coal ash and new pozzolanic materials. *ACI Mater. J.* **2003**, *100*, 208–215.
17. Lachemi, M.; Hossain, K.M.A.; Shehata, M.; Thaha, W. Controlled low strength materials incorporating cement kiln dust from various sources. *Cem. Concr. Compos.* **2008**, *30*, 381–392. [CrossRef]
18. Shon, C.S.; Mukhopadhyay, A.K.; Saylak, D.; Zollinger, D.G.; Mejeoumov, G.G. Potential use of stockpiled circulating fluidized bed combustion ashes in controlled low strength material (CLSM) mixture. *Constr. Build. Mater.* **2010**, *24*, 839–847. [CrossRef]
19. Lin, W.T.; Weng, T.L.; Cheng, A.; Chao, S.J.; Hsu, H.M. Properties of Controlled Low Strength Material with Circulating Fluidized Bed Combustion Ash and Recycled Aggregates. *Materials* **2018**, *11*, 715. [CrossRef] [PubMed]
20. Kuo, W.T.; Gao, Z.C. Engineering Properties of Controlled Low-Strength Materials Containing Bottom Ash of Municipal Solid Waste Incinerator and Water Filter Silt. *Appl. Sci.* **2018**, *8*, 1377. [CrossRef]

21. Wu, H.; Huang, B.; Shu, X.; Yin, J. Utilization of solid wastes/byproducts from paper mills in Controlled Low Strength Material (CLSM). *Constr. Build. Mater.* **2016**, *118*, 155–163. [CrossRef]
22. Park, S.M.; Lee, N.K.; Lee, H.K. Circulating fluidized bed combustion ash as controlled low-strength material (CLSM) by alkaline activation. *Constr. Build. Mater.* **2017**, *156*, 728–738. [CrossRef]
23. Jang, J.G.; Park, S.M.; Chung, S.; Ahn, J.W.; Kim, H.K. Utilization of circulating fluidized bed combustion ash in producing controlled low-strength materials with cement or sodium carbonate as activator. *Constr. Build. Mater.* **2018**, *159*, 642–651. [CrossRef]
24. Kuo, W.T.; Wang, H.Y.; Shu, C.Y.; Su, D.S. Engineering properties of controlled low-strength materials containing waste oyster shells. *Constr. Build. Mater.* **2013**, *46*, 128–133. [CrossRef]
25. Mneina, A.; Solimanb, A.M.; Ahmed, A.; El Naggar, M.H. Engineering properties of controlled low-strength materials containing treated oil sand waste. *Constr. Build. Mater.* **2018**, *159*, 277–285. [CrossRef]
26. Wang, L.; Zou, F.; Fang, X.; Tsang, D.C.W.; Poon, C.S.; Leng, Z.; Baek, K. A novel type of controlled low strength material derived from alum sludge and green materials. *Constr. Build. Mater.* **2018**, *165*, 792–800. [CrossRef]
27. Fang, X.; Wang, L.; Poon, C.S.; Baek, K.; Tsang, D.C.W.; Kwokc, S.K. Transforming waterworks sludge into controlled low-strength material: Bench-scale optimization and field test validation. *J. Environ. Manag.* **2019**, *232*, 254–263. [CrossRef]
28. Tang, C.W.; Cheng, C.K. Partial Replacement of Fine Aggregate using Water Purification Sludge in Producing CLSM. *Sustainability* **2019**, *11*, 1351. [CrossRef]
29. Hung, C.C.; Wang, C.C.; Wang, H.Y. Establishment of the Controlled Low-Strength Desulfurization Slag Prediction Model for Compressive Strength and Surface Resistivity. *Appl. Sci.* **2020**, *10*, 5674. [CrossRef]
30. Park, J.; Hong, G. Strength Characteristics of Controlled Low-Strength Materials with Waste Paper Sludge Ash (WPSA) for Prevention of Sewage Pipe Damage. *Materials* **2020**, *13*, 4238. [CrossRef] [PubMed]
31. El-Hinnawi, E.; Abayazeed, S.D. Characterization of Dimension Stone Sawing Sludge in Egypt. *J. Appl. Sci.* **2011**, *11*. [CrossRef]
32. *Method of Slump Test for Concrete*; Chinese National Standard (CNS) 1176; The Bureau of Standards, Metrology and Inspection (BSMI); The Ministry of Economic Affairs (MOEA): Taipei, Taiwan, 2003.
33. *Method of test for Slump Flow of High Flowing Concrete*; Chinese National Standard (CNS) 14842; The Bureau of Standards, Metrology and Inspection (BSMI); The Ministry of Economic Affairs (MOEA): Taipei, Taiwan, 2004.
34. *Standard Test Method for Time of Setting of Concrete Mixtures by Penetration Resistance*; ASTM C403/C403M-16; ASTM International: West Conshohocken, PA, USA, 2016; Available online: https://www.astm.org/Standards/C403 (accessed on 1 April 2021).
35. *Standard Test. Method for Density (Unit Weight), Yield, Cement Content, and Air Content (Gravimetric) of Controlled Low-Strength Material (CLSM)*; ASTM D6023-16; ASTM International: West Conshohocken, PA, USA, 2016; Available online: https://www.astm.org/Standards/D6023.htm (accessed on 1 April 2021).
36. *Method of Test for Compressive Strength of Cylindrical Concrete Specimens*; Chinese National Standard (CNS) 1232; The Bureau of Standards, Metrology and Inspection (BSMI); The Ministry of Economic Affairs (MOEA): Taipei, Taiwan, 2002.
37. Taguchi, G. *Introduction to Quality Engineering: Designing Quality into Products and Processes*; Asian Productivity Organization: Tokyo, Japan, 1987.
38. Neville, A.M. *Properties of Concrete*; Longman: Harlow, UK, 1994.
39. Tang, C.W. Properties of fired bricks by incorporating TFT-LCD waste glass powder with reservoir sediments. *Sustainability* **2018**, *10*, 2503. [CrossRef]
40. Kaliyavaradhan, S.K.; Ling, T.C.; Guo, M.Z.; Mo, K.H. Waste resources recycling in controlled low-strength material (CLSM): A critical review on plastic properties. *J. Environ. Manag.* **2019**, *241*, 383–396. [CrossRef]
41. Trejo, D.; Folliard, K.J.; Du, L. Sustainable Development Using Controlled Low-Strength Material. In Proceedings of the International Workshop on Sustainable Development and Concrete Technology, Beijing, China, 20–21 May 2004; pp. 231–250.
42. Do, T.M.; Kim, Y.S. Engineering properties of controlled low strength material (CLSM) incorporating red mud. *Geo-Engineering* **2016**, *7*, 1–17. [CrossRef]
43. Razak, H.A.; Naganathan, S.; Hamid, S.N.A. Performance appraisal of industrial waste incineration bottom ash as controlled low-strength material. *J. Hazard. Mater.* **2009**, *172*, 862–867. [CrossRef]
44. Türkel, S. Strength properties of fly ash based controlled low strength materials. *J. Hazard. Mater.* **2007**, *147*, 1015–1019. [CrossRef]
45. Taha, R.A.; Alnuaimi, A.S.; Al-Jabr, K.S.; Al-Harthy, A.S. Evaluation of controlled low strength materials containing industrial by-products. *Build. Environ.* **2007**, *42*, 3366–3372. [CrossRef]

 *sustainability*

Article

Classifying the Level of Bid Price Volatility Based on Machine Learning with Parameters from Bid Documents as Risk Factors

YeEun Jang, JeongWook Son and June-Seong Yi *

Department of Architectural & Urban Systems Engineering, Ewha Womans University, Seoul 03760, Korea; jyee@ewha.ac.kr (Y.J.); jwson@ewha.ac.kr (J.S.)
* Correspondence: jsyi@ewha.ac.kr; Tel.: +82-2-3277-4454

Abstract: The purpose of this study is to classify the bid price volatility level with machine learning and parameters from bid documents as risk factors. To this end, we studied project-oriented risk factors affecting the bid price and pre-bid clarification document as the uncertainty of bid documents through preliminary research. The authors collected Caltrans's bid summary and pre-bid clarification document from 2011–2018 as data samples. To train the classification model, the data were preprocessed to create a final dataset of 269 projects consisting of input and output parameters. The projects in which the bid inquiries were not resolved in the pre-bid clarification had higher bid averages and bid ranges than the risk-resolved projects. Besides this, regarding the two classification models with neural network (NN) algorithms, Model 2, which included the uncertainty in the bid documents as a parameter, predicted the bid average risk and bid range risk more accurately (52.5% and 72.5%, respectively) than Model 1 (26.4% and 23.3%, respectively). The accuracy of Model 2 was verified with 40 verification test datasets.

Keywords: risk management; risk analysis; bid price volatility; uncertainty in bid documents; pre-bid clarification document; machine learning (ML), classification model; public project; sustainable project management

Citation: Jang, Y.; Son, J.; Yi, J.-S. Classifying the Level of Bid Price Volatility Based on Machine Learning with Parameters from Bid Documents as Risk Factors. *Sustainability* **2021**, *13*, 3886. https://doi.org/10.3390/su13073886

Academic Editor: Sunkuk Kim

Received: 28 February 2021
Accepted: 25 March 2021
Published: 1 April 2021

Publisher's Note: MDPI stays neutral with regard to jurisdictional claims in published maps and institutional affiliations.

1. Introduction

Sustainability refers to the whole life cycle from siting to design, construction, operation, maintenance, renovation, and deconstruction [1,2]. Traditional research focused on the design and construction stages to maximize profits has gradually expanded to the entire life cycle of construction projects to realize sustainable development. Accordingly, many researchers have conducted valuable studies to minimize the impact on the environment by improving the energy performance of buildings and reducing waste. As a result, many studies on sustainability have developed remarkably around maintenance and subsequent steps. Recently, this trend has been further expanded to realize the results of many studies conducted so far [3]. Therefore, many studies have refocused on project management, which corresponds to the preceding stages in terms of sustainability [4]. The success or failure of a construction project starts from the initial stage of the project. More precisely, the feasibility at the bid and contract phases, stipulating plans for the future, enables the completion of a sustainable project.

Contracts for construction projects are created based on competitive bids. In general, the bidder who offers the lowest price is selected as the final winner. Therefore, determining the final price is crucial for bidders [5]. It is also difficult because the bid price affects the likelihood of gaining a satisfactory profit and winning the project [6]. The client provides a bid document to the bidders, who then examine it to estimate the bid price. Thus, the bid document plays an essential role in determining the bid price. If the content of the bid document is uncertain, the intention of the construction object may be ambiguous and cause mistakes during the construction phase, which may lead to construction rework and claims [7,8]. Thus, bidders include the cost in the bid price to cover those risks. In general, the bid price can be expressed as follows (Equation (1)):

$$B_i = C_i(1 + M_i),$$ (1)

C_i represents the construction cost and M_i is the markup (i.e., contingency), which means the risk cost due to uncertainty in the bid document [9]. In other words, if the risk cost increases owing to uncertainty in the bid documents, the bid price increases [10].

The uncertainty factor in bid documents that causes the risk cost must be determined and investigated. However, because reviewing all bid documents in a limited time frame is difficult [11], businesses often rely on their experience rather than quantitative uncertainty measurements [12–15]. Because uncertainty in bid documents is affected by complex factors that are difficult to measure quantitatively, most qualitative research studies have been conducted in academia. Therefore, determining the risk cost remains a tough challenge [16–20]. Bid prices that are not adequately set negatively affect bidders, clients, and users alike. Bidders take the risk to a severe degree, which not only does not yield the expected return but can lead to more serious financial difficulties. Simultaneously, such a bid price may increase the cost of completion due to frequent design changes during project execution, increasing the burden on the client. Eventually, the project quality completed by this process could be worse, causing great inconvenience to users.

Although reviewing all documents may be challenging, pre-bid clarification documents contain much more uncertain information than other documents. This document type includes inquiries and answers from bidders and clients about the uncertainty factors in the bid documents; this information can be used as an input parameter for a machine learning-based model to construct a bid price. This study aims to examine whether the uncertainty measured in pre-bid clarification documents affects the bid price. This uncertainty may change the mean value or variance of the bid price. In this paper, these two changes are operatively defined as "bid price volatility."

In this study, a sample of data from the California Department of Transportation (Caltrans) in the US was used to see how uncertainty in the tender document changes bid price volatility. Analyzing the uncertainty of the bidding document is very difficult. In particular, the volume of bidding documents is enormous because of construction project size, making analysis difficult. Crucially, construction project data has an unstructured text format, making quantitative analysis even more difficult. However, the authors solved this problem using the pre-bid clarification document, which inquired about this uncertainty as a proxy, and used it with the bid summary. This study suggests that the uncertainty of the bid document affects the change in the bid price volatility. This allows bidders to execute the project at a reasonable price between earning a profit and winning the project. Further, this reasonable price can improve the project performance and realize the client's satisfaction. More ultimately, it can extend sustainability in terms of the life cycle of the project.

2. Preliminary Research: Project Risk in Bid Phase and Uncertainty in Bid Documents

2.1. Definition of Project Risk in Bid Phase

The definition of risk depends on the subject and purpose in a field. Because risk is a concept defined to quantify the uncertainty regarding danger, it differs from the latter; it is defined as the "possibility of loss or injury" in dictionaries. In academia, the risk is more clearly defined as a factor or condition that can cause loss or injury owing to uncertainty; this definition focuses more on the possibility of risk rather than the risk itself, typically expressed in the following equation [21]:

$$Risk\ Magnitude = Risk\ Severity \times Risk\ Likelihood,$$ (2)

Risk magnitude is one of several attempts to measure risk [16]. It is a useful indicator for determining priorities among various risk factors using "risk" and "uncertainty" as variables. However, this also acts as a limitation because relative comparisons between risk factors are possible, but absolute comparisons are impossible. Therefore, the quantitative relationship with the bidding price resulting from the risk in the actual bidding process is blurred. This suggests that a new indicator that can reflect the risk of bidding price is necessary for at least construction projects, and there have been many studies related to this. Abotaleb and El-Adaway [9] attempted to measure bidding risk as a percentage of markups. Besides the total construction cost, bidders present the total construction cost plus a specific rate as the bid price for pursuing profit while preparing for risks. In addition, a study was conducted to determine whether the successful bid price was a price that had more risk than necessary by using the contrast between the successful bid price and the average bid price [22]. Lee et al. [23] attempted to measure the bid risk by using an equation similar to the equation of Williams [22], but in which the engineer's estimate replaced the successful bid price. However, the previously suggested equations have limitations in that they are challenging to use in this study in the following aspects. First, it is a matter of the possibility of utilizing the markups. It is correct that the contingency is included in the price, but a third party such as researchers other than bidders cannot check from the bid history. This is because the contingency is included in one or more of several bid items of the bill of quantities (BOQ). Second, the successful bid price and the average bid price are values determined after the bidding date. It is difficult to predict similar projects' risks during the bid phase using these values.

In this study, the risk is defined as the quantitative uncertainty regarding risk, and the risk factor represents a factor that causes uncertainty regarding time, cost, and quality risk. We use two metrics that match our definition of risk in Section 4.1.2. The scope of this study covers construction projects, and the project risk corresponds to the uncertainty regarding risks that arise from the characteristics of the construction project. The bid varies according to the project delivery method in actual construction projects. In this study, the bid phase is considered as the period in which the construction bid is made with the design–bid–build method. Uncertainty in a bid document is one of the many project risk factors in the bid phase.

2.2. Project Risk Factors Affecting the Bid Price

Several researchers have studied project risk factors that affect the bid price. Construction projects can be classified into several types depending on the case, and any project type can include risk. Therefore, researchers have analyzed the risk without considering the project type. In this study, risk factors that affect the bid price are extracted from 13 reviews in the field of transportation (Table 1).

The above studies are of great significance in that they have substantially advanced the critical risk identification stage in risk management. Many studies have extracted common factors as considerations when bidding for projects. Existing studies have facilitated more detailed risk management by deriving or breaking down the priorities of risks to be considered when performing projects based on surveys of most experts. On the other hand, some studies have analyzed how the number of bidders affects the bid price and predicts the bid price through simulation using multiple variables instead of one variable. However, there is a limitation in not considering how the project risk is integrated into the project's initial bid price.

Table 1. Project Risk Factors Affecting Bid Price According to Previous Studies.

No.	Author	Year	Risk Factors
1	Ahmad & Minkarah [24]	1988	Degree of difficulty; type of project; client; location; design document quality; size of job; competition; contingency; duration
2	Shash [25]	1993	Number of competitors with bidding experience in such projects; client identity; contract conditions; project type; project size; bid method
3	Chua & Li [26]	2000	Degree of technological difficulty; identity of client/consultant; size of project; completeness of drawings and specifications; consultants' interpretation of the specification; project timescale and penalty for non-completion; time for bid preparation
4	Wanous et al. [27]	2000	Financial capability of client; relationship to and reputation of client; project size; availability of time for bidding; site clearance of obstructions; project duration
5	Han & Diekmann [28]	2001	Geography and climate conditions of country; document issues and contract conditions; project cost uncertainty; project schedule uncertainty
6	Egemen & Mohamed [18]	2007	Project size (total project value); current financial capability of client; type of work; technological difficulty of project
7	Zeng et al. [21]	2007	Communication; layout and space; site constraints; work scheduling; condition
8	Bageis & Fortune [29]	2009	Financial capacity of client; contract conditions; experience with similar projects; size of contract
9	Chan & Au [19]	2009	Poor employer's reputation to honor payment on time; very tight contract period; onerous contract conditions and rigid specifications
10	El-Mashaleh [17]	2012	Project type; project size (contract price); quality of bid documents (e.g., drawings, and specifications); client's reputation
11	Leśniak & Plebankiewicz [30]	2013	Type of work; contract documents; client's reputation; project value; project duration; criteria of bid selection; location of project; time window for bid preparation; degree of complexity of works
12	Ahiaga-Dagbui & Smith [31]	2014	Bid strategy; site information; ground conditions; type of soil; information; scope of project
13	Delaney [32]	2018	Project size; project type; level of competition; contract document quality; project-specific items; unforeseen conditions

2.3. Uncertainty in Bid Document as a Project Risk Factor

Uncertainty in bid documents is one of the most crucial risk factors. The bid phase is the first stage of a project contract. The bidder submits the bid price after reviewing the extensive bid documents, which contain information on the following three aspects: (1) the bid procedure (e.g., the announcement, guide, participation application, participation notice, and bid), (2) contract (e.g., the general and special conditions), and (3) construction (e.g., the drawings, specification, and pre-bid clarification document). Each bid document has a different scope and form. For example, the specification document contains a set of documented requirements and the drawings that present the building requirements. The special conditions are contract clauses that apply only to the project subject to the contract; they are created by changing, adding, or deleting existing content in the General Conditions section. In other words, the bid documents present standards and procedures regarding the design, construction method, materials, and inspection for the completion of the construction object; thus, the bid documents constitute the basis for calculating the bid price. In addition, because bid documents are contract documents, they are the basis for judgment when legal problems arise in the future. Cost overrun can occur if the bidder fails to review the bid document's risk factors in advance [23]. Therefore, analyzing the uncertainty in bid documents is crucial.

Discrepancies, errors, and omissions cause uncertainty in bid documents; these are the leading causes of legal adjustment, arbitration, or litigation regarding the project costs. According to Tanaka [33], 74.4% of construction-related claims in the United States are due to uncertainty in bid documents; Erdis and Ozdemir [34] studied the dispute between a client and bidder, arguing that uncertain expressions in a bid document could lead to construction disputes.

Public projects in the US include a pre-bid clarification procedure that can resolve all uncertainty in bid documents before the bid. If bidders find uncertainty in a bid document during the quotation, they can contact the client, who must respond within the deadline. Relieving all uncertainty in bid documents through this approach helps bidders present the correct bid price [35]. New Work State in the US emphasized that the pre-bid clarification is a significant procedure for the client and bidder [36]. The former can calculate the project cost more accurately with less uncertainty. Pre-bid clarification is an institutional method that helps present accurate project costs and prevents possible future design changes, extensions, additional construction costs, and disputes [37].

2.4. Pre-Bid Clarification Document as a Proxy for Uncertainty in Bid Documents

Uncertainty in bid documents includes (1) unclear communication caused by discrepancies, errors, or the omission of information or (2) unclear requirements regarding the project object. In general, the bid process for a construction project involves many bid documents [23]. Each document may independently contain risk factors; besides this, they may interact with each other, creating risk factors. Hence, all bid documents must be carefully reviewed to determine the uncertainty level. However, it is complicated for bidders to identify all hidden risks within a short bid preparation time [16–20,23]. Therefore, in the actual field and academia, the uncertainty of bid documents has been considered a complex problem to solve [11,38] and risk beyond control [28].

The uncertainty that arises in the pre-bid clarification procedure is caused by factors, which occur in all the bid documents that the bidders read. These documents are incorporated into the pre-bid clarification document, which can serve as a proxy variable that gauges the entire bid document's uncertainty. For example, Daoud and Allouche [39] analyzed pre-bid clarification documents to examine which uncertainty factors occur in the bid documents of construction projects.

2.5. Hypothesis Development

From the literature review, there is a widely believed proposition: uncertain things during the bid phase affect bid price on the theoretical plane. However, the problem is that the factors classified as risks are mixed with what can be measured and what is not, what can be controlled and what is not possible, making quantitative analysis impossible. For this reason, when practitioners calculate prices, these uncertainties are guessed and reflected in prices without a factual basis. We made the following two assumptions to establish the hypothesis: (1) As the uncertainty increases, the bidders will reflect this in their prices, causing an increase in the overall average bid price. (2) The greater the uncertainty, the more significant the difference in prices offered by bidders will also increase, resulting in an increase in the range of bidding prices formed. Under these assumptions, we set up the following two hypotheses on the empirical plane.

Hypothesis 1 (H$_1$). *Factors derived from bid summary and pre-bid clarification document affect $F_1(x)$, representing the volatility of bid price.*

Hypothesis 2 (H$_2$). *Factors derived from bid summary and pre-bid clarification document affect $F_2(x)$, representing the volatility of bid price.*

In H_1 and H_2, the factors consist of seven independent parameters obtained from the bid summary and pre-bid clarification document. Then, $F_1(x)$ of H_1 becomes Bid Average Risk (Equation (3)), and $F_2(x)$ of H_2 becomes Bid Range Risk (Equation (4)) discussed in Section 4.

2.6. Research Gaps and Research Questions

According to the Project Management Body of Knowledge [40], risk management research is based on (1) risk identification, (2) risk assessment, and (3) risk plan and control. The risk assessment, which is a leading step in risk planning and control, quantifies the potential impact of these uncertain factors [11]. However, many variables to be considered and interrelated make the analysis in the actual field and research studies difficult [14,24].

The fundamental reason is risk identification (which is a leading step); the general approach is to subdivide all project risk factors into controllable units based on specific criteria. Analyzing segmented risks can reduce uncertainty in the bid phase [37]. However, this approach has not been both quantitatively and qualitatively studied for risk factors in bid documents [23] because they contain vast amounts of information and differ in content depending on the project. Therefore, uncertainty in bid documents has been classified as an uncontrollable risk [28]. As mentioned in Section 2.2, researchers have only progressed to Level 1 by suggesting that uncertainty in bid documents is a risk factor; there are no sufficient specific studies on Level 2 [23]. In other words, published management studies have mainly focused on high-level risk factors and surveys with expert groups [11,38]. However, these data [41] only serve as references for determining the bid price in the bid phase.

The following factors must be investigated: first, regarding the social background, a reasonable bid price is crucial for establishing a reasonable project budget for the bidder and client [20]. This requires a decision support tool that can be used by practitioners who encounter difficulties in the bid price prediction. In research, a more quantitative study based on actual bid data is required to assess whether uncertainty in bid documents affects bid prices. This study aims to meet both academic and practical needs by analyzing whether uncertainty in bid documents affects the bid price.

Uncertainty in a bid document is expected to have the following effects on the bid price. First, each bidder will represent this risk factor in his/her bid price, thereby increasing the project's overall bid price (i.e., the bid price average). Besides this, the other bidders represent this risk factor in prices, which increases the range of the established bid price bands. In this study, these two X are defined as "bid average risk" and "bid range risk," respectively. Further, the bid price volatility comprises the two types of bid price fluctuations due to uncertainty in a bid document (i.e., the increase in the bid price average and range). This study aims to provide answers to the following two questions:

- *Research Question 1*: Does the uncertainty in the pre-bid clarification document increase the bid average risk and bid range risk?
- *Research Question 2*: Does the parameter of uncertainty in bid documents improve the classification of the bid price volatility?

The results of this study are two types of bid price volatility level classification models:

- *Model 1*: level classification model without uncertainty in the bid documents;
- *Model 2*: level classification model with uncertainty in the bid documents.

In this study, the performance of Model 2 is evaluated to support decision-making about bid prices.

3. Materials and Methods: Modeling Approach

Regarding risk management, this study on assessment is different from risk plan and control, which supports decision-making on participation in the bid phase. This is because the decision-making process regarding bid prices of bidders who have already decided to participate is supported in this study. In general, risk assessment studies can be classified into studies of B_i and studies of M_i (Equation (1)). Because it is difficult to collect and analyze sufficient data, mainly M_i has been studied; in this study, B_i is empirically evaluated based on the actual bid results. In addition, in other published studies, the uncertainty factors of bid documents were analyzed with a proxy (i.e., a pre-bid clarification document).

Uncertainty factors in bid documents are natural phenomena because construction projects are typically one-off projects. Considering the toxin clause that partially exists in the special conditions, an uncertainty factor in bid documents is problematic because the artificial content clearly defines who should be responsible in certain circumstances.

Therefore, the uncertainty factors in bid documents considered in this study are limited to those that occur naturally because of specific characteristics of the construction industry.

3.1. Materials

The data from the bid results regarding the project risk factors discussed in Section 2.2 are the variables of interest. To study their effects, the data in which the influence of other factors can be minimized should be analyzed. The public construction project of Caltrans meets this purpose because of the following reasons: first, the uncertainty in the bid documents can be analyzed. Because Caltrans includes a pre-bid clarification process in the bid phase, the pre-bid clarification document is publicly available. Second, the quantity and quality of available project data are sufficient. Caltrans invests approximately $1.7 billion per year in approximately 450 projects, which is the largest of the 50 US states. The thousands of standardized project datasets of Caltrans have led to large amounts of high-quality data and excellent project management capabilities based on experience. Third, the absence of special conditions reduces influences from other than the variables of interest. Standard contracts used worldwide include the FIDIC (Fédération Internationale Des Ingénieurs-Conseils), JCT (Joint Contracts Tribunal), NEC (New Engineering Contract), and AIA (American Institute of Architects), mainly applied to private projects. By contrast, Caltrans uses federal-aid construction contracts (FHWA-1273) for public projects. In this case, only general conditions without special conditions (unlike private projects) are applied, which means that the projects are relatively standardized. The bid document of a standardized project reduces the influences of numerous external factors; it is considered suitable for observing the effect of uncertainty in bid documents on the bid price because of the absence of special conditions.

Caltrans has published all the bid results online since 2004 (they provide all bid documents, bid summaries, and important information). However, the online services for pre-bid clarification documents have been operated since 2011. The number of projects since the access date was 3584 during 2011–2018. In total, 3578 datasets were collected (six cases were excluded because they could not be accessed owing to system errors).

3.2. Methods

Pre-Data Analysis (Data Preprocessing Based on Bid Summary and Pre-Bid Clarification Document): in this step, information that can be obtained from the bid result is preprocessed into input parameters (IPs) and output parameters (OPs). Caltrans has published a bid summary containing the critical details of the bid results. In this study, the data related to project risk factors affecting the bid price (which was discussed in Section 2.2) are extracted from the bid summary (B. S.) and pre-bid clarification document (P. C. D.). Subsequently, the final dataset is constructed from the raw data.

Data Analysis (Two Classification Models of Bid Price Volatility Based on Machine Learning): Methods of analyzing data can be classified into several categories depending on the purpose of the study and the characteristics of the data. When analyzing data that is large and composed of various factors, techniques such as data mining through machine learning (ML) are mainly used, and the data mining method is actively applied in recent risk analysis studies [42]. Such data mining can be classified mainly into a prediction technique that derives a regression equation, such as statistical analysis, and a classification technique that determines the category of data. Therefore, this study uses a machine learning-based data mining classification technique as a data analysis technique to classify the level of bidding risk with a large amount of data composed of various variables. In this study, machine learning can be used to classify the OPs in data consisting of multiple IPs. In this study, the class of the OP is designated such that the model algorithm learns to classify the levels of bid price volatility with MATLAB. To evaluate the model performance during training and validate it through validation tests in a post-data analysis, the pre-data analysis's final dataset is classified into training and validation data (for the training validation and validation test, respectively). As a result, Model 1 (which does not include the uncertainty in the bid documents in the IPs) and Model 2 (which includes the uncertainty in the bid documents in the IPs) are generated.

Post-Data Analysis (Validation): Models 1 and 2 are tested in a validation test to determine whether the models created in the data analysis step show similar performance characteristics for new data other than the data used for training. The test results are presented in a confusion matrix, which is analyzed and discussed.

4. Model for Classification of Level of Bid Price Volatility

4.1. Pre-Data Analysis: Preprocessing of Data from Bid Summary and Pre-Bid Clarification Document

4.1.1. Input Parameters

Because of the wide range of types and the number of bid documents, Caltrans's bid summary contains significant information about the bid. The pre-bid clarification document contains uncertain details of risk on the bid documents. Based on the factors discussed in Section 2.2, highly relevant parameters to the bid price are extracted from these two documents (Table 2).

Table 2. Metadata of Input Parameters.

Risk Type	No.	Input Parameter	Relevant Risk Factors in Previous Studies	Origin	Type	Unit	Coding
Time	IP-1	Working Days	Duration (No. 1, 2, 4, 7, & 11); project schedule uncertainty (No. 5)	B. S.	Num.	Days	47–1530
Time	IP-2	Project Location	Location of the project (No. 1 & 12)	B. S.	Categ.	County	1–12
Cost	IP-3	Engineer's Estimate	Size of project (No. 1 [1], 2, 3, 4, 7, 10, 12, & 13); size of contract (No. 8); value of project (No. 11)	B. S. [2]	Num. [4]	$	10,000,000–280,000,000
Cost	IP-4	Bid Preparation Days	Available time for bid preparation (No. 3, 4, 9, & 12)	B. S.	Num.	Day	18–237
Cost	IP-5	Number of Bidders	Competition (No. 1 & 13); the number of competitors with bidding experience in such projects (No. 2)	B. S.	Num.	#of Bidders	2–12
Quality	IP-6	Project Type	Project type (No. 1, 2, 10, 11, & 13)	B. S.	Categ. [5]	Road or Bridge	1–2
Quality	IP-7	Uncertainty in Bid Documents	Document quality (No. 1, 3, 5, 8, 9, 10, 11 & 13); communication (No. 7)	P. C. D. [3]	Num.	# of U. B. I. [6]	2–59

[1] Numbers in "Relevant Risk Factors … " match numbers in Table 1; [2] B. S.: bid summary; [3] P. C. D.: pre-bid clarification document; [4] Num.: numeric. [5] Categ.: categorical; [6] U. B. I.: unsolved bid inquiries.

Meanwhile, the information extracted from the bid summary requires preprocessing to be used as a model parameter. Further, text data must be standardized through nominalization, and numeric data must be filtered based on a chosen range such that the outliers do not affect the model performance. The pre-bid clarification document contains the inquiries and answers for the project (Table 3).

Most inquiries aim to accurately estimate the bid prices by resolving uncertainty; because some inquiries do not, they must be preprocessed to include only those related to uncertainty. If these inquiries can be resolved with appropriate answers, they are excluded. The following describes the seven input parameters presented in Table 2 in detail for each risk type.

Table 3. Examples of Inquiry and Answer in Pre-Bid Clarification Document.

No.	Inquiry	Answer
1	(Inquiry #6) *Please check the shoulder backing quantity.* *Please check the tack coat quantity.* *Please check the HMA quantity.*	(Response #1): Unsolved *Bid per current contract documents.*
2	(Inquiry #36) *Bolted Connections and Faying Surfaces:* *1. Are all A325 and A490 HS bolted connections considered slip critical or bearing connections? If so, provide reference in plans and specs.*	(Response #4): Solved *1. Refer to Addendum No 4, dated 13 February 2018. Section 55–1.02E(6)(c) of the Special Provisions has been modified to define the HS bolted joint type.*

Time Risk

IP-1 (Working Days): IP-1 represents the period of completion of the project required by the client. If IP-1 is relatively short considering the size of the project, the bid price may increase owing to required rush or night work. The projects considered in this study have IP-1 values between 47 and 1530.

IP-2 (Project Location): the IP-2 of the raw data is an address close to the construction site. IP-2 is related to the local price index, affecting the bid price. In the US, the price index is generally determined at the state level; however, differences in prices can occur within a single state. Because California is a large state in the US, Caltrans divides its administration into 12 districts separate from their counties. Accordingly, in this study, IP-2 is coded as 1–12 according to the district.

Cost Risk

IP-3 (Engineer's Estimate): the project cost to which bidders can refer for the bid price is IP-3 at the time of the announcement; because the raw data have a too wide IP-1 distribution, the range must be adjusted. In this study, projects in the range of $10,000,000–$280,000,000 are used in the model.

IP-4 (Bid Preparation Days): IP-4 is when bidders have to review the bid document (including the uncertainties); thus, this time affects the accuracy of the bid price. IP-4 is calculated as the period from the bid announcement date to the bid opening date extracted from the raw data (values between 18 and 237).

IP-5 (Number of Bidders): to use IP-5 as an input parameter, it must be checked whether the information is known before the bid opening. Researchers have argued that the variable IP-5 influences the bid price [24,25,32]. When it increases, the bidders deliberately lower the bid price to win the project [43,44]. Thus, the bidders are aware of the number of competitors in advance; Christodoulou [43] studied the optimal M_i (Equation (1)) based on this premise. Therefore, IP-5 is included as an input parameter with values from 2 to 12.

Quality Risk

IP-6 (Project Type): IP-6 can be mainly classified into roads (e.g., highways, freeways, or roadways) and bridges; the numbers "1" and "2" represent a road and bridge, respectively.

IP-7 (Uncertainty of Bid Documents): 3578 raw datasets are screened through Section 4.1.1, which result in 269 final datasets with 6682 bid inquiries. As mentioned in Section 2.4, there are two uncertainty factors in bid documents: unclear communication (BI. 1–3) and unclear requirements (BI. 4–5). Uncertain communication includes discrepancies, errors, and the omission of information in the bid document, each of which has overlapping meanings. For example, omission means that necessary information is missing owing to an error; thus, it can be interpreted as an error itself. Therefore, each term is clearly classified according to the mutually exclusive and collectively exhaustive principle. When certain identical information in various bidding documents causes conflicts, the case corresponds to case BI. 1: discrepancy. The case in which being inquired by an error of single information itself is categorized as case BI. 2: error. Uncertainty due to the lack of specific information is classified as case BI. 3: omission. Furthermore, uncertain requirements are classified into two types that ask for insufficient but non-essential information (BI. 4: insufficient information) or accept alternatives to the existing guidelines (BI. 5: alternative information). That is, only inquiries corresponding to BI. 1-5 among the pre-bid clarification document content are regarded as uncertainty factors. Through this process, 52 bid inquiries are excluded. After excluding the questions, the uncertainty of which has been resolved with appropriate answers (4336), the number of unsolved bid inquiries is 1994 with values between 2 and 59 for each project (i.e., the IP-7).

All coded IPs are used to train the model in the normalized form.

4.1.2. Output Parameters

As stated in Section 2.6, the output parameters of the models are the bid average risk (OP-1) and bid range risk (OP-2), which are based on the bid price of the raw data. The bid average risk is the ratio of the average price to the engineer's estimate (Equation (3)):

$$Bid\ Average\ Risk = \frac{Average\ Bid\ Price}{Engineer's\ Estimate} = F_1(x). \tag{3}$$

For example, if the engineer's estimates of projects A and B are $10 billion and the respective average bid prices are $10 billion and $13 billion, the bid average risks are 1.0 and 1.3, respectively. Thus, it can be assumed that the bidders expect a higher risk for project B. Moreover, the bid range risk (OP-2) refers to the difference between the maximal and minimal bid price concerning the engineer's estimate (Equation (4)):

$$Bid\ Range\ Risk = \frac{Max.\ Bid\ Price - Min.\ Bid\ Price}{Engineer's\ Estimate} = F_2(x). \tag{4}$$

For projects A and B with estimates of $10 billion and $1 billion, respectively, the differences between the maximal and minimal bid prices would be identical ($2 billion); however, the differences between the uncertainties of the two projects cannot be considered identical: project B is riskier than A. Finally, as mentioned in Section 2, the bid average risk and bid range risk defined in this section act as $F_1(x)$ of H_1 and $F_2(x)$ of H_2.

4.1.3. Impact of IP-7 on Bid Price Volatility

The raw data of 3578 projects are preprocessed to create the final dataset of 269 projects. To answer research question 1 in Section 2.6, the final dataset should be classified into groups with and without uncertainty factors. In this study, projects with IP-7 values between 2 and 59 are considered a group with uncertainty factors, and projects with IP-7 values between 0 and 1 are considered a group without uncertainty factors. The OP-1 and OP-2 in each group are presented in Figure 1 and Table 4.

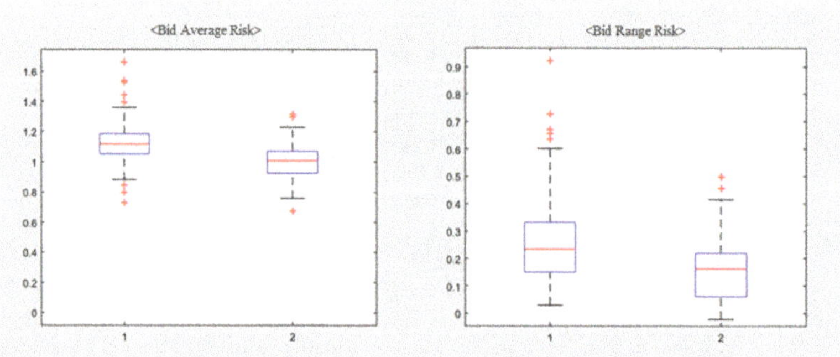

Figure 1. Comparison between OP-1 and OP-2 of Projects with IP-7 Values of (1) 2–59 and (2) 0–1.

Table 4. Comparison between Data Statistics of OP-1 and OP-2 of Projects 1 and 2.

No.	Output Parameter	Data Statistics							
		Project 1				Project 2			
		Med.	Max.	Min.	Mean	Med.	Max.	Min.	Mean
OP-1	Bid Average Risk	1.1172	1.6612	0.7295	1.1303	0.9759	1.2787	0.6482	0.8653
OP-2	Bid Range Risk	0.2339	0.9222	0.0297	0.2684	0.2004	0.5338	0.0174	0.1761

As shown in Figure 1 and Table 4, projects with uncertainty (Project 1) score higher values for the bid average risk and bid range risk than projects without uncertainty (Project 2). In other words, uncertainty in bid documents increases the bid price volatility.

4.2. Data Analysis: Two Classification Models Based on Machine Learning for Bid Price Volatility
4.2.1. Design

In the data analysis stage, the final data in Section 4.1 is used to implement a model for classifying the level of volatility in bid prices. This study presupposes that the uncertainty of the bid document is related to the bid price and ultimately tries to improve the accuracy of the bidding price volatility level classification model by adding the variable of the uncertainty of the bid document to the existing bid-related variables.

SPSS, MATLAB, R, and Python are mainly used for machine learning-based data mining classification. In this study, data analysis and model development were performed using Mathworks' MATLAB software as a data analysis tool.

Because this model predicts the classes of OP-1 and 2 with input parameters, the models are trained through supervised learning after the class designation. A suitable number of classes is significant because too many classes decrease the reliability of the prediction results; too few classes make the interpretation of the results difficult. In this study, four levels according to the bid price volatility distribution are considered.

When the boundary between the classes is set, a natural breakpoint is preferable; the breakpoint is the point at which the distribution of the OP values suddenly breaks. If there is no natural breakpoint, a boundary should be set such that the data of each class is evenly distributed to ensure reliability. Figure 2 shows the distributions of OP-1 and OP-2 of the 269 final projects; the parameters do not have natural breakpoints. The set boundaries between the classes of the models are presented in Table 5. The OP-1 and OP-2 have a total of 4 classes: OP-1 is a "−" class that is much smaller than 0, a slightly smaller class is "-," a slightly larger class is "+," and a much larger class is "++." On the other hand, classes of OP-2 were named with the symbols "+," "++," "+++," and "++++" in the order of close to 0.

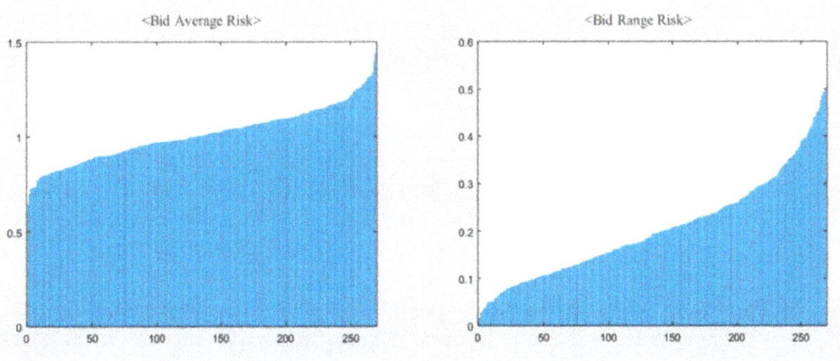

Figure 2. Distribution of Bid Price Volatility of 269 Projects.

Table 5. Class Designation for Output Parameters.

No.	Output Parameter		Class Designation			
		Class	−	-	+	++
OP-1	Bid Average Risk	Range	(min., 0.9)	(0.9, 1.0)	(1.0, 1.1)	(1.1, max.)
		Ratio	22%	26%	27%	25%
		Class	+	++	+++	++++
OP-2	Bid Range Risk	Range	(min., 0.27)	(0.2, 0.27)	(0.27, 0.31)	(0.31, max.)
		Ratio	25%	23%	24%	28%

Not all final datasets are used to train the model. The remaining datasets are used to check whether the classification model performs consistently for new data (typically 30% of the total data). Half these data is used for the validation during training, and the rest is applied in the validation tests (Section 4.3). Accordingly, 269 datasets are allocated to 40 for the training validation, 40 for the validation test, and 189 for the model training with machine learning. There are several machine learning algorithms for training classification models; in this study, neural net (NN), which shows good performance, is used. In addition, three algorithms are combined to evaluate the results derived with neural net. In total, four classification algorithms are used for the models: NN, decision tree (Tree), support vector machine (SVM), and K-nearest neighbor (KNN).

4.2.2. Results

In the bid price volatility level classification based on the model design in Section 4.2.1, Model 1 is trained without IP-7, whereas Model 2 is trained with IP-7. The performance of the model is expressed as the accuracy (i.e., the number of correct classifications compared to the total number of classifications):

$$Accuracy\ (\%) = \frac{Number\ of\ Correct\ Classifications}{Total\ Number\ of\ Classifications}. \tag{5}$$

In this section, the accuracies of Model 1 trained without IP-7 and Model 2 trained with IP-7 are compared to answer research question 2 in Section 2.6 (Table 6).

Table 6. Accuracies of Classification Models.

Algorithm	Accuracy (%)			
	Model 1		Model 2	
	Bid Average Risk	Bid Range Risk	Bid Average Risk	Bid Range Risk
NN	37.5	42.5	63.9	65.8
Tree	31.5	34.3	61.7	60.6
SVM	34.6	40.5	59.5	60.5
KNN	32.8	36.1	58.4	56.1
Average	34.1	38.4	60.9	60.8

According to Table 6, Model 2 performs better than Model 1 for all algorithms, including the NN. Second, the classification Models 1 and 2 result in higher accuracies for OP-2 than for OP-1 in most cases. Third, the NN exhibits the best performance in Models 1 and 2.

Model 1 results in accuracies with the order of SVM > KNN > Tree, and Model 2 results in accuracies with the order of Tree > SVM > KNN. Thus, IP-7 (a parameter of the uncertainty in bid documents) improves the performance of the model that classifies the level of the bid price volatility. Regarding the NN algorithm classifier, the performance scores of Model 1 (without IP-7) are 37.5% (OP-1) and 42.5% (OP-2), whereas those of Model 2 (with IP-7) are 63.9% (OP-1) and 65.8% (OP-2); these results are 26.4% and 23.3% higher than those of Model 1, respectively. This trend is identical for the averages of all algorithms: the accuracies of Model 1 are 34.1% (OP-1) and 38.4% (OP-2), while those of Model 2 are 60.9% (OP-1) and 60.8 % (OP-2).

4.3. Post-Data Analysis: Validation

In the validation test, the accuracy of a classification model is evaluated with data that have not been used for training. In this section, the training performance of Model 2 with IP-7 is validated based on the NN. There are various approaches for validating a machine learning model (e.g., the N-fold cross-validation, bootstrap, and sliding window methods); N-fold cross-validation is the most used approach. The results of the validation test are presented in a confusion matrix, which enables the determination of the accuracy of the model and true positive rate (*TPR*, which represents the recall and sensitivity):

$$True\ Positive\ Rate\ (TPR) = \frac{TP\ (True\ Positive)}{TP\ (True\ Positive) + FN\ (False\ Negative)}, \tag{6}$$

where *"True Positive"* (*TP*) refers to samples in which the positive cases are correctly classified, whereas *"False Negative"* (*FN*) refers to samples in which the negative cases are incorrectly classified. In other words, the TPR is the ratio of correct samples to total samples classified by the model. Figure 3 presents the validation results of Model 2 trained with the NN for the classification of the bid price volatility.

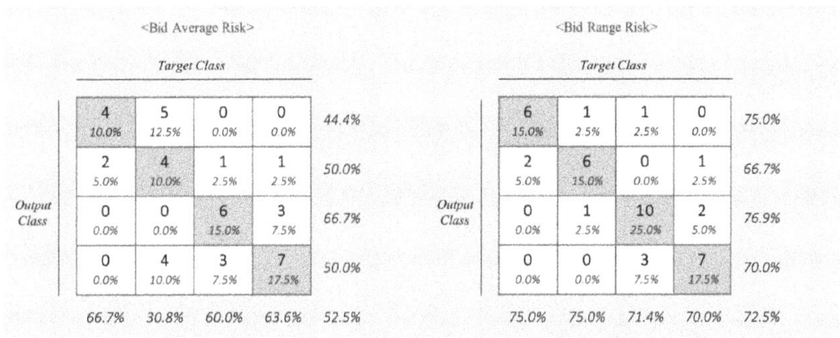

Figure 3. Confusion Matrices of Model 2: Bid Average Risk and Bid Range Risk.

The validation accuracies of Model 2 for OP-1 and OP-2 are 52.5% and 72.5%, respectively; thus, the model correctly classifies the bid price volatility levels of 21 and 29 out of 40 projects. First, the TPR of Model 2 for OP-1 is highest in the "+" class (66.7%) and lowest in the "−"class (44.4%). However, according to the confusion matrix, incorrectly classified samples are classified into groups of similar rather than entirely different classes. This is because successive values are cut by the artificial boundaries of the class designation. Moreover, the accuracy of the model for the test data (52.5%) is slightly lower than the accuracy of the model for the training data (69.3%). Second, the TPR of Model 2 for OP-2 is highest in the "+++" class (76.9%) and lowest in the "++" class (66.7%). For OP-1, the model classifies most of the samples into same or similar classes. In addition, the performance of the model for the validation test data is 72.5%, which is better than the accuracy of the model for the training data (64.6%).

5. Discussion

From a background in Introduction and Research Gaps, a decision support tool for bid price budgeting for the actual field and research studies was created based on quantitative data analyses with project-based bid results from other studies. Based on these requirements, two research questions were established, and Caltrans's bid results were used to select related parameters based on previous studies. Based on the 269 final project datasets obtained after preprocessing, the uncertainty level in bid documents was quantified as the number of inquiries corresponding to BI. 1-5 in the pre-bid clarification document that have not been resolved.

To answer research question 1, the 269 projects were classified into two groups: a group with IP-7 values of 0–1 and another group with IP-7 values of 2–59. The group with uncertainty in the bid documents generally had a higher bid average risk and bid range risk than the group without uncertainty. Thus, uncertainty in bid documents increases the bid price volatility. It is noteworthy that the project with an IP-7 value of 1 was also considered a project without uncertainty in this study because the considered uncertainty level depends on the stakeholders' opinion. For example, for projects with IP-7 values of 1, some bidders may believe that there is little uncertainty, whereas others may believe that the uncertainty level is relatively high. Therefore, the boundary of uncertainty can be expressed as follows: {project without uncertainty: IP-7 value of $x \,|\, x = 0$} $\cup$ {project with uncertainty: IP-7 value of $x \,|\, x > 0$} for risk-averse bidders and as {project without uncertainty: IP-7 value of $x \,|\, 0 \leq x \leq n$, where $n = 1, \dots, 59$} $\cup$ {project with uncertainty: IP-7 value of $x \,|\, x > n$} for risk-takers. In other words, for projects with IP-7 values of n, the level of perceived uncertainty depends on the stakeholder; the case $n = 1$ was presented as an example in this study.

Furthermore, Models 1 and 2 were trained with an NN to examine whether IP-7 affects the bid price volatility classification. The accuracies of Model 1 for the bid average risk and bid range risk were 37.5% and 42.5%, respectively; those of Model 2 were 63.9% and 65.8%, respectively.

The accuracies of Model 1 exceeded 25%, which corresponds to the mathematical probability when one of four classes is randomly selected; nevertheless, the accuracies were too low for decision making. This means that parameters that are not included in the model have a more substantial influence on the bid price. By contrast, the accuracies of Model 2 were better, and similar phenomena were observed for the three other algorithms and NN. In other words, the influence of IP-7 is relatively strong compared to those of the other parameters included in the model. However, the fact that the accuracy of the models has stopped at the current level proves that there are still other parameters that are not included in the models and that significantly affect the bid price. However, this effect can be attributed to the nature of IP-7 itself; IP-7 represents the number of unsolved bid inquiries; this surrogate endpoint is set because it is impossible to cost the situation mentioned in each inquiry. For example, two inquiries might be worth $100 and $10,000, respectively; however, they are treated equally as a value of 1 in the models. This may be why the models score higher accuracies for the bid range risk than for the bid average risk. From a superficial point of view, this implies that the model's application may be limited due to its accuracy. However, the more time it takes to consider accurately the cost of each inquiry, the less time it takes to identify more risks, which in turn increases the uncertainty of the bidding document. In this respect, obtaining such a relatively high accuracy only with the number of unsolved bid inquiries is a good sign.

Moreover, the results were validated with test data, and the answer to research question 2 was found. Bidders should determine the bid price by integrating project-related information. Before this, the bidders should add the cost of project uncertainty in the markups along with Ci (Equation (1)), which is calculated based on construction budgeting. Because "uncertainty" literally means "lack of knowledge," there is no raw data for calculating it; thus, Model 2 can be applied. For example, if the project's bid average risk is classified as "−" the average of the bid price is likely to be much lower than the engineers' estimate. If the currently calculated Ci is high, Mi must be lowered, or Ci must be adjusted to win the bid. If the results are below the lower bound of the expected profit, the bidder may stop participating in the project. Likewise, if the bid range risk is classified as "+++," many bidders may present prices that deviate much from the engineers' estimate; thus, the bidder may use this strategy to adjust Ci and Mi.

Since there are too many uncertainties affecting the bid price of a construction project and many of them are risks, researchers have performed many valuable studies, and practitioners have been through trial and error. However, when a specific point in time and space is determined, many risks are eventually settled. For example, a city's price index itself fluctuates over time, but finally, the city's price index is determined as one at the time and place of the project. So, for bidders, one dilemma between winning bids and project profits is, therefore, whether they should call more or less than expected. These fine-grained adjustments are ultimately determined by project-intrinsic-risks that have not been finalized. The bidding document is a kind of contract document, and all the explicit contents contained in it are directly related to the price. However, there is a problem that uncertain content cannot be dealt with one-to-one with the price. Moreover, since the contents of the bidding document are all different for each project, it was impossible to solve this with the existing linear method. On the other hand, the authors measured the uncertainty of each bid document, not the content of the bid document, and analyzed the risk through the NN algorithm. This study provided an answer to how much the uncertainty in the contents of bid documents increases or decreases the already expected price.

Nevertheless, the method presented in this study has several limitations, which must be considered in future research studies. First, the data are limited. Caltrans's project data were used to minimize the influence of external factors; however, owing to the characteristics of standardized data, the impact of uncertainty in bid documents (which is one of the internal factors) may be relatively weak. In this respect, it might be difficult to generalize the results of this study. Another limitation is related to the preprocessing of IP-7. Because the pre-bid clarification document contains unstructured text data, which is challenging to be computationally processed, there is no automated method for quantifying it; thus, the authors manually analyzed 6682 datasets. Because this process can introduce human error, the authors mutually verified the results. Third, there are obvious limitations in that this paper does not address all of these factors and only deals with the risk of uncertainty in the bid document. The risk of fluctuations in the material and workforce market is a significant factor affecting the bid amount and must be considered in the bidding stage. The bidders scrutinize these through market research. Meanwhile, the risk of fluctuations in the material and labor market varies with time and location, which means they are variables. If they are determined, the risk of volatility could be determined, too. Since the bidders bid simultaneously for work performed at the same location, these costs will reach some agreed value. However, this cannot be confirmed as a single value, so it remains a risk of the bid price. Future studies have to further consider the remaining major fluctuation risks and use them as parameters. In that case, significant improvements are expected to improve model accuracy.

6. Conclusions

In this study, a classification model for the bid price volatility level was developed by analyzing the relationship between the uncertainty in bid documents and bid price based on bid data. The model results in this study reveal that uncertainty in bid documents is causing bid prices to rise or fall more than necessary. Therefore, it is essential to conduct a thorough review of items causing uncertainty before bidding. We present these items as discrepancy, error, omission, insufficient information, and alternative information. Besides this, the inclusion of a pre-bid clarification process allows the price of the project to converge more appropriately if the remaining uncertainties are eliminated at the time of bidding. This study has the following contributions: the first step in quantifying the uncertainty in bid documents. The results of published qualitative studies of risk identification were evaluated with bid result data in this quantitative study. The model proposed in this paper enables risk management at a lower level: in the new approach, the uncertainty in bid documents considered to have an uncontrollable risk is analyzed with a pre-bid clarification document. Through this, the theoretical gaps are closed. The results of this study can help bidders to determine the bid price. According to the results, the pre-bid clarification in the bid phase is an essential process because the resolution of uncertainty in the bid documents can reduce the bid average risk and bid range risk. Accordingly, it is expected that the bid price of the project, the risk of which has been resolved during the pre-bid clarification process, will converge to a more acceptable price. The construction objectives created with this price will improve bidders' profitability and meet the client's expectations, which ensures a successful construction project. The final contribution of this study is that the concept of sustainability has been further expanded in construction projects. In construction, the idea of sustainability is meaningful in that it extends the project management, which focused on the design and construction stages, to the entire life cycle. Until now, researchers have carried out valuable studies related to sustainable energy use, noting the importance of maintenance and operation phase after completion. However, the early stages of the project are also of great importance. This study tried to ensure the success and sustainability of the project through research on a reasonable project price. In the future, the authors will establish a method that comprehensively analyzes the uncertainty in unstructured text data from public projects of various institutions that provide pre-bid clarification documents for the automatic extraction of the IP-7 content. Further, the authors will combine this with the results of this study to establish a more general and highly accurate risk classification model.

Author Contributions: Conceptualization, Y.J. and J.-S.Y.; methodology, Y.J. and J.S.; software, Y.J.; validation, Y.J., J.-S.Y. and J.S.; formal analysis, Y.J.; investigation, Y.J.; resources, Y.J.; data curation, Y.J. and J.-S.Y.; writing—original draft preparation, Y.J., J.-S.Y. and J.S.; writing—review and editing, Y.J., J.-S.Y. and J.S.; visualization, Y.J.; supervision, J.-S.Y. and J.S.; project administration, J.-S.Y. and J.S; funding acquisition, J.-S.Y. All authors have read and agreed to the published version of the manuscript.

Funding: This work is supported by the Korea Agency for Infrastructure Technology Advancement (KAIA) grant funded by the Ministry of Land, Infrastructure and Transport (Grant 21ORPS-B158109-02). This study is also supported by the Ewha Womans University Scholarship of 2019.

Institutional Review Board Statement: Not applicable.

Informed Consent Statement: Not applicable.

Data Availability Statement: Caltrans's raw data (2011–2018) used in this study can be accessed through the following link: http://ppmoe.dot.ca.gov/des/oe/project-bucket.php (accessed on 27 March 2021).

Acknowledgments: The authors would like to thank Lee, Assistant Professor of Department of Civil & Environmental Engineering and Construction of University of Nevada, Las Vegas, for helpful advice on various technical issues examined in this paper.

Conflicts of Interest: The authors declare no conflict of interest.

References

1. Tomczak, M.; Jaśkowski, P. New approach to improve general contractor Crew's work continuity in repetitive construction projects. *J. Constr. Eng. Manag.* **2020**, *146*, 04020043. [CrossRef]
2. Dasović, B.; Galić, M.; Klanšek, U. A survey on integration of optimization and project management tools for sustainable construction scheduling. *Sustainability* **2020**, *12*, 3405. [CrossRef]
3. Czarnecki, L.; Kaproń, M. Definiowanie zrównoważonego budownictwa. Cz. 2. *Mater. Bud.* **2010**, nr2, 46–47.
4. Hermarij, J. *Better Practices of Project Management based on IPMA competences*; Van Haren: Hertogenbosch, The Netherlands, 2013.
5. KPMG International. *Global Construction Survey 2015*; Climbing the curve; KPMG International: Amstelveen, The Netherlands, 2015.
6. Ishii, N.; Takano, Y.; Muraki, M. A bidding price decision process in consideration of cost estimation accuracy and deficit order probability for engineer-to-order manufacturing. In *Technical Report No. 2011-1*; Tokyo Institute of Technology: Tokyo, Japan, 2011.
7. Doloi, H. Cost overruns and failure in project management: Understanding the roles of key stakeholders in construction projects. *J. Constr. Eng. Manag.* **2012**, *139*, 267–279. [CrossRef]
8. Maemura, Y.; Kim, E.; Ozawa, K. Root Causes of Recurring Contractual Conflicts in International Construction Projects: Five Case Studies from Vietnam. *J. Constr. Eng. Manag.* **2018**, *144*, 05018008. [CrossRef]
9. Abotaleb, I.S.; El-Adaway, I.H. Construction bidding markup estimation using a decision theory approach. *J. Constr. Eng. Manag.* **2016**, *143*, 04016079. [CrossRef]
10. Miller, R.; Lessard, D.R. *Evolving Strategy: Risk Management and the Shaping of Mega-Projects*; Edward Elgar: Cheltenham, UK, 2008.

11. Dikmen, I.; Budayan, C.; Talat Birgonul, M.; Hayat, E. Effects of Risk Attitude and Controllability Assumption on Risk Ratings: Observational Study on International Construction Project Risk Assessment. *J. Manag. Eng.* **2018**, *34*, 04018037. [CrossRef]
12. Messner, J.I. An Information Framework for Evaluating International Construction Projects. Ph.D. Thesis, Pensylvania State University, University Park, PA, USA, 1995.
13. Fayek, A.; Ghoshal, I.; AbouRizk, S. A survey of the bidding practices of Canadian civil engineering construction contractors. *Can. J. Civ. Eng.* **1999**, *26*, 13–25. [CrossRef]
14. Chua, D.K.H.; Li, D.Z.; Chan, W.T. Case-based reasoning approach in bid decision making. *J. Constr. Eng. Manag.* **2001**, *127*, 35–45. [CrossRef]
15. El-Mashaleh, M.S. Decision to bid or not to bid: A data envelopment analysis approach. *Can. J. Civ. Eng.* **2010**, *37*, 37–44. [CrossRef]
16. An, M.; Baker, C.; Zeng, J. A fuzzy-logic-based approach to qualitative risk modeling in the construction process. *World J. Eng.* **2005**, *2*, 1–12.
17. El-Mashaleh, M.S. Empirical framework for making the bid/no-bid decision. *J. Manag. Eng.* **2012**, *29*, 200–205. [CrossRef]
18. Egemen, M.; Mohamed, A.N. A framework for contractors to reach strategically correct bid/no bid and mark-up size decisions. *Build. Environ.* **2007**, *42*, 1373–1385. [CrossRef]
19. Chan, E.H.; Au, M.C. Factors influencing building contractors' pricing for time-related risks in bids. *J. Constr. Eng. Manag.* **2009**, *135*, 135–145. [CrossRef]
20. Williams, T.P.; Gong, J. Predicting construction cost overruns using text mining, numerical data and ensemble classifiers. *Autom. Constr.* **2014**, *43*, 23–29. [CrossRef]
21. Zeng, J.; An, M.; Smith, N.J. Application of a fuzzy based decision making methodology to construction project risk assessment. *Int. J. Proj. Manag.* **2007**, *25*, 589–600. [CrossRef]
22. Williams, T.P. Bidding ratios to predict highway project costs. *Eng. Constr. Archit. Manag.* **2005**, *12*, 38–51. [CrossRef]
23. Lee, J.; Yi, J.S. Predicting Project's Uncertainty Risk in the Bid Process by Integrating Unstructured Text Data and Structured Numerical Data Using Text Mining. *Appl. Sci.* **2017**, *7*, 1141. [CrossRef]
24. Ahmad, I.; Minkarah, I. Questionnaire survey on bid in construction. *J. Manag. Eng.* **1988**, *4*, 229–243. [CrossRef]
25. Shash, A.A. Factors considered in tendering decisions by top UK contractors. *Constr. Manag. Econ.* **1993**, *11*, 111–118. [CrossRef]
26. Chua, D.K.H.; Li, D. Key factors in bid reasoning model. *J. Constr. Eng. Manag.* **2000**, *126*, 349–357. [CrossRef]
27. Wanous, M.; Boussabaine, A.A.; Lewis, J. To bid or not to bid: A parametric solution. *Constr. Manag. Econ.* **2000**, *18*, 457–466. [CrossRef]
28. Han, S.H.; Diekmann, J.E. Approaches for making risk-based go/no-go decision for international projects. *J. Constr. Eng. Manag.* **2001**, *127*, 300–308. [CrossRef]
29. Bagies, A.; Fortune, C. Bid/no-bid decision modelling for construction projects. In Proceedings of the 22nd Annual ARCOM Conference, Birmingham, UK, 4–6 September 2006.
30. Leśniak, A.; Plebankiewicz, E. Modeling the decision-making process concerning participation in construction bidding. *J. Manag. Eng.* **2015**, *31*, 04014032. [CrossRef]
31. Ahiaga-Dagbui, D.D.; Smith, S.D. Dealing with construction cost overruns using data mining. *Constr. Manag. Econ.* **2014**, *32*, 682–694. [CrossRef]
32. Delaney, J.W. The Effect of Competition on Bid Quality and Final Results on State DOT Projects. Ph.D. Thesis, State University of New York at Buffalo, Buffalo, NY, USA, 2018.
33. Tanaka, T. Analysis of Claims in Us Construction Projects. Ph.D. Dissertation, Massachusetts Institute of Technology, Cambridge, MA, USA, 1988.
34. Erdis, E.; Ozdemir, S.A. Analysis of technical specification-based disputes in construction industry. *KSCE J. Civ. Eng.* **2013**, *17*, 1541–1550. [CrossRef]
35. Trost, S.M.; Oberlender, G.D. Predicting accuracy of early cost estimates using factor analysis and multivariate regression. *J. Constr. Eng. Manag.* **2003**, *129*, 198–204. [CrossRef]
36. OGS. Pre-Bid Inquiry & Response Policy. Available online: https://online.ogs.ny.gov/dnc/contractorconsultant/esb/prebidinquiryresponsepolicy.asp (accessed on 27 November 2020).
37. Duzkale, A.K.; Lucko, G. Exposing uncertainty in bid preparation of steel construction cost estimating: II. Comparative analysis and quantitative CIVIL classification. *J. Constr. Eng. Manag.* **2016**, *142*, 04016050. [CrossRef]
38. Baloi, D.; Price, A.D. Modelling global risk factors affecting construction cost performance. *Int. J. Proj. Manag.* **2003**, *21*, 261–269. [CrossRef]
39. Daoud, O.E.; Allouche, E.N. Bid queries as a gauge for quality control of documents. In Proceedings of the Canadian Society for Civil Engineering, Moncton, NB, Canada, 4–7 June 2003.
40. Project Management Institute (PMI). *A Guide to the Project Management Body of Knowledge (PMBOK Guide)*, 4th ed.; Project Management Institute: Newtown Square, PA, USA, 2008.
41. Winch, G.M. *Managing Construction Projects*; John Wiley & Sons: Hoboken, NJ, USA, 2010.
42. Katal, A.; Wazid, M.; Goudar, R.H. Big data: Issues, challenges, tools and good practices. In Proceedings of the 2013 Sixth International Conference on Contemporary Computing (IC3), Noida, India, 8–10 August 2013; IEEE: Piscataway, NJ, USA, 2013; pp. 404–409.

43. Christodoulou, S. Optimum bid markup calculation using neurofuzzy systems and multidimensional risk analysis algorithm. *J. Comput. Civ. Eng.* **2004**, *18*, 322–330. [CrossRef]
44. Lo, W.; Lin, C.L.; Yan, M.R. Contractor's opportunistic bid behavior and equilibrium price level in the construction market. *J. Constr. Eng. Manag.* **2007**, *133*, 409–416. [CrossRef]

sustainability

Article

Excavation Method Determination of Earth-Retaining Wall for Sustainable Environment and Economy: Life Cycle Assessment Based on Construction Cases in Korea

Youngman Seol [1], Seungjoo Lee [2], Jong-Young Lee [3], Jung-Geun Han [3,4,*] and Gigwon Hong [5,*]

[1] Department of Civil Engineering, Chung-Ang University, 84 Heukseok-Ro, Dongjak-gu, Seoul 06974, Korea; great@kecgroup.kr

[2] NEO-TRANS Co., Ltd., 33 Daewangpangyo-Ro, 606 Beon-Gil, Bundang-Gu, Seongnam-Si, Gyeonggi-do 13524, Korea; seunglee@doosan.com

[3] School of Civil and Environmental Engineering, Urban Design and Studies, Chung-Ang University, 84 Heukseok-Ro, Dongjak-gu, Seoul 06974, Korea; geoljy@cau.ac.kr

[4] Department of Intelligent Energy and Industry, Chung-Ang University, 84 Heukseok-Ro, Dongjak-gu, Seoul 06974, Korea

[5] Department of Civil and Disaster Prevention Engineering, Halla University, 28 Halladae-gil, Wonju-si, Gangwon-do 26404, Korea

* Correspondence: jghan@cau.ac.kr (J.-G.H.); g.hong@halla.ac.kr (G.H.)

Abstract: This study describes life cycle assessment (LCA) results of the excavation depth and ground condition of a medium-sized excavation ground in order to examine the effect of construction methods on environmental and economic feasibility for an earth-retaining wall. LCA is conducted in consideration of eight environmental impact categories according to the construction stage of the earth-retaining wall. In addition, the environmental cost of construction method for the earth-retaining wall was calculated, and its selection criteria were analyzed based on the calculation results. The evaluation results of the environmental load of construction methods for the earth-retaining wall show that the H-Pile+Earth plate construction method has low economic efficiency because the construction method significantly increased the environmental load due to the increased ecological toxicity. The environmental load characteristics have a greater effect on the selection of construction methods in sandy soil than in composite soil when the excavation depth is the same. The evaluation result of the environmental cost of the construction methods for the earth-retaining wall shows that the environmental cost increased as the excavation depth increased, and the sandy soil conditions have higher environmental costs than complex soil conditions.

Keywords: LCA (life cycle assessment); earth-retaining wall; excavation; environment load; environment cost

Citation: Seol, Y.; Lee, S.; Lee, J.-Y.; Han, J.-G.; Hong, G. Excavation Method Determination of Earth-Retaining Wall for Sustainable Environment and Economy: Life Cycle Assessment Based on Construction Cases in Korea. *Sustainability* **2021**, *13*, 2974. https://doi.org/10.3390/su13052974

Academic Editor: Sunkuk Kim

Received: 21 January 2021
Accepted: 5 March 2021
Published: 9 March 2021

Publisher's Note: MDPI stays neutral with regard to jurisdictional claims in published maps and institutional affiliations.

1. Introduction

LCA, called life cycle assessment or life cycle environmental load assessment, is defined as a technique that identifies life cycle flows, such as raw material and energy input, pollutant occurrence, and recycling in product production, and it assesses potential environmental impacts. That is, it is an evaluation of the environmental impact of the entire process of obtaining raw materials for products, production, application, and disuse, i.e., the entire process from the acquisition of raw materials to the final disposal of the product [1–3]. LCA, an environmental evaluation technique, is actively used as a technology evaluation method to secure source technologies to respond to climate change worldwide. [4–8]. LCA is not limited to assessing greenhouse gas emissions, but they are focused on in the literature review section of this study, because Korea is facing the considerable issue of greenhouse gas emissions in the field of construction.

Large-scale facilities are planned mainly in the construction industry. The application of LCA in this field can sufficiently consider the environmental impact, because there are many types and quantities of materials, and high-energy facilities are applied. In particular, rapid decision support is possible for environmental issues if LCA is performed in the early stages of a project [9,10]. As a result of forecasting greenhouse gas emissions by the industry sector by 2030, Lee [11] predicted that emissions associated with the construction industry will increase by 2.2% by 2030. In 2015, the International Energy Agency (IEA) established a plan to induce and support greenhouse gas reduction activities with the aim of implementing greenhouse gas target management in the construction industry; in Korea, 8.34% and 2.07% reduction targets were established in the building and transportation sectors, respectively. As mentioned above, various studies on environmental impact assessment of greenhouse gases emitted from construction activities have been actively conducted in order to respond to the international situation [12–14].

The Korean construction market is expanding not only to the infrastructure sector but also to the energy and building sectors, mainly in the carbon emission rights market, renewable energy market, and green building market, so it is time to require a more aggressive response and greater investment. Overseas, it is reported that Europe classifies the construction industry as one of the seven major sectors that emit greenhouse gases, and the construction industry accounts for 36% of total industrial carbon emissions and 40% of total energy consumption. It was determined that the cause of these results is closely related to the fuel use of construction equipment and gas emissions due to various construction activities, and studies have been conducted to contribute to reducing the emission of greenhouse gases [15–17].

Research on LCA has actively been carried out abroad for more than two decades. Europe is a leader in the field of LCA research, and many studies have been conducted on methodology, life cycle inventory (LCI) DB (Database) construction, and program development in the field of the environment [18]. Japan is attempting a systematic approach to LCA, and Australia has constructed an LCI DB mainly of infrastructure facilities, such as buildings, raw materials, iron, minerals, and packaging materials. In addition, various case studies have been conducted to evaluate the environmental impact related to greenhouse gases on the foundation work of buildings and residential buildings [19–22]. Moreover, in many advanced countries, evaluation programs that take into account the life cycle of construction materials have been developed and put into use, and they have been set as sustainable development goals to reduce the environmental load in the construction industry [23–25]. Recently, research on LCA has been conducted in various environmental fields in Korea. It has been only 5 years since the study on the field of civil engineering took off in Korea, so the available data related to construction materials and construction are insufficient. Additionally, LCA is partly applied to SOC (social overhead capital) facilities, such as roads, bridges, and tunnels, in which the target facilities are standardized [26,27].

Therefore, this study aims to improve the process by which existing construction methods are selected by additionally applying the results of LCA analysis, such as constructability and economic feasibility, to the way a construction method is selected when considering various soil conditions. To this end, the earth-retaining wall, a representative soil structure, was selected as the target structure, and a case of securing stability through a series of design processes was established for various excavation conditions and construction methods after simplifying the excavation-related ground conditions. Afterward, the environmental loads for the eight major categories in the environmental product declaration (EPD), such as greenhouse gas emissions and energy consumption, which are the main management targets of the Greenhouse Gas and Energy Target Management System, were analyzed and applied to the established case. Based on this analysis, in order to minimize the environmental load when selecting a construction method for an earth-retaining wall, LCA analysis for an earth-retaining wall according to excavation depth and soil conditions was conducted to prepare improvement measures.

2. Theoretical Review of LCA Technique

2.1. Concept of LCA Technique

LCA, also called "life cycle environmental load assessment," is a technique to identify the inputs of raw materials, energy, chemicals, etc. and outputs of wastes, pollutants, recycling, etc. in the life cycle of a product and to evaluate potential environmental impacts (Figure 1).

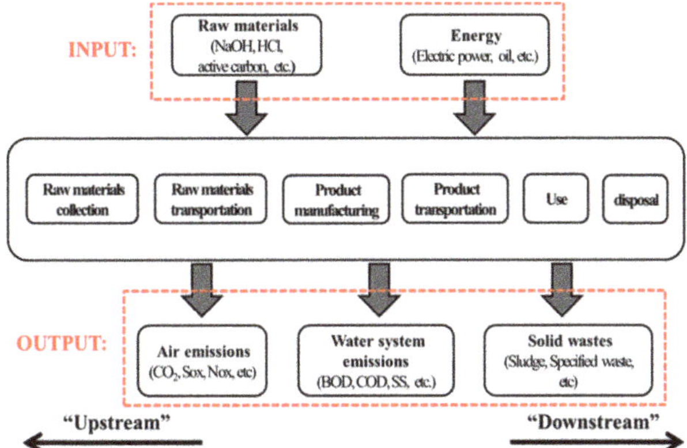

Figure 1. Overall process and input/output of life cycle assessment (LCA).

Raw materials, energy, and utilities are inputs, and air emissions, water system emissions, solid wastes, etc. in the manufacturing process, the use process, and the disposal process are outputs. Early stages of the construction process such as collection and transportation of raw materials are referred to as "upstream," whereas product use and disposal are "downstream."

General guidelines to LCA structures and procedures used to assess environmental performance in a series of processes can be found in ISO standards 14040 and 14044, international standards for environmental management (green management) established by the International Organization for Standardization (ISO) [1,2]. As shown in Figure 2, the LCA consists largely of objectives and scope definitions, inventory analysis (LCI), impact assessment (LCA), and interpretation of results.

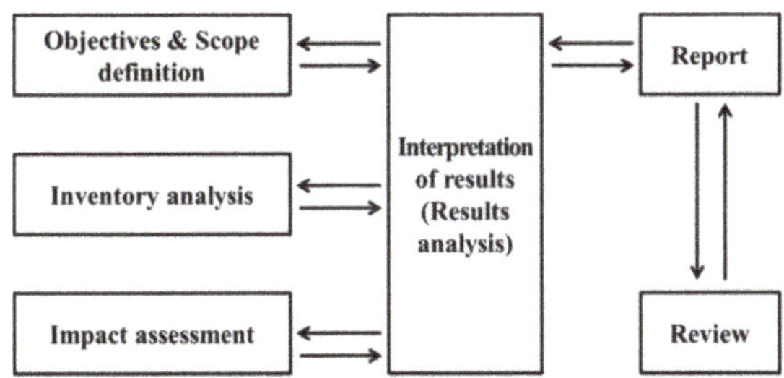

Figure 2. Procedure of life cycle assessment (LCA).

LCA is used to provide a scientific basis for determining which of several processes has a significant environmental impact or which of several products is environmentally friendly. For instance, LCA can be performed to identify which construction method, A or B, has a smaller impact on the environment. This process makes quantitative numerical comparisons possible by collecting data on materials and equipment that are inputted during the construction and maintenance stages of a comparative construction method and by setting inputted material, energy, and resource usage units. LCA has recently been applied to the construction industry internationally to reflect various environmental impact assessments in the planning and design stages, making it possible to design alternatives by taking into account the environmental friendliness, such as the comparison of routes and construction methods. Therefore, as it is necessary to introduce and effectively apply decision-making methods for environmentally friendly development in the construction sector, the LCA, in which environmental performance in terms of construction environment and environmental value through the quantification of environmental load are evaluated, is a significant factor.

2.2. Application of LCA in the Construction Industry

Although there have not yet been many cases in which an LCA evaluation was conducted in Korea, the results of analyzing various cases performed concerning roads/bridges, ports, and railways are as follows: First, the evaluation method was conducted by analyzing material and equipment inputs through information collection and analysis and then evaluating environmental and economic feasibility by calculating the environmental load through LCA evaluation by comparison. In addition, LCA analysis as a comparison method is performed in the application stage of a construction method, and LCA analysis of the basic plan and basic design is conducted after dividing it into the initial construction stage, maintenance stage, and dismantling and disposal stage. In other words, in life cycle cost (LCC) analysis as a comparison method, the environmental loads for eight environmental impact categories (abiotic resource depletion, global warming, ozone depletion, photochemical oxidant creation, acidification, eutrophication, ecotoxicity, and human toxicity) are calculated, and those for key contributors to global warming (carbon dioxide (CO_2), sulfur dioxide (SO_2), nitrogen dioxide (NO_2), and carbon monoxide (CO)) out of all environmental impact categories are calculated and compared to alternatives. After analyzing the effect on the reduction of environmental load of the basic design, reflecting the final LCA evaluation results, with the reduced values of the environmental indices compared to the basic plan, the basic design was presented, which makes environmental economic feasibility or environmentally friendly design possible.

Meanwhile, more work is being done in foreign countries. The Netherlands has been developing LCA evaluation programs for the construction industry since 1994, with work being conducted by major construction-related organizations (e.g., the Ministry of Housing, Spatial Planning, and Environmental), with various types of data now being provided, such as the reliability of LCA. In Finland, LCA of the construction industry is conducted by the VTT Technical Research Centre of Finland; the scope of the LCA is set at each life cycle stage of an individual building, such as material production, transportation, construction, maintenance, and dismantling, and the environmental impact data obtained from these results are used in marketing, product display, system management, and product design [28]. Recently, Han et al. [29] developed a tool that considers cost and environmental impact together by utilizing building information modeling (BIM) based on information and communication technologies (ICTs) to link LCC throughout the life cycle of a building to LCA tools. In order to develop a database that can reflect greenhouse gas reduction, Japan developed an environmental load inventory for individual items by utilizing a method to correct the estimates with inter-industry relational tables based on the detailed DB calculated using the estimation method. By making use of these methods, the environmental load of new materials such as eco-cement to consider the environment can be updated from time to time through the DB, and a basis for conducting evaluations that reflect an environmental load of materials has been prepared. The American Society for Testing and Materials (ASTM) in the United States prepares guidelines for LCA of construction materials, design of green buildings, construction, and operation, and the National Institute of Standards and Technology (NIST) develops Building for Environmental and Economic Sustainability (BEES) to support the selection of economic and environmentally friendly construction materials.

As mentioned above, mainly overseas, evaluation software that considers all aspects of construction materials has been developed and utilized mainly in the construction sector, and various activities have been carried out to reduce the environmental pollution load in the construction sector with the goal of sustainable development. Table 1 shows these research activities by country.

Table 1. Life cycle assessment (LCA) application status by country (Kwon [30]).

Country	Purpose	Project Contents	Research (Managing) Institute
The United States	Decision support for purchasing construction materials with excellent environmental economic feasibility. Developed as part of the US EPA (Environmental Protection Agency) Green Purchasing Program.	Standardize both the LCA as an environmental performance evaluation tool and the LCC as an economic feasibility evaluation tools into ASTM (American Society for Testing and Materials). Development of a methodology and software called BEES (Building for Environmental and Economic Sustainability) to integrate and make a decision.	NIST (National Institute of Standards and Technology) of the United States
Finland	Finding ways to convert construction materials, construction, and construction waste treatment in civil infrastructure projects such as road construction in an environmentally friendly manner.	LCI DB (Life Cycle Inventory DataBase) construction for construction industry. LCA implementation for various construction scenarios. Comparative evaluation by scenario.	Road Corporation VTT
Sweden	Identifying the significance of road maintenance from an LCA perspective.	Identification of environmental impacts throughout the entire process of road construction, maintenance, and disposal, and support of various decision-making processes.	Road Corporation IVL
Netherlands	Identifying environmental impacts on national infrastructure industries, such as sewage facilities, through LCA techniques.	Identifying environmental performance through LCA techniques in constructing various national infrastructures. Support for environmentally friendly design.	Concrete Association Cement Association INTRON BRE
England	Building material certification program	Quantifying the environmental performance of construction materials using LCA techniques.	BRE certification authority
Australia	Transitioning to an environmentally friendly construction industry using LCA technique.	Identifying opportunities for environmental improvement for construction materials and systems through performing LCA.	Ministry of Environment RMIT (Royal Melbourne Institute of Technology)
Japan	Increasing the Recycling Rate of Construction Waste.	Identification of carbon dioxide (CO_2) throughout the entire process of the construction industry	KAJIMA Construction Company

2.3. Application of Similar Techniques for the Selection of Construction Methods on Civil Engineering Structures

Bae [31] suggested a system for selection of construction method by classifying influential factors by applying the analytical hierarchy process (AHP) technique to the selection of construction methods for an underground retaining wall; Han and Lee [32] applied the AHP technique to work conducted by a group of experts in related fields when selecting the reinforcing method for a cut slope. Lee et al. [33] once presented a decision model for selecting soft ground improvement methods using AHP techniques, and Lee and Jeong [34] proposed a decision-making system using the AHP technique and preference function (PF) when selecting the basic construction method for structures.

In order to resolve the inaccuracies intrinsic to the subjective judging process and reduce the uncertainty and ambiguity of the AHP method in bridge construction projects, Pan [35] proposed the fuzzy AHP (FAHP) model by applying triangular and trapezoidal fuzzy numbers and the α-cut concept. Ebrahimian et al. [36] pointed out that application of the existing AHP technique has the drawback that the pairwise comparison required for hierarchy analysis is tedious and time-consuming in the planning phase of a construction project when complex interests are concerned, such as urban construction projects, and suggested a combined model of fuzzy AHP (FAHP) and compromise programming (CP).

Shen et al. [37] introduced text mining case-based reasoning (TM-CBR), which can extract the most similar case from a design by integrating the text mining technique into the CRB system in order to improve the efficiency of decision-making in environmentally friendly design. Lorenz and Jost [38] reported that the system dynamic model is an efficient way to select the best method for a given purpose; Tsai et al. [39] proposed the multiple criteria decision making (MCDM) approach to resolve the impact on the goal of the time, cost and environmental Impacts (TCEI) analysis, the selective issue on how decision-makers determine the most appropriate construction methods.

In order to rationalize selection of construction methods for a retaining wall, Kim et al. [40] used a neural network system to verify the rationality of the selection at approximately 160 sites and showed predictive results of 88% in the selection of a construction method and 90% in the selection of the wall retaining method. Furthermore, the selection of the construction method for a retaining wall has many factors to consider and is based on uncertain information, resulting in frequent design changes and consequent delays in construction and lots of economic loss. To overcome this issue, we highlight the limitation that artificial intelligence (AI) technology is limited to new projects even though it can be used to support complex decision-making processes [40,41]; when selecting tunnel construction methods, Park et al. [42] applied the AHP technique to the existing problems of value engineering (VE), and LCC and proposed the life cycle social cost (LCSC) evaluation method to convert social loss expenditures, which could not be applied in the LCC technique.

However, as mentioned above, in most previous research, several decision-making methods have been adopted to rationally select the construction method for an earth-retaining wall, and most of them suggested only the applicability and rationality of appropriately applied construction methods based on the existing application cases.

That is, in order to select a rational construction method, an evaluation system that considers social loss expenses (environmental factors) and social factors has been used only with improvements. Therefore, there is a limit to using the mechanical relationship among construction methods, soil, material, and environment based on a stability-based design for various soil conditions when selecting construction methods for an existing earth-retaining wall.

Therefore, beyond the selection of a construction method that focuses on the given soil conditions and the usability and stability of the materials in each construction method, a study is needed that addresses how to select a construction method for earth-retaining walls that considers economic feasibility and environmental performance applied the conversion of environmental costs as well as LCA analysis considering environmental performance in the existing method.

3. Selection of Cases and Stability Review for LCA of Earth-Retaining Wall

3.1. Evaluation of Case Selection and Soil Characteristics

In this section, we set the selection criteria with which the construction method for a retaining wall is applied and propose a resultant rational selection method. In this study, rational selection methods are classified by taking excavation size and excavation depth into account based on the "Special Law about Underground Safety Management (2018)" and "Review Guideline on Excavation for Safe Building Construction" of Seoul Metropolitan City, created for special safety management considering the stability of recent ground subsidence. The criterion for excavation depth under the Special Law about Underground Safety Management is 20 m, and the criterion for excavation depth of buildings under architectural design-review in urban areas is 10 m. Therefore, as shown in Table 2, the excavation depth at which the earth-retaining wall was installed was 15 m, the middle value of the two standards. This can be viewed as a criterion considering the fact that various construction methods use a 15 m excavation depth. Additionally, the characteristics of the soil to be installed are mostly distributed from the surface to the topsoil, weathered soil, weathered rock, soft rock, and hard rock, in that order, and the weathered soil is mostly composed of deposits. There is also a composition of the sandy soil layer and soft clay layer on the rock layer of a riverbank or shoreline, and, most commonly, it is to consist of composite stratum (typically weathered soil layers) on the rock layer. Thus, the new construction method can be applied if it is composed of only rock layers, so the general sediment layer consists of the sandy soil layer, soft clay layer (soft clay ground), and the mixed stratum of the sandy soil and soft clay. Therefore, we decided to conduct an analysis based on these soil compositions in this study (Table 3). The applied equipment and the construction management method are different according to the excavation scale and ground conditions in excavation construction. Thus, the excavation scale and the ground conditions were applied as comparative criteria in this study.

Table 2. Excavation conditions and soil conditions in each case.

No	Excavation Area (m²)	Excavation Depth (m)	Soil Conditions	Construction Method for Earth Retaining Wall
Case 01				C.I.P (Cast-In-Placed pile)
Case 02			Composite Soil	S.C.W (Soil Cement Wall)
Case 03				Sheet Pile
Case 04		15 (Shallow Excavation)		H-Pile+Earth Plate
Case 05				C.I.P (Cast-In-Placed pile)
Case 06			Sandy Soil	S.C.W (Soil Cement Wall)
Case 07				Sheet Pile
Case 08				H-Pile + Earth Plate
Case 09	50 m × 50 m (Medium-Scale)		Soft Clay Soil	Sheet Pile
Case 10				C.I.P (Cast-In-Placed pile)
Case 11			Composite Soil	S.C.W (Soil Cement Wall)
Case 12				Sheet Pile
Case 13		40 (Deep Excavation)		H-Pile + Earth Plate
Case 14				C.I.P (Cast-In-Placed pile)
Case 15			Sandy Soil	S.C.W (Soil Cement Wall)
Case 16				Sheet Pile
Case 17				H-Pile+Earth Plate
Case 18			Soft Clay Soil	Sheet Pile

Table 3. Soil properties applied to the case analysis (Excavation area: 50 m $\times$ 50 m).

Excavation Depth (m)	Soil Condition	Depth (m)	γ_t (kN/m^3)	γ_{sat} (kN/m^3)	c (kN/m^2)	φ (deg)	N	Coefficient of Horizontal Subgrade Reaction (kN/m^3)
15	Composite Soil	3	18	19	3	33	20	2200
		8			8	35	25	7200
		13			13	38	30	13,400
		25			35	42	40	18,000
	Sandy Soil	3	18	19	0	33	20	2000
		8				35	25	6000
		13				38	30	12,000
		25				42	40	15,000
	Soft Clay Soil	3	17	18	4	5	4	500
		8			7	10	8	1000
		13			14	15	15	2000
		35			14	15	15	2000
40	Composite Soil	5	18	19	4	33	20	2200
		10			7	35	25	7200
		15			14	38	30	13,400
		60			14	42	40	18,000
	Sandy Soil	5	18	19	0	33	20	2000
		10				35	25	6000
		15				38	30	12,000
		60				42	40	15,000
	Soft Clay Soil	5	17	18	4	4	20	500
		10	17	18	7	8	25	1000
		15	17	18	14	15	30	2000
		60	17	18	15	17	40	2000

The excavation area is medium-sized (50 $\times$ 50 m), and the deepest excavation point (excavation depth: 40 m) was determined to be 40 m, a depth which makes the application of the construction method for a retaining wall clearly distinguished, in consideration of the maximum possible construction depth (less than 50 m allowed).

3.2. Evaluation of Stability in Each Case

The program used in the design case is Midas GeoX V.4.6.0. Earth pressure applied to the retaining wall causes stress and displacement of the structure. The deformation analysis of the retaining wall is generally performed by the elastoplastic analysis, because the stress and displacement of the retaining wall change depending on the excavation stage of ground. Midas GeoX V.4.6.0 allows the elastoplastic analysis considering the excavation stage.

All cases applied to the LCA analysis were assumed to have both internal and external stability at each excavation stage. The assessment of internal stability was conducted by a review of the cross-section of the structure (member), and the structural stability of H-Pile, C.I.P, Sheet Pile, S.C.W, Strut, Wale, etc., which form a wall, was evaluated by construction stage (excavation stage). External stability was evaluated by dividing it into the stability on the earth pressure acting on the retaining wall and the stability on the surrounding ground subsidence, etc. during the excavation stage and final excavation stages. Table 4 summarizes the application method of each item for the evaluation of stability performed for the earth-retaining wall in this study.

Table 4. Method for review of stability by construction method and item.

Classification	Construction Method or Item	Method for Review
Member Sections (Structural Analysis)	H-Pile Sheet Pile C.I.P (Cast-In-Placed pile) S.C.W (Soil Cement Wall)	Review of Bending Safety Review of Shear Review of Axial Force (S.C.W)
	Strut	Review of Applied Load Axial Force against Earth Pressure Axial Force due to Temperature Change Axial Force Applied to the Vertical Load and Auxiliary Reinforcement
	Wale	Buckling Length Section Review
Excavation Face	Embedded Depth	Reviewing after Dividing it into the Final Excavation and Pre-stage of Final Excavation
	Surrounding Subsidence	Final Excavation Stage Review by Caspe (1966) Method
	Boiling (Sandy Soil, Composite Soil)	Final Excavation Stage Terzaghi Critical Hydraulic Gradient
	Heaving (Soft Clay Soil)	Final Excavation Stage Method by Bearing Capacity Formula Terzaghi-Peck (Review by Surcharge Load Strength or Ultimate Bearing Capacity) Bjerrum-O.Eide (Review by Rotational Moment and Resisting Moment)

Figure 3 shows a schematic diagram of the numerical analysis carried out in this study. The underground water level is reflected in the analysis on the premise that it is lowered according to the stage of excavation and lowered to the excavation surface. The review of stability, such as the stability of the embedded unit, the stability of subsidence, and heaving in each case, considered only the impact on excavation depth because it was affected by the increase in stress depending on excavation depth and was independent of the excavation width.

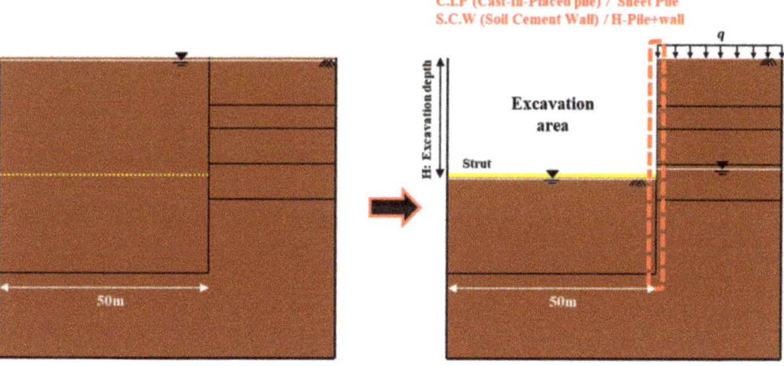

Figure 3. Schematic diagram of the numerical analysis.

Table 5 shows the results of the stability review for each case based on analysis conditions. First, in the stability evaluation of the embedded depth (required safety factor: 1.2) based on the Earth-retaining Wall Design Standard of the Ministry of Land, Infrastructure and Transport in Korea [43] for a shallow excavation depth, the safety factors are in the following orders from high to low: C.I.P, S.C.W, Sheet Pile, and H-Pile+Earth Plate construction method in the composite soil, and C.I.P, H-Pile, S.C.W, and Sheet Pile construction method in the sandy soil. For deep excavations in composite soil, the safety factors are in the following orders from high to low: S.C.W, C.I.P, Sheet Pile, and H-Pile+Earth Plate construction method. For deep excavations in sandy soil, they increase in the following order: C.I.P, S.C.W, Sheet Pile, and H-Pile+Earth Plate construction method. Furthermore, it was confirmed that the deeper the excavation depth, the greater the safety factor in soft clay ground.

Table 5. Stability review results by case.

No		Stage	Safety Factor (=1.2)	Maximum Subsidence around Retaining Wall (m)	Boiling (Safety Factor Criteria = 2.0)		Heaving (Safety Factor Criteria = 1.2)	Soil Condition
					Terzaghi Analysis	Critical Hydraulic Gradient		
Excavation Depth 15 m	Case 01	① ②	9.899 9.010	−0.005	5.400	6.300		Composite Soil
	Case 02	① ②	4.719 6.424	−0.008	5.400	6.300		
	Case 03	① ②	4.719 6.424	−0.008	5.400	6.300		
	Case 04	① ②	4.580 3.820	−0.010	2.700	3.600		
	Case 05	① ②	3.456 3.535	−0.005	5.400	6.300		Sandy Soil
	Case 06	① ②	1.598 2.520	−0.008	5.400	6.300		
	Case 07	① ②	1.598 2.520	−0.010	5.400	6.300		
	Case 08	① ②	3.097 2.893	−0.010	5.400	6.300		
	Case 09	① ②	1.245 2.889	−0.073			2.652	Soft Clay Soil
Excavation Depth 40 m	Case 10	① ②	1.972 8.646	−0.044	5.400	6.300		Composite Soil
	Case 11	① ②	2.499 11.799	−0.039	5.400	9.900		
	Case 12	① ②	1.319 5.368	−0.047	9.000	6.300		
	Case 13	① ②	2.755 4.309	−0.044	5.400	6.300		
	Case 14	① ②	1.284 6.090	−0.067	5.400	11.700		Sandy Soil
	Case 15	① ②	1.346 5.085	−0.048	10.800	15.300		
	Case 16	① ②	1.333 4.510	−0.120	14.400	6.300		
	Case 17	① ②	1.753 2.605	−0.069	5.400	6.300		
	Case 18	① ②	1.696 8.371	−0.256			3.791	Soft Clay Soil

Here, pre-final excavation stage—①, final excavation stage—②.

Caspe [44] estimation of subsidence on the soil was based on a method redefined by Bowles [45], which is relatively consistent with actual data. However, this method has the premise that the displacement (subsidence) due to an increase in effective stress caused by a drop in groundwater level should be calculated separately. As input data for analysis, lateral displacement of the wall by depth, excavation depth, excavation width, and shear resistance angle are required, and for lateral displacement of the wall, computerized analysis data using the beam on elasto-plastic foundation analysis were used.

The deeper the excavation depth, the larger the maximum subsidence, and subsidence occurred more in sandy soil than in composite ground. In addition, in composite soil, when the excavation depth is shallow, the H-Pile+Earth Plate construction method produces the largest amount of subsidence, but the deeper the excavation depth, the greater the subsidence in the Sheet Pile construction method. When the excavation depth is shallow in sandy soil, the Sheet Pile and the H-Pile+Earth Plate construction methods have the largest subsidence, and the C.I.P construction method has the smallest subsidence. When the excavation depth is deep, the Sheet Pile construction method has the largest subsidence, and the S.C.W. construction method has the smallest one. Meanwhile, in soft clay ground, the deeper the excavation depth, the more rapidly the subsidence increases. This result is based on the design of the retaining wall structure with secured stability, so only a very small amount of subsidence occurs; only the tendency of the occurrence of subsidence was analyzed.

Boiling on the bottom of an excavation is generally assessed to increase the safety factor as excavation depth increases, and at this time, the safety factor applied to the boiling judgment was 2.0 [43]. When the excavation depth is shallow, in composite soil, the H-Pile+Earth Plate construction method has a smaller safety factor than do the other construction methods. In sandy soil, as the excavation depth increases, the safety factor increases rapidly, and the safety factor is high in the order of Sheet Pile, S.C.W, C.I.P, and H-Pile+Earth Plate construction method. On the other hand, if pile stiffness and penetration depth are met, a review of heaving is considered in the soft clay layer, so in Sheet Pile application, the deeper the excavation depth, the greater the calculated safety factor necessary to meet the safety factor requirements. The required safety factor was applied to 1.2 [43].

4. Analysis of LCA on Earth-Retaining Wall

4.1. Method and Scope of the Evaluation of Environmental Impact Assessment

LCA analysis was performed on the applicable construction method of an earth-retaining wall by each installation condition, and then the environmental impact characteristics were analyzed. In Korea, the environmental impact assessment of earth-retaining wall is considered as a temporary structure, which reflects only the production and consumption of input resources in the construction stage. Therefore, construction details of material and equipment usage, standards of construction estimates, and energy statistics data of Korea were used to perform inventory analysis on all items applied to the construction of the earth-retaining wall method in this study. In addition, the LCI DB of the Ministry of Environment (MOE) and Ministry of Trade, Industry and Energy (MPTIE) of Korea was used for inventory analysis of the surveyed resources that were required. LCA software (Tool for TypeIII Labeling and LCA, hereinafter referred to as TOTAL) suggested by the Ministry of Environment in Korea was used. The environmental impact assessment was performed on the temporary earth protection facility based on the results after inventory analysis for each case object was performed. Abiotic resource depletion (ARD), global warming (GW), ozone depletion (OD), photochemical oxidant creation (POC), and acidification (AC), eutrophication (EU), ecotoxicity (ET), and human toxicity (HT) were applied as impact categories in order to establish the evaluation comparison criterion. In the environmental load assessment, the construction cost considering the construction method and ground conditions of the earth-retaining wall was applied based on standard of construction estimates in Korea [46].

4.2. LCA Results of the Earth-Retaining Wall According to Excavation Depth

4.2.1. Evaluation Results of Environment Load

Tables 6 and 7 and Figure 4 show the results of identifying and evaluating major environmental impacts through list analysis and impact assessment results for cases where the excavation area is medium scale (50 × 50 m) in shallow excavation (15 m) and deep excavation (40 m) depending on the ground conditions.

Table 6. Results of environmental load (shallow excavation: H = 15 m).

Environmental Impact Factor	Soil Condition	Construction Method			
		C.I.P	S.C.W	Sheet Pile	H-Pile+Earth Plate
Abiotic Resource Depletion (ARD)	Composite Soil	2.50×10^{-5}	2.58×10^{-5}	1.59×10^{-5}	1.59×10^{-5}
	Sandy Soil	2.56×10^{-5}	2.67×10^{-5}	1.79×10^{-5}	1.74×10^{-5}
	Soft Clay Soil			4.01×10^{-5}	
Global Warming (GW)	Composite Soil	5.37×10^{-5}	5.49×10^{-5}	2.70×10^{-5}	5.94×10^{-5}
	Sandy Soil	5.64×10^{-5}	5.80×10^{-5}	3.08×10^{-5}	6.22×10^{-5}
	Soft Clay Soil			6.78×10^{-5}	
Ozone Depletion (OD)	Composite Soil	1.40×10^{-7}	1.41×10^{-7}	1.37×10^{-7}	2.25×10^{-7}
	Sandy Soil	1.36×10^{-7}	1.40×10^{-7}	1.52×10^{-7}	2.38×10^{-7}
	Soft Clay Soil			3.48×10^{-7}	
Photochemical Oxidant Creation (POC)	Composite Soil	2.24×10^{-7}	2.35×10^{-7}	1.32×10^{-7}	3.87×10^{-7}
	Sandy Soil	2.35×10^{-7}	2.48×10^{-7}	1.50×10^{-7}	4.00×10^{-7}
	Soft Clay Soil			3.32×10^{-7}	
Acidification (AC)	Composite Soil	2.00×10^{-6}	1.98×10^{-6}	1.37×10^{-6}	1.33×10^{-6}
	Sandy Soil	2.13×10^{-6}	2.13×10^{-6}	1.57×10^{-6}	1.47×10^{-6}
	Soft Clay Soil			3.45×10^{-6}	
Eutrophication (EU)	Composite Soil	3.47×10^{-9}	3.41×10^{-9}	2.21×10^{-9}	8.20×10^{-9}
	Sandy Soil	3.60×10^{-9}	3.58×10^{-9}	2.50×10^{-9}	8.42×10^{-9}
	Soft Clay Soil			5.58×10^{-9}	
Ecotoxicity (ET)	Composite Soil	6.91×10^{-6}	6.96×10^{-6}	5.16×10^{-6}	5.45×10^{-4}
	Sandy Soil	7.04×10^{-6}	7.19×10^{-6}	5.78×10^{-6}	5.46×10^{-4}
	Soft Clay Soil			1.30×10^{-5}	
Human Toxicity (HT)	Composite Soil	4.53×10^{-6}	4.63×10^{-6}	2.49×10^{-6}	3.79×10^{-6}
	Sandy Soil	4.49×10^{-6}	4.64×10^{-6}	2.76×10^{-6}	4.03×10^{-6}
	Soft Clay Soil			2.76×10^{-6}	
Total	Composite Soil	9.24×10^{-5}	9.47×10^{-5}	5.21×10^{-5}	6.26×10^{-4}
	Sandy Soil	9.60×10^{-5}	9.91×10^{-5}	5.90×10^{-5}	6.32×10^{-4}
	Soft Clay Soil			1.31×10^{-4}	

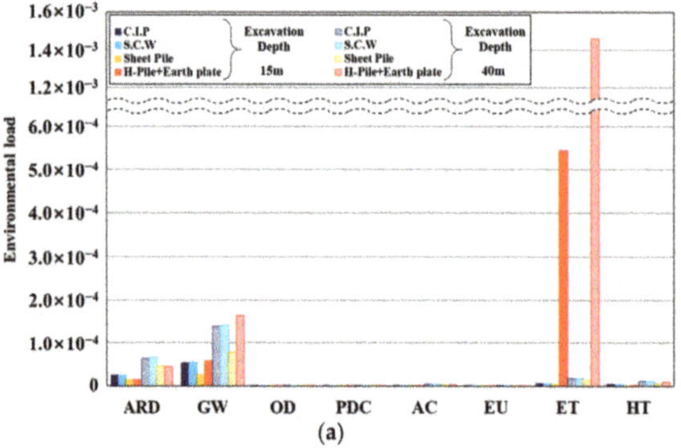

Figure 4. *Cont.*

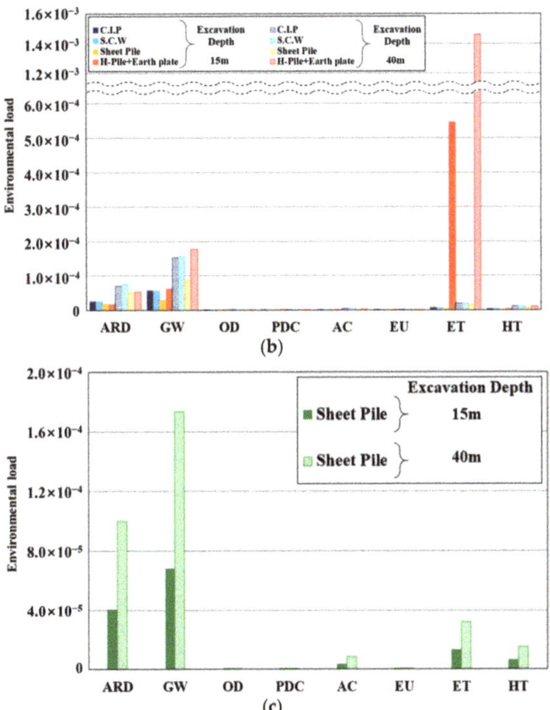

Figure 4. Relationship between environmental impact factor and environmental load by soil condition (excavation depth 15 m, 40 m): (**a**) composite soil; (**b**) sandy soil; (**c**) soft clay soil.

Table 7. Results of environmental load (deep excavation: H = 40 m).

Environmental Impact Factor	Soil Condition	Construction Method			
		C.I.P	S.C.W	Sheet Pile	H-Pile+Earth Plate
Abiotic Resource Depletion (ARD)	Composite Soil	6.45×10^{-5}	6.76×10^{-5}	4.72×10^{-5}	4.58×10^{-5}
	Sandy Soil	7.07×10^{-5}	7.49×10^{-5}	5.19×10^{-5}	5.30×10^{-5}
	Soft Clay Soil			1.00×10^{-4}	
Global Warming (GW)	Composite Soil	1.39×10^{-4}	1.43×10^{-4}	7.94×10^{-5}	1.65×10^{-4}
	Sandy Soil	1.52×10^{-4}	1.58×10^{-4}	8.75×10^{-5}	1.77×10^{-4}
	Soft Clay Soil			1.74×10^{-4}	
Ozone Depletion (OD)	Composite Soil	3.64×10^{-7}	3.77×10^{-7}	4.13×10^{-7}	6.27×10^{-7}
	Sandy Soil	4.02×10^{-7}	4.24×10^{-7}	4.52×10^{-7}	6.88×10^{-7}
	Soft Clay Soil			8.39×10^{-7}	
Photochemical Oxidant Creation (POC)	Composite Soil	5.84×10^{-7}	6.19×10^{-7}	3.89×10^{-7}	1.06×10^{-6}
	Sandy Soil	6.44×10^{-7}	6.87×10^{-7}	4.28×10^{-7}	1.12×10^{-6}
	Soft Clay Soil			8.46×10^{-7}	
Acidification (AC)	Composite Soil	5.32×10^{-6}	5.36×10^{-6}	4.05×10^{-6}	3.88×10^{-6}
	Sandy Soil	5.96×10^{-6}	6.08×10^{-6}	4.46×10^{-6}	4.50×10^{-6}
	Soft Clay Soil			8.86×10^{-6}	
Eutrophication (EU)	Composite Soil	9.04×10^{-9}	9.05×10^{-9}	6.56×10^{-9}	2.24×10^{-8}
	Sandy Soil	9.97×10^{-9}	1.01×10^{-8}	7.22×10^{-9}	2.34×10^{-8}
	Soft Clay Soil			1.41×10^{-8}	

Table 7. *Cont.*

Environmental Impact Factor	Soil Condition	Construction Method			
		C.I.P	S.C.W	Sheet Pile	H-Pile+Earth Plate
Ecotoxicity (ET)	Composite Soil	1.80×10^{-5}	1.85×10^{-5}	1.54×10^{-5}	1.46×10^{-3}
	Sandy Soil	1.99×10^{-5}	2.07×10^{-5}	1.69×10^{-5}	1.46×10^{-3}
	Soft Clay Soil			3.23×10^{-5}	
Human Toxicity (HT)	Composite Soil	1.14×10^{-5}	1.18×10^{-5}	7.47×10^{-6}	1.06×10^{-5}
	Sandy Soil	1.21×10^{-5}	1.27×10^{-5}	8.19×10^{-6}	1.17×10^{-5}
	Soft Clay Soil			1.53×10^{-5}	
Total	Composite Soil	2.39×10^{-4}	2.48×10^{-4}	1.54×10^{-4}	1.68×10^{-3}
	Sandy Soil	2.62×10^{-4}	2.74×10^{-4}	1.70×10^{-4}	1.71×10^{-3}
	Soft Clay Soil			3.33×10^{-4}	

First, in the composite soil condition of shallow excavation (as shown in Table 6), the environmental load of the H-Pile+Earth Plate construction method was the highest, at 6.26×10^{-4}, which shows that the impact of the environmental load was great due to the use of wood. Next, the environmental loads were high in the S.C.W, C.I.P, and Sheet Pile construction methods, in that order. In the environmental impact factor, the H-Pile+Earth Plate construction method showed the highest ecological toxicity, and the other three construction methods (C.I.P, S.C.W, and Sheet Pile) showed the highest environmental load in the order of global warming and resource depletion. In the composite soil condition of deep excavation, (as shown in Table 7), the environmental load of the H-Pile+Earth Plate construction method for the earth-retaining wall was 1.68×10^{-3} (the highest), and the environmental load of the other construction methods was high in the following order: S.C.W, C.I.P and Sheet Pile. Considering the environmental impact factor, the H-Pile+Earth Plate construction method had the largest environmental load for ecotoxicity, and the environmental load of the other three construction methods was high in the order of global warming and resource depletion.

Second, in the sandy soil condition of shallow excavation (as shown in Table 6), out of the four construction methods for the earth-retaining wall, the environmental load of H-Pile+Earth Plate was the highest (6.32×10^{-4}), and the environmental load was high in the order of S.C.W., C.I.P, and Sheet Pile. When compared by environmental impact factor, the H-Pile+Earth Plate construction method had the highest environmental load for ecotoxicity, and the environmental loads of the other three construction methods were high in the order of global warming, resources depletion, and ecotoxicity. The impact of global warming and resource depletion was greater than that of the other environmental impact categories. Moreover, in the sandy soil condition of deep excavation (as shown in Table 7), the environmental load of the H-Pile+Earth Plate construction method out of four construction methods for the earth-retaining wall was 1.71×10^{-3}, followed by the remaining three in the order of S.C.W, C.I.P, and Sheet Pile construction method. According to the environmental impact categories, the environmental load of ecotoxicity in the H-Pile+Earth Plate construction method was the highest, and the environmental load of the other three construction methods was high for global warming, resource depletion, and ecotoxicity, in that order.

Third, in soft clay ground in shallow excavation (as shown in Table 6), the environmental load of the Sheet Pile construction method was 1.31×10^{-4}, and the environmental load was high in the order of global warming and resource depletion among all categories of environmental impact. Additionally, the environmental load of the Sheet Pile construction method in soft clay ground was much higher than that of the other soil conditions, and the worse the condition of the soil, the greater the associated environmental load because of the need for more input resources (e.g., reinforcing materials, etc.). In soft clay ground in deep excavation (as shown in Table 7), the environmental load for the Sheet Pile construction method was 3.33×10^{-4}, and according to the environmental impact factor, the environmental load amount was associated with global warming and resource depletion in that order. Compared to the Sheet Pile construction method in other soil conditions, the Sheet Pile construction method in soft clay soil had a higher environmental load.

4.2.2. Evaluation Results of Environment Cost

Tables 8 and 9 and Figure 5 show the results of identifying and evaluating major environmental impacts through list analysis and impact assessment results following the purpose and scope for cases where the excavation area is of medium scale (50 × 50 m) in shallow excavation (15 m) and deep excavation (40 m) depending on the ground conditions. On this basis, in order to evaluate the environmental and economic impacts of the earth-retaining wall, evaluation of environmental economic feasibility was conducted by applying the environmental cost per unit of pollutants based on the environmental impact factor to the characteristics results of the environmental load amount for the eight categories previously calculated.

Table 8. Results of environmental cost * (shallow excavation: H = 15 m).

Environmental Impact Factor	Soil Condition	Construction Method			
		C.I.P	S.C.W	Sheet Pile	H-Pile+Earth Plate
Abiotic Resource Depletion (ARD)	Composite Soil	0.9	0.9	0.6	0.6
	Sandy Soil	29.7	30.6	16.3	32.9
	Soft Clay Soil			1.4	
Global Warming (GW)	Composite Soil	28.3	29.2	14.3	31.4
	Sandy Soil	28.3	29.2	14.3	31.4
	Soft Clay Soil			36.0	
Ozone Depletion (OD)	Composite Soil	0.0	0.0	0.0	0.0
	Sandy Soil	0.0	0.0	0.0	0.0
	Soft Clay Soil			0.0	
Photochemical Oxidant Creation (POC)	Composite Soil	0.0	0.0	0.0	0.0
	Sandy Soil	0.0	0.0	0.0	0.0
	Soft Clay Soil			0.0	
Acidification (AC)	Composite Soil	0.2	0.2	0.1	0.1
	Sandy Soil	0.2	0.2	0.1	0.1
	Soft Clay Soil			0.3	
Eutrophication (EU)	Composite Soil	0.0	0.0	0.0	0.0
	Sandy Soil	0.0	0.0	0.0	0.0
	Soft Clay Soil			0.0	
Ecotoxicity (ET)	Composite Soil	1.2	1.2	0.9	90.9
	Sandy Soil	1.2	1.2	1.0	90.9
	Soft Clay Soil			2.2	
Human Toxicity (HT)	Composite Soil	6.4	6.5	3.5	5.3
	Sandy Soil	6.3	6.5	3.9	5.7
	Soft Clay Soil			8.9	
Total	Composite Soil	36.9	37.9	19.3	128.3
	Sandy Soil	38.3	39.5	21.9	130.2
	Soft Clay Soil			48.7	

* Environmental cost(E-Cost) unit: KRW 1 million.

Table 9. Results of environmental cost * (deep excavation: H = 40 m).

Environmental Impact Factor	Soil Condition	Construction Method			
		C.I.P	S.C.W	Sheet Pile	H-Pile+Earth Plate
Abiotic Resource Depletion (ARD)	Composite Soil	2.3	2.4	1.7	1.7
	Sandy Soil	2.6	2.7	1.9	1.9
	Soft Clay Soil			3.6	
Global Warming (GW)	Composite Soil	73.4	75.9	41.9	87.2
	Sandy Soil	80.4	83.8	46.4	93.7
	Soft Clay Soil			92.3	

Table 9. *Cont.*

Environmental Impact Factor	Soil Condition	Construction Method			
		C.I.P	S.C.W	Sheet Pile	H-Pile+Earth Plate
Ozone Depletion (OD)	Composite Soil	0.0	0.0	0.0	0.0
	Sandy Soil	0.0	0.0	0.0	0.0
	Soft Clay Soil			0.0	
Photochemical Oxidant Creation (POC)	Composite Soil	0.0	0.0	0.0	0.0
	Sandy Soil	0.0	0.0	0.0	0.0
	Soft Clay Soil			0.0	
Acidification (AC)	Composite Soil	0.4	0.4	0.3	0.3
	Sandy Soil	0.5	0.5	0.3	0.3
	Soft Clay Soil			0.7	
Eutrophication (EU)	Composite Soil	0.0	0.0	0.0	0.0
	Sandy Soil	0.0	0.0	0.0	0.0
	Soft Clay Soil			0.0	
Ecotoxicity (ET)	Composite Soil	3.0	3.1	2.6	242.5
	Sandy Soil	3.3	3.4	2.8	243.4
	Soft Clay Soil			0.0	
Human Toxicity (HT)	Composite Soil	16.0	16.6	10.5	14.9
	Sandy Soil	17.1	17.9	11.6	16.5
	Soft Clay Soil			21.6	
Total	Composite Soil	95.1	98.4	57.0	346.6
	Sandy Soil	103.9	108.4	63.0	355.9
	Soft Clay Soil			123.6	

* Environmental cost(E-Cost) unit: KRW 1 million.

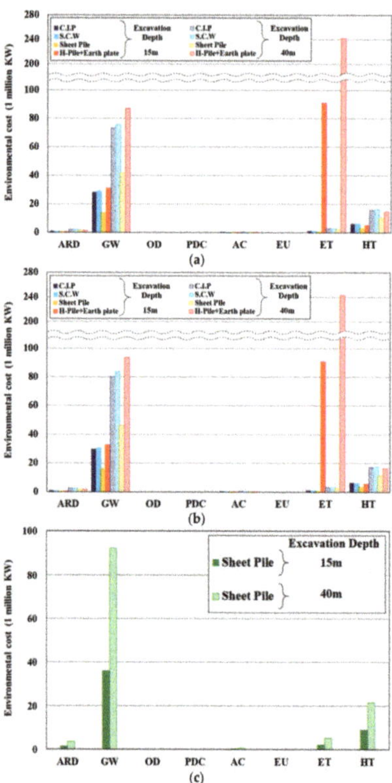

Figure 5. Relationship between environmental impact factor and environmental cost by soil condition (excavation depth 15 m, 40 m): (**a**) composite soil; (**b**) sandy soil; (**c**) soft clay soil.

First, in the composite soil condition of shallow excavation (as shown in Table 8), the H-Pile+Earth Plate construction method showed the highest environmental cost for ecotoxicity at KRW 90.9 million, and the other three construction methods had the largest environmental costs due to global warming. Thus, when it comes to the expected total environmental costs at the construction stages for each installation condition of the construction methods for the earth-retaining wall considering all environmental costs corresponding to the eight environmental impact categories, the total environmental cost of the H-Pile+Earth Plate construction method is the highest (KRW 128.3 million), and the total environmental costs are high in the order of S.C.W, C.I.P, and Sheet Pile construction method. Furthermore, the environmental costs of the S.C.W and C.I.P construction methods are quite similar. In the composite soil condition of deep excavation (as shown in Table 9), in the environmental cost calculation conducted by analyzing the environmental economic feasibility, as was carried out for the shallow excavation, when it comes to the total environmental costs expected in the construction stage of the installation condition for the earth-retaining wall, the H-Pile+Earth Plate construction method had the highest costs (KRW 346.6 million), and the environmental cost associated with ecotoxicity was the highest. For the other three construction methods, excluding the H-Pile+Earth Plate construction method, the highest environmental cost was associated with global warming.

Second, in the sandy soil condition of shallow excavation (as shown in Table 8), when it comes to the expected total environmental cost at the construction stage of the corresponding earth-retaining wall, the H-Pile+Earth Plate construction method had the largest cost (KRW 130.2 million), with the largest share of that cost due to ecotoxicity. Moreover, the three construction methods excluding H-Pile+Earth Plate had the largest environmental costs due to resource depletion and global warming, and it was found that the environmental costs of the C.I.P and S.C.W construction methods are similar. In the sandy soil condition of deep excavation (as shown in Table 9), when it comes to the total environmental cost expected in the construction stage of the earth-retaining wall installation condition, the total environmental cost of H-Pile+Earth Plate was the highest (KRW 355.9 million), and the environmental cost for ecotoxicity was the highest. For the three construction methods, excluding H-Pile+Earth Plate, the environmental cost for global warming was the highest.

Third, in soft clay ground in shallow excavation (as shown in Table 8), the total environmental costs expected in the construction stage of the Sheet Pile installation condition were KRW 48.7 Million, and the total environmental costs in the shallow excavation and medium-sized Sheet Pile installation condition were twice as high as the total environmental costs in other soil conditions. The environmental costs due to global warming account for the largest share. In soft clay ground in deep excavation (as shown in Table 9), the total environmental cost of the Sheet Pile construction method was KRW 123.6 million, which is twice as high as the cost in other soil conditions in a deep and medium-sized excavation (H = 15 m, 50 × 50 m), and the environmental cost associated with global warming was the highest.

4.3. Relationship between Excavation Depth, Total Environmental Load, and Total Environmental Cost by Soil Condition

As shown in Figure 6, the total environmental cost of the H-Pile+Earth Plate construction method was the highest in composite soil, and that cost was higher than the cost associated with the other three construction methods. Moreover, the deeper the excavation depth, the clearer the increase in total environmental cost. We confirmed that the total environmental costs of the C.I.P and S.C.W construction methods were similar, and this tendency remained the same when the excavation depth increased. The total environmental cost in sandy soil was similar to that in composite soil, but the cost in sandy soil was slightly greater. In soft clay soil, the total environmental cost of the Sheet Pile construction method increased as excavation depth increased, and the total environmental cost in soft clay soil was twice as high as that in other soil conditions. Furthermore, the assessments of environmental load and environmental cost were similar.

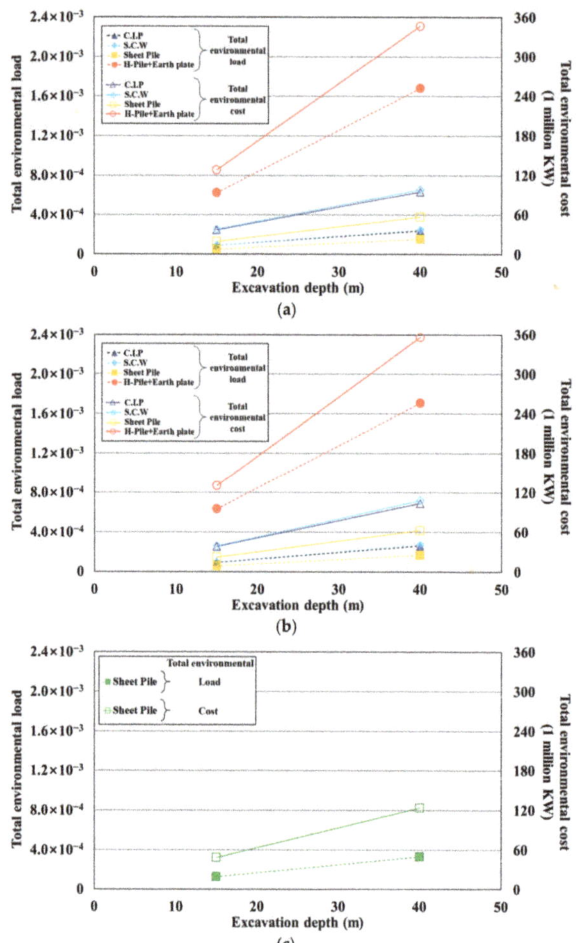

Figure 6. Relationship between total environmental load and total environmental cost by excavation depth and soil condition: (**a**) composite soil; (**b**) sandy soil; (**c**) soft clay soil.

5. Conclusions

This study evaluated the combination of excavation depth and soil condition in medium-sized excavation ground in order to examine the effect of construction methods on environmental economic feasibility for an earth-retaining wall during soil excavation. LCA analysis of the construction stage of the earth-retaining wall was conducted in consideration of eight environmental impact categories, the criteria for selecting the construction method for the earth-retaining wall considering the environmental costs of each construction method were reviewed, and the following conclusions were obtained as a result of this research:

1. If a calculation is conducted after calculating the environmental load by list analysis of the construction stage, this affects the selection of the construction method for the earth-retaining wall, so it is possible to select an optimal construction method for an earth-retaining wall considering stability and economic feasibility in various soil conditions via selection of a construction method that considers environmental loads in line with international trends.

2. Evaluation of the stability of the earth-retaining wall revealed that the C.I.P construction method was the best in terms of stability in both composite soil and sandy soil in the case of a shallow excavation. In terms of stability in the case of deep excavation, the S.C.W construction method was the best in composite soil and the C.I.P construction method was the best in sandy soil. In soft clay soil, the deeper the excavation depth, the greater the safety factor.

3. Evaluation of the environmental load of construction methods for the earth-retaining wall revealed that the H-Pile+Earth Plate construction method had low economic feasibility compared to the other construction methods because the environmental load of the H-Pile+Earth Plate method increased due to an increase in ecotoxicity. Furthermore, at the same excavation depth, the environmental load characteristics had a greater effect on the selection of construction methods in sandy soil than in composite soil.

4. Evaluation of the environmental costs of the construction methods for the earth-retaining wall revealed that the deeper the excavation depth, the greater the environmental cost. For a shallow excavation, in both composite and sandy soil, the H-Pile+Earth Plate construction method had low economic feasibility with the highest environmental cost, and the same is true for a deep excavation. In the case of soft clay soil, the environmental cost of the Sheet Pile construction method was higher than in other soil conditions, and the environmental cost was higher in sandy soil than in composite soil.

This study considered only the environmental effect in the determination of the retaining wall. Therefore, research should be conducted on the effect of various cost conditions on sustainability in order to be applied to the site.

Author Contributions: Conceptualization, Y.S., S.L. and J.-G.H.; methodology, J.-G.H. and G.H.; software, Y.S. and S.L.; validation, Y.S., J.-Y.L. and G.H.; formal analysis, Y.S., J.-G.H. and G.H.; investigation, S.L. and J.-Y.L.; resources, Y.S. and S.L.; data curation, J.-G.H. and G.H.; writing—original draft preparation, Y.S.; writing—review and editing, J.-G.H. and G.H.; visualization, J.-Y.L. and G.H.; supervision, J.-G.H.; project administration, S.L. All authors have read and agreed to the published version of the manuscript.

Funding: This research was supported by the MSIT (Ministry of Science and ICT), Korea, under the ITRC (Information Technology Research Center) support program (IITP-2020-2020-0-01655), the MSIT(NRF-2019R1A2C2088962) and the X-mind Corps program (2017H1D8A1030599) from the National Research Foundation (NRF) of Korea, the Human Resources Development (No.20204030200090) of the Korea Institute of Energy Technology Evaluation and Planning (KETEP) grant funded by the Korean government, and the Korea Agency for Infrastructure Technology Advancement funded by the Ministry of Land, Infrastructure and Transport of the Korean government (19SCIP- B108153-05).

Institutional Review Board Statement: Not applicable.

Informed Consent Statement: Not applicable.

Data Availability Statement: The data presented in this study are available on request to the corresponding author. The data are not publicly available as they form part of an ongoing study.

Conflicts of Interest: The authors declare no conflict of interest.

References

1. ISO. *ISO 14040: 2006-Environmental Management, Life Cycle Assessment, Principles and Framework*; International Organization for Standardization: Geneva, Switzerland, 2006.

2. ISO. *ISO 14044: 2006-Environmental Management, Life Cycle Assessment, Requirement sand Guidelines*; International Organization for Standardization: Geneva, Switzerland, 2006.

3. Guggemos, A.A.; Horvath, A. Comparison of Environmental Effects of Steel- and Concrete-Framed Buildings. *J. Infrastruct. Syst.* **2005**, *11*, 93–101. [CrossRef]

4. Keoleian, G.A.; Kendall, A.; Dettling, J.E.; Vanessa, M.S.; Richard, F.C.; Michael, D.L.; Victor, C.L. Life cycle modeling of concrete bridge design: Comparison of engineered cementitious composite link slabs and conventional steel expansion joints. *J. Infrastruct. Syst.* **2005**, *11*, 51–60. [CrossRef]

5.	Araújo, J.P.C.; Oliveira, J.R.M.; Silva, H.M.R.D. The importance of the use phase on the LCA of environmentally friendly solutions for asphalt road pavements. *Transp. Res. Part D Transp. Environ.* **2014**, *32*, 97–110. [CrossRef]

6.	Society of Environmental Toxicology and Chemistry (SETAC). *Guidelines for Life-Cycle Assessment: A "Code of Practice"*; SETAC Publications: Pensacola, FL, USA, 1993.

7.	China National Institute of Standardization. *GB/T 24040-2008 Environmental Management-Life Cycle Assessment-Principle and Framework*; Standard Press of China: Beijing, China, 2008.

8.	Hui, M.; Zhigang, Z.; Xia, Z.; Shuang, W. A Comparative Life Cycle Assessment (LCA) of Warm Mix Asphalt (WMA) and Hot Mix Asphalt (HMA) Pavement: A Case Study in China. *Adv. Civ. Eng.* **2019**, *19*, 9391857.

9.	Guggemos, A.A. Environmental Impacts of On-Site Construction Processes: Focus on Structural Frames. Ph.D. Thesis, University of California, Berkeley, CA, USA, 2003.

10.	Guggemos, A.A.; Horvath, A. Decision-support tool for assessing the environmental effects of constructing commercial buildings. *J. Arch. Eng.* **2006**, *12*, 187–195. [CrossRef]

11.	Lee, J.H. A Study on GHG Emissions and Mitigation Potentials of the Industrial Sector in Korea. Master's Thesis, Keimyung University, Daegu, Korea, 2010.

12.	Mao, C.; Shen, Q.; Shen, L.; Tang, L. Comparative study of greenhouse gas emissions between off-site prefabrication and conventional construction methods: Two case studies of residential projects. *Energy Build.* **2013**, *66*, 165–176. [CrossRef]

13.	Hong, J.; Shen, G.Q.; Feng, Y.; Lau, W.S.T.; Mao, C. Greenhouse gas emissions during the construction phase of a building: A case study in China. *J. Clean. Prod.* **2015**, *103*, 249–259. [CrossRef]

14.	Luo, W.; Sandanayake, M.; Zhang, G. Direct and indirect carbon emissions in foundation construction—Two case studies of driven precast and cast-in-situ piles. *J. Clean. Prod.* **2019**, *211*, 1517–1526. [CrossRef]

15.	Frey, H.C.; Rasdorf, W.; Lewis, P. Comprehensive field study of fuel use and emissions of nonroad diesel construction equipment. *Transp. Res. Rec. J. Transp. Res. Board* **2010**, *2158*, 69–76. [CrossRef]

16.	Lewis, P. Estimating Fuel Use and Emission Rates of Nonroad Diesel Construction Equipment Performing Representative Duty Cycles. Ph.D. Thesis, North Carolina State University, Raleigh, NC, USA, 2009.

17.	Lewis, P.; Frey, H.C.; Rasdorf, W. Development and use of emissions inventories for construction vehicles. *Transp. Res. Rec. J. Transp. Res. Board* **2009**, *2123*, 46–53. [CrossRef]

18.	Junnila, S.; Horvath, A.; Guggemos, A.A. Life-Cycle Assessment of Office Buildings in Europe and the United States. *J. Infrastruct. Syst.* **2006**, *12*, 10–17. [CrossRef]

19.	Sandanayake, M.; Zhang, G.; Setunge, S. Environmental emissions at foundation construction stage of buildings—Two case studies. *Build. Environ.* **2016**, *95*, 189–198. [CrossRef]

20.	Zhang, G.; Sandanayake, M.; Setunge, S.; Li, C.; Fang, J. Selection of emission factor standards for estimating emissions from diesel construction equipment in building construction in the Australian context. *J. Environ. Manag.* **2017**, *187*, 527–536. [CrossRef]

21.	Sandanayake, M.; Zhang, G.; Setunge, S.; Luo, W. Estimation and comparison of environmental emissions and impacts at foundation and structure construction stages of a building—A case study. *J. Clean. Prod.* **2017**, *151*, 319–329. [CrossRef]

22.	Sandanayake, M.; Zhang, G.; Setunge, S. A comparative method of air emission impact assessment for building construction activities. *Environ. Impact Assess. Rev.* **2018**, *68*, 1–9. [CrossRef]

23.	Santero, N.J. Pavements and the Environment: A Life-Cycle Assessment Approach. Ph.D. Thesis, University of California Berkeley, Berkeley, CA, USA, 2009. Unpublished work.

24.	Bjarne, S.; Jeppe, C.D. CO_2 emission reduction by exploitation of rolling resistance modelling of pavements. *Procedia Soc. Behav. Sci.* **2012**, *48*, 311–320.

25.	Shang, C.; Zhang, Z.; Li, X. Research on energy consumption and emission of life cycle of expressway. *J. Highway Transp. Res. Dev.* **2010**, *27*, 149–154.

26.	Ministry of Land, Infrastructure and Transport of Republic of Korea. *Environmental Load Reduction Type LCA (Life Cycle Assessment) Based Design and Construction Technology Development*; Construction & Transportation R&D Report; Ministry of Land, Infrastructure and Transport: Sejong, Korea, 2017.

27.	Lee, Y.K.; Han, J.G.; Kwon, S.H. A Study on the Evaluation of Environmental Load Based on LCA Using BIM—Focused on the Case of NATM Tunnel. *J. Korean Soc. Civil Eng.* **2018**, *38*, 477–485.

28.	Häkkinen, T.; Mäkelä, K. *Environmental Adaption of Concrete: Environmental Impact of Concrete and Asphalt Pavements*; VTT Technical Research Centre of Finland: Espoo, Finland, 1996.

29.	Han, I.X.; Zhow, W.; Tang, L.C.M. Model-based life cycle cost and assessment tool for sustainable building design decision. In Proceedings of the 4th International Conference on Construction Engineering and Project Management (ICCEPM-20011), Sydney, Australia, 16–18 February 2011.

30.	Kwon, S.H. Development of Assessment Model for Environmental Economics of Construction Projects. Ph.D. Thesis, Chung-Ang University, Seoul, Korea, 2008.

31.	Bae, C.H. A Study on the Selection of Retaining Wall System for the Underground. Master's Thesis, Chonnam National University, Gwangju, Korea, 2010.

32.	Han, J.G.; Lee, J.Y. Case Study on AHP Technique Application for the Reinforcing Method Selection on a Cut-Slope. *J. Korean Geotech. Soc.* **2008**, *24*, 81–88.

33. Lee, H.C.; Woo, S.K.; Kim, O.K. Development of Decision Making Model for Selecting the Soft Foundation Improvement Method Using AHP technique. *J. Korean Soc. Civ. Eng.* **2007**, *27*, 499–506.
34. Lee, C.H.; Jeong, K.C. An Analytic Hierarchy Process based Decision Support System for Selecting Foundation Practice. *Korean J. Constr. Eng. Manag.* **2012**, *13*, 129–139. [CrossRef]
35. Pan, N.F. Fuzzy AHP approach for selecting the suitable bridge construction method. *Autom. Constr.* **2008**, *17*, 958–965. [CrossRef]
36. Ebrahimian, A.; Ardeshir, A.; Radb, I.Z.; Ghodsypour, S.H. Urban stormwater construction method selection using a hybrid multi-criteria approach. *Autom. Constr.* **2015**, *58*, 118–128. [CrossRef]
37. Shen, L.; Yan, H.; Fan, H.; Wu, Y.; Zhang, Y. An integrated system of text mining technique and case-based reasoning (TM-CBR) for supporting green building design. *Build. Environ.* **2017**, *124*, 388–401. [CrossRef]
38. Lorenz, T.; Jost, A. Towards an orientation framework in multi-paradigm modelling: Aligning purpose, object and methodology in System Dynamics, Agent-based Modeling and Discrete-Event-Simulation. In Proceedings of the 24th International Conference of the System Dynamics Society, Nijmegen, The Netherlands, 23–27 July 2006; pp. 2134–2151.
39. Tsai, W.H.; Lin, S.J.; Lee, Y.F.; Chang, Y.C.; Hsu, J.L. Construction method selection for green building projects to improve environmental sustainability by using an MCDM approach. *J. Environ. Plan. Manag.* **2013**, *56*, 1487–1510. [CrossRef]
40. Kim, J.Y.; Park, W.Y.; Kang, K.I. A Study on the Selection System of Retaining Wall Methods Using Neural Network. *J. Archit. Inst. Korea Struct. Constr.* **2002**, *18*, 69–76.
41. Yau, N.J.; Yang, J.B. Applying case-based reasoning technique to retaining wall selection. *Autom. Constr.* **1998**, *7*, 271–283. [CrossRef]
42. Park, J.K.; Jun, S.K.; Kim, Y.K.; Heo, E.N. A study on the improvement in decision making analysis for the selection of tunnel construction method. *J. Korean Tunnel Undergr. Space Assoc.* **2002**, *4*, 261–276.
43. Ministry of Land, Infrastructure and Transport. *KDS 21 30 00: 2016, Korean Design Standard*; Ministry of Land, Infrastructure and Transport: Sejong, Korea, 2016.
44. Caspe, M.S. Surface settlement adjacent to braced open cut. *J. Soil Mech. Found. Div.* **1966**, *92*, 51–59. [CrossRef]
45. Bowles, J.E. *Foundation Design and Analysis*, 4th ed.; McGraw-Hill Book Company Limited: Maidenhead, UK, 1988; pp. 658–661.
46. Ministry of Land, Infrastructure and Transport of Republic of Korea. *Standard of Construction Estimate*; Ministry of Land, Infrastructure and Transport: Sejong, Korea, 2017.

 *sustainability*

Article

Environmental and Economic Optimization of a Conventional Concrete Building Foundation: Selecting the Best of 28 Alternatives by Applying the Pareto Front

Ester Pujadas-Gispert [1,*], Joost G. Vogtländer [2] and S. P. G. (Faas) Moonen [1]

[1] Department of the Built Environment, Eindhoven University of Technology, 5600 MB Eindhoven, The Netherlands; S.P.G.Moonen@tue.nl

[2] Department of Industrial Design Engineering, Delft University of Technology, 2628 CE Delft, The Netherlands; J.G.Vogtlander@tudelft.nl

* Correspondence: e.pujadas.gispert@tue.nl; Tel.: +31-40-247-6047

Abstract: This research optimizes the environmental impact of a conventional building foundation in Northern Europe while considering the economic cost. The foundation is composed of piles and ground beams. Calculations are performed following relevant building Eurocodes and using life cycle assessment methodology. Concrete and steel accounted for the majority of the environmental impact of foundation alternatives; in particular, steel on piles has a significant influence. Selecting small sections of precast piles or low-reinforcement vibro-piles instead of continuous-flight auger piles can reduce the environmental impacts and economic costs of a foundation by 55% and 40%, respectively. However, using precast beams rather than building them on site can increase the global warming potential (GWP) by up to 10%. Increasing the concrete strength in vibro-piles can reduce the eco-costs, ReCiPe indicator, and cumulated energy demand (CED) by up to 30%; the GWP by 25%; and the economic costs by up to 15%. Designing three piles instead of four piles per beam reduces the eco-costs and ReCiPe by 20–30%, the GWP by 15–20%, the CED by 15–25%, and the costs by 12%. A Pareto analysis was used to select the best foundation alternatives in terms of the combination of costs and eco-burdens, which are those with vibro-piles with higher concrete strengths (low reinforcement), cast in situ or prefabricated beams and four piles per beam.

Keywords: ground beam; LCA; prefabrication; vibro-pile; eurocode; precast prestressed concrete pile; continuous flight auger pile; eco-costs; life cycle assessment; economic

Citation: Pujadas-Gispert, E.; Vogtländer, J.G.; Moonen, S.P.G.(. Environmental and Economic Optimization of a Conventional Concrete Building Foundation: Selecting the Best of 28 Alternatives by Applying the Pareto Front. *Sustainability* **2021**, *13*, 1496. https://doi.org/10.3390/su13031496

Academic Editor: Sunkuk Kim
Received: 12 December 2020
Accepted: 20 January 2021
Published: 1 February 2021

Publisher's Note: MDPI stays neutral with regard to jurisdictional claims in published maps and institutional affiliations.

1. Introduction

1.1. Background

To keep the global temperature rise preferably at no more than 1.5 degrees Celsius by the end of the century [1], it is necessary to diminish global emissions by more than 50% by 2030 and work towards carbon neutrality by 2050. The construction sector accounts for 36% of final energy use and 39% of carbon dioxide (CO_2) related to energy and processes [2]. To date, the focus has been on reducing the energy consumed during the use of buildings. However, embodied energy related to the materials, construction, maintenance, and end of life of buildings is becoming increasingly important [3]. Life Cycle Assessment (LCA) has proven to be a suitable tool to reduce the environmental impact of buildings [4,5]. Nevertheless, uncertainties in the LCA calculation must be minimized and reliable benchmarks must be provided for evaluating buildings [6,7]. The LCA can be carried out at various levels of the system, such as for portions, components, or the entire building [8]. Nonetheless, the foundation, which is the lowest part of the building and in contact with the soil, is rarely assessed despite its considerable impact at an aggregate level, leaving ample room for improvement [9,10]. Consequently, rigorous studies are required to optimize foundations, thus reducing the emissions from the construction sector.

1.2. Deep Foundations

A deep foundation is a type of foundation that depends on the deeper layers of the soil, while a shallow foundation is based on the surface layers of the soil [11]. Deep foundations tend to have a greater environmental impact than shallow foundations due to the greater amount of materials used [12,13]. A typical deep foundation for buildings and other civil works consists of piles, which are vertical structural elements driven or drilled deep into the ground at a building site. Piles tend to work in groups braced by ground beams, pile caps, and slabs. A conventional foundation in Northern Europe is composed of piles and ground beams. The foundation can be either built directly on the ground (cast in situ) or built in a factory (prefabricated) and then transported and installed on site [14]. During the installation, the joints between precast components are connected with steel reinforcing bars and sealed with mortar. Cast in situ beams can be built into different types of formwork (e.g., removable or non-removable, different materials and forms, etc.), and a blinding or plastic foil is often required at the bottom. Below are some of the most commonly used types of piles in Northern Europe [11]:

- Precast prestressed concrete (PPC) piles are prefabricated piles that are prestressed and driven in the ground using a diesel or hydraulic hammer (Figure 1). The main advantage of PPC piles over conventional precast piles is that they are more resistant given the same pile cross-section. Therefore, PPC piles are slender and lighter, making them easier to lift and drive. Additionally, the effect of prestressing closes the cracks of concrete caused during handling and driving, which combined with high-quality concrete, extends the durability of the prestressed pile.
- Fundex piles are built by drilling a metallic tube with a tip on the ground (Figure 2). Then, the reinforcement cage is installed and concrete is poured inside the tube. Finally, the metal tube is removed, leaving the sacrificial drill point in the soil, and the pile head is cut to ensure a good connection with the upper structure.
- Continuous flight auger (CFA) piles are drilled and concreted in one continuous operation, reducing installation time compared to bored piles (Figure 3). The reinforcement cage is then placed in the upper meters of the pile when concrete is still wet.
- Vibro-piles are built by driving a metal tube closed at the end by a sacrificial plate (Figure 4). Then, the reinforcement cage is installed, and concrete is poured inside the tube. Finally, an outer ring vibrator is used to extract the tube from the soil, leaving the sacrificial plate in the soil.

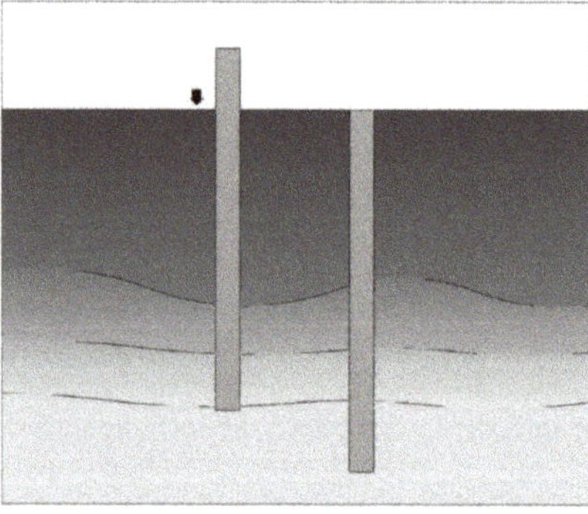

Figure 1. Precast prestressed concrete pile.

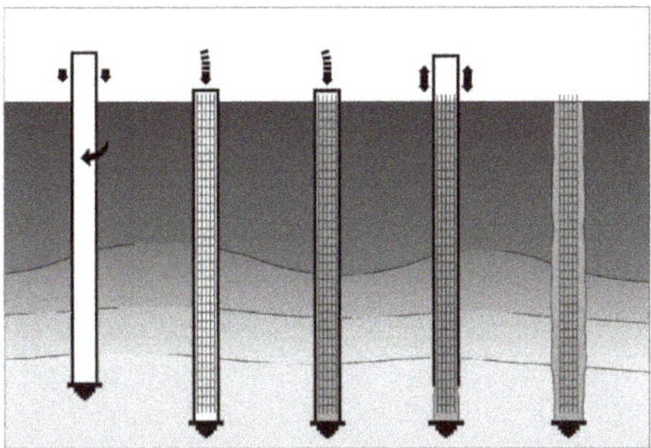

Figure 2. Fundex pile.

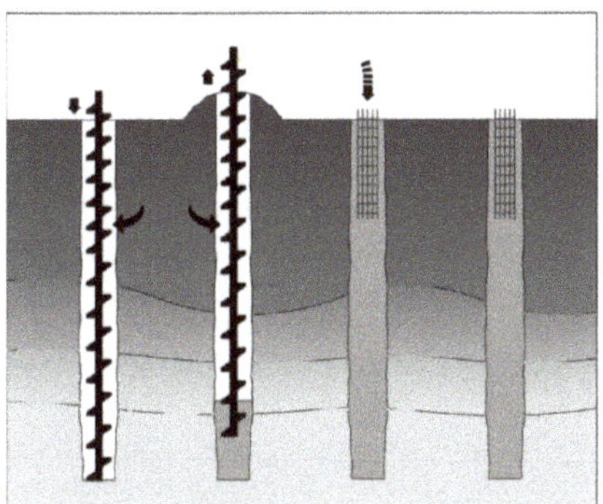

Figure 3. Continuous flight auger (CFA) pile.

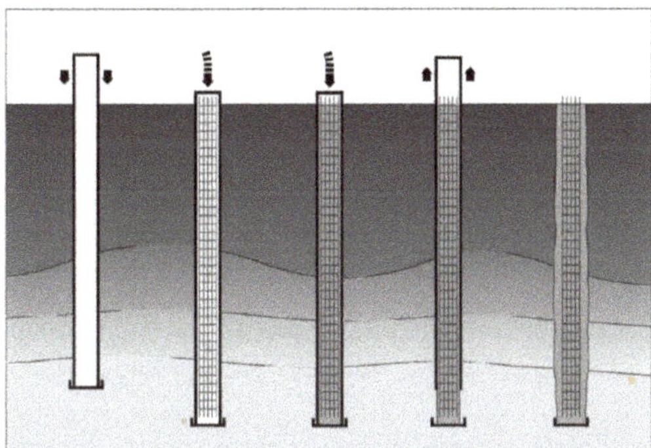

Figure 4. Vibro-pile.

Concretes for precast foundations tend to have higher strengths than cast in situ concrete because precast components are built in a factory where production conditions are more controlled and it is easier to obtain higher strengths. The nomenclature for concrete strength from the Eurocode consists of a capital C followed by two compressive strengths, one measured using concrete cylinder test specimens and the other using concrete cube specimens (e.g., C25/30) [15].

1.3. Previous Literature

Previous studies have reported that concrete and steel account for the highest greenhouse gas (GHG) emissions during the construction of deep concrete foundations (66–90%), followed by equipment usage (12–19%) and transportation (9–16%) [14,16–19]. In addition, piles account for the majority of the environmental impact in a foundation compared to a concrete raft or pile caps. Therefore, it is desirable to reduce the environmental impact of foundation materials, particularly for piles.

The decisions made at the design stage are important because they affect the long-term impact of a foundation [20,21]. For instance, the use of precast concrete piles instead of CFA piles was shown to halve the environmental impact of a foundation, fully counteracting the extra emissions from prefabrication (higher concrete strength, more transport, and installation). The same study also reported that prefabrication may be more expensive than conventionally cast in situ methods, although this is closely related to the nature of each work (e.g., number of units, location). Conversely, Luo et al. [16] reported that precast concrete piles can create 5% more GHG emissions than bored piles, with materials and transportation being points to be optimized. However, the emissions from the construction of precast concrete piles were closely related to the area, cost, and number of piles in a foundation [22].

The combination of variables in the design of a foundation also showed reductions in the GWP. For instance, the selection of the level of prefabrication, concrete strength, type, and calculation codes of a foundation reduced the GWP impact of the foundation by up to 50–60% [14,17]. For this reason, this research also incorporated these variables and others derived from common practice with the intention that the resulting recommendations could be easily implemented in future foundation designs. Furthermore, the study also included some types of piles which are used in Northern Europe (PPC, Fundex, and vibro-piles) that have yet to be environmentally assessed. The economic cost was also included, since the implementation of a new solution in the construction sector depends significantly on its economic cost [16,23,24]. Additionally, a Pareto analysis was performed on building foundations to select the best alternatives in terms of the combination of costs and ecological burdens. Furthermore, the study focuses solely on the foundation and its environmental optimization [9,10], which is normally approached as part of a building [7]. In this respect, prior studies [14,16,17] optimized environmentally part of a foundation (i.e., a pile cap with piles, a footing, and only piles of a foundation); the present study aims to go one step further by considering the entire foundation of the building.

1.4. Objectives

The research is based on a real case and aims to optimize a conventional foundation for Northern Europe in terms of the environment by considering the economic cost and studying the variables of the prefabrication level (fully precast, semi-precast, and cast in situ), concrete strength of cast in situ piles (C20/25, C25/30, C30/37, C40/50), type of pile (precast prestressed concrete pile (PPC), continuous flight auger (CFA), Fundex, and vibro-pile), and the number of piles per ground beam (3 and 4 piles). The specific objectives are (i) to conduct a structural analysis to determine the dimensions of foundation alternatives; (ii) to calculate and analyze the environmental burden using LCA; (iii) to calculate and analyze the economic cost; (iv) to apply a Pareto analysis (the so-called Pareto front) to select the best solution(s), given a combination of the eco-burdens and costs; and (v) to assess the influence of the study variables on the environmental burden and the economic cost of a foundation and, in doing so, define specific design conclusions and recommendations.

2. Materials and Methods

The integrated methodology applied to determine the environmental influence of the study variables includes a selection of equivalent alternatives (Section 2.1), an explanation of a case study (Section 2.2), a definition of a functional unit (FU) (Section 2.3) as well as system boundaries (Section 2.4) and quantitative model (Section 2.5), an explanation of the foundation design (Sections 2.6 and 2.7) and LCA (Section 2.8), and a compilation of the data sources used (Section 2.9).

2.1. Selection of Equivalent Alternatives

The type of foundation and the study variables are common in usual Northern Europe practices. The abbreviations used to designate the study alternatives are shown in Table 1.

Table 1. Abbreviations used in the study.

	Variables	Abbreviations
1	Number of foundation	1–28
2	Type of pile + diameter/side of the pile (mm)	Continuous flight auger pile (C), Vibro-pile (V), Fundex pile (F), Precast prestressed concrete pile (P)
3	Concrete strength	Cast in situ: C20/25 (20), C25/30 (25), C30/37 (30), C40/50 (40). Precast: 35/45 (35)
4	Beam	Cast in situ (I) (concrete is poured on site), Precast (P) (concrete is poured in a specialized facility)
5	Piles per beam	(3) and (4)
6	Width of the beam (mm)	Cast in situ (500, 550, 650) and Precast (300, 350, 400, 450)

Example: 1-C600-20/I3.650 is foundation number 1 and is composed of C600-20 piles and I3.650 beams. The C600-20 pile is a continuous flight auger pile 600 mm in diameter with a concrete strength of C20/25. I3.650 is a cast in situ beam supported by three piles with a width of 650 mm.

2.2. Case Study

The reference project is a neutral energy housing project in Vianen (The Netherlands) (Figure 5) which is composed of 16 buildings with similar characteristics. The foundation of building 13 was selected for assessment (Figure 6). The foundation is composed of precast concrete piles and ground beams (Figure 7). The main characteristics of the foundation of the reference project are shown in Supplementary Material S1. The soil is composed of loose layers of sand that increase in resistance to levels of 8–10 m where the piles are embedded. The reference project has been adapted to better analyze the influence of the variables on the study results. For instance, the prefabricated concrete walls were turned into sand-lime brick and the partition walls into two 120 mm sand-lime bricks. Table 2 shows the building materials and permanent and variable loads considered in the study. Finally, wind load was not considered, given that it does not alter the results of the study.

Figure 5. Aerial photo of the reference project [25].

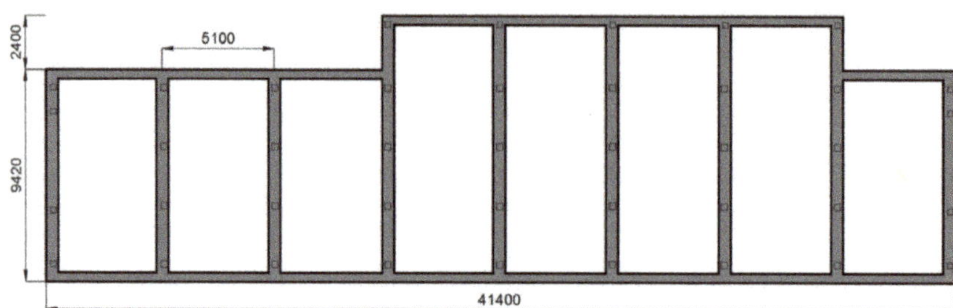

Figure 6. Foundation of building 13 of the reference project in Vianen.

Figure 7. A foundation of the reference project [25].

Table 2. Building materials and permanent loads used in the study.

Element	Construction System	Loads
Ground floor	PS insulation floor	$3.6/2.55\ \text{kN/m}^2$
First and second floors	Concrete floor	$5.7/2.55\ \text{kN/m}^2$
Roof	Timber + tiles	$1.0/0.0\ \text{kN/m}^2$
Facades	Timber frame construction + masonry	$2.5\ \text{kN/m}^2$
Front facade	Sand-lime bricks + masonry	$4.3\ \text{kN/m}^2$
Building wall	2×120 mm sand-lime bricks construction	$4.5\ \text{kN/m}^2$
Extension facade	Sand-lime bricks + masonry	$4.3\ \text{kN/m}^2$
Roof extension	Concrete floor	$5.5/0.0\ \text{kN/m}^2$

2.3. Functional Unit

The functional unit (FU) is a conventional foundation that consists of concrete ground beams and piles considering different levels of prefabrication, types of pile, concrete compressive strength, and number of piles per beam for a useful life of 50 years.

2.4. System Boundaries

Figure 8 shows the phases of the LCA and the elements considered in each of them. The phases comprise steps from the extraction of raw materials to the construction of the foundation on site. The various transports are included in the corresponding phases (e.g., the transport of raw material, products, waste, soil). The levelling of the ground prior to building the foundation was not considered because it is very specific to each work and does not alter the study comparison. Nevertheless, the excavation of each foundation has been considered because it is different depending on the alternative. The pumping of concrete and the transport of the machines to the site were excluded because a preliminary study showed that their environmental impact was low. The use phase was also excluded because well-designed foundations tend to not require maintenance. Similarly, the end of life was not considered because recycling or reusing foundations is not the norm, although there is great potential in this area [7]. The piles are usually left installed in the ground, although precast ground beams, which are a very recent foundation element, have not yet reached the end of their useful life. However, in cases where the precast ground beams were built in temporary buildings (10 and 15 years), which is not the usual case, at the end of their useful life, they were (i) reused in the same building, (ii) reused in another building. or (iii) dismantled and demolished. In cases (i) and (ii), the end of life was considered in the design stage of the beams.

System boundaries

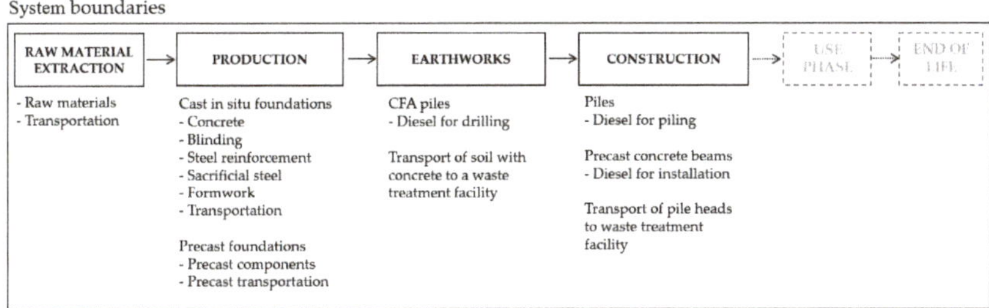

Figure 8. Life cycle diagram and system boundaries of the construction of the foundation alternatives.

2.5. Quantitative Model for Environmental Impact Category Calculation

Below the quantitative model used to calculate the eco-burdens for each category is shown [17].

$$E = \sum_{i=1}^{n} \sum_{j=1}^{m} p_i \, Q_{ij} \tag{1}$$

where E is the single indicator score (i.e., Eco-costs 2017, ReCiPe 2016 Endpoint World (2010) H/A, GWP 100-year 2013 Intergovernmental Panel on Climate Change and cumulated energy demand) of the FU; n is the total number of reference flows; i is the reference flow (i.e., material, diesel, formwork, or transportation); m is the total number of phases in the life cycle (i.e., 4); j is the phase of the life cycle (i.e., raw material extraction, production, earthworks, or construction); p_i is the combined factor for characterization, normalization, and weighting per unit of reference flow (e.g., factor per kg or m^2 material); and Q_{ij} is the quantity of reference flow in an FU phase.

2.6. Structural Design of Ground Beams

Loads were calculated according to NEN-EN 1990+A1+A1/C2:2019/NB:2019 [26] and NEN-EN 1991-1-1+C1+C11:2019/NB:2019 [27]. Reinforced concrete beams (cast in place and precast) were designed following NEN-EN 1992-1-1+C1:2011/NB:2016+A1:2020 [28]. Calculations of the beams were performed with Technosoft Balkroosters [29]. All the beams were designed for an XC3 environmental class (moderate humidity) according to Table 4.1 in [28], with concrete covers of 35 mm for cast in situ beams and 30 mm for precast beams complying with article 4.4.1 in [28]. In addition, cast-in-place beams were designed to be cast in a common removable timber formwork [30] so that all the alternatives could be compared without insulation. Additionally, they were designed to be built over a 50 mm concrete blinding [28]. The formwork was considered to be used 5 times. The concrete strength considered for cast in situ beams was C30/37, that for concrete blinding was C12/15, and that for precast beams was C40/50, which is aligned with normal practices. However, different strengths of concrete for ground beams were not considered because a preliminary study showed that increasing the strength of concrete in beams minimally increases their resistance, because beams work mostly in bending rather than in compression. Ground beams were dimensioned following the usual Dutch practice. The height was set at 500 mm for all the ground beams (cast in situ and precast). For the cast-in-situ ground beams, a minimum width of 500 mm for transverse walls and 400 mm for longitudinal walls was established so that the width of the beams matched the width of the walls. However, for the precast floor beams, this was not necessary because the consoles, which are elements that protrude from both sides of the top of the beam, allow the beam to be adjusted to the width of the wall. It should be mentioned that consoles were not included in the material calculation of the study due to their low impact on the results. Finally, various PPC pile and wide beam foundation options were discarded from the study because they were oversized and unrealistic for reasons of economic cost, transportation, and installation.

2.7. Geotechnical and Structural Design of Piles

The bearing capacity of the piles was calculated following NEN 9997-1+C2:2017 [31]. All the piles were designed for an XC4 environmental class (cyclic wet and dry) according to Table 4.1 in [28]. The normal loads to the piles were 510 kN for the four-pile beams and 760 kN for the three-pile beams. Concrete covers were designed to be 70 mm for CFA piles, 50 mm for Fundex piles, 40 mm for vibro-piles, and 30 mm for PPC piles, which is in line with article 4.4.1 in [28]. The concrete strengths considered for cast in situ piles were C20/25, C25/30, and C30/37, complying with NEN-EN 1536:2010+A1:2015 [32], and C35/45 for PPC piles according to product specifications. In most cases, the heads of the piles were not cut off. However, in Fundex piles, it has been considered that the pile head is cut off (1 m), which aligns with the usual practice. Afterwards, the heads of all types of piles were rebuilt with mortar, and a reinforcing bar was placed on top of each pile to connect them with the precast ground beams, which was not necessary for the cast-in-situ beams, since they were built directly on top of piles.

In terms of reinforcement, all the pile types were reinforced with the minimum amount of reinforcement, which is specified in Table 9.6 from [28], and an additional reinforcement was also calculated to resist the bending moment at the head of the pile (38 kNm in three-beam piles and 25.5 kNm in four-beam piles). This bending moment arises from considering a 50 mm eccentricity in the structural calculations to cover possible unforeseen events, such as construction misalignments and horizontal forces. As a result, a larger pile diameter may be required to arm this bending moment. Furthermore, the minimum diameter considered for the longitudinal reinforcement was 12 mm in all piles [28], while for stirrups it was considered to be 6 mm in CFA piles, 8 mm in vibro-piles, and 5 mm in Fundex piles according to article 10.2.4 in [33]. The reinforcement for PPC piles was retrieved from product specifications. Finally, most of the piles were reinforced at all lengths, except CFA piles, which were reinforced only at the three superior meters [33], and an additional bar of Ø20 mm was arranged in the center of the CFA pile to compensate for the weakness of the superior layers of the soil (article 7.1.7 in [32]). Regarding cast in situ piles, the minimum cement content considered was 375 kg/m^3, complying with Annex D of [15], and a reduction in the pile diameter was applied to cover the uncertainty of building a pile directly on the ground following article 2.3.4.2 in [28].

The higher the concrete strength in the piles, the less steel reinforcement is required. It is worth mentioning that a higher concrete strength was not considered once the cast in situ piles reached the minimum steel reinforcement because it did not add information in the study. As previously mentioned, the environmental impact of the cubic meter of concrete is the same for the different concrete resistances in this study following the Dutch regulations. Therefore, rather than looking for the best option that may depend on each case (loads, soil, etc.), it is intended to detect the influence of the study variables on the environmental and economic results of the alternatives to consider them in future designs and codes of foundations.

2.8. Life Cycle Assessment

The LCA method was applied to determine the environmental impact, as defined in international codes [34,35]. The software SimaPro version 9.2 [36] was used for the calculations. Since LCA is used here to determine whether "system A is better than System B", so-called "single score methods" are applied, as has been recommended by the Society of Environmental Toxicology and Chemistry [37] and the Joint Research Centre of the EU [38]. The calculations were performed for 4 indicator types: (i) carbon footprint, as a "single issue method"; (ii) ReCiPe, as a "damage-based method"; (iii) eco-costs, as a "monetized prevention-based method"; (iv) and embodied energy, as a "single issue method". In their sectors, these four indicators are the most applied methods in science. The eco-costs comply with [39]. Although monetization in LCA is not very common, the advantage of eco-costs is that they are so-called "external costs" (i.e., costs for our society that are not incorporated in the price of a product), so they have a direct meaning to architects, business managers, and governmental policy makers. Recently, there have been increasing applications of eco-costs in the building industry—e.g., for concrete construction [40] and beams [41]. Eco-costs are also applied in full cost accounting (FCA), which is also called true cost accounting (TCA). The basic philosophy behind TCA is that the external costs (=environmental burden) of a product should be added to the economic costs to enable a fair comparison in product benchmarking between a cheap but polluting product and a "clean" product. Another way to address the issue of "ecology versus economy" is to display the external costs and the economic costs in a two-dimensional graph and determine the Pareto front (being the best solution). Section 3.4 explains how such a Pareto analysis works in practice.

2.9. Data Sources

The economic and construction data were mainly provided by leading foundation and concrete companies in the Netherlands. Data from Vroom Funderingstechnieken [25] provided the quantities of diesel needed to install the various types of piles (CFA, Vibro, Prefab, and Fundex) and precast beams and the quantities of sacrificial steel for Fundex and Vibro piles. The specifications for the PPC piles were provided by a Dutch precast concrete company. Mebin B.V. [42] provided concrete dosages for cast in situ foundations. Most of the economic data were obtained from the EcoQuaestor database [43], except for the cast in situ concretes, which were provided by Mebin B.V. The installation and removal of machinery on site were not considered in either the environmental or economic costs due to their marginal influence on the study results.

Environmental data were retrieved from the Ecoinvent v.3.5 [44] and Idemat database [45]. The various piles used different types of steel. Nonetheless, the same steel was considered for all reinforcements (steel reinforcement with a working process) because not all steels were found in the consulted databases. In this regard, the importance of retrieving the data from the same databases should be remarked on to ensure a fair comparison. Please see Supplementary Material S2 for more information on the materials/processes and quantities introduced in SimaPro. The transport distances for materials and components were obtained from the literature and are summarized in Table 3. Note that these distances are for a trip by truck; however, in the study two trips by truck were considered (one to deliver the product and then another to return to the empty truck to the factory).

Table 3. Transport distances used for calculation.

Item	Transportation		Distances (km)	Retrieved from
	From	**To**		
Cement	Place of production	Concrete plant Precast concrete plant	75	[14,17,46]
Aggregates	Place of production	Concrete plant Precast concrete plant	40	
Steel reinforce-ment	Place of production	Construction site Precast concrete plant	130	
Concrete	Place of production	Construction site	30	
Soil	Construction site	Landfill sites	30	
Waste	Construction site	Waste management facility	30	
Sawn timber	Place of production	Construction site	50	[46]
Additives	Place of production	Concrete plant Precast concrete plant	100	[14,17]
Precast units	Precast concrete plant	Construction site	150	[47]

3. Results and Discussion

The results of the research are presented and discussed in the subsections below: structural results of the foundation alternatives (Section 3.1), environmental results of only piles (Section 3.2.1), environmental results of foundation alternatives (piles and beams) (Section 3.2.2), economic results of the foundation alternatives (Section 3.3), and the economic-environmental results of the foundation alternatives (Pareto front) (Section 3.4). Please find absolute values of the environmental and economic results in Supplementary Material S5.

3.1. Structural Results

Table 4 shows the study alternatives along with the main characteristics of the alternatives for later conducting the environmental and economic analysis. Please consult Supplementary Materials S3 and S4 for more information on the structural results.

Table 4. Main characteristics of the foundation alternatives.

Foundation Alternative Code	Number	Piles	Concrete	Steel	Steel Reinforcement for Piles
		u	m^3	kg	
1-C600-20/I3.650	1	32	169	4900	7Ø16 (3 m) + Ø25 (10 m)
2-V305-20/I3.500	2	32	70	11,540	8Ø20 *
3-V305-30/I3.500	3	32	70	9469	6Ø20 *
4-V305-40/I3.500	4	32	70	6568	5Ø16 *
5-V356-20/I3.500	5	32	81	6582	5Ø16 *
6-V356-25/I3.500	6	32	81	5132	5Ø12 *
7-F380-20/I3.500	7	32	83	7438	5Ø16 *
8-F460-20/I3.550	8	32	106	7651	7Ø12 *
9-P250-35/I4.500	9	41	66	2406	4Ø6.9 *
10-P350-35/I3.500	10	32	92	3383	4Ø9.3 *
11-V305-20/P3.400	11	32	65	11,468	8Ø20 *
12-V305-30/P3.400	12	32	65	9397	6Ø20 *

Table 4. *Cont.*

Foundation Alternative Code	Number	Piles	Concrete	Steel	Steel Reinforcement for Piles
		u	m³	kg	
13-V305-40/P3.400	13	32	65	6497	5Ø16 *
14-V356-20/P3.400	14	32	76	6510	5Ø16 *
15-V356-25/P3.400	15	32	76	5060	5Ø12 *
16-P350-35/P3.400	16	32	87	3311	4Ø9.3 *
17-C500-20/I4.550	17	32	151	4622	5Ø16 (3 m) + Ø25 (10 m)
18-V273-30/I4.500	18	41	70	10,765	6Ø20 *
19-V273-40/I4.500	19	41	70	7048	5Ø16 *
20-V305-20/I4.500	20	41	78	7911	6Ø16 *
21-V305-30/I4.500	21	41	78	4727	4Ø12 *
22-F380-20/I4.500	22	41	95	6460	5Ø12 *
23-V273-30/P4.350	23	41	63	10,794	6Ø20 *
24-V273-40/P4.350	24	41	63	7077	5Ø16 *
25-V305-20/P4.350	25	41	71	7940	6Ø16 *
26-V305-30/P4.350	26	41	71	4757	4Ø12 *
27-F380-20/P4.450	27	41	93	6554	5Ø12 *
28-P250-35/P4.300	28	41	64	2509	4Ø6.9 *

Terminology: foundation alternative number − type of pile (continuous flight auger (C), Vibro (V), Fundex (F), precast prestressed concrete (P) pile) + pile diameter/side (mm) − pile concrete strength (C20/25 (20), C25/30 (25), C30/37 (30), C35/45 (35), C40/50 (40))/Cast in situ (I) and precast (P) beams + (3) and (4) piles per beam + width of the beam (mm). * Reinforcement all the length of the pile.

3.2. Environmental Results

3.2.1. Piles

Figure 9 compares the GHG emissions of various piles from alternatives with cast in situ beams and three piles per beam. The piles from foundation alternatives with precast beams and four piles per beam are not shown, as they display a similar trend.

If we observe the environmental impact of each element in the construction of a pile (e.g., reinforcement, transport, etc.), concrete and steel play an important role in the environmental results of all piles, representing 65–95% of the impact. In addition, pile driving (i.e., drilling, driving, etc.) can represent up to 20% of the GWP and CED from pile construction, as is the case with Fundex piles, given that for this type of pile diesel is needed not only for piling but also for an external unit for pumping concrete. Additionally, the sacrificial steel in the Fundex pile type can represent up to 15% of GWP, up to 25% of eco-costs, and 20% of ReCiPe and CED. Transportation in precast piles accounts for 15–20% of the environmental impact. However, the transportation of waste derived from installation in all piles has little impact on the environmental results.

If we compare the environmental impact between the piles, it can be observed that piles with the least amount of reinforcement obtained the best environmental results, namely precast piles with the smallest cross-sections (e.g., P250-35) and vibro-piles with the highest concrete strengths. On the other hand, the types of piles that obtained the worst environmental results were those with large amounts of concrete and/or steel. These included CFA piles (e.g., C600-20), vibro-piles with low concrete strengths and/or large amounts of reinforcement (e.g., V305-20), and Fundex piles with larger cross-sections (e.g., F460-20).

If we analyze the influence of the study variables on the environmental results of piles, we can see that prefabricated piles with the smallest cross-sections obtained the best environmental results (e.g., P250-35). Additionally, the increase in concrete strength in vibro-piles led to reductions in GHG emissions, which is in line with a previous study [17]. Finally, four-pile ground beam piles obtained better results than three-pile ground beam piles because the latter have more steel reinforcement to compensate for higher buckling loads.

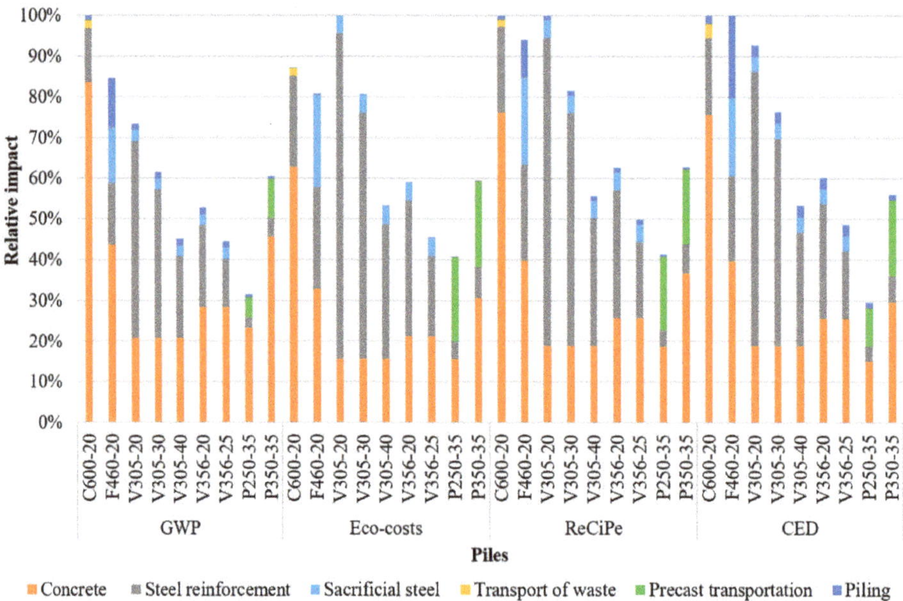

Figure 9. Relative impact of piles from beams with three piles considering the indicators of GWP, Eco-costs, ReCiPe, and CED. Terminology: type of pile (continuous flight auger (C), Vibro (V), Fundex (F), precast prestressed concrete (P) pile) + pile diameter/side (mm) − pile concrete strength (C20/25 (20), C25/30 (25), C30/37 (30), C35/45 (35), C40/50 (40)).

3.2.2. Foundation Alternatives (Piles and Beams)

Materials are the main contributor to environmental impact (85–95%) in all indicators of the foundation alternatives, which is aligned with previous studies [14,17]. Figure 10 shows some of the relevant environmental results to allow for a discussion of the influence of study variables.

In terms of prefabrication, foundation alternatives with small cross-section precast piles (PPCs) obtained the best environmental results in the study (e.g., 9–28). However, the use of prefabricated beams instead of cast in situ beams increased the environmental impact of the foundation alternatives by up to 5% in terms of eco-costs, ReCiPe, and CED and 10% in terms of GWP. This is because concrete in precast beams has a greater impact (more cement) than concrete in cast in situ beams. In this sense, concrete has a special effect on the GWP indicator, while steel is on the eco-cost indicator. Furthermore, precast ground beams require transportation to and installation at the building site, which far outweighs the impact of concrete blinding and the larger volumes of concrete in cast in situ beams. Most likely, as the beam becomes wider (precast or cast in situ), the amount of concrete and steel increases, and consequently, the environmental impact of the beam is higher. However, the prefabrication of ground beams might be interesting from a cradle-to-grave perspective [48], as they can have a prolonged service life (from reuse).

The design of the piles has an important effect on the environmental results of the hole foundation. Nevertheless, the type of pile itself is not a guarantee that the foundation alternative is sustainable, although the reduced amounts of concrete and particularly steel are sustainable. Nevertheless, it should be noted that some types of piles use fewer materials and resources than others to support the same load. The foundation alternatives that resulted in the lowest environmental impact compared to the worst environmental result were those with small cross-section PPC piles (9, 28), which obtained up to 55% lower environmental impacts, and vibro-piles with low amounts of reinforcement, which obtained up to 45% smaller impacts (e.g., 21→4Ø12). Surprisingly, these alternatives have the lowest amounts of steel reinforcement in the piles examined in this study. In contrast, the foundation alternatives that inflicted the greatest environmental impact were those with vibro-piles with large amounts of reinforcement (e.g., 11→8Ø20), CFA piles (e.g., 1→diameter 600) with large amounts of concrete, and Fundex piles with moderately high amounts of concrete and steel, which includes the sacrificial steel (e.g., 8→ diameter 460 + 7Ø12 + 160 kg tip).

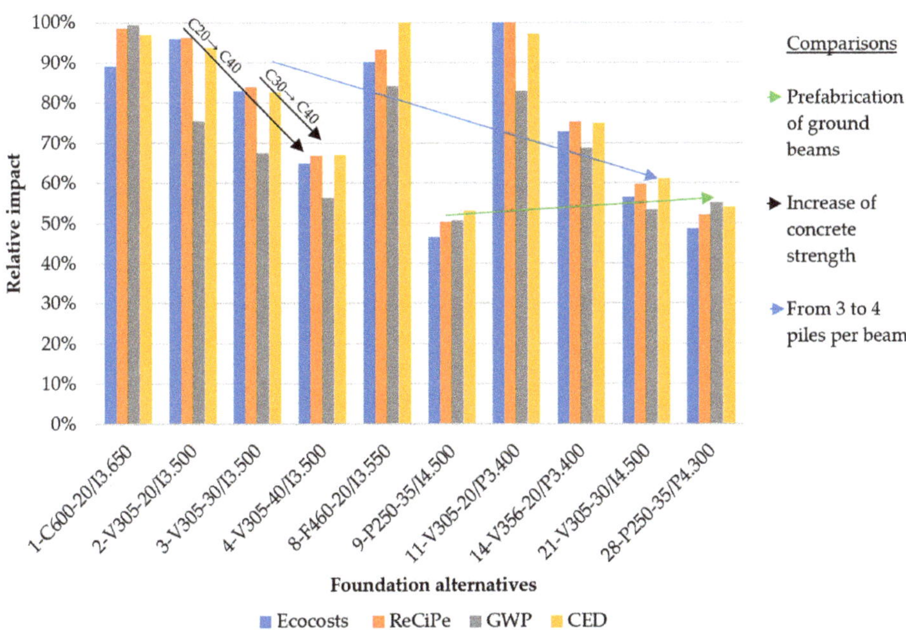

Figure 10. Relative impact of the eco-costs, ReCiPe, GWP, and CED of relevant study foundation alternatives. Terminology: foundation alternative number − type of pile (continuous flight auger (C), Vibro (V), Fundex (F), precast prestressed concrete (P) pile) + pile diameter/side (mm) − pile concrete strength (C20/25 (20), C25/30 (25), C30/37 (30), C35/45 (35), C40/50 (40))/cast in situ (I) and precast (P) beams + (3) and (4) piles per beam + width of the beam (mm).

The increase in concrete strength in vibro-piles from C20/25 to C40/50 reduced the eco-cost, ReCiPe, and CED impacts by up to 30% and GWP by 25% (e.g., 2–4). Similarly, the increase in concrete strength from C30/37 to C40/50 reduced the eco-costs and ReCiPe impacts by up to 25%, and GWP and CED by up to 20% (e.g., 3–4). Surprisingly, vibro-piles that differed only in concrete strength had a similar impact from concrete but a different impact from steel (e.g., 3 and 4). This is explained because the concrete for cast in situ piles must have a minimal cement content of 375 kg/m^3 according to Dutch regulation [15], which assimilates the impact of concrete between cast-in-situ piles. However, the impact of steel between these piles is different because as the higher concrete strength is, the less steel reinforcement is required, since concrete contributes more to resisting the forces.

In terms of the number of piles in beams, four-pile foundations obtained 20–30% fewer impacts in eco-costs and ReCiPe, 15–20% in GWP and 15–25% in CED compared to those with three piles (e.g., 3–21). This is because less steel is required to compensate for the bending moments and shear forces in the beams and buckling forces in the piles.

Given the same beam, alternatives with vibro-piles that fit the beams better (i.e., larger piles) obtained 25–40% lower eco-costs and ReCiPe and up to 30% lower GWPs and CEDs compared to smaller diameter piles (e.g., 11–14). This is because a larger diameter pile has a higher capacity since the axial forces are distributed over a larger surface and therefore the internal forces are smaller, reducing the required steel reinforcement.

3.3. Economic Results of Foundation Alternatives

Piles accounted for 40–60% of the cost of foundation alternatives. Figure 11 shows some of the representative economic results of the study to allow for a discussion of the study variables. Foundation alternatives with piles with larger concrete pile cross-sections incurred the highest economic costs (e.g., 1, 10, 16, 17), while foundations with low-reinforcement vibro-piles with precast beams (e.g., 15, 26) and cast in situ beams (e.g., 6, 21) were the most economical options (up to 40% cheaper). The increase in the strength of concrete in vibro-piles considerably reduced the cost of the foundation by 7–10% from C20 to C30 and from C30 to C40 and up to 15% from C20 to C40. This is because as the strength of the concrete increases, less steel reinforcement is required, although the cost of concrete is slightly higher. Moreover, foundations with four piles per beam were 7–12% more economical than foundations with three piles because the former have less steel reinforcement (e.g., 3–21). Alternatives with vibro-piles that fit the beams better (i.e., larger pile for the same beam) were up to 20% cheaper. Finally, foundations with the same piles and with precast or cast in situ beams obtained similar economic costs (approximately 5% up and down).

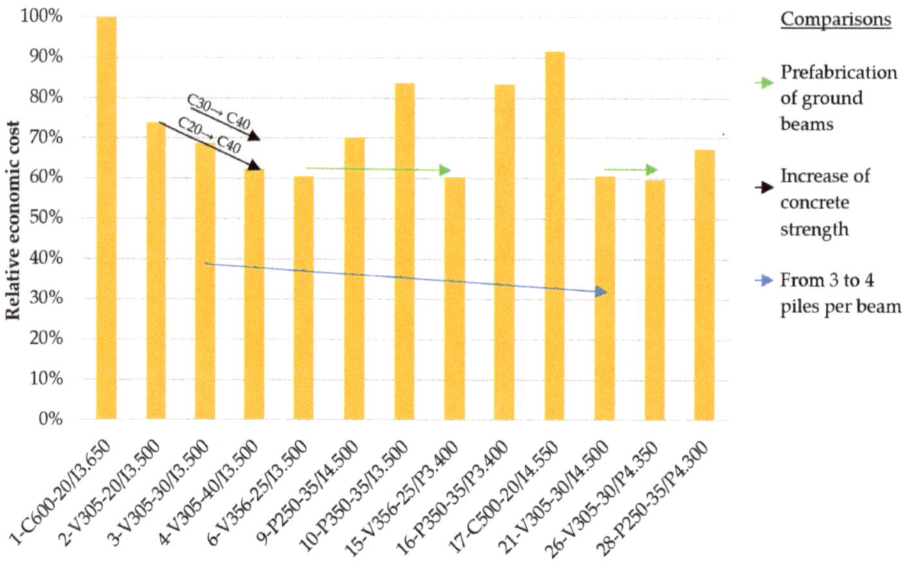

Figure 11. Economic cost of relevant study foundation alternatives. Terminology: foundation alternative number − type of pile (continuous flight auger (C), Vibro (V), Fundex (F), precast prestressed concrete (P) pile) + pile diameter/side (mm) − pile concrete strength (C20/25 (20), C25/30 (25), C30/37 (30), C35/45 (35), C40/50 (40))/cast in situ (I) and precast (P) beams + (3) and (4) piles per beam + width of the beam (mm).

3.4. Environmental-Economic Results of Foundation Alternatives (Pareto Front)

The best choices on the basis of both economy and ecology are shown in Figures 12 and 13. In a one-dimensional system, there is one best choice, but in a two-dimensional system the best choices are given on a line: the so-called Pareto Front. A solution at the Pareto Front has no better alternative in the sense that there are no alternatives that score better on eco-costs and, at the same time, on costs. Such a solution is called Pareto Optimal, or Pareto Efficient [49]. There are many examples of this mathematical concept of multi-objective optimization (MOO), often related to the costs of energy conservation systems in the building industry [50], the refurbishment of buildings [51], and industrial processes [52]. The method is used to select the best solutions out of a cloud of alternatives and has the advantage that such a selection is still free of subjective choices.

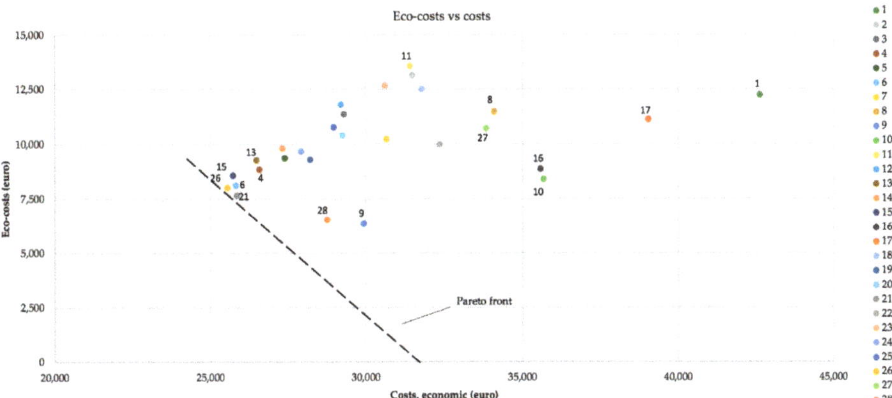

Figure 12. Pareto front of all foundation alternatives, eco-costs vs. costs. Note that the dotted line also depicts true costs = eco-costs + costs = constant (e.g., solutions 26 and 21 have the same true costs).

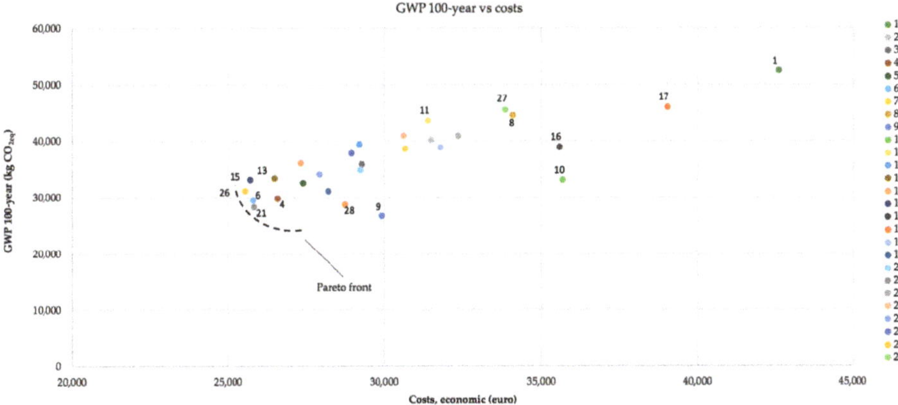

Figure 13. Pareto front of all foundation alternatives, 100-year GWP vs. costs.

A final choice of the best solution on the Pareto front, however, is a matter of a subjective choice—in this case, the relative importance of the eco-costs versus the costs (i.e., "how important is the ecology compared to the economy?"). A recent approach is to minimize the "true costs", where "true costs" = "eco-costs" + "costs" [53], and where "true costs" = "constant" is a straight line in Figure 12. In our case, it is a co-incidence that the Pareto front falls along this straight line. That means that, in this case, an additional subjective criterion must be applied to make a final choice. When the carbon footprint in Figure 13 is expressed in terms of money (e.g., the (future) price of carbon allowances, or the "eco-costs of carbon footprint"), the additional criterion of minimizing the true costs leads to a final choice.

The foundation alternatives that obtained the lowest results from this perspective were those composed of vibro-piles with a medium section (305 mm), low reinforcement (4Ø12), higher compressive concrete strengths (C30), four-pile cast in situ beams (21), and four-pile precast beams (26). It is worth mentioning that the same alternative with greater concrete strength (C40) was not considered because the amount of steel in the pile (21) is already the minimum established by regulations and therefore would have obtained similar environmental results. Second, the best-rated foundation alternatives were those with vibro-piles with a wide cross-section (356 mm), low reinforcement (5Ø12), the highest compressive concrete strength (C25), four-pile precast beams (15) and three-pile cast in situ beams (6). Surprisingly, these alternatives (21, 26, 6, 16) have the lowest amounts of steel reinforcement in cast in situ piles. Prefabricated piles with small cross-sections have even reduced amounts of reinforcement (9, 28). The alternatives with prefabricated piles of small cross-sections obtained the best environmental results (Section 3.2.2.), although from an environmental and economic perspective the aforementioned alternatives with vibro-piles obtained better results because vibro-piles are cheaper (according to the database consulted). In contrast, the alternatives that obtained the worst results were the CFA piles with three and four piles per beam cast in situ (1, 17), as they required the highest amounts of concrete. Second, the worst-rated foundation alternatives were those with Fundex piles (8) and PPC piles (10, 16) with large pile cross-sections and highly reinforced vibro-piles (11). It should be mentioned that the larger the section of a pile was, the greater the width of the beam would be, thus increasing the amount of concrete.

4. Conclusions

An assessment has been presented, from an environmental and economic perspective, of the construction of a conventional building foundation composed of piles and ground beams according to the variables of level of prefabrication (fully precast, semiprecast, and cast in situ), the concrete strength of cast in situ piles (C20/25, C25/30, C30/37, C40/50), the type of pile (precast prestressed concrete (PPC), continuous flight auger (CFA), Fundex, and Vibro piles), and the number of piles per beam (3 and 4 piles). Some of the main conclusions of the study are summarized below.

- None of the study variables guarantees that a foundation is sustainable. However, a combination of selected variables can reduce the environmental impact by up to 55% and the economic costs by up to 40% compared to the worst study alternatives. Materials accounted for 85–95% of the impact of the foundation; therefore, the combination of variables must guarantee a reduction in the impact of materials, particularly of steel on piles.
- From an environmental perspective, it is recommended to use PPC piles with small cross-sections because the environmental impact of the foundation is significantly reduced by up to 45–55% in all environmental indicators. It is also recommended to use vibro-piles with low amounts of reinforcement because the environmental impact can be reduced by up to 40–45%. In contrast, it is not recommended to use CFA piles or highly reinforced vibro-piles in this type of foundation. In addition, the use of prefabricated beams instead of cast in situ beams increases the environmental impact of the foundation by up to 5% in terms of eco-costs, ReCiPe, and CED and up to 10% in GWP over 100 years. Moreover, increasing the compressive strength of concrete in vibro-piles is highly recommended because it reduces the environmental impact of the foundation. For instance, the increase in concrete strength from C20/25 to C40/50 reduces the eco-costs, ReCiPe, and CED by up to 30% and the GWP over 100 years by up to 25%. Finally, the design of four piles per beam instead of three piles per beam can reduce the eco-costs and ReCiPe impact by 20–30%, GWP over 100 years by 15–20%, and CED by 15–25% due to the reduction in steel reinforcing amounts.
- From an economic perspective, it is preferable to select foundations with vibro-piles with high concrete strengths and with four-pile beams either cast in situ or prefabricated. Conversely, it is not recommended to select CPI piles or precast piles with large cross-sections. Additionally, increasing the strength of concrete from C20 to C30 and from C30 to C40 can reduce the cost of the foundation between 7% and 12% and from C20 to C40 by up to 15%. Moreover, designing four piles instead of three can reduce the economic cost of the foundation by up to 12%.

- From an environmental and economic perspective (Pareto front), foundations should have low amounts of concrete and especially steel in the piles. Thus, it is recommended to use piles with reduced cross-sections, as their width also conditions the amount of materials in beams. However, reducing the pile cross-section may increase the amount of steel reinforcement in piles, and then the use of a higher concrete strength can moderate steel amounts. Vibro-piles with higher concrete strengths with cast in situ or prefabricated beams and four piles per beam are the most recommended alternatives from this perspective. Alternatives with CFA piles, Fundex piles, and PPC piles with large cross-sections and three piles per beam are the least recommended.

This paper has highlighted that changing certain common variables in the design of a foundation can significantly reduce the environmental and economic cost of the construction of a foundation. However, it must be considered that each type of pile has its optimal application that depends on many factors, such as the type of soil, loads, regulations, and tradition. Besides this, the economic cost of construction can be variable depending on the location and size of the work and the construction company, among other things. However, an attempt was made to minimize this uncertainty—on the one hand, by starting from a real case that was built, and on the other hand by using the databases and resources conventionally used in real practice. Future research could consider the study variables to optimize other constructive elements (slabs, etc.) and include other variables that lead to a significant reduction in environmental impact and economic cost. Likewise, it is interesting to consider other more sustainable materials (geopolymer concrete, biobased reinforcement, etc.) as well as other optimized designs [48]. This research aims to influence future buildings and codes to contribute to the improvement of environmental sustainability in the construction sector.

Supplementary Materials: The following are available online at https://www.mdpi.com/2071-1050/13/3/1496/s1.

Author Contributions: E.P.-G. was responsible for data curation, investigation, writing—original draft, and writing—review and editing; J.G.V. was responsible for data curation and writing—review and editing; S.P.G.M. was responsible for supervision and writing—review and editing. All authors have read and agreed to the published version of the manuscript.

Funding: This research received no external funding.

Institutional Review Board Statement: Not applicable.

Informed Consent Statement: Not applicable.

Data Availability Statement: Not applicable.

Acknowledgments: We would like to thank the following people for their advice, without whose help this work would never have been possible: Viola Friebel, Justin van der Eerden, Leo Dekker, Tim de Jonge, and Rawaz Kurda. We also want to thank Vroom Funderingstechnieken, Mebin B.V., and EcoQuaestor for the help provided.

Conflicts of Interest: The authors declare no conflict of interest.

References

1. Paris Agreement. Available online: https://ec.europa.eu/clima/policies/international/negotiations/paris_en (accessed on 17 January 2020).
2. IEA. GlobalABC Roadmap for Buildings and Construction 2020–2050—Analysis. Available online: https://www.iea.org/reports/globalabc-roadmap-for-buildings-and-construction-2020-2050 (accessed on 17 January 2020).
3. de Klijn-Chevalerias, M.; Javed, S. The Dutch approach for assessing and reducing environmental impacts of building materials. *Build. Environ.* **2017**, *111*, 147–159. [CrossRef]
4. Demertzi, M.; Silvestre, J.; Garrido, M.; Correia, J.R.; Durão, V.; Proença, M. Life cycle assessment of alternative building floor rehabilitation systems. *Structures* **2020**, *26*, 237–246. [CrossRef]
5. Ingrao, C.; Messineo, A.; Beltramo, R.; Yigitcanlar, T.; Ioppolo, G. How can life cycle thinking support sustainability of buildings? Investigating life cycle assessment applications for energy efficiency and environmental performance. *J. Clean. Prod.* **2018**, *201*, 556–569.

6. Hoxha, E.; Habert, G.; Lasvaux, S.; Chevalier, J.; Le Roy, R. Influence of construction material uncertainties on residential building LCA reliability. *J. Clean. Prod.* **2017**, *144*, 33–47. [CrossRef]

7. Song, X.; Carlsson, C.; Kiilsgaard, R.; Bendz, D.; Kennedy, H. Life Cycle Assessment of Geotechnical Works in Building Construction: A Review and Recommendations. *Sustainability* **2020**, *12*, 8442. [CrossRef]

8. Emami, N.; Heinonen, J.; Marteinsson, B.; Säynäjoki, A.; Junnonen, J.-M.; Laine, J.; Junnila, S. A Life Cycle Assessment of Two Residential Buildings Using Two Different LCA Database-Software Combinations: Recognizing Uniformities and Inconsistencies. *Buildings* **2019**, *9*, 20. [CrossRef]

9. Ondova, M.; Estokova, A. Environmental impact assessment of building foundation in masonry family houses related to the total used building materials. *Environ. Prog. Sustain. Energy* **2016**, *35*, 1113–1120. [CrossRef]

10. Sandanayake, M.; Zhang, G.; Setunge, S. Environmental emissions at foundation construction stage of buildings—Two case studies. *Build. Environ.* **2016**, *95*, 189–198. [CrossRef]

11. Tomlinson, M.J.; Woodward, J. *Pile Design and Construction Practice*, 6th ed.; CRC Press, Taylor & Francis Group: Boca Raton, FL, USA, 2014; ISBN 9781466592636.

12. Ay-Eldeen, M.K.; Negm, A.M. Global Warming Potential impact due to pile foundation construction using life cycle assessment. *Electron. J. Geotech. Eng.* **2015**, *20*, 4413–4421.

13. Bonamente, E.; Cotana, F. Carbon and energy footprints of prefabricated industrial buildings: A systematic life cycle assessment analysis. *Energies* **2015**, *8*, 12685–12701. [CrossRef]

14. Pujadas-Gispert, E.; Sanjuan-Delmás, D.; Josa, A. Environmental analysis of building shallow foundations: The influence of prefabrication, typology, and structural design codes. *J. Clean. Prod.* **2018**, *186*, 407–417. [CrossRef]

15. European Union. *NEN-EN 206+NEN 8005:2017 Beton—Specificatie, Eigenschappen, Vervaardiging en Conformiteit + Nederlandse Invulling van NEN-EN 206*; Nederlands Normalisatie Instituut: Delft, The Netherlands, 2017.

16. Luo, W.; Sandanayake, M.; Zhang, G. Direct and indirect carbon emissions in foundation construction—Two case studies of driven precast and cast-in-situ piles. *J. Clean. Prod.* **2019**, *211*, 1517–1526. [CrossRef]

17. Pujadas-Gispert, E.; Sanjuan-Delmás, D.; de la Fuente, A.; Moonen, S.P.G.; Josa, A. Environmental analysis of concrete deep foundations: Influence of prefabrication, concrete strength, and design codes. *J. Clean. Prod.* **2020**, *244*, 118751. [CrossRef]

18. Sandanayake, M.; Zhang, G.; Setunge, S.; Li, C.-Q.; Fang, J. Models and method for estimation and comparison of direct emissions in building construction in Australia and a case study. *Energy Build.* **2016**, *126*, 128–138. [CrossRef]

19. Zhang, X.; Wang, F. Assessment of embodied carbon emissions for building construction in China: Comparative case studies using alternative methods. *Energy Build.* **2016**, *130*, 330–340. [CrossRef]

20. Lee, M.; Basu, D. Environmental Impacts of Drilled Shafts and Driven Piles in Sand. In Proceedings of the IFCEE 2018, Orlando, FL, USA, 5–10 March 2018; American Society of Civil Engineers: Reston, VA, USA, 2018; pp. 643–652.

21. Lee, M.; Basu, D. Impacts of the Design Methods of Drilled Shafts in Sand on the Environment. In Proceedings of the Geo-Chicago 2016, Chicago, IL, USA, 14–18 August 2016; American Society of Civil Engineers: Reston, VA, USA, 2016; pp. 673–682.

22. Li, X.J.; Zheng, Y.D. Using LCA to research carbon footprint for precast concrete piles during the building construction stage: A China study. *J. Clean. Prod.* **2020**, *245*, 118754. [CrossRef]

23. Pujadas, E.; de Llorens, J.I.; Moonen, S.P.G. Prefabricated Foundations for 3D Modular Housing. In Proceedings of the 39th World Congress on Housing Science: Changing Needs, Adaptive Buildings, Smart Cities (IAHS), Milan, Italy, 17–20 September 2013; ISBN 978-84-16724-93-2.

24. Pujadas Gispert, E. Prefabricated Foundations for Housing Applied to Room Modules. Ph.D. Thesis, Universitat Politècnica de Catalunya, Barcelona, Spain, 2016.

25. Vroom Funderingstechnieken. Available online: https://www.vroom.nl/ (accessed on 17 January 2020).

26. European Union. *NEN-EN 1990+A1+A1/C2:2019/NB:2019 Nationale Bijlage bij NEN-EN 1990+A1:2006+A1:2006/C2:2019 Eurocode: Grondslagen van het Constructief Ontwerp*; Nederlands Normalisatie Instituut: Delft, The Netherlands, 2019.

27. European Union. *NEN-EN 1991-1-1+C1+C11:2019/NB:2019 Nationale Bijlage bij NEN-EN 1991-1-1+C1+C11: Eurocode 1: Belastingen op Constructies—Deel 1-1: Algemene Belastingen—Volumieke Gewichten, Eigen Gewicht en Opgelegde Belastingen voor Gebouwen*; Nederlands Normalisatie Instituut: Delft, The Netherlands, 2019.

28. European Union. *NEN-EN 1992-1-1+C1:2011/NB:2016+A1:2020 Nationale Bijlage bij NEN-EN 1992-1-1+C2 Eurocode 2: Ontwerp en Berekening van Betonconstructies—Deel 1-1: Algemene Regels en Regels voor Gebouwen*; Nederlands Normalisatie Instituut: Delft, The Netherlands, 2020.

29. Technosoft. Balkroosters. Available online: https://www.technosoft.nl/rekensoftware/producten/balkroosters (accessed on 17 January 2020).

30. Bouwbestel. Available online: https://www.bouwbestel.nl/ (accessed on 17 January 2020).

31. European Union. *NEN 9997-1+C2:2017 Geotechnisch Ontwerp van Constructies—Deel 1: Algemene Regels*; Nederlands Normalisatie Instituut: Delft, The Netherlands, 2017.

32. European Union. *NEN-EN 1536:2010+A1:2015 en Uitvoering van Bijzonder Geotechnisch werk—Boorpalen*; Nederlands Normalisatie Instituut: Delft, The Netherlands, 2015.

33. *NVN 6724:2001 Voorschriften Beton—In de Grond Gevormde Funderingselementen van Beton of Mortel*; Nederlands Normalisatie Instituut: Delft, The Netherlands, 2001.

34. ISO. *Environmental Management—Life Cycle Assessment—Principles and Framework*; ISO 14040:2006; International Organization for Standardization: Geneve, Switzerland, 2006; Volume 1997.
35. ISO. Environmental Management—Life Cycle Assessment—Requirements and Guidelines. ISO 14044:2006; International Organization for Standardization: Geneve, Switzerland, 2006.
36. PRé SimaPro 9.2. Available online: https://pre-sustainability.com/ (accessed on 17 January 2020).
37. Kägi, T.; Dinkel, F.; Frischknecht, R.; Humbert, S.; Lindberg, J.; De Mester, S.; Ponsioen, T.; Sala, S.; Schenker, U.W. Session "Midpoint, endpoint or single score for decision-making?"—SETAC Europe 25th Annual Meeting, May 5th, 2015. *Int. J. Life Cycle Assess.* **2016**, *21*, 129–132. [CrossRef]
38. Sala, S.; Cerutti, A.K.; Pant, R. *Development of a Weighting Approach for the Environmental Footprint*; Publications Office of the European Union: Luxembourg, 2018; ISBN 97892796804127.
39. ISO 14008:2019—Monetary Valuation of Environmental Impacts and Related Environmental Aspects. Available online: https://www.iso.org/standard/43243.html (accessed on 17 January 2020).
40. Hrabova, K.; Teply, B.; Vymazal, T. Sustainability assessment of concrete mixes. In Proceedings of the IOP Conference Series: Earth and Environmental Science, Ostrava, Czech Republic, 25–27 November 2019; IOP Publishing: Bristol, UK, 2020.
41. Zula, T.; Kravanja, S. Optimization of the sustainability profit generated by the production of beams. In Proceedings of the 1st International Conference on Technologies & Business Models for Circular Economy, Portorož, Slovenia, 5–7 September 2018.
42. Mebin, B.V. Available online: https://www.mebin.nl/nl (accessed on 17 January 2020).
43. EcoQuaestor. Available online: https://www.ecoquaestor.nl/de-aanpak/ecokosten/ (accessed on 17 January 2020).
44. Ecoinvent. Available online: https://www.ecoinvent.org/ (accessed on 17 January 2020).
45. Idemat. Available online: https://www.ecocostsvalue.com/EVR/model/theory/5-data.html (accessed on 17 January 2020).
46. Kellenberger, D.; Althaus, H.-J. Relevance of simplifications in LCA of building components. *Build. Environ.* **2009**, *44*, 818–825. [CrossRef]
47. Concrete Centre. *The Concrete Centre Sustainability Performance Report*; 1st Report; Concrete Centre: Surrey, UK, 2009.
48. van Loon, R.R.L.; Pujadas-Gispert, E.; Moonen, S.P.G.; Blok, R. Environmental optimization of precast concrete beams using fibre reinforced polymers. *Sustainability* **2019**, *11*, 2174. [CrossRef]
49. Pareto Efficiency Definition. Available online: https://www.investopedia.com/terms/p/pareto-efficiency.asp (accessed on 17 January 2021).
50. Caldas, L.G.; Norford, L.K. Genetic Algorithms for Optimization of Building Envelopes and the Design and Control of HVAC Systems. *J. Sol. Energy Eng.* **2003**, *125*, 343–351. [CrossRef]
51. Ostermeyer, Y.; Wallbaum, H.; Reuter, F. Multidimensional Pareto optimization as an approach for site-specific building refurbishment solutions applicable for life cycle sustainability assessment. *Int. J. Life Cycle Assess.* **2013**, *18*, 1762–1779. [CrossRef]
52. Bernier, E.; Maréchal, F.; Samson, R. Life cycle optimization of energy-intensive processes using eco-costs. *Int. J. Life Cycle Assess.* **2013**, *18*, 1747–1761. [CrossRef]
53. True Cost Economics Definition. Available online: https://www.investopedia.com/terms/t/truecosteconomics.asp (accessed on 17 January 2021).

sustainability

Article

Towards Effective Safety Cost Budgeting for Apartment Construction: A Case Study of Occupational Safety and Health Expenses in South Korea

Kanghyeok Yang [1], Kiltae Kim [2] and Seongseok Go [1,*]

[1] School of Architecture, Chonnam National University, 77 Yongbong-ro, Buk-gu, Gwangju 61186, Korea; kyang@jnu.ac.kr

[2] Goosan Construction, 15-15, Cheonbyeon 3-gil, Damyang-eup, Damyang-gun, Jeollanam-do 57843, Korea; kimkiltaebos0907@naver.com

* Correspondence: ssgo@jnu.ac.kr; Tel.: +82-62-530-1643

Abstract: The construction industry has experienced a lot of occupational accidents, and construction work is considered one of the most dangerous occupations. In order to reduce the number of occupational injuries from construction, the South Korean government legislated the occupational safety and health expense law, requiring companies to reserve a reasonable budget for safety management activities when budgeting for construction projects. However, safety budgets have not been spent based on the risk of accidents, and a large amount of the safety budget is spent either in the beginning or late stages of construction projects. Various accident risk factors, such as activity types, previous accident records, and the number of workers on a construction site, need to be considered when determining the safety budget. To solve such problems, this study investigated the expenditure trends of occupational safety and health expenses for 10 apartment construction projects in South Korea. This study also proposed an accident risk index that can be incorporated with the project costs, schedule, the number of workers, and historical accident records when budgeting for the safety costs. The results from the case study illustrate the limitations of the current planning strategy for safety expenditures and demonstrate the need for effective safety budgeting for accident prevention. The proposed safety cost expenditure guideline helps safety practitioners when budgeting for the occupational safety and health expenses while considering accident risk and the characteristics of safety cost expenditures in practice. The outcome of this research will contribute to the development of regulations for the budgeting of safety costs and help to prevent occupational injuries by providing a reasonable budget for safety management activities in an apartment construction project.

Keywords: occupational safety and health expenses; construction safety; safety cost expenditures; apartment construction

Citation: Yang, K.; Kim, K.; Go, S. Towards Effective Safety Cost Budgeting for Apartment Construction: A Case Study of Occupational Safety and Health Expenses in South Korea. *Sustainability* **2021**, *13*, 1335. https://doi.org/10.3390/su13031335

Academic Editor: Sunkuk Kim

Received: 10 December 2020

Accepted: 19 January 2021

Published: 27 January 2021

Publisher's Note: MDPI stays neutral with regard to jurisdictional claims in published maps and institutional affiliations.

1. Introduction

The construction industry in South Korea has rapidly grown over the last few decades [1], with infrastructure and residential facility constructions to accommodate the rapid expansion of the major cities. However, the construction industry has experienced a lot of occupational injuries, and construction work is considered one of the most dangerous occupations due to the dynamic and temporary nature of the workplace [2–4]. Specifically, most construction work takes place outdoors and work conditions (e.g., temperature, humidity, and light conditions) and the number of required workers frequently changes, which increases the difficulty of safety management. According to the Korea Occupational Safety and Health Agency (KOSHA), fatal injuries of construction workers have been increasing since the year 2000. In 2018, the construction industry experienced the highest number of fatal accidents, accounting for 49.95% of total fatalities in South Korea [5]. The safety of construction workers is a global issue. In 2011, the construction industry employed almost

7% of the world's workforce, while the industry recorded 30–40% of the world's fatal injuries [6]. There are many different contributing factors associated with the occurrence of fatal accidents [7–9], but one of the major issues is the lack of appropriate countermeasures to reduce the risk of accidents in construction environments [10]. The South Korean government legislated the occupational safety and health expense law to ensure companies secured a minimum safety management budget, the size of which depends on the size and type of the construction project. The budget for safety costs for a general construction project must equal or exceed 1.86% of the total material and labor costs. Construction projects that have budgets less than 500 million Korean republic Won (KRW) are required to have a higher ratio (e.g., 2.93%) compared to that of projects with more than 500 million KRW of construction costs, to help protect workers in smaller construction projects.

The safety expense law aims to enhance the safety of construction workplaces and restricts the use of the safety budget to certain types of expenses. Specifically, the safety budget is only available for performing safety management activities such as purchasing personal protective equipment, safety education, safety consulting by experts, and so on. The safety expense law also includes a regulation that requires a construction firm to spend a certain amount of the safety budget according to the progress of the construction project. This regulation is adopted to effectively protect workers from an accident by requiring the firm to spend money on safety, but the requirement is insufficient for accident prevention. The regulation enforces construction firms to spend more than half of the budget before completion of 70% of the construction project. Under these conditions, safety budgets are spent either at the early or the late stages of the construction project. In addition, safety related studies argue that safety-cost planning in practice does not consider the risk of ongoing activities, which is not suitable to effectively prevent accidents at construction sites [11,12]. To reduce the accident risk and protect construction workers from accidents, safety costs should be allocated based on the risk of the on-going construction activity. In addition, the safety budget needs to be allocated while considering various safety-related risk factors such as the number of workers, historical accident records, and other site conditions. In short, there is a definite need to analyze how to effectively use the safety budget to decrease the risk of accidents and to advance the safety of construction sites.

To address current issues in safety-cost planning, this study first investigated the budgeting and execution of safety and health expenses by conducting a case study analysis of 10 apartment construction sites in South Korea. The results from this case study illustrated the current problems in safety-cost budgeting and executions in practice. In addition, data on factors related to accident risk (e.g., cost, schedule, number of workers on-duty) were also collected from case study sites to comprehensively assess the accident risk during construction projects. The accident risk index was lastly proposed to consider the abovementioned risk factors (i.e., construction schedule, construction costs, number of workers, and historical accident records) in the effective planning of budgets for safety costs. The recommendation for the expenditure of the safety budget is presented to facilitate the outcome of this study and help safety practitioners perform effective safety management activities. The remaining sections of the manuscript are organized as follows. The research background reviewed the previous research on accident prevention, safety management activities, and safety cost budgeting. The material and method section describe how to compute the accident risk index and analyze the trend of safety cost budgeting in apartment constructions. The remaining sections explain the results of the analysis and the conclusion of this research.

2. Literature Review

The safety of a construction worker is an important issue in many nations, since construction environments are complex and often unsafe due to their dynamic and labor-intensive characteristics (e.g., largely relying on a worker's labor and heavy equipment) [13]. In addition, construction works are often placed at elevations that could highly increase the risk of accident. The weather conditions are other factors that could adversely affect the safety of a worker on a construction site. As a result, the construction industry has recorded a poor safety performance and experienced a lot of fatal and non-fatal injuries [14]. According to Shafique and Rafiq (2019), the construction industry accounted for around 20% of the occupational fatalities that occurred in Japan, United Kingdom, the United States, and Hong Kong in 2017 [15]. The research from the Workplace Safety and Health institute (WSH) illustrated that construction sites in Asia experienced a greater number of fatal injuries compared to the sites in other continents. In South Korea, a large number of fatal injuries also occurred during construction [5]. Among the various types of accidents that can occur, falling from a great height is the leading cause of fatalities [16] and therefore the prevention of fall accidents is a critical issue for the safety of construction workers [17–20]. The Occupational Safety and Health Organization (OSHA) in the United States forced employers to provide a fall protection system that can prevent fall accidents when the work surface is located over 1.82 m (i.e., 6 feet) above the ground or a lower floor. The guardrail, safety net, and personal fall arrest system are the examples of the fall protection systems and the employees should not start their works before the installation of such fall protection systems in the workplace.

The occurrence of accidents are related to the various factors and previous research emphasized the significance of two accident-related factors, which are unsafe work environments and psychological/behavioral characteristics of an individual worker [21–23]. The above-mentioned fall protection systems are used for improving the safety of the workplace by modifying the work environments. However, an individual's unsafe behavior is a persisting issue, since a large portion of construction works are performed by a worker's hand or by manually using equipment. Behavior-based safety is the one solution that can prevent accidents originating from a worker's unsafe behaviors [24]. Several previous studies indicated that more than 80% of accidents could be attributed to a worker's unsafe behaviors [25–27]. Ascending/descending using stairs without holding a guardrail is an example of a worker employing unsafe behavior. Poor housekeeping in a construction site is a result of unsafe worker behavior that can involve neglecting activities such as storing equipment or cleaning the floor after completing a task. A worker's unsafe behaviors are often triggered by factors such as needing to meet excessive production targets, a competitive atmosphere, a tight construction schedule and a lack of available resource [28]. Also, inappropriate safety management activities conducted by the safety manager could strengthen a worker's attitude toward unsafe behavior during the construction process. In short, the prevention of accidents during construction is a complicated issue and it requires various efforts to be addressed including the improvement of the work environment, safety related education, safety observations, and proper safety interventions. Also, financial resources for safety management activities are vital for the success of accident prevention in construction.

Safety management is an important research topic to decide the proper amount of safety costs and to quantity the risk of accidents occurring for effective safety budget allocations. A study by Pinto et al. (2011) analyzed the financial costs of construction-related accidents. The occupational injuries assessed did not only badly affect the worker's wellbeing but also adversely affected the cost of the construction projects due to requiring high medical costs [29]. According to the analysis from Everett and Frank (1996), the occupational injuries from non-residential construction projects account for 7.9% to 15% of total construction costs [30]. This research illustrated that the prevention of occupational injuries is essential for both a worker's safety and the success of a construction project. The risk of accidents occurring is commonly defined as the significance of these risky events in terms of the occurrence probability and the severity of a potential injury [31–33]. The previous risk assessment studies utilized the analytic hierarchy process (AHP) technique—which is a structured multi-attribute decision method for complex decision making—while maintaining consistency of experts' judgements [34]. The AHP technique has been utilized to rank various safety factors by assessing the severity and the probability of accidents [35] or injuries [36]. Such a risk assessment technique is beneficial for effectivity quantifying the risk level of accident-related factors, but the process largely relies on subjective decisions, which are prone to being biased. Also, assessment results from the previous studies are not suitable for safety cost budgeting at the project level, since they were conducted to rank different types of hazards. Further research on developing a safety risk index that includes the influences of the factors related to the risk of accidents is essential for effective safety cost budgeting at the project level.

Occupational safety and health expenses were legislated by the Korean government under the law to require securing appropriate budgets for safety management activities. The amount of safety costs required are determined based on the type and size of the relevant construction projects. Specifically, the safety budget is a proportion of the total labor and material costs. Table 1 presents the 15 categories for safety cost budgeting, which is classified by the type of works being conducted (i.e., 5 different construction types) and the total amount of construction costs (i.e., 3 different cost ranges). The usage of the safety cost is limited to the (1) labor costs of safety managers, (2) costs for protective equipment, (3) costs for personal protective equipment, (4) external safety inspection or consulting fees, (5) costs for safety education, (6) health care fees for workers, (7) safety technology fees, and (8) costs for the safety organization to be established in the construction headquarters. The safety costs play an important role in enhancing the safety level of the construction site; however, the allocations of the safety budget are still not optimal in terms of the prevention of accidents in the construction industry. For example, the risk levels for construction works are different depending on the stage of the construction project. As previously described, the risk level of falling accidents is not significant at the initial stage of the construction project, since the excavation and the foundation works are the main construction activities being completed at this stage. Also, the most of construction works at the late stage of the construction project are the finishing works, which are generally performed when the structural works of the building are completed. As a result, the risk of falling accidents at the late stage would not be significant compared to during the middle stage of the construction project. A safety cost expenditure guideline is beneficial to effectively allocate the safety budget for accident prevention and prevent occupational injuries. In this context, this study firstly investigated the expenditures of safety budgets using data from apartment construction projects and proposed an accident risk index and safety cost expenditure guideline to enhance the safety of construction workplaces and protect construction workers.

Table 1. Occupational safety and health expense rates by type and size of construction projects.

Construction Types	Sizes of Construction Projects			Projects Required to Hire a Safety Manager **
	Smaller Than 500 million KRW *	Between 500 million KRW * and 5000 million KRW * (Baseline Cost)	More Than 5000 million KRW *	
General Construction (A)	2.93%	1.86% (5.349 million KRW *)	1.97%	2.15%
General Construction (B)	3.09%	1.99% (5.499 million KRW *)	2.10%	2.29%
Heavy Construction	3.43%	2.35% (5.4 million KRW *)	2.44%	2.66%
Railway Construction	2.45%	1.57% (4.411 million KRW *)	1.66%	1.83%
Special Construction	1.85%	1.20% (3.25 million KRW *)	1.27%	1.31%

* KRW: Korean Republic Won. ** defined by the Occupational Safety and Health Act in South Korea.

3. Materials and Methods

3.1. Occupational Safety and Health Expenses in Case Study

This research study analyzed the expenditures on occupational health and safety expenses in apartment construction projects. The number of housing units, total construction periods, and total construction costs were considered during the selection of the construction sites. A total of 10 construction sites with a similar size, number of construction periods, and construction dates (between 2015 and 2017), were selected for the case study analysis (See Table 2 for details) to avoid possible distortionary issues hindering comparisons, such as inflation and temporal material shortages during the construction process.

Table 2. Information on the apartment construction sites for the case study.

	Median	Min	Max
Number of Housing Units (EA)	356	303	410
Construction Periods (Months)	28	24	32
Construction Costs (million KRW)	37,900	32,500	45,000
Completion Date (YYYY-MM)	2016-05	2015-02	2017-11

The 2.29% proportion of material and labor costs for construction projects were applied for the occupational health and safety expenses in construction sites for the case study. In the analysis, safety expenditures were categorized as (1) labor costs for safety professionals, (2) costs for protective devices and facilities for safety activities, (3) costs for personal protective equipment, and (4) costs for other safety related activities (e.g., safety consulting, safety education, and so on). The details of the safety budgeting and expenditures are summarized in Table 3.

Table 3. Budgeting and expenditure for occupational safety and health expenses from 10 case study sites.

Case Study (10 Sites)	Safety Cost Budgeting (Million KRW)			Safety Cost Expenditure Ratio (%)			
	Labor Costs	Material Costs	Safety and Health Expenses	Labor Costs	Protective Device/Safety Facility	Personal Protective Equipment	Other Safety Activities
Site A	12,478	13,212	588.3	48.7	32.6	11.2	7.5
Site B	11,700	11,700	535.9	46.3	35.7	11.9	6.1
Site C	12,639	12,639	622.7	44.9	35.4	13.2	6.5
Site D	15,750	15,750	721.4	50.3	31.8	12.7	5.2
Site E	12,032	12,032	611.3	46.2	34.5	14.5	4.8
Site F	15,365	15,365	723.8	47.4	32.3	12.5	6.8
Site G	11,154	11,154	557.3	51.8	30.6	11.4	6.2
Site H	14,580	14,580	677.1	46.1	33.2	14.6	6.1
Site I	12,578	12,578	560.9	46.5	32.0	13.9	7.6
Site J	14,972	14,972	658.6	50.4	33.6	11.1	4.7

Most of the safety budget was spent on the labor of safety professionals, protective devices, and personal protective equipment. On average, 47.9% of the safety budget was spent on the labor of safety professionals. The costs for protective devices and personal protective equipment were determined to be 33.2% and 12.7%, respectively. The other safety activities were 6.2% of the total safety budgets. As shown in Table 3, the labor costs and protective device/safety facilities were the two major components of the safety budget (81.1%) while other safety activities accounted for only a small portion of the budget (18.9%). Considering the fact that the labor costs would be spent evenly during the entire construction period, the expenditure trends (as shown in Figure 1) revealed that safety management related activities, including installation of the protective devices and purchasing the personal protective equipment, can be performed irregularly and this would increase the risk of accidents occurring on construction sites. The analysis results demonstrate the need for better safety budget planning to enhance the safety level of the construction sites. This study introduced the accident risk index to be incorporated with construction site information (i.e., current progress, the number of workers, and construction costs) and historical accident records (i.e., the number of accidents associated with each activity) in safety cost budgeting. The accident risk index can be calculated in accordance with the construction progress (i.e., 0% to 100%) to determine the appropriate expenditure for occupational safety and health expenses at the project level.

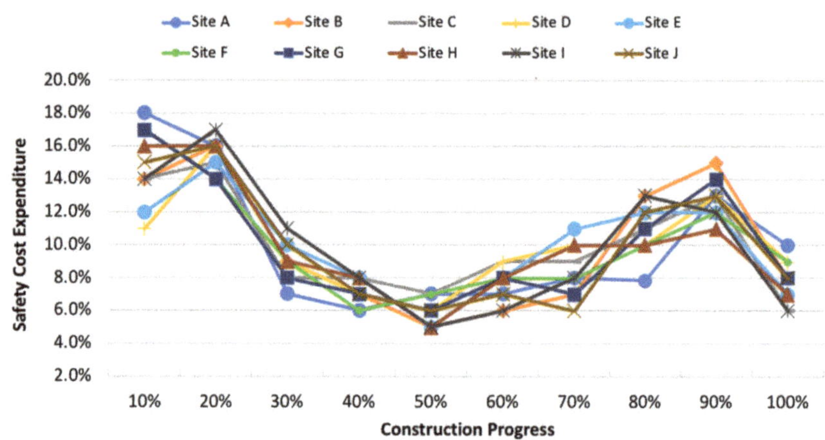

Figure 1. Expenditures on occupational safety and health expenses in apartment construction.

3.2. Accident Risk Index

The proposed index in this study considers: (1) the number of major accidents, (2) the number of workers on duty, (3) the required working time for the construction work, and (4) the amount of progress payments needed to measure the risk of accidents occurring. These factors are decided based on the fact that the risk level of an accident occurring during a certain stage of the project on a construction site is related to the number of workers on-duty, the progress of the project, the amount of payments that have been made linked to the progress of the project, and previous accident records. In fact, the factors related to the occurrence of accidents are numerous and the selection of these factors is highly related to the risk assessment level. Specifically, risk assessment could be performed with the various risk factors being assessed at different management levels (e.g., the task, activity, and project) and the level of risk management would affect the type of risk factors to consider. As examples, the task location, age of the workers, previous injury records, and the levels of experience could be included when managing the risk of accidents occurring at the task level. However, this study specifically aimed to assess risk management at the project level and the corresponding risk factors that are the information available at that level. The research from Gurcanli et al. (2015) utilized the total construction cost, number of required workers, required construction time, and the risk of completing various activities to decide a reasonable safety budget amounts and budget allocation for construction projects [37]. Similar to the previous study, four attributes were utilized in this study (i.e., the progress ratio, cost ratio, worker ratio, and risk ratio) and such attributes are measured by Equations (1)–(4). The progress ratio is the proportion of the required working time for an activity to the total construction time. The cost ratio is the proportion of construction costs for an activity to the total construction costs. The worker ratio is the proportion of the number of required workers for an activity to the total number of construction workers. The risk ratio is the proportion of the number of major accidents while completing an activity to the total number of major accidents in historical accident data. The accident risk index, representing the risk level of accidents at a certain period by combining the four aforementioned attributes, is calculated by Equation (5):

$$\text{Progress Ratio}(i) = \text{Construction Time}(i)/\text{Total Construction Time} \times 100 \qquad (1)$$

$$\text{Cost Ratio}(i) = \text{Construction Cost}(i)/\text{Total Construction Cost} \times 100 \qquad (2)$$

$$\text{Worker Ratio}(i) = \text{Construction Worker}(i)/\text{Total Construction Worker} \times 100 \qquad (3)$$

$$\text{Risk Ratio}(i) = \text{Number of Major Accidents}(i)/\text{Total Number of Major Accidents} \times 100 \qquad (4)$$

$$\text{Accident Risk}(i) = [\text{Progress Ratio}(i) + \text{Cost Ratio}(i) + \text{Worker Ratio}(i)] \times \text{Risk Ratio}(i) \qquad (5)$$

where Construction Time(i), Construction Cost(i), Construction Worker(i), and Number of Major Accidents(i) are the construction time, construction costs, number of workers, and the number of major accidents for each activity or period(i), respectively.

The statistical data published by the KOSHA was utilized to calculate the risk ratio by measuring the number of major accidents that occurred during construction. The major accidents were defined by the KOSHA as accidents that resulted in a fatality or an illness requiring medical care after more than 3 months, or accidents where injuries affected more than 10 workers at once. This study utilized the historical accident data collected between 2014 and 2016 and a total of 118 major accidents were recorded. Laborers were recorded to have the largest number of major injuries (i.e., 27 accidents) while scaffolders and carpenters had the next highest numbers (i.e., 21 and 17 injuries, respectively). This study utilized the progress percentage, representing 10 different levels of construction progress (e.g., 10%, 20%, and 100%) from the beginning to the end of the construction project, to ease the implementation of the proposed accident risk index.

4. Case Study Data Analysis

4.1. Trends of Safety Expenditures and Computations for Progress, Cost, Worker, and Risk Ratios

The expenditure on safety costs was reorganized corresponding to the construction project progress (e.g., every 10% progress) to investigate the problems for the expenditures in practice. The expenditures from 10 case study sites were then analyzed and detailed results are presented in Figure 1 and Table 4. As shown in Figure 1, more than 30% of safety budgets were spent before 20% of the projects' completion had occurred and relatively small amounts of safety budgets were used for 30% to 70% of the projects' progress. However, according to the historical accident data from the KOSHA, more than 50% of the major accidents occurred between 30% and 70% of the projects' completion (as shown in Figure 2). Such facts illustrated the problem of current safety cost expenditures and the need for better safety budget planning while considering the risk of accidents occurring to effectively prevent a major accident on a construction site.

Table 4. Details of safety cost expenditure corresponding to construction progress (Unit: thousand KRW).

Case Study (10 Sites)	Construction Progress									
	10%	20%	30%	40%	50%	60%	70%	80%	90%	100%
Site A	105.89	94.13	41.18	35.30	41.18	41.18	47.06	46.06	76.48	58.83
Site B	75.02	85.74	48.23	37.51	26.79	32.15	37.51	69.66	80.38	42.87
Site C	87.18	93.41	49.82	49.82	43.59	56.04	56.04	68.50	80.95	37.36
Site D	79.35	115.42	64.92	50.49	43.28	64.92	72.14	72.14	93.78	64.92
Site E	73.36	91.70	61.13	48.91	30.57	48.91	67.25	73.36	73.36	42.79
Site F	123.05	101.34	65.14	43.43	50.67	57.91	57.91	72.38	86.86	65.14
Site G	94.74	78.02	44.58	39.01	33.44	44.58	39.01	61.30	78.02	44.58
Site H	108.33	108.33	60.93	54.16	33.85	54.16	67.70	67.70	74.47	47.39
Site I	78.53	95.36	61.70	44.87	28.05	33.65	44.87	72.92	67.31	33.65
Site J	98.80	105.38	65.86	46.11	39.52	46.11	39.52	79.04	85.62	52.69

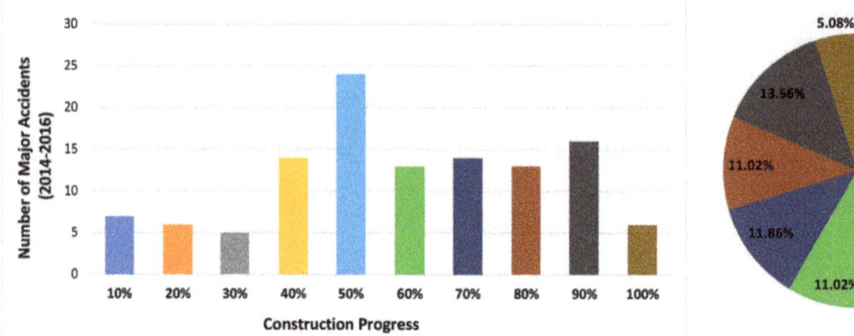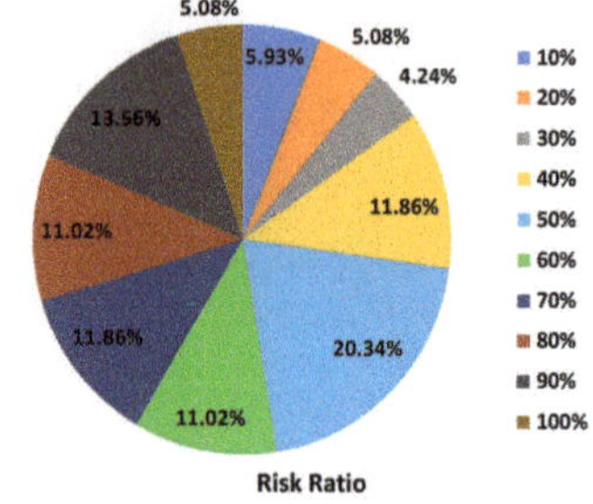

Figure 2. Number (**left**) and proportion (**right**) of major accidents that occurred between 2014 and 2016.

Data for attribute computations from the construction sites were also reorganized. The required time for the completion of an apartment construction was 730 days on average. The largest amount of project time (i.e., 146 days) was spent on the 0% to 10% progress stage, which includes the excavation and foundation works. The second largest time (i.e., 102 days) was required for the 80% to 90% progress stage, which is the period for the fishing works. The cost ratio was analyzed similarly, and the results showed that costs were evenly spent during the whole construction progress. The largest cost ratio is 12.5% at the end of the project (i.e., the 90% to 100% stage) and the lowest cost ratio is 6.0% between 20% and 30% progress.

The worker ratio, which is the ratio of the number of workers on-duty to the total number of workers, was further analyzed to investigate the change of the required workforce corresponding to the construction progress. On average, a total of 726 workers participated in an apartment construction project and the construction progress from 50% to 60% employed the largest number of workers, which was 18.2% of the total number of workers (See Table 5 for details). This construction period was the moment when both structural and finishing works were performed simultaneously. Similar to the historical accident records, construction works between 40% and 80% progress employed 74.1% of the total number of workers but corresponding safety expenditure was only 40.1% (See Figures 1 and 2 for details). This fact might be one of the reasons why the construction industry recorded a high rate of accidents, considering the comparably low safety cost expenditure for these periods.

Table 5. Number of workers on-duty corresponding to construction progress.

Case Study (10 Sites)	Construction Progress									
	10%	20%	30%	40%	50%	60%	70%	80%	90%	100%
Site A	30	45	56	90	110	136	120	91	29	23
Site B	25	33	53	75	89	95	108	92	24	22
Site C	33	50	58	91	113	141	121	95	30	25
Site D	36	51	67	103	119	156	127	101	49	28
Site E	28	42	55	85	108	135	115	92	39	21
Site F	33	47	63	96	101	142	119	97	42	26
Site G	26	38	52	76	94	115	102	83	31	19
Site H	35	51	59	95	117	153	127	98	32	25
Site I	26	36	51	72	91	99	104	90	30	20
Site J	21	49	62	98	115	158	131	99	35	21

The risk ratio was lastly computed from the historical accident records collected between 2014 and 2016. The largest risk ratio is observed during the 40% and 50% construction progress stage that accounts for 20.34% of the major accidents (See Figure 2 for details). The construction period from 40% to 80% showed a higher accident risk (i.e.,

55.08%) compared to other periods but safety cost expenditure during this period (i.e., 29.01%) was relatively small, as previously described. These facts demonstrate the existing safety cost expenditure problem and illustrate the necessity of an accident risk index for better safety budgeting for apartment construction projects.

The computed ratios associated with the accident risk are summarized in Table 6. Also, the ratios and the trend of safety expenditures are visualized in Figure 3. The trend of the safety expenditure seems to be similar to the progress ratio, while the worker ratio shows a similar pattern to the risk ratio. These results would imply that the current expenditure of the safety cost is related to the required construction time, while the accident risk has a relationship with the number of workers on-duty. These results also indicate the need for an accident risk index to comprehensively assess various safety related factors and better safety budgeting for construction safety.

Table 6. Results of the progress ratio, cost ratio, worker ratio, and risk ratio computations.

Attributes		Construction Progress									
		10%	20%	30%	40%	50%	60%	70%	80%	90%	100%
Progress	Avg. (days)	146	37	28	42	62	79	86	95	102	53
	Ratio (%)	20.00	5.07	3.84	5.75	8.49	10.82	11.78	13.01	13.97	7.26
Cost	Avg. (million KRW)	3872	2112	3168	3520	4224	2816	2992	4083	3942	4370
	Ratio (%)	11.03	6.02	9.02	10.03	12.03	8.02	8.52	11.63	11.23	12.48
Worker	Avg. (workers)	29	44	58	88	106	133	117	94	34	23
	Ratio (%)	4.04	6.06	7.95	12.12	14.57	18.20	16.21	13.00	4.68	3.17
Risk	Events	7	6	5	14	24	13	14	13	16	6
	Ratio (%)	5.93	5.08	4.24	11.86	20.34	11.02	11.86	11.02	13.56	5.08

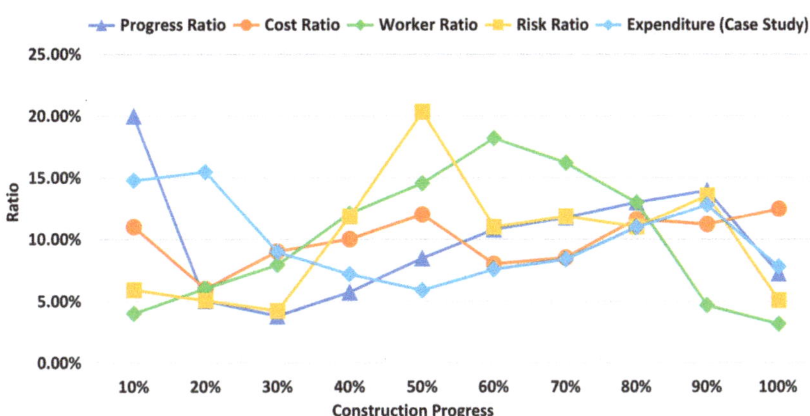

Figure 3. Visualization of the computed attributes and safety cost expenditure from case study sites.

4.2. Accident Risk Index for Safety Cost Budgeting

The accident risk index was computed to determine the construction periods containing a high risk of accidents and to effectively budget safety costs. The computation results illustrated that the 40% to 50% progress stage has the highest accident risk (i.e., 717.73), which is equivalent to 22.27% of the total accident risk. The second largest accident risk was observed at the 60% to 70% progress stage and this period contained the second largest number of workers on-duty. The analysis results indicated that construction periods between the 40% to 90% progress stage accounted for 74.08% of the total accident risks. The results also demonstrated the importance of enhancing the safety management efforts during these periods (See Table 7 for details).

Table 7. Accident risk index and recommended schedule for the safety cost expenditure.

	Construction Progress									
	10%	20%	30%	40%	50%	60%	70%	80%	90%	100%
Accident Risk Index	207.97	87.12	88.23	330.89	713.73	408.18	433.01	414.79	405.17	116.38
Accident Risk Ratio	6.49	2.72	2.75	10.32	22.27	12.73	13.51	12.94	12.64	3.63
Rank	7	10	9	6	1	4	2	3	5	8

The differences between the accident risk index and safety expenditures were investigated to suggest a guideline for better safety budgeting. The government's safety law forced companies to spend more than 50% and 70% of their construction safety budgets before 70% and 90% completion of their construction project, respectively. Considering the computed risk index figures during construction projects, more detailed recommendations are necessary to effectively utilize the safety budget and increase the effectiveness of the safety management activities to avoid major accidents. As shown in Figure 4, a huge gap between the computed accident risk ratio and the safety expenditures from analysis of the case studies was found in terms of safety cost executions. Considering that the accident risk index represents the risk level of construction environments during certain periods, the safety expenditure guideline rate is calculated by finding the average of the accident risk ratio and the expenditure ratio to consider the importance of completing various on-site safety management activities. For example, safety facilities should be built at the initial stage of the construction projects and purchasing protective equipment, including personal protective equipment, should be completed before the start of certain construction works. Therefore, this study proposed a safety cost expenditure guideline (shown in Table 8) that consider both the characteristics of construction projects and the risk of accidents occurring for on-going construction projects. Considering that current legislation forces companies to spend more than 50% of their budgeted safety costs before 70% completion of their construction projects, the recommended guideline could help to budget safety costs for high risk periods by considering the accident related factors on construction sites.

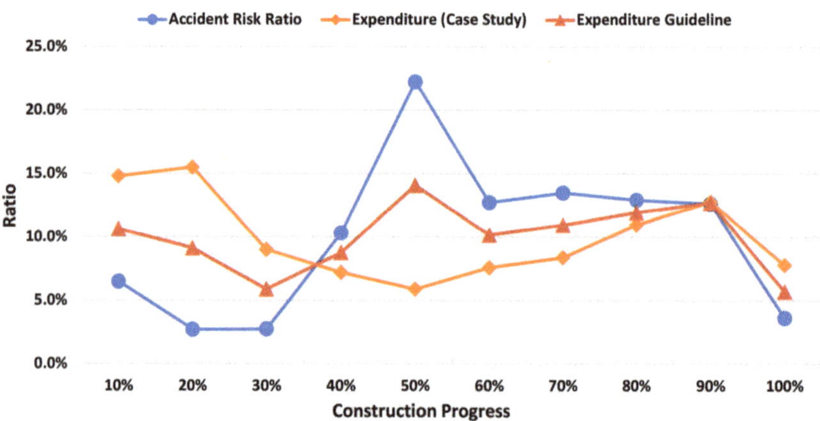

Figure 4. Visualization of the accident risk ratio, safety cost expenditure (Case Study), and safety cost expenditure guideline.

Table 8. Comparison of the accumulated accident risk ratio, accumulated expenditure ratio from the case study analysis, expenditure guideline rate provided by the occupational safety and health expense law, and expenditure guideline rate recommended by this study.

	Construction Progress									
	10%	20%	30%	40%	50%	60%	70%	80%	90%	100%
Accident Risk Ratio	6.49	9.21	11.96	22.28	44.55	57.28	70.79	83.73	96.37	100
Expenditure Ratio (Case Study)	14.80	30.31	39.31	46.51	52.41	60.01	68.41	79.40	92.20	100
Expenditure Guideline (Law)					50	50	70	70	90	
Expenditure Guideline (Recommended)	10.65	19.76	25.63	34.39	48.48	58.65	69.60	81.56	94.28	100

5. Conclusions

The safety of construction workers is one of the important management factors for ensuring the success of the construction projects. This study investigated the expenditure for safety and health expenses by conducting a case study with 10 apartment construction projects in South Korea. The safety expenditures derived from the case study revealed that most of the mandated safety costs were spent during the initial and the last stages of the assessed construction projects and these expenditures did not correlate well with the risk level of accidents occurring during different stages of apartment construction. At the 50% progress stage, the highest accident rate and the lowest safety expenditure rate were observed, and this is a significant problem for the prevention of accidents on construction sites. In addition, the Occupational Health and Safety Expense Law does not have a detailed expenditure guideline covering the 0% to 40% construction progress stage, although almost 30% of the major accidents occur during this stage. To address these problems, this study proposed an accident risk index that can incorporate the construction schedule, construction costs, the number of workers on-duty, and historical accident records in the safety cost budgeting. The safety cost expenditure guideline was also developed by combining the characteristics of the safety cost expenditure in practice and the risk level of accidents occurring corresponding to the construction schedule. The proposed accident risk index would offer information about the risk level of on-going construction activities. The recommended expenditure guideline helps to understand the required safety management efforts for the accident preventions corresponding to the construction schedule. Considering that the legal expenditure guideline would not provide any information about the risk of accidents occurring, the proposed guideline in this study will help safety practitioners to perform effective safety cost budgeting while considering the accident risks and enhancing the level of safety management for apartment constructions.

However, several limitations remained to be assessed by future research. There are many contributing factors (e.g., the construction methods, size of construction projects) to the occurrences of accidents on construction sites, but this study utilized only small sets of attributes for the computation of the accident risk index and the development of the safety cost expenditure guideline. Future research would be essential to investigate the relationships between the occurrence of accidents and various accident-related factors in apartment construction. In addition, this study performed safety cost budgeting corresponding to the construction schedule. Another important issue is how to utilize the safety costs for enhancing the safety of construction sites. The effectiveness of safety management activities needs to be further investigated to find the optimal use of safety budgets to increase construction safety.

Author Contributions: Conceptualization, K.Y., K.K., and S.G.; data curation, K.Y., K.K., and S.G.; writing original draft, K.Y., K.K., and S.G.; writing review and editing, K.Y., K.K., and S.G. All authors have read and agreed to the published version of the manuscript.

Funding: This work was supported by the National Research Foundation of Korea (NRF) grant funded by the Korea government (MSIT) (No. 2019R1G1A1100365).

Data Availability Statement: The data presented in this study are available upon request.

Conflicts of Interest: The authors declare no conflict of interest.

References

1. Choi, S.D.; Guo, L.; Kim, J.; Xiong, S. Comparison of Fatal Occupational Injuries in Construction Industry in the United States, South Korea, and China. *Int. J. Ind. Erg.* **2019**, *71*, 64–74. [CrossRef]
2. Fang, D.; Wu, H. Development of a Safety Culture Interaction (SCI) Model for Construction Projects. *Saf. Sci.* **2013**, *57*, 138–149. [CrossRef]
3. Wanberg, J.; Harper, C.; Hallowell, M.; Rajendran, S. Relationship between Construction Safety and Quality Performance. *J. Constr. Eng. Manag.* **2013**, *139*, 04013003. [CrossRef]
4. Fang, D.; Wu, C.; Wu, H. Impact of the Supervisor on Worker Safety Behavior in Construction Projects. *J. Manag. Eng.* **2015**, *31*, 04015001. [CrossRef]
5. Korea Occupational Safety and Health Agency Statistics of Occupational Accidents in 2018. Available online: http://www.kosha.or.kr/kosha/data/industrialAccidentStatus.do?mode=view&articleNo=410303&article.offset=0&articleLimit=10 (accessed on 19 October 2020).
6. Sunindijo, R.Y.; Zou, P.X. Political Skill for Developing Construction Safety Climate. *J. Constr. Eng. Manag.* **2012**, *138*, 605–612. [CrossRef]
7. Chang, D.-S.; Tsai, Y.-C. Investigating the Long-Term Change of Injury Pattern on Severity, Accident Types and Sources of Injury in Taiwan's Manufacturing Sector between 1996 and 2012. *Saf. Sci.* **2014**, *68*, 231–242. [CrossRef]
8. Fabiano, B.; Parentini, I.; Ferraiolo, A.; Pastorino, R. A Century of Accidents in the Italian Industry: Relationship with the Production Cycle. *Saf. Sci.* **1995**, *21*, 65–74. [CrossRef]
9. Dong, X.S.; Choi, S.D.; Borchardt, J.G.; Wang, X.; Largay, J.A. Fatal Falls from Roofs among US Construction Workers. *J. Saf. Res.* **2013**, *44*, 17–24. [CrossRef]
10. Sawacha, E.; Naoum, S.; Fong, D. Factors Affecting Safety Performance on Construction Sites. *Int. J. Proj. Manag.* **1999**, *17*, 309–315. [CrossRef]
11. Yi, K.-J. Preventive Occupational Health and Safety Expense Estimation Method Based on Fatality Statistics and Progress Model. *J. Korea Inst. Build. Constr.* **2017**, *17*, 191–197. [CrossRef]
12. Baek, Y.; Wee, K.; Baek, I.; Kim, J. A Study on Improvement of Occupational Safety and Health Management Cost Accounting Standards. *Korean J. Constr. Eng. Manag.* **2020**, *21*, 39–46.
13. Durdyev, S.; Mohamed, S.; Lay, M.L.; Ismail, S. Key Factors Affecting Construction Safety Performance in Developing Countries: Evidence from Cambodia. *Constr. Econ. Build.* **2017**, *17*, 48. [CrossRef]
14. Mohamed, S. Empirical Investigation of Construction Safety Management Activities and Performance in Australia. *Saf. Sci.* **1999**, *33*, 129–142. [CrossRef]
15. Shafique, M.; Rafiq, M. An Overview of Construction Occupational Accidents in Hong Kong: A Recent Trend and Future Perspectives. *Appl. Sci.* **2019**, *9*, 2069. [CrossRef]
16. Jo, B.W.; Lee, Y.S.; Kim, J.H.; Khan, R.M.A. Trend Analysis of Construction Industrial Accidents in Korea from 2011 to 2015. *Sustainability* **2017**, *9*, 1297. [CrossRef]
17. Yang, K.; Ahn, C.R.; Vuran, M.C.; Aria, S.S. Semi-Supervised near-Miss Fall Detection for Ironworkers with a Wearable Inertial Measurement Unit. *Autom. Constr.* **2016**, *68*, 194–202. [CrossRef]
18. Yang, K.; Ahn, C.R.; Vuran, M.C.; Kim, H. Collective Sensing of Workers' Gait Patterns to Identify Fall Hazards in Construction. *Autom. Constr.* **2017**, *82*, 166–178. [CrossRef]
19. Kim, H.; Ahn, C.R.; Yang, K. Identifying Safety Hazards Using Collective Bodily Responses of Workers. *J. Constr. Eng. Manag.* **2017**, *143*, 04016090. [CrossRef]
20. Jebelli, H.; Ahn, C.R.; Stentz, T.L. Fall Risk Analysis of Construction Workers Using Inertial Measurement Units: Validating the Usefulness of the Postural Stability Metrics in Construction. *Saf. Sci.* **2016**, *84*, 161–170. [CrossRef]
21. Iverson, R.D.; Erwin, P.J. Predicting Occupational Injury: The Role of Affectivity. *J. Occup. Organ. Psychol.* **1997**, *70*, 113–128. [CrossRef]
22. Sheehy, N.P.; Chapman, A.J. Industrial accidents. In *International Review of Industrial and Organizational Psychology 1987*; John Wiley & Sons: Oxford, UK, 1987; pp. 201–227.
23. Oliver, A.; Cheyne, A.; Tomas, J.M.; Cox, S. The Effects of Organizational and Individual Factors on Occupational Accidents. *J. Occup. Organ. Psychol.* **2002**, *75*, 473–488. [CrossRef]
24. Choudhry, R.M.; Fang, D. Why Operatives Engage in Unsafe Work Behavior: Investigating Factors on Construction Sites. *Saf. Sci.* **2008**, *46*, 566–584. [CrossRef]
25. Heinrich, H.W. *Industrial Accident Prevention. A Scientific Approach*; McGraw-Hill Book Company Inc.: New York, NY, USA; London, UK, 1941.

26. Blackmon, R.B.; Gramopadhye, A.K. Improving Construction Safety by Providing Positive Feedback on Backup Alarms. *J. Constr. Eng. Manag.* **1995**, *121*, 166–171. [CrossRef]
27. Fleming, M.; Lardner, R. *Strategies to Promote Safe Behaviour as Part of a Health and Safety Management System*; HSE Books: London, UK, 2002.
28. Choudhry, R.M. Behavior-Based Safety on Construction Sites: A Case Study. *Accid. Anal. Prev.* **2014**, *70*, 14–23. [CrossRef] [PubMed]
29. Pinto, A.; Nunes, I.L.; Ribeiro, R.A. Occupational Risk Assessment in Construction Industry–Overview and Reflection. *Saf. Sci.* **2011**, *49*, 616–624. [CrossRef]
30. Everett, J.G.; Frank, P.B., Jr. Costs of Accidents and Injuries to the Construction Industry. *J. Constr. Eng. Manag.* **1996**, *122*, 158–164. [CrossRef]
31. Aminbakhsh, S.; Gunduz, M.; Sonmez, R. Safety Risk Assessment Using Analytic Hierarchy Process (AHP) during Planning and Budgeting of Construction Projects. *J. Saf. Res.* **2013**, *46*, 99–105. [CrossRef]
32. Hariri-Ardebili, M.A. Risk, Reliability, Resilience (R3) and beyond in Dam Engineering: A State-of-the-Art Review. *Int. J. Disaster Risk Reduct.* **2018**, *31*, 806–831. [CrossRef]
33. Zacchei, E.; Molina, J.L. Reviewing Arch-Dams' Building Risk Reduction Through a Sustainability–Safety Management Approach. *Sustainability* **2020**, *12*, 392. [CrossRef]
34. Saaty, T.L. How to Make a Decision: The Analytic Hierarchy Process. *Eur. J. Oper. Res.* **1990**, *48*, 9–26. [CrossRef]
35. Badri, A.; Nadeau, S.; Gbodossou, A. Proposal of a Risk-Factor-Based Analytical Approach for Integrating Occupational Health and Safety into Project Risk Evaluation. *Accid. Anal. Prev.* **2012**, *48*, 223–234. [CrossRef] [PubMed]
36. Padma, T.; Balasubramanie, P. Knowledge Based Decision Support System to Assist Work-Related Risk Analysis in Musculoskeletal Disorder. *Knowl. Based Syst.* **2009**, *22*, 72–78. [CrossRef]
37. Gurcanli, G.E.; Bilir, S.; Sevim, M. Activity Based Risk Assessment and Safety Cost Estimation for Residential Building Construction Projects. *Saf. Sci.* **2015**, *80*, 1–12. [CrossRef]

 *sustainability*

Article

Using Recycled Material from the Paper Industry as a Backfill Material for Retaining Walls near Railway Lines

Karmen Fifer Bizjak *, Barbara Likar and Stanislav Lenart

Department of Geotechnics and Traffic Infrastructure, Slovenian Building and Civil Engineering Institute, ZAG, 1000 Ljubljana, Slovenia; barbara.likar@zag.si (B.L.); stanislav.lenart@zag.si (S.L.)
* Correspondence: karmen.fifer@zag.si; Tel.: +386-41-39-55-51

Abstract: The construction industry uses a large amount of natural virgin material for different geotechnical structures. In Europe alone, 11 million tonnes of solid waste is generated per year as a result of the production of almost 100 million tonnes of paper. The objective of this research is to develop a new geotechnical composite from residues of the deinking paper industry and to present its practical application, e.g., as a backfill material behind a retaining structure. After different mixtures were tested in a laboratory, the technology was validated by building a pilot retaining wall structure in a landslide region near a railway line. It was confirmed that a composite with 30% deinking sludge and 70% deinking sludge ash had a high enough strength but experienced some deformations before failure. Special attention was paid to the impact of transport, which, due to the time lag between the mixing and installation of the composite, significantly reduced its strength. The pilot retaining wall structure promotes the use of recycled materials with a sustainable design, while adhering to government-mandated measures.

Keywords: paper sludge ash; deinking sludge; paper industry; backfill material; retaining wall

Citation: Bizjak, K.F.; Likar, B.; Lenart, S. Using Recycled Material from the Paper Industry as a Backfill Material for Retaining Walls near Railway Lines. *Sustainability* **2021**, *13*, 979. https://doi.org/10.3390/su13020979

Received: 2 December 2020
Accepted: 14 January 2021
Published: 19 January 2021

Publisher's Note: MDPI stays neutral with regard to jurisdictional claims in published maps and institutional affiliations.

1. Introduction

According to information published by the European Aggregate Association, the demand for European aggregates is 3 billion tonnes annually [1]. About half of natural (virgin) material is consumed by the construction industry, which also generates a large amount of waste material [2]. Undoubtedly, virgin material can partially be replaced by other materials, such as recycled industrial material, including material made from paper industry waste. This recycled waste can be substituted for virgin aggregates that are used in various applications in the building sector in huge quantities, especially for roads and earthworks [3]. Of course, the mechanical and environment criteria for recycled materials according to the national legislation must be satisfied.

Globally, 420 million tonnes of paper and paperboard are produced annually [4], and production is growing. The production processes result in significant waste generation; 11 million tonnes of solid waste are generated per year in Europe [5]. Approximately 70% of this waste is from paper recycling, for example, deinking sludge [6]. According to the Integrated Pollution Prevention and Control Directive 1996/61/CE [7], the paper industry is required to minimize the amount of waste and develop more sustainable technologies for waste treatment. There are also EU waste management legislative measures and policies [8] implementing a waste hierarchy, with landfilling being the least desirable option and recycling the most, supported by increased taxes for landfilling. Recycled paper residues are a potential material that could be substituted for virgin raw materials from a technical and economical point of view [9,10].

Examples of pulp and paper industry residue implementation have been presented by other authors [11–14], but in general, most paper industry waste is burned in power plant boilers or landfilled. The production process with different fillers, pigments, and coagulates influences the type of paper ash. Also, the technology and temperature in the boilers have an effect [15,16].

If paper sludge ash were to be used only as a binder in the construction industry, some problems due to the presence of lime would be observed, but it could be very useful for the stabilization of road structures or as a backfill material [17]. Different mixtures of paper ash and paper sludge have been tested on a laboratory scale. A mixture of sand, paper fly ash, paper sludge, and cement has been used in laboratory research [18]. The mixture reached a compressive strength of 0.8 MPa, which is high enough for use as a backfill material for a foundation structure, a structural fill, or a hydraulically bound layer in a road structure. If the paper sludge ash is mixed with recycled concrete aggregate (RCA), the mechanical properties are improved, especially the resistance to acid and sulfate attacks [19]. Highly plastic clay soil was stabilized with paper sludge ash [20–22] and the compressive strength increased enough (0.7, 1 MPa) for the mixture to be used without any other additives for a pozzolanic reaction. For mining backfill material, a mixture of paper sludge ash and sewage sludge ash [23] was prepared. Both materials were mixed and calcinated at high temperatures. In addition to paper sludge ash, paper sludge can be used. Paper sludge was used with marine-washed sand, aggregate, and Portland cement [24]. With this mixture, a compressive strength of 8 MPa was achieved. Remediation of contaminated soil by red mud was used with paper ash as a binder material [25].

Most of the research is related to laboratory tests, but some field results have also been published. A road subgrade was stabilized in a length of 250 m with a mixture of paper sludge and cement in Portugal [26]. The installed mixture achieved an unconfined compressive strength of 4.5 MPa. Paper fly ash was used for gravel road stabilization of a hydraulically bound layer [27] in Spain. The hydraulically bound layer reached an unconfined compressive strength of 5 MPa.

Paper sludge ash is also used in the cement industry as supplementary cementitious material in mortar [28,29], concrete manufacturing [30,31], and the brick industry [32].

When using ash as a building material, particular attention should be paid to the impact on the environment. Studies have demonstrated the wide applicability of ash, but it is necessary to carefully investigate the potential environmental impact and use technology that is appropriate for individual recycled materials [33,34].

Investors and designers find it difficult to decide to use recycled material in construction due to a lack of knowledge about the material, technology of installation, high cost of production, and often a negative attitude towards all new materials [1]. The objective of this research is to develop a new backfill material (composite) used behind retaining walls from the residue of paper industry production and promote the use of the recycled material with a pilot structure. Especially in mountainous regions, landslides represent a threat to roads and railways, which must be reduced by slope stabilization with different retaining walls. A new composite must have high enough unconfined compressive strength and shear properties, but at the same time has to allow elastic deformation before cracking. Until now, paper sludge ash and deinking sludge have been used in different mixtures, usually in mixtures with soil and other binders. At present, deinking sludge ash and deinking sludge are mixed together as a new composite in a precise ratio and compacted under strict conditions behind a retaining wall. This type of composite has not been tested in the laboratory so far, nor has its use been validated in field tests.

None of the studies to date have dealt with changing the strength characteristics of the material during mixing and installation. Here, the time of transport of the material from the place of mixing to installation is crucial. In the study, we found that, over time, the strength properties decrease significantly, which may be crucial for the stability of the retaining wall. The study notes that the materials in the laboratory must also be tested in terms of installation time in order to provide the designer with relevant data regarding the geomechanical characteristics of the composite.

The pilot retaining wall structure promotes the concept of a circular economy from idea to laboratory tests, installing the structure, and monitoring it over a long period of time. The new composite and the technology were tested at a construction site and then monitored over a longer period of time. This gives us information about the details of construction and proves that the structure is stable, usable, and meets all the technical and environmental standards available to investors, designers, and contractors. Pilot structures could help suppliers, investors, designers, and contractors identify the factors hampering the use of recycled materials in the construction sector as well as provide strategies that can be adopted to form an economical and sustainable product.

2. Materials and Methods

2.1. Material Used

Deinking sludge ash (DSA) and deinking sludge (DS), used in this study as raw materials, represent the main waste from recycled deinking paper pulp production at a paper industry company, VIPAP Videm Krško d.d., in Slovenia. The DSA is a combustion residue formed in a steam boiler during the incineration of DS. It consists of a mixture of bottom ash (approx. 90 wt%) and fly ash (approx. 10 wt%). VIPAP recycles around 600 tonnes of paper daily. Annually, 25,000 tonnes of DSA and 67,000 tonnes of DS are produced. According to the European Waste Catalogue (EWC), DSA is classified as 10 01 01, while DS is classified as 03 03 05 [35].

2.2. Methods

2.2.1. Testing of Raw Materials

The bulk chemical composition of the DSA and DS was determined by a Wavelength Dispersive X-ray Fluorescence (WD XRF), using a Thermo Scientific ARL PERFORM'X Spectrometer (Waltham, MA, USA). Analysis of loss on ignition (LOI) and the total chloride content of DSA was performed according to SIST EN 196-2 [36]. The physical and mechanical properties of the raw materials were tested according to the standards in Table 1.

Table 1. Physical and mechanical properties of raw materials.

Property	STANDARD
Initial Moisture Content (w) (%)	SIST EN ISO 17892-1:2015
Specific Gravity (γ_s) (Mg/m^3)	SIST EN ISO 17892-3:2016
Optimum Water Content (w_{opt}) (%)	SIST EN 13286-2:2010/AC:2013
Maximum Dry Density ($\gamma_{d,max}$) (Mg/m^3)	SIST EN 13286-2:2010/AC:2013
Unconfined composite strength after compaction (q_u) (MPa)	SIST EN 13286-41:2004
Liquid Limit (LL) (%)	SIST-TS CEN ISO/TS 17892-12:2004
Plastic Limit (PL) (%)	SIST-TS CEN ISO/TS 17892-12:2004
Particle Size Distribution	
Particle (>2.5 mm) (%)	SIST EN 933-1:2012
Particle (0.063–2.5 mm) (%)	SIST EN 933-1:2012
Particle (0.002–0.063 mm) (%)	SIST EN ISO 17892-4:2017
Particle (<0.002 mm) (%)	SIST EN ISO 17892-4:2017

Particle size distribution of DSA was measured by laser diffraction analysis (particles < 400 μm) using a CILAS 920 Particle Size Analyser (Cilas, Orléans, France).

2.2.2. Preliminary Laboratory Tests of Composite

In order to design a backfill material for a retaining wall structure for the stabilization of a landslide near a railway line, several mixtures consisting of different ratios of DSA and DS were tested. Among them, two mixtures (Table 2) with sufficiently good geomechanical characteristics and suitable properties for compaction and installation were tested in detail.

Table 2. Mixing proportions of the investigated mixture composites and their designations.

Designation of the Composites	Mixing Ratios (% Dry Mass)	
	DSA	**DS**
D80/20	80	20
D70/30	70	30

DSA: deinking sludge ash; DS: deinking sludge.

Geomechanical tests were performed in an accredited geomechanical laboratory, according to SIST EN ISO/IEC 17025 [37].

The components were mixed in a 20 L planetary mixer. Two kilograms were mixed for 2 min until a homogeneous mixture was obtained. Mixtures were compacted at the maximum dry density according to the SIST EN 13286-2 [38]. In order to prevent evaporation, the composites were stored and cured in a climatic chamber at 90% RH and 22 °C.

Compressive strength was tested according to SIST EN 13286-41 [39] immediately after compaction and after one, four, seven, 28, and 50 days of curing.

Freezing/thawing tests were performed according to Slovenian technical specification TSC 06.320 [40]. According to the specification, the composites were exposed to 12 cycles of freezing at −23 °C and thawing at 20 °C in a climate chamber.

The shear characteristics of the composites were tested according to the SIST EN ISO 17892-10 [41] directly after compaction and after seven days of curing. The permeability of the composite was tested in a triaxial cell, according to SIST EN ISO 17892:11 [42] under a pressure of 50 kPa.

In order to investigate the impact of the transport time to the construction site, the time delay between mixing and compacting was taken into account. The tests were performed with two different testing procedures:

- The material was moistened to the maximum water content (w_{max}) and cured in the open air,
- The material was moistened to the optimal water content (w_{opt}) and cured in closed boxes.

After moistening and mixing, the mixtures were compacted in the following time intervals: immediately, and after 4, 8, and 24 h. After seven days, an unconfined compressive strength test was performed on each specimen.

2.2.3. Test of Pilot Structure

In the region between Ljubljana and Novo Mesto, there is a landslide risk that endangers the safety of a railway line. The instability of the slope was already evident from the geological and geomechanical mapping of the site [43]. Subsequently, after carrying out a detailed geomechanical investigation of the railway zone, the final pilot structure location was selected in the exact location shown in Figure 1.

Construction work started in August 2018. The length of the structure was 50 m, with a height of 1.5 m. The retaining wall was made of gabions and backfill material with composite. The width of the backfill material between the gabions and landslide slope was between 2 and 3.5 m.

The mixture was prepared at the VIPAP facility from DPA and DS and transported to the construction site, located 70 km from VIPAP's facility. Mixing was performed by a stirrer (TERREX, Norwalk, CA, USA) for at least 15 min. The composite was transported to the construction site and subsequently compacted to the required density (Figure 2) in a 30 cm thick layer by a compactor (BOMAG, Boppard, Germany). For the whole structure, 100 t of the mixture was compacted into nine layers (Figure 3).

Figure 1. Location of the pilot structure.

Figure 2. Compacted layer of the composite.

Figure 3. Compacting the last layer of the composite.

Special attention was paid to the drainage system since the composite is impermeable. A drainage system was installed under the backfill and at the contact point between the backfill and the landslide slope.

Each compacted layer was tested at the construction site for dry density and moisture content (γ_d, w) according to TSC 06.711 [44] by a nuclear soil moisture density gauge (TROXLER, Ettenheim, Germany) and for dynamic deformation (E_{vd}) modules according to TSC 06.720 [45] (2003), by a dynamic plate (ZORN, Stendal, Germany). In order to perform the leaching procedure, according to SIST EN 1744-3 [46], samples were collected from the last layer. The leachates were analyzed by inductively coupled plasma mass spectrometry (ICP-MS) and UV-Vis spectrophotometry. Samples were also taken to verify the shear characteristics of the composite according to the standard SIST EN ISO 17892-10 [41].

2.2.4. Long-Term Monitoring

For long-term monitoring, an automatic station was placed at the construction site (Figure 4). It collects measurements from the weather station, inclinometers, piezometers from boreholes, and moisture and temperature probes installed in the backfill material behind the retaining wall made from gabions.

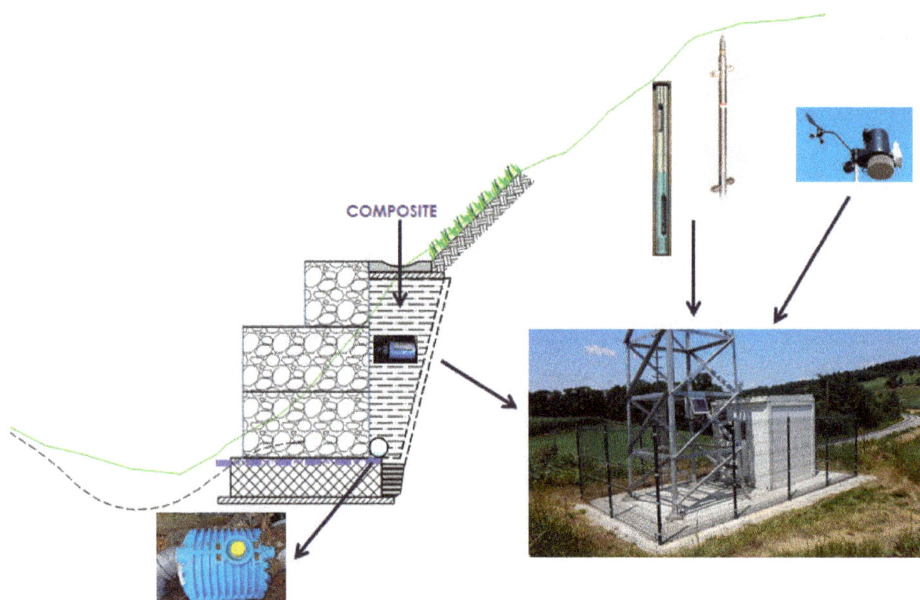

Figure 4. Monitoring system.

A soil moisture probe (Imko, Ettlingen, Germany) was installed 90 cm below the surface, and temperature sensors were installed 30, 60, and 90 cm below the surface of the composite.

For environmental monitoring, a water tank was built at the end of the retaining wall. The water flowing from the drainage system between the gabions and the composite (backfill material) was collected in a plastic tank and taken for chemical analyses.

The displacements that can occur above the retaining wall structure were monitored with a manual inclinometer (Interfels, Ulm, Germany), while those of the entire structure (specifically those of the gabions and the foundation) were monitored by a laser scanner (Leica, Aarau, Switzerland).

3. Results and Discussion

Preliminary investigations of two types of composites were performed in the laboratory, and the composite with the most suitable properties was selected for implementation.

3.1. Raw Material

Chemical analyses showed that the major oxide of DSA (Table 3) was CaO, accounting for almost 50 wt%. More than a quarter of DSA mass was lost after the LOI treatment. The most abundant oxides of DS were CaO (accounted for almost one-third) and the measured LOI was more than 50 wt%. High values of LOI were due to high fibre or other organic compounds. Elements such as Si, Al, and Mg were in concentrations below 0.25 wt%. Higher CaO content in DSA and DS was due to the use of fillers and coatings in the deinking process of the production of new paper from used paper. Also, higher CaO led to the higher pozzolanic reactivity of the composite made from DSA and DS.

Table 3. Chemical composition of raw materials.

Raw Material	Parameter (wt%)											
	SiO_2	Al_2O_3	Fe_2O_3	CaO	P_2O_5	MgO	K_2O	Na_2O	TiO_2	SO_3	LOI	Total
DSA	11.37	8.26	0.39	47.94	0.17	1.70	0.27	0.17	0.18	0.31	27.26	98.02
DS	7.27	5.56	0.24	31.78	0.06	1.03	0.16	0.14	0.11	0.06	53.41	99.82

LOI: loss on ignition.

The physical and mechanical properties of DSA and DS are presented in Table 4. DSA is a dry material, while the water content of DS ranges between 45% and 50%. In comparison with DS, the specific gravity of DSA was higher by about 20%. A standard Proctor test (SPP) showed that the optimal water content (w_{opt}) and maximal dry density ($\gamma_{d,max}$) were higher for DS. The unconfined compressive strength of DSA is between 300 and 500 kPa, which is in the range of very stiff soil according to the criteria for virgin materials. DS is a softer material in the range of stiff soil. Both materials were nonplastic.

Table 4. Physical and mechanical properties of raw materials.

Property	DSA	DS
Initial Moisture Content (w) (%)	0	45–50
Specific Gravity (γ_s) (Mg/m^3)	2.64	2.15
Optimum Water Content (w_{opt}) (%)	51	56.5
Maximum Dry density ($\gamma_{d,max}$) (Mg/m^3)	0.99	0.89
Unconfined compressive strength after compaction (q_u) (MPa)	0.3–0.5	0.22
Liquid Limit (LL) (%)		
Plastic Limit (PL) (%)	Nonplastic	Nonplastic
Particle Size Distribution		
Particle (>2.5 mm) (%)	0	-
Particle (0.063–2.5 mm) (%)	13.3	-
Particle (0.002–0.063 mm) (%)	75.59	-
Particle (<0.002 mm) (%)	11.11	-
D_{10} (mm)	0.002	-
D_{50} (mm)	0.02–0.06	-
D_{90} (mm)	0.4–0.8	-

3.2. Results of Preliminary Laboratory Tests

3.2.1. Unconfined Compressive Strength—q_u

The unconfined compressive strength decreased (Figure 5) with a higher quantity of DS in the composites and increased with curing time. The tests performed immediately after compaction showed relatively similar values of q_u (0.2–0.3 MPa), independent of the composition. After one day of curing, composites with higher percentages of DSA showed higher q_u values, accounting for the more intense hydration process in those composites.

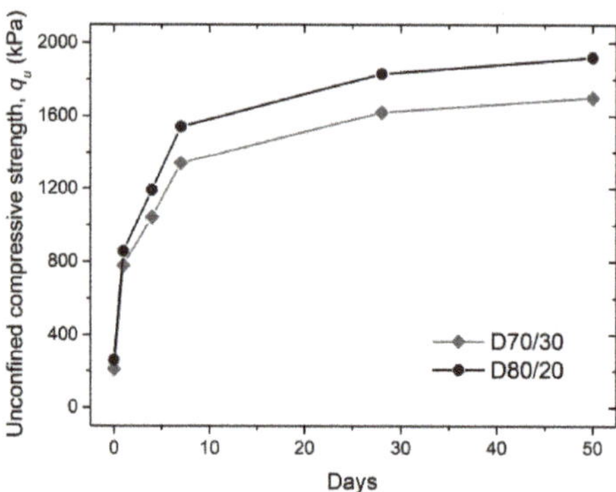

Figure 5. Uniaxial compressive tests results.

The results of the vertical deformation at the sample failure during the uniaxial compressive test are shown in Figure 6. With an increasing content of DS in the composite, the vertical deformations increased. This is reflected in the rapid breaking and very small elastic deformations, i.e., the brittle deformation behavior of the D80/20. By contrast, ductile behavior was observed in the D70/30.

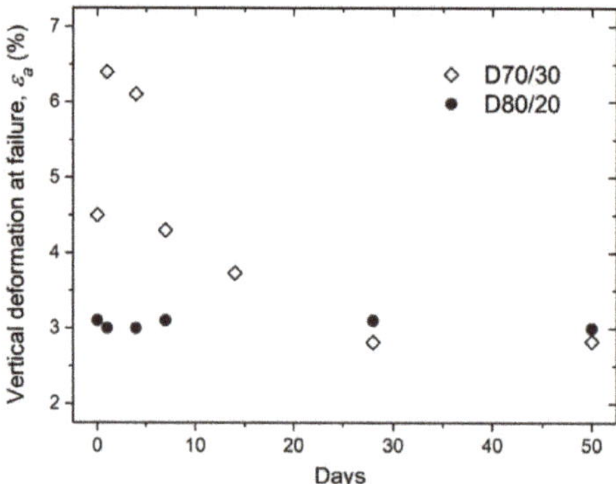

Figure 6. Deformations after uniaxial compressive tests.

The curing time has a strong influence on the vertical deformation (ε_A) at failure. While this is not as obvious with D80/20, in the case of D70/30, the ε_A at failure increased rapidly in one day and then decreased over time. After 28 days, the ε_A was almost the same for both composites. This points to the fact that the failure mode of mixtures changes from ductile to brittle between the first and fourth days of curing.

Even after a retaining wall is built, some deformations can still occur in the structure because of the stabilization processes in the landslide mass, especially in the early days of construction. Because of this, it is desirable for a backfill material to possess higher elasticity and, at the same time, have enough strength to prevent landslides.

3.2.2. Time Effect

The results of the investigation of the impact of the delay between mixing and compacting (Figure 7) showed that q_u decreased with the increased time between them. The highest values of q_u were seen in samples prepared at the w_{opt} (Figure 8) and compacted immediately after mixing. For samples compacted 4 h after mixing, higher q_u values were measured for composites prepared at the w_{max}. The hydration process dried the mixtures and after 4 h resulted in a lower q_u for mixtures with lower water content. After 24 h, no difference could be observed in q_u. Based on the results, it can be concluded that the mixture compacted immediately must be moistened to the w_{opt}. For a mixture that will be compacted after 4 h due to transport, it is more appropriate to moisten it to the w_{max}.

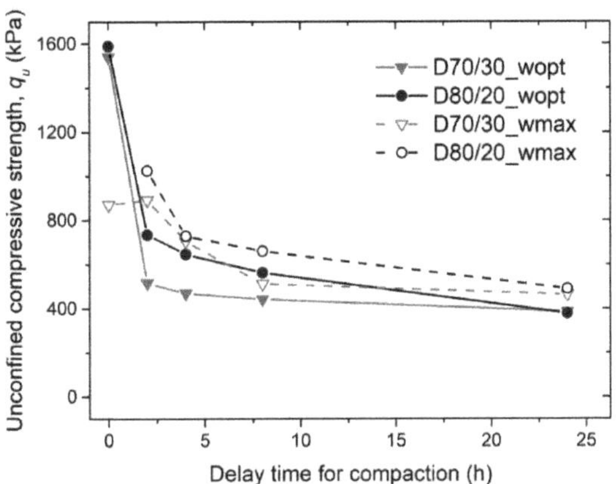

Figure 7. Unconfined compressive strength of samples with delay in compaction time. wopt: optimal water content; wmax: maximal water content.

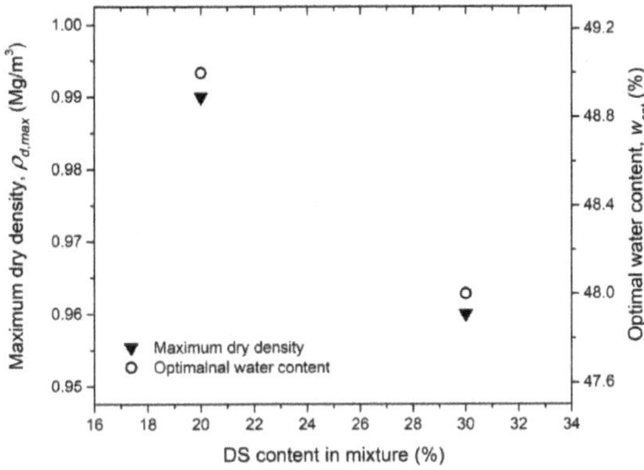

Figure 8. Results of standard proctor tests (SPP) for the investigated composites.

The results of these tests are essential to determine the methods of mixing, transport, and installation of the mixture on the construction site and to assess the maximum distance between the composite production site and the composite construction site. The results from the laboratory and field tests showed that the mixture has to be moistened above the w_{opt} before transport. It is a very important requirement for the installation procedure. If the mixture at the construction site does not have a high enough water content, it will not be properly compacted, and the strength of the installed layer will be lower than it has to be, according to the legislation. Three to five percent of the water is consumed during the hydration process of DSA.

3.2.3. Standard Proctor Tests (SPP)

The results of the SPP showed that the optimal water content (w_{opt}) increases slightly with a higher percent of DSA in composites (Figure 8). A higher percentage of DSA required more water for the hydration process in the composite. A similar trend was observed for maximal dry density as it increased from 0.96 Mg/m^3 in D70/30 to 0.99 Mg/m^3 in D80/20.

The main difference between the investigated composites and the natural gravel material is the density. The $\gamma_{d,max}$ for a natural gravel material is 2.3 Mg/m^3 on average, while the investigated composites show a much lower $\gamma_{d,max}$. This physical characteristic enables the use of the investigated composites in low-bearing-capacity foundation soils. Such a backfill material is light and does not cause large settlement, as does the heavy virgin gravel material. This is one of the important advantages of using the investigated composites as a backfill material for geotechnical structures in regions with difficult geotechnical conditions.

3.2.4. Frost Resistance

According to the Slovenian National Technical Specifications for roads, TSC 06.320 [40] on freeze/thaw resistance, the freeze ratio has to be more than 0.7 (the ratio between the unconfined compressive strength of exposed samples and curing samples at atmosphere-controlled conditions).

Both composites are above the limit value of 0.7. Based on these results, it can be concluded that this material could be used as a backfill material even in the top meter below the surface. That is the depth of freezing at the pilot structure's location.

3.2.5. Shear Properties

The results of a friction angle (f) and cohesion (c) analysis showed that shear characteristics increase with the decrease in DS content of composites, but after 28 days, the differences were not so obvious (Figure 9). The shear characteristics are higher in comparison with the natural gravel backfill material conventionally used for retaining wall structures. The friction angle for gravel is around 30° with no cohesion. The high shear properties of the composite allow for the construction of a thinner retaining wall than would be needed if virgin gravel material was used. These characteristics are very important, especially for retaining structures near railway lines, where there is generally not enough space for large geotechnical structures.

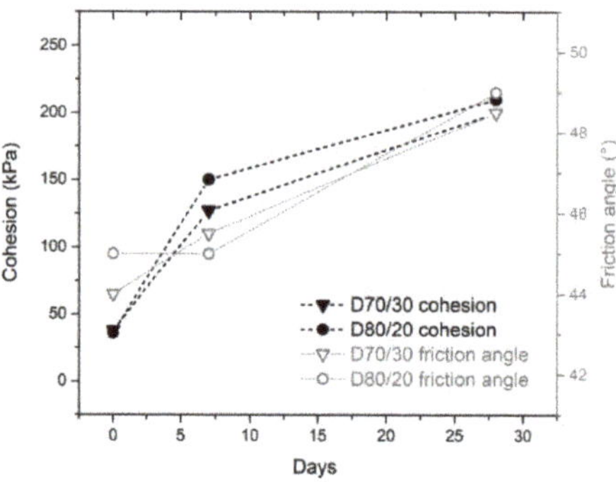

Figure 9. Angles of friction and cohesion for the mixtures.

3.2.6. Composite Used for Backfill Material

Based on the results of the laboratory tests, the composite D70/30 was chosen for use as a backfill material behind the retaining wall structure. The composite has a high enough q_u even when compacted 4 h after mixing, and at the same time allows small deformations before cracking. It is ductile enough to sustain small deformations behind the retaining wall in case of landslides. The composite has high enough freeze/thaw resistance to withstand extreme weather conditions. Shear properties are higher in comparison with virgin material (gravel). Because of these properties, the retaining wall structure can be lower and cheaper. Instead of 2 m, the gabion structure that was built had to be only 1.5 m high.

3.3. Results from the Pilot Structure

3.3.1. Field Tests

According to the technical specification TSC 06.711 [44], the average compaction of the last layer has to be at least 95% $\gamma_{d,max}$. The results of the compaction tests of the installed composite D70/30 showed that only the first three layers were compacted below 95% $\gamma_{d,max}$ (Figure 10). The reason for the lower compactness in the first layers is the uncompacted drainage layer under the composite. In all other layers, a higher compaction degree was measured. Measurements confirmed that the quality of the installed composite was within the requirements of the technical specifications.

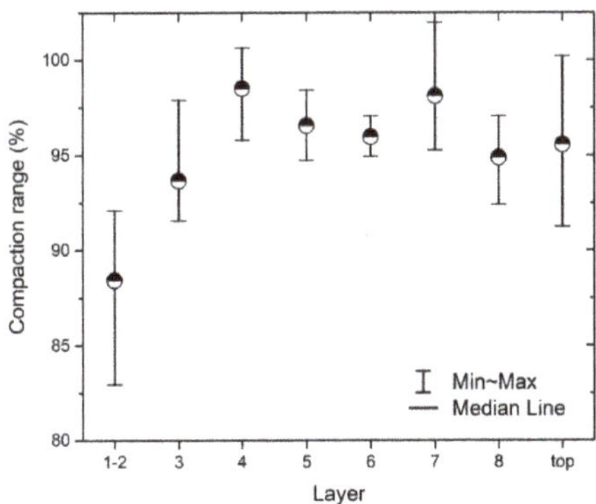

Figure 10. Results of the measurements with the neutron probe.

3.3.2. Laboratory Tests of Samples from the Pilot Retaining Wall

Shear tests of the samples taken from the composite at the construction site reach high enough shear properties. The results are similar to the results from the preliminary laboratory investigation; even cohesion increases significantly after 28 days. In the design project [43] for the stability analysis, the design parameters were lower, especially for cohesion, which means that the safety factor of the structure was higher. Shear characteristic values obtained from the in-built composite are gathered in Table 5.

Table 5. Shear properties of the compacted composite.

Shear Properties	Laboratory Results		Demo Field			Design Parameter
	Immediately	After Seven Days	Immediately	After Seven Days	After 28 Days	
F (°)	44	45	37	40	45	40
c (kPa)	12	42	35	45	200	40

The new composite had significantly higher shear characteristics compared to a natural gravel backfill material. The shear properties of the composite are similar to the shear properties of soft rock. If virgin gravel material is used as the backfill, a larger retaining wall structure should be designed to prevent landslide movement, as was confirmed in the project design [43].

A leaching test of the samples taken from the structure showed (Table 6) that none of the components in the water exceeded the limits established by Slovenian legislation

(UL RS, No. 10/14, 22 February 2014). The installed mixture does not have an adverse environmental impact because it is impermeable, and water from the composite does not leach hazardous substances. For the composite, very low permeability (only 2.2×10^{-10} under the load of 50 kPa) was measured in the laboratory.

Table 6. Results from the leaching tests.

Component	Limit	Sample One after Two Days	Sample Two after 28 Days
		(mg/L)	
As	0.5	0.003	0.0033
Ba	20	16.04	8.82
Cd	0.04	<0.002	<0.002
Cr total	0.5	0.033	<0.002
Cu	2	1.866	0.61
Hg	0.01	0.005	<0.001
Mo	0.5	0.092	0.097
Ni	0.4	0.021	0.0064
Pb	0.5	0.005	<0.005
Sb	0.06	<0.001	<0.001
Se	0.1	0.003	<0.003
Zn	4	0.035	<0.005
Chlorides	800	29.1	13.5
Fluorides	10	4.01	3.6
Sulfates	1000	<10	<10

3.4. Results of the Long-Term Monitoring

3.4.1. Stability of the Landslide

To ensure the safety of passengers and cargo, the landslide was stabilized with a retaining wall structure (Figure 11). Deformations of the slope above the railway line were measured with an inclinometer. The measurements did not indicate any displacements because the retaining wall successfully stopped the landslide (Figure 12).

For the retaining wall structure, measurements were taken by a laser scanner. The measurements showed that the foundation of the retaining wall was stable. Both the measurements of the retaining wall and the measurements of the landslide deformations showed that the landslide was successfully stabilized.

Figure 11. Retaining wall with gabions and the composite as the backfill material.

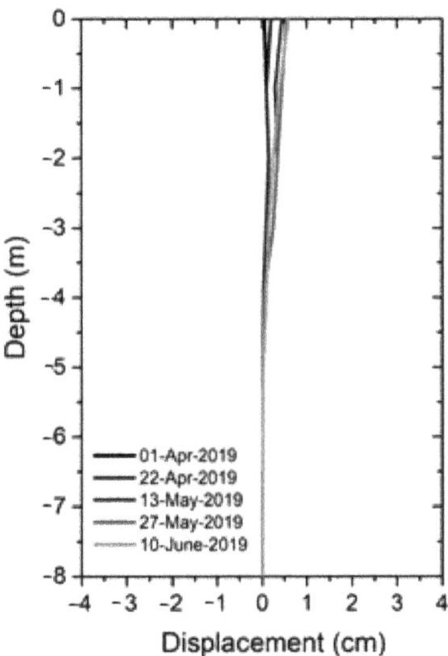

Figure 12. Horizontal displacements in borehole above the retaining wall.

The composite used has to be sustainable and resistant to weather changes. The results of the temperature measurements showed that the weather does not influence the temperature of the composite significantly (Figure 13). Some correlation with the outside temperature (T) can be observed in the sensor that is 30 cm below the surface (T1), while the temperatures at 60 cm (T2) and 90 cm (T3) below the surface remained more or less stable despite the fluctuation of the outside temperature.

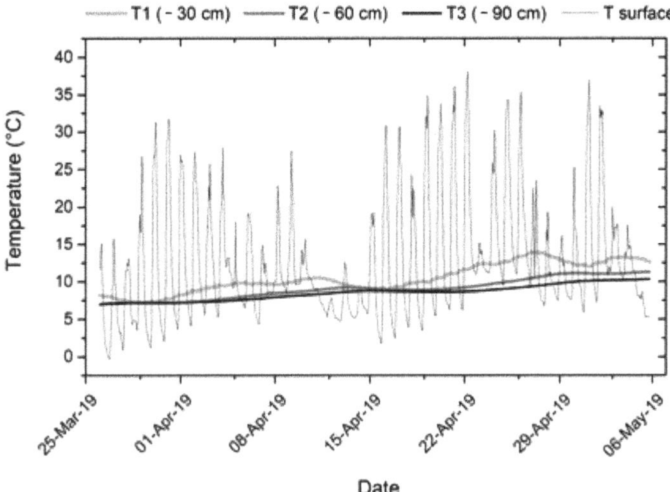

Figure 13. The temperature outside and in the backfill composite in the retaining wall structure (26 March 2019–10 May 2019).

The water content of the composite at the time of compaction was 57% and is currently around 47%. Comparison with the precipitation showed that the composite is impermeable because there was no correlation between the water content and precipitation (Figure 14).

The temperature and water content sensors showed that the weather does not have an influence on the composite quality.

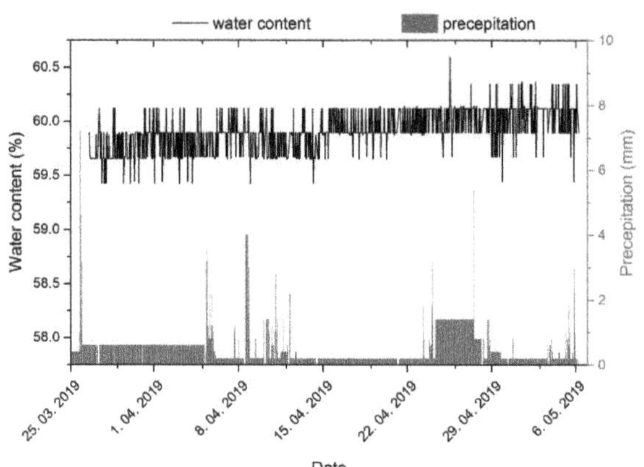

Figure 14. Water content in backfill material in the retaining structure vs. precipitation (26 March 2019–10 May 2019).

3.4.2. Environmental Monitoring

Samples were taken from the water tank and tested in an accredited chemical laboratory. The results show that the composite did not have a negative influence on the water quality since the concentration of none of the tested parameters exceeded the limits set by Slovenian legislation UR RS No. 98/15 (Table 7).

Table 7. Results of the chemical analysis of the water from the drainage system.

Component	Limit	Water Sample April 2019	Water Sample May 2019	Water Sample October 2019
		(mg/L)		
As	0.1	0.0019	0.0006	0.0017
Ba	5	0.0080	0.0073	0.122
Cd	0.025	<0.0002	<0.0002	<0.0002
Cr total	0.5	0.0031	0.0010	0.011
Cu	0.5	0.042	0.011	0.013
Hg	0.005	<0.0001	<0.0001	<0.0001
Mo	1	0.018	0.0024	0.0028
Ni	0.5	0.0011	0.0015	0.0024
Pb	0.5	0.0005	0.0006	<0.0005
Sb	0.3	0.0039	0.0011	0.0017
Se	0.6	0.0005	<0.0003	0.0003
Zn	2	<0.0005	0.0016	0.0005
Chlorides	800	5.17	1.52	2.19
Fluorides	10	0.264	<0.10	0.204
Sulfates	1000	19	2	14

4. Conclusions

A new composite from paper residues was developed as a backfill material for retaining wall structures. Preliminary tests were performed in an accredited geomechanical laboratory, and the results were later verified with field and laboratory measurements on a pilot structure.

A composite has to have geomechanical properties that are high enough to be used as a backfill material behind a retaining wall structure in an unstable area near a railway line threatened by landslides. Several mixtures with different contents of DS and DSA were initially tested to choose the optimal composite with the proper geomechanical properties. Two of them were studied in detail within the present research to estimate the effect of their composition upon particular basic characteristics.

No studies to date have dealt with the changing strength characteristics of material during mixing and installation. The time of transport is an important parameter for using this mixture as a backfill material. Results from the laboratory showed that the mixture has to be moistened above the w_{opt} before being transported and used at the construction site within 4 h.

Based on the laboratory results, it was decided to use composite D70/30 as the backfill material for the retaining wall. The composite is a new mixture that has not yet been tested in a laboratory or installed in the retaining wall structure at the construction site. The composite with 30% DS had a high enough q_u and shear strength but allowed small deformations before failure. The high shear characteristics of the composite allowed for a slimmer retaining structure to support the unstable slope behind it. At the same time, the DS in the composite enabled ductile behavior of the structure and prevented it from brittle failure.

In 2018, the composite D70/30 was used as a backfill material of the retaining wall structure built by gabions in the south part of Slovenia, near the railway line, for landslide stabilization. All laboratory and field tests confirmed the physical characteristics measured in the research phase and confirmed that the environmental requirements are being reached.

At the construction site, the material was installed in 30 cm layers. Each layer was compacted and controlled to reach the optimal moisture and maximum density. Before, during, and after the construction, landslide stability was assessed, and environmental monitoring was performed. With the test results from the pilot structure presented in this paper, the technology of mixing and compacting was improved. Mixing is usually not a problem in the laboratory, but at a construction site, a large quantity of material has to be mixed with the proper technology. The monitoring system confirmed that the retaining wall with the composite stabilizes the landslide near the railway. The composite was an impermeable material, and precipitation did not influence its stability. As the area covered with such an impermeable layer was not very wide, the impact on the groundwater recharge was limited. However, one should consider such a structure as a groundwater barrier that disturbs shallow groundwater flows. Thus, the use of this kind of structure should be combined with a properly designed drainage system, which minimizes the effect of groundwater disturbance. In the case of the retaining structure presented within this paper, vertical and horizontal drainage were employed to enable effective drainage of water behind the retaining structure.

On the other hand, the low density of the composite from paper residues also has several advantages. There is great potential for its use in construction on soft ground. In such a case, the utilization of very light materials, like the new composite, could prevent large settlement.

The pilot retaining wall structure represents a practical case for future investors, designers, and construction companies to encourage them to use recycled materials from the paper industry instead of virgin materials. The new composite and its installation technology were successfully tested. The presented circular case between the paper industry and the construction sector shows the advantages for both sides. Instead of disposing of the waste in a landfill for a very high price (in Slovenia, 80–150 €/tonne), the paper manufacturer processes the waste into a composite for a retaining wall structure. Contractors, meanwhile, get a composite that is cheap and has better deformation properties than the virgin material.

5. Patents

The Slovenian Technical Approval STS—1870011 was granted for this product.

Author Contributions: Conceptualization, K.F.B. and S.L.; methodology, B.L., K.F.B. and S.L.; formal analysis, B.L. and K.F.B.; investigation, K.F.B., S.L. and B.L.; resources, K.F.B., S.L. and B.L.; writing—original draft preparation, K.F.B.; writing—review and editing, S.L. and B.L.; supervisor, K.F.B., funding acquisition, K.F.B. All authors have read and agreed to the published version of the manuscript.

Funding: This research was funded by the European Union, co-financed by the Horizon 2020 Research and Innovation Programme under grant agreement no. 730305, project Paperchain and the ARRS financial support from the research core funding No. P2-0273, Building structures and materials.

Institutional Review Board Statement: Not applicable.

Informed Consent Statement: Not applicable.

Data Availability Statement: Data available in a publicly accessible repository.

Conflicts of Interest: The authors declare no conflict of interest.

References

1. UEPG. *Annual Review 2017–2018*; UEPG: Brussels, Belgium, 2018.
2. Oyedele, L.O.; Ajayi, S.O.; Kadiri, K.O. Use of recycled products in UK construction industry: An empirical investigation into critical imepiments and strategies for improvement. *Resour. Conserv. Recycl.* **2014**, *93*, 23–31. [CrossRef]
3. Poulikakos, L.D.; Papadaskalopoulou, C.; Hofko, B.; Gschosser, F.; Falchetto, A.C.; Bueno, M.; Arraigada, M.; Sousa, J.; Ruiz, R.; Petit, C.; et al. harvesting the unexplred potential of European waste material for road construction. *Resour. Conserv. Recycl.* **2017**, *116*, 32–44. [CrossRef]
4. Garside, M. Paper Industry—Statistics & Facts. Statista. 2019. Available online: https://www.statista.com/topics/1701/paper-industry/Published (accessed on 4 January 2021).
5. Zhang, C.; Jiang, G.; Liu, X.; Wang, Z. Lateral displacement of silty clay under cement−fly ash−gravel pile−supported embankments: Analytical consideration and field evidence. *J. Cent. South Univ.* **2015**, *22*, 1477–1489. [CrossRef]
6. Monte, M.C.; Fuente, E.; Blanco, A.; Negro, C. Waste management from pulp and paper production in the European Union. *Waste Manag.* **2009**, *29*, 293–308. [CrossRef]
7. IPPC. Council Directive 96/61/EC of 24 September 1996 concerning Integrated Pollution Prevention and Control (IPPC). *Off. J.* **1996**, *257*, 10.
8. EAA. *The Roadmap to a Resource Efficient Europe (EU COM 2011/571)*; European Commission: Brussels, Belgium, 2011.
9. Fischer, C.; Lehner, M.; McKinnon, D.L. Overview of the Use of Landfill Taxes in Europe. 2012. Available online: https://ec.europa.eu/environment/waste/pdf/strategy/3.%20%20Christian%20Fischer%20EEA%20Landfill%20taxes.pdf (accessed on 4 January 2021).
10. Watkins, E.; Hogg, D.; Mitsios, A.; Mudgal, S.; Neubauer, A.; Reisinger, H.; Troeltzsch, J.; Van Acoleyen, M. *Use of Economic Instruments and Waste Management Performances*; Bio Intelligence Service S.A.S: Paris, France, 2012.
11. Bajpai, P. *Recycling and Deinking of Recovered Paper*; Elsevier: Amsterdam, The Netherlands, 2012.
12. Saeli, M.; Senff, L.; Tobaldi, D.M.; La Scalia, G.; Seabra, M.P.; Labrincha, J.A. Innovative Recycling of Lime Slaker Grits from Paper-Pulp Industry Reused as Aggregate in Ambient Cured Biomass Fly Ash-Based Geopolymers for Sustainable Construction Material. *Sustainability* **2019**, *11*, 3481. [CrossRef]
13. Ferreira, I.A.; Fraga, M.C.; Godina, R.; Barreiros, M.S.; Carvalho, H. A Proposed Index of the Implementation and Maturity of Circular Economy Practices—The Case of the Pulp and Paper Industries of Portugal and Spain. *Sustainability* **2019**, *11*, 1722. [CrossRef]
14. Gabriel, M.; Schöggl, J.P.; Posch, A. Early Front-End Innovation Decisions for Self-Organized Industrial Symbiosis Dynamics—A Case Study on Lignin Utilization. *Sustainability* **2017**, *9*, 515. [CrossRef]
15. García, R.; de la Villa, R.V.; Rodríguez, O. Mineral phases formation on the pozzolan/lime/water system. *Appl. Clay Sci.* **2009**, *43*, 331–335. [CrossRef]
16. Fernández, R.; Nebreda, B.; de la Villa, R.V.; García, R.; Frías, M. Mineralogical and chemical evolution of hydrated phases in the pozzolanic reaction of calcined paper sludge. *Cem. Concr. Compos.* **2010**, *32*, 775–782. [CrossRef]
17. Segui, P.; Aubert, J.E.; Husson, B.; Measson, M. Characterization of wastepaper sludge ash for its valorization as a component of hydraulic binders. *Appl. Clay Sci.* **2012**, *57*, 79–85. [CrossRef]
18. Wu, H.; Yin, J.; Bai, S. Experimental investigation of utilizing industrial waste and byproduct material in controlled low strength materials (CLMS). *Adv. Mater. Res.* **2013**, *639–640*, 299–303. [CrossRef]
19. Bui, K.N.; Satomi, T.; Takahashi, H. Influence of industrial by-products and waste paper sludge ash on properties of recycled aggregate concrete. *J. Clean. Prod.* **2019**, *214*, 403–418. [CrossRef]
20. Khalid, N.; Mukri, M.; Kamarudin, F.; Arshad, M.F. Clay Soil Stabilized Using Waste Paper Sludge Ash (WPSA) Mixtures. *EJGE* **2012**, *12*, 1215–1225.

21. Mavroulidou, M. Use of waste paper sludge ash as a calcium-based stabiliser for clay soils. *Waste Manag. Res.* **2017**, *36*, 1066–1072. [CrossRef] [PubMed]
22. Dharan, R.B. Effect of Waste Paper Sludge Ash on Engineering Behaviors of Black Cotton Soils. *Int. J. Earth Sci. Eng.* **2016**, *9*, 188–191.
23. Lu, Q.M.; Zhang, Y.X.; Zhang, R.L.; Sun, W.B. Analysis of mechanical properties and microstructure of sludge-ash and cement cementitious system. *J. Saf. Environ.* **2019**, *19*, 308–311.
24. Ahmadi, B.; Al-Khaja, W. Utilization of paper waste sludge in the building construction industry. *Resour. Conserv. Recycl.* **2001**, *21*, 105–113. [CrossRef]
25. Oprčkal, P.; Mladenović, A.; Zupančič, N.; Ščančar, J.; Milačič, R.; Zala Serjun, V. Remediation of contaminated soil by red mud and paper ash. *J. Clean. Prod.* **2020**, *256*, 120440. [CrossRef]
26. Lisbona, A.; Vegas, I.; Ainchil, J.; Riso, C. Soil Stabilization with Calcined Paper Sludge: Laboratory and Field Tests. *J. Mater. Civ. Eng.* **2012**, *24*, 666–673. [CrossRef]
27. Vestin, J.; Arm, M.; Nordmark, D.; Lagerkvist, A.; Hallgren, P.; Lind, B. Fly ash as a road construction material. In *WASCON*; Arm, M., Vandecasteele, C., Heynen, J., Suer, P., Lind, B., Eds.; SGI: Gothenburg, Sweden, 2012; pp. 1–8.
28. Vegas, I.; Urreta, J.; Frias, M.; Garcia, R. Freeze–thaw resistance of blended cements containing calcined paper sludge. *Constr. Build. Mater.* **2009**, *23*, 2862–2868. [CrossRef]
29. Sadique, M.; Al-Nageim, H.; Atherton, W.; Seton, L.; Dempster, N. Analytical investigation of hydration mechanism of a non-Portland binder with waste paper sludge ash. *Constr. Build. Mater.* **2019**, *211*, 80–87. [CrossRef]
30. Vashistha, P.; Kumar, V.; Singh, S.K.; Dutt, D.; Tomar, G.; Yadav, P. Valorization of paper mill lime sludge via application in building construction materials: A review. *Constr. Build. Mater.* **2019**, *211*, 371–382. [CrossRef]
31. Fava, G.; Ruello, M.L.; Corinaldesi, V. The properties of wastepaper sludge ash and its generic applications. *J. Mater. Civ. Eng.* **2011**, *23*, 772–776. [CrossRef]
32. Singh, D.; Kumar, A. Performance Evaluation and Geo-Characterization of Municipal Solid Waste Incineration Ash Material Amended with Cement and Fibre. *Int. J. Geosynth. Ground Eng.* **2017**, *3*, 16. [CrossRef]
33. Mauko, P.A.; Oprčkal, P.; Mladenović, A.; Zapušek, P.; Urleb, M.; Turk, J. Comparative Life Cycle Assessment of possible methods for the treatment of contaminated soil at an environmentally degraded site. *J. Environ. Manag.* **2016**, *218*, 497–508. [CrossRef]
34. Kinnarinen, T.; Golmaei, M.; Jernström, E.; Häkkinen, A. Removal of hazardous trace elements from recovery boiler fly ash with an ash dissolution method. *J. Clean. Prod.* **2019**, *209*, 1264–1273. [CrossRef]
35. European Commission. Commission Decision of 3 May 2000 Replacing Decision 94/3/EC Establishing a List of Wastes Pursuant to Article 1(a) of Council Directive 75/442/EEC on Waste and Council Decision 94/904/EC Establishing a List of Hazardous Waste Pursuant to Article 1(4) of Council Directive 91/689/EEC on Hazardous Waste (Notified under Document Number C(2000) 1147). 2000. Available online: https://eur-lex.europa.eu/legal-content/EN/TXT/?uri=CELEX:02000D0532-20150601 (accessed on 4 January 2021).
36. SIST EN 196-2:2013. *Method of Testing Cement—Part 2: Chemical Analysis of Cement*; SIST: Ljubljana, Slovenia, 2013.
37. SIST EN ISO/IEC 17025:2017. *General Requirements for the Competence of Testing and Calibration Laboratories*; SIST: Ljubljana, Slovenia, 2017.
38. SIST EN 13286-2:2010/AC:2013. *Unbound and Hydraulically Bound. Mixtures—Part 2: Test. Methods for Laboratory Reference Densityand Water Content—Proctor Compaction*; SIST: Ljubljana, Slovenia, 2013.
39. SIST EN 13286-41:2004. *Unbound and Hydraulically Bound. Mixtures. Test. Method for Determination of the Compressive Strength of Hydraulically Bound. Mixtures*; SIST: Ljubljana, Slovenia, 2004.
40. TSC 06.320:2001. *Vezane Spodnje Nosilne Plasti s Hidravličnimi Vezivi*; DRSI: Ljubljana, Slovenia, 2001.
41. SIST EN ISO 17892-10. *Geotechnical Investigation and Testing—Laboratory Testing of Soil—Part 10: Direct Shear Tests*; SIST: Ljubljana, Slovenia, 2019.
42. SIST EN ISO 17892:11. *Geotechnical Investigation and Testing—Laboratory Testing of Soil—Part 11: Permeability Tests*; SIST: Ljubljana, Slovenia, 2019.
43. Fifer, K.B. *Geološko Geotehnično Poročilo*; ZAG: Ljubljana, Slovenia, 2018.
44. TSC 06.711:2001. *Meritev Gostote in Vlage. Postopek z Izotopskim Merilnikom*; DRSI: Ljubljana, Slovenia, 2001.
45. TSC 06.720:2003. *Meritve in Preiskave. Deformacijski Moduli Vgrajenih Materialov*; DRSI: Ljubljana, Slovenia, 2003.
46. SIST EN 1744-3:2002. *Tests for Chemical Properties of Aggregates—Part 3: Preparation of Eluates by Leaching of Aggregates*; SIST: Ljubljana, Slovenia, 2002.

 *sustainability*

Article

Modeling Building Stock Development

Antti Kurvinen [1,*], **Arto Saari** [1], **Juhani Heljo** [1] **and Eero Nippala** [2]

1. Faculty of Built Environment, Tampere University, Korkeakoulunkatu 5, FI-33720 Tampere, Finland; arto.saari@tuni.fi (A.S.); juhani.heljo@tuni.fi (J.H.)
2. School of Built Environment and Bioeconomy, Tampere University of Applied Sciences, Kuntokatu 3, FI-33520 Tampere, Finland; eero.nippala@tuni.fi
* Correspondence: antti.kurvinen@tuni.fi

Abstract: It is widely agreed that dynamics of building stocks are relatively poorly known even if it is recognized to be an important research topic. Better understanding of building stock dynamics and future development is crucial, e.g., for sustainable management of the built environment as various analyses require long-term projections of building stock development. Recognizing the uncertainty in relation to long-term modeling, we propose a transparent calculation-based QuantiSTOCK model for modeling building stock development. Our approach not only provides a tangible tool for understanding development when selected assumptions are valid but also, most importantly, allows for studying the sensitivity of results to alternative developments of the key variables. Therefore, this relatively simple modeling approach provides fruitful grounds for understanding the impact of different key variables, which is needed to facilitate meaningful debate on different housing, land use, and environment-related policies. The QuantiSTOCK model may be extended in numerous ways and lays the groundwork for modeling the future developments of building stocks. The presented model may be used in a wide range of analyses ranging from assessing housing demand at the regional level to providing input for defining sustainable pathways towards climate targets. Due to the availability of high-quality data, the Finnish building stock provided a great test arena for the model development.

Keywords: modeling; building stock development; mortality of building stock; residential buildings; public buildings; commercial buildings

Citation: Kurvinen, A.; Saari, A.; Heljo, J.; Nippala, E. Modeling Building Stock Development. *Sustainability* **2021**, *13*, 723. https://doi.org/10.3390/su13020723

Received: 7 November 2020
Accepted: 11 January 2021
Published: 13 January 2021

Publisher's Note: MDPI stays neutral with regard to jurisdictional claims in published maps and institutional affiliations.

1. Introduction

It is widely agreed that dynamics of building stocks are relatively poorly known even if it is recognized to be an important research topic. Better understanding of building stock dynamics and future development is crucial, e.g., for sustainable management of the built environment [1]. More advanced and transparent modeling of building stocks also contributes to improving analyses that lean on building stock data. The research fields that are in need of improved information on building stocks include but are not limited to land use planning, energy analysis, life cycle assessment, life cycle costing, mass flow analysis, calculation of green gross domestic product, service life estimation of components, simulation of maintenance and refurbishment, cultural heritage protection, comfort and public health, and resilience. As the fields for which building stock information is relevant are various, different levels of detail in building stock development may be significant for analysis within these fields. For example, depending on the intended use, in some cases, information related to buildings or dwellings is of interest while in others, a further subdivision according to building types or age bands is relevant. Thus, a simple and transparent modeling approach that is modifiable to several purposes should serve the needs of these various fields.

Research on built environments often has important policy implications, such as contributing to strategies to achieve the goals set by the EU (e.g., energy performance of buildings [2,3], low-carbon economy [4,5], and no net land take by 2050 [6]). Despite the importance, many analyses still seem to rely on vague assumptions about the development of building stock. In sustainability-related research, the description of the stock's current state is usually based on approaches using constructed building archetypes (e.g., [7]) or sample buildings (e.g., [8]). As an alternative to those, Nägeli et al. [9] introduced an approach where the idea is to create synthetic microdata on building stocks to describe individual buildings and their usage instead of using aggregate average archetype buildings. Yet another recent approach by Nägeli et al. [10] is an agent-based building stock model which combines a bottom-up building stock model (BSM) with agent-based modeling (ABM) to incorporate the interaction between building owners' decision making and relevant influencing factors. Moreover, previous literature on building stock dynamics covers topics like reconstitution of building stock dynamics [1], mortality of building stock [11], statistical analysis on demolished buildings [12], and vacancy of residential buildings [13].

Even if the abovementioned approaches in the previous literature are enough to provide a relatively good understanding of the current states of building stocks, options for modeling long-term future development of building stocks are limited. To forecast construction demand, researchers have used, e.g., multiple regression analysis [14], a panel vector error correction approach [15], a combination of neural networks and genetic algorithms [16], grey forecasting [17], and Box–Jenkins model [18]. However, these modeling approaches tend to be better suited for predicting short- or medium-term development than for long-term projections. In some of the most closely related studies, dynamic material flow analysis has been applied to model housing stock long-term in the Netherlands [19] and in Norway [20,21]. However, those focus on housing stock alone while our approach also pays attention to other building types. Moreover, there are established practices to assess long-term housing needs in many countries. These include but are not limited to Finland [22,23], Sweden [24,25], Norway [26], Denmark [27], England [28], the US [29], and Australia [30].

To create feasible strategies towards sustainability targets, a better understanding of the development of building stocks is urgently needed. Specifically, various analyses on built environments require long-term projections of building stock development. However, long-term forecasts even at their best include a great amount of uncertainty. Recognizing this inherent uncertainty, we propose a transparent calculation model for modeling building stock development, QuantiSTOCK. Our approach not only provides a tangible tool for understanding the development when selected assumptions are valid but also, most importantly, allows for studying the sensitivity of results to alternative developments of the key variables. Thus, this relatively simple modeling approach provides fruitful grounds for understanding the impact of different key variables, which is needed to facilitate meaningful debate on different housing, land use, and environment-related policies.

The model is particularly developed for modeling the development of Finnish building stock but may also be widely applied to other geographic locations when fitted for location-specific data. The developed QuantiSTOCK model may be extended in numerous ways and lays the groundwork for modeling the future developments of building stocks. The presented model may be used to a wide range of analyses ranging from assessing housing demand at the regional level to providing input for defining sustainable pathways towards climate targets. Thus, the results should be of interest to a wide range of researchers, policymakers, and community stakeholders who contribute to housing and land use policies.

This work is divided as follows: Section 2 outlines the modeling approach, which is followed by the Section 3 that explains the modeling procedure. In Section 4, the results of the analysis are presented, and thereafter, Section 5 provides the discussion. Finally, concluding remarks are presented in Section 6.

2. Modeling Approach

The quantitative building stock model (QuantiSTOCK) provides a relatively straightforward calculation-based approach to model future development of building stocks. The basic assumptions for modeling are that, logically, (1) population change and (2) mortality of existing buildings are the main drivers for quantitative changes in the building stock. Moreover, (3) gross floor area per capita ratio is an important modeling attribute that captures many overlapping processes, including but not limited to changes in residential density and distribution of housing types, and potential excess of new construction.

Two distinct advantages of the selected method are its comprehensibility and transparency. Comprehensibility refers to the modeling procedure being based on logical attributes that are suggested by common sense, while transparency, in this context, means that the modeling is based on publicly available data and the modeling procedure is clearly described in contrast to various black-box models; thus, the reader can understand how the model is constructed and what the role of the different modeling attributes is.

Relying on publicly available statistics, the model is easy to update when new statistics become available. Moreover, application of the model only requires relatively little effort in comparison to more complicated simulation approaches. Recognizing the fact that publicly available data from Statistics Finland is of an exceptionally high quality and well documented, which is not always the case, it is important that the applicability of publicly available data is evaluated case-by-case. When the reliability of public data is low, it is necessary that the user compiles a consistent dataset for modeling.

As projections of needed new construction cannot be considered an exact science, there is no such model that would produce an exact number of future needs [25]. Thus, it is critical to understand that, due to the great uncertainty about the predictor attributes in the long-term, it is by no means self-evident that the prediction accuracy of more complicated approaches would be any better than the outcome from this stripped-down model. For example, According to Boverket [25], Schmuecker [31] obtains similar results using a more simple method compared to the results from more advanced approaches used in England. Another important aspect of using a relatively simple approach is that it allows transparently, putting into perspective which factors are important relative to the big picture. In contrast, complicated modeling approaches may focus on complicated descriptions of the modeling procedure while the understanding of the critical factors may be blurred.

Here, the selected modeling approach does not aim to produce any exact numbers of future development as it would not be meaningful in terms of long-term projections, but the objective is to picture the potential pathways of future development and help to understand the impacts of these alternative scenarios. Such an approach helps us to understand the relationships between the key attributes and building stock development. Even if a great amount of uncertainty is still present in the selected modeling approach, this strategy combined with relevant sensitivity analyses provides a tangible tool for understanding the boundaries within which future development will fall into.

The structure of the QuantiSTOCK modeling approach is presented in Figure 1. The first step is to define the situation of the building stock at the beginning of the modeling period and then to define the modeling attributes for the future projection. As the modeling parameters are uncertain estimates, sensitivity to their changes is also necessary to be modeled to better understand the boundaries for the actual development. In this study, the sensitivity of building stock development to following key variables is modeled: mortality rate (low, historical, and high), population change (a decrease or an increase of five percent relative to the official population projection), and residential gross floor area per capita ratio (an annual decrease or increase of 0.5 percent relative to that in 2020). The lower and upper limits for the sensitivity analysis are defined by the research group members so that the values should present plausible boundaries for the fluctuation range of the variable. These steps are explained more precisely in the following Section 3, where a detailed description of the modeling procedure is provided.

As the urbanization trend still strongly affects the development of building stocks, allowing regional level heterogeneity is important. However, due to the high uncertainty of modeling attributes, a too fine-grained modeling approach is not meaningful either. In this study, this has been addressed by grouping the Finnish cities into three groups. The first group only includes the fast-growing Helsinki region, while the second group contains other Finnish cities that are growing but still at a slower pace than the capital region. Those include the regions of Tampere, Turku, Oulu, and Jyväskylä. The third group contains the rest of the cities that are, based on the official population projection, non-growing or declining in the study period. This grouping simplifies the modeling but still allows heterogeneity between the regions that are on different development paths.

As an outcome, the QuantiSTOCK model produces a projection of the building stock development in the study period. In this paper, the projection is presented for the above described grouping of cities: (i) fast-growing Helsinki region, (ii) growing regions, and (iii) zero-growth or declining regions. Moreover, an aggregation of the development of the entire Finnish building stock is presented. In addition to the overall development of the building stock, the outcome includes the projected demand for new construction, volume of building stock mortality, and distribution between the existing and new building stock in the study period of 2020–2050.

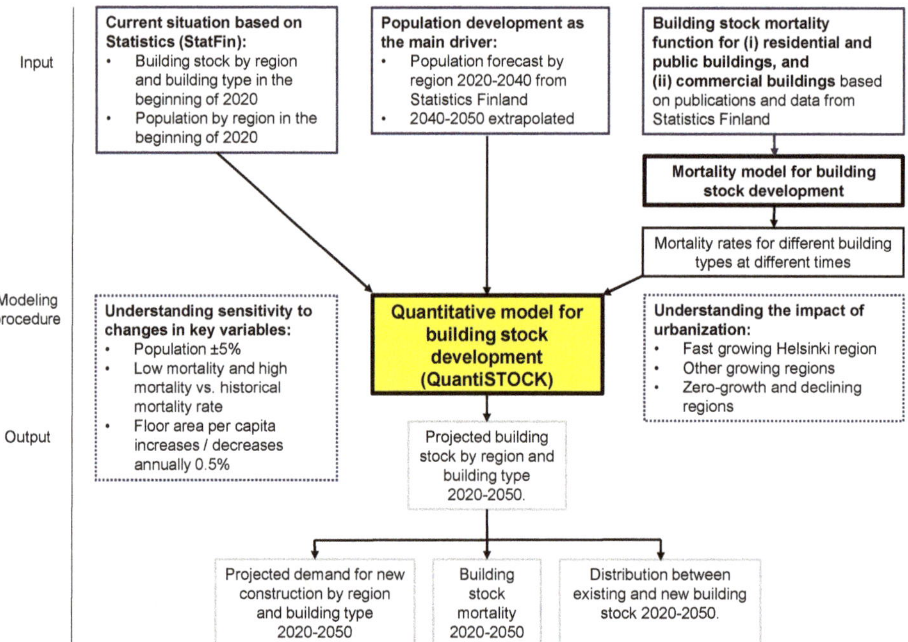

Figure 1. A diagram of the QuantiSTOCK model.

3. Modeling Procedure

The main steps of the QuantiSTOCK modeling approach are illustrated in Figure 1 in the previous section and now follows a more detailed description of the modeling procedure. The starting point for the modeling is the current state of the building stock. In this study, that is the existing building stock in Finland at the beginning of 2020. The building stock data follow the classification of buildings by Statistics Finland [32]. However, industrial and agricultural buildings are excluded from the model, as due to their heterogeneous nature, modeling attempts would not be meaningful in this context. Other important inputs for the QuantiSTOCK model are the regional distribution of population at the beginning of the modeling period and the regional population projection, which in this study, was available for the period of 2020–2040. To cover the entire study period, the official population projection was extrapolated to reach the end of 2050. The used population projection is described in more detail in Section 3.1. Moreover, gross floor area per capita ratios are calculated based on the situation at the beginning of the modeling period as more precisely described in Section 3.1. The raw data for all the above mentioned input data sets were acquired from the StatFin database [33]. The fourth required input data for the QuantiSTOCK model are mortality rates for different building types, which are defined based on mortality functions that lean on history statistics. As mortality data are not directly available from the statistics, the definition of mortality and creation process of mortality rates are explained in detail in Section 3.2.

When all input data is available, the modeling procedure may start to model the development of building stock in the study period. First, demand for annual new construction is modeled based on the parameters in Equations (1) and (2):

- When demand at the beginning of the year < stock at the beginning of the year:

$$
\begin{aligned}
&\textit{Demand for new construction} \\
&= \textit{Annual change in demand} + \textit{Mortality} + \textit{New non-permanently occupied floor area}
\end{aligned}
\tag{1}
$$

- When demand at the beginning of the year > stock at the beginning of the year:

$$
\begin{aligned}
&\textit{Demand for new construction} \\
&= (\textit{Demand at the beginning of the year} - \textit{Stock at the beginning of the year}) \\
&+ \textit{Annual change in demand} + \textit{Mortality} + \textit{New non-permanently occupied floor area}
\end{aligned}
\tag{2}
$$

where annual change in demand = gross floor area per capita ratio × annual population growth + annual change in gross floor area per capita ratio × total population.

Equation (1) is applied when the demand at the beginning of the year is less than the size of existing stock, while Equation (2) is used when the demand at the beginning of the year is greater than the stock at the beginning of the year. Demand is modeled separately for each building type and separately for the three region groups, including (i) the fast-growing Helsinki region, (ii) growing regions, and (iii) zero-growth and declining regions. Furthermore, the regional analysis is also aggregated to describe the development of the entire Finnish building stock.

The next step is to model the size of the building stock at the beginning of next year, which is performed based on the parameters in Equations (3) and (4):

- When demand for new housing construction < 0

$$
\begin{aligned}
&\textit{Stock atthe beginning of the year} \\
&= \textit{Stock at the beginning of the previous year} - \textit{Mortality}
\end{aligned}
\tag{3}
$$

- When demand for new housing construction > 0

$$
\begin{aligned}
&\textit{Stock at the beginning of the year} \\
&= \textit{Stock at the beginning of the previous year} - \textit{Mortality} \\
&+ \textit{Demand for new construction}
\end{aligned}
\tag{4}
$$

Equation (3) is used when there is no demand for new construction, while Equation (4) is applied when demand for new construction occurs. Again, modeling is performed separately for each building type and separately for the three region groups and, in the final stage, the regional results are aggregated to describe the development of the entire Finnish building stock. Next, a more detailed description of the definitions of the modeling attributes follows.

3.1. Population Growth and Gross Floor Area per Capita

In the QuantiSTOCK model, population growth is a main modeling attribute for the demand of residential building stock. Figure 2 depicts the population projection for the study period of 2020–2050. From 2020 to 2040, it follows the official regional population projection from Statistics Finland while, for the period of 2040–2050, the assumed population growth is extrapolated from the official projection. The impact of urbanization is easy to see in Figure 2. In 2020, the fast-growing Helsinki region, growing regions, and zero-growth and declining regions accommodate 29%, 21%, and 50% of the population, respectively, while, in 2050, it is projected that the respective proportions are 34%, 23%, and 43%. This should notably affect building stock dynamics within these region groups. However, it is important to notice that, even in declining regions, new construction is needed as the existing buildings do not meet all the demand and migration within regions also occurs. In particular, the ageing demographic structure causes moves from more distant rural locations to more attractive locations that are close to district centers and better services.

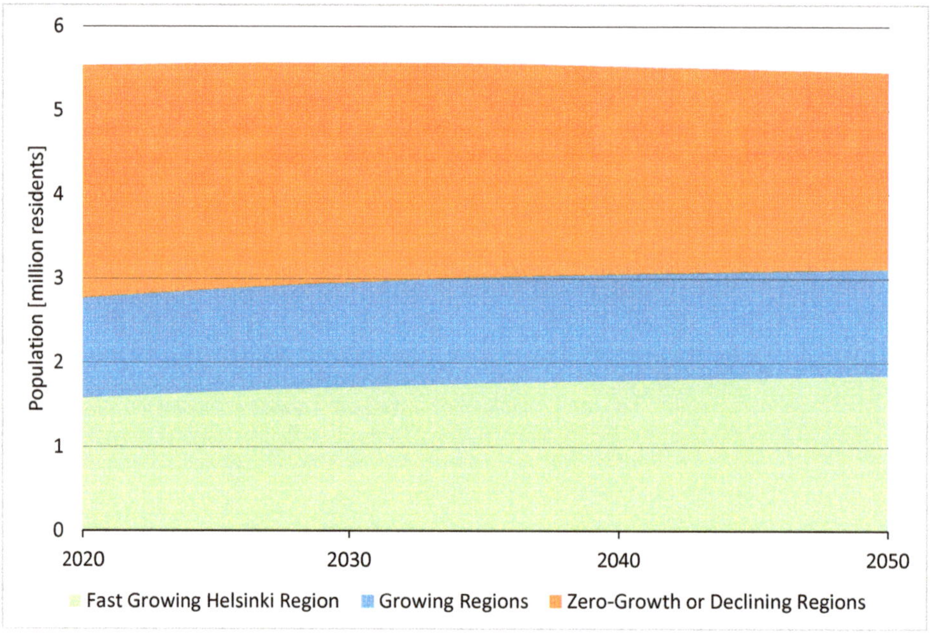

Figure 2. Population projection by region type in Finland for the period of 2020–2050.

Moreover, gross floor area per capita ratio is used to assess how many square meters of each building type are needed. The QuantiSTOCK model operates with gross floor areas instead of more detailed descriptions of building stock units. In the interest of simplicity, this allows for a straightforward approach that still provides important information for multiple purposes. Still, an indicative distribution between (i) detached houses, (ii) semi-detached and row houses, and (iii) apartment buildings is reported based on official statistics, with an assumption that the proportions of these different residential building types are assumed to remain at the same level throughout the study period. However, these distributions are reported for information purposes only and they do not affect the modeling procedure where the possible variation in different types of housing units is included in gross floor per capita ratio. Even if this is the case in the base version of the QuantiSTOCK model, alternative approaches such as using headship rate method (e.g., [23,25]) may be incorporated in the model if relevant.

It is also good to notice that, as dwelling densities tend to vary with various factors, such as residential building type and location, we instead use gross floor area ratio per capita as an input in the QuantiSTOCK model. This allows for including uncertainties about various factors into one modeling attribute. Those include changes in residential density and changes in proportions of different residential building types and even potential excess of new construction. In this study, the gross floor area per capita ratio is specified based on the official statistics at the beginning of the study period. For residential buildings, the proportion of gross floor area that is reported to be "non-permanently occupied" is excluded from the ratio. As ratios differ between different regions, a separate ratio is defined for each of the three region types. Furthermore, the ratio is assumed to remain at the same level throughout the study period.

For non-residential buildings, including public buildings and commercial buildings, the gross floor area per capita ratio is specified in the same way as for housing but, with the exception, that the "non-permanently occupied" floor area of public or commercial buildings cannot be distinguished. In the QuantiSTOCK model, different building types are categorized based on the classification of buildings by Statistics Finland [32]. However, we also include office buildings in the group of commercial buildings while our specification of public buildings includes transport and communication buildings, buildings for institutional care, assembly buildings, and educational buildings.

3.2. Mortality of Building Stock

Mortality rates of the existing buildings are yet another central input for the QuantiSTOCK model. As there are different types of mortality, including (A) demolition, (B) alterations to purpose of use, and (C) merger of spaces, it is important to explain what mortality of building stock means in the context of this paper. As the QuantiSTOCK model operates with gross floor areas, the types of mortality that are included are limited to types A and B. This outline is due to data limitations, as demolition of buildings and alterations to purpose of use are visible in building stock statistics while merger of spaces is not. It is also important to notice that types A and B cannot be separated from each other as only the total changes are reported in the official statistics.

To predict the mortality of the existing building stock, mortality functions were constructed based on the official statistics from Statistics Finland: more precisely, Population and Housing Census reports with ten-year intervals between 1950 and 2000 that were acquired from the Doria repository of Statistics Finland [34] that is maintained by National Library of Finland, while the latest cross sections for the years 2010 and 2018 were acquired from the StatFIN database [33]. The collected data account for the size of the stock for different types of buildings by year of building at different cross-sectional years, allowing for construction of separate mortality functions for each classified purpose of use by completion decades. These mortality functions describe the proportion of buildings from their respective decades that still exist at different cross sections of time.

Second, an integrated mortality function for each purpose of use was constructed based on the mortality functions that depict buildings from different decades separately. The first two steps provided information on the differences and similarities between the mortality of different purpose of use classes, allowing further integration of the mortality functions for similarly behaving purposes of use classes. The final integration resulted in two different mortality functions for the entire building stock, including (A) mortality of residential and public buildings, and (B) mortality of commercial buildings (Figure 3). The figure reveals a faster mortality of commercial buildings relative to residential and public buildings. More precisely, the pace of mortality of commercial buildings rapidly increases after the age of 40, and by the age of 70, the majority of commercial buildings do not exist anymore. At the same time, the lifecycle of residential and public buildings is notably longer.

The middle line (black) in Figure 3 depicts the statistics-based mortality of buildings. However, there is no guarantee that the mortality rate of building stock in the future should follow this history, which makes sensitivity analysis for alternative development paths necessary. To make the sensitivity analysis meaningful, plausible lower and upper limits for mortality development were defined by the members of the research group. In Figure 3, the upper line (green) depicts a low mortality scenario and the lower line (red) denotes a fast mortality scenario. By studying these three alternatives, an adequate understanding of the impact of changes within realistic boundaries should be achieved.

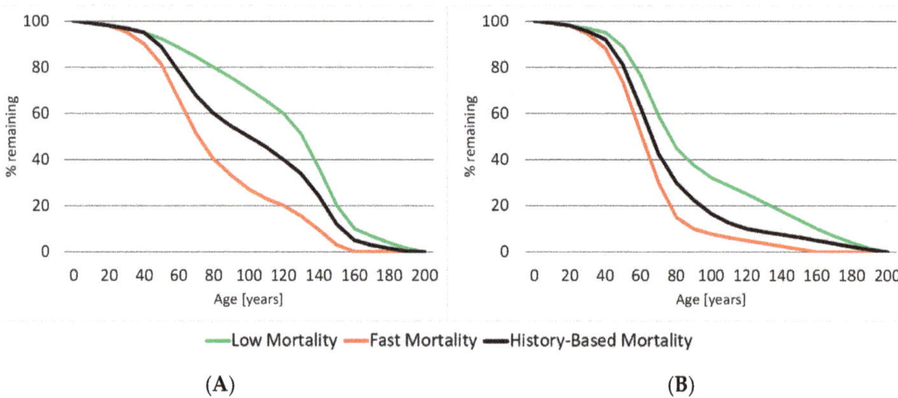

Figure 3. Three scenarios for the remaining proportion of the building stock by building types as a function of time: (**A**) residential and public buildings; (**B**) commercial buildings.

The final step is to convert the mortality functions into a usable form in terms of the QuantiSTOCK model. This is done by using a mortality sub-model, where the different variations of mortality functions are combined with the data on the current building stock. To define mortality rates for different building types at different cross sections in the future, the mortality sub-model also incorporates the needed new construction over time. These mortality rates, which are used as an input for the QuantiSTOCK model, are defined separately for (A) residential buildings and public buildings, and (B) commercial buildings. The rates vary between ten-year periods.

3.3. Validation of the Modeling Procedure

To validate the modeling procedure, development of the Finnish building stock in a past period from 2006 to 2019 was modeled using the QuantiSTOCK model. Then, the modeled results were compared to the building stock statistics for the same period to prove that the outcomes are in the expected range. In the test period, the gross floor area per capita ratio annually increased by 0.8% for residential buildings while the yearly increase of gross floor area for non-residential buildings was 1.7%, which was taken into account in the modeling attributes. Figure 4 shows that the calculation-based results from the QuantiSTOCK model seem to correspond well with the actual development of the building stock. This proves that the model is capable of producing accurate results, if the modeling attributes are in line with the actual development. However, the challenge here is to be able to assess the real development of modeling attributes. Given that uncertainty is always present in these assessments, the importance of sensitivity checks should be emphasized in an attempt to find the boundaries for real future development.

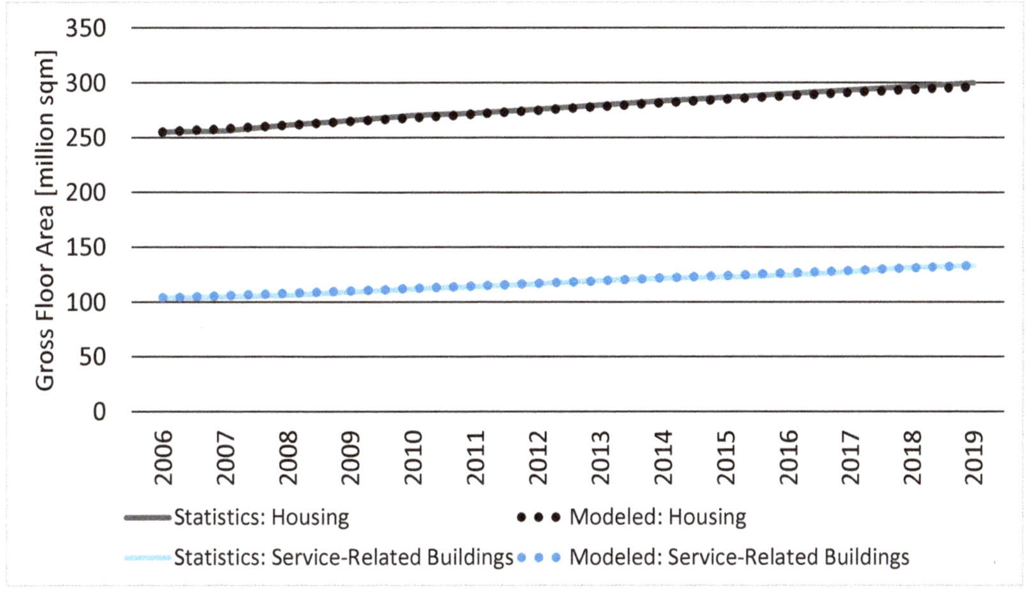

Figure 4. Modeled development vs. statistics in the period of 2006–2019.

4. Results

In this section, the modeling results for the Finnish building stock in the period of 2020–2050 are presented. The results are provided for the entire building stock, and regional differences are reported in accordance with (i) the fast-growing Helsinki region, (ii) growing regions, and (iii) zero-growth and declining regions. Moreover, sensitivity to changes in the mortality rate, population growth, and floor area per capita ratio is illustrated.

4.1. Modeled Development of Finnish Building Stock

Figure 5 depicts the modeled development of the Finnish building stock in the period of 2020–2050. The results suggest that, in 2050, less than 25 percent of the building stock is built after 2020. This finding confirms the importance of addressing the existing building stocks in strategies to achieve the EU's carbon neutral targets by 2050. Another interesting observation based on the results is that the total size of the Finnish building stock will not increase if the population development is in line with the 2019 population projection from Statistics Finland. Instead, a decrease of two percent is modeled relative to the building stock size in 2020.

In Figure 6, the focus is on a need for new construction over the study period and its distribution into residential buildings and non-residential buildings, including public and commercial buildings. In the study period of 2020–2050, 65 percent of the need for new construction is modeled to be residential buildings, equaling 65 million square meters of gross floor area (if evenly distributed, some 49,000 housing units annually). At the same time, non-residential buildings represent 35 percent of the need for new construction, equaling 35 million square meters of gross floor area.

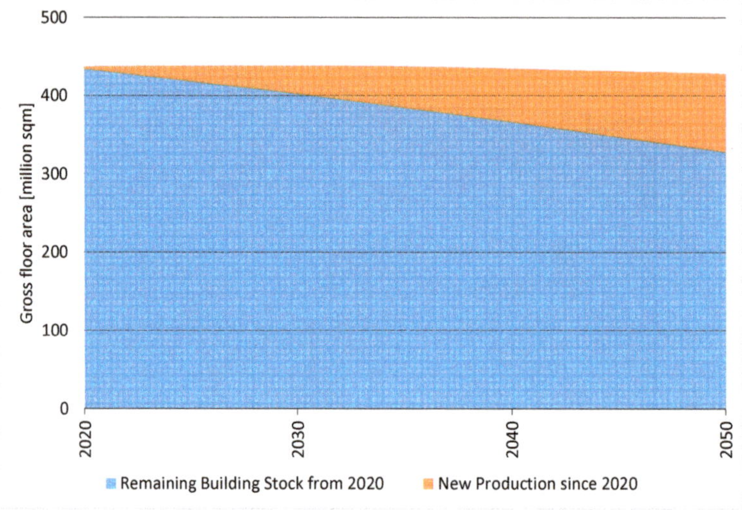

Figure 5. Modeled building stock development 2020–2050.

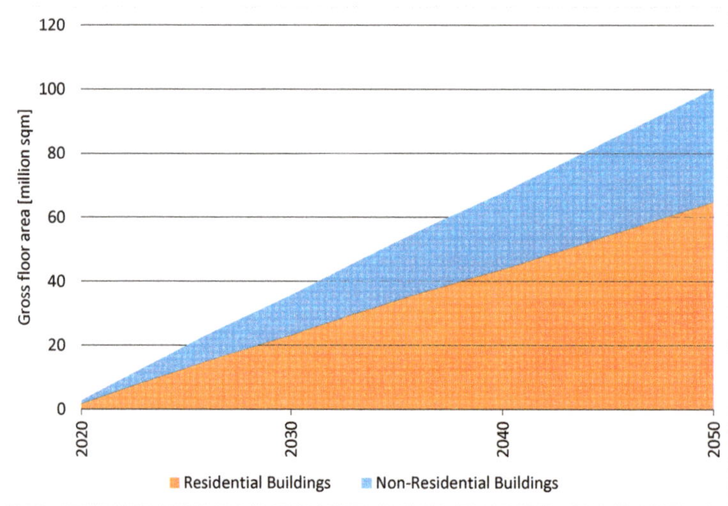

Figure 6. Modeled cumulative building production in the study period of 2020–2050: residential and non-residential buildings.

4.2. Regional Differences

In Figure 7, the modeled development is presented separately for the three different region groups, including (a) the fast-growing Helsinki region, (b) growing regions, and (c) zero-growth and declining regions. This more fine-grained representation reveals clear differences between the region groups, making the interpretations more meaningful. The modeling reveals that building stock is growing by 17 percent (19 million square meters of gross floor area) in the fast-growing Helsinki region and by 6 percent (5.5 million square meters) in other growing regions, while in zero-growth and declining regions, the total stock decreases 14 percent (33 million square meters). Despite the decreasing total stock, new construction is also needed in the zero-growth and declining regions due to migration from more distant locations to district centers, where the ageing population has better access to services, increasing demand for housing in regionally central locations.

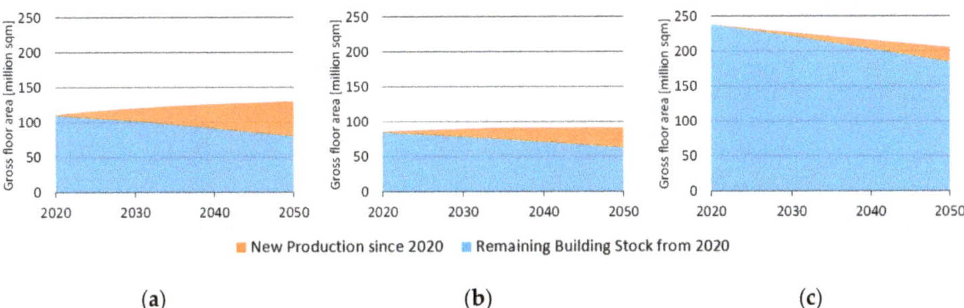

(a)   (b)   (c)

Figure 7. Modeled building stock development in different groups of regions: (**a**) fast-growing Helsinki region; (**b**) growing regions; and (**c**) zero-growth and declining regions.

Figure 8 reveals the distribution of the modeled new production into residential and non-residential buildings. In the fast-growing Helsinki region, the proportion of residential new production is 65 percent, equaling 33 million square meters of gross floor area (some 25,000 housing units annually). In the group of other growing regions, the proportion of new residential building production is 68 percent, equaling 19 million square meters of gross floor area (some 14,000 new housing units annually). In the zero-growth and declining regions, the proportion of residential buildings is 62 percent, equaling 13 million square meters (some 10,000 new housing units annually).

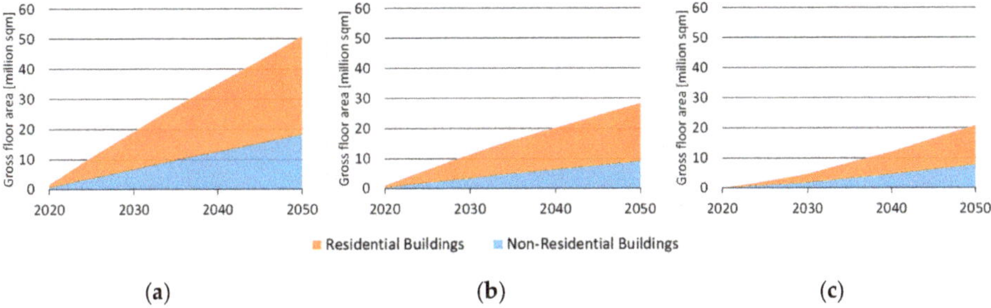

(a)   (b)   (c)

Figure 8. Modeled cumulative need for new construction in different regions: (**a**) fast-growing Helsinki region; (**b**) growing regions; and (**c**) zero-growth and declining regions.

4.3. Sensitivity Analysis

As modeling of future development always leans on assumptions, it is important to recognize the key variables and explore how the modeling results are affected if values of these variables vary. This kind of sensitivity analysis allows a better understanding of the boundaries for actual development as modeling only one potential development scenario could result in faulty conclusions. Below, the sensitivity of building stock development to mortality rate, population change, and residential gross floor area per capita ratio is illustrated, ceteris paribus.

4.3.1. Sensitivity to Mortality Rate

Figure 9 depicts the impact of mortality rate on the development of building stock. The panels reveal that mortality rate has a notable impact on the structure of the future building stock. Specifically, in the low mortality scenario, the proportion of new construction since 2020 is 13 percent; in the history-based scenario, it is 23 percent; and in the high mortality scenario, it is 35 percent of the stock. However, based on the results, changes in the mortality rate alone do not seem to affect the total size of the building stock but the size of the stock is the same in all three scenarios at the end of the study period. This is because of the assumption that all mortality has to be replaced with new production within the region. The reasoning behind this is that, if this proportion of the building stock was an essential part of accommodating residents in the region before its mortality, these people need a place to live after the mortality as well if population growth is not negative. Whereas the total size of the building stock starts decreasing, if the population growth turns negative. Then, only the proportion of mortality that has a demand in the market is replaced with new production.

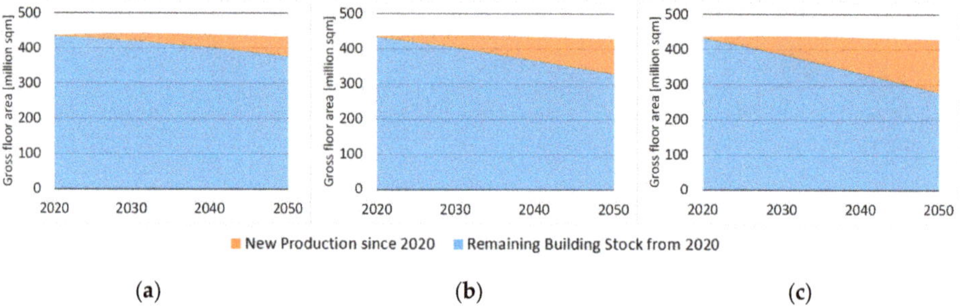

Figure 9. Modeled development of building stock with different mortality rates: (**a**) low mortality rate; (**b**) history-based mortality rate; and (**c**) high mortality rate.

4.3.2. Sensitivity to Population Growth

In Figure 10, the impact of changes in population growth is illustrated. The middle scenario (b) is based on the official population projection from StatFIN database [33], and sensitivities to a decrease and an increase of 5 percent (approximately 270,000 residents, equaling to some 9000 residents annually if evenly distributed over the study period) are modeled in scenarios (a) and (c). The figure reveals that the impact is notable on both new production and the total size of the building stock. The total size of the stock is 6 percent (27 million square meters) smaller in 2050 than in 2020 in scenario (a), a little less than 2 percent (8 million square meters) greater in scenario (b), and over 2 percent (11 million square meters) greater in scenario (c). At the same time, the respective proportions of new production since 2020 at the end of the study period are 19 percent, 23 percent, and 28 percent, of which the proportion of housing is some 65 percent in all scenarios.

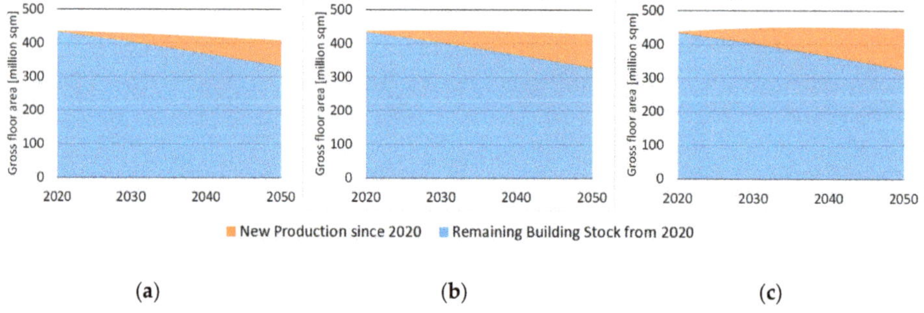

Figure 10. Modeled development of building stock with different population changes: (**a**) the population in 2050 at 5% less than in the official population projection; (**b**) the population in 2050 in line with the official population projection; and (**c**) the population in 2050 at 5% more than in the official population projection.

4.3.3. Sensitivity to Residential Gross Floor Area per Capita

Finally, Figure 11 illustrates the sensitivity to changes in residential gross floor area per capita ratio that captures several overlapping processes, including but not limited to changes in residential density and distribution of housing types, and potential excess of new construction. In scenario (a), the residential gross floor area per capita annually decreases by 0.5 percent; in scenario (b), the ratio remains the same as is in 2020 throughout the modeling period; and in scenario (c), the residential gross floor area per capita ratio annually increases by 0.5 percent. The figure reveals that the impact is notable on both new production and the total size of the stock. The total size of the stock in 2050 is modeled to be 9 percent (41 million square meters) smaller, 2 percent (8 million square meters) smaller, and 8 percent (36 million square meters) greater than the total stock size in 2020 in scenarios (a), (b), and (c), respectively. At the same time, the proportion of new production since 2020 in the respective scenarios is 15 percent, 23 percent, and 32 percent, of which the proportion of residential buildings is 38 percent, 65 percent, and 77 percent in scenarios (a), (b), and (c), respectively. The proportion of residential buildings since 2020 varies notably between the scenarios, as only residential gross floor area per capita ratio is changed, while the ratio for non-residential buildings remains the same. Changing gross floor area per capita ratios for both residential and non-residential buildings at the same time would not be meaningful, as the ratios may develop towards opposite directions, making scrutinizing only one change at a time more informative.

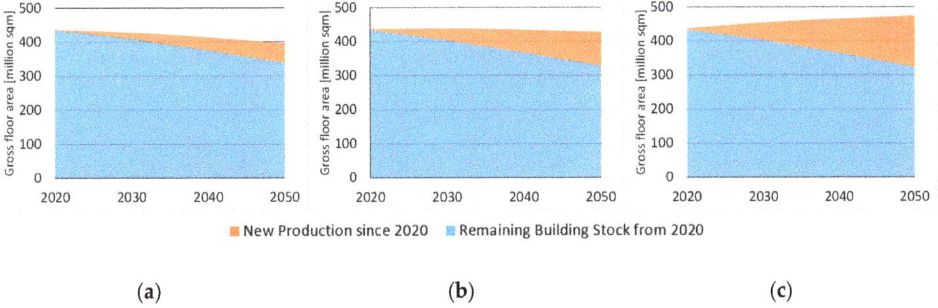

Figure 11. Modeled development of building stock with different gross floor area per capita ratios: (**a**) gross floor area per capita decreases annually by 0.5%; (**b**) gross floor area ratio through the modeling period as in 2020; and (**c**) gross floor area per capita grows annually by 0.5%.

5. Discussion

In this paper, we introduced a calculation-based model for modeling the quantitative future development of building stocks in the long-term. The previous literature covers various ways to describe the current state of the building stock, including approaches that are based on constructed building archetypes [7], sample buildings [8], synthetic microdata [9], and agent-based building stock model [10]. Moreover, there are numerous papers where advanced forecasting approaches have been applied to produce short- or medium-term forecasts, for example, on construction demand. These approaches include but are not limited to multiple regression analysis [14], a panel vector error correction model [15], a combination of neural networks and genetic algorithms [16], grey forecasting [17], and Box–Jenkins model [18]. As our focus is on long-term modeling, the most closely related studies lean on material flow analysis [19–21] and various assessment approaches to forecast long-term housing needs [22–30]. However, these most closely related studies are usually limited to projections of housing stocks while other building types are excluded from the analysis. The presented QuantiSTOCK model provides a novel contribution to sustainable management of building stocks by combining approaches akin to what has been presented on dynamic material flow analysis in [21], on the assessment of housing needs in [23,25], and on mortality analysis in [11]. In addition to residential buildings, as opposed to the previous literature, the QuantiSTOCK model also covers other building types. The only excluded building categories are industrial and agricultural buildings as their heterogenic nature would make modeling highly uncertain.

The base version of the QuantiSTOCK model operates in gross floor area units and provides a relatively straightforward calculation-based approach to model future development of building stocks, where (1) population growth, (2) mortality of existing buildings, and (3) gross floor area per capita ratio are the three main drivers for quantitative changes in the building stock. The population growth input is directly based on the official population projection from the StatFIN database [33]. Regarding mortality, the compiled mortality functions are akin to the survival functions in [11], but in our simplified approach, we do not apply any mathematical equations but the curves are rather visually fitted based on the points from history statistics. Next, curves that are relatively similar to each other were merged, which in this study resulted in separate mortality curves for (i) residential and public buildings, and (ii) commercial buildings. Finally, taking also into account the current state of building stock and cumulative need for new construction, the mortality curves are translated into mortality rates. These mortality rates vary with time as the building stock evolves. These rates provide a statistics-based base scenario for the analysis, which is complemented with low and high scenarios that provide boundaries for the range of variation.

In terms of gross floor per capita ratio, it is important to notice that this ratio is a multifaceted modeling attribute that captures many overlapping processes, including but not limited to changes in residential density and distribution of housing types, and potential excess of new construction. Thus, using gross floor area per capita ratio differs from using residential density as a modeling attribute instead. For example, urbanization trends contribute to an increasing proportion of apartment buildings where residential densities tend to be lower than in single-family houses. Additionally, as a result of increased housing prices in urban centers, more and more people may still prefer good locations but choose to consume less floor area, which also leads to lower gross floor area per capita ratio. On the other hand, if excessive new construction occurs in an area, it may seem that residential density has increased. However, in such a case, the actual reason for the higher gross floor area per capita ratio—or at least part of it—would be that more new construction has been delivered to the market in relation to the number of new residents who have moved in. Thus, it is important to properly consider which factors may affect gross floor area per capita ratio in each case. Relative to the methods in assessing housing needs in [23,25], our base modeling approach is a simplified version as QuantiSTOCK does not take into account headship rates for different groups. However, we recognize that this may be necessary for some analyses, for example, if the data allows a more detailed analysis for different age cohorts or the number of housing units is of a particular interest. To address these potential needs, the QuantiSTOCK model is easily modifiable to include such an alternative modeling approach. Another distinct advantage of the QuantiSTOCK approach is that it relies on publicly available statistics, making it easy to update the model when new statistics becomes available. Of course, this is limited to locations where public high-quality data are available. Otherwise, it is advisable to use self-compiled data that is tailored for the analysis. In cases where high-quality data are easily accessible, application of the model only requires relatively little effort, as opposed to more complicated simulation approaches. It is also critical to understand that, because of the great uncertainty about the predictor attributes in the long-term, it is not self-evident that application of more complicated approaches would results in more accurate outcomes than the outcome from this stripped-down model. This is supported by Boverket [25] who reports that Schmuecker [31] obtained similar results using a simpler method compared to more advanced approaches used in England. Another important aspect in using a relatively simple approach is that it allows one to transparently put into perspective which factors are important relative to the big picture. In contrast, complicated modeling approaches may focus on complicated descriptions of the modeling procedure, leaving an actual understanding of the critical factors blurred.

The QuantiSTOCK model seeks to provide a simplified and transparent modeling approach that is easy to understand. In this paper, we use "demand for new construction" and "need for new construction" as synonyms. The demand is considered here as objectively assessed need for new construction that is required to address changes in demographics and building stocks. In other words, the definition for demand is broader than the strict traditional definition in economics. As long-term projections even at the best include a great amount of uncertainty, the focus here is rather on major lines than in trifling matters. It is still good to bear in mind that a simplified approach always requires choices and approximations that may also hide some critical aspects. Therefore, it is critical to understand the impact of the incorporated assumptions as well as to perform adequate sensitivity checks for the development of key modeling attributes. In this study, the research group members defined together the sensitivity checks for our analysis on the Finnish building stock. The challenge in this strategy is to be able to define lower and upper limits for the modeling attributes so that they provide proper boundaries for actual development. In order to succeed in setting these proper limits, expertise in the field of a built environment is necessary and the modeling results are only reliable if the interpreters understand the underlying assumptions. Still, any "black box" approaches do not solve this problem either, as they only tend to increase the risk of unexplainable and unreliable modeling results. However, we recognize that, in some contexts, it may be necessary to increase the degree of complexity in the QuantiSTOCK modeling procedure, for example, to better understand the impacts of changing demographics. It is also important to understand that economic conditions and restrictions from land use planning notably affect the volume and structure of new construction that occurs in real world.

6. Conclusions

The QuantiSTOCK model is particularly developed for modeling the development of the Finnish building stock, which was used as a development and test arena in this study due to the good availability of high-quality data. However, this relatively simple modeling approach may also be widely applied to other geographic locations when fitted for location-specific data. By being transparent, the model provides fruitful grounds for understanding the impact of different key variables. This is necessary to allow more reliable analyses on the built environment and to facilitate meaningful debate on different housing, land use, and environment-related policies. As the proposed model is relatively easily modifiable, we consider it to have a great potential to be widely applied in various fields.

The modeling using Finnish data revealed that, in the study period of 2020–2050, the total size of the Finnish building stock will not be growing and the size of the stock may even slightly decrease if the population growth is in line with the official population projection; 65 percent of the new production in the study period is modeled to be residential buildings. However, the proportion of buildings that are built since 2020 is less than 25 percent in 2050, which once again is a reminder that measures addressing the existing building stock are critical in an attempt to achieve the EU's carbon neutral targets.

Further examination of the modeling results by region type reveals notable regional differences in the building stock development, confirming the high impact of urbanization. In the fast-growing Helsinki region, other fast-growing regions, and zero-growth and declining regions, the percentage changes of the total building stock size in the period of 2020–2050 are 17 percent, 6 percent, and −14 percent, respectively. However, the decreasing stock size does not directly mean that there is no need for new construction but that migration within these regions from more distant locations to district centers increases the demand in regionally central locations even when the total size of building stock in the region is decreasing.

Due to the uncertainty in modeling future development, sensitivity checks to changes in key modeling attributes are necessary to understand the boundaries of actual building stock development. The sensitivity analysis demonstrated that mortality rate has a notable impact on the structure of the future building stock, as proportions of new construction since 2020 varied from 13 percent to 35 percent. Also, a decrease or an increase of 5 percent (some 270,000 people) in the projected total population for 2050 had a clear impact on building stock development as the proportion of new production in 2050 varied between 19 and 28 percent and the size of the total stock varied between −6 and 2 percent. Furthermore, residential gross floor area per capita ratio was observed to have a high impact on the modeling outcome both in terms of new production and the total size of the stock. An annual decrease and increase of 0.5% in residential gross floor area per capita ratio resulted in the total size of the building stock in 2050 varying between −9 percent and 8 percent relative to the beginning of the study period. At the same time, the proportions of new construction since 2020 varied between 15 and 32 percent.

The introduced QuantiSTOCK model may be extended in numerous ways, and it lays the groundwork for modeling the future developments of building stocks. Some potentially fruitful strands for future work that could help develop the QuantiSTOCK model should be, for example, (i) a more detailed study on the dynamics of building stock development at different scales, (ii) the impact of local conditions on building stock development, (iii) the impact of municipal land use planning on building stock development, (iv) a further and more robust validation of the model using longer time horizon and data from other countries, (v) mortality differences between owner-occupied and rental buildings, and (vi) recognizing if construction techniques and materials have an impact on mortality. Nevertheless, already, today's version of the QuantiSTOCK model may be used in a wide range of analyses ranging from assessing housing demand at the regional level to providing input for defining sustainable pathways towards climate and land use targets. Thus, the results should be of interest to a wide range of researchers, policymakers, and community stakeholders who contribute to creating better built environments.

Author Contributions: Conceptualization, A.K., A.S., J.H. and E.N.; methodology, A.K., A.S., J.H. and E.N.; validation, A.K., A.S., J.H. and E.N.; formal analysis, A.K., A.S., J.H. and E.N.; investigation, A.K., A.S., J.H. and E.N.; resources, A.K., A.S., J.H. and E.N.; data curation, A.K., J.H. and E.N.; writing—original draft preparation, A.K.; writing—review and editing, A.K., A.S., J.H. and E.N.; visualization, A.K.; supervision, A.S.; project administration, A.S.; funding acquisition, A.S., J.H. and A.K. All authors have read and agreed to the published version of the manuscript.

Funding: This research was funded by the Academy of Finland (decision number: 309,067, project: RenewFIN—Optimal transformation pathway towards the 2050 low-carbon target: integrated buildings, grids, and national energy system for the case of Finland). The main author would also like to thank the Finnish Foundation for Technology Promotion for funding.

Institutional Review Board Statement: Not applicable.

Informed Consent Statement: Not applicable.

Data Availability Statement: Publicly available datasets were analyzed in this study. This data can be found here: [https://www.stat.fi/tup/statfin/index_en.html] and [https://www.doria.fi/handle/10024/67150].

Conflicts of Interest: The authors declare no conflict of interest.

References

1. Aksözen, M.; Hassler, U.; Kohler, N. Reconstitution of the dynamics of an urban building stock. *Build. Res. Inf.* **2017**, *45*, 239–258. [CrossRef]
2. *Directive 2010/31/EU of the European Parliament and of the Council of 19 May 2010 on the Energy Performance of Buildings*; The Publication Office of the European Union (Publications Office): Luxembourg, 2010.
3. *Directive 2012/27/EU of the European Parliament and of the Council of 25 October 2012 on Energy Efficiency, Amending Directives 2009/125/EC and 2010/30/EU and Repealing Directives 2004/8/EC and 2006/32/EC Text with EEA Relevance*; The Publication Office of the European Union (Publications Office): Luxembourg, 2012.
4. *Long-Term Low Greenhouse gas Emission Development Strategy of the European Union and its Member States-Submission to the UNFCCC on Behalf of the European Union and its Member States*; Council of the European Union: Zagreb, Croatia, 2020.

5. European Parliament. *Communication from the Commission to the European Parliament, the Council, the European Economic and Social Committee and the Committee of the Regions A Renovation Wave for Europe-Greening our Buildings, Creating Jobs, Improving Lives*; COM(2020) 662 Final; The Publications Office of the European Union (Publications Office): Luxembourg, 2020.
6. European Parliament. *Communication from the Commission to the European Parliament, the Council, the European Economic and Social Committee and the Committee of the Regions. Roadmap to a Resource Efficient Europe*; COM(2011) 571 Final; The Publications Office of the European Union (Publications Office): Luxembourg, 2011.
7. Swan, L.G.; Ugursal, V.I. Modeling of end-use energy consumption in the residential sector: A review of modeling techniques. *Renew. Sustain. Energy Rev.* **2009**, *13*, 1819–1835. [CrossRef]
8. Mata, É.; Sasic Kalagasidis, A.; Johnsson, F. Building-stock aggregation through archetype buildings: France, Germany, Spain and the UK. *Build. Environ.* **2014**, *81*, 270–282. [CrossRef]
9. Nägeli, C.; Camarasa, C.; Jakob, M.; Catenazzi, G.; Ostermeyer, Y. Synthetic building stocks as a way to assess the energy demand and greenhouse gas emissions of national building stocks. *Energy Build.* **2018**, *173*, 443–460. [CrossRef]
10. Nägeli, C.; Jakob, M.; Catenazzi, G.; Ostermeyer, Y. Towards agent-based building stock modeling: Bottom-up modeling of long-term stock dynamics affecting the energy and climate impact of building stocks. *Energy Build.* **2020**, *211*, 109763. [CrossRef]
11. Aksözen, M.; Hassler, U.; Rivallain, M.; Kohler, N. Mortality analysis of an urban building stock. *Build. Res. Inf.* **2017**, *45*, 259–277. [CrossRef]
12. Huuhka, S.; Lahdensivu, J. Statistical and geographical study on demolished buildings. *Build. Res. Inf.* **2016**, *44*, 73–96. [CrossRef]
13. Huuhka, S. Vacant residential buildings as potential reserves: A geographical and statistical study. *Build. Res. Inf.* **2016**, *44*, 816–839. [CrossRef]
14. Bee-Hua, G. An evaluation of the accuracy of the multiple regression approach in forecasting sectoral construction demand in Singapore. *Constr. Manag. Econ.* **1999**, *17*, 231–241. [CrossRef]
15. Jiang, H.; Liu, C. A panel vector error correction approach to forecasting demand in regional construction markets. *Constr. Manag. Econ.* **2014**, *32*, 1205–1221. [CrossRef]
16. Bee-Hua, G. Evaluating the performance of combining neural networks and genetic algorithms to forecast construction demand: The case of the Singapore residential sector. *Constr. Manag. Econ.* **2000**, *18*, 209–217. [CrossRef]
17. Tan, Y.; Langston, C.; Wu, M.; Ochoa, J.J. Grey Forecasting of Construction Demand in Hong Kong over the Next Ten Years. *Int. J. Constr. Manag.* **2015**, *15*, 219–228. [CrossRef]
18. Fan, R.Y.C.; Ng, S.T.; Wong, J.M.W. Reliability of the Box–Jenkins model for forecasting construction demand covering times of economic austerity. *Constr. Manag. Econ.* **2010**, *28*, 241–254. [CrossRef]
19. Müller, D.B. Stock dynamics for forecasting material flows—Case study for housing in The Netherlands. *Ecol. Econ.* **2006**, *59*, 142–156.
20. Bergsdal, H.; Brattebø, H.; Bohne, R.A.; Müller, D.B. Dynamic material flow analysis for Norway's dwelling stock. *Build. Res. Inf.* **2007**, *35*, 557–570. [CrossRef]
21. Sartori, I.; Bergsdal, H.; Müller, D.B.; Brattebø, H. Towards modelling of construction, renovation and demolition activities: Norway's dwelling stock, 1900–2100. *Build. Res. Inf.* **2008**, *36*, 412–425. [CrossRef]
22. Vainio, T. *Asuntotuotantotarve 2020–2040*; VTT Technical Research Centre of Finland: Espoo, Finland, 2020; ISBN 978-951-38-8735-3.
23. Lankinen, M. *Asuntorakentamisen Ennakointi*; Oy Edita Ab, Suomen Ympäristö: Helsinki, Finland, 1996; Volume 43, ISBN 952-11-0062-1.
24. Boverket. *Behov av Nya Bostäder 2018–2025*; Boverket: Karlskrona, Sweden, 2018.
25. Boverket. *En Metod för Bedömning av Bostadsbyggnadsbehovet*; Boverket: Karlskrona, Sweden, 2016.
26. Ruud, M.E.; Barlindhaug, R.; Nørve, S. *Fremtidige Boligbehov*; NIBR-Rapport; Norsk Institutt for By-og Regionforskning: Oslo, Norway, 2013.
27. Socialministeriet. Den Almene Boligsektors Fremtid. Rapport Fra Arbejdsgruppen Vedrørende Fremtidsperspektiver for en Mere Selvbærende Almen Sektor. Socialminsteriet: København, Denmark, 2006.
28. Ministry of Housing, Communities & Local Government Housing and Economic Needs Assessment. Available online: https://www.gov.uk/guidance/housing-and-economic-development-needs-assessments (accessed on 9 December 2020).
29. Landis, J.D. *Raising the Roof: California Housing Development Projections and Constraints, 1997–2020*; California Department of Housing and Community Development: Sacramento, CA, USA, 2000.
30. Rowley, S.; Leishman, C.; Baker, E.; Bentley, R.; Lester, L. Modelling Housing Need in Australia to 2025. Available online: https://www.ahuri.edu.au/research/final-reports/287 (accessed on 13 December 2020).
31. Schmuecker, K. *The Good, the Bad and the Ugly. Housing Demand 2025*; Institute for Public Policy Research: London, UK, 2011; p. 27.
32. Statistics Finland. Classification of Buildings 2018. Available online: https://www.stat.fi/en/luokitukset/rakennus/rakennus_1_20180712/ (accessed on 21 October 2020).
33. Statistics Finland. StatFin Database. Available online: https://www.stat.fi/tup/statfin/index_en.html (accessed on 21 October 2020).
34. Statistics Finland. Doria Repository of Statistics Finland. Available online: https://www.doria.fi/handle/10024/67150 (accessed on 5 November 2020).

 *sustainability*

Article

The Contribution of Peripheral Large Scientific Infrastructures to Sustainable Development from a Global and Territorial Perspective: The Case of IFMIF-DONES

Virginia Fernández-Pérez [1] and Antonio Peña-García [2,*]

1 Department of Organización de Empresas I, University of Granada, 18071 Granada, Spain; vfperez@ugr.es
2 Department of Civil Engineering & PI UGR "DONES Preparatory Phase" (CE Ref. 870186),
   University of Granada, 18071 Granada, Spain
* Correspondence: pgarcia@ugr.es

Abstract: Large scientific infrastructures are a major focus of progress. They have a big impact on the economic and social development of their surroundings. Departing from these well-known facts, it is not trivial to affirm whether the global contribution to Sustainable Development (SD) is higher when they are built in peripheral and not highly developed provinces instead of capitals and rich areas. Besides the economic impact on depressed areas, other SD-related parameters like the attachment of young and skilled people to their homeland, the avoidance of uncontrolled migrations from rural to dense urban zones, the growth of new focuses of knowledge independent from the lines of research established in the universities of the capitals, the indirect impact of auxiliary infrastructures and others must be analyzed. Concerning the next implementation of the "International Fusion Materials Irradiation Facility—Demo Oriented Neutron Source" (IFMIF-DONES) project in Granada (Spain), one depressed and tourism-dependent zone, an analysis and comparison with similar infrastructures were done and presented.

Keywords: sustainable development; global sustainability; scientific infrastructures; Post-COVID-19 Scenario

Citation: Fernández-Pérez, V.; Peña-García, A. The Contribution of Peripheral Large Scientific Infrastructures to Sustainable Development from a Global and Territorial Perspective: The Case of IFMIF-DONES. *Sustainability* **2021**, *13*, 454. https://doi.org/10.3390/su13020454

Received: 16 December 2020
Accepted: 2 January 2021
Published: 6 January 2021

Publisher's Note: MDPI stays neutral with regard to jurisdictional claims in published maps and institutional affiliations.

1. Introduction

Science and Technology have been acknowledged for a long time as key motors to foster progress and well-being. However, the way scientific and especially technological development has been focused has evolved from an absolutely results-oriented philosophy to a more global perspective including not immediate benefits. The first philosophy lasted from the Industrial Revolution in the early 19th century until the third quarter of the 20th century, whereas the new perspective taking Environment and Sustainability as the main factors has been a constant trend for about 40 years.

Although there has been no single inflection point from the first to the current philosophy, we might be able to find one remarkably important milestone in the so-called Brundtland Report: "Sustainable development must meet the needs of current generations without compromising the ability of future generations to meet their own needs" [1].

The recent compilation of the Sustainable Development Goals (SDGs) [2], which summarizes and attempts to channel the efforts toward a better world, is fully applicable to the technological development and the infrastructures built to get it. Thus, besides the expected scientific and technical advances, the actions and projects with a strong technological background should bring highly qualified employment, investments and economic prosperity for the territory contributing to the achievements of the SDGs. Table 1 summarizes the SDGs and their official labels.

Table 1. United Nations Sustainable Development Goals (SDGs) [2].

	SDG
1	No Poverty
2	Zero Hunger
3	Good Health and Well-being
4	Quality Education
5	Gender Equality
6	Clean Water and Sanitation
7	Affordable and Clean Energy
8	Decent Work and Economic Growth
9	Industry, Innovation and Infrastructure
10	Reduced Inequality
11	Sustainable Cities and Communities
12	Responsible Consumption and Production
13	Climate Action
14	Life Below Water
15	Life on Land
16	Peace and Justice Strong Institutions
17	Partnerships to Achieve the Goal

In this scenario, it is really important to compile information and foresee the economic and social consequences of each project in order to support or disregard those decisions dealing with their implementation and the way to do it, including its location, timetable, public investment and other factors.

Thus, the role and responsibility of governments and public administrations have remarkably changed because now it is clearly understood and accepted that [3,4]:

(1) The environmental impacts and the demand for natural resources must be limited.
(2) Productive models must be based on technology allowing sustainable processes.
(3) Sustainable Development requires organizational structures that must be adapted accordingly.

The so-called "Circular Economy", more and more deeply implemented in Engineering processes going from design to decommissioning [5,6], is a good example.

Attending to its significance and dimension (environmental, social and economic) [7], it could seem that Sustainable Development is a holistic concept, and technology focused on the SDGs cannot be understood out of a global framework. Certainly, SDGs cannot be achieved from isolated perspectives, and some transversal branches of Science and Technology closely related to them are claiming for global visions [8]. However, it is important to keep in mind that a global perspective of technology is fully compatible and even necessary for territorial sustainable development exclusively focused on concrete areas.

This research presents and analyses the situation and potential effects of critical scientific and technological infrastructures on their near geographical framework from the perspective of Global Sustainability and SDGs. The particular case of the "International Fusion Materials Irradiation Facility—Demo Oriented Neutron Source" (IFMIF-DONES) is taken as an example and guide. Besides the main target of this project, that is, providing some key milestones from other perspectives parallel to the path toward clean and almost unlimited energy from nuclear fusion, it seeks to contribute to sustainable regional development in the abovementioned wide contexts of the Brundtland Report and SDGs (see Figure 1).

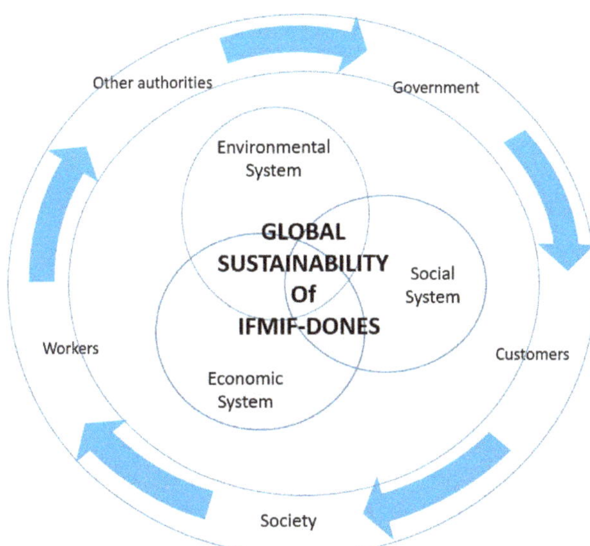

Figure 1. "International Fusion Materials Irradiation Facility—Demo Oriented Neutron Source" (IFMIF-DONES) from a global and territorial perspective. Sustainable energy economically and socially viable attending territorial necessities.

The next section will briefly present the IFMIF-DONES project and its objectives.

2. IFMIF-DONES: The Key Milestone toward the Use of Fusion Energy

2.1. The Situation with Nuclear Power

The energy inside the matter can be obtained by breaking heavy and unstable nuclei like Uranium-235 of Plutonium-239 (nuclear fission) or by fusing light nuclei like hydrogen into heavier ones (nuclear fusion). The fundamentals of both, fission and fusion, are not complex and have been understood since the 1930s. However, the technology to actually obtain energy from the nuclei is extremely complex. In the case of fission, big and complex centers have been built since the 1940s, and it is nowadays an essential energy source. In Spain, 20% of the produced energy is achieved with just seven reactors that started to work between 1983 and 1988 [9]. This 20% is not enough and additional energy from nuclear origin must be bought from other countries. However, nuclear fusion is a technical challenge, and its control to produce useful energy is still far away. This is a problem because, whilst fission requires the use, management and storage of hazardous materials, fusion is expected to be a cleaner, safer and much more efficient source of energy.

In this general framework where mankind needs more and more energy to ensure economic growth and resources for everyone without harming the environment, both the scientific community and governments around the world are working hard to control nuclear fusion and thus profit from its huge benefits.

As mentioned above, the control of fusion is very difficult from a technical point of view. There are two main limitations: (1) fusing light elements requires working in extreme conditions around 100 million °C; and (2) the neutrons arising from the nuclear reaction are extremely difficult to control, and they would quickly damage the whole installation.

Several experimental installations have tried to demonstrate the feasibility of fusion by solving the abovementioned problems one way or another. Among them, the "International Thermonuclear Experimental Reactor" (ITER, www.iter.org) is the most ambitious project up to date. It is currently under construction in Cadarache (France) and, according to the official sources of the Project, it is expected to maintain fusion for long periods of time and test the integrated technologies, materials and physics regimes necessary for the commercial production of fusion-based electricity [10]. The fact that the ITER members are China, the European Union, India, Japan, Korea, Russia and the United States shows the high strategic interest of this experimental infrastructure and the geopolitical importance of fusion control. Figure 2 shows the state of part of the installation in December 2020.

Figure 2. ITER under construction. Credit © ITER Organization, http://www.iter.org/.

ITER Organization retains copyright in the pictures and videos. Download and use of the image do not amount to a transfer of intellectual property (https://www.iter.org/media/www/downloads/av_terms_of_use.pdf).

In parallel to ITER, the "International Fusion Materials Irradiation Facility—DEMO Oriented Neutron Source", known as IFMIF-DONES (www.ifmifdones.org), is the key to solve the second shortcoming from a more specific perspective.

2.2. IFMIF-DONES: What It Is and How It Will Work

Although the target of this work is not an exhaustive technical description of IFMIF-DONES, this section will briefly describe it so that its potential impact can be situated in the right framework.

The target of IFMIF-DONES is to obtain neutrons like those to be produced in real fusion reactions without reaching the extremely high temperatures of these reactions. In other words, it is an installation where the neutrons that we want to learn how to stop will be produced. The study of their effects on different materials will be studied, and the materials that support that flux of neutrons in the best way will be selected to build the fusion reactors in the future.

To achieve that key milestone in the race to fusion, IFMIF-DONES will have three main parts:

(1) Particle accelerator, where deuterons (nuclei of hydrogen with one proton and one neutron) will be accelerated to very high speed. In the second phase, two parallel accelerators will work simultaneously.
(2) Liquid lithium target, where the accelerated deuterons will impact producing nuclear reactions whose product are neutrons with the same energy as those created in fusion.
(3) Test module, where different materials will be placed and irradiated with the neutrons.

2.3. The Situation of IFMIF-DONES: Current Development of the Project

IFMIF-DONES is expected to be installed in Escúzar, a small town in the Province of Granada (South of Spain). This is a rather economically depressed zone where most people work in agriculture or low-qualified services.

The project is expected to cost about EUR 700 million, which makes it the largest scientific infrastructure ever built in Spain. The costs will be funded by the Central Government of Spain and the Regional Government of Andalusia, the Spanish region where it will be located. Most of these national funds will come from the European Regional Development Fund, which finances lots of infrastructures in poorly developed European countries. Other funds will come from organizations like EUROfusion and other European programs.

In the first phase, already running in November 2020, the first budget of EUR 32.6 million has been transferred to the main implementing institutions of IFMIF-DONES: CIEMAT (Centro de Investigaciones Energéticas, Medioambientales y Tecnológicas) and the University of Granada, both in Spain. These EUR 32.6 million are being used to start the construction and launch the first actions. In addition, EUR 4 million more have been provided by the European Strategy Forum on Research Infrastructures (ESFRI) in this first phase.

It is interesting to remark that, together with Escúzar, there were other candidates in the European Union to host IFMIF-DONES. They were in Poland and Croatia, two countries whose economic level is lower than the European average. It is also the case in Spain. Finally, Escúzar was appointed as an EU candidate to host IFMIF-DONES, but the three potential locations considered by the EU are a good indication of the potential of critical technological infrastructures as a motor for regional development.

3. Effects of Peripheral Large Scientific Infrastructures on Social and Territorial Sustainability

Large scientific infrastructures connect with the terms of territorial social responsibility, sustainable development, social and economic sustainability and even education for sustainability [11]. Thereby, this type of project has impact and potential enough to contribute to change the territory in which they are inserted, and even far beyond.

According to Torjman [12], "human well-being cannot be sustained without a healthy environment and is equally unlikely in the absence of a vibrant economy". In this sense, the total sum of the investment planned for the first phase of construction and IFMIF-DONES will have a very positive socioeconomic and environmental impact because its implementation amounts to EUR 729.57 million, which will be executed over approximately nine years. Thus, the quality management of large projects is linked to sustainability and the development of a society and vice versa [13].

There is a growing recognition that sustainability is more than just the "green agenda", and it is considered that social practices are increasingly more important [14,15]. Nonetheless, integrated approaches to sustainability have failed to examine social sustainability in adequate detail [16]. For this reason, the scope of this research is the social impact of IFMIF-DONES and its consequences on the surroundings of the future installation. We start by defining social sustainability and how the project contributes to it.

There are several definitions of social sustainability with different ranges. From a wider perspective, the Western Australia Council of Social Services considers that "social sustainability occurs when the formal and informal processes, systems, structures, and relationships actively support the capacity of current and future generations to create healthy and livable communities" [16]. Secondly, social sustainability of a city (territorial approach) "is defined as development that is compatible with harmonious evolution of civil society, fostering an environment conducive to the compatible cohabitation of culturally and socially diverse groups and encouraging social integration, with improvements in the quality of life for all segments of the population" [17]. Lastly, regarding businesses, social sustainability is understood more generally as a business that influences individuals' or society's well-being [18,19] or, in other words, a system that meets the expectations of stakeholders without causing harm to the well-being of society and its members [20]. Here, the idea of social sustainability is commonly interpreted as the ability to continue to stay in business through good relations with stakeholders [21].

All these definitions reflect the development that social sustainability covers the broadest aspects of public and private activities and the effects that they have on employees, customers, suppliers, investors, and local and global communities, thus protecting all stakeholders and fairly respecting social diversity [22].

On the other hand, the key to achieving the sustainable development of a territory (territorial sustainability) is an understanding and modeling of its identity [23,24]. Sustainable territorial development can refer to different levels, and in particular to regional and local or relative to a country, continent or the entire world economy. Many researchers of social sustainability have focused mainly on urban studies from both academic and policy perspectives [25,26].

In this work, we review social sustainability from the perspectives of peripheral development. Ezcúzar and its province, Granada, have quite specific socioeconomic indicators as shown in the next sections. Therefore, it is necessary to work in this dependent context due to the influence of the relevance of the project through the perspective of social and territorial sustainability. Table 2 compares both cities in terms of population and economy.

Table 2. Comparison between the village of Escúzar and the city of Granada (Province Capital). Source: Instituto de Estadística y Cartografía de Andalucía 2019 https://www.juntadeandalucia.es/institutodeestadisticaycartografia/sima.

	Granada	Escúzar
Total population	232,462	791
Population under 20 years old (%)	18.3	17.7
Relative increase in population in 10 years	−0.8	−4.7
Public infrastructures	155 schools, 10 libraries, 21 medical centers, 1 public university.	1 school, 2 nurseries, 1 medical center.
Economy	Wide variety of services: 22,917 businesses with economic activity and 13,406 places to host (tourism). Some agriculture and industry.	Mainly agriculture. Recently, some industry.
Rate of unemployment	23.2	23.8

In the next sections, the specific foreseen impact of IFMIF-DONES on the territory and its contribution to social and territorial sustainability, as well as the duly relationship with the SDGs, and the interconnection between all are analyzed.

4. The Overall Impact of IFMIF-DONES at Different Levels

The implementation of a scientific infrastructure like IFMIF-DONES, the largest in Spain ever, has effects at various levels: worldwide, European, national, regional and local.

(1) Global level: Given that the expected output of IFMIF-DONES will be the decision about the most suitable materials to construct the future fusion reactors, its impact at the global level will be to permit the development of fusion energy, which is expected to have a very high efficiency producing energy from non-hazardous fuels.

(2) European Union level: The EU is one of the parts of the Broader Approach with Japan. This treaty is the framework toward fusion energy, where the EU has put a large amount of funds over the years. Furthermore, the major part of the funds financing IFMIF-DONES come from European programs like EUROfusion, ESFRI and, mainly, the European Regional Development Fund (EFDR). Therefore, the successful

construction and operation of IFMIF-DONES will be a success for the European scientific policy.

(3)   National level: The expected location for IFMIF-DONES is in the South of Spain. The Spanish Government has managed all the different eventualities in order to successfully become the European candidate to host the Project. Furthermore, the Spanish Government is dedicating a remarkable part of its EFDR to fund IFMIF-DONES, which demonstrates a clear willingness. In this scenario, the increase in the number of contracts and the attraction of highly specialized construction and technological enterprises are a very worthy output of the project for Spain. Furthermore, the rise of Spanish research centers in important indicators and rankings like Shangai and others is another attractor and motor for progress.

(4)   Regional level: The Regional Government of Andalusia is carrying out the first investments of the project implementation at 50% with the Central Government. The origin of these funds is mainly the EFDR with some additional investment from its own budget. The expected impact will be also financial and scientific, but there is one particularity: the IFMIF-DONES project can become the needed boost for the shift of a mainly tourism-based economy to a knowledge-based one.

(5)   Local impact: Given the particularities of the territory where IFMIF-DONES is expected to be built and developed, it is necessary to enter in more details.

The Province of Granada and its surroundings are characterized by:

(1)   High dependence on agriculture, tourism, construction and its related areas (hotels, bars, restaurants, etc.) (REF: Instituto de Estadística y Cartografía de Andalucía. Explotación de la Encuesta de Población Activa del INE.2020-3er trimestre. Thousands people).

(2)   Low rate of industrialization and big companies (industrial production index 90,1 and yearly variation −14,1).

(3)   Big cultural gap between young and older people. Whilst most young people have studied at a university, most people above 60 have not even got undergraduate formation.

(4)   An outstanding university, the fourth largest in Spain, with more than 60,000 students and 7000 professors, researchers and administrative staff [10]. It is at the top of Shangai ranking in some disciplines [12]. In addition, there are a few top research centers mainly linked to the university, but few job opportunities for their graduates.

(5)   Few but good hospitals. Some of them at the highest national level in some specialties.

(6)   High rates of unemployment (Granada: 108,100 unemployed, unemployment rate 25.94%; Andalusia: 932,300 unemployed, unemployment rate 23.80%).

The consequence of this configuration in the Province of Granada, where IFMIF-DONES is expected to be built, is a high rate of public/private employees, most of the employees of private companies working in tourism-related jobs, and a socioeconomic and cultural framework where the University of Granada is, by far, the main actor. Other provinces around Granada, like Almería, Jaén or Málaga, have also a big dependence on tourism, but their industry and/or agriculture has more relative weight.

In this particular territorial framework, the potential territorial yields of IFMIF-DONES are:

(1)   Implementation of new high-technology private companies and stimulation of existing ones to work as suppliers during construction and operation of the facility.

(2)   Stimulation of regional construction companies.

(3)   Creation of highly qualified jobs for local graduates and additional jobs of services for non-graduate local workers currently in unemployment.

(4)   Creation of new lines of research complementary to the ones of the University and the research centers in the area.

(5)   Improvement of auxiliary infrastructures.

In spite of the good perspectives, the desired yields of the installation of IFMIF-DONES in the Province of Granada is not immediate and, obviously, not direct for these companies and citizens that are not prepared to offer the needed services. In addition, given that the vast majority of the contracts will be decided after public concurrence, enterprises and workers from other parts of Spain and even from all over the world could be hired. Table 3 below shows a schematic analysis of positive, negative and uncertain points regarding the implementation of IFMIF-DONES in the site of Escúzar.

Table 3. Threats, opportunities, weaknesses and strengths regarding the implementation of IFMIF-DONES in the site of Escúzar.

THREATS	OPPORTUNITIES
- Regulatory changes impacting the installation - Local reluctance from the side of neighbors - Low utilization - Difficulty to find suppliers nearby	- International awareness of the necessity of clean and sustainable energy - Compromise of financing from Central and Regional Governments - New lines of research for the local University - Investments in the area (auxiliary infrastructures, etc.)
WEAKNESS	**STRENGTHS**
- Difficulty attracting highly qualified human resources - Technical difficulties inherent to such complex and large Project - High budget demands. Very high budget needed - Potential high dependence on certain people whose know-how is essential	- Optimal location of the facility - The goal of the project is to facilitate the control of clean and sustainable energy, which is in line with the SDGs - Application of previous know-how that can be used now - Possibility of extending the research of IFMIF-DONES to other sectors like health, communications, etc.

In summary, the yields of the installation of a critical scientific and technological facility like IFMIF-DONES are expected to be very high, but its contribution to really sustainable development and the relevant SDGs are not free of risks and uncertainties and will need big efforts and careful preparation at all the levels.

In the next section, the potential contributions of IFMIF-DONES to the SDGs are presented and analyzed.

5. The Impact of IFMIF-DONES on Social Sustainability and SDGs

Beyond the impact of IFMIF-DONES from the pure perspective of financial profitability and yields, it is interesting to focus on aspects not frequently included when studying large scientific and technological infrastructures. In this sense, the analysis from the perspective of social sustainability is more frequent in other frameworks more related to actions to eliminate poverty, inequality, etc.

Theoretically, social sustainability as a concept covers broad societal issues [27] and has various interpretations in different fields [28]. Laureate Amartya and Sen identified five dimensions in social sustainability—equity, diversity, social cohesion, quality of life and democracy and governance—which have been considered in determining if a business or a project is socially sustainable. However, these have been extended in many studies. It has also been acknowledged that the social dimension of the SDGs needs further development [15].

In this section it is shown that large scientific and technological infrastructures are an efficient instrument for social sustainability through their direct impact on specific SDGs, using Khan's classification [29]. As it is evident from his table, numerous authors point toward similar themes as they remain the primary constituents of social sustainability.

Table 5 below summarizes the main impacts.

Table 4. Impact of IFMIF-DONES implementation on socioeconomic indicators and specific SDGs. Adapted from Khan [29].

Topics on Social Sustainability	Promotion of Social Sustainability IFMIF-DONES	Promotion of Sustainable Development Goals (SDGs)
Human health and well-being/well-being of generations	The ultimate goal of IFMIF-DONES is to provide the keys to build fusion reactors that will produce clean energy, avoiding the emission of pollutants and greenhouse gasses and their negative effects on the health and well-being of present and future generations.	Goal 3
Basic needs and quality of life	The expected shift from the primary sector to highly qualified jobs will mean higher incomes and better heritage in the economy and better quality of life.	Goals 1, 2, 3, 6, 7 and 8
Social Coherence	IFMIF-DONES is in agreement with the recent attempt from regional, national and European levels to transform the region into a digital green-economy-based one.	Goals 1, 4, 5 and 10
Social justice and equity		Goals 1, 2, 3, 4 and 5
Democratic/engaged government and democratic society		
Human rights		
Social inclusion	More opportunities for everyone. Access to more opportunities for disadvantaged classes.	Goals 1, 4 and 5
Diversity	The region of Escúzar will pass from a closed structure formed by several families (around 700 inhabitants) to a more heterogeneous one due to the move of foreign engineers and workers of the facility.	Goals 5 and 10
Decline of poverty		
Social infrastructure	The infrastructures complementary to IFMIF-DONES (roads, telecom, new schools) are expected to be remarkable.	Goal 9
Social capital	The social capital of the original population of Escúzar will be enriched. More open-minded perspectives of life and formation will be brought with new visitors and inhabitants.	Goal 11
Behavioral changes	The shift to a new productive framework and the arrival of people from other countries to the region will bring behavioral changes.	Goals 5, 10, 12 and 13
Preservation of socio-cultural patterns and practices	The abovementioned changes can in no way cause the disappearance of the socio-cultural patterns and practices of the original population. The necessary bodies and practices for cultural preservation must be created and fostered.	Goals 4, 10 and 11

Table 5. Impact of IFMIF-DONES implementation on socioeconomic indicators and specific SDGs. Adapted from Khan [29].

Topics on Social Sustainability	Promotion of Social Sustainability IFMIF-DONES	Promotion of Sustainable Development Goals (SDGs)
Participation (including stakeholder participation)	The local and regional institutions will have a strong and permanent presence and participation in all the aspects of the DONES environment.	Goal 16
Human dignity		
Safety and security	Better infrastructures, safer roads, etc.	Goal 9
Sense of place and belonging	Regional pride of one infrastructure that will be a key milestone toward nuclear fusion will undoubtedly reinforce the sense of place and belonging.	
Education and training	More schools and more international students will be a challenge for education and training, which will be improved.	Goal 4
Employment	Many workers, qualified and not qualified, will be needed and hired during several decades.	Goal 8
Community involvement and development, community resilience	One of the main targets of the University of Granada, as an implementing body of the project, is community involvement. For this reason, continuous activities to communicate to the community what is being done are being carried out.	Goal 16
Fair operating practices	The final target is the decrease in uncontrolled energy consumption and emission of hazardous substances. Therefore, IFMIF-DONES will result in fair operating practices in industry, transport and other key activities.	Goals 12, 13, 14 and 15
Capacity for learning	The activities of the University of Granada result in deeper know-how in scientific and operational terms based on continuous retrofit.	Goal 4
No structural obstacles (to health, influence, competence, impartiality and meaning-making)		

6. Conclusions

Large scientific and technological infrastructures have impacts on many aspects of human communities. These aspects go from international to territorial levels and also from economic to social perspectives, including a wide variety of intermediate levels.

These impacts are even greater when the infrastructures have deep implications as experimental nuclear fusion facilities. The necessity of cleaner, safer and more efficient energy from the atomic nuclei is a key milestone in the long way to really sustainable development. The current rate of development requires huge amounts of energy whilst the worrying situation of the environment, the negative effects on human health and the lack of more and more natural resources leaves the question of energy as one of the main challenges for the upcoming decades.

In this framework, the "International Fusion Materials Irradiation Facility—Demo Oriented Neutron Source" (IFMIF-DONES) is a unique project to test different material candidates for the construction of future fusion reactors. In other words, without the results of IFMIF-DONES, access to fusion energy will be impossible during this century.

Departing from this evidence and trying to enlarge the spectrum of positive effects at all levels, large projects like IFMIF-DONES should incorporate targets and strategies beyond the classic social responsibility in order to satisfy the necessities of the territories. Thus it is essential to create added economic and environmental value whereas, in parallel, creating social value [29]. This ensures sustainable economic success in the territory.

There is a mutual relationship between territory and project because the project prospers when the society prospers and vice versa: new enterprises, better houses, schools, businesses, etc., that equally generate new projects. This is one of the reasons why all big projects should foster the achievement of the relevant Sustainable Development Goals (SDGs) in a broad and integral perspective.

In this work, based on the forecasts and current advances in the implementation of IFMIF-DONES in a rural zone of the South of Spain, we highlighted that, besides the well-known financial outputs and profits, large scientific and technological infrastructures are an efficient instrument for social sustainability through their direct impact on specific SDGs.

Furthermore, the asymmetrical social and economic impact of large scientific and technological infrastructures is also a matter of major concern. Whilst life conditions of people are not highly improved in big and developed regions, in small peripheral provinces with low average incomes, the effect on social sustainability, directly contributing to the achievement of certain SDGs, is very remarkable. Thus, people with a medium-to-high socioeconomic level and a wide variety of job opportunities may have some improvements in their conditions, but not great changes. On the other hand, unemployed people without a high level of studies and/or very limited options to change their jobs can experience big changes that will be transmitted to the rest of society through their big change in incomes, way of life and others.

Urban centers could be spaces where the sources of employment are larger than in rural spaces. As technology develops, the workforce has to be more specialized and therefore a greater gap will be generated between regions. Therefore, the migration for years of the rural population toward the cities is narrowing the labor market of these areas dedicated to agriculture, which is reducing the difference in income and opportunities between the rural and urban population. Thus, land attachment could also be favored with this kind of project that produces high returns to the territorial area in which it would be implanted.

As regards asymmetrical impact, institutions should treat laws and regulations of the projects as opportunities for improvement, development and sustainability, insisting on ethical behavior in their interactions with stakeholders [30]. However, trust and collective action are core topics because sustainability derives from accessible and inclusive processes. An integrated perspective on sustainability is thus implicated in more effective social sustainability, which in turn relies upon attention to local contexts and ideas.

Furthermore, the development of a territorial identity seems basic not only for avoiding offshoring risks but also because new implementations require specific characteristics and demanding rules for service quality in the territory. The regional and local identity among citizens, politicians and society in general allows an integrated approach to environmental and social sustainability that represents an added value when considering the attraction of future investments.

In summary, besides its important technical implications and its contribution to the control of fusion energy, IFMIF-DONES is expected to become an engine fostering the development of a depressed area thus demonstrating some hypotheses and becoming a great field experiment in social and economic sustainable development.

Author Contributions: Formal analysis, A.P.-G. and V.F.-P.; Funding acquisition, A.P.-G.; Investigation, A.P.-G. and V.F.-P.; Methodology, A.P.-G. and V.F.-P.; Project administration, A.P.-G.; Resources, A.P.-G.; Visualization, V.F.-P.; Writing—original draft, A.P.-G. and V.F.-P. All authors have read and agreed to the published version of the manuscript.

Funding: This work was supported by the European Commission as part of the project "DONES Preparatory Phase" (Ref. 870186).

Institutional Review Board Statement: Not relevant.

Informed Consent Statement: Not relevant.

Data Availability Statement: Not relevant.

Conflicts of Interest: The authors declare no conflict of interest.

References

1. World Commission on Environment and Development. *Our Common Future*; Oxford University Press: Oxford, UK, 1987.
2. United Nations Division for Sustainable Development Goals (DSDG). Available online: https://sustainabledevelopment.un.org/ (accessed on 30 December 2020).
3. Henriques, A.; Richardson, J. *The Triple Bottom Line: Does It All Add up? Assessing the Sustainability of Business and CSR*; Earthscan: London, UK, 2004.
4. Cardillo, E.; Longo, M.C. Managerial Reporting Tools for Social Sustainability: Insights from a Local Government Experience. *Sustainability* **2020**, *12*, 3675. [CrossRef]
5. Molina-Moreno, V.; Leyva-Díaz, J.C.; Sánchez-Molina, J.; Peña-García, A. Proposal to foster sustainability through circular economy-based engineering: A profitable chain from waste management to tunnel lighting. *Sustainability* **2017**, *9*, 2229. [CrossRef]
6. Hermoso-Orzáez, M.J.; Lozano-Miralles, J.A.; Lopez-Garcia, R.; Brito, P. Environmental criteria for assessing the competitiveness of public tenders with the replacement of large-scale LEDs in the outdoor lighting of cities as a key element for sustainable development: A case study applied with PROMETHEE methodology. *Sustainability* **2019**, *11*, 5982. [CrossRef]
7. Elkington, J. Accounting for the triple bottom line. *Meas. Bus. Excell.* **1998**, *2*, 18–22. [CrossRef]
8. Peña-García, A.; Salata, F. The perspective of Total Lighting as a key factor to increase the Sustainability of strategic activities. *Sustainability* **2020**, *12*, 2751. [CrossRef]
9. Spanish Ministry for Ecological Transition and Demographic Challenge. Available online: https://energia.gob.es/nuclear/Centrales/Espana/Paginas/CentralesEspana.aspx (accessed on 29 November 2020).
10. Available online: https://www.iter.org/proj/inafewlines (accessed on 30 December 2020).
11. Levy, N. Taking Responsibility for Responsibility. *Public Health Ethics* **2019**, *12*, 103–113. [CrossRef] [PubMed]
12. Torjman, S. *The Social Dimension of Sustainable Development, a Paper Prepared for Commissioner of Environment and Sustainable Development*; Caledon Institute of Social Policy: Ottawa, ON, Canada, 2000.
13. Hall, J.; Matos, S.; Sheehan, L.; Silvestre, B. Entrepreneurship and Innovation at the Base of the Pyramid: A Recipe for Inclusive Growth or Social Exclusion? *J. Manag. Stud.* **2012**, *49*, 785–812. [CrossRef]
14. Zink, K.J. Designing sustainable work systems: The need for a systems approach. *Appl. Ergon.* **2014**, *45*, 126–132. [CrossRef]
15. Missimer, M.; Robèrt, K.-H.; Broman, G. A Strategic Approach to Social Sustainability—Part 2: A Principle-Based Definition. *J. Clean. Prod.* **2016**, *140*, 42–52. [CrossRef]
16. McKenzie, S. *Social Sustainability: Towards Some Definitions*; Hawke Research Institute Working Paper Series No 27; University of South Australia: Magill, Australia, 2004.
17. Polese, M.; Stren, R. *The Social Sustainability of Cities: Diversity and the Management of Change*; University of Toronto Press: Toronto, ON, Canada, 2000.
18. Geibler, J.; Liedtke, C.; Wallbaum, H.; Schaller, S. Accounting for the Social Dimension of Sustainability: Experiences from the Biotechnology Industry. *Bus. Strategy Environ.* **2006**, *15*, 334–346. [CrossRef]
19. Huq, F.A.; Stevenson, M.; Zorzini, M. Social sustainability in developing country suppliers. *Int. J. Oper. Prod. Manag.* **2014**, *34*, 610–638.
20. Lindgreen, A.; Antioco, M.; Harness, D.; van Sloot, R. Purchasing and Marketing of Social and Environmental Sustainability for High-Tech Medical Equipment. *J. Bus. Ethics* **2009**, *85*, 445–462. [CrossRef]
21. Brown, D.; Dillard, J.; Marshall, R.S. *Triple Bottom Line: A Business Metaphor for a Social Construct*; School of Business Administration, Portland State University: Portland, OR, USA, 2006.
22. Vavik, T.; Keitsch, M. Exploring relationships between Universal Design and Social Sustainable Development: Some Methodological Aspects to the Debate on the Sciences of Sustainability. *Sustain. Dev.* **2010**, *18*, 295–305. [CrossRef]
23. Carta, M. Reimagining urbanism. In *Creative, Smart and Green Cities for the Changing Times*; List Lab: Trento, Italy, 2014.
24. Shao, G.; Li, F.; Tang, L. Multidisciplinary perspectives on sustainable development. *Int. J. Sustain. Dev. World Ecol.* **2011**, *18*, 187–189. [CrossRef]
25. Ghahramanpouri, A.; Lamit, H.; Sedaghatnia, S. Urban Social Sustainability Trends in Research Literature. *Asian Soc. Sci.* **2013**, *9*, 185–193. [CrossRef]
26. Weingaertner, C.; Moberg, Å. Exploring social sustainability: Learning from perspectives on urban development and companies and products. *Sustain. Dev.* **2014**, *22*, 122–133. [CrossRef]
27. Suopajärvi, L.; Poelzer, G.A.; Ejdemo, T.; Klyuchnikova, E.; Korchak, E.; Nygaard, V. Social sustainability innorthern mining communities: A study of the European North and Northwest Russia. *Resour. Policy* **2016**, *47*, 61–68. [CrossRef]

28. Vasquez, R.V.; Klotz, L.E. Social Sustainability Considerations during Planning and Design: Framework of Processes for Construction Projects. *J. Constr. Eng. Manag.* **2013**, *139*, 80–89. [CrossRef]
29. Khan, R. How Frugal Innovation Promotes Social Sustainability. *Sustainability* **2016**, *8*, 1034. [CrossRef]
30. Wolsko, C.; Marino, E.; Doherty, T.; Fisher, S.; Goodwin, B.; Green, A.; Reese, R.; Wirth, A. Systems of Access: A Multidisciplinary Strategy for Assessing the Social Dimensions of Sustainability. *Sustain. Sci. Pract. Policy* **2016**, *12*, 88–100. [CrossRef]

sustainability

Article

Priority of Accident Cause Based on Tower Crane Type for the Realization of Sustainable Management at Korean Construction Sites

Ju Yong Kim [1], Don Soo Lee [2], Jin Dong Kim [3] and Gwang Hee Kim [1,*]

1   Department of Architectural Engineering, Kyonggi University, Yeongtong-Gu, Suwon 16227, Korea; ju2020@kyonggi.ac.kr
2   DAELIM Construction, 14, Miraero, Namdong-Gu, Incheon 21556, Korea; donandon14@hanmail.net
3   Department of Architecture, Yeonsung University, Manan-Gu, Anyang-Si 14011, Korea; kjd@yeonsung.ac.kr
*   Correspondence: ghkim@kyonggi.ac.kr; Tel.: +82-31-249-9757

Abstract: Construction safety is a key factor among the many factors related to the sustainable management of construction sites. Although research is underway to reduce potential accidents in the construction industry in Korea, the number of tower crane (T/C) accidents is consistently increasing based on the increased use of such cranes. In this study, the priorities of accident causes for each T/C type were derived and utilized for the sustainable management of construction sites. An analytic hierarchy process (AHP) questionnaire was completed by experts such as construction engineers, construction managers, safety engineers, and T/C operators with more than ten years of field experience. The results of the AHP questionnaire revealed that the leading cause of cab-control T/C accidents is poor operator visibility, while the leading cause of accidents related to remote-control T/Cs is the poor management of lifting objects and control of surroundings. The high-ranking causes derived in this study should be managed and priority measures should be implemented to reduce the number of T/C accidents.

Keywords: sustainable construction management; tower crane accident reduction; priority of tower crane accident causes

Citation: Kim, J.Y.; Lee, D.S.; Kim, J.D.; Kim, G.H. Priority of Accident Cause Based on Tower Crane Type for the Realization of Sustainable Management at Korean Construction Sites. *Sustainability* **2021**, *13*, 242. https://doi.org/10.3390/su13010242

Received: 14 November 2020
Accepted: 25 December 2020
Published: 29 December 2020

Publisher's Note: MDPI stays neutral with regard to jurisdictional claims in published maps and institutional affiliations.

1. Introduction

The crane has become a major symbol of building construction sites and is often the most prominent piece of equipment at a building construction site based on its size and the key role that it plays at many construction sites [1]. The use of tower cranes (T/Cs) at construction sites has consistently increased since their introduction into the Korean construction industry in the 1980s. According to statistics on construction machinery statuses from the Ministry of Land, Infrastructure, and Transport, in 2015, Korea contained 3408 cab-control (CC) T/Cs and 272 remote-control (RC) T/Cs. Generally, CC T/Cs are used for lifting objects weighing three tons or more and RC T/Cs are used for lifting objects weighing less than three tons. In 2019, the number of CC T/Cs increased by 22% to 4385 and the number of RC T/Cs increased by over 85% to 1845 [2].

One of the major causes of fatalities is the usage of cranes during lifting operations in the construction phase of the construction project lifecycle [3]. As the number of T/Cs used at construction sites has increased steadily, there has been an increase in fatalities and accidents because T/Cs are relatively dangerous and various risk factors are inherent to assembly, lifting, and disassembly works [4]. As accidents at construction sites are closely related to construction time, cost, scope, and company reputation [5], and because construction workplace safety and health are essential elements of sustainable construction management [6], construction accidents must be reduced for the sustainable management of construction sites. Reyes et al. [7] stated that when quantifying the sustainable value of a construction project, health and safety indexes should be considered. Therefore,

173

the government, academia, and practitioners in Korea have made various efforts to reduce T/C-related accidents. To reduce T/C accidents, the Korean government revised the enforcement regulations of the Construction Equipment Management Act in October of 2019, subdividing the safety training programs for construction equipment operators into 19 types and shortening the training cycle of RC T/C operators. Members of academia have also conducted research [8–10] on various causes of accident occurrence to reduce T/C accidents.

Although many efforts have been made to prevent accidents related to CC T/Cs, research on RC T/Cs, which are becoming increasingly common at Korean construction sites, is insufficient, leading to many accidents at construction sites. Every year, the number of T/C-related accidents at Korean construction sites continues to increase. The numbers of fatalities related to T/Cs were nine in 2016, seven in 2017, six in 2018, and eight in 2019 [11]. T/C accidents were officially announced during the first quarter of 2020, when five casualties had already occurred. Figure 1 presents an image from January 2 of 2020, where a 30 m T/C collapsed at a construction site in Incheon, Korea. This accident caused two fatalities and one injury. Kim [12] analyzed T/Cs at Korean construction sites and proposed the following main accident causes: (1) In the case of CC T/Cs, the main causes are equipment age, insufficient work management, violation of work guidelines and safety rules, and lack of communication. (2) In the case of RC T/Cs, the main causes are a lack of knowledge regarding work manuals for installation workers, insufficient checking of the cables used for fixing lifted objects, a lack of simultaneous checking of camera feeds during tying and lifting work, and insufficient checking of the specifications of heavy objects.

Figure 1. A 30 m T/C (tower crane) collapse at a construction site in Incheon, Korea [5].

Therefore, this study aimed to identify whether ranking can be utilized for the accident causes' management of T/C types by analyzing the importance of accident causes for each crane type. The results of this study can contribute to reducing construction accidents by identifying management causes for T/Cs that should be considered during the planning process for construction accident prevention and safety management activities.

2. Literature Review

2.1. Previous Research

Since the 2000s, various studies related to T/C accidents have been conducted around the world. These studies can be classified into three major categories: (1) risk analysis for analyzing T/C accidents [13,14], (2) development of management goals or plans by analyzing T/C accidents [14–16], (3) derivation of the major accident causes related to T/C accidents [17–21], and (4) presenting measures for preventing tower crane accidents [22]. Thus far, most studies have focused on the causes or risks of T/C accidents based on specific causes and direct management. Especially, Fang et al. [22] developed a framework for real-time pro-active safety assistance (RPSA) for mobile crane lifting operation, and Zhang et al. [20] and Zhou et al. [21] presented the tower-crane accident cause system (TCACS) model, which was a quite logical model through system thinking and case analysis to quantify the tower crane accident causes. However, there was a limit to revealing the direct cause of T/C accidents. Therefore, to reduce T/C accidents, it is necessary to manage various accident causes that cause accidents comprehensively. In other words, it is necessary to manage the accident causes that cause accidents with a high frequency or high probability more intensively. Additionally, in Korea, the use of RC T/Cs has increased based on pressure from the T/C union and the revision of the labor laws that limit working hours. Thus far, most research has focused on CC T/Cs, but there is a need to proceed with research to reduce all types of T/C accidents, including RC T/C accidents.

2.2. Construction Safety in Sustainability

The area for sustainability appears to be focusing on limiting environmental impact, reducing energy, and incorporating less harmful material. Additionally, sustainability takes into account the environmental, economic, social and resource impacts of construction as well as the implementation of its principles throughout the lifecycle of building. However, Chandra [23] insisted that sustainable construction safety and health are an integral part of sustainable and environmentally friendly construction efforts. In addition, sustainable construction is defined as the creation of a healthy construction environment and responsible management based on resource efficiency and ecological principles. Rajendran et al. [24] recommended for research investigating the impact of green design and construction on worker safety and health, taking into account the safety and health of workers as well as the safety and health of the end user. Especially, the Leadership in Energy and Environmental Design (LEED) is designed to define eco-friendly buildings by establishing a common standard for measurement or rating systems and to achieve three main objectives: market innovation, design integration, and education on sustainable principles and sharing ideas [6]. Hinze et al. [6] presented the concept that worker safety and concern belong to education—the third plan of the LEED objectives. As mentioned, the sustainability certification, LEED, also includes worker safety in the construction process, and construction safety has become an indispensable item in sustainability.

2.3. T/Cs at Construction Sites

Cranes mainly used in the construction industry are classified into two equipment types: tower cranes and mobile cranes [1]. Mobile cranes can be classified as truck-mounted mobile crane and crawler crane. The basic truck-mounted crane configuration is a "boom truck" featuring a rear-mounted rotary telescopic boom crane mounted on a commercial truck. The crawler crane is boomed on a vehicle with a crawler track set that provides both stability and mobility. For many years, particularly in Korea, T/Cs have been widely used in all types of building construction projects in both urban and rural areas. In high-rise construction, T/Cs are a key type of equipment for moving materials, building elements, and form work components horizontally and vertically [1]. As shown in Figure 2, the major parts of the T/C are the mast, main jib, and counter jib. The mast, which is a steel structure that serves as a pillar supporting the T/C, is constructed on the upper part of the basic mast, which is connected to the mounting configuration. Masts are available as rail-mounted units, stationary units, climbing units, and mobile units. Depending on whether the mast is fixed or slewing, a T/C can be classified as a fixed or a slewing T/C. Additionally, T/Cs can be classified as top-slewing and bottom-slewing T/Cs according to the T/C rotation position. The T/C cabin is attached to the crane structure or installed at a remote location. The cabin of a top-slewing crane is almost always at the top of the mast, often at a significant distance from the work area. In this type of crane, it is important to improve the quality of operator visibility. Bottom-slewing cranes do not have an operator cab attached to the crane structure. The main types of jibs on T/Cs are saddle jibs and luffing jibs. A saddle jib is supported by a pendant in a horizontal or near-horizontal position and the load hook changes the hook radius by moving along the jib on a trolley. A luffing jib rotates on the jib foot and is supported by a luffing cable. The load-bearing hoist rope typically passes through the sheave of the jib head and changes the hook radius by changing the inclination angle of the jib. A saddle jib typically has a smaller minimum working radius than an equivalent luffing jib, so it can handle loads closer to the crane tower. Many luffing jib cranes have very short counter jibs, which can be advantageous when a crane is operating near obstacles such as other cranes or adjacent buildings. In the case of Korean construction sites, CC T/Cs have dominated in the past, but RC T/Cs have been increasing in use rapidly in recent years. This trend appears to have been partially influenced by pressure from the T/C operator union and construction labor union.

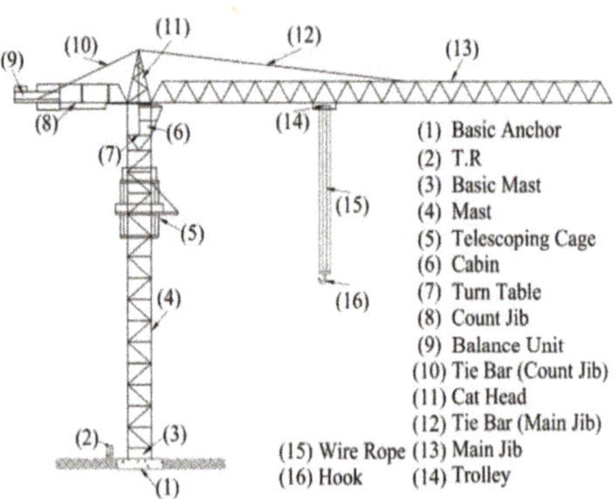

Figure 2. T/C configuration.

2.4. Types of T/C Accidents

The 40 tower crane accidents that occurred in Korea from 2015 to October 2020 were analyzed and classified into three categories such as the processes of erection, dismantling, and operation and management. The accident cases based on the analysis of T/C accident reports published from Korea Occupational Safety and Health Agency are the result of analyzing 40 accident cases—the most frequently occurred in the operation and management stage, 11 cases occurred in the erection stage, 7 cases in the dismantling stage, and 22 cases in the operation and management stage as shown in Table 1. Representative examples of each stage are as follows. In the erection stage, accidents related to eccentricity occur based on the operation of a T/C in a state where the telescopic component, mast, and other components are not fully fixed. Telescopic accidents break the balance of the jib while the telescopic mast is lifting or moving in an unstable state because the lower part of the turntable and upper pin of the telescopic cage are not fastened [25]. Collapse accidents occur because the member of the telescopic cage buckles based on improper use or non-use of a balance weight for the bidirectional balance of the jib during telescopic work [26]. Accidents in the dismantling stage are caused by the mast losing balance or the basic anchor being damaged. The position of the bolt hole for fixing the mast may be misaligned, so when an operator attempts to adjust the hole position, the crane may lose its balance and collapse. There can also be deviation between the cage roller and mast, causing a dismantling worker to disable the interlock function that stops crane operation. Forcible manipulation in this scenario can result in an accident [25]. Some accidents are also caused by the inadequate selection of standard lifting positions for dismantling, which causes the connecting part of the jib to split. During the process of T/C operation and management, the causes of accidents are mainly non-compliance with the work procedures suggested by manufacturers and the negligence of management in terms of safety inspections and education.

Table 1. T/C accident cases by construction phase/task.

Year	2017	2018	2019	2020.10	Total
Erection	2	3	3	3	11
Dismantling	2	1	3	1	7
Operation and management	5	5	6	6	22
Total	9	9	12	10	40

3. Methods

This study aimed to derive the importance of T/C accident causes and the potential for T/C accidents by quantifying the experiences of experts related to construction accidents. Intensively managing such factors should aid in reducing T/C accidents. The analytic hierarchy process (AHP) technique was adopted to quantify the experiences of experts related to construction accidents in the field. The AHP is a structured decision-making technique developed by Saaty in the early 1970s. It can reflect the knowledge, experience, and intuition of respondents in pair-wise comparisons based on the elements of the hierarchy of decision-making [27]. The experts related to construction accidents for the AHP are construction engineers, construction managers, safety managers, and T/C operators that have more than 10 years of field experience.

As shown in Figure 3, this research process can be divided into the following 4 steps. (1) Identify major T/C accidents through a literature review. (2) Extract first-level phase/task and second-level structure accident causes for the AHP through interviews with related experts. After extracting 11 phases/tasks from the previous literature [12,14,15,18,19,21,28], 5 items were selected by integrating 11 items through interviews with the related experts. For the second-level accident causes, 15 items were selected by conducting a preliminary survey of accident causes extracted from the previous literature (refer to Table 2) as a Likert scale to the related experts, and the results are presented in Table 3. (3) Provide an AHP questionnaire to a total of 44 related experts, such as construction managers, 10 safety managers, 14 construction engineers, and 10 T/C operators. (4) Follow the AHP to derive priority management accident causes for reducing T/C accidents.

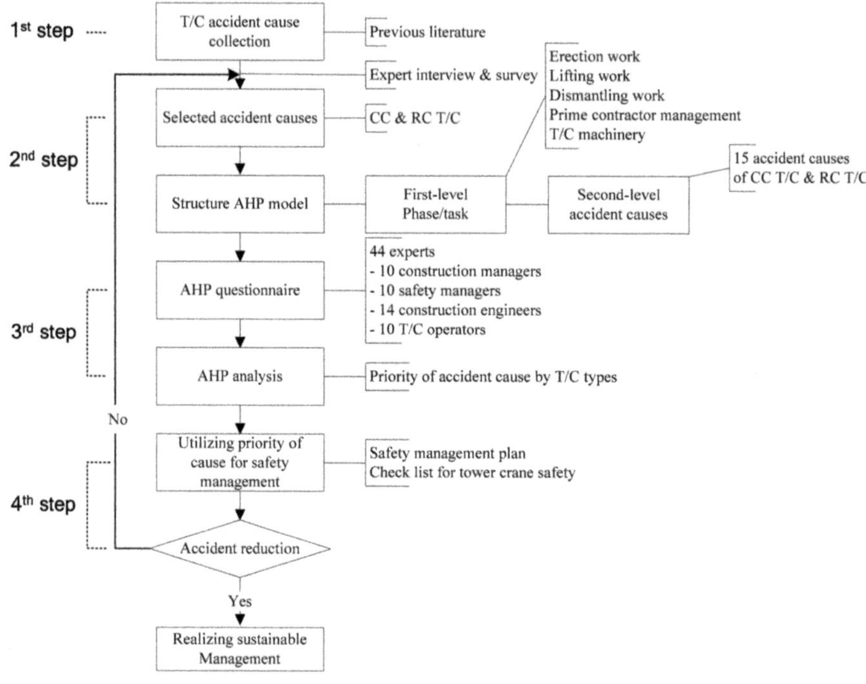

Figure 3. Developed model and research process.

Table 2. Causes of the AHP (analytic hierarchy process) model for T/C accidents.

First Level Phase/Task	Second-Level (Sub) Accident Causes of the AHP Structure for T/C Accidents	Reference
Erection work	Improper bolting on the brace/mast/telescopic element	Kwon (2015), Cho (2017), Wei et al. (2018)
	Failure to comply with safety rules and work guidelines for erection work	Kwon (2015), Cho (2017), Song (2018), in (2018)
	Insufficient worker skill for erection work	In (2018) and Song (2018)
Dismantling work	Poor understanding of accident risk during the dismantling process	Vivian and Ivan (2011), Heng et al. (2012), Kwon (2015), Song (2018), in (2018)
	Improper tightening of bolts	Cho (2017), Song (2018), in (2018)
	Low skill level of dismantling workers	Vivian and Ivan (2011), Heng et al. (2012), Kwon (2015), Cho (2017), Song (2018)
Lifting work	Poor management of lifting objects and control of surroundings	Kwon (2015), Cho (2017), Song (2018)
	Improper tying of the sling leg	Kwon (2015), Cho (2017), Song (2018)
	Poor operator visibility	Cho (2017) and Kim (2018)
	Operator unable to check the weight and specifications of the lifted objects	Kim (2018)
	Operator cannot check the wire and sling leg	Kim (2018)

Table 2. *Cont.*

First Level Phase/Task	Second-Level (Sub) Accident Causes of the AHP Structure for T/C Accidents	Reference
Prime contractor management	Inappropriate personnel and equipment placement	Kwon (2015) and Cho (2017)
	Poor subcontracting technology management process	Vivian and Ivan (2011), Kwon (2015), Cho (2017), Wei et al. (2018)
	Insufficient safety management of equipment and personnel	Kwon (2015), Cho (2017), Wei et al. (2018)
T/C machinery	Equipment aging	Cho (2017) and Kim (2018)
	Crane operation error or failure	Cho (2017) and In (2018)
	Problems with overseas parts procurement	Cho (2017)

Table 3. Preliminary survey results of accident causes.

Figure		Second-Level (Sub) Causes for T/C Accidents	Ranking
Erection work		Failure to comply with safety rules and work guidelines for erection work	1
		Insufficient worker skill for erection work	2
		Bad bolting brace/mast/telescopic element	3
		Not used seat belt during erection work	4
		Asymmetric load due to deflected foundation anchor	5
Dismantling work		Poor understanding of accident risk during the dismantling process	1
		Low skill level of dismantling workers	2
		Poor tightening of bolts	3
		Jib imbalance or damage	4
		Non-compliance with manufacturer's instruction and work order	5
Lifting work	CC T/C	Poor operator visibility	1
		Poor management of lifting objects and control of surroundings	2
		Poor tying of the sling leg	3
		Bad signal	4
		Equipment defect	5
	RC T/C	Operator cannot check the wire and sling leg	1
		Operator unable to check the weight and specifications of the lifted objects	2
		Poor management of lifting objects and control of surroundings	3
		Bad signal	4
		Equipment defect	5
Prime contractor management		Inappropriate personnel and equipment placement	1
		Insufficient safety management of equipment and personnel	2
		Poor subcontracting technology management process	3
		Improper risk assessment	4
		Equipment maintenance history unconfirmed	5
T/C machinery		Aging equipment	1
		Problems with overseas parts procurement	2
		Crane operation error or failure	3
		Design and construction errors	4
		Insufficient electrical insulation	5

4. AHP Model for T/C Accident Factors

4.1. Extract T/C Accident Factors and Structure of the AHP Model

Major causes were extracted by interviewing related experts, such as construction managers, safety managers, and construction engineers, after arranging the causes of

T/C accidents discussed in the previous literature. Expert interviews for classifying the extracted major phases/tasks and causes, and identifying first-level phases/tasks and second-level causes were conducted to structure the hierarchy of the AHP model. The first-level phase/task in the AHP model are five categories of dismantling work, lifting work, erection work, prime contractor management, and T/C machinery (Figure 4). Table 3 lists a total of five first-level phases/tasks and second-level accident causes that were extracted from the previous literature based on expert interviews and preliminary surveys. These accident causes were used in our AHP model for T/C accidents.

4.2. Structure of the AHP Model

Figure 4 presents an AHP model for CC T/Cs and RC T/Cs. F1 to F5 in Figure 4 represent the first-level phase/task. Among the second-level accident causes for lifting management, F21 is applicable to both types of cranes, whereas F22 and F23 are relevant to CC T/Cs, and F24 and F25 are relevant to RC T/Cs.

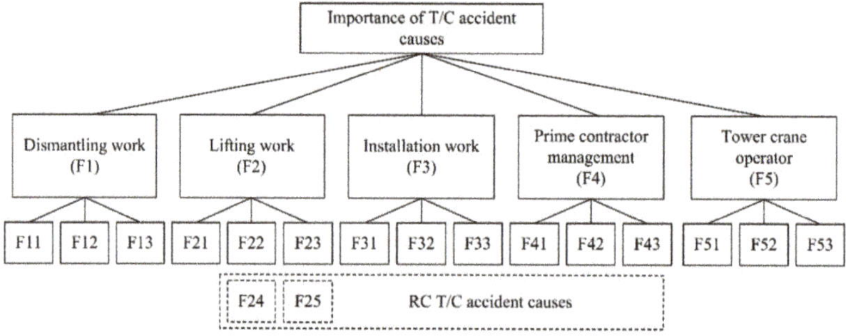

Figure 4. AHP structure for T/C accidents.

4.3. AHP Survey

The analysis results for the AHP questionnaire are presented in Table 4. The weights of the first-level phase/task indicate that erection work is the most important phase/task with a value of 0.226, followed by T/C machinery (0.216), lifting work (0.214), and prime contractor management (0.175), while dismantling work is the least important with a value of 0.170 for CC T/Cs. In the case of RC T/Cs, lifting work is the most important phase/task with a value of 0.264, followed by erection work (0.254), dismantling work (0.170), and T/C machinery (0.167). Prime contractor management has the lowest value of 0.146.

Table 4. Weight ranking of first-level phase/task in the AHP model.

	Phase/Task	Construction Engineer		Construction Manager		Safety Engineer		T/C Operator		Total	
		Wt *		R **		Wt		R		Wt	
	Erection work	0.156	3	0.205	3	0.219	3	0.322	1	0.226	1
	Dismantling work	0.391	1	0.090	5	0.100	5	0.099	5	0.170	5
CC T/C	Lifting work	0.294	2	0.293	1	0.121	4	0.146	4	0.214	3
	Prime contractor management	0.087	4	0.164	4	0.254	2	0.194	3	0.175	4
	T/C machinery	0.072	5	0.248	2	0.305	1	0.239	2	0.216	2
	Erection work	0.155	3	0.279	2	0.181	2	0.399	1	0.254	2
	Dismantling work	0.390	1	0.116	4	0.123	3	0.050	5	0.170	3
RC T/C	Lifting work	0.296	2	0.094	5	0.577	1	0.091	4	0.264	1
	Prime contractor management	0.086	4	0.219	3	0.092	4	0.185	3	0.146	5
	T/C machinery	0.073	5	0.292	1	0.027	5	0.275	2	0.167	4

* Wt: Weight; ** R: Ranking.

The analysis results for the AHP questionnaire on the second-level causes of T/C accidents are presented in Table 5. The weights of the second-level causes for CC T/Cs appear from largest to smallest as follows: poor operator visibility, improper bolting of the brace/mast/telescopic element, poor subcontracting technology management process, failure to comply with safety rules and work guidelines for erection work, problems with overseas parts procurement, and poor understanding of risk factors during the dismantling process. As shown in Table 5, in the case of RC T/Cs, poor management of lifted objects and control of surroundings are the most important causes, followed by the operator being unable to check the weight and specifications of the lifted objects, failure to comply with safety rules and work guidelines for erection work, insufficient worker skill for erection work, and improper bolting of the brace/mast/telescopic element.

Table 5. Weight ranking of second-level accident causes in the AHP model.

First Level	Second Level	Construction Engineer		Construction Manager		Safety Engineer		T/C Operator		Total	
		Wt.	R	Wt.	R	Wt.	R	Wt.	R	Wt.	R
CC T/C											
Erection work	Improper bolting on the brace/mast/telescopic element	0.081	5	0.122	2	0.126	2	0.116	3	0.098	2
	Failure to comply with safety rules and work guidelines for erection work	0.040	9	0.057	7	0.054	9	0.180	1	0.080	4
	Insufficient worker skill for erection work	0.036	10	0.027	14	0.040	11	0.085	6	0.046	13
Dismantling work	Poor understanding of accident risk during dismantling process	0.204	1	0.039	10	0.028	12	0.048	8	0.073	6
	Improper tightening of bolts	0.066	6	0.030	13	0.052	10	0.011	15	0.048	12
	Low skill level of dismantling workers	0.121	3	0.021	15	0.020	13	0.040	10	0.049	11
Lifting work	Poor management of lifting objects and control of surroundings	0.042	8	0.117	4	0.015	14	0.096	4	0.071	8
	Improper tying of the sling leg	0.107	4	0.041	9	0.012	15	0.011	14	0.037	15
	Poor operator visibility	0.144	2	0.135	1	0.093	4	0.039	11	0.107	1
Prime contractor management	Inappropriate personnel and equipment placement	0.018	14	0.039	11	0.067	7	0.044	9	0.042	14
	Poor subcontracting technology management process	0.047	7	0.036	12	0.123	3	0.131	2	0.084	3
	Insufficient safety management of equipment and personnel	0.022	13	0.089	5	0.064	8	0.019	13	0.050	10
T/C machine	Equipment aging	0.023	12	0.119	3	0.079	5	0.079	7	0.064	9
	Crane operation error or failure	0.018	15	0.083	6	0.075	6	0.026	12	0.071	7
	Problems with overseas parts procurement	0.031	11	0.047	8	0.152	1	0.090	5	0.075	5

Table 5. *Cont.*

First Level	Second Level	Construction Engineer		Construction Manager		Safety Engineer		T/C Operator		Total	
		Wt.	R	Wt.	R	Wt.	R	Wt.	R	Wt.	R
RC T/C	Erection work										
	Improper bolting on the brace/mast/telescopic element	0.080	5	0.107	4	0.048	8	0.061	6	0.084	5
	Failure to comply with safety rules and work guidelines for erection work	0.039	8	0.127	3	0.058	5	0.134	3	0.086	3
	Insufficient worker skill for erection work	0.036	10	0.046	6	0.075	4	0.204	1	0.084	4
	Dismantling work										
	Poor understanding of accident risk during dismantling process	0.186	1	0.015	15	0.039	9	0.009	14	0.047	12
	Improper tightening of bolts	0.082	4	0.044	11	0.032	10	0.007	15	0.042	13
	Low skill level of dismantling workers	0.122	3	0.057	5	0.051	7	0.034	10	0.081	7
	Lifting work										
	Poor management of lifting objects and control of surroundings	0.178	2	0.033	13	0.311	1	0.010	13	0.105	1
	Operator unable to check the weight and specifications of the lifted objects	0.062	6	0.040	12	0.082	3	0.071	5	0.103	2
	Operator cannot check the wire and sling leg	0.057	7	0.022	14	0.185	2	0.011	12	0.057	9
	Prime contractor management										
	Inappropriate personnel and equipment placement	0.019	14	0.045	9	0.020	11	0.037	9	0.031	15
	Poor subcontracting technology management process	0.036	11	0.045	10	0.052	6	0.111	4	0.065	8
	Insufficient safety management of equipment and personnel	0.031	12	0.130	2	0.020	12	0.037	8	0.050	11
	T/C machine										
	Equipment aging	0.010	15	0.200	1	0.005	14	0.061	7	0.052	10
	Crane operation error or failure	0.024	13	0.046	7	0.005	15	0.025	11	0.032	14
	Problems with overseas parts procurement	0.039	9	0.046	8	0.017	13	0.189	2	0.083	6

5. Case Study

Since it is difficult to measure the rate of reduction in accident causes by applying the results of this AHP analysis to actual construction sites, we propose the reduction rate of accident causes through the scenario of installing a camera and wireless transmitter on the trolley of tower cranes and also placing a safety manager of tower crane (refer to Figure 5). In this scenario, it is assumed that the tower crane operator (both CC T/C and RC T/C), the safety manager of tower crane, and the field office have installed monitors that can check the lifting work of the tower crane. As shown in Table 6, the weight calculated in AHP analysis was converted into the probability of accident causes in order to calculate the reduction rate of accident causes. The probability of erection phase is 22% for CC T/C and 25% for RC T/C. Dismantling phase is 17% for both T/C and Lifting work; is 22% for CC T/C and 27% for RC T/C and prime contractor management; is 17% for CC

T/C and 15% for RC T/C and T/C machinery; is 22% for CC T/C and 16% for RC T/C. As shown in Table 6, in the case of installing the camera to tower crane for tower crane operator, the safety manager of tower crane and field office, most of the accident causes are eliminated, so that the reduction probability for CC T/C is 29% and RC T/C is 30%. In the case of placing the safety manager of tower crane, the reduction probability for CC T/C is 55% and RC T/C is 49%. Although this reduction probability is not the result obtained after applying two cases to the actual construction site, it is believed that the camera attached to tower crane and safety manager of tower crane can eliminate most of the actual causes of tower crane accidents.

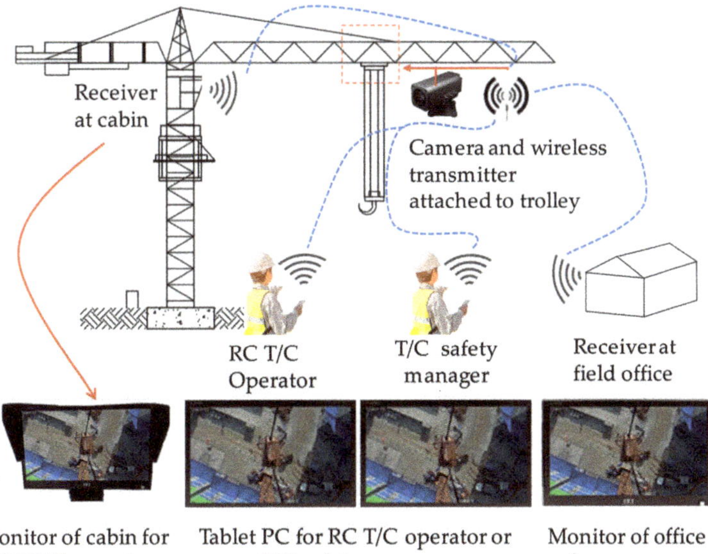

Figure 5. Safety camera and wireless transmission for preventing tower crane accidents.

Table 6. Accident probability and reduction probability applied to T/C camera and manager.

First Level Phase/Task	Second-Level (Sub) Accident Causes of the AHP Structure for T/C Accidents	Weights by AHP Analysis		Probability by AHP Analysis		Reduction Probability			
						Safety Camera		T/C Manager	
		CC T/C	RC T/C	CC T/C	RC T/C	CC T/C	RC T/C	CC T/C	RC T/C
Erection work	Improper bolting on the brace/mast/telescopic element	0.098	0.084	10	8	0	0	10	8
	Failure to comply with safety rules and work guidelines for erection work	0.080	0.086	8	9	0	0	8	9
	Insufficient worker skill for erection work	0.046	0.084	4	8	0	0	0	0
	Subtotal			22	25	0	0	18	17
Dismantling work	Poor understanding of accident risk during the dismantling process	0.073	0.047	7	5	0	0	7	5
	Improper tightening of bolts	0.048	0.042	5	4	0	0	5	4
	Low skill level of dismantling workers	0.049	0.081	5	8	0	0	0	0
	Subtotal			17	17	0	0	12	9

Table 6. *Cont.*

First Level Phase/Task	Second-Level (Sub) Accident Causes of the AHP Structure for T/C Accidents	Weights by AHP Analysis		Probability by AHP Analysis		Reduction Probability			
						Safety Camera		T/C Manager	
		CC T/C	RC T/C	CC T/C	RC T/C	CC T/C	RC T/C	CC T/C	RC T/C
Lifting work	Poor management of lifting objects and control of surroundings	0.0707	0.105	7	11	7	11	0	0
	Improper tying of the sling leg	0.037	-	4	-	4	-	0	-
	Poor operator visibility	0.107	-	11	-	11	-	0	-
	Operator unable to check the weight and specifications of the lifted objects	-	0.103	-	10	-	10	-	0
	Operator cannot check the wire and sling leg	-	0.057	-	6	-	6	-	0
	Subtotal			22	27	22	27	0	0
Prime contractor management	Inappropriate personnel and equipment placement	0.042	0.031	4	3	0	0	4	3
	Poor subcontracting technology management process	0.084	0.065	8	7	0	0	8	7
	Insufficient safety management of equipment and personnel	0.050	0.050	5	5	0	0	5	5
	Subtotal			17	15	0	0	17	15
T/C machinery	Equipment aging	0.064	0.052	7	5	0	0	0	0
	Crane operation error or failure	0.071	0.032	7	3	7	3	0	0
	Problems with overseas parts procurement	0.075	0.083	8	8	0	0	8	8
	Subtotal			22	16	7	3	8	8
	Total			100	100	29	30	55	49

6. Discussion and Conclusions

Despite various efforts to reduce accidents related to T/Cs, the number of accidents is still increasing. This study was conducted to help prevent T/C-related accidents by ranking the accident causes related to CC T/Cs and RC T/Cs, as well as the weights of each cause to be used as a reference for management. As shown in the results of our AHP questionnaire, various accidents occur when an operator cannot directly check the status or ties when a lifting object is hidden by other structures or objects, and is largely dependent on the signals and radio communications of other workers. The result of this study is "operator visibility" as first ranking cause in case of CC T/C, and is "operator unable to check the weight and specification of the lifting objects" as second ranking cause in case of RC T/C. In previous study [17], the operator's impact such as "operator proficiency", "operator character", and "employment source (operator)" was dominated. The collective weight of these causes is nearly 24%. The cause of the tower crane "operator proficiency" was suggested as the second-level accident cause. Especially, the comprehensive cause of operator impact was suggested. The operator proficiency is an ambiguous cause of tower crane accidents. In other words, the way to solve this cause is ambiguous. To solve this cause, the operator training cycle in Korea has recently been shortened. In this study, operator visibility and to check lifting objects are very specific causes that can be managed and are suggested solutions at construction sites. Therefore, one could prevent tower accidents by attaching a device like RPSA to the jib or hook for both CC T/Cs and RC T/Cs. Such device could help T/C operators monitor their work from the cabin and make decisions based on signals from other workers and their own judgment. Additionally, it is necessary to train managers and workers continuously to help them maintain and comply with guidelines and manuals related to erection work, lifting work, and operation.

The major accident cause for CC T/Cs were ranked in descending order of "poor operator visibility", "bad bolting on the brace/mast/telescopic element", and "poor subcontractor technology". Such causes like "bad bolting on the brace/mast/telescopic element" and "poor subcontractor technology" are combined various factors such as the management problem of the prime contractor, the management problem of the subcontractor, and the skill of erecting/dismantling worker. The major accident cause for RC T/Cs were ranked in descending order of "poor management of lifting objects and control of surroundings", "operator unable to check the weight and specifications of lifting objects", and "failure to comply with safety rules and work guidelines". In particular, in the case of poor management of lifting objects and control of surroundings, there is a problem in the function of properly controlling and managing T/Cs because cranes are controlled and managed by numerous operators who have completed the required training for each type of construction work. Therefore, it is recommended to have a separate manager in charge of supervising work using T/Cs at a construction site. Furthermore, the second major accident cause of checking the weight and specifications of lifting objects can be addressed by attaching RPSA to help operators to make informed decisions. This situation is similar to that of a CC T/C.

In previous study [17], the "site-level safety management" is the highest weight cause affecting safety on construction sites with tower crane. Especially, the superintendent effect has "superintendent character" and "site-level safety management" to influence the safety of the crane-related site, a total of over 23%. In other previous study [20], as a result of analyzing the previous literature, the causal factors belonging to "site personnel management" are also very important as the frequency occupies the first to third place. In this study, most accident causes are related to situation/surrounding control using T/Cs and management issues that require workers to comply with work guidelines and rules of erection, dismantling, and lifting works that reflect reality. Therefore, it is important to designate a safety manager of tower crane with sufficient experience and knowledge regarding T/Cs to train T/C operators, engineers, and managers, and to revise, supplement, and manage various instructions and manuals. Additionally, the safety manager of tower crane is expected to provide sufficient help in terms of reducing accidents through consistent safety-based management and exercising practical control authority over T/C erection, telescoping, and lifting operations. In the case of an RC T/C, plans for supplemental education should be prepared as it becomes increasingly easy to obtain an operating license. In the field, it is necessary to establish a reinforced training plan for safety rules during erection, operation, and dismantling.

To prevent T/C accidents, we derived the priorities of accident causes for different T/C types using an AHP questionnaire. It is crucial to reduce accidents for sustainable management at construction sites. If the priorities of T/C accident causes presented in this paper are utilized in various checklists or management plans, and construction site management is conducted accordingly, then sustainable construction management can be realized. In the future, if additional research is performed by narrowing the scope of AHP questionnaire respondents to workers who directly use T/Cs and experts familiar with T/Cs, more realistic results can be derived, which will further reduce T/C accidents.

7. Limitations and Future Research

This research was conducted to derive the priority of tower crane accident causes based on the experience of the related experts of construction sites with tower cranes. However, the effect was not verified by actual application to safety management plans or tower crane checklists at the real construction sites. Therefore, in the future research, it is necessary to apply the priority of accident causes based on expert experience to the practical example in accordance with the T/C configuration to identify the actual accident reduction level.

Author Contributions: J.Y.K. reviewed the existing literature, completed the AHP questionnaire, and conducted the result data coding of AHP questionnaire and analysis.; D.S.L. analyzed the characteristics of each tower crane type and suggested a new tower crane safety model for application of the research results.; J.D.K. selected the subjects of the AHP questionnaire and conducted a AHP survey; G.H.K. conceived the whole this study and conducted a review of the research results. All authors have read and agreed to the published version of the manuscript.

Funding: No funding.

Institutional Review Board Statement: Not applicable.

Informed Consent Statement: Not applicable.

Data Availability Statement: The data presented in this study are available on request from the corresponding author. The data are not publicly available due to restrictions on right of privacy.

Conflicts of Interest: The authors declare no conflict of interest.

References

1. Shapira, A.; Lucko, G.; Schexnayder, C.J. Cranes for Building Construction Projects. *J. Constr. Eng. Manag.* **2007**, *133*, 690–700. [CrossRef]
2. Ministry of Land, Infrastructure and Transport. Available online: http://kosis.kr (accessed on 27 July 2020).
3. Beavers, J.E.; Moore, J.R.; Rinehart, R.; Schriver, W.R. Crane-Related Fatalities in the Construction Industry. *J. Constr. Eng. Manag.* **2006**, *132*, 901–910. [CrossRef]
4. Choi, C.H. A Study on the Risk Analysis and Measures of Reduction through Tower Crane Accidents Cases. Master's Thesis, Hanyang University, Seoul, Korea, 2017.
5. The Korea Times. Available online: http://www.koreatimes.co.kr/www/nation/2020/02/281_281363.html (accessed on 3 November 2020).
6. Hinze, J.; Godfrey, R.; Sullivan, J. Integration of Construction Worker Safety and Health in Assessment of Sustainable Construction. *J. Constr. Eng. Manag.* **2013**, *139*, 594–600. [CrossRef]
7. Asanka, W.A.; Ranasinghe, M. Study on the impact of accidents on construction project. In Proceedings of the 6th International Conference on Structural Engineering and Construction Management 2015, Kandy, Sri Lanka, 11–13 December 2015; pp. 58–67.
8. Lee, H.S. A study on the Accident Analysis and the Control of Tower Cranes. Master's Thesis, Seoul National University of Science and Technology, Seoul, Korea, 2014.
9. Choi, S.-Y.; Cho, K.-H.; Park, D.-H.; Gil Choi, B. A study on the work environment and accident exposure status of Tower Crane workers. *J. Korea Saf. Manag. Sci.* **2015**, *17*, 115–123. [CrossRef]
10. Yun, D.H.; Jong, Y.P.; Jung, H.K. Measures to reduce tower crane accidents during operation by improving signal system and education for signalmen. *J. Korea Saf. Manag. Sci.* **2019**, *34*, 59–75. [CrossRef]
11. Ministry of Land, Infrastructure and Transport. Available online: http://www.molit.go.kr (accessed on 27 July 2020).
12. Kim, Y.U. Problems and Improvement Schemes for Unmanned Tower Crane Accident through Case Analysis at Construction site. Master's Thesis, Chung-Ang University, Seoul, Korea, 2018.
13. Aneziris, O.; Papazoglou, I.; Mud, M.; Damen, M.; Kuiper, J.; Baksteen, H.; Ale, B.; Bellamy, L.; Hale, A.; Bloemhoff, A.; et al. Towards risk assessment for crane activities. *Saf. Sci.* **2008**, *46*, 872–884. [CrossRef]
14. Song, P.Y. A Study on Improvement of Safety Management through Analysis of Tower Crane Disaster at Construction Site. Master's Thesis, Pukyong University Safety Engineering, Seoul, Korea, 2018.
15. Kwon, O.M. Deduction of Accident Cause for Tower-Crane Using FMEA Method. Master's Thesis, Hanyang University, Seoul, Korea, 2015.
16. Richard, L.N.; Noah, S.S.; Kyle, K.R. A Review of crane safety in the construction industry. *Appl. Occup. Environ. Mental Hyg.* **2001**, *16*, 1106–1117. [CrossRef]
17. Shapira, A.; Simcha, M. AHP-Based Weighting of Factors Affecting Safety on Construction Sites with Tower Cranes. *J. Constr. Eng. Manag.* **2009**, *135*, 307–318. [CrossRef]
18. Cho, Y.R. Importance Evaluation of the Safety Accident Factors of Tower Crane Using Analytic Hierarchy Process. Master's Thesis, Kyonggi University, Suwon, Korea, 2017.
19. In, J.S. Factors that affect safety of tower crane installation/dismantling in construction industry. *Saf. Sci.* **2014**, *72*, 379–390. [CrossRef]
20. Zhang, X.; Zhang, W.; Jiang, L.; Zhao, T. Identification of Critical Causes of Tower-Crane Accidents through System Thinking and Case Analysis. *J. Constr. Eng. Manag.* **2020**, *146*, 04020071. [CrossRef]
21. Zhou, W.; Zhao, T.; Liu, W.; Tang, J. Tower crane safety on construction sites: A complex sociotechnical system perspective. *Saf. Sci.* **2018**, *109*, 95–108. [CrossRef]

22. Fang, Y.; Cho, Y.K.; Chen, J. A framework for real-time pro-active safety assistance for mobile crane lifting operations. *Autom. Constr.* **2016**, *72*, 367–379. [CrossRef]
23. Chandra, H.P. Initial Investigation for Potential Motivators to Achieve Sustainable Construction Safety and Health. *Procedia Eng.* **2015**, *125*, 103–108. [CrossRef]
24. Rajendran, S.; Gambatese, J.; Behm, M. Impact of Green Building Design and Construction on Worker Safety and Health. *J. Constr. Eng. Manag.* **2009**, *135*, 1058–1066. [CrossRef]
25. Korea Occupational Safety and Health Agency. Available online: http://www.kosha.or.kr/kosha/data/construction.do (accessed on 27 July 2020).
26. Korea Occupational Safety and Health Agency. *Safety Work Guide to Prevent Death of Construction Machinery and Equipment*; Korea Occupational Safety and Health Agency: Seoul, Korea, 2018; pp. 1–97.
27. Cho, G.T.; Cho, Y.G.; Kang, H.S. *The Analytic Hierarchy Process*; Donghun: Seoul, Korea, 2003; pp. 1–311.
28. Tam, V.W.; Fung, I.W. Tower crane safety in the construction industry: A Hong Kong study. *Saf. Sci.* **2011**, *49*, 208–215. [CrossRef]

Article

The Transition from Traditional Infrastructure to Living SOC and Its Effectiveness for Community Sustainability: The Case of South Korea

Yeonsoo Kim, Jooseok Oh and Seiyong Kim *

Department of Architecture, Korea University, Seoul 02841, Korea; yeonsooclarekim@gmail.com (Y.K.); ohjooseok@korea.ac.kr (J.O.)
* Correspondence: kksy@korea.ac.kr; Tel.: +(02)-3290-3914

Received: 23 October 2020; Accepted: 5 December 2020; Published: 8 December 2020

Abstract: In 2018, the South Korean government began promoting a "livelihood-improving" social overhead capital policy based on the concepts of an inclusive city, smart shrinkage, and the balanced development of metropolitan and provincial cities. Based on a review of the extant literature and relevant policies from South Korea, this study explores this policy's implementation and makes some suggestions for its sustainability. This study compares the current state of South Korea's urban facilities' and the balance of their supply between metropolitan and provincial cities. To discern which type of facility central and local governments should prioritize, this study conducts a stepwise regression analysis and identifies which preexisting facilities influence the facility type proposed by the current policy. Results show that South Korea's living infrastructure is well distributed among metropolitan and provincial cities. However, urban planning shows little consideration for minimizing the distance between facilities and residential zones. In terms of facility types, the supply of education and local community facilities was adequate throughout the country, while culture and art facilities were inadequate. In metropolitan cities, the supply of sports and leisure facilities was insufficient.

Keywords: social capital; living environment; living infrastructure; soft infrastructure; living social overhead capital; inclusive growth; inclusive city

---

1. Introduction

First presented by the United Nations as a theme of its Global Campaign on Urban Governance in 1999 [1], the concept of city inclusion comprises three dimensions: social, economic, and spatial. Of these, spatial inclusion refers to equal accessibility to living infrastructure and public services [2], mainly because people with limited access to living infrastructure and public services experience social exclusion from various social opportunities [3]. Accordingly, the global community has sought sustainable, inclusive growth by ensuring universal access to living infrastructure and public services [4]. In addition to the concept of inclusive growth, there is a growing emphasis on nonphysical infrastructure (or soft infrastructure) beyond the conventional concept of living infrastructure and public services [5]. Unlike physical infrastructure (or hard infrastructure), such as roads and ports, nonphysical infrastructure refers to all services essential for maintaining a nation's economy, health, and culture. From the late twentieth century, European and North American scholars have defined parks, green areas, and community sports facilities as social infrastructure and have actively participated in policymaking and research related to social infrastructure [6–9].

In 2018, South Korea proposed a "livelihood-improving" social overhead capital policy (hereinafter, Living SOC) as a practical alternative to realizing "spatial inclusiveness". Defined as "a small-scale living infrastructure easily accessible to people in the community", the Living SOC reflects the paradigm of global change while promoting balanced development between South Korea's regions and cities and creating more equitable living standards [10]. While the policy draws on the concept of the "living infrastructure" insofar as it includes the same type of facilities, it also reflects the concepts of "spatial equality" and "equal accessibility" [11]. The Living SOC policy was motivated by South Korea's 2018 Gini coefficient, which showed that the regional disparity between metropolitan and provincial cities in the Living SOC supply across the country was worse than the income disparity between individuals [12]. In general, metropolitan cities refer to large central cities with a population of more than 50,000. They serve as hubs for social and economic activities in the surrounding area, while provincial cities refer to other areas [13]. However, following Article 175 of the Local Autonomy Act, South Korea has defined metropolitan cities as specific cities with a population of more than 500,000 and eligible for special treatment, and this study follows this definition. With the national budget increasing by approximately KRW 10.4 billion per annum [14], the government allocated funds for a national project for balanced development based on the Living SOC—including some KRW 500 billion for startup expenses alone [15]. As such, policies related to Living SOC are increasing in significance.

However, despite relatively refined standards, the Living SOC policy faces the criticism that it is not much different from the previous policy of supplying "living infrastructure" in terms of its exhaustive list of facilities [16]. Urban shrinkage is becoming common in cities worldwide, including over 20 cities in South Korea [17]. Here, urban shrinkage is a concept established through the 2002 in German miniature city project [18,19]. Urban shrinkage does not mean that a city's physical size is getting smaller, but rather an urban phenomenon in which boundaries and infrastructure remain the same, while the population and economy decline [20,21]. Provincial cities experiencing urban shrinkage may suffer from various problems, including poor usability due to a superfluous supply of facilities and their deteriorating conditions, and the difficulty of procuring the financial resources necessary for the upkeep of facilities. Nonetheless, with the area of various convenience facilities intended to sustain people's lives predicted to increase from 2792 km^2 in 2015 to 3842 km^2 in 2040 [22], the significance of adequately supplying such infrastructure has also been emphasized.

As such, South Korea is facing the complicated task of supplying Living SOC equitably and sustainably to cope with urban change and resolve spatial inequality. Accordingly, it is necessary to analyze the characteristics of cities to address urban shrinkage and identify the type of infrastructure. In particular, as the current Living SOC policy overlaps with preexisting living infrastructure in urban areas, it is essential to determine whether the current supply of living infrastructure overlaps with the facilities proposed by the new policy and whether the pertinent facilities are distributed equitably. Extant studies from South Korea [23–25] are limited insofar as they primarily focus on examining the condition of major living infrastructure and strategies for improving accessibility. Moreover, investigating and analyzing the physical and demographic conditions of each city, as well as the current condition of the major urban facilities that can be categorized as Living SOC, will improve the implementation of relevant policies.

Considering the foregoing, this study examines the implementation and sustainability of South Korea's Living SOC policy. To overcome the limitation in the extant research and relevant policies, this study examines the current condition of the living infrastructure established in South Korea before 2018, when the Living SOC was introduced, from the perspective of urban planning and land use. Per the concept of balanced development, this study examines the distribution of the preexisting facilities and analyzes whether the pertinent facilities are evenly distributed between metropolitan and provincial cities. By discerning which types of preexisting facility influence that proposed in the Living SOC policy, this study identifies which type of facilities the central and local governments should prioritize in terms of supply. The findings of this study can facilitate economic stability and sustainability by improving the implementation of relevant policies going forward.

2. Literature Review

2.1. Social, Soft, and Living Infrastructure

This study understands the concept of Living SOC to be similar to that of an inclusive city, which was proposed in the 1990s as a solution to a regional imbalance between cities [10]. The concept also draws on practical land and urban planning strategies and policies proposed by countries in Europe, North America, and East Asia [12].

In this respect, the concept of Living SOC is the closest to that of social infrastructure—that is, a composite of resources and facilities—including spaces, services, and networks—that vitalize the local community [26] and preserve the happiness and quality of life of community residents [27]. The concept of social infrastructure is generally contrasted by physical or economic infrastructure. While the physical infrastructure directly supports economic growth, social infrastructure aims to help build the community by providing the necessary social services [28] and improving residents' quality of life [29]. Social infrastructure also contributes to economic development by ensuring the effective utilization of human resources [30]. Social infrastructure can be defined as the physical environment that determines the successful development of social capital [31].

Compared to social infrastructure, living-related infrastructure refers to more specific physical facilities that community members need for daily life, such as houses, parks or green areas, water facilities, parking lots, and hospitals [32]. Meanwhile, living infrastructure refers to physically alive and easy infrastructure for community members to access and utilize [33]. Living infrastructure is similar to the living-related infrastructure insofar as it is defined in the scope of the preexisting infrastructure in close relation to daily life from the perspective of social and natural science. However, in contrast to living-related infrastructure, the concept of living infrastructure emphasizes the sustainability of the local and urban residents by adding to it "being alive" [34].

Infrastructure can also be divided into hard and soft infrastructure which provide both physical and social services. Hard infrastructure is a new category of large-scale infrastructure, comprising the basic urban structures such as roads, ports, electric/energy plants, water supply, and sewage systems [35]. In contrast, soft infrastructure refers to the necessary services for maintaining a community's economy, health, and culture [5].

As such, new definitions of infrastructure transcend the traditional definition of infrastructure as the physical and essential facilities for constructing and operating cities to include those intended to improve community sustainability and improve residents' quality of life. Such a perspective of infrastructure is widely accepted by developed and advanced countries seeking to ensure cities' sustainability and their inclusive growth. Certainly, South Korea's latest policy adopts a social infrastructure perspective—recognizing that improving quality of life by providing facilities and services supporting people's daily lives will positively affect local production. Living SOC differs from conventional SOC (or social infrastructure) in that it tries to provide equal access to essential living services [36]. In this respect, South Korea's current Living SOC policy seeks to minimize the physical distance between residential zones and the daily living services, ultimately making the routes of urban residents more compact.

2.2. Living SOC as a Community-Supportive Infrastructure

Since the late twentieth century, policies similar to South Korea's Living SOC policy have been established and implemented in several countries in Europe and East Asia [6–9]. To understand similar policies, first, it is necessary to understand the concept of Smart Shrinkage and Compact City, which is one way to achieve an inclusive city [37]. Smart Shrinkage and Compact City is an urban regeneration method that focuses on improving the quality of life of existing urban residents while reducing population and building land use [38]. This concept differs from existing urban regeneration in that it improves the quality of life rather than inducing population inflow and employment growth [39]. Smart Shrinkage and Compact City can be a strategy to prevent the vicious cycle of decline by reorganizing the urban infrastructure to fit a new level of population, such as returning the abandoned neighborhoods of the city to nature, increasing the walking space, and fitting housing prices [40]. Poppers defined this as less planning, less population, fewer buildings, and less land use, and argued that small could be beautiful [41].

To address the population decline and a worsening local economy, Japan implemented the concept of a Compact City in 2014, placing residential zones in proximity to public transportation and necessary service facilities [42]. To be specific, in July 2014, the Ministry of Land, Infrastructure, and Transport and Tourism (MLIT) unveiled the "Grand Design of National Spatial Development towards 2050", with Aggressive Smart Shrinkage as an alternative to population decline [43]. The concept of Aggressive Smart Shrinkage is not a defensive response that prevents the city from shrinking and disappearing if the population decline is inevitable; it is a reduction plan to proactively continue urban function even if the population decreases by predicting the reduction mode. "Grand Design of National Spatial Development towards 2050" proposes a connection between Compact and Network to maintain urban functions even in the face of a declining population [44]. It is a strategy to prevent urban functions' departure by spatially integrating urbanized areas and resolving insufficient urban functions through the interaction of surrounding areas by reinforcing public transportation. Expressly, in a city with a population of 100,000 or more, a 1-km grid range of reach within an hour is set as an urban area, and a high-level regional urban association is established so that the urban area can sustain a population of 300,000 or more. An institutional response that applies the concept of a Compact City is a plan to appropriate its location. It will seek a network that allows access to medical, welfare, and business facilities through public transportation in areas where population reduction is expected lead to a failure in meeting the minimum residential standards. More specific measures include inducing urban functional facilities, such as medical care and welfare in the hub area, inducing residences in areas with public transportation connection, overhauling the walking and vehicle environments in the center, and introducing community buses. Besides, urban function inducement zones were established to enable urban function services to ease regulations and provide subsidies to induce urban function concentration rather than coercion. To ensure the smooth execution of the policy, the Japanese government implemented a "city function initiation zone" initiative by designating residential zones and collecting feedback from citizens to maintain an optimal population density [45]. The city function initiation zone promotes healthcare, business, education, and basic service facilities, thereby providing optimal services for urban residents. Regarding administration, the Japanese government monitors current convenience facilities and provides support in policy and finance for pertinent facilities when there is a shortage of individual convenience facilities.

In a similar context, Germany is implementing a policy to ensure equal living conditions based on the constitution, which guarantees "living conditions with equal value", and the 1965 Federal Space Planning Act [46]. As the industrial structure changed in Germany, manufacturing declined, and suburbanization increased, resulting in a decline of cities. In particular, in the former East German region after reunification, urban shrinkage became more severe as people who lost their jobs moved to the former West Germany or surrounding large cities [47]. Cities such as Dresden, Leipzig, and Cottbus are typical examples. House remain unoccupied in both the old and newly redeveloped areas of some cities. Therefore, the German government has adopted and implemented a strategic plan for Smart Shrinkage at the local level [48]. INSEK, an integrated strategic plan to respond to the smart-reduction problem, is the basis for allocating all subsidies to promote urban regeneration. In Germany, the government has stipulated that subsidies should be made only after establishing INSEK after 2002. INSEK is an integrated plan for urban development, established by each local government, to review the development priorities to designate areas subject to maintenance and areas of focus. The overall direction of urban development is thus set, and specific plans are flexibly adjusted according to the circumstances. More specifically, the law mandates that all 38 provinces identify a hub city with a population of over 100,000 people and equally distribute various living facilities for each hub city so that each region can enjoy similar living standards [46]. The guidelines are divided into the social infrastructure category, which involves service facilities such as daycares and hospitals, and the technical infrastructure category, including water supply and treatment facilities. The German government has also focused on creating a universal living environment by monitoring changes in regional characteristics [49].

Numerous scholars have researched balanced urban development with the land as the spatial background. For example, Peters et al. [50] examined communities' social infrastructure to determine the degree of smart shrinkage in small towns based on population, land use, and transportation. Similarly, Chang and Liao [51] identified strategies for improving accessibility to urban parks and balanced distribution while highlighting public facilities' spatial equity. Examining the size and shape of a city and the attributes of urban planning, Hodge and Gatrell [52] highlighted the significance of determining the service area and argued that the attributes of urban planning could be a constraint on the related activities of the political, economic, and social systems. Regarding well-balanced development—the ultimate goal of South Korea's Living SOC policy—the extant research demonstrates the significance of the following: the establishment of relevant strategies prioritizing the investigation of urban land use status [53], an appropriate supply of facilities and commercial zones [54], the equal supply of education facilities [55], and hard and soft infrastructure [56]. Moreover, as the supply should meet the demand in terms of the accessibility and availability of these facilities [57], identifying the demand and the current condition of relevant facilities can positively affect the optimal supply of facilities when implementing relevant policy initiatives.

3. Methodology

3.1. Research Model

Like other countries, South Korea defines the Living SOC's scope as facilities assisting citizens in their daily lives, including those related to education, healthcare, welfare, transportation, culture, sports, and parks. South Korea also emphasizes the accessibility of these facilities. In addition to promoting the development of different regions, South Korea seeks to develop land equitably and improve citizens' lives. Therefore, this study examines the entire territory of South Korea to derive implications for the Living SOC policy. Data collection, analysis, and interpretation were conducted in two stages, as follows.

First, this study compares metropolitan and provincial cities in terms of population, the average age of urban residents, urbanization ratio, and the area size of each zone according to South Korea's land-use planning (i.e., residential, commercial, industrial, and green zones, respectively), and the total number of major facilities in the category of the Living SOC policy presented by the government. In doing so, this study examines the type of imbalance between metropolitan and provincial cities using independent sample t-tests. Various studies have verified and highlighted the validity of the variables mentioned above [58] and the use of independent t-test to compare regional differences [59,60].

Second, this study examines how the amount of basic service, convenience, and cultural and sports facilities in each city influences the number of Living SOC. There is already a wide variety of preexisting basic service, convenience, and cultural and sports facilities throughout the country, many of which are included in the Living SOC policy. Accordingly, if the research model is statistically significant, the type of facilities closely related to the current Living SOC policy is already sufficiently distributed from the policy's perspective when the number of the Living SOC is set as the dependent variable and other facilities as independent variables. Therefore, they are relatively unimportant. By analyzing each city throughout the country, this study derives implications for improving the implementation and efficacy of future policy initiatives. To identify the determined balanced development, we divided the cities into metropolitan cities and provincial cities, conducted two rounds of regression analysis, and compared the results. Various studies have highlighted the importance of conducting regression analysis in analyzing the current condition of facilities in a specific area [61–63]. Figure 1 illustrates this study's research process and methodology.

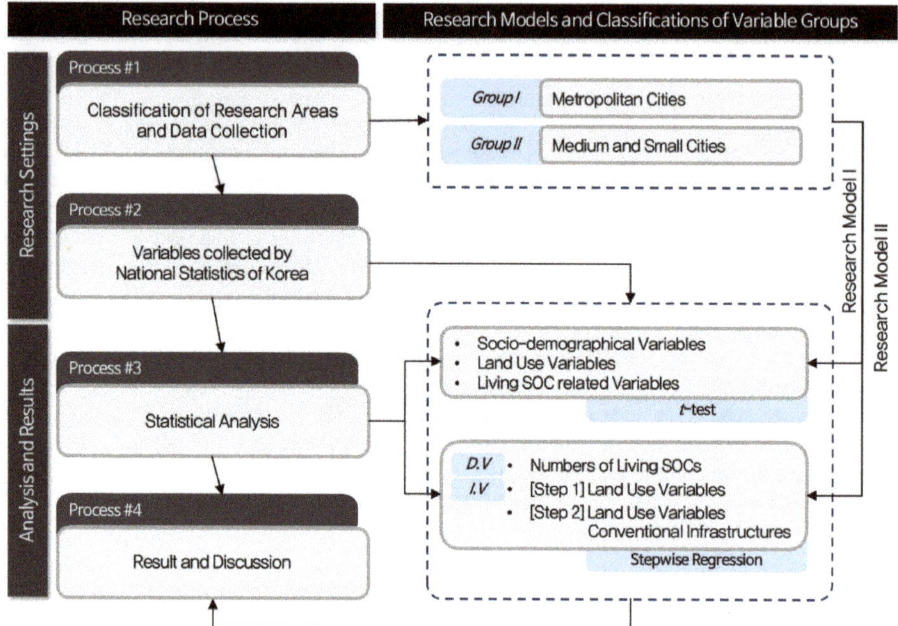

Figure 1. Research process and methodology.

3.2. Study Areas and Variables

As noted, the spatial scope of this study is the entire territory of South Korea. All data used in this study are based on the 2020 administrative division of South Korea. A total of 229 regions were used as samples, including 69 autonomous districts, 75 autonomous cities, 82 counties, one special self-governing city, and eight provinces (Figure 2). These samples are the minimum unit of all data used for statistics and include the entire are of South Korea.

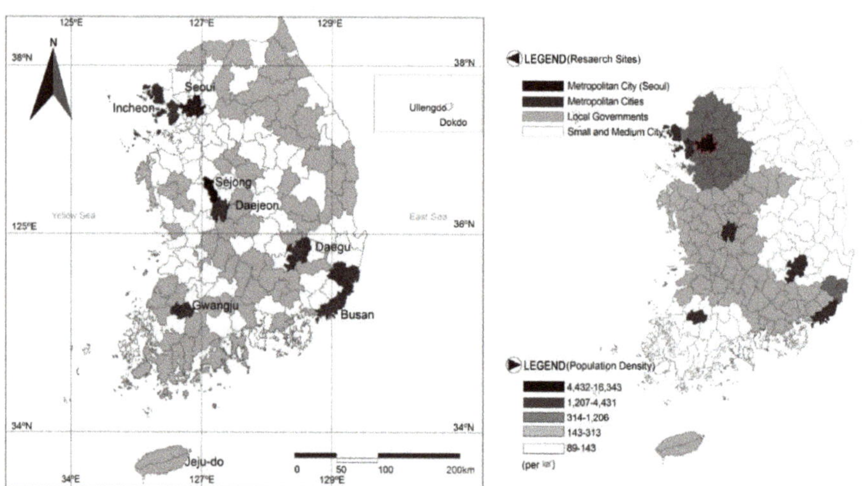

Figure 2. Classification of research areas and population density (from Statistics Korea).

For the two rounds of regression analysis, we categorized cities into metropolitan cities and provincial cities based on the relevant statutes, resulting in 74 self-governing cities (or self-governing areas) and 155 provincial cities. In this categorization process, we collected data to compare the two categories of cities in terms of population, average age, number of residents per city, and the ratio of residential/commercial/industrial/green zones. This study used open-access data from the Korea Statistical Information Service (KOSIS).

This study used the following two-step method to obtain and classify convenience facilities. First, according to the government's proposal, Living SOC facilities are intended to enhance the convenience of people's lives and encompass culture, sports, education, healthcare, welfare, and park facilities [13]. Specific facilities include community sports centers; outdoor sports facilities, such as baseball parks, soccer fields, gate ball courts, and artificial rock-climbing walls; and cultural and educational facilities, such as libraries, museums, art galleries, parking lots, daycare centers, kindergartens, elementary schools, welfare facilities for elderly, hospitals, highway rest areas for safe traffic, fire or disaster safety facilities, forests, recreation forests, campsites, and urban parks. Among the listed facilities, the supply of Living SOC facilities is provided by the public sector, and statistics are officially totaled by the central government. There are seven types of Living SOC facilities in total: elementary schools, job training schools, libraries, culture centers, post offices, police stations, and fire stations. As the supply of Living SOC facilities is provided by the central and provincial government, other private facilities were not included as variables. However, there is a possibility that, at the local level, critical private facilities were excluded from this process.

Second, the preexisting living infrastructure used as an independent variable in the regression analysis includes public and private facilities proposed in the pertinent policy and statues. Among over 30 facilities, we selected nine facility types with available open-access data from KOSIS as follows: elementary schools, job training schools, libraries, culture and art facilities such as museums, art galleries, and culture centers; sports facilities such as baseball parks, basketball courts, soccer fields, and gyms; and local community facilities such as recreation forests, campsites, fields, and urban parks. There is a possibility that major facilities may have been excluded during the process of limiting the variables to those for which national data exist. This study recategorized the nine selected facility types into facilities with similar functions to ensure commonality between the variables, resulting in the following four categories of facilities: (1) education and empowerment facilities, (2) culture and art facilities, (3) sports and leisure facilities, and (4) local community facilities—Table 1 details the content of each category. The three-year Living SOC plan of the South Korean Government was referred to for the classification of categories.

Table 1. Specific types of facilities.

	Facilities
Living SOC	Total number of Living SOC facilities.
Education and Empowerment	Total number of elementary schools, job training schools, and libraries.
Culture and Arts	Total number of museums and galleries.
Sports and Leisure	Total number of sports facilities (gyms, pools, bowling alleys, golf clubs, tennis courts, etc.).
Local Community	Total number of community sports facilities and urban parks.

4. Results

4.1. Comparison between Metropolitan and Provincial Cities

South Korea's Living SOC policy aims to supply facilities equitably among metropolitan cities and provincial cities. Using the aforementioned methodology, this study quantitatively compares metropolitan and provincial cities' conditions by conducting independent sample *t*-tests. Through *t*-tests, this study examines the difference between the two types of cities to identify the significant differences in the major variables reflecting regional attributes, such as population, average age of residents, urbanization ratio, the level of Living SOC, and the area ratio of residential, commercial, industrial, and green zones in South Korean urban planning-Table 2 details the results of the analysis.

Table 2. Comparison of the key variables by region.

Dependent Variables	Group	Numbers	Mean	Std. Deviation	t	p
Population	Metropolitan	74	300,907.72	147,607.04	4.239	0.000 ***
	Provincial	155	177,991.13	227,457.26		
Ave. Age	Metropolitan	74	41.45	2.65	−7.305	0.000 ***
	Provincial	155	45.21	5.14		
Urbanization Ratio	Metropolitan	74	95.43	18.28	19.315	0.000 ***
	Provincial	155	32.58	30.69		
Residential zone	Metropolitan	74	34.76	21.73	7.421	0.000 ***
	Provincial	155	15.57	6.86		
Commercial zone	Metropolitan	74	5.27	8.42	3.329	0.001 **
	Provincial	155	2.00	1.13		
Industrial zone	Metropolitan	74	7.12	10.78	0.316	0.753
	Provincial	155	6.68	7.48		
Green Area	Metropolitan	74	47.07	24.84	−8.509	0.000 ***
	Provincial	155	73.19	12.92		
Living SOC	Metropolitan	74	192.78	87.17	2.078	0.039 *
	Provincial	155	164.21	115.72		

* $p < 0.05$, ** $p < 0.01$, *** $p < 0.001$. t = t-value; p = p-value.

The results of the analysis show that the two groups differ significantly in terms of population, average age, urbanization ratio, and the ratio of residential, commercial, green, and Living SOC. With regard to population, metropolitan cities were found to have a larger population than provincial cities ($t = 4.239$). With regard to age, the average age of the residents in provincial cities was found to be higher than that in metropolitan cities ($t = −7.305$). With regard to urbanization ratio, metropolitan cities were found to have a higher urbanization ratio than provincial cities ($t = 19.315$). With regard to the ratio of residential zones, metropolitan cities were found to have a higher ratio of residential zones than provincial cities ($t = 7.421$). With regard to the ratio of commercial zones, metropolitan cities were found to have a higher ratio of commercial zones than provincial cities ($t = 3.329$). With regard to the ratio of green areas, provincial cities were found to have a higher ratio of green zones than metropolitan cities ($t = −8.509$). With regard to the Living SOC, metropolitan cities were found to have a higher level of Living SOC than provincial cities ($t = 2.078$). However, with regard to the ratio of industrial zones, there was no significant difference between the metropolitan and provincial cities.

The findings can be summarized as follows. Compared to metropolitan cities, provincial cities had a smaller population, a lower urbanization rate, and a lower ratio of residential and commercial zones. Meanwhile, the average age of the residents and the ratio of green zones in provincial cities were greater than those in metropolitan cities—a common difference between metropolitan and provincial cities. Moreover, with respect to Living SOC facilities, metropolitan cities were found to have a higher number of Living SOC facilities. However, the maximum capacity per facility in metropolitan cities was 1560, whereas the maximum capacity per facility in provincial cities was 1083. This shows that public-initiated Living SOC facilities are successfully supplied in provincial cities throughout the country.

4.2. Results of the Regression Analysis of the Living SOC Perspective

This study conducted a regression analysis to test the independent variables' effect on the current Living SOC of South Korea's metropolitan cities. The independent variables included the supply of education, culture and arts, sports and leisure, and local community facilities, classified based on population, average age, urban planning attributes, the urbanization ratio, and preexisting convenience facilities. Between rounds of analysis, this study employed stepwise regression analysis to derive the results. This method has the advantage of showing an increase in the explanatory power in accordance with the inclusion of each independent variable group by stage. This study examined the effect of the urban planning perspective on Living SOC as well as its implications by applying population, age, and land-use status to Model I. In contrast, in Model II, this study applied all the variables to examine the related facilities' overall effect and compared the results for metropolitan and provincial cities.

Table 3 presents the results of the regression analysis in metropolitan cities. The analysis of the results showed that the regression model was statistically significant in Stage 1 (F = 13.423, $p < 0.001$) and Stage 2 (F = 46.709, $p < 0.001$). Based on the adjusted R2, the explanatory power was 57.7% in Stage 1 and increased to 88.3% in Model II, indicating a high explanatory power. The Durbin–Watson statistic was 1.731, producing an approximate value of 2. This indicates that the residuals can be presumed to be independent. The variance inflation factor (VIF) was also found to be below 10, indicating no problems with the correlation of variables. Accordingly, the majority of current facilities in metropolitan cities belong to the category covered by the Living SOC policy.

In Model I, the population was found to have a significant effect on the dependent variable ($t = 7.348$, $p = 0.000$). This result indicates that cities with large populations are the main recipients of Living SOC. In contrast, in Model II, which includes all the variables, the size of the population was not significant, while the average age of the residents was found to influence the dependent variable. In both Models I and II, land use was found to be insignificant, indicating a need for revision of the current government policy of creating a dense assortment of Living SOC facilities around the residential zones. Of the four independent variables (education, culture and arts, sports and leisure, and local community facilities), education ($t = 7.327$, $p = 0.000$) and local community facilities ($t = 9.870$, $p = 0.000$) were found to affect the dependent variable. In other words, in metropolitan cities with dense populations, the supply of Living SOC increases in areas with a sufficient supply of education and local community facilities.

Using the same method, this study performed a stepwise regression analysis on provincial cities, the results of which are presented in Table 4. The results show that the regression model was statistically significant in both Stage 1 (F = 13.423, $p < 0.001$) and Stage 2 (F = 46.709, $p < 0.001$). Based on the adjusted R2, the explanatory power was 20.9% in Model I and increased to 72.5% in Model II, indicating a relatively high explanatory power. The Durbin–Watson statistic was 2.120, showing no problem with presuming the independence of the residuals. The VIF was also found to be below 10, indicating no issue with the correlation of variables. As such, similar to metropolitan cities, the majority of the current facilities in provincial cities also influence the Living SOC.

Table 3. Results of the regression analysis of metropolitan cities ($n = 74$).

Class.	Independent Variables	Model I					Model II				
		B	β	t	p	VIF	B	β	t	p	VIF
-	(Constant)	−2.869		−1.219	0.227		−4.397		−3.528	0.001 ***	
Demographical Variables	Population	0.000	0.755	7.348 ***	0.000	1.821	−0.000	−0.014	−0.167	0.868	4.592
	Average Age	0.016	0.042	0.312	0.756	3.156	0.105	0.279	3.789	0.000 ***	3.367
Land-Use Variables	Residential Zone	0.005	0.115	0.829	0.410	3.338	0.003	0.063	0.845	0.401	3.421
	Commercial Zone	0.005	0.044	0.435	0.665	1.751	−0.002	−0.020	−0.375	0.709	1.791
	Industrial Area	−0.002	−0.022	−0.209	0.835	1.954	−0.006	−0.068	−1.201	0.235	1.994
	Green Area	0.010	0.249	1.690	0.096	3.737	0.000	0.010	0.128	0.899	4.013
	Urban Area	0.090	0.090	0.694	0.490	2.893	0.002	0.002	0.030	0.976	3.221
	Non-urban Area	0.060	0.060	0.568	0.572	1.956	−0.094	−0.094	−1.569	0.122	2.248
Facility Variables	Education	-	-	-	-		0.491	0.491	7.327	0.000 ***	2.795
	Culture and Arts	-	-	-	-		0.034	0.034	0.762	0.449	1.209
	Sports and Leisure	-	-	-	-		0.064	0.064	1.107	0.273	2.076
	Local Community	-	-	-	-		0.676	0.676	9.870	0.000 ***	2.916
F		13.423 ($p < 0.001$)					46.709 ($p < 0.001$)				
R^2		0.623					0.902				
${}_{adj}R^2$		0.577					0.883				

*** $p < 0.001$. B = Unstandardized Coefficients; β = Standardized Coefficients; t = t-value; p = p-value.

Table 4. Results of the regression analysis of provincial cities ($n = 155$).

Class.	Independent Variables	Model I					Model II				
		B	β	t	p	VIF	B	β	t	p	VIF
-	(Constant)	−0.263		−0.196	0.845		−1.091		−1.378	0.171	
Demographical Variables	Population	0.000	0.458	4.656	0.000 ***	1.883	−0.000	−0.212	−1.551	0.123	8.502
	Average Age	−0.007	−0.037	−0.370	0.712	1.980	0.013	0.067	1.110	0.269	2.064
Land-Use Variables	Residential Zone	0.010	0.066	0.537	0.592	2.906	0.005	0.036	0.484	0.629	3.084
	Commercial Zone	−0.119	−0.135	−1.103	0.272	2.912	0.005	0.006	0.084	0.933	2.977
	Industrial Area	0.007	0.054	0.464	0.643	2.611	0.005	0.039	0.561	0.576	2.677
	Green Area	0.004	0.048	0.353	0.725	3.661	0.007	0.095	1.144	0.254	3.858
	Urban Area	0.119	0.119	1.535	0.127	1.160	−0.076	−0.076	−1.300	0.196	1.892
	Non-urban Area	0.201	0.201	2.487	0.014 *	1.273	0.053	0.053	1.093	0.276	1.323
Facility Variables	Education						0.406	0.406	2.828	0.005 **	9.539
	Culture and Arts						0.073	0.073	1.146	0.254	2.251
	Sports and Leisure						0.311	0.311	6.307	0.000 ***	1.358
	Local Community						0.613	0.613	11.067	0.000 ***	1.721
F		6.074($p < 0.001$)					34.827($p < 0.001$)				
R2		0.250					0.746				
adjR2		0.209					0.725				

* $p < 0.05$, ** $p < 0.01$, *** $p < 0.001$. B = Unstandardized Coefficients; β = Standardized Coefficients; t = t-value; p = p-value.

Testing the significance of the regression coefficients, this study found that population and the non-urbanization ratio were significant and positive (+) in Model I. Specifically, population ($t = 4.656$, $p = 0.000$) and the non-urbanization ratio ($t = 2.487$, $p = 0.000$) affected the supply of Living SOC. This indicates that, in provincial cities, Living SOC is generally established in non-urban areas rather than residential zones. In contrast, the aforementioned independent variables were not statistically significant in Model II. Of the four independent variables (education, culture and arts, sports and leisure, and local community facilities), all except for culture and art facilities were found to affect the dependent variable in Model II. In this respect, the supply of Living SOC was found to be sufficient in terms of education facilities ($t = 2.828$, $p = 0.005$), sports and leisure facilities ($t = 6.307$, $p = 0.000$), and local community facilities ($t = 11.067$, $p = 0.000$).

5. Discussion

The extant literature reflects the growing significance of the inclusive city concept—a global trend in the development of Living SOC policy, particularly with respect to improving the spatial equality and quality of life in communities. In this regard, this study aligns with extant urban planning theories, including notions of Compact City and Smart Shrinkage. Examining current urban planning in South Korea, this study found that various living infrastructures lack a connection to residential zones, irrespective of city type (i.e., metropolitan or provincial cities). In terms of facility type, this study found that the supply of facilities related to local communities and education is adequate. The supply of culture and art facilities is insufficient throughout the country, and the supply of sports and leisure facilities in provincial cities is relatively adequate. Based on these findings, this study identified the Living SOC's scope in terms of land planning to ensure the balanced development of metropolitan and provincial cities. This study also discerns the actual disparity between metropolitan and provincial cities and how this might be resolved. The results of this study can be summarized as follows.

First, establishing necessary community support facilities that are easily accessible by foot—as advanced by notions of Compact City and Living SOC—is a significant government policy in many countries, including South Korea. In Germany, the federal government operated a support program to solve urban shrinkage and established a plan (INSEK) to cope with various problems caused by urban reduction, such as vacancies in residential areas. In Japan's case, it has recognized the problems of public infrastructure operations due to aging and the occurrence of vacant homes. However, since many large residential areas had already been built in suburban areas with a population density below the stipulated number, they chose centralization as the solution. In this respect, particular emphasis is placed on the need to establish such facilities in residential zones. This study shows that the independent variables related to current land use—namely, the supply of education, culture and arts, sports and leisure, and local community facilities—did not significantly influence the dependent variable, Living SOC. Accordingly, the future policy needs to provide Living SOC centering on the residential zone(s) within a city.

Second, an analysis of South Korea's metropolitan and provincial cities reveals that population size significantly impacts Living SOC. In other words, the population was the most significant sociodemographic factor impacting the supply of preexisting education, culture, sports, and community facilities. This finding is consistent with the country's current Living SOC policy.

Third, it is crucial to increase the supply of facilities related to culture and the arts and sports and leisure facilities. Regression analysis results show that these two types of facilities did not significantly impact the dependent variable in provincial cities or metropolitan cities with dense populations. This indicates the need to ensure the sustainability of the Living SOC policy in South Korea by supplying related facilities in the future. Given the non-significant impact of such facilities in provincial cities, establishing culture and art facilities may be critical to the successful realization of the Living SOC policy in terms of providing equitable supply and access to facilities in provincial and metropolitan cities.

Fourth, it is essential to perform a business feasibility review regarding the sustainability of the pertinent facilities, particularly as South Korea's Living SOC policy is heavily funded by the central and local government and reliant on the financial soundness of local governments. For instance, despite the result of the analysis showing the lack of sports facilities in metropolitan cities, policyholders should have a lengthy deliberation when establishing large-scale sports facilities in a metropolitan city. Because most local governments of South Korea have insufficient financial independence, consideration of efficiency should come first.

Finally, the supply of Living SOC is based on the current condition of provincial cities. South Korea's land-use system consists of urban areas (16.6% or 17,614 km^2), comprising residential, commercial, industrial, and green zones; and non-urban areas (83.4% or 88,448 km^2), comprising management (areas requiring systematic management), farming, and natural environment protection zones [64]. It is worth noting that the non-urban areas of provincial cities are some five to nine times larger than those of metropolitan cities, and that such non-urban areas house many residential buildings. Therefore, future policy should direct local government to establish facilities based on accessible "distance" by considering the land use and the location and density of the residential buildings. However, achieving this requires conducting micro-level research of each city or region rather than the entire country.

6. Conclusions

Establishing and executing the Living SOC policy in terms of land and urban planning is key to improving South Korea and its cities' sustainability. Policies related to cities serve as the guidelines in establishing ordinances and the duties of the local government. Therefore, to ensure the policy's sustainability, its development and implementation must reflect the local community's attributes from the perspective of balanced land development. This study is significant in that it examines the entire territory of South Korea, but further and more localized research is necessary.

Using currently available data, this study focused on current urban planning conditions and the impact on Living SOC policy implementation—a hitherto unexamined topic.

However, this study has limitations that need to be considered. First, the analysis of facilities was hindered by the difficulty of obtaining data due to data source limitations. This study focused on quantitative aspects, such as the number of infrastructures. Thus, this study excluded some qualitative aspects regarding quality of life. Further research based on a survey, which includes qualitative aspects, is necessary. Second, this study failed to consider the disparity among cities regarding urban planning, population, and social and economic status. In the same context, this study failed to consider accessibility, which is a crucial Living SOC concept. Accordingly, future studies should examine the Living SOC of each region, considering each facilities' accessibility.

Nonetheless, it is expected that the findings on South Korean policy, including those of this study, will prove useful to the global community and provide baseline data for follow-up studies in South Korea. It is expected to be a barometer to prevent budget waste through the reckless introduction of Living SOC facilities in the future. In particular, this study used quantitative research models to ensure objectivity in assessing policies, which would help the Korean government when they supply Living SOC. Indeed, it conforms to the "benefit responsiveness" section of the Nakamura and Smallwood polities, widely used in policy evaluation. Therefore, it can be used to determine how much Living SOC has been beneficial in improving the residents' quality of life in the area.

Author Contributions: Conceptualization, Y.K. and J.O.; methodology, Y.K.; software, Y.K.; formal analysis, Y.K.; data curation, Y.K.; writing—original draft preparation, Y.K.; writing—review and editing, J.O.; visualization, J.O.; supervision, S.K.; All authors have read and agreed to the published version of the manuscript.

Funding: This research received no external funding.

Conflicts of Interest: The authors declare no conflict of interest.

References

1. UN-Habitat. *The Global Campaign on Urban Governance*; UN-Habitat: Nairobi, Kenya, 2002.

2.   Armendaris, F. *World—Inclusive Cities Approach Paper*; World Bank Group: Washington, DC, USA, 2015.

3.   Cass, N.; Shove, E.; Urry, J. Social exclusion, mobility and access1. *Sociol. Rev.* **2005**, *53*, 539–555. [CrossRef]

4.   The United Nations Conference on Housing and Sustainable Urban Development. Adopted Draft of the New Urban Agenda. In Proceedings of the Habitat III Conference, Quito, Ecuador, 17–20 October 2016.

5.   Evans, B.; O'Brien, M. Local Governance and Soft Infrastructure for Sustainability and Resilience. In *Risk Governance: The Articulation of Hazard, Politics and Ecology*; Fra.Paleo, U., Ed.; Springer: Dordrecht, The Netherlands, 2015; pp. 77–97. [CrossRef]

6.   Putnam, R.D. *Bowling Alone: The Collapse and Revival of American Community*; Simon & Schuster: New York, NY, USA, 2000.

7.   Wolf, J.; Adger, W.N.; Lorenzoni, I.; Abrahamson, V.; Raine, R. Social capital, individual responses to heat waves and climate change adaptation: An empirical study of two UK cities. *Glob. Environ. Chang.* **2010**, *20*, 44–52. [CrossRef]

8.   Frank, L.D.; Sallis, J.F.; Saelens, B.E.; Leary, L.; Cain, K.; Conway, T.L.; Hess, P.M. The development of a walkability index: Application to the Neighborhood Quality of Life Study. *Br. J. Sports Med.* **2010**, *44*, 924–933. [CrossRef]

9.   Lotfi, S.; Koohsari, M.J. Neighborhood Walkability in a City within a Developing Country. *J. Urban Plan. Dev.* **2011**, *137*, 402–408. [CrossRef]

10.   Kim, Y.-G. Development of Index of Park Derivation to Promote Inclusive Living SOC Policy. *J. Korean Inst. Landsc. Archit.* **2019**, *47*, 28–40.

11.   Jo, H.-E.; Nam, J.-H. A Study on the Scope of Life SOC and Characteristics of Facilities by Type and Region—Focusing on the Facilities Status of Life SOC in Gyeonggi-do and its Improvement Plan. *J. Urban Des. Inst. Korea Urban Des.* **2019**, *20*, 33–52. [CrossRef]

12.   Koo, H.; Lee, D.; Park, J. *A Strategy for the Provision of Social Infrastructure at the Regional Level*; Korea Research Institute for Human Settlements: Seoul, Korea, 2019; Volume 19–23.

13.   Katz, B.; Lang, R.E. *Redefining Urban and Suburban America: Evidence from Census 2000*; Brookings Institution Press: Washington, DC, USA, 2004.

14.   Office for Government Policy Coordination. *Living SOC Three-Year Plan*; Office for Government Policy Coordination: Seoul, Korea, 2019.

15.   Ministry of Economy and Finance. *The Government Budget for 2021*; Ministry of Economy and Finance: Seoul, Korea, 2020.

16.   KIm, Y. *A Study on the Mixed-Use Living SOC for the Role of Urban Regeneration Centers: An Analysis of Accessibility by Applying National Minimum Standards by Facilities*; Korea Research Institute for Human Settlements: Sejong, Korea, 2020.

17.   Koo, H.; Kim, T.; Lee, S. *Urban Shrinkage in Korea: Current Status and Policy Implications*; Korea Research Institute for Human Settlements: Sejong, Korea, 2016.

18.   Haase, A.; Rink, D.; Grossmann, K.; Bernt, M.; Mykhnenko, V. Conceptualizing Urban Shrinkage. *Environ. Plan. A Econ. Space* **2014**, *46*, 1519–1534. [CrossRef]

19.   Hollstein, L.M. Planning Decision for Vacant Lots in the Context of Shrinking Cities: A Survey and Comparison of Practices in the United States. Ph.D. Thesis, The University of Texas at Austin, Austin, TX, USA, 2014.

20.   Luescher, A.; Shetty, S. *An Introductory Review to the Special Issue: Shrinking Cities and Towns: Challenge and Responses*; Springer: Berlin/Heidelberg, Germany, 2013.

21.   Pallagst, K. Shrinking cities: Planning challenges from an international perspective. *Cities Grow. Smaller* **2008**, *10*, 5–16.

22.   Lim, E.; Lee, Y.; Hwang, M.; Oh, C. *Direction and Challenges of National Land Policy in the Big Data Era*; Korea Research Institute For Human Settlements: Sejong, Korea, 2018.

23.   Jo, P.; Min, B.; Son, K.; Park, S.; Kim, S.; Kim, J.; Kim, D.; Suh, M.; Lee, S.; Park, G.; et al. *Places for the People: How Social Infrastructure Can Help Fight Inequality, Polarization, and the Decline of Civic Life*; Korea Research Institute for Human Settlements: Sejong, Korea, 2013; Volume 21.

24.   Seong, E.; Lim, Y.; Lim, H. *Accessibility and Availability od Neighborhood Facilities in Ild Residential Area*; Architecture & Urban Research Institute: Sejong, Korea, 2013.

25.   Lee, S.; Park, J. *A Study on Supply of Public Facilities Considering Regional Demand*; Korea Research Institute for Local Administration: Sejong, Korea, 2014.

26. Aldrich, D.P. Social, not physical, infrastructure: The critical role of civil society after the 1923 Tokyo earthquake. *Disasters* **2012**, *36*, 398–419. [CrossRef]

27. Australia, I. *Australian Infrastructure Audit 2019*; Infrastructure Australia: Sydney, Australia, 2019.

28. Latham, A.; Layton, J. Social infrastructure and the public life of cities: Studying urban sociality and public spaces. *Geogr. Compass* **2019**, *13*, e12444. [CrossRef]

29. Kumari, A.; Sharma, A.K. Physical & social infrastructure in India & its relationship with economic development. *World Dev. Perspect.* **2017**, *5*, 30–33. [CrossRef]

30. Dash, R.K.; Sahoo, P. Economic growth in India: The role of physical and social infrastructure. *J. Econ. Policy Reform* **2010**, *13*, 373–385. [CrossRef]

31. Klinenberg, E. *Palaces for the People: How Social Infrastructure Can Help Fight Inequality, Polarization, and the Decline of Civic Life*; Crown: New York, NY, USA, 2018.

32. Ogawa, T., III. Social Overhead Capital. *Jpn. Econ. Stud.* **1976**, *5*, 32–42. [CrossRef]

33. Reimers, K.; Johnston, R.B. Living Infrastructure. In *Collaboration in the Digital Age: How Technology Enables Individuals, Teams and Businesses*; Riemer, K., Schellhammer, S., Meinert, M., Eds.; Springer International Publishing: Cham, Switzerland, 2019; pp. 249–267. [CrossRef]

34. Alexandra, J.; Norman, B.; Steffen, W.; Maher, W. *Planning and Implementing Living Infrastructure in the Australian Capital Territory—Final Report*; Canberra Urban and Regional Futures, University of Canberra: Canberra, Australia, 2017.

35. Mehmood, R.; Katib, S.S.I.; Chlamtac, I. *Smart Infrastructure and Applications*; Springer: Cham, Switzerland, 2020.

36. Ministry of Economy and Finance. *Act on Public-Private Partnerships in Infrastructure*; Ministry of Economy and Finance: Seoul, Korea, 2018.

37. Panagopoulos, T.; Barreira, A.P. Determinants and shrink smart strategies for the municipalities of Portugal. In Proceedings of the Artigo Apresentado na Conferência Shrinkage in Europe: Causes, Effects and Policy Strategies, Amsterdam, The Netherlands, 16–17 February 2011; pp. 16–17.

38. Oswalt, P. *Shrinking Cities, Volume 2: Interventions*; Hatje Cantz Verlag: Ostfildern, Germany, 2006.

39. Wiechmann, T. Errors expected—Aligning urban strategy with demographic uncertainty in shrinking cities. *Int. Plan. Stud.* **2008**, *13*, 431–446. [CrossRef]

40. Hollander, J.B.; Németh, J. The bounds of smart decline: A foundational theory for planning shrinking cities. *Hous. Policy Debate* **2011**, *21*, 349–367. [CrossRef]

41. Popper, D.E.; Popper, F.J. Small can be beautiful. *Planning* **2002**, *68*, 20.

42. Ministry of Land, Infrastructure, Transport and Tourism. *Guidelines for Effective Utilization of Public Real Estate (PRE) for Division Town Development*; Ministry of Land, Infrastructure, Transport and Tourism: Tokyo, Japan, 2014.

43. Ministry of Land, Infrastructure, Transport and Tourism. *Grand Design 2050*; Ministry of Land, Infrastructure, Transport and Tourism: Tokyo, Japan, 2014.

44. Yokobari, M. Regarding the revised Act on Special Measures Concerning Urban Revitalization and the Plan for Proper Location (Special Feature: Consider future land issues). *Land Compr. Res.* **2015**, *23*, 33–36.

45. Ministry of Land, Infrastructure, Transport and Tourism. Status of Efforts to Prepare a Site Rationalization Plan. Available online: https://www.mlit.go.jp/en/toshi/city_plan/compactcity_network.html (accessed on 17 November 2020).

46. The Federal Office for Building and Regional Planning. *Regionalstrategie Daseinsvorsorge–Denkanstöße für die Praxis*; The Federal Office for Building and Regional Planning: Berlin, Germany, 2011.

47. Wiechmann, T.; Pallagst, K.M. Urban shrinkage in Germany and the USA: A comparison of transformation patterns and local strategies. *Int. J. Urban Reg. Res.* **2012**, *36*, 261–280. [CrossRef] [PubMed]

48. Pallagst, K.; Fleschurz, R.; Said, S. What drives planning in a shrinking city? Tales from two German and two American cases. *Town Plan. Rev.* **2017**, *88*, 15–29. [CrossRef]

49. Haase, D.; Haase, A.; Kabisch, N.; Kabisch, S.; Rink, D. Actors and factors in land-use simulation: The challenge of urban shrinkage. *Environ. Model. Softw.* **2012**, *35*, 92–103. [CrossRef]

50. Peters, D.J.; Hamideh, S.; Zarecor, K.E.; Ghandour, M. Using entrepreneurial social infrastructure to understand smart shrinkage in small towns. *J. Rural Stud.* **2018**, *64*, 39–49. [CrossRef]

51. Chang, H.-S.; Liao, C.-H. Exploring an integrated method for measuring the relative spatial equity in public facilities in the context of urban parks. *Cities* **2011**, *28*, 361–371. [CrossRef]

52. Hodge, D.; Gatrell, A. Spatial Constraint and the Location of Urban Public Facilities. *Environ. Plan. A Econ. Space* **1976**, *8*, 215–230. [CrossRef]

53. Liu, Y.; Luo, T.; Liu, Z.; Kong, X.; Li, J.; Tan, R. A comparative analysis of urban and rural construction land use change and driving forces: Implications for urban–rural coordination development in Wuhan, Central China. *Habitat Int.* **2015**, *47*, 113–125. [CrossRef]

54. Shizhong, C. Empirical Research on a Balanced Development of Urbanization, Industrialization and Agricultural Modernization in the Building of Central Economic Region. *Manag. Agric. Sci. Technol.* **2012**, *3*, 14–16.

55. Wei, B.; Li, S. Theoretical Thinking on Balanced Development of Primary and Secondary School Sports in Our Country's Urban and Rural Area. *J. Sports Sci.* **2010**, *3*, 99–103.

56. Gu, Q. Integrating soft and hard infrastructures for inclusive development. *J. Infrastruct. Policy Dev.* **2017**, *1*, 1–3. [CrossRef]

57. Thapa, B. Soft-infrastructure in tourism development in developing countries. *Ann. Tour. Res.* **2012**, *39*, 1705–1710. [CrossRef]

58. Langford, M.; Higgs, G.; Radcliffe, J.; White, S. Urban population distribution models and service accessibility estimation. *Comput. Environ. Urban Syst.* **2008**, *32*, 66–80. [CrossRef]

59. Naseem, K.; Khurshid, S.; Khan, S.F.; Moeen, A.; Farooq, M.U.; Sheikh, S.; Bajwa, S.; Tariq, N.; Yawar, A. Health related quality of life in pregnant women: A comparison between urban and rural populations. *JPMA J. Pak. Med. Assoc.* **2011**, *61*, 308–312.

60. Pajonk, F.-G.; Schmitt, P.; Biedler, A.; Richter, J.C.; Meyer, W.; Luiz, T.; Madler, C. Psychiatric emergencies in prehospital emergency medical systems: A prospective comparison of two urban settings. *Gen. Hosp. Psychiatry* **2008**, *30*, 360–366. [CrossRef] [PubMed]

61. Nilsson, P. Natural amenities in urban space–A geographically weighted regression approach. *Landsc. Urban Plan.* **2014**, *121*, 45–54. [CrossRef]

62. Yuan, F.; Wei, Y.D.; Wu, J. Amenity effects of urban facilities on housing prices in China: Accessibility, scarcity, and urban spaces. *Cities* **2020**, *96*, 102433. [CrossRef]

63. Buys, L.; Miller, E. Residential satisfaction in inner urban higher-density Brisbane, Australia: Role of dwelling design, neighbourhood and neighbours. *J. Environ. Plan. Manag.* **2012**, *55*, 319–338. [CrossRef]

64. Ministry of Land, Infrastructure and Transport. *A Study on the Improvement of National Land and Urban Planning Management System in Response to the Change of Conditions*; Ministry of Land, Infrastructure and Transport: Seoul, Korea, 2017.

Publisher's Note: MDPI stays neutral with regard to jurisdictional claims in published maps and institutional affiliations.

Article

Energy and CO$_2$ Reduction of Aluminum Powder Molds for Producing Free-Form Concrete Panels

Donghoon Lee [1] and Sunkuk Kim [2,*]

[1] Department of Architectural Engineering, Hanbat National University, Daejeon 34158, South Chungcheong, Korea; donghoon@hanbat.ac.kr

[2] Department of Architectural Engineering, Kyung Hee University, Yongin-si 17104, Gyeonggi-do, Korea

* Correspondence: kimskuk@khu.ac.kr; Tel.: +82-31-201-3685

Received: 27 October 2020; Accepted: 14 November 2020; Published: 18 November 2020

Abstract: Free-form design may enhance the architectural value of buildings in terms of aesthetic and symbolic effects. However, it is difficult to reuse the mold of free-form concrete segments, so they are manufactured for single use. Manufacturing these molds is a time-consuming process that requires a lot of manpower. To solve these problems, there have been numerous studies on the use of phase change materials (PCMs) to make the molds. PCM molds represent a new technique of producing free-form panels using a computerized numeric control (CNC) machine that employs low-cost material to produce free-form concrete panels. However, PCM molds require a substantial amount of time and energy during fabrication because repeated heating and cooling cycles are required during panel production, and this process increases the CO$_2$ emissions. Thus, the purposes of this study were to develop composite molds using aluminum powder to improve PCM mold performance and to conduct experiments to quantify the reduction of energy use and CO$_2$ emissions. As a result of cooling experiments, it was found that the aluminum powder mold had an energy reduction effect of 14.3% against the PCM mold that had been produced only with paraffin wax, and CO$_2$ reduction effect of more than 50% against the conventional mold.

Keywords: free-form building; free-form concrete panel; aluminum powder; composite PCM mold; CO$_2$ emission reduction

1. Introduction

Free-form design is being increasingly adopted in monumental buildings to improve aesthetic and symbolic effects. However, the molds used for the production of free-form panels are 3–10 times more expensive than conventional molds [1]. This is because free-form concrete segments are not produced in fixed shapes, which make it difficult to produce a mold using conventional materials like metal, wood, and synthetic resins; it is also impossible to reuse the produced mold [2]. Presently, molds for the production of free-form concrete segments (FCS) are only used one time. Due to this, the construction of free-form buildings requires longer construction times, is significantly more expensive, and also produces more CO$_2$ emissions compared to conventional construction. However, there have been no studies on the development of practical production technologies for free-form concrete segments to realize the environmentally-friendly construction of free-form buildings [3].

Since free-forms are composed of various irregular curved surfaces, they require more construction time and resources than ordinary molds. There has been a wide range of technologies applied to the production of free-form concrete segments [4–6], but the production of free-form concrete segments is not fit for sale [2,6].

Recently developed phase change material (PCM) molds can be semi-permanently reused, and it is easy and quick to produce free-form concrete segments (FCSs) with this new technique [7]. Here, PCM refers to a material that changes its phase from liquid to solid and vice-versa, depending on the temperature. PCM molds are used in the solid-state for FCS production, and it changes the mold back to a liquid state for reuse. However, heating and cooling must be repeated every time concrete segments are produced, so there is great concern about the environmental impact, including energy consumption and CO_2 emissions arising from energy consumption. It is absolutely necessary to evaluate energy consumption as well as CO_2 emissions when developing new molding materials [8]. Thus, the purposes of this study were to develop aluminum powder molds to improve PCM mold performance and to verify the reduction in energy consumption and CO_2 emissions achieved using these molds. This study is limited to frequently used conventional PCM molds, plywood (wooden) molds, and the newly developed aluminum powder molds. We also analyzed the effect of these techniques on energy consumption and CO_2 emissions. This study was conducted in five steps, as described below:

(1) Consideration of previous studies
(2) Development of aluminum powder molds
(3) Analysis of the characteristics of aluminum powder molds
(4) Analysis of energy consumption
(5) Analysis of CO_2 emissions

2. Consideration of Previous Studies

Steel, wood, expandable polystyrene (EPS), and plastic were used to produce free-form segments in many of the free-form buildings that have recently been completed [6,9–12]. Latorre [13] developed a "pneumatic system" to produce free-form domes, yet each material had to be individually produced, and it was difficult to produce shapes other than domes. Verhaegh [6] developed a "fabric formwork" using fabric forms, however many of the molds were used for shape constraints, and a great deal of manufacturing time was needed. This method failed to improve construction duration and cost when compared to conventional methods. Mandl et al. [14] and Lindsey and Gehry [15] conducted studies on the formwork made with EPS using computerized numeric control (CNC) technology and Toyo Ito and Associates [16] developed a wooden system form. Franken and ABB [17] used digital forms with CNC and acrylic glass to produce free-form concrete. However, these molds cannot be reused, generate a lot of waste, and require an extremely long manufacturing time. Researchers at the University of Southern California hoped to come up with a building process using a robot automation system, but additional studies are required due to limitations in the production of large segments. To address the problems with free-form molds, CRAFT (Center for Rapid Automated Fabrication Technologies) [18] are conducting a study on the 3D printing of buildings. However, the concrete deposited using a 3D printer nozzle needs time for curing, and it is not yet ready for commercialization because the size of the 3D printer is limited.

The need for developing economical, variable mold, and free-form concrete production technologies is on the rise and has focused on the reuse of molds and the reduction of production time [3]. Recognizing this need, Oesterle et al. [19] developed a reusable wax mold. PCM changes from solid to liquid and vice-versa depending on the type of external stimulation (temperature, electric current, etc.). Therefore, PCM molds can be used to make shapes and can be reused to reduce waste generation. Paraffin, one of the PCMs used in this study, is an alkane hydrocarbon with a chemical formula expressed as C_nH_{2n+2} (n $\geq$ 19). It does not dissolve in water, but it does dissolve in ether, benzene, or ester. Paraffin is a mixture of hydrocarbon molecules comprised of 20 or 40 carbon molecules. It is a soft, white solid without color that is derived from petroleum, coal, or oil shale. The melting point of paraffin ranges from 47 °C to 64 °C, and the density is 0.9 g/cm^3 [20–22].

However, Oesterle et al. failed to analyze and discover solutions to address the increased energy consumption, increased CO_2 emission arising from the energy consumption, long freezing times, crystallization effects, strength issues, solidification shrinkage, and cracking. The study focused only on the realization of wax mold shapes and did not comprehensively explain the device or technology. In particular, the study had some limits in that it lacked specific verification of the shrinkage in wax molds, freezing time, and energy consumption.

Lee et al. [4] developed PCM molds using a CNC machine and suggested a production process of FCP (Free-form Concrete Panel). The production process was composed of a total of nine stages, as illustrated in Figure 1. In the stage wherein PCM molds were produced, liquid PCM was first filled into a CNC machine to produce a shape. After shaping, the hot liquid PCM was cooled to solidify the PCM. The solid PCM was then separated from the CNC machine and was attached to the frame on the side to produce a PCM mold. Production of the FCP involved pouring concrete into the PCM mold, followed by curing to produce FCPs [7]. However, the study failed to suggest solutions to various problems, including the extensive cooling energy of the PCM mold and the effects of crystallization and cooling time.

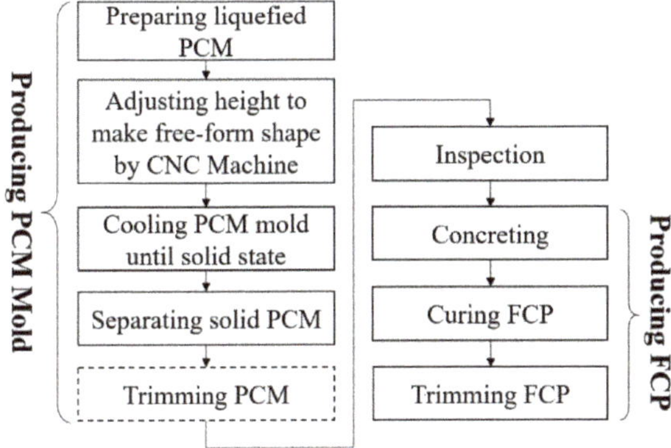

Figure 1. FCP (Free-form Concrete Panel) production process with phase change material (PCM) mold (Lee et al., 2015).

3. Aluminum Powder Mold

As examined in Section 2, there have been great efforts to develop free-form concrete panel molds. In addition, there have been some studies on the development of PCM molds that can be used to make free shapes. However, there are no case studies on the production of concrete panels for free-form buildings using novel PCM molds. This is because newly developed molds may be less practical or have some problems. A wide range of shapes can be produced with PCM molds, yet two problems still exist. First, PCM cooling takes a long time and requires a lot of energy. Second, the low strength of the PCM may result in crushing or sinking of the mold when concrete is poured into it. A light aluminum with outstanding thermal conduction was chosen as a material to improve the performance of the PCM molds in this study. The aluminum powder mold (c) shown in Figure 2 is a free-form mold mixed with aluminum powder (a) and paraffin wax (b); this mixture is produced using a CNC machine.

Aluminum, the main ingredient of the mold, is the most common chemical element found in the earth after oxygen and silicon. The heat of fusion of aluminum is 94.6 cal/g, and it has a melting point of 660 °C. As listed in Table 1, aluminum has a specific gravity of 2.71, and it is fairly light, only 1/3 of the weight of copper or iron. In addition, its thermal conductivity is 204 W/m·h·K, which is more than three times higher than that of iron (67 W/m·h·K) [23,24]. In this study, we developed free-form molds by mixing the aluminum powder with paraffin wax. The aluminum powder mold could be used to produce any shape in the heated, liquid state using a CNC machine. Then, it was cooled and used as a robust mold in the solid-state. The thermal conductivity of the aluminum powder mold was high, which solved previous problems related to the high consumption of cooling energy and long cooling times in other PCM molds. When the aluminum/paraffin wax PCM mold was cooled, the surface was first cooled and then frozen, letting down the energy efficiency. However, the thermal conductivity of the aluminum powder mold could solve the problem.

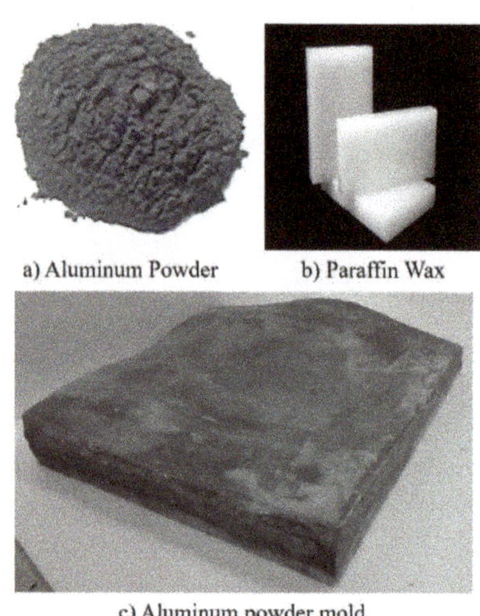

a) Aluminum Powder　　　　b) Paraffin Wax

c) Aluminum powder mold

Figure 2. Ingredients of the aluminum powder mold.

Table 1. Specific heat and thermal conductivity of main metal and paraffin wax [25].

Ingredient	Specific Heat (cal/g·°C)	Thermal Conductivity (W/m·h·°C)	Specific Gravity
Al	0.215	204	2.71
Fe	0.107	67	7.86
Cu	0.0924	386	8.93
Paraffin wax	0.7	0.25	0.9
Water	1	0.016	1

Powder molds have been used for casting metal molds for a long time. They are reusable and are able to produce a wide range of shapes. However, the shape can be easily crushed or deformed. When paraffin wax is added to these powder molds, there is no change to the powder, which is densely packed after freezing. The molds are movable, and they are highly resistant to compression. As shown in Table 2 these types of molds can be reused semi-permanently, yet melting energy and cooling energy must be added during each reuse. Aluminum powder with high thermal conductivity is a suitable material for solving this problem.

Table 2. Characteristics of the mold materials.

Materials	Characteristics
Paraffin wax	- Owing to the crystallization effect, energy consumption, and cooling time increased upon cooling. - Deformation occurs due to compression caused by pouring concrete. - Any shape can be produced, and it can be moved after freezing.
Powder (sand, minerals, etc.)	- There is almost no energy consumption since it requires no melting or cooling. - The mold material can be reused semi-permanently. - Shapes are easily crushed or deformed. - The mold is not movable.
Powder + Paraffin wax	- The mold material can be reused semi-permanently. - Any shape can be produced, and it can be moved after freezing. - There is no deformation due to compression after addition of concrete. - Energy is needed for melting and cooling.

4. Analysis of Energy Consumption and CO_2 Emission

4.1. Analysis of Energy Consumption

Since aluminum powder molds and PCM molds are plate-type molds, they release heat mostly through the upper and lower parts. We used the experimental apparatus shown in Figure 3 to study the heat release of molds for the analysis of energy consumption. A movable heating device was used for heating and cooling of the materials, and a cooler along with other auxiliary devices, were used to maintain a constant cooling temperature. Furthermore, the surface upper and lower parts were exposed so that the heat was released from those parts, and the sides were insulated. The volume of the mixture used in the experiments was 600 mL, and the mixture was sufficiently heated to 90 °C or above and then air-cooled to 20 °C. Aluminum powder (3 μm) was mixed with commercial-grade paraffin wax. After melting the paraffin wax and mixing it with the aluminum powder, we found that it became saturated at a volume ratio of approximately 30:17. At this ratio, the ingredients were mixed so that the paraffin wax filled in the small spaces between the aluminum particles.

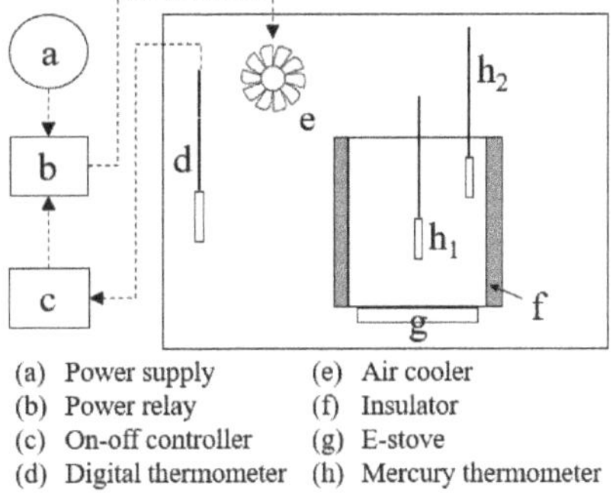

Figure 3. Schematic diagram of the experimental apparatus.

There was a significant difference in temperature profile during the cooling of paraffin wax molds and aluminum powder molds, as shown in Figure 4. It was instructive to compare the cooling times from a heated state of 60 °C to the temperature where the paraffin wax congealed (40 °C). This cooling step took 50 min for the Al powder-based material and took 255 min for the paraffin wax, a more than five-fold difference. This result implied that a CNC machine that produces shapes of melted molds could make five times more molds in the same amount of time.

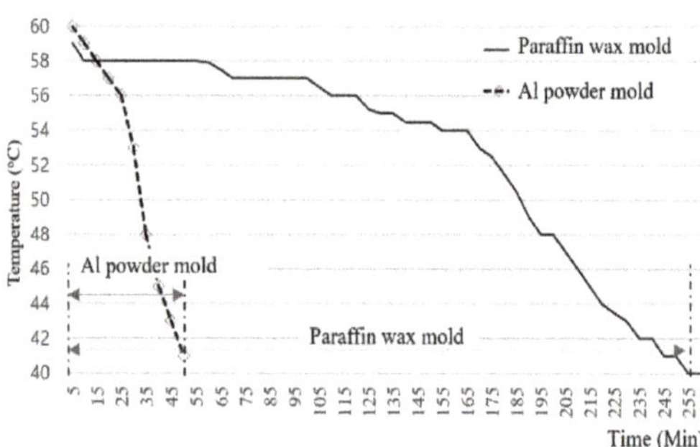

Figure 4. Temperature variation of the molds.

The temperature variation in the liquid state was determined by the removal of sensible heat. When the upper/lower cooling surfaces were exposed to air that was below the phase change temperature, the liquid started to solidify. At this point, natural convection began to be restricted. There was no difference in the output of sensible heat through the thin solid surface layers between the two mixtures in the early stages of cooling. However, as shown in Table 3 and Figure 5, as the solid layers thicken, the transfer of sensible heat inside the PCM molds dropped significantly as the liquid near the cooling surface solidified. The PCM mold must be completely frozen to be used as a mold, so long cooling times are required to freeze all of the paraffin in a solid-state.

Table 3. Temperature variation and heat output of the PCM mold.

Thickness of Solid Layer (mm)	10	15	20	25	30	35	40
Temperature of the Surface (°C)	39	38	37.5	36	35	32.5	31
Temperature of the Inner Part (°C)	55	54.5	54	53	50.5	45	42
Cooling Temperature (°C)	20	20	20	20	20	20	20
Heat Output per Min. (kcal)	9.92	6.52	4.82	3.74	2.88	2.02	1.56

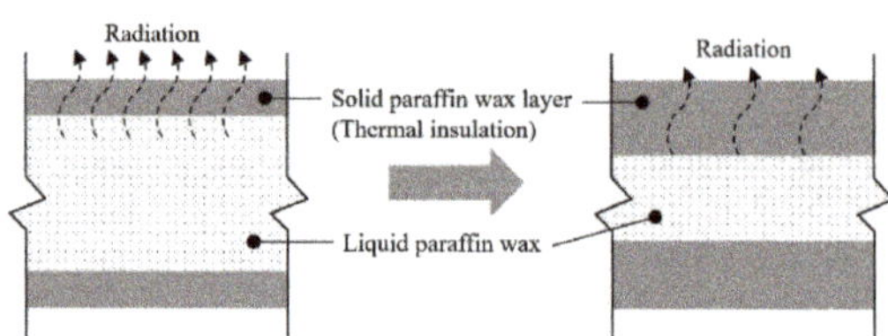

Figure 5. Reduced heat output owing to the solid layer.

This phenomenon did not occur in the case of the aluminum powder molds because the inner thermal energy was quickly transferred. Figure 6 shows the temperature variation inside the surface of the molds. The freezing temperature of both materials was about 40 °C. The difference between the time required for the inner surface to freeze was about 10 min for the aluminum powder mold and more than 110 min for the paraffin wax mold. As shown in Figure 6a,b, the temperature difference between the surface and inner part of the PCM mold after 160 min of cooling was 15 °C or above. After 260 min, the temperature difference in the PCM mold was 10 °C or above. In contrast, the aluminum powder mold showed a temperature difference between the surface and inner part of around 10 °C in the earlier stage, as shown in Figure 6c, but the temperature difference drastically declined after 40 min of cooling, as shown in Figure 6d. This was because the inner heat was immediately transferred to the air since the heat transfer efficiency of the material itself was high.

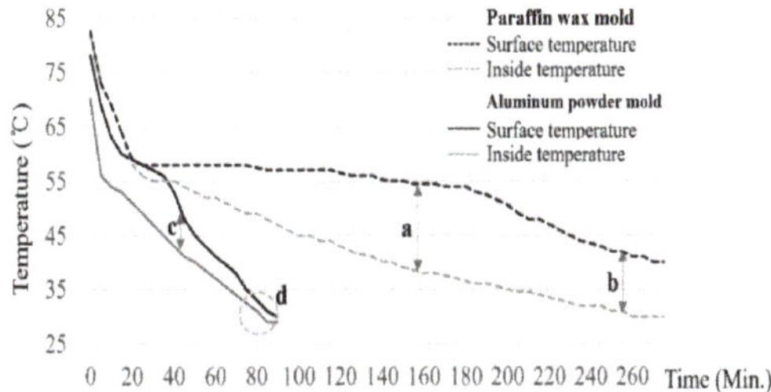

Figure 6. Temperature variation on the inside and on the surface of the molds.

In order to freeze and cool 1 L of hot paraffin wax from 80 °C to 40 °C, 56.7 kcal must be removed from the system, as shown in Table 4. This estimation was reached by adding the 25.2 kcal required to cool the wax and lower the temperature and the 31.5 kcal required for the phase change (freezing). A total of 49.125 kcal was needed to cool 1 L of the aluminum powder mold. The specific heat of this material was 0.215 kcal/kg, which was significantly lower than that of the paraffin wax (0.7 kcal/kg), and there was no phase change between 80 °C and 40 °C, which implied that there was no need to account for latent heat. Although the specific gravity of aluminum powder mold was bigger, it had a 14.3% lower required cooling energy.

Table 4. Heat consumption of molds (based on 1 L).

Type	Weight (kg)	Cooling Heat (kcal)	Remarks
Paraffin wax mold	0.9	56.7	
- Paraffin wax	0.9	25.2 31.5 (latent heat of fusion)	0.9 kg · 0.7 kcal/kg · 40 °C 0.9 kg · 35 kcal/kg
Aluminum powder mold	1.316	49.125	
- Aluminum	0.621	5.34	0.621 kg · 0.215 kcal/kg · 40 °C
- Paraffin wax	0.695	19.46 24.325 (latent heat of fusion)	0.695 kg · 0.7 kcal/kg · 40 °C 0.695 kg · 35 kcal/kg

The PCM molds required around 150 L of paraffin wax to produce 1 m^2 of concrete panels; 8505 kcal of heat was required to melt this amount of paraffin at 80 °C. As shown in Table 5 one kWh of electric power could generate 2150 kcal of heat, so 3.96 kWh was needed to heat 1 m^2 of the PCM mold. Assuming that the efficiency of the cooler was 67% and its heat loss is 50%, this process required 11.88 kWh of electric power to cool 150 L of the PCM mold at 40 °C. Using the same method, the energies required for heating and cooling the aluminum powder mold (1 m^2) were 3.43 kWh and 13.71 kWh, respectively. We, therefore, concluded that there was an energy reduction effect of around 2.1 kWh as compared with the PCM mold.

Table 5. Power consumption of PCM mold and aluminum powder mold.

	PCM Mold	**Aluminum Powder Mold**
Heating	3.96 kWh (56.7 kcal × 150 L/2150 kcal/kw)	3.43 kWh (49.13 kcal × 150 L/2150 kcal/kw)
Cooling	11.88 kWh (3.96 kWh/0.67/0.5)	10.28 kWh (3.43 kWh/0.67/0.5)
Total	15.84 kWh	13.71 kWh

4.2. Analysis of CO_2 Emissions

The molds that are used in conventional free-form building projects have different curvatures, sizes, and shapes, and it is therefore not possible to reuse them. In addition, they are frequently twisted and damaged during the separation of the mold. The increased use of temporary materials leads to increased CO_2 emissions and costs. To compare the energy consumption of conventional methods where numerous temporary wooden boards were used with the use of PCM and aluminum powder molds, we evaluated CO_2 emissions from the manufacture of the materials to the completion of the molds. As shown in Table 6, we found that there was a great difference in CO_2 emissions between the aluminum powder mold that can be reused semi-permanently and with conventional plywood.

Table 6. Comparison of total CO_2 emissions.

Type	Wooden Mold	PCM Mold	Aluminum Powder Mold
Material use	Plywood (12 mm, 7 ply) 105 m^2	Paraffin wax 135 kg	Aluminum powder 93.15 kg Paraffin wax 104.25 kg
	2226 kg-CO_2 (105 m^2 × 21.2 kg-CO_2/m^2)	398 kg-CO_2 (135 kg × 2.95 kg-CO_2/kg)	484.53 kg-CO_2 (93.15 kg × 1.9 kg-CO_2/kg + 104.25 kg × 2.95 kg-CO_2/kg)
Electricity use	-	1584 kWh (Heating: 396 kWh, Cooling: 1188 kWh)	1371 kWh (Heating: 343 kWh, Cooling: 1028 kWh)
	-	671.6 kg-CO_2 (1584 kWh × 0.424 kg-CO_2/kWh)	581.3 kg-CO_2 (1371 kWh × 0.424 kg-CO_2/kWh)
Total CO_2 emissions	2226 kg-CO_2	1069.6 kg-CO_2	1065.83 kg-CO_2

The energy consumption and CO_2 emissions required to produce each material can be estimated using interindustry analysis. Interindustry analysis uses an interindustry transaction matrix to construct an input coefficient matrix that can be used to estimate the total energy input for manufacturing the materials. This study applied this method to calculate all CO_2 generated associated with panel production. Lee et al. [26] used interindustry analysis to analyze the energy needed for plywood manufacturing, and they found that 3.37 kg-CO_2/kg of plywood was emitted. Considering the unit weight of plywood, 21.2 kg-CO_2 is emitted per m^2. Here, a small amount of electric power was used for assembling the mold, but this was not included for calculation since it was extremely small.

Since PCM molds can be reused semi-permanently, CO_2 emissions were estimated based on 150 liters of paraffin wax (the amount needed to produce 1 m^2 of concrete panels). In addition, heating to 80 °C and cooling to 40 °C were repeated with 150 L of paraffin wax whenever 1 m^2 of concrete panels were produced. The total electric energy input during this process was analyzed. In the case of PCM molds, 1584 kWh of electric power in total was used. In the case of the aluminum powder molds, 1371 kWh of electric power was used. A coefficient of 0.424 kg-CO_2/kWh [27,28] was used to estimate the total CO_2 emissions, which were 1069.6 kg-CO_2 and 1065.83 kg-CO_2, respectively, for PCM molds and aluminum powder molds. Therefore, compared with CO_2 emission of conventional wooden molds (2226 kg-CO_2), PCM and aluminum powder molds could be used to reduce the emission to less than 50%. Taking into consideration the waste generated after removal of a single-use wooden mold, there should be a greater reduction of CO_2 emissions from aluminum powder molds that do not generate any waste.

5. Conclusions

In this study, we developed an aluminum powder mold to improve energy consumption and reduce the time required for heating and cooling PCM molds. Our mold utilized the effect of phase change, so it was a type of PCM mold. Since heating and cooling should be repeated for the production of concrete panels using PCM molds, there was a negative impact on the environment, including increased energy consumption and the corresponding higher CO_2 emissions.

Our aluminum powder molds had a high thermal conductivity, which allowed them to reduce the required cooling energy and reduced the cooling time. The surface of the aluminum powder mold was first cooled and then congealed to prevent heat blockage, which reduced the consumption of cooling energy. A mold made only of paraffin wax required 1584 kWh of electric power in total, but the aluminum mold required 1371 kWh of electric power, reducing power consumption by 14.3%. In addition, CO_2 emissions that result from using the PCM mold and the aluminum powder mold were 1069.6 kg-CO_2 and 1065.83 kg-CO_2, respectively. These represent a reduction of 50% compared to the use of conventional wooden molds (2226 kg-CO_2). Considering the waste generated after the removal of single-use wooden molds, we believe that there was an even greater reduction in CO_2 emissions from aluminum powder molds that do not generate any waste.

Aluminum powder molds could be used to easily produce any shapes, just like conventional PCM molds. The aluminum powder molds could also reduce energy consumption and CO_2 emissions relative to conventional molds. Our aluminum powder-based molds could be used to improve the economic feasibility of the construction of free-form concrete buildings and reduce energy consumption and CO_2 emissions, which would be helpful in realizing various building designs and creating value. Additional studies related to composite molds mixed with a wide range of materials are needed.

Author Contributions: Conceptualization, D.L. and S.K.; writing—original draft preparation, D.L.; writing—review and editing, S.K.; All authors have read and agreed to the published version of the manuscript.

Funding: This work is supported by the Korea Agency for Infrastructure Technology Advancement (KAIA) grant, funded by the Ministry of Land, Infrastructure and Transport (Grant 20CTAP-C151959-02).

Acknowledgments: This work is supported by the Korea Agency for Infrastructure Technology Advancement (KAIA) grant, funded by the Ministry of Land, Infrastructure and Transport (Grant 20CTAP-C151959-02).

Conflicts of Interest: The authors declare no conflict of interest.

References

1. Kim, K.; Son, K.; Kim, E.D.; Kim, S. Current trends and future directions of free-form building technology. *Archit. Sci. Rev.* **2015**, *58*, 230–243. [CrossRef]
2. Lee, D.; Jang, D.B.; Kim, S. *Production Technology of Free-Form Concrete Segments Using Phase Change Material*; CSE: Kuala Lumpur, Malaysia, 2014.
3. Lee, D. A Study of Construction and Management Technology of Free-Form Buildings. Ph. D. Thesis, Kyung Hee University, Seoul, Korea, 2015.
4. Lee, D.; Hong, W.K.; Kim, J.T.; Kim, S. Conceptual Study of Production Technology of Free-Form Concrete Segments. *Int. J. Eng. Technol.* **2015**, *7*, 321. [CrossRef]
5. Payne, J. The Sydney Opera House, Inside and Out. 2003, pp. 247–277. Available online: https://ses.library.usyd.edu.au/bitstream/handle/2123/1415/08chapter7.pdf;jsessionid=A02ECB8BE3B55C6E3CD3746D1B9B9738?sequence=8 (accessed on 10 October 2020).
6. Verhaegh, R.W.A. Free Forms in Concrete Fabric. Master's Thesis, Eindhoven University of Technology, Eindhoven, The Netherlands, 2010.
7. Lee, D.; Kim, S. Development of PCM-enabled atypical concrete segment production process. *J. Archit. Inst. Korea* **2015**, *17*, 219–224.

8.   Lee, D.; Lim, C.; Kim, S. CO_2 emission reduction effects of an innovative composite precast concrete structure applied to heavy loaded and long span buildings. *Energy Build.* **2016**, *126*, 36–43. [CrossRef]

9.   Schipper, H.R.; Janssen, B. Manufacturing Double-Curved Elements in Precast Concrete Using a Flexible Mould First Experimental Results. Fib Symposium PRAGUE. 2011. Available online: https://repository.tudelft.nl/islandora/object/uuid:fc718916-cecf-42a8-a3d9-0bb8c2c8fe02 (accessed on 10 October 2020).

10. Nedcam. Available online: http://www.nedcam.nl (accessed on 10 October 2020).

11. Fuehrer, G. Available online: http://www.yatzer.com/First-photos-of-Spencer-Dock-Bridge-in-Dublin (accessed on 10 October 2020).

12. Peri. Available online: http://www.peri.de/ww/en/projects.cfm/fuseaction/diashow/reference_ID/581/currentimage/15/referencecategory_ID/17.cfm (accessed on 10 October 2020).

13. Latorre, J.I.P. Construction Method to Build Ice Shells with Pneumatic Formwork. Master's Thesis, University of Technology Faculty of Civil Engineering, Vienna, Austria, 2010.

14. Mandl, P.; Winter, P.; Schmid, V. Freeforms in Composite Constructions. The new House of Music and Music Theatre "MUMUTH" in Graz, EUROSTEEL, Austria, Volume 9(c). 2008. Available online: https://structurae.net/en/literature/conference-paper/free-forms-in-composite-constructions-the-new-house-of-music-and-music-theatre-in-graz (accessed on 10 October 2020).

15. Lindsey, B.; Gehry, F.O. Englische Ausgabe.: Material Resistance Digital Construction. Springer Science & Business Media: Berlin, Germany, 2001.

16. Toyo Ito and Associates, Meiso no Mori Crematorium Gifu, Kakamigahara Japan. 2006. Available online: http://www.toyo-ito.co.jp/WWW/index/index_en.html (accessed on 10 October 2020).

17. Architeckten, F.; Architeckten, A. The Bubble. 1999. Available online: https://www.german-architects.com/en/franken-architekten-frankfurt-am-main/project/bubble (accessed on 10 October 2020).

18. Craft-usc.com. Available online: http://www.craft-usc.com/ (accessed on 10 October 2020).

19. Oesterle, S.; Vansteenkiste, A.; Mirjan, A. Zero Waste Free-Form Formwork. In Proceedings of the Second International Conference on Flexible Formwork, ICFF, Bath, UK, 27–29 June 2012.

20. Britannica Korea, Wax. Available online: http://www.britannica.co.kr/ (accessed on 10 October 2020).

21. Karim, L.; Barbeon, F.; Gegout, P.; Bontemps, A.; Royon, L. New phase-change material components for thermal management of the light weight envelope of buildings. *Energy Build.* **2014**, *68*, 703–706. [CrossRef]

22. Li, D.; Zheng, Y.; Li, Z.; Qi, H. Optical properties of a liquid paraffin-filled double glazing unit. *Energy Build.* **2015**, *108*, 381–386. [CrossRef]

23. Lai, C.; Shuichi, H. Thermal performance of an aluminum honeycomb wallboard incorporating microencapsulated PCM. *Energy Build.* **2014**, *73*, 37–47. [CrossRef]

24. Asdrubali, F.; Giorgio, B.; Francesco, B. Influence of cavities geometric and emissivity properties on the overall thermal performance of aluminum frames for windows. *Energy Build.* **2013**, *60*, 298–309. [CrossRef]

25. Engineeringtoolbox.com, Material Properties. Available online: www.engineeringtoolbox.com (accessed on 10 October 2020).

26. Lee, K.H.; Yang, J.H. A Study on the Junctional Unit Estimation of Energy Consumption and Carbon Dioxide Emission in the Construction Materials. *J. Arch. Inst. Korea* **2009**, *25*, 43–50.

27. Korea Environmental Industry and Technology Institute (KEITI), Carbon Dioxide Emission Factor. Available online: http://www.keiti.re.kr/en/index.do (accessed on 10 October 2020).

28. Intergovernmental Panel on Climate Change, 2006 IPCC Guidelines for National Greenhouse Gas Inventories. Intergovernmental Panel on Climate Change. 2016. Available online: https://www.ipcc.ch/ (accessed on 10 October 2020).

Publisher's Note: MDPI stays neutral with regard to jurisdictional claims in published maps and institutional affiliations.

 *sustainability*

Article

Construction Management Solutions to Mitigate Elevator Noise and Vibration of High-Rise Residential Buildings

Yangki Oh [1], Minwoo Kang [1], Kwangchae Lee [2] and Sunkuk Kim [2,*]

[1] Department of Architecture, Mokpo National University, Muan-gun, Jeollanam-do 58554, Korea; oh.duoh@gmail.com (Y.O.); drminuby@naver.com (M.K.)
[2] Department of Architectural Engineering, Kyung Hee University, Yongin-si, Gyeonggi-do 17104, Korea; woojiri@gmail.com
*   Correspondence: kimskuk@khu.ac.kr; Tel.: +82-31-201-2922

Received: 11 September 2020; Accepted: 19 October 2020; Published: 27 October 2020

Abstract: In high-rise residential buildings (HRBs), elevators run at a high speed, which causes problems such as change of atmospheric pressure, noise, and vibration. Elevator noise and vibration (ENV) of HRBs causes both mental anxiety and a consistently negative effect for promoting a comfortable residential area. Therefore, a solution for alleviating the ENV of HRBs is essential. To date, studies related to ENV have been mostly conducted in the approach of mechanical and electric aspects. There have been few cases conducted from the perspective of construction management (CM), which integrates design and construction. Therefore, the aim of this study is to propose CM solutions to mitigate the ENV of HRB. For this study, the CM solution is presented after identifying the ENV problems of HRBs through documented research and case measurement. By measuring the noise of HRB that the solution was applied to, the noise level, especially in a range of >125 Hz, was extensively reduced. The result of this study will be used as sustainable guidelines that alleviate ENV problems in the process of design and construction of HRB elevators. It is expected that studies for improving ENV problems that occur in high-rise elevators will increase on the basis of the results of this study.

Keywords: elevator; noise; vibration; construction management; high-rise residential building

1. Introduction

A high-rise residential building (HRB) is a type of housing that has multi-dwelling units built on the same land. This housing has become popular in urban areas because of the increase in land cost [1]. Efficient vertical mobility is a critical component of developing and constructing tall buildings [2]. Advances in elevators over the past 20 years are probably the greatest advances we have seen in tall buildings [3].

However, the elevators of HRB operated at a high-speed cause problems such as changes in atmospheric pressure inside and outside a lift car, noise, and vibration [4–13]. In particular, elevator noise and vibration (ENV) cause both mental anxiety for passengers and a consistent negative effect on promoting a comfortable residential area close to the elevator shaft [11,14–21]. To secure a sustainable living environment, the impacts can be significant issues related to sound quality, sleeping conditions, and enjoyment within residences [14,16–20,22–25].

To solve these problems, multiple studies have been conducted, which mostly focused on mechanical and electric aspects [6,7,10,26–36]. However, it is not easy to solve ENV problems only with machinery solutions for HRB elevators that run at a speed of >90 m/min [21,37–39]. The reason is that not only do noise and vibrations occur in the elevating machinery itself, but there are architectural problems of elevator shafts or air turbulence because HRB elevators run at a high speed [4,14,37–39]. Thus, identifying solutions after analyzing such causes is necessary. Solutions for ENV problems should be provided in detail in the design phase primarily, and construction should be followed according to the design details. However, we have confirmed that ENV problems continue to occur due to errors or mistakes in the design and construction phase through research and site surveys over the past several years.

This is because there is no CM solution that integrates design and construction for ENV problems. As shown in routines (1) and (2) in Figure 1, even if ENV solutions are provided with the documents including drawings and specifications in the design phase, the details including the precision, quality, and tolerance of construction suitable for the site condition must be determined in the construction phase after design review. As shown in routines (3), (4), and (5) in Figure 1, if design errors or omissions are confirmed after design review, construction details are determined after performing supplementary design. In other words, additional designs can be carried out according to the site situation at the construction phase. Solutions applied to the design and construction phases and integrated management of them are defined as CM solutions in this paper. The aim of this study is to propose CM solutions to mitigate the ENV of HRB.

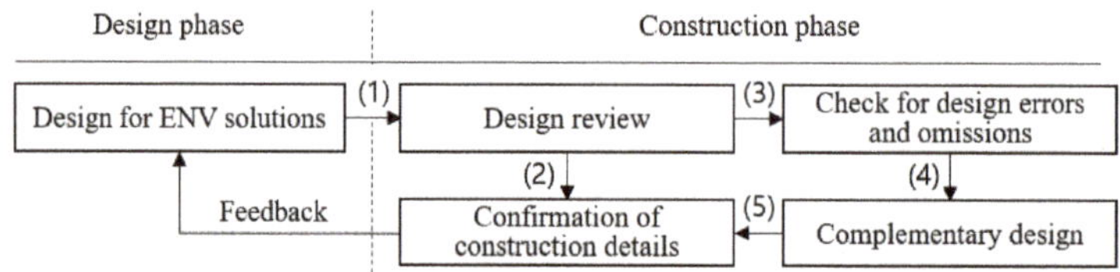

Figure 1. Design and construction integrated construction management (CM) solutions for elevator noise and vibration (ENV) problems.

Figure 2 shows the methods and procedures of this study. First, review the definition of HRB and elevator and ENV sources and transmissions by surveying the design guidelines. Second, analyze the ENV sources and the transmissions of HRBs through documents and examine the site, and then measure ENV as a designated HRB that CM solutions have not been applied to as a case study. Third, propose design and construction solutions that have been confirmed using multiple documents that have been presented and studies that have been previously conducted. Moreover, confirm the effectiveness of CM solutions proposed in this study after measuring ENV of a case HRB that these solutions are applied with. Fourth, discuss the consistent improvement of problems that occur in high-rise elevators as per the results of this study and then describe results in the conclusion.

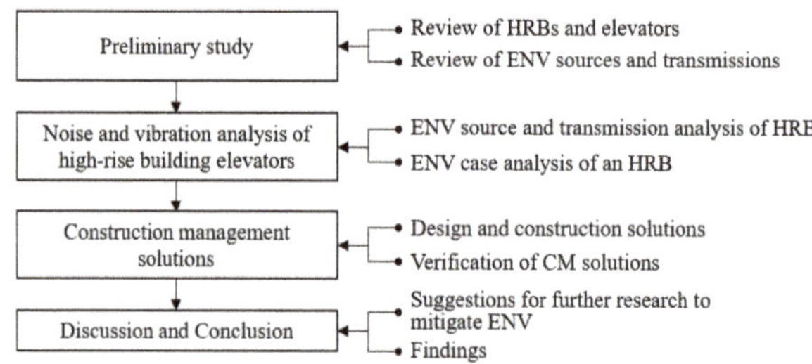

Figure 2. Research process and methodology. HRB = high-rise residential buildings.

2. ENV of HRBs

2.1. Review of HRBs and Elevators

Emporis Standards defines a high-rise building as a multi-story structure between 35 and 100 m or a building of unknown height from 12 to 39 floors [40]. Korea Land and Housing Corporation (KLHC), Korea, which develops public apartments, classifies low-rise buildings as those with 5–6 floors, mid-rise as those with 7–15 floors, high-rise as those with 16–20 floors, and super high-rise as those with 21–49 floors as per residential building guidelines [41]. For construction at a relatively large scale, residential building projects comprise many buildings with different numbers of floors. Depending on the number of floors and households per floor, the capacity and speed of elevators in each building varies [41]. As shown in Table 1, the capacity comprises many passengers, loading capacity, box size, size of exits that are available, and elevator speed regulated by KLHC guidelines comprising six levels with the range from 60 m/min as a minimum to 180 m/min as a maximum as per the number of floors in a building. For the reference, HRBs in Korea have mostly 12–40 floors, and there is recently a case of a building with >40 floors. This study is precisely processed on HRBs with >12 floors and elevators at a speed of >90 m/min.

Table 1. Elevator speed by operating floor [42].

No. of Floors	Elevator Speed	Remarks
Under 10th floor	60 m/min	
10 to 14th floor	90 m/min	
15 to 25th floor	105 m/min	Applied to elevators installed after June 2011
26 to 30th floor	120 m/min	
31 to 40th floor	150 m/min	
Over 40th floor	180 m/min	

The change of air pressure, noise, and vibration generated while HRB elevators are being operated cause passengers to be discomforted and has a negative effect on the residential environment of nearby residents [4,8,11–13]. As per the building code of many countries, the characteristic noise level, because of a life within an apartment building, should not exceed 30 dB(A) in any bedroom or living room of apartments [15,43–45]. However, for many HRBs, there are multiple cases that noise level exceeds 30 dB(A) in the apartments around elevators [15,19,21]. Many researchers state that any noise problem may be described in terms of a source, a transmission path, and a receiver; furthermore, noise control may take the form of altering any one or all elements. The noise source is the one in which the vibratory mechanical energy originates because of a physical phenomenon such as mechanical shock, impacts, friction, or turbulent airflow [14,46,47].

Both noise and vibration that occur in HRB elevators are said to be noise sources; moreover, the parts of elevating system and structure of building are transmission paths and residents are considered as receivers. Therefore, it is primary solutions in response with the location of the source that reduce mechanical shock, impacts, and friction sources occurring while elevator machinery is being operated, or alleviate the occurrence of turbulent air noise generated while an elevator car is on the move. The secondary solution is controlling transmission paths in which elevators are arranged in isolation in the housing units of HRB, or they are designed to have a buffer space between them. Although many studies have been focusing on primary solutions, the ENV problems of HRBs have not been sufficiently solved [14,15,24,25,27,36]. Therefore, in-depth studies on primary and secondary solutions are required and then applied. The reason is that it is difficult to reduce elevator noise under 30 dB(A) caused at a high speed of it, even if noise and vibration can be partly alleviated by primary solution in the case of HRB elevators. This is because the damage largely escalates if ENV problems occur because the number of HRB residents is less.

2.2. Review of ENV Sources

Fullerton [14] and Ingold [23] specifically described sources, transmission paths, and control of ENV. Torres and Haugen [27] reported a case study regarding noise and vibration because of machine room-less (MRL) elevators of apartment buildings and proposed an approach for alleviation [27]. Based on multiple studies, MRL elevators as gearless synchronous machines reduce electric energy by 50% compared to the geared traction elevators [27,48,49].

Moreover, many researchers published studies on ENV sources and transmissions [14,16,18–21,35]. The noise source is the one in which the vibratory mechanical energy originates because of a physical phenomenon such as mechanical shock, impacts, friction, tonal sound, or turbulent airflow [14,27,34,50–52]. Especially, interactions between the pulley and the cables that suspend the elevator cab show the potential for a tonal sound. These sounds vary in loudness and frequency with the system's speed, with the loudest airborne sounds being attributed to the fastest speed of operation [14,19]. The elevators of HRBs runs at a high speed from 90 to 180 m/min, which caused tonal sound at a considerable level.

In many studies, although ENV sources were introduced, most of them were from the perspective of a mechanical system and little was covered in terms of CM solutions for HRB [14,15,26–36,48]. Thus, in this study, ENV sources and transmission paths of HRB are analyzed. Confirming clearly airborne and structure-borne transmission paths through a study is to obtain clues to CM solutions to alleviate the ENV of HRBs. Several researchers, including Fullerton [14], Ingold [23], and Torres and Haugen [27], have partially presented ENV solutions from a design and construction perspective. However, they did not proceed from the perspective of a CM solution that integrates the design and construction phases as introduced in Figure 1.

3. Noise and Vibration Analysis of HRBs

3.1. ENV Source and Transmission Analysis of HRBs

Figure 3 shows the paths on which noise and vibration generated in the traction elevators of HRBs are transmitted to a residence. ENV occurs only when the machinery primarily runs in a machine room and a hoist way. Noise caused from various sources comprises airborne noise transmitted in the form of soundwaves through air particles, as shown in Figure 3, as well as structure-borne noise transmitted through the slabs, walls, and ceilings of buildings [7,14,17,18,21,27]. These two types of noises are delivered to a space that requires quietness, such as bedrooms and living rooms of residences, and consistently hinders living comfort.

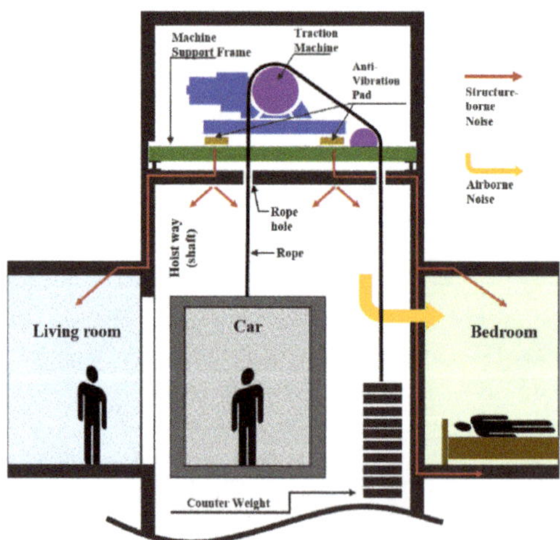

Figure 3. The transmission paths of traction ENV [17,18,21].

By considering various studies [14,16,18–21,27,34,35,50–53] and the research of previous years, the sources, causes, and types of traction elevators have been identified, which are organized in Table 2. For machine rooms, noise is generated by operating multiple machine parts along with the high-frequency rotation of motors, meshing frequency of the gear system, on-and-off brakes, and electrical contact switches to control the elevator. For the hoist way, the sources are noise caused in the process of operation of door parts and by the misalignment of the door and the noise caused when an elevator car goes up and down as per the rails and elevator car guide, and from activities such as the rotation of bearings and rollers, passing through rail joints, and interactions between pulleys and cables.

As shown in Table 1, elevators at a high speed of >90 m/min are used in HRBs. At this time, inrush, air friction, and puff noises generated by elevator cars work as sources [8,9,38,39]. Moreover, the operation of devices such as machine cooling fans, car ventilation fans, car arrival signals, and friction and impact of machine parts work as sources. As shown in Table 2, various forms of noises comprise mechanical shock, impacts, friction, tonal sound, or turbulent airflow in a physical sense.

Table 2. Noise sources, causes, and types of traction elevators.

Sources			Causes		Types of Noises
Machine Room	Motor	-	High frequency rotation	-	Tonal sound
	Gear and brake	-	Meshing frequency of the gear system	-	Mechanical shock, Impact, Friction
		-	On-and-off the brakes	-	Impact
	Control panel	-	Electrical contact switches to control the elevator	-	Impact
Hoist Way	Door	-	Operation of door parts	-	Friction
		-	Misalignment		
	Rails and elevator car guide	-	Rotation of bearings	-	Tonal sound
		-	Rotation of rollers	-	Tonal sound, Friction
		-	Pass through rail joints	-	Mechanical shock, Impact
		-	Interactions between pulley and cables	-	Tonal sound, Friction
	Car	-	Inrush, air friction, and narrow–section pass of car	-	Turbulent airflow

Table 3 shows the specific analysis of paths in which noises generated from various sources, as introduced in Figure 3, are transformed into structure-borne and airborne noises and transmitted to the residence. For structure-borne noise, vibratory noise generated by elevating machinery in the machine room is transmitted to residences via an anti-vibration pad, a machine support frame, a machine room slab, and a hoist way wall. Recently, it is common that an anti-vibration pad is designed double-layered between traction machine and support frame and is designed as a vibration transfer area and the size of the machine support frame are reduced, which leads to the considerable reduction of structure-borne noise transmitted from the machine room compared to the past. For structure-borne noise, vibratory noise generated by an elevator car as well as the ascent and descent of counter weight is transmitted to residences via an elevator car guide, rails, rail brackets, and hoist way.

Table 3. Transmission paths of ENV [14,21,39].

Description	Sources	Transmission Paths
Structure–borne noises	Machine room	Machinery → Anti-vibration pad → Machine support frame → Machine room slab → Hoist way wall → Residences
	Hoist way	Elevator car guide → Rails → Rail brackets → Hoist way wall → Residences
Airborne noises	Machine room	Rope hole → Hoist way wall → Residences Outside ventilation openings → Neighboring residence windows → Residences
	Hoist way	Hoist way → Hoist way wall → Residence

Airborne noise produced in a machine room is transferred via a rope hole because Figure 3 shows passing by a hoist wall to residences; moreover, airborne noise is transferred to residences past ventilation openings and neighboring residence windows of the machine. Airborne noise produced in the hoist way is transferred to residences through a hoist way wall. Moreover, there could be airborne noise produced in the hoist way, which can be easily controlled by a >200-mm-thick reinforced concrete (RC) wall and tightly closed concrete placement without any cracks.

3.2. ENV Case Analysis of an HRB

To present CM solutions, it is necessary to analyze ENV problems and the actual condition of the ENV of HRB. We performed measurements of 2 buildings, as Figure 4 shows, as designating one area of apartment complex 'W' located in Seoul known for ENV problems. As shown in Figure 4, the elevator is adjacent to the bedroom or living room and is a traction-type, having a machine room. Figure 4a shows a case that the elevator of Building 'A' is adjacent to a living room of a nearby residence, whereas Figure 4b is a case that the elevator of Building 'B' is adjacent to a bedroom of a nearby residence. As shown in Figure 4c, the vibration was measured with four sensors attached to the wall, and another four sensors were then installed on the upper part of a room floor to measure noise. The measurement time was 50 seconds, and the elevator speed was 60 m/min as the lowest speed of HRB classified in Table 1. Table 4 shows the measurement system used to measure noise and vibrations produced by the elevator with a capacity of 550 kg, i.e., 8 persons.

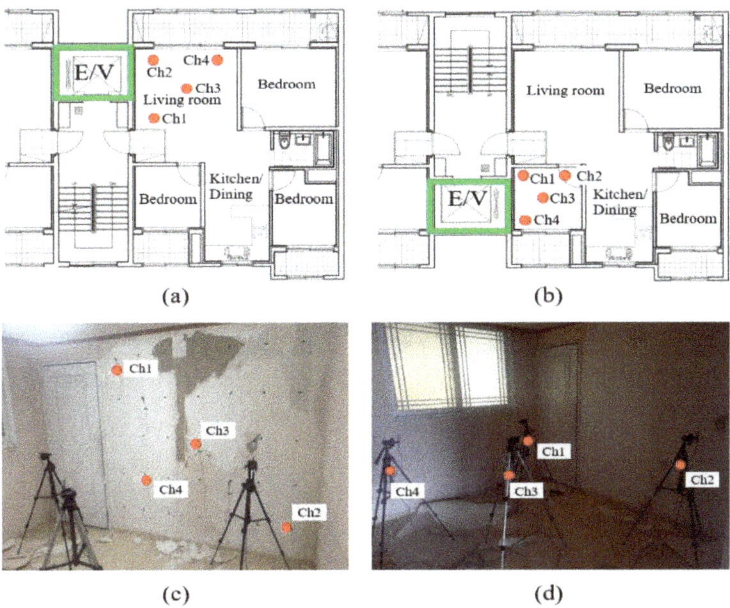

Figure 4. ENV measurement of HRB 'W': (**a**) The living room in Building 'A'; (**b**) the bedroom in Building 'B'; (**c**) the vibration measurement sensors on the living room or bedroom wall; (**d**) the noise measurement sensors in the living room or bedroom space.

Table 4. Measurement system.

	Description	Model	Manufacturer	Details
Noise	Analyzer	Apollo	SINUS	Bandwidth: DC~80 kHz/Dynamic range: 120 dB
	Microphone	CLASS 0 (LEMO)	G.R.A.S	Bandwidth: 3.15~20 kHz/Dynamic range: 135 dB
Vibration	Analyzer	SA-01	RION	Bandwidth: 0.5~20 kHz/Dynamic range: 140 dB
	Accelerometer	Single Axis Accelerometer (SW)	B.S.W.A	Bandwidth: 0.5~14 kHz/Maximum Acceleration: 0.0002 g rms

Figure 5 shows a graph in which noise and the result of vibratory measurements are simultaneously written; it is a case in which a living room is adjacent to the elevator-like Figure 4a. Figure 5b shows a case in which the bedroom is adjacent to the elevator-like Figure 4b. To date, elevator noise has shown a sound pressure level of 35 dBA on an average when reviewing the result of the noise measurement of the elevator in an apartment building in Korea. Although this noise level is quieter than the sound level criteria in a library, many residents face discomfort because the noise in low-frequency bands has a considerable influence on them. Note that the energy distribution of the general environment noise ranges from 125 Hz to 4 kHz; however, as shown in Figure 5, elevator noise is distributed up to 63 Hz or even up to 32 Hz. In other words, the noise accompanying vibrations is sensed with considerable discomfort even if it is low. The common tendency identified in both graphs is the increase in value of contrast measurement data of the background noise (BGN), and the background vibration (BGV) appears to be high within the range of low-frequency bands of <500 Hz. This corresponds to a figure reported in the guidelines of many previous published studies [16,54,55].

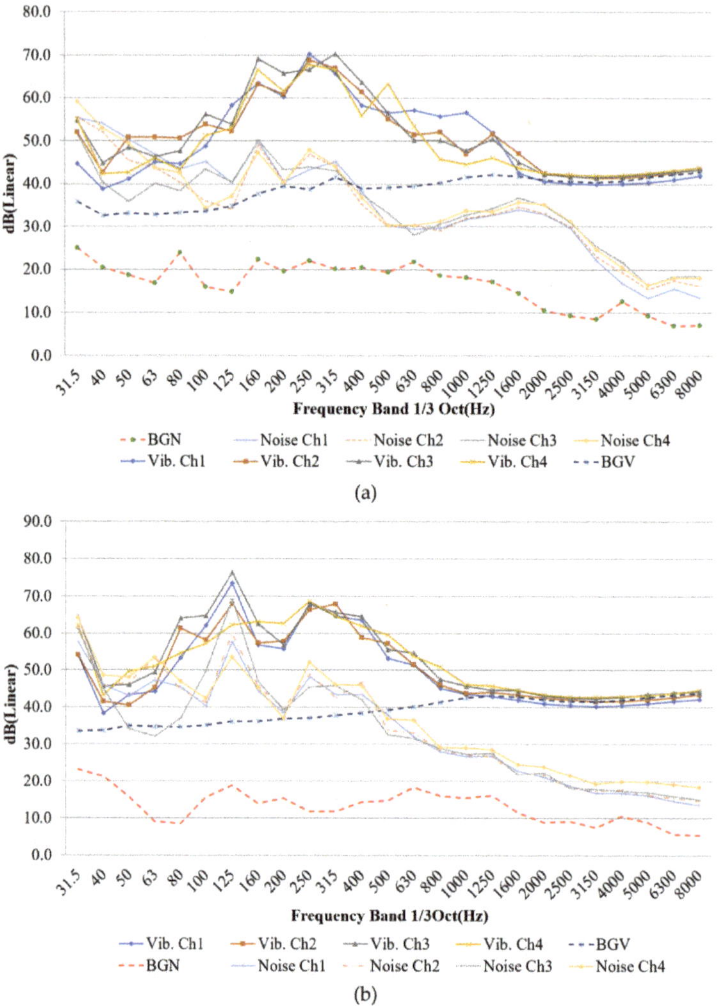

Figure 5. ENV measurement results of HRB 'W': (**a**) the ENV measurement results of living room; (**b**) the ENV measurement results of bedroom.

The measurement results in Figure 5 confirmed that the noise in the low-frequency band among noises transferred in the form of structure-borne noise and produced by the operation of the elevator affected the living comfort of residents. For measurements in the living room, the effect appeared to be the most in the range of 250 Hz. Moreover, for measurements in the bedroom, the effect appeared to be considerable in the range of 125 Hz. Regarding noise and vibration, the same tendency was confirmed in both the living room and bedroom. Based on these results, solutions should be presented in low-frequency bands in which noise and vibration produced during the operation has considerable influence on residences so as to reduce the elevator noise.

4. CM Solutions

4.1. Design Solutions

4.1.1. Separation of Residences and Elevators

In the ground of residents, to relieve ENV damages of HRBs, solutions applied in the design phase comprise two approaches. First, to separate space that is sensitive to noise such as bedrooms and living rooms from hoist ways or elevator shafts. Second, to arrange a buffer space between the rooms that are adjacent to hoist ways to increase the thickness of machine room slabs and hoist way walls, and to supplement partly building components similar to the detailed changes of hoist ways.

The most efficient approach to relieve the effects of noise and vibration of HRB elevators that run at a high speed is to separate elevators from residences and arrange them. By comparing the results of noise measurement from the case when an elevator is attached to a living room or bedroom, like Figure 4, with the results of noise measurement from the case when an elevator is separated from residence, the effect can be confirmed. In particular, it is confirmed that noise became as low as 10–20 dB in the band of >125 Hz where the elevator is separated from residences. However, in a limited building area, residence design with high density should be high and if considering design-constraint conditions, such as daylight, view, and space function, there is a case that space that is sensitive to noise like living rooms and bedrooms cannot be separated and arranged. In this case, the second design solutions should be identified.

4.1.2. Buffer Space Design

As shown in Figure 6a, because the elevator hoist way is adjacent to bedrooms, noise and vibration are directly transferred if the elevator is operated. The case of toilet has a reduced impact of ENV; the case of a bedroom has a bad influence on rest and sleeping time. For HRBs in Korea, because a hoist way is designed with a 200-mm-thick bearing walls, airborne noise is blocked. However, the shield effect of structure-borne noise is not high. Thus, to relieve the transmission of structure-borne noise, residents feel it can be reduced if a wardrobe is placed similar to Figure 6b. To check the effect of it, one case of noise measurement in an apartment bedroom where there was a wardrobe sharing the wall with the elevator shaft was measured in comparison with another case without the wardrobe in the same condition. In Figure 6b, similar to the case of the installed wardrobe, the installation of wardrobe reduced noise level by ~7.6 dB, although Figure 6b is in the same plane compared to the general rectangular shape bedroom of Figure 6a. Such noise reduction was basically made possible by the effect of sound absorption or sound insulation using sound-absorbing materials such as blanket and clothing, both of which are filled at both sides of the wardrobe and inside of it. Moreover, the installation of the wardrobe made it possible to control the phenomenon of noise in a low-frequency band, known as room mode, which breaks the rectangular shape of a bedroom.

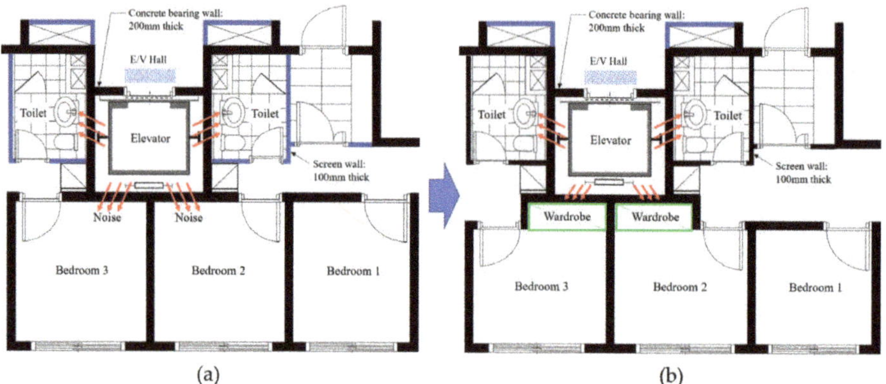

(a)   (b)

Figure 6. Noise reduction design using a wardrobe: (**a**) A floor plan without a wardrobe; (**b**) a floor plan with a wardrobe [17,21].

4.1.3. Change of Hoist Way Details

The elevator that runs at 120 m/min should be installed in HRBs with >26 floors (Table 1), and for HRBs with >40 floors, the elevator should run at a high speed of 180 m/min. For elevators that run at a high speed, turbulent airflow is produced, which leads to air friction noise, puff noise, inrush noise, and draft noise [38,39,56]. Design solutions regarding airborne noises are attributed to such a turbulent airflow.

(**1**) Air Friction Noise

As an elevator runs at a high speed in a hoist way (Figure 7a), air is compressed in a car-heading direction and elevator piston effect that increases air pressure [4,8,9,37–39,57,58]. Compressed air travels through the cramped gap between car and hoist way to the upper part of the car, which creates noise known as an air friction noise. The air friction noise gets louder as the area of the gap gets smaller compared to the area of the car with a faster speed of the elevator. This occurs as either a single elevator or double elevators run at a high speed in a hoist way. When two elevators run at the same time and in the same direction, the noise gets even louder. There is not an issue in the case that the elevator runs at 120 m/min in the hoist way for a single elevator, and the elevator runs at 180 m/min in the hoist way for double elevators. Note that the area of the hoist way should be designed to be bigger than normal by 40% at a higher speed such as Figure 7b [38,39].

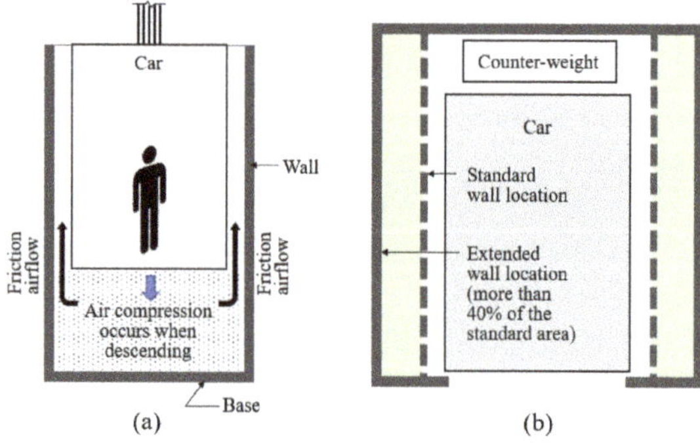

(a)   (b)

Figure 7. Air friction noise and design solution: (**a**) Air friction noise; (**b**) a design solution for air friction noise.

(2) Narrow-Section Passing Noise

Narrow-section passing noise, called puff noise [38], is a noise that is produced when an elevator goes through protruding steel installed as Figure 8a because of the structural mechanism or the installation of a guide rail. The elevator goes through an RC beam, which creates wind pressure that contributes to the noise. For HRBs, such protruding parts are reported per floor; therefore, the noise repeatedly occurs. To prevent this, architectural designing should be processed without the protruding parts inside the hoist way. However, in cases where it is unavoidable to have protruding parts, it is appropriate to install an air sliding panel upon and underneath the protruding parts to relieve the wind pressure that occurs in that area (Figure 8b). The appropriate angle of the air sliding panel is 4–8° [38,39]. This solution is required in the hoist way for a single elevator that runs at a speed of >150 m/min; in other words, with >31 floors in HRBs. The same solution is to be considered in the case that the car speed is >180 m/min in a hoist way for double elevators.

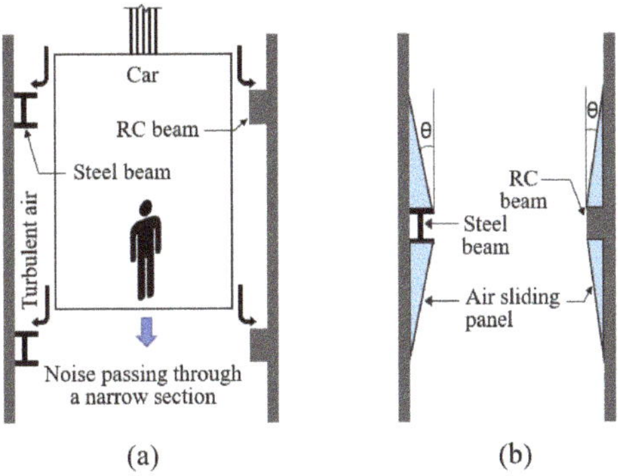

Figure 8. Narrow-section passing noise and design solution: (**a**) Noise generation; (**b**) the design solution.

(3) Inrush Noise

It is common to arrange more than two elevators in HRB where there are >20 floors and four householders on each floor. When only one elevator of multiple elevators enters in the single hoist way (Figure 9a), noise is produced by compressed air flow, which is called inrush noise [38]. It is common that the car vibration occurs along with noise when the inrush noise by the high speed is considerable. This is preventable because the cause of inrush noise is rapid air compression when an elevator rushes in. The best approach is not to have the single hoist way; however, in the case where it is unavoidable to have one because of structural constraints, two solutions can be considered. First, similar to Figure 9b, ventilation openings can be placed either on one side or both sides of the single hoist way wall. The size of the opening varies depending on the size and speed of the car; however, there is no trouble if it is designed within 1.5 m^2. If it is difficult to place the ventilation opening, the wall is expanded such that the floor area of the hoist way is increased by >40% (Figure 9c) [38,39].

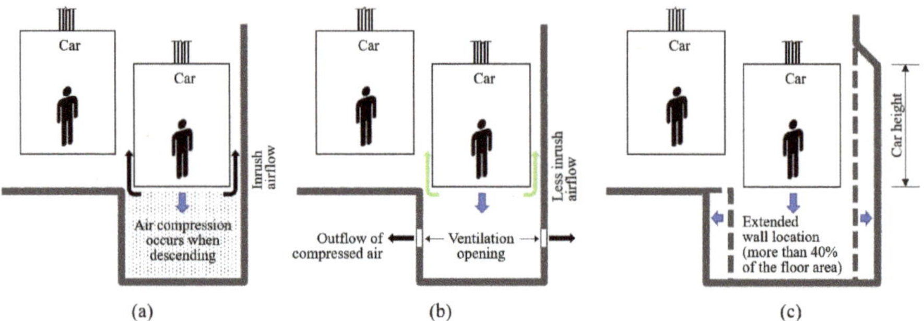

Figure 9. Inrush noise and design solution: (**a**) Inrush noise; (**b**) a ventilation opening arrangement; (**c**) expansion of the hoist way floor area.

(4) Draft Noise

Generally, in a high-lift elevator, a pressure difference is generated between the landing and hoist way. Therefore, air flows into and out of the hoist way from the gap around the landing door; at that time, a blowing sound called draft noise is generated [39,56]. In particular, this noise gets worse when the air produced by a heater in winter flows in the hoist way and the elevator ascends. The solution to this is to design an entrance door that is double-layered or a revolving door such that cold air outside does not enter within HRB. Moreover, it should be designed with air-shielding details between and around elevator opening frame of the landing floor such as a landing door, jamb, and sill [56].

4.1.4. Other Design Solutions

Note that design solutions should be considered, in addition to design solutions that are suggested, so as to alleviate the ENV that residents of HRBs feel.

- As shown in Figure 3, a machine support frame should be suspended from a machine room floor slab.
- The thickness of the machine room floor is designed to be possibly >350 mm, including a 200-mm-thick RC slab and 150-mm-sized light-weight concrete.
- The thickness of the wall in the machine room is designed to be >150 mm, whereas the sound absorption layer is placed within the wall.
- The thickness of the hoist way wall is designed to be >200 mm.
- If the hoist way space is sufficiently big, separated beams are supplemented. Moreover, the installation of rail brackets reduces noise a lot while an elevator is running.
- The counter-weight of the elevator is planned to be installed on the staircase or external wall rather than adjacent residences.

4.2. Construction Solutions

Construction solutions are the ones to be applied to at the construction phase and are about the selection of elevating machinery, the location, and the method of installation, in addition to design solutions that present solutions to ENV problems at a planning phase. Similar to Table 5, the problems are classified into structure-borne and airborne noises, and again classified into the machine room and hoist way (shaft), and then construction solutions are analyzed and presented. They are subdivided into either controlling elevator vibration or noise sources and controlling transmission paths.

Table 5. Construction solutions to mitigate ENV.

Description	Sources	Solutions
Structure-Borne Noises	Machine room	Measures for vibration sources - Use of high-quality motors - Precise balance of brakes, gears, and elevator car Measures for anti-vibration - Isolation of the traction machine from the support frame - Isolation of the switchgear cabinet from the machine room slab
	Hoist way	Measures for vibration sources - Improvement of guide rail machining accuracy: Machining error within ±2 mm/5 m - Improvement of guide rail installation accuracy: Seam surface difference between rails within ±0.05 mm Measures for anti-vibration - Fastening rail brackets at the edge of slab - Isolation of guide rails from hoist way wall
Airborne Noises	Machine room	Measures for noise sources - Use of high-quality motors - Precise machining and fabrication of brakes, gears, and coupling parts - Use of low-noise cooling fan or self-cooling system Measures for anti-noise - Installation of sound insulation cover over the rope hole - Isolation of airborne noise inside machine room
	Hoist way	- Precise installation of elevator doors for silent operation - Volume adjustment of door enunciator

Table 5 shows the construction solutions to mitigate ENV that are obtained through the studies for years after contemplating various research documents [14,18–21,27,38,39,50–52]. Structure-borne noise generated in the machine room primarily occurs via the vibration of the elevating machinery. To alleviate this, elevating motor with power filters with high quality basically are used and brakes, gears, and elevator car should be installed to maintain the precise balance of elevator machinery. Moreover, vibration isolation pads that have high damping performance such as neoprene or rubber should be inserted at fixing points such that the vibration is generated in the traction machine and switchgear cabinet [14,18–21,50–52]. When anti-vibration pads are installed in the traction machine, the first natural vibration frequency of the whole elevator should be maintained at a less than audible frequency band. For the traction machine to work precisely aligned with the height between elevator car and hall landing, elevator machinery and anti-vibration pads should be installed, thus maintaining balance to minimize isolation deflection. For this purpose, close cooperation with an elevator manufacturer and installer is required at the installation phase.

To alleviate structure-borne noise generated in the hoist way or elevator shaft, the machining errors of guide rails are to be maintained within ±2 mm/5 m, and seam surface difference between guide rails of the guide rails should be installed within ±0.05 mm, which alleviates vibration because of the movement of rollers. Vibration-borne noise produced in the hoist way occurs between the guide rollers of the car and rails. To prevent this, brackets should be installed next to the floor slab to fixate the rail (Figure 10). The floor slab edges are inherently stiffer than the shaft walls and will limit the transmission of rail and roller guide interactions from generating the structure-borne noise in adjacent spaces [14,17,21]. Furthermore, anti-vibration pads should be installed between brackets and floor slab edges. For reference, depending on the location of the rail brackets, it is confirmed that the noise difference by ~4 dB on an average occurs because of the measurement of noise while an elevator is running [21].

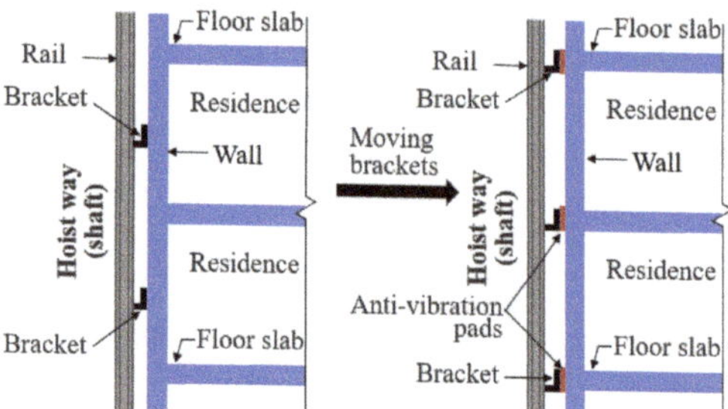

Figure 10. Fastening rail brackets at the edge of a slab.

Measures in correspondence with the construction phase should be considered because airborne noises generated in the machine room primarily occur by the tonal sound of the elevating machinery. It is essential to select motors of good quality to encounter noise sources and precise machining; moreover, the fabrication of brakes, gears, and coupling parts should be performed.

A low-noise cooling fan or self-cooling system in the machine room should be used. As shown in Figure 3, sound insulation cover should be installed around the rope hole as transmission paths of airborne noises in the machine room. Moreover, soundproofing materials should be used to encounter isolation of airborne noise on the interior wall and the ceiling in the machine room. After glass wool as thick as 50 mm is placed on the wall and ceiling in the machine room, it is confirmed that the level of noise generated in the elevating machinery is reduced by 6.2 dB on average. In particular, it is confirmed that the noise level in the range of 1 kHz frequency is reduced by 9.4 dB at the maximum.

Airborne noises generated in the hoist way are primarily related to the elevator door. The primary noise sources of the door are its misalignments or improper installation and alarm sound for arrival on each floor. To solve such problems, the precise installation of doors is required, and the volume of the door enunciator is to be adjusted to <60 dB(A).

4.3. Verification of CM Solutions

In this study, the suggested CM solutions are applied and how much ENV is alleviated should be confirmed. For this purpose, after selecting the HRB project with a 25-story building as a case, the most effective elevators of CM solutions and design solutions, such as separation of residences and elevators as well as the hoist way wall as thick as 200 mm, were applied. One noticeable aspect is that it is designed using MRL elevators [49] that are generally used for buildings of <20 floors. Moreover, after selecting an OTIS elevating machine of good quality at the construction phase, the construction solutions suggested in Table 5 were mostly applied. The case project was completed in 2020, whereas the elevator noise of two buildings was measured in the same method of Figure 4d. The vibration measurement with the sensors attached on the wall was not done without the agreement of the residents (Figure 4c).

As shown in Figure 11a, the staircase was placed between the elevator and the bedroom of residence. Two elevators were placed as per design guidelines because there were four householders on each floor in Building 'A' [41,42]. Figure 11b shows that the elevator is separated from residence using the elevator hall. For this measurement, it was conditional that the elevator runs departing from the second floor as the lowest floor and stopped on the 25th floor. The measurement was made in bedrooms on the 25th floor as the highest floor. Moreover, the measurement time was 90 seconds, and the elevator speed was 105 m/mm, which is the speed of the 25-story HRB classified in Table 1. The same measuring system as in Table 5 was used for measuring noise generated by the elevator with a capacity of 15 persons and 1150 kg.

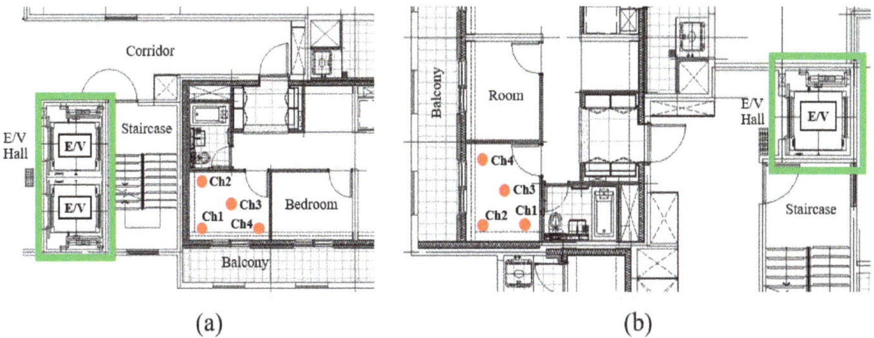

Figure 11. Elevator noise measurement of HRB 'D': (**a**) The 25th floor plan of Building 'A'; (**b**) the 25th floor plan of Building 'B'.

Figure 12 shows the result of noise measurement for the case that elevators and residences were isolated, as seen in Figure 11a,b. Similar to Figure 5, the effect of noise measurements compared to BGN in the range of <500 Hz appears to be extensive. Note that regardless of different conditions, it shows that it was a low-frequency band that had an influence on elevator noise. In particular, the effect of the two cases seemed to be extensive at 63 Hz; moreover, the common aspect is that the effect of low-frequency bands at the central measuring point appeared to be less. The reason is that the low-pitched superposition phenomenon by the room mode was most remarkably noticeable at the corners of rectangular bedrooms. Figure 13 shows the distribution of the sound pressure level by the overlapping and offsetting of sound energy per frequency band in the rectangular room. The low sound energy at 63 and 125 Hz certainly demonstrated the phenomenon of remarkable overlap at the corners of the room [59].

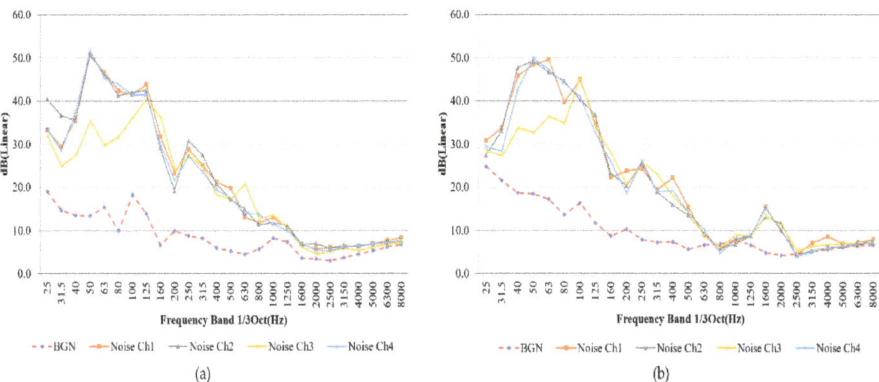

Figure 12. Elevator noise measurement results of HRB 'D': (**a**) The measurement results of 'Building A'; (**b**) the measurement results of 'Building B'.

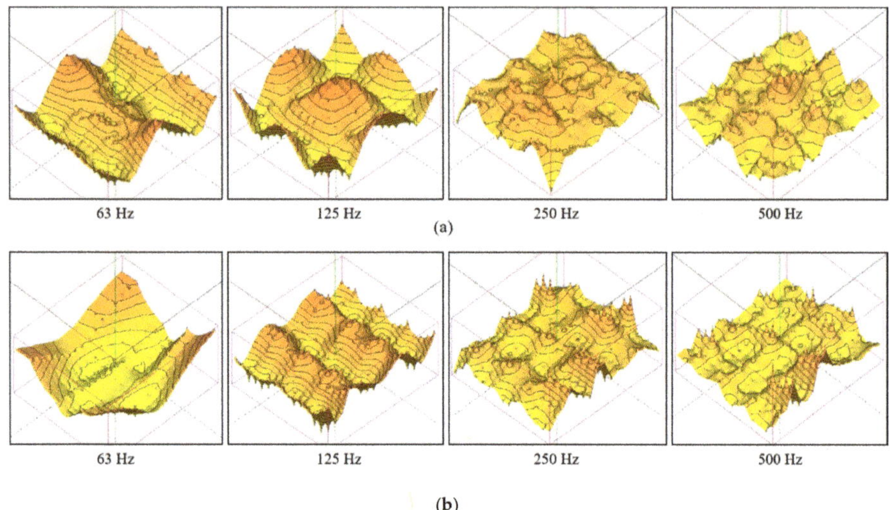

(a)

(b)

Figure 13. Different patterns of spatial distribution of sound pressure level in a room according to the frequency bands [59]: (**a**) The noise measurement results at the center; (**b**) the noise measurement results at the corners.

Figure 14 shows the analysis of the level of noise reduction and noise characteristics as per the frequency band after the average conversion of the measurement results of Figures 5 and 12. The measurement results of Figure 12 show that, in this study, CM solutions that were applied as suggested demonstrate the remarkable reduction of noise in the frequency band of >100 Hz. Figure 12 shows that there was no considerable difference at frequencies of <125 Hz; however, it was confirmed that noise of HRB 'D', in which elevators and residences are isolated in the frequency of >125 Hz, was measured at a noise level lower than that of HRB 'W'. As shown in Figure 13, for HRB 'W', while the average value of the noise level compared to BGN makes a difference within 26 dB(A), the case of HRB 'D' maintains the level of 5 dB(A). This is explained to be the alleviating noise that residents feel with CM solutions suggested in this study. However, regardless of the isolation of elevators and residences in the low-frequency band of <125 Hz, the noise level compared to BGN seems to be high. The structure-borne noise generated by the operation of the elevator is analyzed to be high in the low-frequency band.

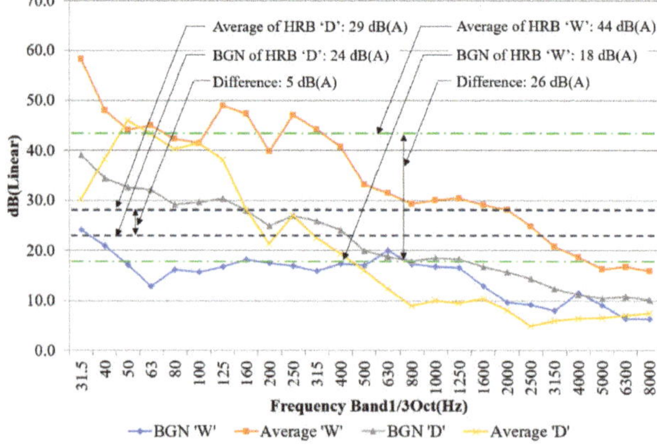

Figure 14. Noise analysis by frequency band according to the location of elevators and residences.

5. Discussion

To solve ENV problems of a building, multiple studies regarding mechanical solutions have been conducted; in certain studies, solutions were presented from the perspective of design and construction, which are only for residential buildings of less than ten floors. Thus, to alleviate the ENV of HRBs that run at >90 m/min, solutions that consider the characteristics of high-speed elevators and HRB floor plans on the basis of knowledge obtained from the results of their study are necessary. We analyzed the characteristics of noises and vibrations produced in machine rooms and hoist ways through ENV-related studies of HRBs for the past few years and sources vs. transmission paths of airborne and structure-borne noises, and then presented solutions that control the stage of design and construction. A case study was processed to confirm the given solutions. CM solutions applied to the case of HRB 'D', including isolation arrangement of elevators and residences, 200-mm-think RC wall design of hoist way and most other solutions are presented in Table 4; they were then applied to the construction phase. We were then able to confirm considerable improvement compared to HRB 'W' case because of the measurement on noise of HRB 'D'. Vibration measurement was impossible because the residents opposed to sensors being installed on the wall; however, it is assumed that there may have been improvement of vibration corresponding to that of noise considering the HRB 'W' case.

Studies on identifying the effectiveness of alleviating noise should continue with thorough experimentation in each item to clarify the utility of the CM solutions given in this study. The extra application of researches of CM solutions should be processed by the subject area of HRBs presented in Table 1.

6. Conclusions

There are have been many studies regarding ENV worldwide, which is proof that the more sensitive humans have been to noise, the more comfortable life environment they have demanded. In particular, for HRBs built in many downtown areas where there are external noise sources, residents demand more static environments, while noise is more likely to occur because of the use of high-speed elevators. To solve such as dilemma, we contemplated multiple documents and presented CM solutions to alleviate ENV of HRBs in this thesis by means of the study of years. An HRB project with ENV problems and an HRB project that uses solutions given in this study were processed as a case study, and its effectiveness was verified, with the results as following.

First, BGN and BGV appear to be relatively high in the range of low frequency of <500 Hz if an elevator hoist way is adjacent to the living room or bedroom. It was confirmed that the noise produced at low-frequency bands of 250 Hz for a living room and those of 125 Hz for a bedroom had a considerable impact on the living comfort of residents.

Second, in the case of HRBs to which CM solutions are applied, the effect of the measured noise compared to BGN is within the range of <500 Hz. This shows that what affects elevator noise were low-frequency bands of <500 Hz. CM solutions application effects appeared to be the most in the range of 63 Hz; moreover, a common aspect is the effects of low-frequency band turning out to be little at the central measurement point Ch3. The reason is because it is analyzed that the distribution of sound pressure level by overlapping and offsetting of sound energy per frequency band in the rectangular room was remarkably noticeable.

Third, compared to the measurement results of the two case projects, the elevator noise of HRBs that CM solutions were applied to in the frequency band of >100 Hz are confirmed to remarkably reduce noise. Especially, it is confirmed that the CM solutions-applied noise of HRB 'D' was measured to be relatively low, in the frequency band of >125 Hz. While the average value of noise level compared to BGN in the case of HRB 'W' with ENV problems makes a difference of 26 dB(A), the case of HRB 'D' that the CM solutions were applied to maintain the level of 5 dB(A). Thus, it is comprehended that, in this study, the CM solutions suggested largely alleviated noise.

Author Contributions: Conceptualization, Y.O. and S.K.; methodology, S.K.; validation, Y.O., M.K., and S.K.; formal analysis, Y.O. and M.K.; measurement and data curation, Y.O. and M.K.; writing—original draft preparation, Y.O. and K.L.; writing—review and editing, S.K.; visualization, K.L.; supervision, S.K.; project administration, S.K.; funding acquisition, Y.O. All authors have read and agreed to the published version of the manuscript.

Funding: This research was supported by the grant (20RERP-B082204-07) from Residential Environment Research Program funded by Ministry of Land, Infrastructure and Transport of Korean government.

Conflicts of Interest: The authors declare no conflict of interest. The funders had no role in the design of the study; in the collection, analyses, or interpretation of data; in the writing of the manuscript, or in the decision to publish the results.

References

1. Rashidah, S.; Che-Ani, A.I.; Sairi, A.; Tawil, N.M.; Razak, M. *Classification of High-Rise Residential Building Facilities: A Descriptive Survey on 170 Housing Scheme in Klang Valley, MATEC Web of Conferences 66*; Guéhot, S., Ed.; EDP Sciences: Les Ulis, France, 2016; Volume 00103, pp. 1–4. [CrossRef]
2. Al-Kodmany, K. Tall buildings and elevators: A review of recent technological advances. *Buildings* **2015**, *5*, 1070–1104. [CrossRef]
3. Wood, A.; Henry, S.; Safarik, D. 2014 Best Tall Buildings. In *Proceedings of the CTBUH Award Ceremony, 2 November 2014*; Illinois Institute of Technology (IIT): Chicago, IL, USA, 2014.
4. Liu, J.; Zhang, R.; He, Q.; Zhang, Q. Study on horizontal vibration characteristics of high-speed elevator with airflow pressure disturbance and guiding system excitation. *Mech. Ind.* **2019**, *20*, 305. [CrossRef]
5. Qiu, L.; Wang, Z.; Zhang, S.; Zhang, L.; Chen, J. A Vibration-Related Design Parameter Optimization Method for High-Speed Elevator Horizontal Vibration Reduction. *Shock Vib.* **2020**. [CrossRef]
6. Kawasaki, R.; Hironaka, Y.; Nishimura, M. Noise and Vibration Analysis of Elevator Traction Machine. In Proceedings of the Inter-noise 2010, Noise and Sustainability, Lisbon, Portugal, 13–16 June 2010; pp. 1–9.
7. Noda, S.; Mizuno, S.; Kamijo, Y.; Matsushita, M. Prediction of Room Noise Caused by Vibration of High Power Elevator Traction Machine. In Proceedings of the 2013 International Conference on Energy, Environment, Ecosystems and Development, Lemesos, Cyprus, 21–23 March 2013; pp. 130–134. Available online: http://www.inase.org/library/2013/rhodes/bypaper/EEED/EEED-16.pdf (accessed on 25 June 2020).
8. Klote, J.H. Elevator pressurization in tall buildings. *Int. J. High-Rise Build.* **2013**, *2*, 341–344. [CrossRef]
9. Klote, J.H. An analysis of the influence of piston effect on elevator smoke control. In *NIST Interagency/Internal Report (NISTIR)—88-3751*; NIST Pubs: Gaithersburg, MD, USA, 1988. [CrossRef]
10. Mutoh, N.; Kagomiya, K.; Kurosawa, T.; Konya, M.; Andoh, T. Horizontal vibration suppression method suitable for super-high-speed elevators. *Electr. Eng. Japan* **1999**, *129*, 59–73. [CrossRef]
11. Szydlo, K.; Wolszczak, P.; Longwic, R.; Litak, G.; Dziubinski, M.; Drozd, A. assessment of lift passenger comfort by the Hilbert–Huang transform. *J. Vib. Eng. Technol.* **2020**, *8*, 373–380. [CrossRef]
12. Szydlo, K.; Maciag, P.; Longwic, R.; Lotko, M. Analysis of vibroacoustic signals recorded in the passenger lift cabin. *Adv. Sci. Technol. Res. J.* **2016**, *10*, 193–201. [CrossRef]
13. Zhang, Y.; Sun, X.; Zhao, Z.; Su, W. Elevator ride comfort monitoring and evaluation using smartphones. *Mech. Syst. Signal Process.* **2018**, *105*, 377–390. [CrossRef]
14. Fullerton, J.L. Review of elevator noise and vibration criteria, sources and control for multifamily residential buildings. In Proceedings of the Inter-Noise and Noise-Con Congress and Conference Proceedings, Honolulu, HI, USA, 3–6 December 2006; Volume 6, pp. 1230–1237.
15. Kalkman Ir, C.; Buijsing, J.H.N. Noise levels in apartment block caused by lift; what can be done in order to reduce complaints. In Proceedings of the 2001 International Congress and Exhibition on Noise Control Engineering, Hague, The Netherlands, 27–30 August 2001. Available online: https://www.peutz.nl/sites/peutz.nl/files/publicaties/Peutz_Publicatie_KK_Internoise_08-2001.pdf (accessed on 25 June 2020).
16. Jeong, A.Y.; Kim, K.W.; Shin, H.K.; Yang, K.S. Criteria and Characteristics of Elevator Noise in Apartments. *Appl. Mech. Mater.* **2017**, *873*, 231–236. [CrossRef]
17. Lee, K.W. A Study on the Measures to Reduce Elevator Noises. Master's Thesis, Catholic Kwandong University, Gangneung, Korea, 2013; pp. 9–35. (In Korean)

18. Kim, H.S.; Kim, H.G.; Kim, M.J.; Oh, Y.I. Reduction Methods of the Elevator-Operating Noise in Apartment Housings. In Proceedings of the International Conference, Korean Society for Noise and Vibration Engineering (KSNVE), Seoul, Korea, 9–12 February 1994; pp. 619–626. Available online: http://www.dbpia.co.kr/journal/articleDetail?nodeId=NODE02420216 (accessed on 23 July 2020).

19. Evans, J.B. Elevator equipment noise mitigation for high-rise residential condominium. *J. Acoust. Soc. Am.* **2012**, *131*, 3262. [CrossRef]

20. Kim, M.J.; Kim, H.G.; Kim, H.S. Research on the Elevator-operating Noise and Vibration in Apartment Buildings. In Proceedings of the Korean Society for Noise and Vibration Engineering (KSNVE), Pyeongchang, Korea, 14–16 November 2001; pp. 488–493. Available online: http://www.dbpia.co.kr/journal/articleDetail?nodeId=NODE02451458 (accessed on 23 July 2020). (In Korean)

21. Kim, H.G.; Kim, M.J. *Reduction Method of Elevator-operating Noise and Vibration in Apartment Housings*; Land and Housing Institute, Korea Land and Housing Corporation: Seongnam, Korea, 2000; pp. 1–128. Available online: https://dl.nanet.go.kr/SearchDetailView.do?cn=MONO1200102249 (accessed on 22 October 2020). (In Korean)

22. Lee, S.; Kim, J.; Kim, D. A Study on the Cause of Noise and Vibration of Elevators. In Proceedings of the Korean Society for Noise and Vibration Engineering (KSNVE), Seoul, Korea, 18–21 May 1994; pp. 94–99. Available online: http://www.auric.or.kr/User/Rdoc/DocRdoc.aspx?returnVal=RD_R&dn=114296#.X03YAcgzZnI (accessed on 27 July 2020). (In Korean)

23. Ingold, D. Mitigating Elevator Noise in Multifamily Residential Buildings. Available online: http://buildipedia.com/aec-pros/construction-materials-and-methods/mitigating-elevator-noise-in-multifamily-residential-buildings?print=1&tmpl=component (accessed on 2 September 2020).

24. Zhou, Y.Q. In-car noise reduction for a newly developed home elevator. *J. Acoust. Soc. Am.* **1997**, *101*, 3018. [CrossRef]

25. Jeon, E.S.; Cho, B.H. The countermeasure which reduces the noise and vibration of the building elevator. *J. Korean Digit. Archit. Inter. Assoc.* **2005**, *5*, 35–42. (In Korean)

26. Watanabe, S.; Yumura, T.; Kariya, Y.; Hoshinoo, T. The Brake Noise Reduction Method for the Elevator Traction Machine. In Proceedings of the Transportation and Logistics Conference, Kawasaki, Japan, 8 December 2003; pp. 133–134. [CrossRef]

27. Torres, J.; Haugen, K. Noise Produced by Lift In Multi-Story Apartment Building, Case Study. In Proceedings of the Inter-noise 2019, Noise Control for a Better Environment, Madrid, Spain, 16–19 June 2019.

28. Kawasaki, R.; Hironaka, Y.; Tanaka, T.; Daikoku, A. Noise and vibration analysis of elevator traction machine. In *Nippon Kikai Gakkai Ronbunshu, C Hen/Transactions of the Japan Society of Mechanical Engineers, Part C*; Japan Society of Mechanical Engineers: Tokyo, Japan, 2010; Volume 76, pp. 2032–2038. [CrossRef]

29. Arrasate, X.; McCloskey, A.; Hernández, X.; Telleria, A. Optimum Design of Traction Electrical Machines in Lift Installations. In Proceedings of the Symposium on Lift & Escalator Technologies, Northampton, UK, 23–24 September 2015; Volume 5. Available online: https://www.researchgate.net/publication/282607214 (accessed on 28 July 2020).

30. Wang, X.W.; Yu, Y.J.; Zhang, R.J.; Wang, S.C.; Tian, Y. The summary research on the noise of high-speed traction elevators. *Appl. Mech. Mater.* **2014**, *541–542*, 716–721. [CrossRef]

31. Landaluze, J.; Portilla, I.; Cabezón, N.; Martínez, A.; Reyero, R. Application of active noise control to an elevator cabin. *Control Eng. Pract.* **2003**, *11*, 1423–1431. [CrossRef]

32. Yu, S.; Pan, P.; Wang, H.; Chen, L.; Tang, R. Investigation on noise and vibration origin in permanent magnet electrical machine for elevator. In Proceedings of the ICEMS 2005 Proceedings of the Eighth International Conference on Electrical Machines and Systems, Nanjing, China, 27–29 September 2005; Volume 1, pp. 330–333. [CrossRef]

33. Shi, L.Q.; Liu, Y.Z.; Jin, S.Y.; Cao, Z.M. Numerical simulation of unsteady turbulent flow induced by two-dimensional elevator car and counter weight system. *J. Hydrodyn.* **2007**, *19*, 720–725. [CrossRef]

34. Salmon, J.K.; Yoo, Y.S. Reduction of noise and vibration in an elevator car by selectively reducing air turbulence. *J. Acoust. Soc. Am.* **1991**, *90*, 3387–3388. [CrossRef]

35. Yang, I.H.; Jeong, J.E.; Jeong, U.C.; Kim, J.S.; Oh, J.E. Improvement of noise reduction performance for a high-speed elevator using modified active noise control. *Appl. Acoust.* **2014**, *79*, 58–68. [CrossRef]

36. Tray Edmonds, P.E.; Fullerton, J.L. Noise and vibration control of an offset traction elevator system. In Proceedings of the INTER-NOISE 2015—44th International Congress and Exposition on Noise Control Engineering, San Francisco, CA, USA, 9–12 August 2015. Available online: https://scholar.google.com/scholar?hl=ko&as_sdt=0%2C5&q=Noise+and+vibration+control+of+an+offset+traction+elevator+system&btnG= (accessed on 25 August 2020).

37. Park, C. *Elevator Noises and Vibrations. Research on Construction Technology*; Ssangyong Construction Co. Ltd.: Seoul, Korea, 2005; Volume 34, pp. 50–57. Available online: http://www.auric.or.kr/User/Rdoc/DocRdoc.aspx?returnVal=RD_R&dn=172387#.X02lu8gzZnI (accessed on 5 August 2020). (In Korean)

38. Song, K.H. Solutions to mitigate elevator noise and vibration. *J. ACRS Korean Assoc. Air Cond. Refrig. Sanit. Eng.* **1999**, *16*, 65–76. (In Korean)

39. Nam, K. *Solutions to Mitigate High-Speed Elevator Noise and Vibration, Construction Management and Technology*; Samsung Construction Co. Ltd.: Seoul, Korea, 1997; Volume 44, pp. 32–41. (In Korean)

40. Emporis. Emporis Standards: High-Rise Building (ESN 18727). Available online: https://www.emporis.com/building/standard/3/high-rise-building (accessed on 5 August 2020).

41. Korea Land and Housing Corporation, Design Guidelines (Architecture), No. 1988. 2018, p. 3. Available online: https://blog.naver.com/pvc1120/221835099946 (accessed on 28 July 2020).

42. Korea Land and Housing Corporation, Design Guidelines (Electrical and Telecommunication Facilities), No. 1990. 2018, p. 138. Available online: https://blog.naver.com/PostView.nhn?blogId=pvc1120&logNo=221824014189&redirect=Dlog&widgetTypeCall=true&directAccess=false (accessed on 28 July 2020).

43. ASHRAE. *2019 ASHRAE Handbook—HVAC Applications: Chapter 49 Noise and Vibration Control (TC 2.6, Sound and Vibration Control)*; American Society of Heating, Refrigerating and Air-Conditioning Engineers: Atlanta, GA, USA, 2019.

44. National Elevator Industry, Inc. Building Transportation Standards and Guidelines NEII-1, NEII. Available online: https://nationalelevatorindustry.org/wp-content/uploads/2019/09/neii-1.pdf (accessed on 19 August 2020).

45. BSI. *BS 8233:2014 Guidance on Sound Insulation and Noise Reduction for Buildings*; The British Standards Institution: London, UK, 2014.

46. Hansen, C.H.; Goelzer, B. Engineering Noise Control. *J. Acoust. Soc. Am.* **1996**, *100*, 1279.

47. Bies, D.A.; Hansen, C.H.; Howard, C.Q. *Engineering Noise Control*, 5th ed.; CRC Press; Taylor & Francis Group: Boca Raton, FL, USA, 2018; pp. 3–7.

48. National Elevator Industry, Inc. *The Benefits of Machine Room Less Elevators, The Insider*; NEII: New York, NY, USA, November 2015. Available online: http://www.neii.org/insider/editions/20151112.pdf (accessed on 3 August 2020).

49. Elevatorpedia. Machine Room Less Elevator. Available online: https://elevation.fandom.com/wiki/Machine_room_less_elevator (accessed on 20 August 2020).

50. Crocker, M.J. *General Introduction to Noise and Vibration Transducers, Measuring Equipment, Measurements, Signal Acquisition, and Processing, Handbook of Noise and Vibration Control*; Crocker, M.J., Ed.; John Wiley & Sons, Inc.: Hoboken, NJ, USA, 2007; Chapter 35, pp. 417–434, ISBN 978-0-471-39599-7.

51. Ojanen, T. Aero-vibro Acoustic Simulation of an Ultrahigh-Speed Elevator. Master's Thesis, Tampere University of Technology, Tampere, Finland, 2016. Available online: https://trepo.tuni.fi/handle/123456789/23851 (accessed on 20 August 2020).

52. Foulkes, T.J. Dynamic absorbers to control elevator noise in buildings—A case study. *J. Acoust. Soc. Am.* **1992**, *91*, 2350. [CrossRef]

53. Research Team, Noise Control in Architecture, Science Teaching Kit for Senior Secondary Curriculum, Hong Kong Institute of Architects. 2012, p. 11. Available online: https://minisite.proj.hkedcity.net/hkiakit/getResources.html?id=4049 (accessed on 27 August 2020).

54. The Engineering ToolBox, NC-Noise Criterion. Available online: https://www.engineeringtoolbox.com/nc-noise-criterion-25.html#:~{}:text=%20Sound%20pressure%20levels%20are%20measured%20for%20different,8%208000%20Hz%20%3A%2045%20dB%20More%20 (accessed on 27 August 2020).

55. The Engineering ToolBox, NR—Noise Rating Curve, An introduction to the Noise Rating—NR—Curve developed by the International Organization for Standardization (ISO). Available online: https://www.engineeringtoolbox.com/nr-noise-rating-d_60.html (accessed on 27 August 2020).

56. Patent. WO2012176297A1—Elevator Landing Door Device—Google Patents. Available online: https: //patents.google.com/patent/WO2012176297A1/en (accessed on 21 August 2020).
57. Miller, R.S. Elevator shaft pressurization for smoke control in tall buildings. *Build. Environ.* **2011**, *46*, 2247–2254. [CrossRef]
58. Tamura, T.; Itoh, Y. Unstable aerodynamic phenomena of a rectangular cylinder with critical section. *J. Wind Eng. Ind. Aerodyn.* **1999**, *83*, 121–133. [CrossRef]
59. Oh, Y.; Joo, M.K.J.; Park, J.Y.; Kim, H.G.; Yang, K.S. Deviation of Heavy-Weight Floor Impact Sound Levels According to Measurement Positions. *J. Acoust. Soc. Korea* **2006**, *25*, 49–55. Available online: https://www.koreascience.or.kr/article/JAKO200614222983617.page (accessed on 28 August 2020). (In Korean)

Publisher's Note: MDPI stays neutral with regard to jurisdictional claims in published maps and institutional affiliations.

 *sustainability*

Article

Dynamic Optimization Model for Estimating In-Situ Production Quantity of PC Members to Minimize Environmental Loads

Jeeyoung Lim and Joseph J. Kim *

Department of Civil Engineering and Construction Engineering Management, Green BIM Laboratory, California State University Long Beach, Long Beach, CA 90840, USA; jeeyoung.lim@csulb.edu
* Correspondence: Joseph.Kim@csulb.edu; Tel.: +1-562-985-1679

Received: 20 August 2020; Accepted: 1 October 2020; Published: 5 October 2020

Abstract: CO_2 emissions account for 80% of greenhouse gases, which lead to the largest contributions to climate change. As the problem of CO_2 emission becomes more and more prominent, research on sustainable technologies to reduce CO_2 emission among environmental loads is continuously being conducted. In-situ production of precast concrete members has advantages over in-plant production in reducing costs, securing equal or enhanced quality under equal conditions, and reducing CO_2 emission. When applying in-situ production to real projects, it is vital to calculate the optimal quantity. This paper presents a dynamic optimization model for estimating in-situ production quantity of precast concrete members subjected to environmental loads. After defining various factors and deriving the objective function, an optimization model is developed using system dynamics. As a result of optimizing the quantity by applying it to the case project, it was confirmed that the optimal case can save 7557 t-CO_2 in CO_2 emissions and 6,966,000 USD in cost, which resulted in 14.58% and 10.53% for environmental loads and cost, respectively. The model developed here can be used to calculate the quantity of in-situ production quickly and easily in consideration of dynamically changing field conditions.

Keywords: in-situ production; environmental loads; CO_2 emission reduction; life cycle assessment; optimization model; system dynamics

---

1. Introduction

Due to climate change, problems such as droughts, heatwaves, and rising sea levels are globally occurring [1]. One of the biggest causes of climate change is greenhouse gas [2], and international regulations on greenhouse gas emissions are being strengthened [3]. CO_2 accounts for 80% of greenhouse gases, and the problem of CO_2 emission becomes more and more prominent [4]. In particular, research on sustainable technologies to reduce CO_2 emission among environmental loads is continuously being conducted [5–9]. The construction industry is recognized as a major cause of environmental pollution [10], and it is important to quantify and evaluate the environmental load.

In studies related to the calculation of environmental loads of construction, Tae et al. evaluated the CO_2 generated during the life cycle of a building and its economic efficiency to assess the environmental loads and costs of buildings that use plaster board drywall [11]. Priatla et al. proposed a methodology to quantify CO_2 emissions by life cycle in water supply construction projects [12]. Park et al. studied correlation analysis between the environmental load computed through life cycle assessment (LCA) using the database of national highway construction cases and the inventory of available information that can be extracted in the road planning stage [13]. Lee et al. developed and validated an environmental load estimating model for the New Austrian Tunneling Method (NATM) tunnel based on the standard quantity of major works in the early design phase [10]. These studies were conducted to develop decision-support tools using quantified environmental loads, or to evaluate environmental loads by applying life cycle cost (LCC) or environmental valuation methodologies.

In-situ production of precast concrete (PC) members not only reduces the cost by 14.5–39.4% compared to in-plant production, but can also result in equal or enhanced quality under equal conditions [14–18]. Through an experiment in which the amount of CO_2 emission reduction was analyzed according to the increase in quantity, it was proved that in-situ production is an eco-friendly technology with a high CO_2 reduction effect [19]. However, in order to apply in-situ production to a project, additional research is needed to calculate the optimal quantity considering the environmental load.

However, it is difficult to produce all quantities in situ due to various field constraints, as well as the given time. The in-situ production quantity is affected by various factors, such as lead-time, number of molds, and number of cranes [17,20]. Although quantity is an important factor that determines the in-situ production scale, it is difficult to estimate it because it is indirectly affected by most of the influencing factors. Therefore, a simple method is necessary to calculate the optimal quantity to apply in-situ production in real projects.

Therefore, the objective of this paper is to develop a dynamic optimization model for estimating the in-situ production quantity of PC members subjected to environmental load. By defining various factors and deriving the objective function, the optimization model is developed using system dynamics and then applied to a case project for verification. The model developed here can be used to calculate the possible quantities of in-situ production quickly and easily in consideration of dynamically changing field conditions.

This study is carried out as follows.

(1)   The influencing factors of in-situ production for CO_2 emission are defined.
(2)   The development process for the optimization model is explained.
(3)   The objective function to minimize CO_2 emission is derived.
(4)   A dynamic optimization model reflecting the objective function is developed.
(5)   The environmental loads for the members applied to the actual in-situ production are calculated.
(6)   The optimization model is simulated using Monte Carlo simulation.
(7)   The control range of each influencing factor is derived using system dynamics.
(8)   Dynamic optimization model is applied in the case project for verification.

2. Preliminary Study

Under equal production conditions, it was verified through experimental studies that in-situ production of PC members secures equivalent or enhanced quality compared to in-plant production while significantly reducing costs [17,18,20,21]. Lee studied the management model and necessary conditions for in-situ production of composite PC members, and suggested cost, quality, process, resources, and safety management as management factors [22]. Park et al. studied the manufacturing technology for in-situ production of ultra-high-strength PC piles and suggested optimal production conditions through experimental production [23]. Won et al. studied the energy efficiency of in-situ production of PC members using steam curing and suggested necessary equipment and production concepts [24]. Lim et al. introduced a detailed management process through a case study of in-situ production and showed that the cost effectiveness ratio increased as the quantity increased [18]. Kim et al. presented the embedded energy efficiency of their proposed precast concrete frames based on the material, structure, and construction characteristics [25]. These studies analyzed specific plans for in-situ production of PC members and items to be considered when planning and showed that they were advantageous in terms of quality and cost of in-situ production.

Recently, CO_2 has been one of the most important causes of global climate change [2], and a method that can apply in-situ production of PC members is needed. Several studies were conducted to reduce CO_2 emissions of the PC method. Dong et al. showed that the PC method is a way to reduce carbon emissions by comparing to the cast in-situ construction method through experiments using life cycle assessments [26]. Yepes et al. proposed a methodology to optimize CO_2 emissions when designing a PC road and developed a hybrid glowworm swarm optimization algorithm [27]. Kim and Chae presented that a considerable amount of CO_2 is emitted during the steam curing stage when making PCs, and proposed a method of evaluating CO_2 emissions throughout the PC life cycle using life cycle assessment [28]. Lim et al. calculated CO_2 emissions with the LCA method through an experimental study of in-situ production and determined that the amount of CO_2 emission reduction increases as the quantity of in-situ production increases [19]. These studies were reviewed from the eco-friendly perspective for field application through optimization and evaluation methods for the existing PC method's CO_2 emissions. In addition, it was confirmed that it is advantageous to apply in-situ production in terms of CO_2 emissions, but more research on minimizing CO_2 emissions is needed.

Several studies used system dynamics in the production and installation of PC members to analyze influencing factors. Tan et al. applied the Pull-Driven Scheduling (PDS) simulation technique for the Singapore Light Rail Transit (LRT) project to produce PC members. The installation process was dynamically analyzed [29]. Ballard et al. conducted a study to improve the productivity of in-plant production for PC members using dynamic analysis techniques based on the measured data [30]. Cho et al. expressed the production, transportation, and construction process for PC-structured apartment houses as an Entity–Relationship Diagram (ERD) [31]. Lim et al. analyzed the factors that influence the calculation of the in-situ production volume and applied the developed simulation model to six scenarios to derive the in-situ production volume [20]. These papers showed that the dynamic relationships of various influence factors have been considered, and more research using them is being conducted. Therefore, in order to minimize CO_2 emissions, this paper aims to calculate the optimal quantity considering the dynamic relationship of various factors such as lead-time, number of molds, and number of cranes.

3. Methodology for the Optimization Model

The authors describe system dynamics and the Monte Carlo simulation method as applied techniques for developing an optimization model for estimating CO_2 emissions, followed by the definition of influence factors for in-situ production of PC members with details in step-by-step processes. Then, a dynamic optimization model is developed to minimize environmental loads in the sequence of a generation model, simulation model, and optimization model.

3.1. Applied Techniques for Simulation

3.1.1. System Dynamics

In carrying out in-situ production of PC materials, there is a limit to the complexity of the relationship between these influencing factors, which is why it is difficult to clarify this with general static analysis [17]. Static analysis is used because the one-way independent variable affects the dependent variable, it expresses the causes and effects of temporary events, and it views things from a partial perspective. Therefore, a system dynamics technique is needed as a means to grasp and quantify the dynamic relationship between influencing factors.

System dynamics can be defined as follows: (1) Rather than obtaining estimated values of a one-time event or variable, more attention is paid to what kind of dynamic change tends to occur over time in the variable of interest. (2) All phenomena are viewed from the perspective of an internal and cyclical closed-loop thinking, and are understood to be caused by circular dynamic interactions [32]. (3) The research focuses on the process of change and how it is actually happening. In other words, system dynamics is a methodology for understanding complex systems, and is a research methodology that explains the changes in the system over time, focusing on the causal relationship and the feedback relationship [33,34].

3.1.2. Monte Carlo Simulation

The dynamic optimization model was developed using the Powersim Studio 10 Expert program and the simulation was run by utilizing influencing factors such as in-situ production quantity, lead-time, number of molds, and number of cranes. Many values for factors must be derived through simulation, and control ranges for the values need to be set. A random number that follows the probability distribution for each factor is generated using Monte Carlo simulation. Using a computer program is the most effective way to generate a series of random numbers [35]. Monte Carlo simulation is performed by generating 100,000 random numbers, assuming a deviation of 0.1. Various values are presented by random numbers for each variable generated through Monte Carlo simulation. In other words, it attempted to overcome the mathematical limitations of the deterministic method by using the probabilistic method [36].

3.2. Definition of Influencing Factors for In-Situ Production

In-situ production of PC members is carried out in the same manner as in-plant production—applying demolding oil, pouring concrete, finishing the surface of the member, curing, demolding, and yard stocking [21]. In-situ production can be carried out at the same level as the factory by placing the assembled reinforcing bars in the steel formwork, as shown in Figure 1a. As shown in Figure 1b, steam curing is performed using a boiler, and the cured PC member is stacked. All processes are accomplished by establishing a production plan that can be supplied in the just-in-time delivery method of PC members according to the installation plan.

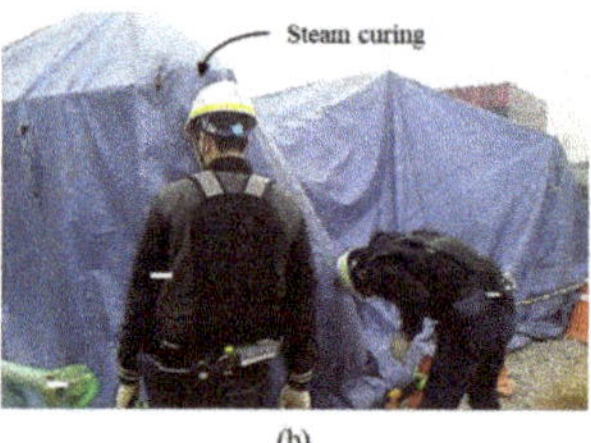

(a)   (b)

Figure 1. Major process of in-situ production: (**a**) manufactured steel mold [20]; (**b**) steam curing [18].

A survey conducted by Lim found that quantity was the most important influence factor for in-situ production [17]. The importance of influence factors was in the following order: number of cranes, number of molds, lead-time, yard stock area, production area, production cycle, erection cycle, material and traffic control, and crane location. In the study, only the factors that directly affect the calculation of the in-situ production quantity were considered by reflecting the results of the existing studies and the field application cases, and five influencing factors were selected, including cost, quantity, lead-time, number of molds, and number of cranes. Figure 2 is a causal loop diagram illustrating in-situ production cost, quantity, lead-time, number of molds, and number of cranes using system dynamics.

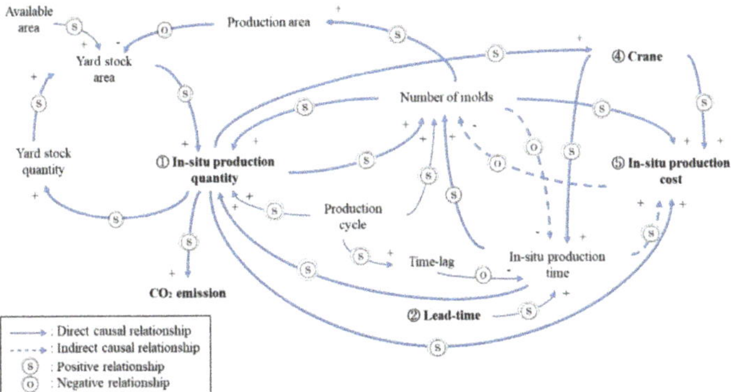

Figure 2. Causal loop diagram for estimating in-situ production quantity.

Details for each influencing factor are as follows.

1. In-situ production quantity: Since CO_2 emissions can be calculated only by the quantity, quantity is the factor that has the greatest influence on CO_2 emissions, and is a key influence factor and result of this study. Equation (1) shows how the quantity can be calculated using the in-situ production time, number of molds, and production cycle. In Figure 2, the in-situ production quantity affects the stock volume, and the stock quantity is determined by the difference between the accumulated production quantity and the accumulated installation quantity. As the stock volume increases, the yard stock area increases, and as the yard stock area increases, the in-situ production quantity can be increased.

$$Q_{SITU} = \frac{T_{SITU} \times N_{MOLD}}{T_{PC}} \tag{1}$$

where Q_{SITU}: in-situ production quantity (unit); T_{SITU}: in-situ production time (day); N_{MOLD}: number of molds (unit); T_{PC}: production cycle time (day).

2. Lead-time: When applying in-situ production, a separate process plan for the site is required. Figure 3 shows that the lead-time is the time of in-situ production in advance before the PC member is installed and is the period from the start of production of the PC member to the start of installation. Considering the curing period, not all PC members can be produced during the installation period, so members must be produced in advance [17,20]. As the in-situ production time increases, the amount of in-situ production available increases, so it is important to secure lead-time, as shown in Figure 2. The lead-time can be calculated using the production cycle, quantity during lead-time, and number of molds, as shown in Equation (2).

$$T_{LEAD} = (T_{PC} \times Q_{SLi}) / N_{MTi} \tag{2}$$

where T_{LEAD}: lead-time (day); T_{PC}: production cycle time (day); Q_{SLi}: in-situ production quantity during lead-time (unit); N_{MTi}: number of mold types (unit); i: number of mold types $(1, \dots , n)$.

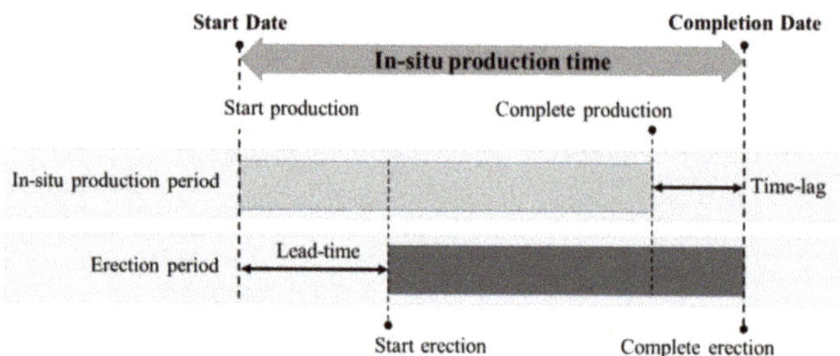

Figure 3. Calculation of in-situ production time [20].

3. Number of molds: As the number of molds increases, the quantity of production per unit of time is increased, resulting in a strong effect of shortening the time, while the cost of in-situ production increases rapidly. The reason is that the steel mold manufacturing cost is high. PC members of various sizes cannot be produced with molds of the same size. Table 1 shows five mold types classified according to the size of the member. Figure 2 shows that as the in-situ production quantity increases, the number of molds increases, and the number of molds affects the production area. The number of molds also affects the in-situ production quantity, so they affect each other. Since the in-situ production cost indirectly increases as the amount of time increases, it is necessary to calculate an appropriate number of molds through a feedback routine. The number of molds is a key influence factor in calculating the quantity of in-situ production, considering the time and cost. In Equation (3), the number of molds can be calculated by using in-situ production quantity, production cycle, and time.

$$N_{mold} = \sum_{i=1}^{n} \frac{Q_{MOLDi} \times T_{PC}}{T_{SITU}}, \tag{3}$$

where N_{MOLD}: number of molds (unit); Q_{MOLDi}: in-situ production quantity of each mold type (unit); T_{PC}: production cycle time (day); T_{SITU}: in-situ production time (day); i: number of mold types $(1, \ldots, n)$.

Table 1. The shapes of precast concrete (PC) members.

PC Type	Mold Type	Size (m)
Column	Type 1	0.8×0.8–1.0×9.2
	Type 2	1.0×1.0–1.7×9.2
	Type 1	0.5×1.0–1.3×11.0
Beam	Type 2	$0.8{\sim}1.3 \times 2.0$–2.5×11.0
	Type 3 (I-girder)	0.6×2.0–$2.3 \times 18.0{\sim}23.0$

4. Number of cranes: The large-scale building covered in this paper has a large floor area but not a high number of floors, making it difficult to use a tower crane, so a mobile crane was used. A crane is used to move the module and to lift and install the PC members. Equation (4) shows that the number of cranes can be calculated by dividing the installation time by the product of the unit usage time per member and the number of installation members, and the number of cranes are an integer equal to or greater than 1.

$$N_{crane} = \frac{(T_{UE} \times Q_{SITU})}{T_{erec}} \tag{4}$$

$$\text{Subject to } N_{crane} \geq 1, \ integer$$

where N_{crane}: number of cranes (unit); T_{UE}: unit erection time (day); Q_{SITU}: in-situ production quantity (unit); T_{erec}: erection time (day).

5. In-situ production cost: The cost is less than the in-plant production cost and is proportional to the quantity and number of molds. If the cost is not satisfied, in-situ production cannot be applied, so it is a limiting condition for minimizing CO_2 emission. If the total production cost of PC components is high, the in-situ production volume, which is lower than the in-plant production cost, is increased [16]. Figure 2 shows that all influence factors affect cost and can finally be collected. Equation (5) shows that cost can be calculated by the number of mold types, unit mold production cost for mold type, in-situ production quantity, and unit PC member production cost.

$$C_{SITU} = \sum_{i=1}^{n} N_{MOLDi} \times C_{MOLDi} + \sum_{i=1}^{n} Q_{MOLDi} \times C_{PRODi} \tag{5}$$

where C_{SITU}: in-situ production cost (USD); N_{MOLDi}: number of mold types (unit); C_{MOLDi}: unit mold production cost for mold type (unit); Q_{MOLDi}: in-situ production quantity for mold type (unit); C_{PRODi}: unit PC member production cost for mold type (USD); i: number of mold types (1, ..., n).

3.3. Dynamic Optimization Model to Minimize Environmental Loads

The dynamic optimization model was developed sequentially through the generation model and simulation model. The generation model determines one case, and the simulation model derives the control range through simulation. The optimization model is used to find the most suitable value for the field condition among simulated values. This development process is shown in Figure 4.

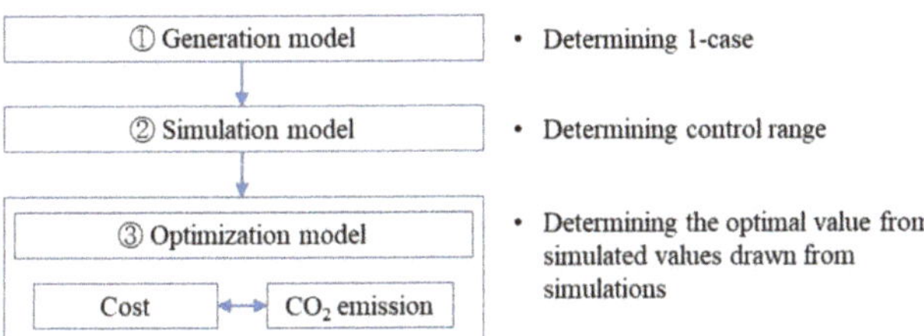

Figure 4. Development process for the optimization model.

3.3.1. Generation Model

The generation model determines one case by deriving the in-situ production cost and time for various influence factors. For instance, the 72 columns of in-situ production conducted in this experimental study are a generation model. Using the quantity calculated based on the design drawings, the CO_2 emissions for the 72 columns generated during in-situ production of total PC members were calculated. Influencing factors each have one value, and Figure 5 shows the mathematical connection of the relationship between influencing factors and CO_2 emissions. That is, an equation defines five influencing factors of in-situ production cost, quantity, lead-time, number of molds, and number of cranes, and calculates CO_2 emissions by substituting the influence factors into each equation. This generation model can be expressed as shown in Figure 5. All the influence factors have a dynamic relationship with each other, as shown in Figure 2.

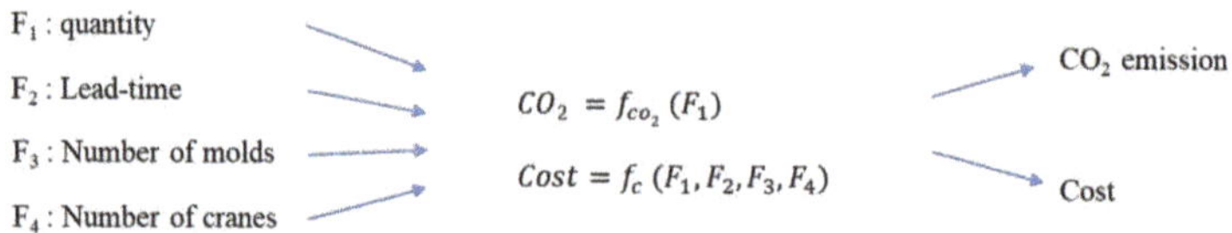

Figure 5. Concept of the generation model.

In this study, CO_2 emission sources, such as material, oil, electricity, and transportation, are classified and calculated for the analysis of the CO_2 emission reduction effect on in-situ production. After defining basic units of CO_2 emission or estimation equations for each source in advance, the quantity of CO_2 emissions generated by in-plant production is calculated using the quantity of sources. Next, the CO_2 emissions by in-situ production are calculated and compared with in-plant production. In order to calculate the material use, each basic unit of CO_2 emission per material quantity is used [37]. Concrete 140 kg-CO_2/m^3 and Steel 3500 kg-CO_2/t can be used to calculate CO_2 emissions. In the studies of Hong et al., Lee et al. and Lim et al., the CO_2 emissions generated in the construction stage of the building were calculated in the same way [19,38–42].

For the CO_2 emissions of the erection process, CO_2 emissions according to the use of oil and electricity must be calculated. Kim et al. analyzed using the LCA technique and proposed a CO_2 emission regression equation at the construction stage [37]. Since this regression equation has a gross floor area as a variable, it is easy to estimate CO_2 emission according to oil use and power consumption. First, the equation for calculating the CO_2 emissions of the construction work according to the use of oil in the construction stage is the same as Equation (7), which can be calculated using Equation (6), an equation for calculating energy consumption [37]. The oil use of in-situ and in-plant production was the same with the same production conditions. The equation for calculating CO_2 emissions at the construction stage according to power consumption is the same as Equation (9), which can be calculated using Equation (8), an equation for calculating energy consumption. CO_2 emissions by transportation equipment use are only applicable when moving PC members manufactured at the plant to the construction site. Unlike in-situ production, in-plant production requires transport of members, and basic units of CO_2 emission are 0.464 kg-CO_2/ton·km and 31.080 kg-CO_2/number of units of equipment [37].

$$E_{CO} = 0.0017 \times A_f + 37.5 \tag{6}$$

$$Q_{CO_2O} = E_{co} \times 3.06 \tag{7}$$

$$E_{CE} = 0.0247 \times A_f^{0.79} \tag{8}$$

$$Q_{CO_2E} = E_{ce} \times 1.64 \tag{9}$$

where E_{CO}: energy (oil) consumption during the construction stage; A_f: total floor area (m^2); Q_{CO2O}: CO_2 emissions based on oil use in the construction stage (t-CO_2); E_{CE}: power consumption in the construction stage; Q_{CO2E}: CO_2 emissions based on power consumption in the construction stage (t-CO_2).

3.3.2. Simulation Model

The simulation model was created based on the generation model [43]. The five influencing factors, such as in-situ production quantity, lead-time, number of molds, and number of cranes, can derive a range value according to the site conditions. The dynamic optimization model was developed using the Powersim Studio 10 Expert program in this paper and can be simulated using the derived influencing factors. Various values are presented by random numbers for each variable generated through the Monte Carlo simulation by using a number of cases as a result of the influencing factor, cost, and the minimum value, which is the control range of CO_2 emissions. The minimum value (Min) and maximum value (Max) can be derived, and Figure 6 is a schematic diagram.

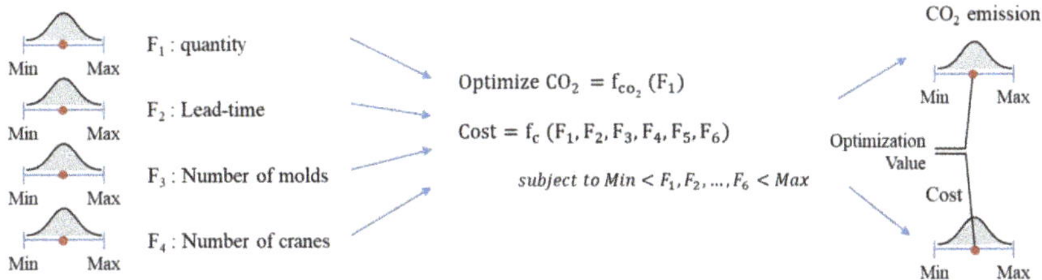

Figure 6. Concept of the simulation model.

3.3.3. Optimization Model

When the simulation model is developed, an optimization model is created using the derived values. The optimization model calculates one most appropriate value from the Min–Max of CO_2 emission, which is obtained from the results of the simulation model. It is possible to derive appropriate values of in-situ production cost, quantity, lead-time, number of molds, and number of cranes, which are influencing factors corresponding to optimal CO_2 emissions. Figure 7 is a schematic of the optimization model, and it can be derived using the Powersim Studio 10 Expert program.

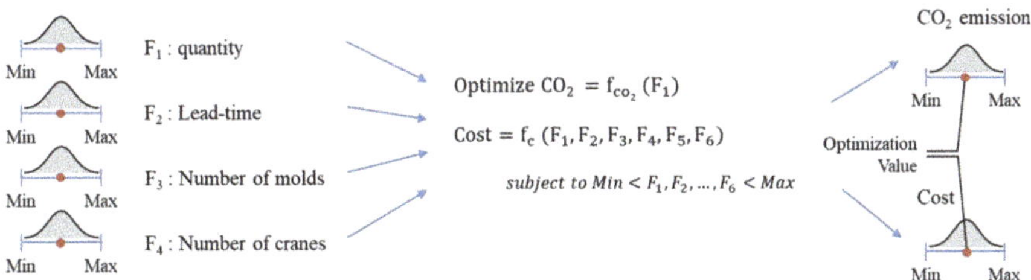

Figure 7. Concept of the optimization model.

Environmental load assessment measures whether CO_2 emission is minimized within the range possible for in-situ production. Environmental loads can be calculated for in-situ and in-plant production. The larger the difference between these values, the more CO_2 emissions are minimized. The cost, time, and yard stock area are within the allowable range. Equation (10) is an objective function and boundary condition to minimize environmental loads. Among the various values generated through the Monte Carlo simulation within the range satisfying these constraints, the maximum difference between in-plant and in-situ production for environmental loads is derived. Equations (11) and (12) are methods of estimating environmental loads for in-situ and in-plant production, respectively. They can be calculated by multiplying the quantity by the sum of the values for each item of material use, oil use, electronic use, and transportation equipment use. Transportation equipment use is calculated only in the case of in-plant production, and these values can be accumulated by member type.

$$Maximize\ f_{co_2}(Q_i) = Q_{CO_2}P - Q_{CO_2}S$$
$$Subject\ to\ C_{req} \leq C_{avail}$$
$$T_{req} \leq T_{avail} \tag{10}$$
$$A_{req} \leq A_{aval}$$

$$Q_{CO_2P} = \sum_{i=1}^{n} \left[Q_{Pi} \times \left(Q_{CO_2M} + Q_{CO_2O} + Q_{CO_2E} + Q_{CO_2T} \right) \right] \tag{11}$$

$$Q_{CO_2S} = \sum_{i=1}^{n} \left[Q_{Si} \times \left(Q_{CO_2M} + Q_{CO_2O} + Q_{CO_2E} \right) \right] \tag{12}$$

where Q_{CO2P}: CO$_2$ emissions of in-plant production (t-CO$_2$); Q_{CO2S}: CO$_2$ emissions of in-situ production (t-CO$_2$); C_{req}: required cost (USD); C_{avail}: available cost (USD); T_{req}: required time (month); T_{avail}: available time (month); A_{req}: required area (m^2); A_{avail}: available area (m^2); Q_{Pi}: in-plant production quantity of mold types (t-CO$_2$); Q_{CO2M}: CO$_2$ emissions of material use (t-CO$_2$); Q_{CO2O}: CO$_2$ emissions of oil use (t-CO$_2$); Q_{CO2E}: CO$_2$ emissions of power consumption (t-CO$_2$); Q_{CO2T}: CO$_2$ emissions of transport equipment use (t-CO$_2$); Q_{Si}: in-situ production quantity of mold types (t-CO$_2$); i: number of mold types (1, … , n).

4. Development of the Dynamic Optimization Model

Based on the previously mentioned causal loop diagram and the dynamic optimization model, the cost and CO$_2$ emission simulation models were created using the Powersim Studio 10 Expert program. The mold type derived based on the previously analyzed PC column and PC beam shape was applied.

4.1. Cost Model

The cost simulation model created for one mold type is shown in Figure 8. From the cost model, the in-situ production time is calculated using time-lag, lead-time, installation start date, and installation completion date (A). By entering the production cycle of two days and five working days per week, the number of parts per week is calculated, and the number of units of production is applied to the calculation of the number of molds together with the quantity and the in-situ production time (B). Material cost, labor cost, and equipment cost can be calculated using these calculated values.

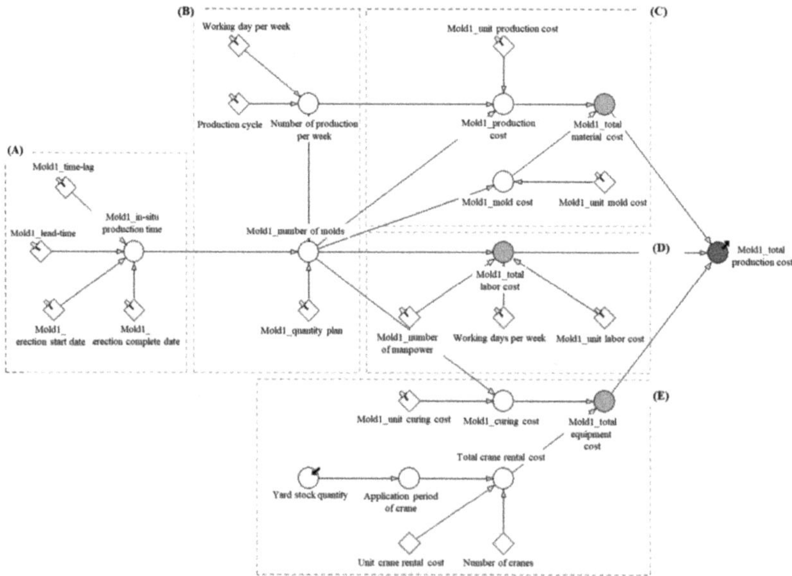

Figure 8. Cost estimation simulation model (mold type 1 for in-situ production); in-situ production time (A), production time (B), total material cost (C), total labor cost (D), total equipment cost (E).

Material cost is calculated by dividing into PC member production cost and mold cost (C). PC member production cost is calculated using the material cost for the production of one member composed of concrete and rebar, the number of parts produced per week, and the quantity of production, and the mold cost can be calculated using the number of molds and the unit cost of the mold. Labor cost is calculated as the input manpower during the period in which the mold was used. The labor cost is calculated using the number of molds, the amount of manpower, the number of working days per week, and the labor unit cost (D). Equipment cost is classified into curing cost and crane rental cost (E). Curing cost is calculated using the number of molds and curing unit cost. Since the stack quantity is determined by the difference between the production quantity and the installation quantity occurring over time, the crane utilization period is calculated based on the quantity of the yard. The crane rental cost is determined by the crane utilization period, unit rental cost, and number of units. The total production cost for one mold type is calculated by summing the material cost, labor cost, and equipment cost. For the production of PC members, the total in-situ production cost is calculated by summing the production cost for each mold type. The calculated in-situ production cost is then compared with in-plant production to determine whether the cost is reduced.

4.2. Environmental Load Model

The environmental load simulation model developed for one mold type is shown in Figure 9. From the perspective of in-plant production, environmental load is calculated by dividing into material use (B), oil use, electric use (C), and transport equipment use (D). First, in (A) of Figure 9, material use is calculated by summing reinforcement works of high-tensile deformed-bar (HD) 13, HD 16, HD 19, super-high-tensile deformed-bar (SHD) 22, SHD 32, SHD 35, embedded steel, etc. by checking the drawing to calculate the rebar weight of the member corresponding to mold type 1. Material use is calculated using this rebar weight, and the amount of concrete is calculated by checking the drawing, the actual steel form used, and the CO_2 base unit of steel and concrete. Oil use and electric use are calculated using the total floor area and number of total members, while transport equipment use is calculated using trailer size, distance from site to factory, CO_2 emission base unit of distance, and transport equipment use.

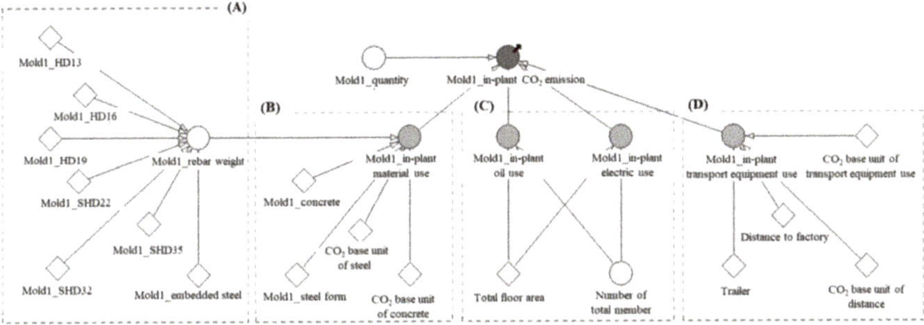

Figure 9. Environmental load simulation model (mold type 1 for in-plant production); material use (A), in-plant production material use (B), oil and electric use (C), transport equipment use (D)

Total CO_2 emissions are calculated by summing material use, oil use, electric use, and transport equipment use. If the total CO_2 emissions for one mold type of in-plant production are calculated, the total in-plant production CO_2 emissions are calculated by summing the production cost for each mold type. CO_2 emissions for in-situ production are calculated excluding the use of transport equipment in all processes. The calculated in-situ production CO_2 emissions can be compared with the in-plant production to determine the degree of CO_2 emission reduction.

5. Application of the Dynamic Optimization Model

5.1. Estimation of Environmental Loads for the Case Study

5.1.1. Estimation of In-Situ Production Quantity

A case study was selected to apply the developed dynamic optimization model. The case project is located in Cheonan-si, Chungcheongnam-do, Republic of Korea, and has a site area of 53,055.60 m^2, building area of 42,406.07 m^2 (246 m long × 178 m width), and total floor area of 167,614.82 m^2. The case project has a scale of four stories above the ground, with 2–4 stories above the ground with a PC structure, a core structure of reinforced concrete, and a roof structure of steel. Therefore, this study targets the 2–4 floors above ground because they are built with a PC structure. A total of 72 members are produced in-situ in the case project. The members to be constructed are columns, girders, and slabs. However, the members capable of in-situ production are limited to columns and girders that require a small production area. The columns and girders are thin and long, so the production space is not wide, but slabs require a large space, making in-situ production difficult in a limited area.

The number of PC members capable of in-situ production at the case site is 1,004 columns and 1252 girders, a total of 2256 members. Table 2 shows that the quantity per column and the 72 columns, as well as the quantity per girder, are calculated. As the resource input for column production is the same in in-situ and in-plant production, the quantity is calculated the same way. Since the material to be actually put in is the same, in-situ and in-plant production were calculated equally. It was confirmed that each quantity was almost proportional to the quantity of each column, as well as the number of columns. The reason is that all 72 columns have similar size and rebar details.

Table 2. Quantity of each column and the 72 columns.

Work	Item	Unit	Each Girder	Each Column	72 Columns
Concrete work	Concrete	m^3	20,957	6.444	464.0
	HD13	t	0.171	0.302	21.7
	HD16	t	0.314	0.104	7.5
	HD19	t	0.971	0.050	3.6
	SHD32	t		0.175	12.6
Reinforcement work	SHD35	t	-	1.301	93.7
	UHD25	t	0.569	-	-
	Embedded steel	t	0.007	0.060	4.3
	Sub-total	t	2.033	1.992	143.4
Form work	Steel form	t	0.010	0.016	1.1

Note. UHD: Ultra-high-tensile deformed-bar.

The material of the mold used in the case site is the same as in the steel mold used in the factory. As shown in Figure 1a, the steel mold applied in the in-situ production is the same as in the in-plant production, and the same specification was ordered for manufacturers and suppliers to the plant. The steel form is not used once, but is reused at least 50 times. Therefore, the amount of CO$_2$ generated during one column's production is calculated and reflected. Steel molds have high durability and high cost. Therefore, if the number of uses is small and reuse is possible, resale is possible. In this paper, each mold was used 36 times and then resold. The quantity of steel forms per column can be calculated as follows. For 72 columns' in-situ production, the purchase cost of two molds was 24,942 USD, and after production, they were resold for 14,000 USD. In other words, the mold cost actually used as input for 36 columns' in-situ production was 5471 USD, and it was determined that 82 columns could be produced when converted. When the total weight of the steel form was 1.297 t and divided by the number of reuses (82 times), the steel form input for production of one column was calculated as 0.016 t.

5.1.2. Estimation of Environmental Loads (CO_2 Emission)

The CO_2 emissions for 72 columns, 1004 columns, and total members were calculated, as shown in Table 3, by using the previously calculated quantities for each column and each beam, as well as using the CO_2 emission calculation formula. For 72 columns, they were calculated as 65,155 kg-CO_2 for concrete, 505,947 kg-CO_2 for steel, and 571 t-CO_2 in total. Here, since the same amount of material was input by applying column members with the same size and reinforcement details, in-situ and in-plant production were calculated equally. In the case of in-situ production, if the total floor area of 167,615 m^2 and Equations (6) and (7) were substituted, the CO_2 emissions from oil use of total members were calculated as 987 t-CO_2. When the areas occupied by the production of members and the mold area were converted into the sum of 72 columns, the emissions were calculated as 147 t-CO_2. If they were calculated by substituting the same total floor area and using Equations (8) and (9), the CO_2 emissions from electric use of total members were calculated as 543 t-CO_2, and if 72 columns were calculated, they were calculated as 40 t-CO_2.

Table 3. CO_2 emission comparison of 72 columns, 1,004 columns, and total members (unit: t-CO_2).

Classification	In-Situ			In-Plant		
	72 Columns	1004 Columns	Total Members	72 Columns	1004 Columns	Total Members
Material use	571	12,940	32,169	571	12,940	32,169
Oil use	147	987	987	147	987	987
Electricity use	40	543	543	40	543	543
Transport equipment use	-	-	-	178	2421	5397
Total	758	14,470	33,699	936	16,891	39,095

In the case of in-plant production, overhead and gantry cranes installed in the factory were used, and an area of 41,292 m^2 used in the production of the members was applied. The CO_2 emissions from transportation equipment use were applied only in the case of in-plant production; the equipment used for transportation is a 25 t trailer, and the distance from the factory to the site is 97.55 km. That is, in the case of 72 columns, CO_2 emissions from in-situ production were calculated as 758 t-CO_2, and from in-plant production, they were 936 t-CO_2, so CO_2 emissions from in-situ production were reduced by 178 t-CO_2 compared to in-plant production.

The CO_2 emissions from 1004 columns were calculated as 14,470 t-CO_2 for in-situ production and 16,891 t-CO_2 for in-plant production, so in-situ production decreased emissions by 2421 t-CO_2 compared to in-plant production. The quantity of 1004 columns and CO_2 emissions are not 1004 times those of one column. The reason is that 72 columns were selected as the largest number of members, because the size of these members and the amount of reinforcement are smaller than those of other columns. As a result of calculating the CO_2 emissions of total columns and girders (1004 columns, 1252 girders), CO_2 emissions from in-situ production were calculated as 33,699 t-CO_2, and from in-plant production, they were calculated as 39,095 t-CO_2. CO_2 emissions from in-situ production was reduced by 5397 t-CO_2 compared to in-plant production, and it was confirmed that CO_2 emissions increased as the quantity of in-situ production increased. The CO_2 emissions of in-situ production without PC were reduced by more than 13.8% compared to in-plant production. Material use accounts for more than 61.0% of the total CO_2 emissions, so the quantity has the greatest effect on the CO_2 emissions. Among them, the environmental loads of the members increased as the amounts of reinforcing bars with large basic units of CO_2 emissions increased.

Table 4 shows that the costs of in-plant and in-situ production with 72 columns applied were calculated by dividing them into material cost, labor cost, equipment cost, transport cost, and overhead and profit (O&P). The cost applied in this paper was the unit price contracted with the PC company and the cost input for actual in-situ production. Service costs were used for transportation costs, and data from PC factories were used for O&P. When calculating the cost, only direct costs, excluding overhead costs, are calculated. The reason for this is that even if the PC member is produced in-plant, as in in-situ production, on-site management costs and on-site land costs are required, so there is no need to additionally calculate additional overhead costs. In-plant production cost is 200,648 USD and in-situ production cost is 160,544 USD, which reduces construction cost by 40,104 USD for in-situ production compared to in-plant production. When 72 PC columns were produced in in-situ production, the cost of each column was reduced by 20% compared to in-plant production.

Table 4. Production cost analysis of in-plant and in-situ production.

Description	In-Plant Production (USD)	In-Situ Production (USD)
Material cost	100,587	100,587
Labor cost	49,939	59,756
Equipment fee	3150	201
Transport cost	15,168	-
Overhead and Profit	31,804	-
Total	200,648	160,544

5.2. Derivation of Control Range (Min–Max) by Factors

A Monte Carlo simulation was performed using the dynamic optimization model developed in this paper. Through the simulation, the number of possible occurrences for factors such as in-situ production quantity, lead time, number of molds, and number of cranes was derived, and the Min and Max for each factor were derived. It was assumed that the construction time was within the allowable range, and this was considered to comply with the 18 months required by the owner and aimed to reduce environmental loads by 10% within the range of cost reduction by more than 10%.

As a result of deriving the management range of Figure 10a, for in-situ production quantity, the number of members was calculated as Min 1757 members and Max 2256. Column type 1 is Min 704 and Max 736, column type 2 is Min 239 and Max 268, beam type 1 is Min 60 and Max 96, beam type 2 is Min 637 and Max 960, and Beam Type 3 is Min 117 and Max 196. The reason for why the number of members does not decrease below a certain number is that the cost reduction ratio increases as the number of members increases. In Figure 10b, as a result of deriving the Min–Max range of lead-time, Min was set to 3.8 weeks and Max was set to 8 weeks. The reason for why the lead-time does not decrease to less than 3.8 weeks is that only PC members that have been produced can be installed, so it is necessary to secure the production time of the members.

As a result of deriving the Min–Max range of Figure 10c, the number of molds was set to Min 37 units and Max 70 units. Column type 1 is Min 11 and Max 20, column type 2 is Min 5 and Max 10, beam type 1 is Min 2 and Max 7, beam type 2 is Min 16 and Max 26, and beam Type 3 is 3 Min and 7 Max. The reason that the number of individual molds does not decrease below a certain number is that if the number of molds decreases, the timely completion of the in-situ production is impossible. In Figure 10d, as a result of deriving the Min-Max range of the crane, Min was set to two units and Max was set to four units. The number of cranes is the result of calculating the in-situ production cost by rounding up the decimal point of the value derived by the simulation. If a crane is added during construction, it is advantageous to shorten the construction time, but this did not increase above a certain level because the cost and CO_2 emissions increase. Since the crane was partially added during construction, the number after the decimal point was derived.

In Figure 11, as a result of deriving the management range of in-situ production cost, Min was set to 615 USD and Max was set to 775 USD. In the case of CO_2 emission reduction, Min was set to 4557 t-CO_2 and Max was set to 5622 t-CO_2. That is, it is possible to save more than 14.38%. These values are the results derived for Min 1757 members and Max 2256 members. As a result of analyzing the management scope as described above, the cost and CO_2 emissions according to the fluctuations of each factor are generally proportional to each other; thus, a graph of similar shape was found. Using simulation, project participants can predict the management range for in-situ production volume, number of molds, lead-time, and number of cranes under various conditions. In addition, the management scope can be changed according to the site conditions.

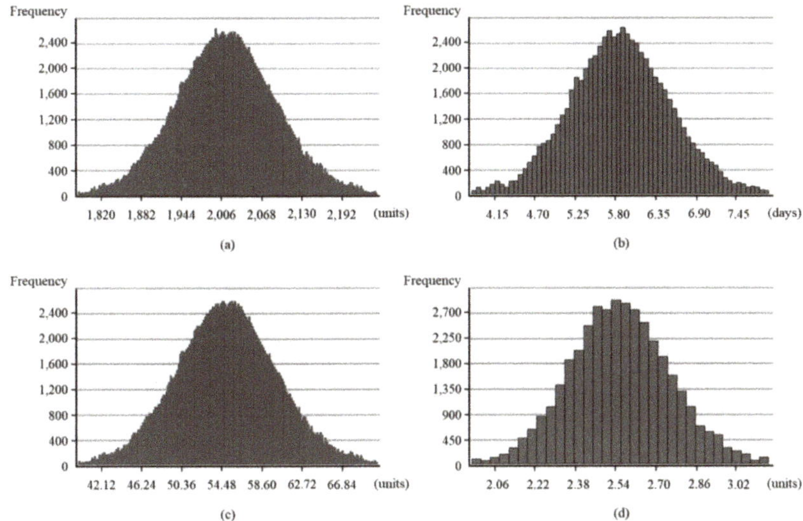

Figure 10. Control ranges of influencing factors: (**a**) in-situ production quantity; (**b**) lead-time; (**c**) number of molds; (**d**) number of cranes.

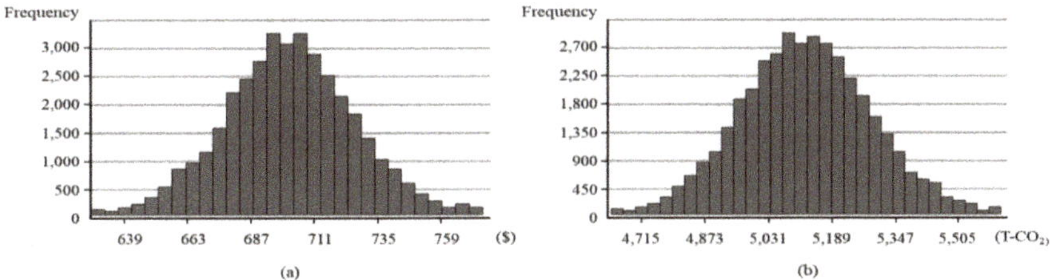

Figure 11. Control range of cost and CO_2 emissions: (**a**) in-situ production cost; (**b**) CO_2 emission reduction.

5.3. Optimization for Estimating In-Situ Production Quantity

The optimal case is derived by utilizing each management range, such as in-situ production volume, lead-time, mold number, and crane number, which were derived through the Monte Carlo simulation. Lim et al. explained six assumptions about the available area, which is selected as the main influencing factor, and derived the highest cost reduction rate among the scenarios applicable to the case project [20]. However, in this study, factors influencing CO_2 emissions are selected and the Min–Max of the influencing factor is derived through simulation. From the derived control range, an optimal value is derived to reduce environmental loads by 10% within a range that can reduce costs by 10% or more.

For simulation, it is assumed that the 18 months required by the client are observed. Table 5 shows the values of the influence factors corresponding to the highest value of the CO_2 emission reduction ratio of 14.58%. The quantity is 1,757 members, accounting for 78% of the total quantity, lead-time is 3.8 weeks, number of molds is 37 units, and number of cranes is two units. In-situ production reduced costs by 6,966,000 USD and CO_2 emission by 7557 t-CO_2 compared to in-plant production. In-situ production costs can be reduced by 10.53% compared to in-plant production. The proposed dynamic optimization model can derive an optimal case in consideration of the CO_2 emission reduction ratio and the cost reduction ratio. In addition, it can support decision-making on whether or not in-situ production can be applied to the field.

Table 5. Optimization result of estimating in-situ production quantity.

Item		Unit	Value
Quantity	Total quantity	ea	1757
	Ratio of quantity	%	78
Lead-time		weeks	3.8
Number of molds		ea	37
Cranes		ea	2
Production cost	In-plant	1000 USD	7752
	In-situ		6966
	Cost reduction ratio	%	10.53
CO_2 emission	In-plant	t-CO_2	23,698
	In-situ		31,255
	CO_2 emission reduction ratio	%	14.58

5.4. Discussion

The construction industry has the potential to improve productivity by applying prefabrication for building [44]. However, automation and robotics are still not common in the construction field [44,45]. Prefabrication has been plagued by dependence on conventional methods [46], complex interfacing [47], scheduling complexity [48], underutilization of factory space [44], cost barriers [49], fragmented information [50], and inconsistent quality [51].

The in-situ production proceeds in the order of installation of rebar setting, pouring concrete, curing, and stacking [16], and is performed at the same level as in-plant production. After stacking the cured PC members, quality checks and finishing work are done by the manpower if there is partial damage to the surface. In other words, PC members are manufactured at the factory, but the production process is mostly performed by manpower rather than automation or robotics. In the case of in-plant production, there is no actual mechanization except for the use of overhead and gantry cranes in the factory. It was assumed that the idle time of the crane used for erection of PC members is used during in-situ production in this study. In general, although the same sizes of PC columns and beams are designed, the rebar arrangements are not designed with the same PC members. In order for PC members to be manufactured, checked, and installed in the same manner as in the design drawing, work by manpower is required. Therefore, to apply in-situ production in the future, it is required to design PC members with the same size and reinforcement details.

Some of the members that have been produced in-plant have had to be reproduced due to issues with cracks, size, and breakage, as well as rebar arrangements that are different from the drawings. According to an interview with a PC factory official, it was confirmed that if the factory owner does not obtain more than 20% of the production cost as a profit, the official does not contract, because this does not cover the factory management overhead [17]. Through several studies, it has been possible to secure the quality and economic feasibility of in-situ production [14–17,19]. This study examined in-situ production at a construction site that complied with the PC production and installation guidelines of the KCI (Korea Concrete Institute); it was confirmed that there were no problems with cracks, breakages, size, or strength quality standards.

This study focused only on the factors of in-situ production quantity, lead-time, number of molds, and number of cranes with respect to CO_2 emissions. It was assumed that the production time and area were sufficient. However, if various field conditions, such as time and production, are additionally considered, different results will be derived. In-situ production can be applied only to sites where the production area is secured due to concerns about interference with ongoing construction, interference around the site, and safety issues.

In this study, a simulation for in-situ production was performed only for PC columns and beams. This means that slender parts, such as columns and beams, can be produced on-site. PC slabs, such as double-T and rib-plus slabs, occupy a large space during production, so in-plant production is more advantageous than in-situ production [18]. That is, in-plant production can be more advantageous depending on the type and number of members. Long-span and heavily loaded buildings, such as the case project examined in this study, can easily secure a production area due to the long distance between columns. However, different results from this study can be derived depending on the site characteristics and the building size.

6. Conclusions

This paper presented a dynamic optimization model that can dynamically predict, control, monitor, and manage five major influencing factors that affect environmental loads during in-situ production. In order to minimize environmental loads, the in-situ production quantity can be adjusted within the range of fluctuations in influence factors. The effect of the model was verified through the simulation results, and the findings are as follows.

First, the optimization model easily and quickly calculated the range of changes in environmental loads by analyzing the dynamic relationship of influencing factors. According to the simulation of the optimization model, in-situ production can reduce a Min of 4557 t-CO_2 and Max of 5622 t-CO_2 compared to in-plant production, which resulted in saving more than 14.58%. Second, the optimization model uses simulation results to create the control range of each influence factor to achieve the target environmental load reduction rate within the target cost reduction range. In the case project, the control range of the in-situ production quantity was 1757–2256 members to achieve the target environmental load reduction rate of 10% or more. In other words, project participants can predict the extent to which influencing factors are managed under various conditions. Third, the proposed dynamic model supports decision-making as to whether in-situ production can be applied in the field. The derived optimal case to minimize the environmental loads from the case project resulted in a quantity of 1757 members through the simulation model. In-situ production was able to reduce environmental loads by 14.58% and cost by 10.53% compared to in-plant production. Fourth, material use accounts for more than 61.0% of total CO_2 emissions, so its quantity has the greatest effect on CO_2 emissions. As the amount of reinforcing bar with a large basic unit of CO_2 emissions increases, the environmental loads of the member increase. Overdesign should be avoided.

The model developed in this paper can control the influence factors of in-situ production and can easily and quickly simulate the influences of factors that change during project execution to derive the optimum value according to the situation. When applying in-situ production using this model, environmental loads of CO_2 emissions can be calculated at the design stage. Furthermore, it is possible to evaluate whether it is applicable at the initial stage of the project in order to establish and review a construction plan. All these values can be used to analyze economic and environmental feasibility. In addition, they can be used to develop a data management system and build a risk management model for analyzing environmental loads of in-situ production in the future. Further research is needed to calculate the optimal in-situ production quantity considering environmental load through more field applications.

Author Contributions: Conceptualization, J.L.; methodology, J.L.; validation, J.L.; formal analysis, J.L.; investigation, J.L.; data curation, J.L.; writing—original draft preparation, J.L. and J.J.K.; writing—review and editing, J.L. and J.J.K.; visualization, J.L.; supervision, J.J.K.; project administration, J.L and J.J.K.; funding acquisition, J.L. All authors have read and agreed to the published version of the manuscript.

Funding: This work was supported by the National Research Foundation of Korea (NRF) grant funded by the Korea government (MOE) (No. NRF-2019R1A6A3A12032427).

Conflicts of Interest: The authors declare no conflict of interest. The funders had no role in the design of the study; in the collection, analyses, or interpretation of data; in the writing of the manuscript, or in the decision to publish the results.

References

1. Intergovernmental Panel on Climate Change (IPCC). *Working Group I Contribution to the IPCC Fifth Assessment Report Climate Change 2013*; Cambridge University Press: Cambridge, UK, 2013.
2. Jung, K.O.; Chung, Y.K. The pollution and economic growth based on the multi-country comparative analysis. *J. Ind. Econ. Bus.* **2004**, *17*, 1077–1098.
3. Sartori, I.; Hestnes, A.G. Energy use in the lifecycle of conventional and low energy buildings: A review article. *Energy Build.* **2007**, *39*, 249–257. [CrossRef]
4. Giesekam, J.; Barrett, J.; Taylor, P.; Owen, A. The greenhouse gas emissions and mitigation options for materials used in UK construction. *Energy Build.* **2014**, *78*, 202–214. [CrossRef]
5. Ecoinvent. Ecoinvent Database. Available online: http://www.ecoinvent.org/ (accessed on 1 August 2019).
6. Giama, E. Life cycle versus carbon footprint analysis for construction materials. In *Energy Performance of Buildings*; Springer: Cham, Switzerland, 2016; pp. 95–106. [CrossRef]
7. Jeong, K.; Hong, T.; Kim, J. Development of a CO_2 emission benchmark for achieving the national CO_2 emission reduction target by 2030. *Energy Build.* **2018**, *158*, 86–90. [CrossRef]
8. Lee, M.C. Reducing CO_2 emissions in the individual hot water circulation piping system. *Energy Build.* **2014**, *84*, 475–482. [CrossRef]
9. Mergos, P.E. Seismic design of reinforced concrete frames for minimum embodied CO_2 emissions. *Energy Build.* **2018**, *162*, 177–186. [CrossRef]
10. Lee, J.H.; Kim, K.; Kim, H. Environmental load estimating model of NATM tunnel based on standard quantity of major works in the early design phase. *KSCE J. Civ. Eng.* **2018**, *22*, 1040–1051. [CrossRef]
11. Tae, S.; Shin, S.; Kim, H.; Ha, S.; Lee, J.; Han, S.; Rhee, J. Life cycle environmental loads and economic efficiencies of apartment buildings built with plaster board drywall. *Renew. Sustain. Energy Rev.* **2011**, *15*, 4145–4155. [CrossRef]
12. Priatla, K.; Ariaratnam, S.; Cohen, A. Estimation of CO_2 emissions from the life cycle of a potable water pipeline project. *J. Manag. Eng.* **2012**, *28*, 22–30. [CrossRef]
13. Park, J.Y.; Lee, D.E.; Kim, B.S. A study on analysis of the environmental load impact factors in the planning stage for highway project. *KSCE J. Civ. Eng.* **2016**, *20*, 2162–2169. [CrossRef]
14. Hong, W.K.; Lee, G.; Lee, S.; Kim, S. Algorithms for in-situ production layout of composite precast concrete members. *Autom. Constr.* **2014**, *41*, 50–59. [CrossRef]
15. Lim, C. Construction Planning Model for In-situ Production and Installation of Composite Precast Concrete Frame. Ph.D. Thesis, Kyung Hee University, Seoul, Korea, 2016.
16. Oh, O.J. A Model for Production and Erection Integration Management of Large Scale PC Structures Using System Dynamics. Ph.D. Thesis, Kyung Hee University, Seoul, Korea, 2017.
17. Lim, J. A Risk Management Model for In-situ Production of Precast Concrete Members Focused on Time and Cost Using System Dynamics. Ph.D. Thesis, Kyung Hee University, Seoul, Korea, 2018.
18. Lim, J.; Park, K.; Son, S.; Kim, S. Cost reduction effects of in-situ PC production for heavily loaded long-span buildings. *J. Asian Archit. Build. Eng.* **2020**, *19*, 242–253. [CrossRef]
19. Lim, J.; Kim, S. Evaluation of CO_2 emission reduction effect using in-situ production of precast concrete components. *J. Asian Archit. Build. Eng.* **2020**, *19*, 176–186. [CrossRef]
20. Lim, J.; Kim, S.; Kim, J.J. Dynamic simulation model for estimating in-situ production quantity of PC members. *Int. J. Civ. Eng.* **2020**, *18*, 935–950. [CrossRef]
21. Na, Y.J.; Kim, S.K. A process for the efficient in-situ production of precast concrete members. *J. Reg. Assoc. Archit. Inst. Korea* **2017**, *19*, 153–161.
22. Lee, G.J. A Study of In-situ Production Management Model of Composite Precast Concrete Members. Ph.D. Thesis, Kyung Hee University, Seoul, Korea, 2012.
23. Park, S.H.; Lim, C.Y.; Lee, W.J.; Kim, D.S.; Jung, Y.S. The experimental study on concrete manufacturing technologies for ultra high strength concrete pile. In Proceeding of the 2013 Spring Annual Conference of the Korea Concrete Institute, Seoul, Korea, 8–10 May 2013; Volume 25, pp. 67–68.
24. Won, I.; Na, Y.; Kim, J.T.; Kim, S. Energy-efficient algorithms of the steam curing for the in situ production of precast concrete members. *Energy Build.* **2013**, *64*, 275–284. [CrossRef]

25. Kim, S.; Hong, W.K.; Kim, J.H.; Kim, J.T. The development of modularized construction of enhanced precast composite structural systems (Smart Green frame) and its embedded energy efficiency. *Energy Build.* **2013**, *66*, 16–21. [CrossRef]

26. Dong, Y.H.; Jaillon, L.; Chu, P.; Poon, C.S. Comparing carbon emissions of precast and cast-in-situ construction methods—A case study of high-rise private building. *Constr. Build. Mater.* **2015**, *99*, 39–53. [CrossRef]

27. Yepes, V.; Martí, J.V.; García-Segura, T. Cost and CO_2 emission optimization of precast–prestressed concrete U-beam road bridges by a hybrid glowworm swarm algorithm. *Autom. Constr.* **2015**, *49*, 123–134. [CrossRef]

28. Kim, T.; Chae, C.U. Evaluation analysis of the CO_2 emission and absorption life cycle for precast concrete in Korea. *Sustainability* **2016**, *8*, 663. [CrossRef]

29. Tan, B.; Huat, D.C.K.; Messner, J.I.; Horman, M.J. Using simulation for pull-driven scheduling with buffer for precast concrete component fabrication and erection. In Proceedings of the 16th CIB World Building Congress, Rotterdam, The Netherlands, 1–7 May 2004; pp. 10–21.

30. Ballard, G.; Harper, N.; Zabelle, T. Learning to see work flow: An application of lean concepts to precast concrete fabrication. *Eng. Constr. Archit. Manag.* **2003**, *10*, 6–14. [CrossRef]

31. Cho, G.H.; Kim, J.J. Integrated management of the production, transportation and installation of precast concrete panels. *J. Archit. Inst. Korea* **1996**, *12*, 185–193.

32. Karnopp, D.C.; Margolis, D.L.; Rosenberg, R.C. *System Dynamics: Modeling, Simulation, and Control of Mechatronic Systems*, 5th ed.; John Wiley & Sons: New Jersey, NJ, USA, 2012; ISBN 9781118160077.

33. Kim, D.; Moon, T.; Kim, D. *Chap. System Dynamics*, 1st ed.; Daeyoung Book: Seoul, Korea, 1999; ISBN 9788976440600.

34. Forrester, J.W. Lessons from system dynamics modeling. *Syst. Dyn. Rev.* **1987**, *3*, 136–149. [CrossRef]

35. Kim, I. *Risk Management in the Construction Projects*, 1st ed.; Kimoondang: Seoul, Korea, 2001; ISBN 9788970864235.

36. Maio, C.; Schexnayder, C.; Knitson, K.; Weber, S. Probability distribution function for construction simulation. *J. Constr. Eng. Manag.* **2000**, *126*, 285–292. [CrossRef]

37. Korea Institute of Construction Technology (KICT). The Environmental Load Unit Composition and Program Development for LCA of Building, The Second Annual Report of the Construction Technology R&D Program. 2004. Available online: http://www.ndsl.kr/ndsl/search/detail/report/reportSearchResultDetail.do?cn=TRKO201000018952 (accessed on 20 May 2019).

38. Hong, W.K.; Kim, J.M.; Park, S.C.; Lee, S.G.; Kim, S.I.; Yoon, K.J.; Kim, H.C.; Kim, J.T. A new apartment construction technology with effective CO_2 emission reduction capabilities with effective CO_2 emission reduction capabilities. *Energy* **2009**, *35*, 2639–2646. [CrossRef]

39. Hong, W.K.; Park, S.C.; Kim, M.M.; Kim, S.I.; Lee, S.G.; Yun, D.Y.; Yoon, T.H.; Ryoo, B.Y. Development of structural composite hybrid systems and their application with regard to the reduction of CO_2 emissions. *Indoor Built Environ.* **2010**, *19*, 151–162. [CrossRef]

40. Lee, D.; Lim, C.; Kim, S. CO_2 emission reduction effects of an innovative composite precast concrete structure applied to heavy loaded and long span buildings. *Energy Build.* **2016**, *126*, 36–43. [CrossRef]

41. Lee, S.H.; Joo, J.K.; Kim, J.T.; Kim, S.K. An analysis of the CO_2 reduction effect of a column-beam structure using composite precast concrete members. *Indoor Built Environ.* **2012**, *21*, 150–162. [CrossRef]

42. Lim, C.; Lee, S.; Kim, S. Embodied Energy and CO_2 emission reduction of a column-beam structure with enhanced composite precast concrete members. *J. Asian Archit. Build. Eng.* **2015**, *14*, 593–600. [CrossRef]

43. Lee, K.; Son, S.; Kim, D.; Kim, S. A dynamic simulation model for economic feasibility of apartment development projects. *Int. J. Strateg. Prop. Manag.* **2019**, *23*, 305–316. [CrossRef]

44. Pan, M.; Pan, W. Determinants of adoption of robotics in precast concrete production for buildings. *J. Manag. Eng.* **2019**, *35*. [CrossRef]

45. Pan, M.; Pan, W. Advancing formwork systems for the production of precast concrete building elements: From manual to robotic. In Proceedings of the 2016 Modular and Offsite Construction (MOC) Summit, Banniff, AB, Canada, 29 September–1 October 2016; pp. 2–9.

46. Mao, C.; Shen, Q.; Pan, W.; Ye, K. Major barriers to off-site construction: The developer's perspective in China. *J. Manag. Eng.* **2015**, *31*. [CrossRef]

47. Pan, W.; Gibb, A.G.F.; Dainty, A.R.J. Strategies for integrating the use of off-site production technologies in house building. *J. Constr. Eng. Manag.* **2012**, *138*, 1331–1340. [CrossRef]

48. Arashpour, M.; Wakefield, R.; Abbasi, B.; Lee, E.W.M.; Minas, J. Off-site construction optimization: Sequencing multiple job classes with time constraints. *Autom. Constr.* **2016**, *71*, 262–270. [CrossRef]
49. Pan, M.; Linner, T.; Pan, W.; Cheng, H.; Bock, T. A framework of indicators for assessing construction automation and robotics in the sustainability context. *J. Clean. Prod.* **2018**, *182*, 82–95. [CrossRef]
50. Li, X.; Shen, G.Q.; Wu, P.; Fan, H.; Wu, H.; Teng, Y. RBL-PHP: Simulation of lean construction and information technologies for prefabrication housing production. *J. Manag. Eng.* **2018**, *34*. [CrossRef]
51. Hwang, B.G.; Shan, M.; Looi, K.Y. Key constraints and mitigation strategies for prefabricated prefinished volumetric construction. *J. Clean. Prod.* **2018**, *183*, 183–193. [CrossRef]

Article

Identifying Risk Indicators for Natural Hazard-Related Power Outages as a Component of Risk Assessment: An Analysis Using Power Outage Data from Hurricane Irma

Sang-Guk Yum [1], Kiyoung Son [2], Seunghyun Son [3] and Ji-Myong Kim [4],*

[1] Department of Civil Engineering and Engineering Mechanics, Columbia University, New York, NY 10027, USA; sy2509@columbia.edu
[2] School of Architectural Engineering, Ulsan University, Ulsan 44610, Korea; sky9852111@ulsan.ac.kr
[3] Department of Architectural Engineering, Kyung Hee University, Suwon 17104, Korea; seunghyun@khu.ac.kr
[4] Department of Architectural Engineering, Mokpo National University, Mokpo 58554, Korea
* Correspondence: jimy6180@gmail.com

Received: 4 August 2020; Accepted: 16 September 2020; Published: 17 September 2020

Abstract: Extensive use has been made of lifecycle-cost assessment to enhance the cost-effectiveness and resilience of facilities management. However, if such assessments are to be truly effective, supplemental information will be needed on the major costs to be expected over buildings' entire lives. Electricity generation and distribution systems, for example, are absolutely indispensable to industry and human society, not least in the operation of buildings and other infrastructure as networks. The widespread disruption that ensues when such power systems are damaged often carries considerable repair costs. Natural disasters likewise can cause extensive societal, economic, and environmental damage. Such damage is often associated with lengthy power outages that, as well as being directly harmful, can hinder emergency response and recovery. Accordingly, the present study investigated the correlations of natural hazard indicators such as wind speed and rainfall, along with environmental data regarding the power failure in Florida caused by Hurricane Irma in 2017 utilizing multiple regression analysis. The environmental data in question, selected on the basis of a thorough literature review, was tree density. Our analysis indicated that the independent variables, maximum wind speed, total rainfall, and tree density, were all significantly correlated with the dependent variable, power failure. Among these, rainfall was the least significant. Despite there being only three independent variables in the model, its adjusted coefficient of determination (0.512) indicated its effectiveness as a predictor of the power outages caused by Hurricane Irma. As such, our results can serve the construction industry's establishment of advanced safety guidelines and structural designs power transmission systems in regions at risk of hurricanes and typhoons. Additionally, insurance companies' loss-assessment modeling for power-system facilities would benefit from incorporating the three identified risk indicators. Finally, our findings can serve as a useful reference to policymakers tasked with mitigating power outages' effects on infrastructure in hurricane-prone areas. It is hoped that this work will be extended, facilitating infrastructure restoration planning and making societies and economies more sustainable.

Keywords: natural hazard; risk management; power system failure; disaster management

---

1. Introduction

Hurricanes, typhoons, and tropical cyclones have been occurring more frequently and increasing in intensity because of global warming [1–4]. Climate change also appears to affect the tracks of these extreme weather events, and thus the amount of damage they cause via high winds and flooding [5]. According to Kreimer and Amond [6] and Chang [7], this has increased both the direct and indirect costs of natural-disaster damage as a whole. A Munich Re report [8] revealed that the 2017 damage caused to the United States by hurricanes—around \$220 billion—was the highest ever recorded, due to the trio of Hurricanes Harvey, Irma, and Maria. Such events can cause severe economic, environmental, and societal damage, not least through lengthy interruptions to the supply of electric power, one of the most important of which is increases to the lifecycle costs of buildings and other infrastructure. In the U.S., the costs of damage specifically due to power outages have been increasing [9,10], and are closely associated with hurricanes. Hurricane Irene, for example, deprived 6.5 million people of power in 2011, while the parallel figure for Hurricane Sandy the following year was 8.5 million [9].

Many studies have investigated the increased damage caused by power outages linked to natural hazards and the ensuing social and economic problems [10–13]. Other research has emphasized that, because the severity of natural hazards is likely to increase in the future, making power systems resilient to such hazards will only become more difficult [14–16]. Some recent work on hurricane damage has focused not only on direct losses, but also on the enormous cost of restoration, which was \$150 billion in the U.S. between January 2004 and December 2005 [17]. Hurricane damage is an inescapable consequence of hurricanes striking built-up areas. Nevertheless, the creation of a prediction model for such damage can form an important part of planning for, and effective responses to, such events. Electric power failure has multiple socio-economic impacts, and can cause serious problems for hospitals, schools, and other critical public infrastructure. However, if utility companies and governments can understand the critical correlations between natural hazard variables and power outages, and design their emergency-response and mitigation plans accordingly, their localities' ability to respond to natural hazards will be greatly improved. Therefore, using multiple regression analysis of a U.S. electric power company's outage data from Hurricane Irma, the present study seeks to identify which hurricane variables are most closely related to power outages. This is a necessary first step toward the better emergency-response plans and prompt restoration of the power facilities mentioned above.

1.1. Research Background

Lifecycle assessment studies are frequently conducted, often focusing on buildings' design stages as a means of making their management more cost-effective. Sometimes, they look at facility management as an aspect of asset management. However, most lifecycle-cost studies pay relatively little attention to buildings' operation and management stages. Though natural hazards remain very difficult to predict, the frequency of severe ones has increased [18]. Moreover, Hurricanes Katrina in 2005, Ike in 2008, and Sandy in 2012 caused damage worth \$108 billion, \$29.5 billion, and \$71.4 billion, respectively [19,20]. It is well known that power outages have significant economic impacts on the operation of facilities essential to society. The negative effects of significant damage to power-infrastructure systems induced by hurricanes and other storms proliferate outward to households, healthcare providers, schools, government facilities, and whole communities, economies, and societies [21]. Protection of these vulnerable lifelines, or—where protection is impossible—their rapid restoration, is essential to reducing risks to life and property [22]. Therefore, assessment of the likely impact of extreme natural hazards would seem critical to effective assessment of the costs of operating and managing buildings and infrastructure over their entire lifecycles, as well as to decision-making for utility companies and emergency managers.

By the same token, blackouts are an important metric of hurricanes' impacts on particular areas. The present research addresses hurricane-related power outages' importance to the operation and management stage of building and infrastructure system lifecycles from a sustainability viewpoint, utilizing actual power-failure data from when Hurricane Irma struck Florida.

1.2. Research Objective

The principal goal of this research is to assess the significance of specific hurricane and power outage variables through multiple regression modeling of historical damage data. First, based on a review of the relevant literature, it will attempt to capture the major expected factors in hurricane-related power. Second, it will use multiple-regression analysis to reveal the relationship between 1326 power outages (the dependent variable) and various characteristics of Hurricane Irma and other aspects of the natural environment (the independent variables). Specifically, this power outage data consists of relative outage frequency (ROF), meaning the number of power outage events per 0.1 million people. The natural hazard indicators include wind speed and rainfall. Tree density will also be used, since the immediate cause of power outages during natural disasters is often fallen trees [23].

2. Literature Review

According to Sissine [24], population growth and rapid urbanization have rendered high-performance buildings an important possible solution for sustainability. If the need for such buildings is to be met, however, their long-term environmental and economic sustainability should be considered simultaneously [25]. The U.S. Green Building Council [26] has proposed that the three main sustainability factors for quality of life are the present and future harmony of society, the economy, and the environment. Thus, balanced long-term pursuit of sustainable development should take account of all three of these factors across buildings' entire lifecycles. To date, however, scholarship on buildings' sustainability performance has focused on short-term social impacts such as fatalities and short-term economic ones such as repair costs, while largely ignoring long-term impacts, especially on the environment [27–30]. Wei et al. [31] suggested that environmental factors have received less attention than economic and social ones due to lack of consensus about what environmental criteria are important, and how they should be defined and measured. Among those studies that have investigated the environmental impacts of natural disasters that strike buildings and other built infrastructure, Chang and Shinozuka's [32] work on seismic risk is perhaps especially important. It yielded an expanded lifecycle-cost framework using a hypothetical water delivery system, with performance-level definitions and criteria for minimizing the lifecycle costs related to the repair of pipeline systems after earthquakes. Shinozuka et al. [33] subsequently proposed a lifecycle-cost estimation framework for bridges at risk of being struck by earthquakes, which includes not only initial bridge construction costs, but also damage repair costs. Similarly, Wei et al. [34] proposed that the estimation of buildings' true lifecycle costs for the purposes of sustainable development should include the potential costs associated with seismic events. Post-event recovery's impacts on systems' lifecycles has also recently been emphasized, with such discussions covering energy demand, building performance metrics, and sustainability rating systems [31,35–40].

According to Sinisuka and Nugraha [41], power generation's role in lifecycle cost includes both deterministic costs, such as asset acquisition and operation/management costs, and probabilistic ones, such as the costs of failures, repairs, and gross-loss margins. The same study reported that probabilistic costs were mainly associated with power systems' maintenance. Power outages related to natural hazards have also been recognized as a huge societal problem, in the first instance via research on earthquakes. For instance, Shinozuka et al. [33,42,43] led an investigation into power outages caused by earthquakes for the U.S.'s Multidisciplinary Center for Earthquake Engineering Research, mainly focusing on the restoration of power in the wake of extreme events. Power systems' fragility curves have also been investigated as a means of assessing damage from power outages before such events have taken place [44,45]. According to Guikema and Nateghi [46], many electric power companies use the Outage Management System (OMS) to record power outages, and historical data gathered from this system have contributed to effective prediction of the magnitude of power outages caused by natural hazards such as hurricanes. However, it remains difficult to forecast power outages using OMS, since extreme climatological events caused by global warming are more likely to happen in the future than they were in the past.

Hurricane-induced power outages cause considerable direct and indirect damage both to utility companies and to end-users of electricity, and thus impose significant restoration and recovery costs on whole affected regions [23]. They are caused by several factors, most notably high winds and flooding [46]. However, most of the relevant studies have focused on direct wind-induced damage to elements of power systems such as poles and power lines, as well as indirect wind damage, e.g., similar damage caused by fallen trees. Davison et al. [47], for example, concluded that wind gusts were correlated with power outages in a certain region, and that such damage could be mitigated by different types of vegetation cover; however, they did not consider other variables such as flooding and precipitation. Liu et al. [22] extended Davison et al.'s [47] work, but only by adding information such as the duration of the maximum wind speed and soil status. Additionally, Reed [48] made novel use of Geographic Information System data to study major storms in the northwestern U.S., but again concluded that storm-induced wind speed was the most impactful factor for power outages there. Nevertheless, Hurricane Sandy caused huge power outages in Manhattan due to inundation of the power system [46], and such possibilities deserve more scholarly attention than they have been given.

Among studies not narrowly focused on wind, Han et al. [49] introduced hurricane intensity and size as factors in power outages' spatial distribution on the U.S. Gulf Coast. The same study also linked physical damage to power systems to the time that elapsed between one hurricane and another. Guikema et al. [50] and Nateghi et al. [51] developed a model for predicting power outages prior to hurricanes, with more accuracy than previous studies of the same type that had relied solely on publicly available data. Even though such prior studies' datasets had been large, they arguably did not take account of wide enough arrays of natural hazard indicators. Balijepalli et al. [52] developed a promising method of estimating power outages related to lightning storms, using a bootstrap method and Monte Carlo simulations. Quiring et al. [53], on the other hand, looked at seldom-studied power outage variables such as soil moisture. Even though they concluded that soil did not have a significant impact on outages, their work suggested that vegetation cover might be a useful indicator of soil stability, and thus of how likely power poles are to fall.

Very recently, lifecycle-cost assessments that take account of natural hazards have become more common. However, most such studies continue to include only the costs of repairing damage in the immediate aftermath of natural disasters [54]. Due to global warming and rapid urbanization, the frequency and intensity of such disasters are both increasing [25]. To supplement and support the findings of the prior studies reviewed above, therefore, new, more holistic approaches will be needed. The current study helps to fill that gap.

3. Research Methods

3.1. Case Study Approach

Irma, a category 5 hurricane on the Saffir–Simpson Scale, devastated the southeastern part of North America, especially Florida, between 30 August and 14 September 2017 (Figure 1). In all, the storm caused $77.16 billion in property damage and more than 100 fatalities. Its 10 min maximum sustained wind speed was 209 kph. The present research targeted an area in Florida covered by one of that state's largest electric system companies, which operated 18 centers, serving 35 counties. The recorded ROF data were limited to the dates in early September during which Hurricane Irma passed through that area.

3.2. Dependent Variable

As the dependent variable for multiple regression analysis, this study used data on 825 ROF that were received by the target system's 18 operations centers during Hurricane Irma's landfall in Florida.

3.3. Independent Variables

Meteorological wind parameters such as maximum wind speed, forward wind motion speed, and radius have all been utilized as key indicators of hurricane damage levels [55,56]. However, according to Burton [55], Watson and Johnson [57], and Vickery et al. [58], maximum wind speed is the best of these. In the current study, therefore, it was used as the independent variable relating to wind. However, following Choi and Fisher's [59] and Brody et al.'s [60] proposals that rainfall could be the dominant indicator of the extent of damage from hurricanes, total rainfall during Hurricane Irma was also adopted as an independent variables.

Aspects of the built environment can also serve as useful indicators of regions' vulnerability to hurricanes, and for estimating their damage and losses [61]. Mitchell [62] reported that tree falls often led to power line failures, especially when wind loading and soil movement caused by hurricanes caused whole trees to be uprooted. However, fallen tree branches can also cause power system failures [63]. In this study, therefore, tree-density data for the target area were collected from the U.S. Federal Government. These data recorded the number of trees per 1000 m^2 in the areas where power outages were reported to the local utility company, with trees being defined for this purpose as woody plants greater than 2.5 cm in diameter and extending more than 1.38 m above the ground surface.

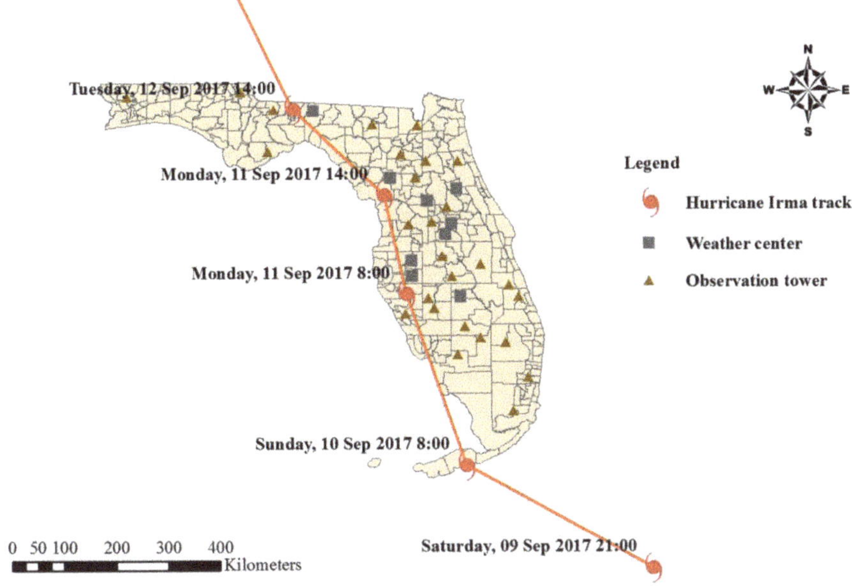

Figure 1. Track of Hurricane Irma, with locations of weather centers and observation towers.

3.4. Data Collection and Management

Table 1 presents further details of the present study's dependent and independent variables. Data on the regional hurricane-induced power outages caused by Hurricane Irma in 2017 were collected and used as the dependent variable. The focal utility company, which provided that data, served approximately 3.1 million customers in the north central, north coastal, south central, and south coastal areas of Florida. The outage data were collected from all 18 power company operations centers within 12 counties in the target area. The 10 min maximum wind speed and total rainfall data were collected from nine weather centers located close to those operations centers. Those weather centers monitor hurricanes and their meteorological characteristics, i.e., wind speed, movement, rainfall, pressure, radius, and intensity. Information about each city with an operations center and/or weather center is shown in Table 1. The numbers by county of reported power outage incidents during Hurricane Irma are shown in Table 2.

Table 1. Data types and sources.

Variable	Description	Unit	Data Source
ROF	Number of power outages per 100,000 people	Number of power outages/100,000	Power company operations centers ($n = 18$) (Apopka, Deland, Jamestown, Longwood, Inverness, Monticello, Ocala, Buena Vista, Clermont, Se Orlando, Highlands, Lake Wales, Winter Garden, Clearwater, Seven Springs, St. Petersburg, Walsingham, Zephyrhills Bay)
Maximum sustained wind speed	10 min maximum sustained wind speed (based on the weather station closest to the point of the outage)	m/s	Weather stations ($n = 9$) (Dover, Balm, Apopka, Bronson, Ocklawaha, Sebring, Monticello, Pierson, Avalon)
Total rainfall	Total amount of rainfall per day	cm	
Tree density	Number of trees per 1000 m^2 where power outage data were reported	Number of trees/1000 m^2	Florida Geographic Data Library

Table 2. Reported number of power outages from Hurricane Irma per county.

County	N
Lake	163
Jefferson	120
Wakulla	86
Polk	82
Citrus	78
Pinellas	78
Highlands	75
Walton	62
Seminole	42
Orange	39
Total	825

The raw power outage data initially comprised more than 1500 incidents. However, the focal weather centers missed some of the meteorological data pertaining to Hurricane Irma, due to being unable to cope with the speed of its unexpected wind gusts. Therefore, only 825 power outage incidents from the power company's dataset had weather center meteorological data corresponding to them. Thus, only those 825 incidents were retained for further analysis. Tree-density information, pertaining specifically to the areas where those same 825 reported power outages occurred, was provided by the Florida Geographic Data Library.

3.5. Multiple Linear Regression Analysis

Regression analysis is a statistical method for predicting the tendencies of collected data. Multiple regression analysis is used to show the linear relationship between more than two independent variables and a dependent variable. The adjusted coefficient of determination (R^2) is used to establish the variances and correlations among the independent variables selected for use in the linear regression model. In addition, R^2 plays the important role of checking the multiple regression models' goodness of fit; and the total variation in the dependent variable (i.e., ROF in this case) is explained as a proportion of the value of R^2 regarding the variability of the independent variables (i.e., here, maximum wind speed, total rainfall, and tree density).

As such, R^2 can be interpreted as approximately how well the linear prediction model can explain the real data. It ranges from 0 to 1, and the closer it is to 1, the better the model's predictive power. It can be estimated by the sum of squares of residual and regression, and the total sum of squares. Those values are also used for estimating F in the analysis. R^2 has a tendency to increase as the numbers of independent variables are increased. To improve this inherent drawback of the parameter, adjusted R^2 is usually used for multiple linear regression analysis. Additionally, analysis of variance (ANOVA) is used for confirming the significance of the regression model. When the significance value produced by an ANOVA is less than 0.05, the regression model is regarded as significant.

Additionally, the present study conducted a normality test for its regression analysis. Such a test can be used for validating that the residuals of the dependent variable are normally distributed. In practice, a p-value larger than 0.000 indicates that the residuals of the variable are normally distributed. In the present study, the normality test was conducted to check the normality of the ROF data before multiple regression analysis was conducted. After verifying that the dependent variable was normally distributed, multiple regression analysis investigated the relationship between ROF and the independent variables. That regression model yielded the predicted trend of the analyzed data, as shown in Equation (1). The straight linear relationship of the function indicates the relationship between the three independent variables and the dependent variable. Specifically, the equation shows that the ROF can be determined along with the independent variables (i.e., X_1, X_2, and X_3) and the regression coefficients (i.e., α, β, γ, and ω) of each independent variable. Thus,

$$\text{ROF} = \alpha + \beta^* X_1 + \gamma^* X_2 + \omega^* X_3 \tag{1}$$

where α is a constant; β is the slope of tree density; γ is the slope of total rainfall, and ω is the slope of maximum wind speed.

From the multiple linear regression analysis results, one can estimate the variance of inflation factor (VIF). VIF can be used as an indicator of multicollinearity among the independent variables. From VIF, the degree of correlations among independent variables can be determined, with larger figures indicating that an independent variable may depend on other independent variables: indicating that it is not, in fact, meaningfully independent. Therefore, analyses that include independent variables with large VIF values can have biased and unpredictable results; and specifically, when VIF is greater than 10, the researcher should consider removing it from the model. A beta (standardized) coefficient is applied to determine the degree of impact of independent variables on the dependent variable in multiple regression analysis, with larger beta coefficients indicating that such impacts are stronger.

4. Results

Table 3 presents descriptive statistics for the dependent variable and independent variables used in this study's model. N stands for the total number of ROF data points during Hurricane Irma's passage through Florida in September 2017.

Table 3. Descriptive statistics of variables.

Category	Reported Cases	Mean	SD
Dependent Variable			
ROF	825	1.955	2.001
Independent Variables			
Maximum sustained wind speed (m/s)	825	15.563	12.821
Total rainfall (cm)	825	0.291	1.051
Tree density (number of trees/1000 m^2)	825	3.134	1.432

Being larger than 0.05, our normality test's significance level of 0.24 indicated that the dependent variables were normally distributed. The multiple regression analysis results are shown in Figures 2 and 3. The *P–P* plot and histogram of the standardized residual (Figure 2) indicate that the residuals in the regression model were normally distributed, while the scatter plot (Figure 3) indicates that the residual's variance was both constant and randomly distributed, i.e., homoscedastic.

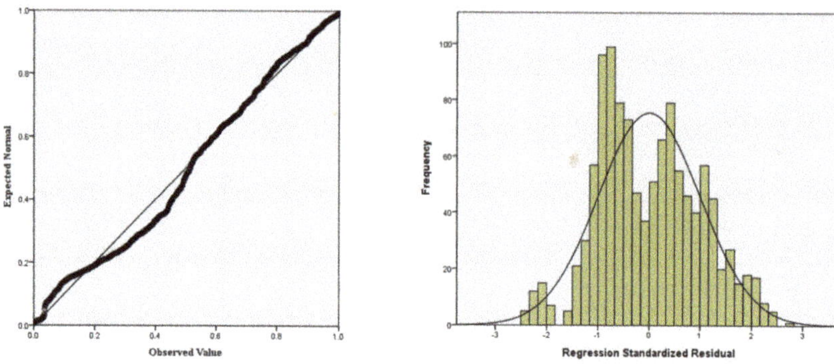

Figure 2. *P–P* plot and histogram of standardized residual from regression analysis.

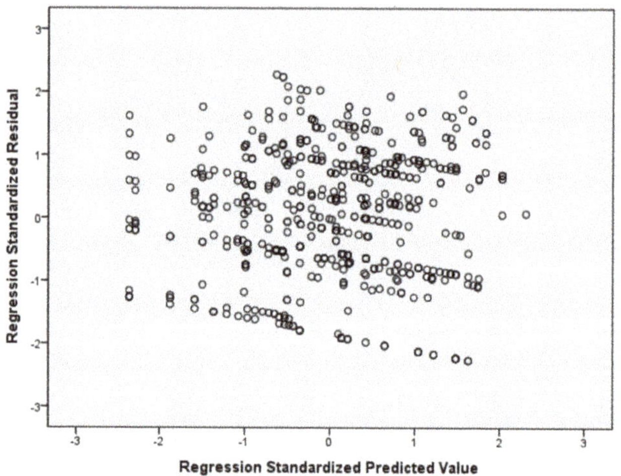

Figure 3. Scatter residual plot from regression analysis.

Analysis of variance (ANOVA) confirmed the above findings. When ANOVA yields a significance level smaller than 0.05, it indicates that the regression model is significant. In this case, it was 0.00, showing that the sampled power outages were related to the independent variables linearly. As shown in Table 4, adjusted R^2 was 0.512, and the value of F was very high (202.768). This meant that 51.2% of the variance in the dependent variable could be explained by all of the independent variables collectively.

Table 4. Summary of ANOVA and multiple regression modeling.

	Sum of Squares	df	Mean Square	F	Sig.	Adjusted R^2
Regression	454.200	5	90.84	202.768	0.000	0.512
Residual	411.712	919	0.448			
Total	895.912	924				
	(2)					

The regression analysis results presented in Table 4 were then checked for statistical relations among the variables. This revealed that maximum sustained wind speed, total rainfall, and tree density were all significant factors, at a confidence interval (CI) of 95% ($p < 0.05$). Additionally, all VIF values were below 10, meaning that there was no multicollinearity among them. The standardized coefficients in the regression analysis indicate the relative strengths of the independent variables' impact on the dependent variable. Here, as shown in Table 5, tree density (0.211) had the largest such impact, followed by rainfall (0.118) and wind speed (0.092).

Table 5. Coefficients of multiple regression analysis.

Variable	Non-Standardized Coefficient	Standardized Coefficient	Significance Probability (p-Value)	Collinearity Statistics (VIF)
(Constant)	0.129	0.259		
Maximum sustained wind speed (m/s)	0.112	0.092	0.024 *	1.283
Total rainfall (cm)	0.227	0.118	0.001 *	1.309
Tree density (number of tree/1000 m^2)	0.094	0.211	0.025 *	1.016

Note. * $p < 0.05$; VIF = variance of inflation factor.

5. Discussion

Natural disasters such as landslides, earthquakes, floods, hurricanes, and wildfires are serious threats to sustainable development. As well as direct losses to the natural capital, they have profound impacts on whole countries' economic growth and sustainability [64]. In part, this is because they can cause or exacerbate imbalances in the supply and demand of social resources [65]. According to Fang et al. [64], due to recent more frequent and intense natural disasters, sustainable development has become a more urgent priority for many societies. Therefore, advanced decision-making processes in the context of natural hazards is increasingly being recognized as critically important.

To achieve more robust estimation of whether construction projects are economically sustainable, scholars have recommended that risk modeling incorporate additional factors such as natural disasters and special vulnerabilities of the local built environment, rather than continuing to focus narrowly on accidents at construction sites [66,67]. Sustainable risk management also needs plentiful information on potential risks if they are to be effectively managed. Especially in the construction industry, such management needs to take more account of risks to the environment, society, and the wider economy than it currently does. Therefore, sustainable construction should focus on balanced development across social, economic, and environmental needs [68,69]. Within the field of sustainable risk management more generally, the most-used method is the management of potential risk factors and uncertainty, which are inherently difficult to predict [70]. For effective decision-making about natural hazards, identifying risk indicators for those hazards is a key first step.

The present study's proposed damage prediction model utilized natural hazard indicators from when Hurricane Irma struck Florida. Its multiple regression analysis showed that the meteorological variables and one topographical variable were all significantly correlated with power outages during Irma. Our normality test's significance level of 0.24 showed that the dependent variable was normally distributed, while our ANOVA's significance level of 0.00 indicates that the selected independent variables had a linear relationship with the dependent one. Together, based on the adjusted R^2 of 0.512, these natural hazard variables explained 51.2% of the variation in the sampled power outages. The other 48.8% of such variation was not explained by the selected variables. Furthermore, the *p*-values of the coefficients of the independent variables in the regression model all indicate significance. Thus, the three selected independent variables—wind speed, rainfall, and tree density—can be used effectively in the prediction of power outages caused by natural hazards such as hurricanes.

These findings support those of various previous studies mentioned in the foregoing literature review. Specifically, wind speed's statistically significant relation to power outages supports the prior conclusions of Burton [55], Watson and Johnson [57], and Vickery et al. [58]. Similarly, the meaningful relation between rainfall and power outages identified in the present study supports previous findings by Choi and Fisher [59] and Brody et al. [60]. Lastly, tree density was revealed to be significantly correlated with power outages during hurricanes, supporting prior results by Mitchell [62] regarding tree- and tree-branch-related power outages.

Additionally, the regression model in Equation (1) explained the straight-line relationship between the dependent and independent variables in the present study. The non-standardized regression coefficients of the independent variables, 0.112, 0.227, and 0.094, were utilized for estimating the trend of ROF corresponding to these natural hazard indicators. Using the regression model and the coefficients, the present study revealed the following. First, when the maximum wind speed increased by 1 m/s, the ROF increased by 11.2%. Second, when total rainfall increased by 1 cm, the ROF increased by 22.7%. And third, when tree density increased by 1 tree/1000 m^2, the ROF increased by 9.4%. The total number of ROF was 3129 when Hurricane Irma made landfall in Florida, and our regression analysis estimated such ROF as 3284, i.e., within a ±95% interval of the actual ROF. As such, these non-standardized coefficients can be used effectively for estimating ROF. Since the actual cost of damage from the power outages caused by Hurricane Irma was not utilized in the present study, due to lack of information (an important limitation), the similarity of the estimated and actual ROFs mean that even more accurate estimation may be possible in future research.

The above findings are also expected to help power system companies improve their preparation for extreme natural hazards in their service areas, including via the adoption of underground cable systems where appropriate. Our results also afford an opportunity to improve traditional lifecycle-cost estimation, by shifting its focus away from the design phase and onto operation and management costs—not only of electric power systems, but also of buildings and coastal infrastructure, in areas prone to extreme weather events.

Insurance companies and construction companies will likely find it advantageous to modify their business plans and construction techniques to reflect natural hazards' impacts on their risk exposure, maximum loss, and so forth. Specifically, in the construction industry, the findings of the present study can be used to help estimate potential losses from power outages by line of business, i.e., commercial, residential, and industrial. Based on estimated damage cost, constructors would be able to give reasoned consideration to the use the underground power system networks as a means of reducing losses from power outages caused by hurricanes (among other, more rational methods of construction). The insurance industry, for its part, can benefit from using the natural hazard indicators we identified, for risk-management and risk-mitigation of insured properties or infrastructure facilities, and thus maximizing their profits. Furthermore, policymakers can refer to the results of the present research when, for example, forecasting the risk factors in hurricane-prone regions and estimating direct and indirect losses to infrastructure facilities and commercial properties. This information, in turn, could be utilized for establishing effective restoration plans that mitigate business interruption; and specifically, tree-related power outages could be mitigated through more careful government-led management of vegetation distribution and tree species in hurricane-prone areas. In the long term, our findings regarding power outage risks can assist advanced risk assessment responses to climate change, by facilitating better-informed asset-management decision making.

Power transmission systems are vulnerable to extreme wind hazards, and the prompt restoration of such systems is critical to industries, utility systems, households, hospitals, and an array of other components of complex societies. Hurricane-induced power outages cause considerable direct damage such as repair costs, and significant indirect ones such as business interruption, in the impacted regions. A number of previous studies have emphasized the long-term repair costs, in time as well as money, of restoring power systems in the wake of disasters [71]. While restoration of power facilities is important, however, it is also very valuable to predict the damage that natural hazards are likely to cause before they occur. Developing vulnerability curves for natural disaster indicators caused by extreme events may be useful to estimating potential damage, as well as the places where power systems are most vulnerable to hurricanes. The natural hazard variables for power failure that we identified above, as well as others that may yet be identified, will be vital to the effective development of such curves.

The findings of the present study can also usefully be extended to the quantification of risk indicators for electric power transmission facilities construction, including socio-economic vulnerabilities arising from power outages, and not just those outages per se. Moreover, our findings regarding some key risk indicators for power failure can contribute to reducing operation and management costs, which are the greatest component of lifecycle cost [41]. Better monitoring of the facilities that may be vulnerable to those key indicators could promote investment in power outage risk mitigation in hurricane-prone areas.

In short, the present study has revealed that sustainable management of lifecycle costs, especially in the operation and management stages, can be enhanced by recognizing that power outages during hurricanes have an inevitable relationship with natural hazard indicators. It has also demonstrated that state-wide power outage data from past hurricanes can be used to identify loss correlations for future ones.

6. Conclusions

Due to its lengthy coastline and geographical placement, Florida has suffered disproportionate financial losses and infrastructural damage from hurricanes as compared to other U.S. states. Hence, it is a site of great demand for hurricane damage prediction models. The results of the present multiple-regression investigation of the relationship between natural hazard factors and power outages during Hurricane Irma in 2017 can help us better understand electrical power systems' vulnerability to hurricane damage in lifecycle-cost terms, despite the limited sizes of the study area and data sample, and the relatively small number of natural hazard variables that were used.

An integrated consideration of social, economic, and environmental impacts should be required to improve lifecycle cost as part of the broader goal of long-term sustainability. As this research did not have an opportunity to consider a full range of geographical, climatological, and economic factors, future research devoted to advanced lifecycle assessment incorporating natural hazards should give due consideration to such factors. Furthermore, such future research should take account of geographically weighted factors such as coastline length and the spatial distribution of damage, and collect data from a wider array of weather centers. This is because the relationship between power outages and natural hazards could differ considerably across geographical locations and their local weather patterns. Other natural hazard indicators such as wind-speed motion and movement direction, along with earthquakes and other environmental factors such as proximity to waterways, should also be considered when expanding this study's model to analysis of the long-term sustainability of buildings and other infrastructure in a wide variety of locations and levels of hazard exposure.

Lastly, a vulnerability function can be estimated if the approximate damage costs arising from power outages can be provided as a function of wind speed or rainfall. Such estimation can help with advance prediction of losses caused by hurricanes, according to the intensity of the natural hazard indicators. In the future, if electric power systems' initial times of failure and restoration time become available, an integrated response and restoration strategy can also be developed to minimize losses to such systems in hurricane prone-areas.

Author Contributions: Conceptualization, S.-G.Y.; Data curation, S.-G.Y., S.S., and J.-M.K.; Funding acquisition, J.-M.K.; Investigation, S.-G.Y. and K.S.; Methodology, S.-G.Y. and J.-M.K.; Project administration, J.-M.K.; Software, J.-M.K.; Supervision, J.-M.K.; Validation, S.-G.Y. and J.-M.K.; Resources, J.-M.K.; Writing—original draft, J.-M.K.; Writing—review and editing, S.-G.Y. and K.S. All authors have read and agreed to the final version of the manuscript.

Funding: This research was funded by the Basic Science Research Program of the National Research Foundation of Korea (NRF-2020R1F1A1048304).

Conflicts of Interest: The authors declare no conflict of interest.

References

1. Intergovernmental Panel on Climate Change (IPCC). *Contribution of Working Group I to the Fourth Assessment Report of the Intergovernmental Panel on Climate Change*; Cambridge University Press: Cambridge, UK, 2007.

2. Kim, J.-M.; Son, S.; Lee, S.; Son, K. Cost of Climate Change: Risk of Building Loss from Typhoon in South Korea. *Sustainability* **2020**, *12*, 7107. [CrossRef]

3. Jongman, B.; Ward, P.J.; Aerts, J.C. Global exposure to river and coastal flooding: Long term trends and changes. *Glob. Environ. Chang.* **2012**, *22*, 823–835. [CrossRef]

4. Jufri, F.H.; Widiputra, V.; Jung, J. State-of-the-art review on power grid resilience to extreme weather events: Definitions, frameworks, quantitative assessment methodologies, and enhancement strategies. *Appl. Energy* **2019**, *239*, 1049–1065. [CrossRef]

5. Emanuel, K. Will Global Warming Make Hurricane Forecasting More Difficult? *Am. Meteorol. Soc.* **2016**, *98*, 294–501. [CrossRef]

6. Kreimer, A.; Arnold, M. World Bank's Role in Reducing Impacts of Disasters. *Nat. Hazards Rev.* **2000**, *1*, 37–42. [CrossRef]

7. Chang, S.E. Evaluating Disaster Mitigations: Methodology for Urban Infrastructure Systems. *Nat. Hazards Rev.* **2003**, *4*, 186–196. [CrossRef]

8. Münchener Rückversicherungs-Gesellschaft (Munich RE). *Topics Geo: Natural Catastrophes 2017: Analysis, Assessment*; Munich RE: München, Germany, 2018.

9. Executive Office of the President. *Economic Benefits of Increasing Electric Grid Resilience to Weather Outages*; The White House: Washington, DC, USA, 2013.

10. Kenward, A.; Raja, U. *Blackout: Extreme Weather, Climate Change and Power Outages*; Climate Central: Princeton, NJ, USA, 2014.

11. Campbell, R.J. *Weather-Related Power Outages and Electric System Resiliency*; Library of Congress Washington DC Congressional Research Service: Washington, DC, USA, 2012.

12. Larsen, H.L.; LaCommare, K.H.; Eto, J.H.; Sweeney, J.L. *Assessing Changes in the Reliability of the U.S. Electric Power Systems*; Lawrence Berkeley National Laboratory: Berkeley, CA, USA, 2015.

13. Nateghi, R.; Guikema, S.D.; Wu, Y.; Bruss, C.B. Critical Assessment of the Foundations of Power Transmission and Distribution Reliability Metrics and Standards. *Risk Anal.* **2015**, *36*, 4–15. [CrossRef] [PubMed]

14. Coumou, D.; Robinson, A. Historic and future increase in the global land area affected by monthly heat extremes. *Environ. Res. Lett.* **2013**, *8*, 034018. [CrossRef]

15. Peterson, T.C.; Hoerling, M.P.; Stott, P.A.; Herring, S.C. Explaining extreme events of 2012 from a climate perspective. *Bull. Am. Meteorol. Soc.* **2013**, *94*, S1–S74. [CrossRef]

16. Webster, P.J.; Holland, G.J.; Curry, J.A.; Chang, H.-R. Changes in Tropical Cyclone Number, Duration, and Intensity in a Warming Environment. *Science* **2005**, *309*, 1844–1846. [CrossRef]

17. Pielke, R.; Gratz, J.; Landsea, C.W.; Collins, D.; Saunders, M.A.; Musulin, R. Normalized Hurricane Damage in the United States: 1900–2005. *Nat. Hazards Rev.* **2008**, *9*, 29–42. [CrossRef]

18. *Natural Catastrophes and Man-Made Disasters in 2014: Convective and Winter Storms Generate Most Losses*; Swiss Re Ltd.: Zurich, Switzerland, 2015.

19. Blake, E.S.; Kimberlain, T.B.; Berg, R.J.; Cangialosi, J.P.; Beven II, J.L. *Tropical Cyclone Report: Hurricane Sandy*; National Hurricane Center: Miami, FL, USA, 2013.

20. Blake, E.S.; Rappaport, E.N.; Landsea, C.W. *The Deadliest, Costliest, and Most Intense United States Tropical Cyclones from 1851 to 2006 (And Other Frequently Requested Hurricane Facts)*; NOAA/National Weather Service: Miami, FL, USA, 2007.

21. Xu, L.; Brown, R.E. Hurricane simulation for Florida utility damage assessment. In Proceedings of the 2008 IEEE Power and Energy Society General Meeting—Conversion and Delivery of Electrical Energy in the 21st Century, Pittsburgh, PA, USA, 20–24 July 2008.

22. Liu, H.; Davidson, R.A.; Rosowsky, D.; Stedinger, J.R. Negative Binomial Regression of Electric Power Outages in Hurricanes. *J. Infrastruct. Syst.* **2005**, *11*, 258–267. [CrossRef]

23. Han, S.-R.; Rosowsky, D.; Guikema, S.D. Integrating Models and Data to Estimate the Structural Reliability of Utility Poles during Hurricanes. *Risk Anal.* **2013**, *34*, 1079–1094. [CrossRef] [PubMed]

24. Sissine, F. *Energy Independence and Security Act of 2007: A Summary of Major Provisions*; Library of Congress Washington DC Congressional Research Service: Washington, DC, USA, 2007.

25. Yum, S.-G.; Kim, J.-M.; Son, K. Natural Hazard Influence Model of Maintenance and Repair Cost for Sustainable Accommodation Facilities. *Sustainability* **2020**, *12*, 4994. [CrossRef]

26. U.S. Green Building Council (U.S. GBC). *LEED Reference Guide for Green Building Design and Construction*; U.S. Green Building Council: Washington, DC, USA, 2009.

27. Federal Emergency Management Agency (FEMA). *HAZUS-MH Estimated Annualized Earthquake Losses for the United States*; FEMA: Washington, DC, USA, 2008.

28. Tantala, M.W.; Nordenson, G.J.; Deodatis, G.; Jacob, K. Earthquake loss estimation for the New York City Metropolitan Region. *Soil Dyn. Earthq. Eng.* **2008**, *28*, 812–835. [CrossRef]

29. Remo, J.W.; Pinter, N. Hazus-MH earthquake modeling in the central USA. *Nat. Hazards* **2012**, *63*, 1055–1081. [CrossRef]

30. Rein, A.; Corotis, R.B. An overview approach to seismic awareness for a "quiescent" region. *Nat. Hazards* **2013**, *67*, 335–363. [CrossRef]

31. Wei, H.-H.; Skibniewski, M.J.; Shohet, I.M.; Yao, X. Life cycle environmental performance of natural hazard mitigation for buildings. *J. Perform. Constr. Facil.* **2015**, *30*, 1–13.

32. Chang, S.E.; Shinozuka, M. Life-Cycle Cost Analysis with Natural Hazard Risk. *J. Infrastruct. Syst.* **1996**, *2*, 118–126. [CrossRef]

33. Shinozuka, M.; Chang, S.E.; Cheng, T.C.; Feng, M.; O'Rourke, T.D.; Saadeghvaziri, M.A.; Dong, X.; Jin, X.; Wang, Y.; Shi, P. *Resilience of Integrated Power and Water Systems*; MCEER: Buffalo, NY, USA, 2004.

34. Wei, H.-H.; Shohet, I.M.; Skibniewski, M.J.; Shapira, S.; Yao, X. Assessing the Lifecycle Sustainability Costs and Benefits of Seismic Mitigation Designs for Buildings. *J. Arch. Eng.* **2016**, *22*, 04015011. [CrossRef]

35. Federal Emergency Management Agency (FEMA). *Next-Generation Performance-Based Seismic Design Guidelines*; FEMA: Washington, DC, USA, 2006.

36. Padgett, J.E.; Tapia, C. Sustainability of Natural Hazard Risk Mitigation: Life Cycle Analysis of Environmental Indicators for Bridge Infrastructure. *J. Infrastruct. Syst.* **2013**, *19*, 395–408. [CrossRef]

37. Hamburger, R.; Rojahn, C.; Heintz, J.; Mahoney, M. FEMA P58: Next-generation building seismic performance assessment methodology. In Proceedings of the 15th World Conference on Earthquake Engineering, Lisbon, Portugal, 24–28 September 2012; International Association for Earthquake Engineering: Tokyo, Japan, 2012.

38. Comber, M.; Poland, C. Disaster resilience and sustainable design: Quantifying the benefits of a holistic design approach. In Proceedings of the Structures Congress, Pittsburgh, PA, USA, 2–4 May 2013.

39. Feese, C.; Li, Y.; Bulleit, W.M. Assessment of Seismic Damage of Buildings and Related Environmental Impacts. *J. Perform. Constr. Facil.* **2015**, *29*, 04014106. [CrossRef]

40. Hossain, K.A.; Gencturk, B. Life-Cycle Environmental Impact Assessment of Reinforced Concrete Buildings Subjected to Natural Hazards. *J. Arch. Eng.* **2016**, *22*, 22. [CrossRef]

41. Sinisuka, N.I.; Nugraha, H. Life cycle cost analysis on the operation of power generation. *J. Qual. Maint. Eng.* **2013**, *19*, 5–24. [CrossRef]

42. Shinozuka, M.; Rose, A.; Eguchi, R. *Engineering and Socioeconomic Impacts of Earthquakes*; MCEER: Buffalo, NY, USA, 1998.

43. Shinozuka, M.; Cheng, T.C.; Feng, M.; Mau, S.T. *Seismic Performance Analysis of Electric Power Systems*; MCEER: Buffalo, NY, USA, 1999.

44. Cagnan, Z.; Davidson, R.A.; Guikema, S.D. Post-Earthquake Restoration Planning for Los Angeles Electric Power. *Earthq. Spectra* **2006**, *22*, 589–608. [CrossRef]

45. Xu, N.; Guikema, S.D.; Davidson, R.; Nozick, L.; Cagnan, Z.; Vaziri, K. Optimizing scheduling of post-earthquake electric power restoration tasks. *Earthq. Eng. Struct. Dyn.* **2007**, *36*, 265–284. [CrossRef]

46. Guikema, S.D.; Nateghi, R. *Modeling Power Outage Risk from Natural Hazards*; Oxford Research Encyclopedia of Natural Hazard Science: Oxford, UK, 2018.

47. Davidson, R.A.; Liu, H.; Sarpong, I.K.; Sparks, P.; Rosowsky, D.V. Electric Power Distribution System Performance in Carolina Hurricanes. *Nat. Hazards Rev.* **2003**, *4*, 36–45. [CrossRef]

48. Reed, D.A. Electric utility distribution analysis for extreme winds. *J. Wind. Eng. Ind. Aerodyn.* **2008**, *96*, 123–140. [CrossRef]

49. Han, S.-R.; Guikema, S.D.; Quiring, S.M. Improving the Predictive Accuracy of Hurricane Power Outage Forecasts Using Generalized Additive Models. *Risk Anal.* **2009**, *29*, 1443–1453. [CrossRef]

50. Guikema, S.D.; Nateghi, R.; Quiring, S.M.; Staid, A.; Reilly, A.; Gao, M. Predicting Hurricane Power Outages to Support Storm Response Planning. *IEEE Access* **2014**, *2*, 1364–1373. [CrossRef]

51. Nateghi, R.; Guikema, S.D.; Quiring, S.M. Power Outage Estimation for Tropical Cyclones: Improved Accuracy with Simpler Models. *Risk Anal.* **2013**, *34*, 1069–1078. [CrossRef] [PubMed]

52. Balijepalli, N.; Venkata, S.; Richter, C.; Christie, R.; Longo, V. Distribution System Reliability Assessment Due to Lightning Storms. *IEEE Trans. Power Deliv.* **2005**, *20*, 2153–2159. [CrossRef]

53. Quiring, S.M.; Zhu, L.; Guikema, S.D. Importance of soil and elevation characteristics for modeling hurricane-induced power outages. *Nat. Hazards* **2010**, *58*, 365–390. [CrossRef]

54. Noshadravan, A.; Miller, T.R.; Gregory, J. A Lifecycle Cost Analysis of Residential Buildings including Natural Hazard Risk. *J. Constr. Eng. Manag.* **2017**, *143*, 04017017. [CrossRef]

55. Burton, C.G. Social Vulnerability and Hurricane Impact Modeling. *Nat. Hazards Rev.* **2010**, *11*, 58–68. [CrossRef]

56. Dunion, J.; Landsea, C.W.; Houston, S.H.; Powell, M.D. A Reanalysis of the Surface Winds for Hurricane Donna of 1960. *Mon. Weather. Rev.* **2003**, *131*, 1992–2011. [CrossRef]

57. Watson, C.C.; Johnson, M.E. Hurricane Loss Estimation Models: Opportunities for Improving the State of the Art. *Bull. Am. Meteorol. Soc.* **2004**, *85*, 1713–1726. [CrossRef]

58. Vickery, P.J.; Skerlj, P.F.; Lin, J.; Twisdale, L.A.; Young, M.A.; Lavelle, F.M. HAZUS-MH Hurricane Model Methodology. II: Damage and Loss Estimation. *Nat. Hazards Rev.* **2006**, *7*, 94–103. [CrossRef]

59. Choi, O.; Fisher, A. The Impacts of Socioeconomic Development and Climate Change on Severe Weather Catastrophe Losses: Mid-Atlantic Region (MAR) and the U.S. *Clim. Chang.* **2003**, *58*, 149–170. [CrossRef]

60. Brody, S.D.; Zahran, S.; Highfield, W.E.; Grover, H.; Vedlitz, A. Identifying the impact of the built environment on flood damage in Texas. *Disasters* **2008**, *32*, 1–18. [CrossRef]

61. Khanduri, A.; Morrow, G. Vulnerability of buildings to windstorms and insurance loss estimation. *J. Wind. Eng. Ind. Aerodyn.* **2003**, *91*, 455–467. [CrossRef]

62. Mitchell, S.J. Stem growth responses in Douglas-fir and Sitka spruce following thinning: Implications for assessing wind-firmness. *For. Ecol. Manag.* **2000**, *135*, 105–114. [CrossRef]

63. Appelt, P.J.; Goodfellow, J.W. Research on how trees cause interruptions—Applications to vegetation management. In Proceedings of the Rural Electric Power Conference, Scottsdale, AZ, USA, 25 May 2004; IEEE: Piscataway, NJ, USA, 2004.

64. Fang, J.; Lau, C.K.M.; Lu, Z.; Wu, W.; Zhu, L. Natural disasters, climate change, and their impact on inclusive wealth in G20 countries. *Environ. Sci. Pollut. Res.* **2018**, *26*, 1455–1463. [CrossRef] [PubMed]

65. Zhou, L.; Wu, X.; Xu, Z.; Fujita, H. Emergency decision making for natural disasters: An overview. *Int. J. Disaster Risk Reduct.* **2018**, *27*, 567–576. [CrossRef]

66. Majdalani, Z.; Ajam, M.; Mezher, T. Sustainability in the construction industry: A Lebanese case study. *Constr. Innov.* **2006**, *6*, 33–46. [CrossRef]

67. Brockett, P.L.; Golden, L.L.; Betak, J. Different Market Methods for Transferring Financial Risks in Construction. In *Risk Management in Construction Projects*; IntechOpen: London, UK, 2019.

68. Bal, M.; Bryde, D.; Fearon, D.; Ochieng, E. Stakeholder Engagement: Achieving Sustainability in the Construction Sector. *Sustainability* **2013**, *5*, 695–710. [CrossRef]

69. Shen, L.; Wu, Y.; Zhang, X. Key Assessment Indicators for the Sustainability of Infrastructure Projects. *J. Constr. Eng. Manag.* **2011**, *137*, 441–451. [CrossRef]

70. Wilderer, P.A.; Renn, O.; Grambow, M.; Molls, M.; Mainzer, K. *Sustainable Risk Management*; Springer International Publishing: Cham, Switzerland, 2018.

71. Duffey, R.B. Power Restoration Prediction Following Extreme Events and Disasters. *Int. J. Disaster Risk Sci.* **2018**, *10*, 134–148. [CrossRef]

Article

Cataloguing of the Defects Existing in Aluminium Window Frames and Their Recurrence According to Pluvio-Climatic Zones

Manuel J. Carretero-Ayuso [1], **Carlos E. Rodríguez-Jiménez** [2,*], **David Bienvenido-Huertas** [2] and **Juan Moyano** [3]

[1] Musaat Foundation and Polytechnic College, University of Extremadura, 10004 Cáceres, Spain; carreteroayuso@yahoo.es
[2] Department of Building Construction II, University of Seville, 41012 Seville, Spain; jbienvenido@us.es
[3] Department of Graphical Expression and Building Engineering, University of Seville, 41012 Seville, Spain; jmoyano@us.es
* Correspondence: ceugenio@us.es

Received: 5 August 2020; Accepted: 6 September 2020; Published: 9 September 2020

Abstract: The sustainability of building envelopes is affected by its windows, since these establish the connection/separation between the indoor rooms and the external environment. They can also lead to problems if they do not offer sufficient protection against external agents. The data source in this research is unprecedented, as it is based on records of court sentences. There is a significant number of cases (1615), which provides high representativeness for the functional reality of windows. The methodology that was developed classifies the defects and the causes that were found, also analysing correspondence with their recurrence according to aspects of climatological location. In the results, the cases pertaining to water infiltration, air permeability and humidity by condensation are highlighted. This study provides a vision that categorizes problems related to aluminium windows that may be useful for future interventions by agents participating in the construction process.

Keywords: air permeability; watertightness; airtightness; infiltration; aluminium window frames

1. Introduction

Windows are indispensable construction units for building facades, given their implications for basic aspects related to habitability and comfort such as watertightness [1], air permeability [2], lighting [3], etc. At the same time, these openings interrupt the continuity of external walls and amount to numerous singularities in the construction solution that still need to meet all requirements of the envelope, without decreasing the performance of the whole. In fact, when characterising windows, one must note both their own elements (the frame and the glazing) and all the perimeter elements that are part of the opening (external windowsill, lintels, jambs and blinds, as well as the assembly and sealants) that share the same functions. As such, windows constitute one of the most problematic elements in the study of the envelopes of buildings, these being determinant for attaining suitable parameters for their performance [4].

In this way, the interrelation between the sustainability of the property and that of its windows is quite significant [5]. Windows exert an influence on basic aspects of building functionality, both at the level of internal comfort and with regard to maintenance, construction quality and repairs throughout buildings' service lives. This is shown by the direct effects that dysfunctions related to windows have on the increase in the work and maintenance costs of the building—all key factors in their sustainability [6,7]. Equally, in the construction phase for buildings, windows are decisive for both quality and costs, featuring among some of the common non-conformities that arise during execution, and lead to significant deviations from the initial budgets for the works [8]. Windows are also usually a part of renovations that take place during the service lives of buildings and are key to evaluating the perception of the quality of houses, as can be observed in the efficiency indicators established subsequent to repair and rehabilitation works [9].

In the studies on the openings of windows, there are numerous lines of enquiry, with a recent focus on energy efficiency, which constitutes one of the critical points in the analysis that was carried out. As for energy savings, windows have different considerations. On the one hand, when aspects related to the air permeability of the thermal envelope are addressed, these are responsible for most of the volume infiltrated through the envelope [10,11]. On the other hand, one must remember that the transmittance of windows is one of the main points of the facades through which internal heat and cold are lost. This is the case because their values are usually clearly lower toward the opaque parts—due to the characteristic values of the materials usually employed in the frames and in the glazing [12,13], as well as to the probability of the appearance of thermal bridges in the most common construction solutions [14]. In this line of research, there are numerous publications that examine the insulation provided by windows, its impact on building energy demands and interventions for improving it [15,16]. From this energy perspective, one should also note the effect of solar radiation entering the building through the glazing of windows, the consequences of which are not always easy to mitigate [17].

Construction aspects are also relevant to the study of windows, since these influence numerous building defects. The problems related to humidity make up a large portion of the damage related to traditional facades [18]. Based on this assumption, the existence of numerous joints and operable parts leads to windows and the contours of their respective openings being especially vulnerable to rainwater infiltration [19]. Equally, the flows of temperature and moist air that go through windows are reflected in the appearance of humidity by condensation [20]. Additionally, despite its lesser impact, one must also point to the roles played by windows and their perimeter elements in the causes of some problems related to fissurations and to the stability walls, especially those made with bricks [21].

Given the amplitude of this set of factors related to defects in windows and their environment, inspection and diagnosis activities that can effectively detect problems and dysfunctions in these construction elements become very important. In the last few years, there has been, in this field, an important development in the methods used for the testing of windows and other elements of the building envelope, largely due to the needs of energy evaluations for buildings [22]. In this way, permeability measurement techniques such as the test blower door [23,24], thermometric tests for the measurement of the transmittance [25], infrared thermography [26] and tests of watertightness against the effects of wind-driven rain [27], among others, are increasingly present (both in the scientific literature and in their spread in the industry).

As a result, the knowledge as to construction defects in windows shows a panorama with multiple variables [28]. In the face of this situation, it is especially useful to have carefully categorised databases for a clear definition of critical points to be able to avoid the repetition of problems in future construction interventions, actively contributing to their sustainability. It is not very common to find publications that incorporate any type of cataloguing of construction problems. Nevertheless, certain examples are found to be focused on the typologies of defects related to the building in general [29] and on the reasons behind repairs and renovations [30]. Of particular interest to this research is the work focusing on window frames, including studies on the influence of frame components [31,32], on degradation [33] and on modelling the inspection for window defects [34] or on predicting their service lives [35,36].

The current study provides a novel analysis of defects related to windows (as well as their causes), since it was carried out on a large database of judicial complaints filed by users. It is thus intended to provide a classification and grading of defects that were actually complained about by building users, as well as evaluating their recurrence according to climatic location. In this regard, the influence of rainfall, climate and latitude on the appearance of defects was also examined.

2. Methodology

2.1. Scope of Study

The data for this research were obtained from the records of complaints for damages of the civil responsibility insurance company of building engineers in Spain [37]. Each record originated from the observation of a construction defect between the years 2008 and 2017 [38] that was subsequently the target of a judicial complaint and was definitively resolved before its inclusion in this paper.

These records (that were handled directly by the authors) contain not only data on the contract between the insurance company and the insured party but also the sentences of the courts of law, according to the complaints filed by the users of the buildings in which the construction defects appeared.

These sentences, issued after the verification by the courts of law that the defects in question were indeed present, detailed the characteristics of the defects, indicating that the problems in question should be resolved. The initial sentences could be appealed to higher courts a number of times until they were no longer appealable and were considered final. That is the point at which the authors proceeded with including the data as part of this research. A copy of all those judicial documents is held by the insurance company of the participating building engineers. It was necessary, as such, to review thousands of pages in these records to extract the technical data and separate them from administrative or contractual information.

No precedents were found for research carried out by other authors on window defects based on this type of database, nor was a relevant data set with such a large number of cases (1615) found.

2.2. Characterisation

The parameters characterised in the research were the following— "effects" (d) and "originating cause" (OC)—both for windows (W), as the construction unit under study. Windows are classified into two types: normal windows (NW) and bay windows (BW); all of the windows had aluminium frames.

Table 1 shows the different types of defects and originating causes, as well as the codes assigned to them (this is the material that is most predominant in Spain, whereas other materials, such as wood, are seldom used in window frames).

Table 1. Types of defects and originating causes that were analysed.

Parameter	Concept	Code
	Water infiltration	dWI
	Air permeability	dAP
Defects	Humidity by condensation	dHC
	Oxidation or corrosion	dOC
	Absence/deficiency of sealant	ocAS
	Inadequate construction material and/or placement	ocIC
Originating causes	Existence of thermal bridges	ocTB
	Incorrect assembly process	ocIA

Likewise, the 'construction typology' of the windows with defects was also studied, being divided into 'flats', 'houses' and 'other buildings'. Below are shown some photographic examples of some of the defects (Figure 1).

Infiltration of rainwater through a window   Humidities on the inside of a wall and two windows

Figure 1. Photographic examples of two types of defects studied.

2.3. Factors of Study

In addition to the parameters indicated above, in this research were also used other concepts that enabled ascribing each one of the defects to a climatological factor. There were three factors, indicated in Table 2. This table also indicates the categories into which each factor is broken down.

Table 2. Types of factors and their categories, as included in the research.

Factor			Categories		
Factor 1:	Rainfall	High	Medium	Low	—
Factor 2:	Climate	Oceanic	Continental	Mediterranean	Subtropical
Factor 3:	Latitude	North	Central	South	—

The information related to the factors established for each geographical area of the territory in question (Spain) is referred to in this study as 'location strips'. Their classification was carried out based on the indications of the Spanish Meteorology Agency [39]:

- Rainfall. It consists of three categories: 'High' (>700 mm/m^2), 'Medium' (between 450 and 700 mm/m^2) and 'Low' ($\leq$450 mm/m^2), according to the available meteorological data.
- Climate. It is made up of four categories: 'Oceanic', 'Continental', 'Mediterranean' and 'Subtropical'. Small sub-areas, with a reduced extension, were not considered in this factor.
- Latitude. It consists of three categories: 'North', 'Central' and 'South'. It corresponds to the geographical areas according to their situation within Spain.

Using said factors, a percentage study was carried out for the recurrence of all defects, applying each factor individually, combining two at a time, and combining the three factors together. In this way, one obtains the 'percentage of number of cases' (%NC), which corresponds to the different cases (location strips) when applying the categories into which Factors 1, 2 and 3 are sub-divided.

Subsequently, using these %NC values and simultaneously applying the three factors (a closer approximation of reality), the percentages are sorted to subsequently establish the 'ranks of concentration of defects'.

In order to homogenise the values obtained, the following was carried out:

- The percentages of the 'location strips' that were closer are added up to one another and sorted from largest to smallest, the value thus obtained being referred to as 'total %NC'.
- The aforementioned value was divided by the number of homes in Spain [40] for that set of location strips, obtaining an dimensionless value that could be compared (relative frequency).

- With the objective of better handling and visualising these last values, the relative frequencies were divided by the largest value, thus obtaining a 'normalised relative frequency' for each pluvio-climatic zone.

The eleven results by 'location strips' and by three factors were characterised to obtain the 'ranks of concentration of defects' according to the percentages obtained. From the values of these ranks, 'zones pluvio-climatic' were configured in such a way that all those eleven combinations were simplified into just three zones (ZONE A, ZONE B and ZONE C).

3. Results

3.1. Results by Type of Element and Defect

Figure 2a shows the number of cases by type of element. Normal windows amount to 95% of the total (NW = 1538 cases), while bay windows constitute the remaining 5% (BW = 77 cases). In turn, Figure 2b represents the percentage distribution according to the types of defects, it being shown that 6 out of every 10 cases belong to 'water infiltrations' (dWI = 60.8%). The following defects—'air permeability' (dAP) and 'humidities by condensation' (dHC)—are practically the same.

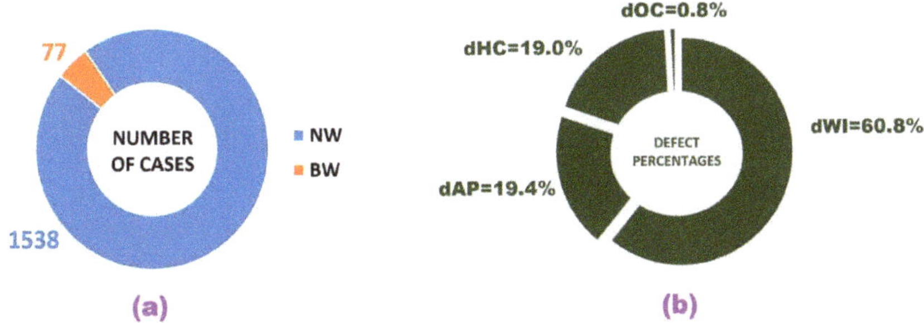

Figure 2. Numbers of cases in the research (**a**) and percentages by types of defects (**b**).

If we break down the number of cases for each type of defect and according to the element in which it occurs (normal window or bay window), we obtain the values that are shown in Figure 3.

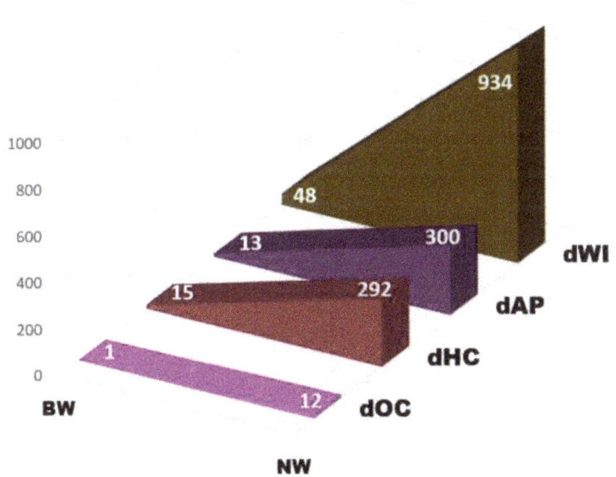

Figure 3. Number of cases by type of defect and by type of element.

Based on Figures 2 and 3, it is evident that problems related to the performance of windows in terms of hermeticity (in terms of permeability to air as well as water) represent nearly the totality of the impact on building users. This implies that the construction configuration and the commissioning of windows must mainly satisfy the requirements on watertightness (absence of infiltrations) and air permeability.

3.2. Results by Type of Originating Cause

As indicated in the methodology section, four different types of originating causes were found in this research. It can be noted that the 'absence/deficiency of sealant' occurs practically 2/3 of the time (ocAS = 66.4%). The second position (ocIC = inadequate construction material and/or placement) and the third position (ocTB = existence of thermal bridges) have quite similar values, differing by less than one percentage point.

Consequently, with the distribution of the originating causes shown in Figure 4, one can observe a majority of issues pertaining to sealing and to the placement during construction. These aspects are directly related to water infiltration and air permeability (dWI and dAP). Complaints related to thermal parameters (ocTB and ocIA) appear to be far less relevant.

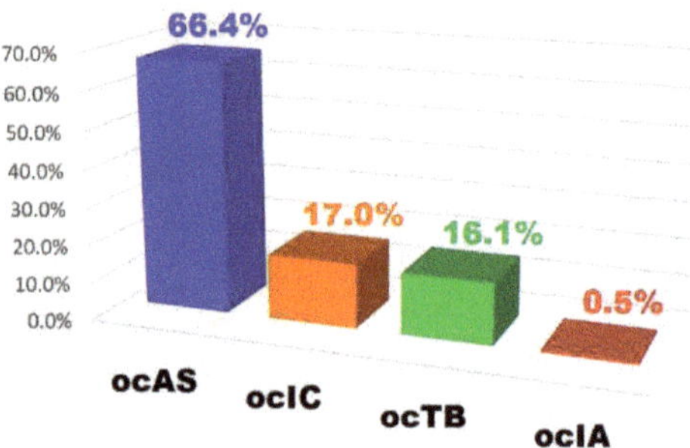

Figure 4. Percentage of recurrence according to the type of originating cause.

3.3. Determination of the Pathology Trinomial Sets

The term 'pathology trinomial set' will be used to refer to construction interrelations that lead a certain type of originating cause to produce a type of defect in one of the two types of windows studied. The data analysed yield 16 different combinations.

Figure 5 shows the number of cases for each one of the 16 pathology trinomial sets found. For a simpler conceptual association, the colours employed for the originating causes are the same as those employed in Figure 4, and the identifying colours for the defects are those used in Figure 3. In addition, Figure 5 indicates the number of cases for each one of the types of defects and originating causes so that they can be evaluated as part of the whole set of the results found.

The most frequent trinomial set is 'absence/deficiency of sealant' that leads to 'water infiltrations' in 'normal windows' (ocAS-dWI-NW = 779). It is followed in second place by the 'existence of thermal bridges' that leads to 'humidity by condensation' in 'normal windows' (ocTB-dHC-NW = 247). In third place is the 'absence/deficiency of sealant' that leads to 'air permeability' in 'normal windows' (ocAS-dAP-NW = 246). The fourth position is held by 'inadequate construction material and/or placement' that leads to 'water infiltrations' in 'normal windows' (ocIC-dWI-NW = 155).

The sum of cases of these first four pathology trinomial sets (from among the 16 sets) equals 1427 cases, or 93% of all the cases. As such, there is a Pareto relation of 25–93; 25% of the pathology trinomial sets lead to 93% of the cases researched.

Some other aspects can also be highlighted: defect dOC is only originated by cause ocIC, cause ocTB only leads to defect dHC, cause ocIA only leads to defect dAP, cause ocAS leads to two defects, and cause ocIC leads to all the types of existing defects.

These pathology trinomial sets confirm that, to reduce most of the problems found and to reduce most of the problems found during the service life, windows should be constructed with the utmost care for their hermeticity (air permeability and watertightness). It is highly recommended to carry out verifications in situ ofn these aspects before the commissioning of the building.

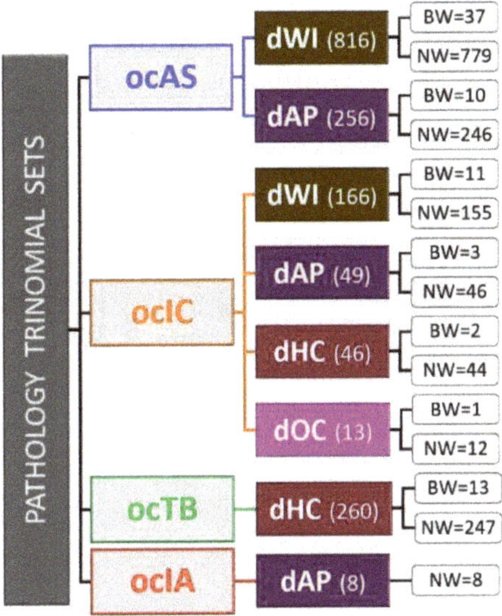

Figure 5. Number of cases according to the type of pathology trinomial set found.

3.4. Results by Construction Typologies

As indicated, during the process of data collection, the typology construction wherein each of the defects occurred was noted. The higher percentages were found in 'flats' (63.03%). See Figure 6.

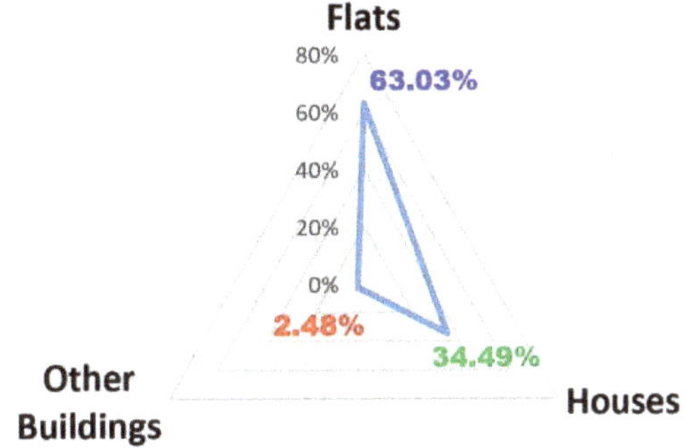

Figure 6. Percentage of defects according to the construction type.

From this distribution, it can be deduced that the greater height of multi-storey buildings, which increases their exposure to weather, may explain the frequency of complaints in this type of building.

3.5. Pluvio-Climatic Study of Defects

Given that the defects collected in the database are mainly related to environmental parameters (humidity, air infiltration, etc.), their relationship with the factors of the different climatic areas will be of interest.

The analysis that is carried out in this section explores the recurrence of these problems according to the pluvio-climatic location of the cases in a way that establishes a quantitative association between defects and weather conditions.

3.5.1. Determination of the Location Strips

Table 2 establishes the three main factors of the environment: rainfall, climate and latitude, as well as their respective categories. Their details are shown in Table 3, which contains the different combinations of these categories, resulting in 'location strips'.

Table 3. Percentages of recurrence of the defects according to each of the factors analysed.

Factor		Location Strip	%NC	% Total
Single Factor	Rainfall	High	28.05%	Factor 1
		Medium	54.86%	
		Low	17.09%	
	Climate	Oceanic	28.92%	Factor 2
		Continental	30.46%	
		Mediterranean	37.15%	
		Subtropical	3.47%	
	Latitude	North	51.27%	Factor 3
		Central	29.47%	
		South	19.26%	
2 Factor Combination	Rainfall and Climate	High-Oceanic	28.92%	Factor 1 + Factor 2
		Medium-Continental	22.72%	
		Medium-Mediterranean	31.27%	
		Low-Continental	7.74%	
		Low-Mediterranean	5.88%	
		Low-Subtropical	3.47%	
	Rainfall and Latitude	High-North	28.92%	Factor 1 + Factor 3
		Medium-North	19.20%	
		Medium-Central	24.89%	
		Medium-South	9.90%	
		Low-North	3.16%	
		Low-Central	4.58%	
		Low-South	9.35%	
	Climate and Latitude	Oceanic-North	28.92%	Factor 2 + Factor 3
		Continental-North	10.40%	
		Continental-Central	17.71%	
		Continental-South	2.35%	
		Mediterranean-North	11.95%	
		Mediterranean-Central	11.76%	
		Mediterranean-South	13.44%	
		Subtropical-South	3.47%	
3 Factor Combination	High	High-Oceanic-North	28.92%	Factor 1 + Factor 2 + Factor 3
	Medium	Medium-Continental-North	7.25%	
		Medium-Continental-Central	13.13%	
		Medium-Continental-South	2.35%	
		Medium-Mediterranean-North	11.95%	
		Medium-Mediterranean-Central	11.76%	
		Medium-Mediterranean-South	7.55%	
	Low	Low-Continental-North	3.16%	
		Low-Continental-Central	4.58%	
		Low-Mediterranean-South	5.88%	
		Low-Subtropical-South	3.47%	

%NC = percentage of the number of cases.

As shown in the upper section of Table 3, the location strip with the highest percentage by rainfall is the 'medium' (54.86%); by climate, it is the 'Mediterranean' (37.15%), and by latitude, it is the 'north' (51.27%).

As indicated in the middle section of Table 3, it can be noted that there is one location strip that shows a higher result than the others by the percentage of recurrence of construction defects: 'medium-Mediterranean' (31.27%). Following it, there is a triple tie between three location strips: 'high-oceanic', 'high-north' and 'oceanic-north'.

When combining Factors 1, 2 and 3 simultaneously (lower section of the same table), the location strips with the most cases are 'high-oceanic-north' (28.92%) and 'medium-continental-central' (13.13%).

3.5.2. Determination of Zones by Ranks of Normalized Frequencies

Table 4 establishes a quadrant in which 'pluvio-climatic zones' are defined according to the 'rank of concentration of defects'. Thus, according to the percentage of the number of cases (%NC), if it is greater than 15, one will be in ZONE A; if %NC is between 7 and 15, one will be in ZONE B; and if %NC is less than or equal to 7, one will be in ZONE C.

Table 4. Classification by ranks of the pluvio-climatic zones sorted by intensity of recurrence.

	Location Strip (Rainfall-Climate-Latitude)	%NC	Total %NC	Relative Frequency ($\times 10^{-5}$)	Normalised Relative Frequency	Pluvio-Climatic Zone	Rank of Concentration of Defects
1	High-Oceanic-North	28.92%	28.92%	7.05	1.00	A	%NC > 15
2	Medium-Continental-Central	13.13%					
3	Medium-Mediterranean-North	11.95%					
4	Medium-Mediterranean-Central	11.76%	51.64%	3.03	0.43	B	$7 < $ %NC ≤ 15
5	Medium-Mediterranean-South	7.55%					
6	Medium-Continental-North	7.25%					
7	Low-Mediterranean-South	5.88%					
8	Low-Continental-Central	4.58%					
9	Low-Subtropical-South	3.47%	19.44%	2.51	0.36	C	%NC ≤ 7
10	Low-Continental-North	3.16%					
11	Medium-Continental-South	2.35%					

The above-mentioned Table 4 also has an intermediate column (relative frequency) that includes the percentages of defects as a function of the numbers of homes in Spain in each of the location strips. Given that the relative frequencies were low, they were standardised: the highest relative frequency was assigned a value of 100%. The remaining values referred to this normalised value (column named 'normalised relative frequency').

Based on Table 4, it can be noted that the regions of Spain situated to the North, with an oceanic climate and high rainfall, show higher standardised proportions of defects. In other words, windows located in buildings in these areas have a much greater risk of having defects that are complained about judicially. This is explained by the fact that these zones have a very moist environment and infiltration defects lead to frequent complaints. The exception is the last row of Table 4 (*'Medium-Continental-South'* strip) that possesses considerable rainfall and a low percentage. In this case, however, one must note the influence of the location to the South and the continental climate, which produces high temperatures, limiting the effect of humidity.

3.5.3. Individual Analysis According to Each Factor and Type of Defect

An analysis was carried to determine in which climate location strips each of the four types of defects found in windows occurs with a higher percentage of recurrence.

- Water infiltrations (dWI)

  This defect occurs more with medium rainfall (54.28%) than with high (29.12%) or low rainfall (16.60%). It was shown that there is a greater presence in the Mediterranean climate (39.82%) than in the oceanic (29.94%) or continental (26.17%) climates. In addition, this defect occurs quite rarely in the subtropical climate (4.07%). As for latitude, it is more common in the North (48.37%) than in the Central part (28.41%) or in the South (23.22%).

- Air permeability (dAP)

  This type of defect occurs more with medium rainfall (61.98%) than with high (23.96%) or low rainfall (14.06%). It was shown to occur with similar frequency in the Mediterranean (37.08%) and continental climates (36.42%), while in third place is the oceanic climate (24.90%), and in last position is the subtropical climate (1.60%). As for latitude, more than half of the time, dAP occurs in the North (54.95%), occurring in the Central part almost one third of the time (31.31%), while the South comes last (13.74%).

- Humidity by condensation (dHC)

  There is medium rainfall on approximately half of the occasions (49.19%), followed by high (28.34%) and low rainfall (22.47%). Presence in the continental climate (39.82%) was observed in the majority of cases, followed closely by the oceanic (29.32%) and the Mediterranean (28.01%) climates. As with previous defects, the subtropical climate was the least recurrent (3.58%). As for latitude, the North clearly remains as the climate strip with the most cases (56.68%), followed by the Central part (30.29%) and the South (13.03%).

- Oxidation or corrosion (dOC)

  There are no cases of this type of defect with low rainfall, while more than one third of the time, it occurs with high rainfall (38.46%), and nearly two thirds of the time, it occurs with medium rainfall (61.54%). This defect also does not occur in the subtropical climate and has quite a low presence in the continental climate (7.69%). It appears occasionally in the oceanic climate (38.51%) and frequently in the Mediterranean climate (53.80%). Lastly, this type of problem does not occur in the strips to the South of the country, and there is not a significant difference between the Central part (46.15%) and the North (53.85%).

The individual analysis by each factor and type of defect confirms the information presented in Table 4, in the sense that the most adverse climatological locations exhibited higher shares of defects. It can be highlighted that, considering only the phenomenon of rainfall, the places with medium rainfall have more cases than those with high rainfall. This must be explained by the higher performance of windows situated in areas where heavy rain leads to more careful construction practices.

3.5.4. Breakdown of Defects in the Strips with Higher %NC

To have a more detailed perspective of how each of the types of defects were distributed in the most problematic location strips, Table 5 was produced. The criterion used was to select the four strips with %NC > 10%.

Table 5. Number of cases according to each type of defect in the four most problematic location strips.

	Location Strip	%NC	dWI	dAP	dHC	dOC	Total of Cases
1	High-Oceanic-North	28.92%	286	75	87	5	453
2	Medium-Continental-Central	13.13%	125	48	38	1	212
3	Medium-Mediterranean-North	11.95%	104	48	39	2	193
4	Medium-Mediterranean-Central	11.76%	123	38	24	5	190

If said table is analysed, it can be seen that the distributions for each type of defect are quite similar with respect to the percentage of cases they represent in each of the location strips. Thus, dWI has a range of presence between 54% and 65% (with 60% as the average), dAP has a range of presence between 17% and 25% (with 21% as the average), dHC has a range of presence between 15% and 20% (with 17% as the average) and dOC has a range of presence between 0.5% and 3% (with 1% as the average).

4. Discussion

4.1. Reflections on the Database

Based on Section 2 above, the method used in this research is not a 'survey' and is certainly not an 'experiment'. It is not a 'simulation', and it is also not a 'case study' (it does not focus on a specific case of a building or situation, or on a limited and unique area with previously selected characteristics). The analysis was carried out over the 'general census of cases' (taking into account and studying the global record of existing judicial records) for the entirety of a country, under the principles of generality and simultaneity.

As such, all the cases analysed correspond to the total set existing in Spain in the period of the study—in other words, 100% of all the emerging cases were collected herein (no case was left out). As such, they do not only constitute a more-or-less characteristic sample whose representativeness needs verifying; rather, they encompass the totality of all the defects found between the years 2008 and 2017 (there is no error or uncertainty, as it is not a partial sample). This constitutes another novel contribution of this research, given that it is generally not possible to collect the entirety of relevant data in a nation.

It should be pointed out that the set that was researched is quite homogeneous regarding building age. All the buildings, and all the windows analysed, had been built relatively recently. Spanish legislation envisions a three-year warranty period for this type of construction element, starting from the time the works are completed. As such, if any type of defect exists, owners should file judicial complaints in this period. This is why there is no dispersion or distortion of the data resulting from the possible deterioration of window frames caused by the passage of time.

As can be seen, the methodology did not include a number of other objectives that could also have been of interest but are out of the scope and fundamental ideas of this study. These could include trying to ascertain the possible responsibilities of the different participants. This aspect was explicitly left out of the permission given to the authors for accessing the data source, for which reason this aspect could not be assessed.

In future studies, it would be of interest to try to correlate defects with other climatological aspects. This analysis must be based on the exact location of each building—an aspect that was not possible to obtain in the data collection process, given the limitation that was imposed to maintain the confidentiality and protection of certain information. Should this sensitive information be accessible, one would correlate it with other external parameters not contained in the judicial process. All these parameters should be processed for each of the 1615 cases in question. This gives an idea of the difficulty of obtaining and quantifying all these variables.

4.2. Several Considerations

The knowledge of the most recurrent construction defects in different construction units is very important for trying to minimise them in subsequent technical interventions, be it during the design [41] or execution stages. There are also two additional advantages: possessing a good maintenance strategy during the service life and having a governmental public catalogue of construction defects. In order to achieve this goal, it would be necessary for a prior compilation to be produced of the most problematic construction points, with the highest numbers of cases, based on (for example) the actual/judicial records, as presented in this research. Furthermore, national regulations could include specific construction details on how these problematic points could be more effectively addressed. This aspect would enable collaboration to minimise future construction costs and improve maintenance during the use phase—a matter, indeed, that is being given significant attention by a number of publications [42–44].

4.3. Teaching These Critical Aspects in Universities

Students of architecture and engineering should have a greater understanding of building defects, so as to have knowledge on the main critical points and avoid making the same mistakes in the future (both in design and execution stages). This aspect is often not given sufficient attention in university degrees.

As a way of facilitating their understanding, and in order to deliver this knowledge in an eminently visual manner (given that the generation they are a part of often works with visual information), the authors have produced a number of infographics on different construction points (starting from the facade openings where windows are inserted). Based on their teaching experience, a number of encouraging results are being obtained. One example of these infographics is shown in Figure 7.

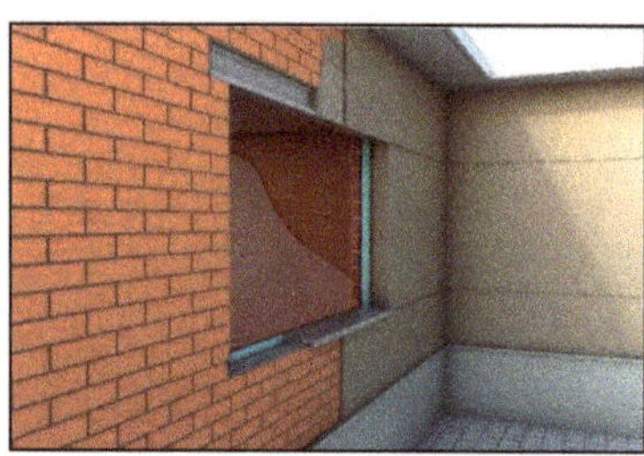

Figure 7. Examples of infographics of the singularities between facades and windows as a learning tool in university classes.

5. Conclusions

The current study considers a database of 1615 cases of judicial complaints filed in Spain regarding defects in windows. In the literature review that was carried out, there were no references to similar work based on this type of source data, the extension of which provides quite a complete view on the reality of the problems that users complain about in these construction units (100% of the cases existing in Spain over 10 years).

By the analysis of said data, it can be noted that there is a clear majority of complaints related to 'water infiltrations' (dWI = 60.8%). Far behind, as to their frequency, and with nearly identical percentages between them are the defects related to 'air permeability' (dAP = 19.4%) and 'humidity by condensation' (dAP = 19.0%). As for the originating causes that lead to the defects, the 'absence/deficiency of sealant' and 'inadequate construction material and/or placement' together result in 83.4% of the cases, it being thus shown that problems are usually focused on aspects related to the placement of windows.

The influence of three climatological factors on the presence of the defects was also studied: rainfall, climate and latitude. As to the influence of the last (latitude), it is clear that there is a greater percentage in location strips situated in the Central part and North of the country (80.75% of cases according to the sum of the %NC values of Rows 1, 2, 3, 4, 6, 8 and 10 of Table 4). From the ranks of the concentrations of the defects analysed, three zones (A, B and C) were created/catalogued, it being noted that there is a greater recurrence in Zone A: locations with high rainfall/oceanic climate/North latitude (in which is obtained a relative frequency of 7.05×10^{-5}).

The results presented herein can be of great interest for researchers of other countries who wish to know the probability of judicial complaints by building users. The values that were handled are highly representative, since they correspond to the totality of cases complained about in the span of time that was indicated.

This information can also be of great assistance for reducing the impact of low-quality processes [45]. The defects and the originating causes found in this research enable the different agents participating in the construction process to possess specific and highly useful information to minimise errors in the design and execution stages [41]. This will significantly lower conflicts and will make the construction process more sustainable, given that there will be less litigation during the stage of the use of the buildings, and repair costs will be significantly reduced.

Author Contributions: Conceptualization, M.J.C.-A.; data curation, M.J.C.-A.; formal analysis, C.E.R.-J. and D.B.-H.; investigation, M.J.C.-A.; methodology, M.J.C.-A.; validation, J.M.; writing—original draft, M.J.C.-A. and C.E.R.-J.; writing—review and editing, D.B.-H. and J.M. All authors have read and agreed to the published version of the manuscript.

Funding: This research received no external funding.

Acknowledgments: The current work was carried out within the MUSAAT Foundation's Action Plan, which envisaged carrying out national research on anomalies in buildings [46].

Conflicts of Interest: The authors declare no conflict of interest.

Nomenclature

dWI—Water infiltration	Entrance of water at weak points of a construction element leading to dripping and/or visible loss of this liquid.
dAP—Air permeability	Penetration of wind through a window, be it through the profiles of the frame or between these and the opening of the facade wall.
dHC—Humidity by condensation	Physical phenomenon through which the environmental humidity liquefies in contact with a cold wall.
dOC—Oxidation or corrosion	Deterioration of a metallic material due to lack of protection against electro-chemical attack.
ocAS—Absence/deficiency of sealant	This action was incorrectly carried out, failing to guarantee watertightness.
ocIC—Inadequate construction material and/or placement	Characteristics of the material and/or the system of placement that were foreseen are insufficient.
ocTB—Existence of thermal bridges	Presence in a specific location of the envelope of a point with different thermal conductivity between different points due to a change in the thickness of the materials, to a lack of continuity between them or to their characteristics.
ocIA—Incorrect assembly process	The process of fitting and joining materials is not correctly carried out.

References

1. Lopez, C.; Masters, F.J.; Bolton, S. Water penetration resistance of residential window and wall systems subjected to steady and unsteady wind loading. *Build. Environ.* **2011**, *46*, 1329–1342. [CrossRef]

2. Almeida, R.M.S.F.; Ramos, N.M.M.; Pereira, P.F. A contribution for the quantification of the influence of windows on the airtightness of Southern European buildings. *Energy Build.* **2017**, *139*, 174–185. [CrossRef]

3. Ramírez, E.; Pujols, W.C.; Casares, F.J.; Ramírez-Faz, J.; López-Luque, R. Development of a suitable synthetic projection to simultaneously study solar exposure and natural lighting in building windows. *Energy Build.* **2013**, *65*, 391–397. [CrossRef]

4. Carretero-Ayuso, M.J.; García-Sanz-Calcedo, J.; Rodríguez-Jiménez, C.E. Characterization and Appraisal of Technical Specifications in Brick Façade Projects in Spain. *J. Perform. Constr. Facil.* **2018**, *32*, 4018012. [CrossRef]

5. Intini, F.; Kühtz, S.; Milano, P.; Dassisti, M. Analysis of sustainability assessment of building windows for italian residential market: Life cycle analysis and LEED. *Procedia Environ. Sci. Eng. Manag.* **2015**, *2*, 239–247.

6. Mills, A.; Love, P.E.; Williams, P. Defect costs in residential construction. *J. Constr. Eng. Manag.* **2009**, *135*, 12–16. [CrossRef]

7. Madureira, S.; Flores-Colen, I.; de Brito, J.; Pereira, C. Maintenance planning of facades in current buildings. *Constr. Build. Mater.* **2017**, *147*, 790–802. [CrossRef]

8. Love, P.E.D.; Smith, J.; Ackermann, F.; Irani, Z.; Teo, P. The costs of rework: Insights from construction and opportunities for learning. *Prod. Plan. Control* **2018**, *29*, 1082–1095. [CrossRef]

9. Vilutiene, T.; Ignatavičius, Č. Towards sustainable renovation: Key performance indicators for quality monitoring. *Sustainability* **2018**, *10*, 1840. [CrossRef]

10. Van Den Bossche, N.; Janssens, A. Airtightness and watertightness of window frames: Comparison of performance and requirements. *Build. Environ.* **2016**, *110*, 129–139. [CrossRef]

11. d'Ambrosio Alfano, F.R.; Dell'Isola, M.; Ficco, G.; Palella, B.I.; Riccio, G. Experimental air-tightness analysis in mediterranean buildings after windows retrofit. *Sustainability* **2016**, *8*, 991. [CrossRef]

12. Baiburin, A.K.; Rybakov, M.M.; Vatin, N.I. Heat loss through the window frames of buildings. *Mag. Civ. Eng.* **2019**, *85*. [CrossRef]

13. Hee, W.J.; Alghoul, M.A.; Bakhtyar, B.; Elayeb, O.; Shameri, M.A.; Alrubaih, M.S.; Sopian, K. The role of window glazing on daylighting and energy saving in buildings. *Renew. Sustain. Energy Rev.* **2015**, *42*, 323–343. [CrossRef]

14. Cappelletti, F.; Gasparella, A.; Romagnoni, P.; Baggio, P. Analysis of the influence of installation thermal bridges on windows performance: The case of clay block walls. *Energy Build.* **2011**, *43*, 1435–1442. [CrossRef]

15. Kim, S.H.; Jeong, H.; Cho, S. A study on changes of window thermal performance by analysis of physical test results in Korea. *Energies* **2019**, *12*, 3822. [CrossRef]

16. Gruner, M.; Matusiak, B.S. A novel dynamic insulation system for windows. *Sustainability* **2018**, *10*, 2907. [CrossRef]

17. Koo, B.; Lee, K.; An, Y.; Lee, K. Solar heat gain reduction of ventilated double skin windows without a shading device. *Sustainability* **2017**, *10*, 64. [CrossRef]

18. Pereira, C.; de Brito, J.; Silvestre, J.D. Contribution of humidity to the degradation of façade claddings in current buildings. *Eng. Fail. Anal.* **2018**, *90*, 103–115. [CrossRef]

19. Van Den Bossche, N.; Huyghe, W.; Moens, J.; Janssens, A.; Depaepe, M. Airtightness of the window-wall interface in cavity brick walls. *Energy Build.* **2012**, *45*, 32–42. [CrossRef]

20. Park, S.; Song, S.-Y. Case study on the inspection and repair of window condensation problems in a new apartment complex. *J. Perform. Constr. Facil.* **2018**, *32*, 4018071. [CrossRef]

21. Calderón, S.; Sandoval, C.; Inzunza, E.; Cruz-Noguez, C.; Rahim, A.B.; Vargas, L. Influence of a window-type opening on the shear response of partially-grouted masonry shear walls. *Eng. Struct.* **2019**, *201*, 109783. [CrossRef]

22. Soares, N.; Martins, C.; Gonçalves, M.; Santos, P.; da Silva, L.S.; Costa, J.J. Laboratory and in-situ non-destructive methods to evaluate the thermal transmittance and behavior of walls, windows, and construction elements with innovative materials: A review. *Energy Build.* **2019**, *182*, 88–110. [CrossRef]

23. Meiss, A.; Feijó-Muñoz, J. The energy impact of infiltration: A study on buildings located in north central Spain. *Energy Effic.* **2015**, *8*, 51–64. [CrossRef]

24. Prignon, M.; Van Moeseke, G. Factors influencing airtightness and airtightness predictive models: A literature review. *Energy Build.* **2017**, *146*, 87–97. [CrossRef]

25. Bienvenido-Huertas, D.; Moyano, J.; Marín, D.; Fresco-Contreras, R. Review of in situ methods for assessing the thermal transmittance of walls. *Renew. Sustain. Energy Rev.* **2019**, *102*, 356–371. [CrossRef]

26. Lucchi, E. Applications of the infrared thermography in the energy audit of buildings: A review. *Renew. Sustain. Energy Rev.* **2018**, *82*, 3077–3090. [CrossRef]

27. Rodríguez-Jiménez, C.E.; Moyano, J.; Carretero-Ayuso, M.J.; Guillén-Lupiáñez, M.I. Methodological proposal for on-site watertightness testing with wind pressure on facade windows. *J. Perform. Constr. Facil.* **2018**, *32*, 4017139. [CrossRef]

28. Santos, A.; Vicente, M.; de Brito, J.; Flores-Colen, I.; Castelo, A. Analysis of the inspection, diagnosis, and repair of external door and window frames. *J. Perform. Constr. Facil.* **2017**, *31*, 4017098. [CrossRef]

29. Siddiqui, A.; Biswas, A.P. Defects a critical issue in construction. *Int. J. Sci. Technol. Res.* **2019**, *8*, 147–150.

30. Love, P.E.D.; Edwards, D.J. Forensic project management: The underlying causes of rework in construction projects. *Civ. Eng. Environ. Syst.* **2004**, *21*, 207–228. [CrossRef]

31. Asif, M.; Muneer, T.; Kubie, J. Sustainability analysis of window frames. *Build. Serv. Eng. Res. Technol.* **2005**, *26*, 71–87. [CrossRef]

32. Carlisle, S.; Friedlander, E. The influence of durability and recycling on life cycle impacts of window frame assemblies. *Int. J. Life Cycle Assess.* **2016**, *21*, 1645–1657. [CrossRef]

33. Ferreira, C.; Silva, A.; de Brito, J.; Dias, I.S.; Flores-Colen, I. The impact of imperfect maintenance actions on the degradation of buildings' envelope components. *J. Build. Eng.* **2020**, *33*, 101571. [CrossRef]

34. Santos, A.; Vicente, M.; De Brito, J.; Flores-Colen, I.; Castelo, A. Inspection, diagnosis, and rehabilitation system of door and window frames. *J. Perform. Constr. Facil.* **2017**, *31*, 4016118. [CrossRef]

35. Fernandes, D.; de Brito, J.; Silva, A. Methodology for service life prediction of window frames. *Can. J. Civ. Eng.* **2019**, *46*, 1010–1020. [CrossRef]

36. Maia, M.; Morais, R.; Silva, A. Application of the factor method to the service life prediction of window frames. *Eng. Fail. Anal.* **2020**, *109*, 104245. [CrossRef]

37. MUSAAT. Expert Records and Reports if Accidents in Spain. In *Mútua de Aparejadores y Arquitectos Técnicos*; MUSAAT: Madrid, Spain, 2017.

38. SERJUTECA. Reports and Documents on Accidents Involving Professional Civil Liability of Building Surveyors and Technical Architects in Spain. In *Servicios Jurídicos Técnicos Aseguradores*; SERJUTECA: Madrid, Spain, 2017.

39. AEMET. Climate and Meteorological Data of Spain. Available online: http://www.aemet.es (accessed on 1 February 2020).

40. Spainish Institute of Statistics. *Population and Housing Census*; Spainish Institute of Statistics: Madrid, Spain, 2019.

41. Carretero-Ayuso, M.J.; García-Sanz-Calcedo, J. Analytical study on design deficiencies in the envelope projects of healthcare buildings in spain. *Sustain. Cities Soc.* **2018**, *42*, 139–147. [CrossRef]

42. Plebankiewicz, E.; Meszek, W.; Zima, K.; Wieczorek, D. Probabilistic and fuzzy approaches for estimating the life cycle costs of buildings under conditions of exposure to risk. *Sustainability* **2020**, *12*, 226. [CrossRef]

43. Farahani, A.; Wallbaum, H.; Dalenbäck, J.-O. Cost-Optimal Maintenance and Renovation Planning in Multifamily Buildings with Annual Budget Constraints. *J. Constr. Eng. Manag.* **2020**, *146*, 4020009. [CrossRef]

44. Andújar-Montoya, M.D.; Galiano-Garrigós, A.; Echarri-Iribarren, V.; Rizo-Maestre, C. BIM-LEAN as a methodology to save execution costs in building construction-An experience under the spanish framework. *Appl. Sci.* **2020**, *10*, 1913. [CrossRef]

45. Forcada, N.; Macarulla, M.; Gangolells, M.; Casals, M.; Fuertes, A.; Roca, X. Posthandover housing defects: Sources and origins. *J. Perform. Constr. Facil.* **2013**, *27*, 756–762. [CrossRef]

46. Carretero-Ayuso, M.J.; Moreno-Cansado, A. *National Statistical Analysis on Construction Pathologies*; MUSAAT Foundation: Madrid, Spain, 2019.

Article

Cost of Climate Change: Risk of Building Loss from Typhoon in South Korea

Ji-Myong Kim [1], Seunghyun Son [2], Sungho Lee [3] and Kiyoung Son [4,*]

1   Department of Architectural Engineering, Mokpo National University, Mokpo 58554, Korea; jimy6180@gmail.com
2   Department of Architectural Engineering, Kyung Hee University, 1732 Deogyeong-Daero, Giheung-Gu, Yongin-Si, Gyeonggi-do 17104, Korea; seunghyun@khu.ac.kr
3   Financial Management Division, Kyung Hee University, 26 Kyungheedae-ro, Dongdaemun-gu, Seoul 02447, Korea; khlsh@khu.ac.kr
4   School of Architectural Engineering, University of Ulsan, 93 Daehak-Ro, Ulsan 44610, Korea
*   Correspondence: sky9852111@ulsan.ac.kr; Tel.: +82-52-259-2788

Received: 30 June 2020; Accepted: 26 August 2020; Published: 31 August 2020

Abstract: In recent years, natural disasters and climate abnormalities have increased worldwide. The Fifth Assessment Report (2014) of the Intergovernmental Panel on Climate Change warned of extreme rainfall events, warming and acidification, global mean temperature rises, and average sea level rises. In many countries, changes in weather disaster patterns, such as typhoons and heavy rains, have already led to increased damage to buildings. However, the empirical quantification of typhoon risk and building damage due to climate change is insufficient. The purpose of this study was to quantify the risk of building loss from typhoon pattern change caused by climate change. To this end, the intensity and frequency of typhoons affecting Korea were analyzed to examine typhoon patterns. In addition, typhoon risk was quantified using the Korean typhoon vulnerability function utilized by insurers, reinsurers, and vendors, the major users of catastrophe modeling. Hence, through this study, it is possible to generate various risk management strategies, which can be used by governments when establishing climate change policies and help insurers to improve their business models through climate risk assessment based on reasonable quantitative typhoon damage scenarios.

Keywords: climate change; typhoon; catastrophe model; typhoon vulnerability function; risk analysis

1. Introduction

Climate change is expected to have serious consequences in a wide range of areas. It is expected to affect extreme weather events in the short term, as well as generating long-term effects such as disease spread and rising sea levels. Extreme weather events could include heat waves, cold waves, windstorms such as hurricanes, heavy rains, floods, a lack of precipitation, and drought. Many regions have suffered from the fatal effects of recent extreme weather events. Such extreme weather events have, of course, always been part of human history. However, recent extreme weather events have become greater in frequency and intensity than those in the past, and the potential for damage has increased rapidly.

Additionally, the current pattern of tropic cyclones is so different from past patterns that they are called super typhoons or super hurricanes. For instance, Typhoon Haiyan occurred in 2013 and became known as Super Typhoon Yolanda, as it was the most extreme tropical cyclone recorded on land. Its severe rain and winds made it difficult for South Asian nations to recover from the shattering damage of about USD 300 billion [1]. In the United States in 2017, three powerful hurricanes (Hurricanes Harvey, Maria, and Irma) caused tremendous damage. The total damage from these hurricanes was about USD 293 billion, with Harvey causing USD 125 billion in damage, Maria causing

USD 90 billion, and Irma USD 77.6 billion worth of damage [2]. In addition, Hurricane Katrina, which occurred in 2005, was one of the most damaging natural disasters in United States history. The heavy rain and strong winds generated by hurricanes have caused US Gulf Coast cities to suffer about USD 180 billion in direct and indirect damage [2]. In Europe, economic losses of around EUR 13 billion were incurred in 1999 due to the record rain and winds of the European storms Anatol, Lothar, and Martin [3].

However, despite these historic events and record damage, there are still debates about climate change and tropical cyclone patterns. Even though many studies have argued that climate change has affected tropical cyclones, other research argues that the evidence for this is poor. For instance, though some have asserted that the intensity of tropical cyclones gradually increases as the climate warms up [4–6], others argue that this increase is within the natural range of fluctuation in the frequency or severity of tropical cyclones in long-term climate observations [7]. Depending on the region, long-term climate observations may not be of sufficient duration to determine how climate change affects tropical cyclones, or the effects may not be clear. It is also difficult to predict how future activities will impact climate change. However, there is evidence that the damage caused by extreme weather events, especially tropical cyclones, is increasing every year [8]. Other studies have shown that this trend is damaging more people and assets, and the damage will be even greater given the high coastal population and property density of many cities, and reduced woodlands [9,10]. While these studies do not adequately rule out damage due to increased social vulnerability (e.g., income and population), it is difficult to view this as only an increase in extreme weather events and tropical cyclones. The trend is clear [11]. Therefore, we will analyze the intensity and frequency of typhoons that have affected Korea for a scientific and quantitative examination of the impact of climate change on typhoons. In addition, this study will assess the risk of building loss to quantify the damage caused by changes in typhoon patterns due to climate change.

2. Literature Review

2.1. Climate Change and Economic Impact

The United Nations Intergovernmental Panel on Climate Change (IPCC) warns against climate change in its 5th Assessment Report (AR5). Compared to pre-industrial levels, the report estimates that global average temperatures will rise by more than 1.5 °C in all scenarios by 2100. In addition, warming will continue as greenhouse gas emissions continue, and moreover, it is likely to exceed 2.0 °C in many scenarios. Additionally, the World Bank (2014) has a similar outlook. Global warming is inevitable due to greenhouse gases in the Earth's atmosphere, and the temperature will be 1.5 °C higher than before industrialization. Without reasonable steps to reduce greenhouse gas emissions, the planet is expected to warm up by up to 2 °C by the middle of the century and up to 4 °C by the end of the century [12]. Furthermore, Stern (2006) reported that in the absence of measures to reduce emissions, greenhouse gas concentrations would reach twice the pre-industrial levels in early 2035, raising the Earth's temperature by nearly 2 °C. This warming is expected to change the water cycle around the world, increasing the difference between wet and dry regions. As the heat expands into the deeper oceans, the ocean's circulation pattern will change and continue to warm, and the Earth's glaciers will decrease. Due to the reduced glaciers, the global average sea level is likely to rise more quickly than the rate of rise over the last 40 years (IPCC 2004). As mentioned above, many studies and research papers show that global climate change is certain and will increase, and warn against the side effects of warming [13].

The literature on the economic impact of climate change is as follows. The IPCC Fifth Assessment (2014) reports that increasing warming above 3 °C will result in a loss of 0.2% to 2.0% of annual GDP (gross domestic product), although estimates of damage vary widely from country to country [14]. The IPCC expects further acceleration of the occurrence of damage if warming exceeds 2 °C, but these effects will be difficult to realize over the next 30 years. Moreover, if warming exceeds 2 °C, negative returns are expected from various portfolios [15]. Dietz and Stern (2014) estimate that when global warming reaches the 4 °C level, annual economic output will decrease by 50% compared to that without warming. They estimated a warming of around 3.5 °C by 2100 [16]. Stern (2006) estimated that over the next two centuries, global warming scenarios between 2.4 and 5.8 °C would result in a mean loss of about 5% (up to 20% in some regions) of global yearly GDP by 2100. These calculations indicate that no action has been taken on global warming, and the costs are expected to increase by more than 20% of GDP, given the wide range of risks and impacts. In addition, with simple extrapolation, it was estimated that extreme weather could damage 0.5% to 1% of global GDP by the middle of the century [13]. Mendelsohn et al. (2000) studied the potential damage using a global warming scenario (an increase of 2.5 °C by 2010), and estimated that the total market impact cost would not exceed 0.1% of GDP in 2100. Market impact may vary based on latitude. For example, in low latitude countries, warming increases damage. On the other hand, income is expected to increase at higher latitudes. However, if global warming is above 2.0 °C, it is expected that the benefits will decrease and the damage will increase. They also found that damage in a global warming scenario (2.0 °C increase by 2060) would be expected to have an aggregate impact of 0.3% damage to GDP in 2060. The study estimated that with warming of 2.0 °C by 2060, most of the damage would occur in agriculture, and the damage would vary widely from country to country [17]. As these studies show, climate change is predicted to have a significant impact on future economic growth and living standards. Losses may vary by region, but the damage is expected to increase globally. In addition, severe weather phenomena are expected to add to the damage.

2.2. Climate Change and Losses from Natural Disasters

The increase in damage caused by natural disasters is closely related to the growth of population and wealth. This is because the world's population is increasing every year, and wealth is also growing. The annual damage caused by natural disasters may be linked to these increases in wealth and population. Therefore, to objectively quantify climate change and the increase in damage from natural disasters, increases in wealth and population must also be considered [18]. To this end, many studies have examined climate change and damage after normalization for population and wealth changes. Nordhaus (2010) argued that since 1900, losses from hurricanes in the United States have increased significantly according to revised data only for GDP [19]. Changnon (2009) argued that insurance losses from hurricanes in the United States increased between 1952 and 2006 but that the growth was concentrated in the western United States and is believed to be due to recent increases in population and wealth in this region [20]. He also examined a study of insurance losses due to hail in the United States since 1992. The amount of insurance losses due to hail has increased, but this is attributed to increased exposure and vulnerability to hail due to the expansion of urban areas. There was no change in the frequency of major hail storms [9]. Chang et al. (2009) detailed a rise in flood loss through a flood loss survey (since 1971) in six cities in Korea. The cause of the increase in flood damage was found to be related to the increase in population, as well as summer rainfall and deforestation [10]. Schmid et al. (2009) found that there was a clear trend in US hurricane losses. However, this trend appeared after 1970 (it was not seen in the entire record dating back to 1950) and was found only after adjustments for wealth and population [21]. Fengqing et al. (2005) investigated flood damage in the Xinjiang Autonomous Region of China and determined that flood damage had increased since 1987. However, he pointed out that reserves and flood control structures, not heavy rains caused by climate change, were responsible for the increase in flood damage [22]. Changnon (2001) reported increased damage according to normalized data due to strong winds, rainfall, lightning, hail, and tornadoes since 1974 in the western United States. Nevertheless, the study also showed increased losses according to normalized data even in areas with reduced thunderstorm activity, suggesting that socioeconomic factors contributed to this trend [23]. Miller et al. (2008) analyzed the loss data for climate disasters around the world after revising them, taking into account wealth and population growth. Their main findings were that since 1970, losses from climate disasters have increased, but this trend does not extend back to 1950. In addition, the authors believe that the increase in losses from climate disasters is due to the hurricane damage in the United States in 2004 and 2005 [24].

As with previous studies, wealth and population are important considerations in the study of the relationship between climate change and losses from natural disasters. This is because increases in population and wealth are important contributors to increased natural disaster damage. It is true that studies of natural disaster damage caused by climate change are difficult due to the close relationship between wealth, population, and natural disaster damage. Therefore, in this study, the vulnerability function was used to exclude the interference of wealth and population for the quantitative study of climate change and losses from natural disasters only. In addition, the existing studies judge the increase and decrease in damage amounts from natural disasters, and thus, it is difficult to quantitatively study building damage due to climate change. Therefore, this study divided buildings into three groups by occupancy (commercial, industrial, and residential) and considered the risk of building loss due to climate change for each category.

3. Framework of Study

The purpose of this study was to quantitatively prove climate change and typhoon changes. To achieve this, the study consisted of two parts. First, this study investigated typhoons that have affected Korea and analyzed the intensity and frequency of typhoons and changes in typhoon patterns.

Second, this study quantified the risk of building loss due to changes in typhoon patterns as a result of climate change. To quantify typhoon risk, this study used the Korean typhoon vulnerability function of major users of catastrophic (CAT) modeling: insurers, reinsurers, and suppliers. The buildings were divided into commercial, industrial, and residential types for analysis. As shown in Figure 1, this study was limited to typhoons that affected S. Korea from the 1970s to 2010s. Therefore, the research scope of this study was limited to S. Korea and reflected the architectural design standards and planning strategies of S. Korea. The results may be different in countries with different geographic or architectural design standards and planning strategies from S. Korea.

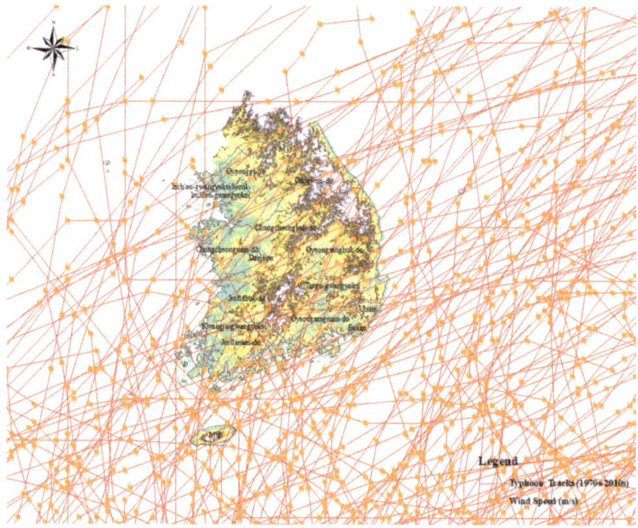

Figure 1. The typhoon that affected S. Korea (1970s to 2010s).

4. Typhoon Patterns

This part will discuss the frequency and severity of typhoons. Since the amount of risk is determined by the product of frequency and severity, both frequency and severity play an important role in risk determination. Therefore, frequency and severity were examined separately for detailed investigation.

Data on typhoons were obtained from the Korea Meteorological Administration (KMA, Seoul, Korea). The KMA was established in 1949 as a Korean government agency providing meteorological services and is responsible for monitoring the weather system and distributing and storing related information. This study collected data on the number of typhoons and the wind speeds that affected Korea between 1973 and 2019 from the KMA.

Frequency and Severity of Typhoons

The frequency of typhoons by year is shown in Figure 2. Korea was affected by an average of 3.3 typhoons annually, with a standard deviation of 1.5. The minimum number of typhoons generated during the survey was zero, and the largest number of typhoons affecting Korea was seven. The linear regression model of typhoon frequency is $y = 0.0036 \times \text{Year} - 3.8831$. The R^2 value is 0.0012. The slope of this model shows that the relationship between the year and number of typhoons is positive, indicating that the number of typhoons has increased slightly each year. However, the R^2 value shows that the relationship between the year and number of typhoons is very weak, since the value is extremely low. Hence, it would be difficult to conclude that climate change has a clear impact on the number of typhoons.

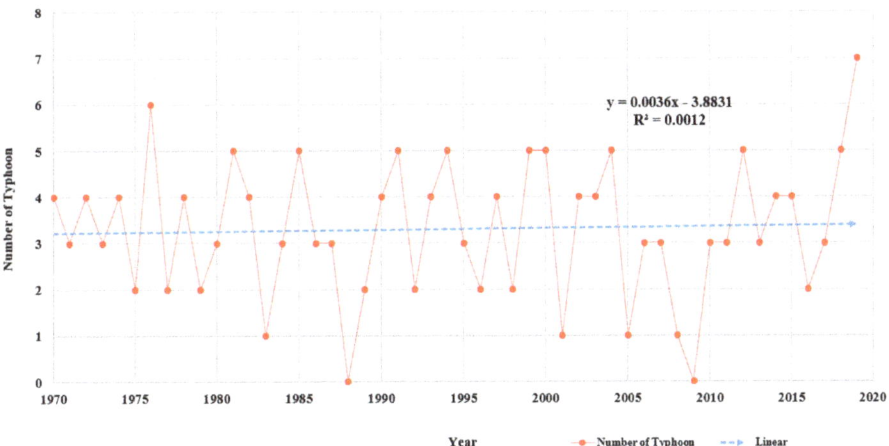

Figure 2. Number of typhoons by year.

The maximum wind speed of typhoons by year is shown in Figure 3. The wind speed was based on 10 min sustained wind speed. When a typhoon affected Korea, the highest wind speed recorded by 96 meteorological stations was considered as the maximum wind speed. After collecting the maximum wind speed by typhoon, the maximum wind speed was determined by year. The average maximum wind speed was 27.6 m/s, with a standard deviation of 8.4 m/s. The highest maximum wind speed was 51.1 m/s, and the lowest was 17.3 m/s.

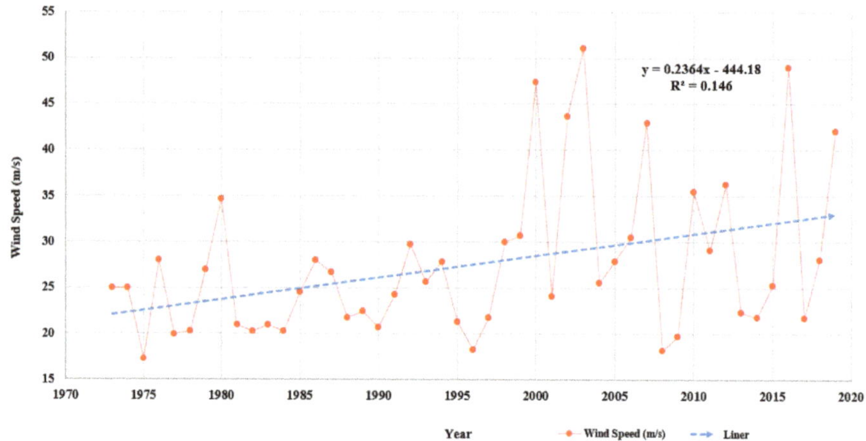

Figure 3. Maximum wind speed of typhoons by year.

The linear regression model of typhoon severity is y = 0.02364 × Year − 444.18. The R^2 value is 0.146. The slope of this model shows that the relationship between the year and maximum wind speed is positive, indicating that the maximum wind speed has increased each year. Furthermore, the R^2 value shows that the year and the maximum wind speed have a weak quantitative linear relationship.

Based on the typhoon data from the KMA, this study investigated the frequency and severity (max. wind speed) of typhoons by year. The frequency of typhoons was found to increase only slightly from year to year, but due to the low R^2 value, the explanatory power is low and not significant. However, the severity of typhoons appears to have increased year by year. Although the R^2 value is relatively small, it is a weak positive relationship and is sufficient to show a potential trend. This suggests that the frequency of typhoons does not increase every year, but the severity does. Thus, it is possible that the risk from typhoons has increased due to increasing severity.

However, the data of the KMA used in this study were recorded for about 50 years from 1973 to 2019. Although this period is a period for which the KMA's current national and regional data are available, it is considered to be a short period for concluding that the severity of typhoons is gradually increasing. Therefore, it is necessary to keep an eye on the trend through additional data collection.

5. Calculation of Increased Typhoon Risk

This section quantifies the risk of building loss due to changes in typhoon patterns resulting from climate change. This study adopted the Korean typhoon vulnerability function used by insurers, reinsurers, and vendors to quantify typhoon risk. The assessment of vulnerability to typhoons is a significant part of the typhoon risk assessment model. It is the vulnerability curve or vulnerability function that is used for this vulnerability assessment. The vulnerability function is expressed by quantifying the vulnerability of the building. The vulnerability function for typhoons explains the correlation between the average loss ratio, wind speed, and inventory information of various buildings, and determines the loss scale. The average damage ratio is the total amount of damage incurred by a building due to a typhoon divided by the total cost of the building. Therefore, the average damage ratio is used as a measure of a building's vulnerability to typhoons. For example, a high average damage ratio indicates that the damage is large due to a high vulnerability to typhoons. The vulnerability function was used because it quantifies the loss ratio for typhoons through various damage indicators such as the inventory information of buildings and wind speed, thus preventing the distortion of damage by wealth and population found in previous studies [25]. The catastrophic (CAT) model has been developed and used by a number of initiatives, global administrations, and private interests as a risk assessment tool for scientifically assessing, responding to, or mitigating natural disaster risk. For example, public models include HAZUS Multi-Hazard in the United States, RiskScape in New Zealand, New Multi-Hazard and Multi-Risk Assessment Method (MATRIX) in Europe, and Central America Probabilistic Risk Assessment in South America. Vendor models include Risk Management Solutions, Applied Insurance Research, and Risk Quantification and Engineering, which develop and use models for natural disasters and other risks as business models. Primary and reinsurance companies quantify risks from natural disasters by actively using in-house or vendor models. They use the CAT model to manage their portfolios, capital, business preferences, and holding strategies and for capacity monitoring based on the quantified risks of natural disasters [25,26]. The CAT model generally consists of four parts (i.e., a hazard module, exposure module, vulnerability module, and financial module). Each module has an independent function, and the operation of the modules proceeds sequentially. First, the hazard module generates events and calculates local intensity to physically define the events and describe the severity and frequency of natural disasters. Second, the exposure module embodies inventory and geographic information for a building. Third, the vulnerability module provides the loss ratio based on the vulnerability function, which determines the average loss ratio based on wind speed and building inventory information. Lastly, the financial module applies certain insurance factors, such as deductibles and liability limits, to calculate financial losses [25]. For instance, in the hazard module, the severity and frequency of typhoons are defined through simulation according to the characteristics of past typhoons in a specific area. In the exposure module, wind speed is determined according to the inventory characteristics and geographic characteristics of buildings in that specific area. Depending on the determined wind speed, the vulnerability module computes the loss amount through the vulnerability function. The calculated loss amount is calculated by considering insurance conditions in the financial module.

Figure 4 illustrates the vulnerability functions for each model. This study used the vulnerability function of loss ratio for wind speed and occupancy. Occupancy, a representative relative vulnerability factor, was used to reflect the vulnerability of building inventory. Occupancy is used in risk management and risk assessment models among other building inventory information. Occupancy also refers to a similar accounting policy in insurance that categorizes buildings as industrial, residential, and commercial. This classification of buildings according to occupancy refers to building units with similar physical and financial characteristics. This study also adopted the occupancy classification and divided buildings into industrial, residential, and commercial groups.

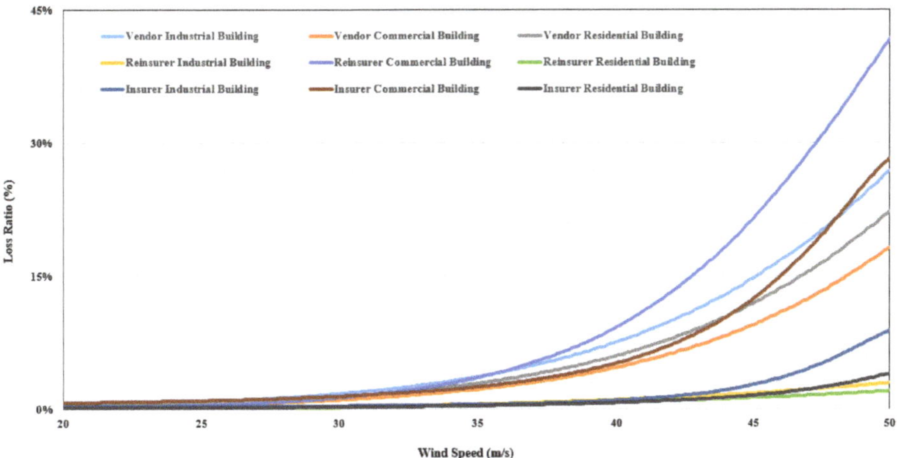

Figure 4. Vulnerability functions for each model.

Results of Analysis

The analysis results are shown in Tables 1–3. To clearly show the increase rate, the five decades from the 1970s, when typhoon data began to be recorded, until the recent 2010s were compared. Additionally, as seen in Equation (1), each decade showed a numerical increase or decrease compared with the 1970s. The period in Equation (1) refers to the 1980s, 1990s, 2000s, or 2010s.

$$\text{The rate of increase each decade (\%)} = (\text{period} - 1970\text{s})/1970\text{s} \tag{1}$$

Table 1. Result (average) summary of each model for industrial buildings.

Period	Vendor		Reinsurer		Insurer		Average		Increase Rate	CV
	Ave.	ST.	Ave.	STD.	Ave.	STD.	Ave.	STD.		
1970s	0.78%	0.35%	0.12%	0.07%	0.35%	0.02%	0.42%	0.15%	-	0.4
1980s	0.97%	0.92%	0.15%	0.15%	0.36%	0.06%	0.50%	0.38%	+19%	0.8
1990s	1.04%	0.58%	0.16%	0.11%	0.37%	0.03%	0.52%	0.24%	+26%	0.5
2000s	7.66%	9.11%	0.94%	1.05%	1.99%	2.70%	3.53%	4.29%	+750%	1.2
2010s	4.80%	6.98%	0.63%	0.81%	1.22%	2.07%	2.22%	3.29%	+434%	1.5

Table 2. Result (average) summary of each model for commercial buildings.

Period	Vendor		Reinsurer		Insurer		Average		Increase Rate	CV
	Ave.	STD.	Ave.	STD.	Ave.	STD.	Ave.	STD.		
1970s	0.61%	0.28%	0.08%	0.05%	0.17%	0.03%	0.29%	0.12%	-	0.4
1980s	0.76%	0.72%	0.10%	0.11%	0.18%	0.07%	0.35%	0.30%	+22%	0.9
1990s	0.81%	0.46%	0.12%	0.07%	0.19%	0.04%	0.37%	0.19%	+30%	0.5
2000s	6.22%	7.50%	0.66%	0.74%	0.99%	1.24%	2.63%	3.16%	+819%	1.2
2010s	3.87%	5.73%	0.45%	0.57%	0.62%	0.92%	1.65%	2.41%	+476%	1.5

Table 3. Result (average) summary of each model for residential buildings.

Period	Vendor		Reinsurer		Insurer		Average		Increase Rate	CV
	Ave.	STD.	Ave.	STD.	Ave.	STD.	Ave.	STD.		
1970s	0.47%	0.22%	0.45%	0.27%	0.95%	0.18%	0.63%	0.22%	-	0.4
1980s	0.60%	0.56%	0.68%	0.93%	1.06%	0.49%	0.78%	0.66%	+24%	0.9
1990s	0.64%	0.36%	0.69%	0.50%	1.09%	0.30%	0.80%	0.38%	+28%	0.5
2000s	5.02%	6.11%	10.91%	14.27%	7.23%	9.09%	7.72%	9.82%	+1133%	1.3
2010s	3.10%	4.67%	6.27%	11.03%	4.43%	7.13%	4.60%	7.61%	+634%	1.7

The differences in the average loss rates for each model ranged from 0.38% to 6.72%. The average loss ratio for each model is as follows. In the case of the Vendor model, industrial buildings have 3.05%, commercial buildings have 1.96%, and residential buildings have 2.45%. In the case of the Reinsurer model, they are 0.40% for industrial buildings, 3.80% for commercial buildings, and 0.28% for residential buildings. In the case of the Insurer model, they are 0.86% for industrial buildings, 2.95% for commercial buildings, and 0.43% for residential buildings.

The vulnerability function is generally developed based on the available or actual losses of the model developers. However, losses may differ due to the differing capital, business preferences, and portfolios of each model developer. These factors will make a difference, even if Korea's vulnerability function is the same. To compensate, the insurance industry typically compares and validates the results of two or more models. Therefore, this study improved the reliability of the results by comparing multiple models from three different fields. This study averaged the results of three models and compared the increase and decrease by decade. Table 1 describes the result (average) summary of each model for industrial buildings. The increase rate has grown gradually over the decades compared with the 1970s, increasing 307% on average and 434% in the recent 2010s. In the 2000s, it ballooned to 750%. This was due to Typhoons Maemi (2001) and Rusa (2002), which caused the most damage in history, with the highest wind speeds in Korea. The coefficient of variation (CV) is used to compare data with different units of measurement. The CV is the standard deviation divided by the arithmetic mean. The larger the CV, the larger the relative difference. The CV also steadily increased every 10 years, indicating that the difference in the intensity of the typhoons (maximum wind speed) increased. Compared to the past, typhoons of various intensity are occurring now, indicating that the current typhoons can exhibit more diverse damage categories than the past typhoons. These values (i.e., average, increase rate, and CV) prove that the loss incurred by buildings is greater than that in the 1970s, and this proves that typhoons are more serious than in the past. Table 2 represents the result (average) summary of each model for commercial buildings. The increase rate has grown progressively over the decades since the 1970s, reaching 455% on average and 634% in the recent 2010s. During the 2000s, it intensified to 1133% due to Typhoons Maemi (2001) and Rusa (2002). The CV rose steadily every period and was projected to vary as the maximum wind speed escalated each decade. This shows that current typhoons produce more varied intensity than past typhoons. Table 3 shows the result (average) summary of each model for residential buildings. The increase rate has grown over the decades since the 1970s, reaching 337% on average and 476% in the recent 2010s. During the 2000s, it increased dramatically to 819% owing to Typhoons Maemi (2001) and Rusa (2002). The CV rose increasingly each decade and was predicted to vary as the maximum wind speed rose. These demonstrate that current typhoons are more severe than past typhoons.

6. Discussion

This study was a quantitative study of the changes in tropical cyclones caused by climate change. It investigated the typhoons affecting Korea in the past and analyzed their intensity and frequency, as well as changes in typhoon patterns. The analysis results showed that the frequency of typhoons did not increase, but their severity did increase. This indicates that the risk from typhoons is growing due to their increasing severity.

To quantify the increased typhoon risk, this study used the vulnerability function of the CAT model, which is not affected by wealth and population. As a result, it was shown that the risk from typhoons increased gradually during the 1970s, 1980s, 1990s, 2000s, and 2010s. On average, the risk to industrial buildings increased by 307%, the risk to commercial buildings by 455%, and the risk to industrial buildings by 337%. The increase rate for commercial buildings was the largest, which is attributed to the fact that commercial buildings consist of more diverse buildings than other building types. In addition, the 2000s displayed the largest increase rate due to the influence of Typhoons Maemi and Rusa. These two typhoons were the largest typhoons in Korea, but they were included in the analysis because they are generally considered to be 15–30 return period typhoons. The analysis results show that the risk from typhoons has increased significantly each year.

For this reason, new strategies are required to respond to changing circumstances and increasing risks. The insurance industry is critically sensitive to hurricanes. For example, 11 insurers in the US were bankrupted by Hurricane Andrew (1992). Therefore, a review of pricing, policy conditions, and reinsurance for increased loss risk is essential. In terms of pricing, there is a need to raise current premiums and modify the current probable maximum loss and limit of liability. Moreover, changes in acquisition, retention, and accumulation management strategies are inevitable due to the changed pricing. Policy conditions require a review of the scope of coverage of existing insurance policies. In facultative and treaty reinsurance arrangements, new strategies for excess loss and layering for enlarged risk are needed. Furthermore, it is crucial to calculate premiums through accurate quantification of the weighted risks, which can be performed through appropriate CAT models. Increased risk, on the other hand, may also be a new opportunity because it requires active risk transfer from the government or private sector, leading to a boom in insurance coverage. The introduction of CAT bonds is also desirable to hedge losses from catastrophic events in developing countries such as Korea. CAT bonds are used in developed countries such as the US and Germany to distribute reinsurance functions through the generation of bonds when the risk for insurance companies exceeds their acquisition capacity.

Governments also need to strengthen architectural design standards and codes to create a sustainable building environment that can withstand extreme weather disasters. Additionally, a separate management guide is required for older buildings constructed with past building codes. Maintaining infrastructure, lifelines, and transportation systems during hurricanes is critical to reducing human and property damage, requiring a more advanced management system. Storm and flood hazard areas should also be mandated to actively transfer the risk of loss through mandatory subscription to storm and flood insurance. This should enable local communities and residents to respond appropriately to changing circumstances.

7. Conclusions

Many historical event and damage analysis studies continue to debate climate change and its effect on tropical cyclones. Therefore, this study aimed to analyze the intensity and frequency of typhoons by examining typhoons affecting Korea for scientific and quantitative research on climate change and typhoon changes and to quantify the damage caused by these changes. In addition, the risk of building loss was quantified. The results show that the severity of typhoons increases year by year, and thus the risk from typhoons also increases year by year. This suggests that climate change is affecting typhoons. Hence, it is necessary to consider government and industry responses to climate change and risk reduction.

On the other hand, since this study only considers wind speed and occupancy, the results may vary when additional inventory information is used. Further research using inventory information from multiple buildings is needed. Because different results can be obtained for different geographic regions of the Korean Peninsula, further research is needed to analyze the results. In particular, the results will differ in the southern part of the main typhoon route. Moreover, this study may not represent other countries, as the research area is limited to Korea. Countries with more coastal development than Korea may be more vulnerable to typhoons, and countries with greater building capacity or maintenance management systems may be better able to adapt to storms. Adaptation policies for climate change and risk reduction have not been modeled. Damage can be reduced through active climate change adaptation policies and programs by individuals or institutions, and comprehensive research that includes this is needed.

Author Contributions: Conceptualization, J.-M.K.; Data curation, J.-M.K.; Funding acquisition, K.S.; Investigation, S.S., S.L.; Methodology, J.-M.K.; Software, S.S., S.L.; Validation, J.-M.K., K.S.; Writing original draft, J.-M.K.; Writing review and editing, J.-M.K., K.S., S.S., and S.L. All authors have read and agreed to the published version of the manuscript.

Funding: This work was supported by the 2020 Research Fund of University of Ulsan.

Conflicts of Interest: The authors declare no conflict of interest.

References

1. Del Rosario; Eduardo, D. *Final Report Effects of Typhoon YOLANDA (HAIYAN(Report)*; National Disaster Risk Reduction and Management Council: Quezon City, Philippines, 2014. Available online: http://www.ndrrmc.gov.ph/attachments/article/1329/FINAL_REPORT_re_E_ects_of_Typhoon_YOLANDA_HAIYAN_06-09NOV2013.pdf (accessed on 20 April 2019).
2. United States National Hurricane Center. *Costliest, U.S. Tropical Cyclones Tables Update*; United States National Hurricane Center: Miami, FL, USA, 2018. Available online: https://www.nhc.noaa.gov/news/UpdatedCostliest.pdf (accessed on 31 May 2020).
3. Ulbrich, U.; Fink, A.H.; Klawa, M.; Pinto, J.G. Three extreme storms over Europe in December 1999. *Weather* **2001**, *56*, 70–80. [CrossRef]
4. Emanuel, K.; Sundararajan, R.; William, J. Tropical cyclones and global warming: Results from downscaling IPCC AR4 simulations. *Bull. Am. Meteorol. Soc.* **2008**, *89*, 347–367. [CrossRef]
5. Webster, P.J.; Holland, G.J.; Curry, J.A.; Chang, H.R. Changes in tropical cyclone number, duration, and intensity in a warming environment. *Science* **2005**, *309*, 1844–1846. [CrossRef] [PubMed]

6. IPCC. *Climate Change 2007: The Physical Science Basis. Contribution of Working Group I to the Fourth Assessment Report of the Intergovernmental Panel on Climate Change*; IPCC: Cambridge, UK, 2007; p. 784.
7. Landsea, C.W.; Harper, B.A.; Hoarau, K.; Knaff, J.A. Can we detect trends in extreme tropical cyclones? *Science* **2006**, *313*, 452–454. [CrossRef] [PubMed]
8. IPCC. *Climate Change 2001: Synthesis Report*; IPCC: Cambridge, UK, 2001; p. 398.
9. Changnon, S.A. Temporal and spatial distributions of wind storm damages in the United States. *Clim. Chang.* **2009**, *94*, 473–482. [CrossRef]
10. Chang, H.; Franczyk, J.; Kim, C.H. What is responsible for increasing flood risks? The case of Gangwon Province, Korea. *Nat. Hazard.* **2009**, *48*, 339–354. [CrossRef]
11. Pielke, R.A., Jr.; Gratz, J.; Landsea, C.W.; Collins, D.; Saunders, M.A.; Musulin, R. Normalized tropical cyclone damage in the United States: 1900_2005. *Nat. Hazard. Rev.* **2008**, *9*, 1–29.
12. The World Bank. *Turn Down the Heat: Confronting the New Climate Normal*; License: Creative Commons Attribution—NonCommercial—No Derivatives 3.0 IGO (CC BY-NC-ND 3.0 IGO); World Bank: Washington, DC, USA, 2013; p. 275.
13. Stern, N. *Stern Review on The Economics of Climate Change, PART II: The Impacts of Climate Change on Growth and Development*; Cabinet Office-HM Treasury: London, UK, 2006; p. 2006.
14. IPCC. Summary for Policymakers. In *Climate Change 2014: Impacts, Adaptation, and Vulnerability. Part A: Global and Sectoral Aspects. Contribution of Working Group II to the Fifth Assessment Report of the Intergovernmental Panel on Climate Change*; IPCC: Cambridge, UK, 2014; p. 1820.
15. Mercer, L.L.C. *Investing in a Time of Climate Change*; International Finance Corporation and The UK Department for International Development: London, UK, 2015. Available online: http://www.mercer.com/services/investments/investment/opportunities/responsible-investment/investing-in-a-time-of-climate-changereport-2015.html (accessed on 20 April 2019).
16. Dietz, S.; Stern, N. Endogenous growth, convexity of damages and climate risk: How Nordhaus' framework supports deep cuts in carbon emissions. *Econ. J.* **2015**, *125*, 574–620. [CrossRef]
17. Mendelsohn, R.; Schlesinger, M.; Williams, L. Comparing Impacts Across Climate Models. *Integr. Assess.* **2000**, *1*, 37–48. [CrossRef]
18. Bouwer, L.M. Have disaster losses increased due to anthropogenic climate change. *Bull. Am. Meteorol. Soc.* **2011**, *92*, 39–46. [CrossRef]
19. Nordhaus, W.D. The economics of hurricanes and implications of global warming. *Clim. Chang. Econ.* **2010**, *1*, 1–20. [CrossRef]
20. Changnon, S.A. Increasing major hail losses in the U.S. *Climat. Chang.* **2009**, *96*, 161–166. [CrossRef]
21. Schmidt, S.; Kemfert, C.; Hoppe, P. Tropical cyclone losses in the USA and the impact of climate change: A trend analysis based on data from a new approach to adjusting storm losses. *Environ. Impact Assess. Rev.* **2009**, *29*, 359–369. [CrossRef]
22. Fengqing, J.; Cheng, Z.; Guijin, M.; Ruji, H.; Qingxia, M. Magnification of flood disasters and its relation to regional precipitation and local human activities since the 1980s in Xinxiang, Northwestern China. *Nat. Hazard.* **2005**, *36*, 307–330. [CrossRef]
23. Changnon, S.A. Damaging thunderstorm activity in the United States. *Bull. Am. Meteorol. Soc.* **2001**, *82*, 597–608. [CrossRef]
24. Miller, S.; Muir-Wood, R.; Boissonnade, A. An exploration of trends in normalized weather-related catastrophe losses. In *Climate Extremes and Society*; Diaz, H.F., Murnane, R.J., Eds.; Cambridge University Press: Cambridge, UK, 2008; pp. 225–347.
25. Kim, J.M.; Kim, T.; Son, K. Revealing building vulnerability to windstorms through an insurance claim payout prediction model: A case study in South Korea. *Geomat. Nat. Hazard. Risk* **2017**, *8*, 1333–1341. [CrossRef]
26. Sanders, D.E.A.; Brix, A.; Duffy, P.; Forster, W.; Hartington, T.; Jones, G.; Levi, C.; Paddam, P.; Papachristou, D.; Perry, G.; et al. The management of losses arising from extreme events. In Proceedings of the Convention General Insurance Study Group GIRO, London, UK, 24–26 September 2019; pp. 1–261.

Article

Development of Design Considerations as a Sustainability Approach for Military Protective Structures: A Case Study of Artillery Fighting Position in South Korea

Kukjoo Kim [1,2] and Youngjun Park [1,*]

[1] Department of Civil Engineering and Environmental Sciences, Korea Military Academy, Seoul 01805, Korea; klauskim@ufl.edu

[2] Nuclear·WMD Protection Research Center, Korea Military Academy, Seoul 01805, Korea

[*] Correspondence: parky@mnd.go.kr

Received: 29 June 2020; Accepted: 5 August 2020; Published: 11 August 2020

Abstract: Republic of Korea (ROK) military installations are scattered across South Korea, but there is a higher concentration of fortifications in the demilitarized zone (DMZ) and eastern and western coastlines. These facilities range from relatively small structures, such as individual and artillery fighting positions, to large buildings, such as ammunition depots and command posts. These military installations have a significant thickness of concrete members to provide a high degree of protection against bombs and projectiles. The Korean military will carry out the integration and dismantling of these protection facilities over the next ten years through the Army transformation plan. Such large-scale construction projects have an impact on the environment in terms of the carbon footprint, because building construction and operations account for 36% of the world's energy use and 40% of energy-related carbon dioxide (CO_2) emissions. It is very important to reduce the concrete materials and reinforcement steel during protective structure construction near the DMZ, which is now recognized as one of the most well-preserved areas in the world. In this study, new sustainable design considerations that allow elasto-plastic or plastic design of concrete elements were evaluated using a case study of an artillery fighting position. The new sustainable design considerations were developed on the basis of mission, enemy, terrain and weather, troops and support available, time and civil considerations (METT + TC) within the context of the current battle situation, as well as protection against near misses. From this study, it was found that new sustainable design considerations provide a reasonable degree of protection that permits good construction practices and maximum structural stability with minimum amount of materials. It was also found that if the new design procedure is used to replace 1000 artillery positions through the Army transformation plan, the CO_2 emissions can be reduced by 476,582.4 tons and the cost reduced by USD 23,829,120.

Keywords: degree of protection; impact damage; blast wave; sustainable design consideration; elasto-plastic design; CO_2 emission

---

1. Introduction

1.1. Background

The design of protective structures is an important factor not only for military construction but also for civilian sectors. As the threat of enemy's weapons of mass destruction increase, protective structure design becomes a common problem for military, civil, and industrial facilities. Currently, there is little information (including experimental data regarding bombs, projectiles, and atomic bomb blasts, etc.) and design procedures available to serve as a design guideline for such protection structures. In conventional works, the maximum degree of protection has been used on the basis of a 00-pound GP bomb detonating at a distance of 00 m (restriction on disclosure due to military secrets). These design criteria produce a structure which is able to sustain a given loading condition within the limits of elastic strain, which requires a significant amount of concrete. Reducing the amount of concrete used in a construction project is very important in terms of sustainability awareness and green planning [1]. The International Energy Agency and United Nations (UN) Environment Programme stated that building construction and operations accounted for 36% of the world's energy use and 40% of energy-related carbon dioxide (CO_2) emissions in 2017 [2].

More specifically, Pacheco-Torgal et al. described that concrete and reinforcement steel account for about 65% of building greenhouse gas (GHG) emissions, 40% of which is CO_2 emissions from concrete [3]. It is noted that the mean embodied carbon dioxide (ECO_2) for building is 340 kg-CO_2/m^2, of which the structure accounts for about 60% [4]. This means that reducing the ECO_2 in the structure frame directly reduces the GHG emissions. Additionally, in terms of the carbon footprint, it is very important to reduce the concrete materials and reinforcement steel during construction projects [5–9].

Recently, the Korean military has formulated plans to carry out the integration and dismantling of these protection facilities through the Army transformation plan over the next ten years. As military protective structures are concentrated at the border, these enormous concrete structures adversely affect the environment, particularly in the demilitarized zone (DMZ), which is now recognized as one of the most well-preserved areas in the world.

To identify the appropriate degree of protection, a design process must consider the weapon effects and dynamic factors pertaining to mission, enemy, terrain and weather, troops and support available, time and civil considerations (METT + TC) [10] within the context of the current battle situation. It must be considered that structure members can resist dynamic loads under relatively large plastic deformation. Such local overstresses in the member, or even some failures, should not seriously impair the overall structure. Some protective structures, such as artillery fighting positions, require protective ability only once. If the protective structure design process ignores the METT + TC factors, it produces structural members with massive thickness. As a large amount of concrete materials is consumed, the construction of these structures has a direct impact on the natural environment. Therefore, it is important to reduce the use of concrete and non-renewable materials during construction works.

In this study, new protective structure design considerations were developed to improve the resistance of structure members, resulting in large plastic deformation. The new design considerations evaluate the amount of concrete that was saved while providing an appropriate degree of protection by using finite element (FE) analysis as a case study.

1.2. Objectives and Scope

The primary objective of this study was to develop new sustainable design considerations for protective structures, using the Delphi technique on the basis of METT + TC factors within the context of the current battle situation. Then, after applying the proposed design consideration to the case project, the CO_2 emission and cost reduction corresponding to the concrete savings were analyzed. To do this, a three-dimensional FE analysis was conducted to assess the potential performance of the artillery fighting position as a case study in South Korea.

2. Protection against Conventional Weapons

For the purpose of protection against weapons, the protective structures may be classified into two general groups: those which provide protection against (1) the impact of a weapon's penetration and (2) the blast of a weapon's explosion. Penetration is caused by weapons such as projectiles fired from guns, conventional bombs with a charge-to-weight ratio smaller than 20%, rockets, and guided missiles. Explosion blast is caused by weapons such as high explosive or conventional bombs with a charge-to-weight ratio higher than 20%. For the purpose of structural analysis, a weapon's impact causes severe local damage, while the weapon's blast causes overall damage of relatively less severity [11,12].

When a bomb or projectile strikes a concrete member, there is the formation of an irregularly shaped crater and considerable cracking in the opposite side of the slab. The severity of such cracking decreases as the concrete thickness increases. Because of the inherent low tensile strength of concrete, both faces of the slab tend to rupture with the reflected shock wave in the impact face and the propagated wave in the opposite face. Design information about the weapon system and condition of protection to be provided is necessary. In most cases, the desired level of protection of a structure differs [13–15]. For example, if the building is to be located near the border between nations, within the range of army artillery, the required protection of the exposed walls would be the loading due to an armor piercing (A.P.) type projectile. In contrast, the design of the roof of the building would consider the loading of and A.P. type bomb released from a carrier plane.

In many cases, the functional importance of a protective building, its size and the thickness of structural members is larger so as to provide a certain degree of lateral and overhead protection against blast and fragments of a bomb. In the South Korea Army, a reasonable degree of protection has been developed on the basis of a 00-pound GP bomb detonating at a distance of 00 m. The thickness of structural members resulting from this consideration only permits the induced stresses of structure elements to remain in the elastic range. However, the blast loading on a protective building caused by a high explosive detonation in a bomb depends on the peak pressure and the impulse of the incident and dynamic pressures. For the analysis of structures under dynamic loading, such as blast loading, the analysis of inertial force and kinetic energy is required, as the applied load changes rapidly with time, as shown in Figure 1 [16].

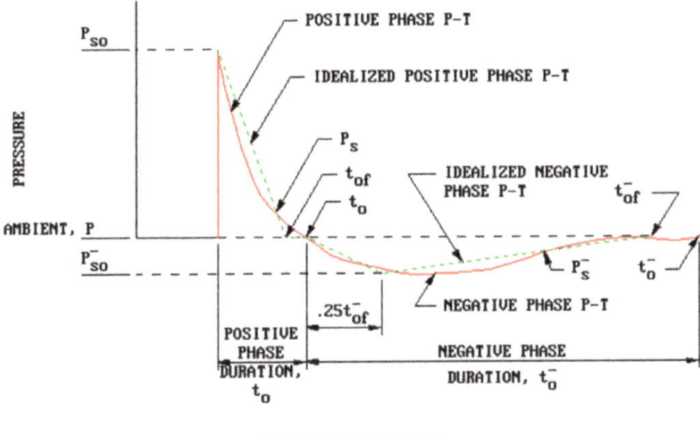

Figure 1. Idealized pressure-time curve of a blast wave.

For design purposes, the effects of the inertial force in the equation of dynamic equilibrium and kinetic energy in the equation of energy conservation related to the mass of the structure must be considered. The response of a concrete element in a protective structure can be defined as ductile or brittle structural behavior. In the ductile mode of response, large inelastic deflections without complete collapse occur in the structure element, while partial failure or total collapse of the element occurs in the brittle mode [17]. If the ductile behavior is selected for an element of protective structures in the current design consideration, there can be savings in concrete materials within the desired level of protection. The flexural action of a reinforced concrete member was demonstrated by the resistance–deflection curve shown in Figure 2 [16].

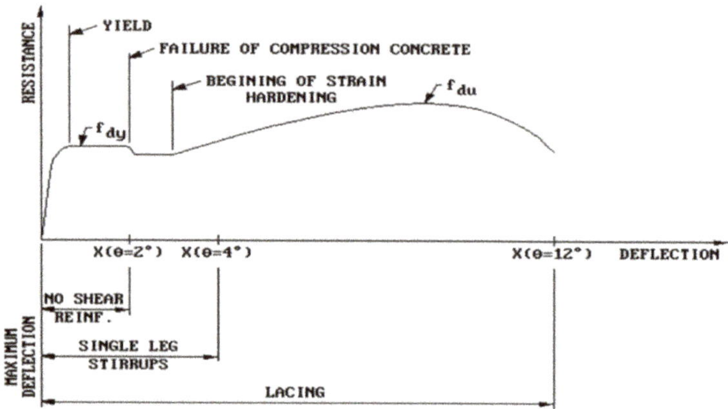

Figure 2. Resistance-deflection curve for flexural response of concrete elements.

The magnitude of stresses produced in the protective structure responding in the plastic range cannot be directly related to the strain. The average stress over portions of the plastic range can be determined by relating this average stress to the deflection of the element defined in terms of the angular rotation at the supports. Therefore, the elasto-plastic or plastic design considerations of concrete protective structures must be considered [18] in the current Army protective structure design standard to use maximum protection capability of concrete elements, and sustainable and economical design approaches.

3. Development of New Design Considerations as a Sustainable Approach

The protection standard of the Republic of Korea (ROK) armed forces comprises four stages, each of which is determined based on comprehensive considerations for the threat of the enemy and the protective capability from the aspect of military operations and for the purposes from the aspect of facilities. Once a degree of protection is set, the corresponding protection level of a structure is determined based on blast loads. In general, protection levels according to protection degrees represent the thresholds of the displacement ductility factor and rotation angle that are proposed by the Unified Facility Consideration (UFC) 3-340-02 [16]. Table 1 presents the permissible limits of rotation angle for brittle materials such as concrete at each protection level.

Table 1. Design consideration of the protective facilities in UFC 3-340-02.

Protection Level	Construction Method	Damage Aspect	Maximum Support Rotation Angle
A	Elastic design (Working stress design)	Microcrack	0–2°
B	Carbon design	Protection for human lives (Crack, crush)	2–6°
C	Plastic design (limit design)	Severe collapse (Separation of concrete from the reinforcement bar)	6–12°

The protection levels of Table 1 are different concepts from the degrees of protection. The protection levels are distinguished based on design concepts. In the case of the ROK armed forces, when the protection degree of a protective facility is determined, an elastic design corresponding to the protection level A is adopted.

For the design of a protective structure, it is necessary not only to analyze severe dynamic loads comprising blended impacts of blast waves and fragments, but also to examine various and complex battlefield conditions where projectiles might directly blast and penetrate the structure. However, the protective degrees currently in use in the ROK armed forces are still grounded on the dated concept of protection focusing on the thickness of heavy-weight structures.

This study aims to propose guidelines for determining bullet/explosion-proof degree of protection, whose application ranges from high-tech precision guided weapons to the artillery strength for pinpoint strike, and to examine guidelines for future revisions in the area of protection in the standard of defense and military facilities. To achieve these goals, the Delphi technique was used to accurately reflect objective opinions of experts from government, military, and private sectors. Based on the opinions collected, the guidelines for determining degrees of protection were derived by horizontally and vertically synthesizing key words that were extracted from the Korea Army innovation assessments and the innovation school of the Korea Army Research Center for Future and Innovation (KARCFI). To achieve a fair and even distribution, a group of 21 experts (7 civilian experts, 7 government officials, and 7 servicemen) was organized. All the experts were experienced in defense and military facilities.

After organizing the expert group, we conducted several rounds of survey (first round with open-ended questionnaires and second to fourth rounds with closed-ended questionnaires). Based on the survey, we derived considerations for setting protection degrees. In particular, the Shapiro–Wilk normality test was performed to quantify the agreement between each panel during the second to fourth rounds. Then, a factor analysis was performed to identify common features of considerations in each factor [19–21]. The result was summarized into the five tactical considerations (METT + TC). Then, the innovation school and assessments for future battlefield environment led by KARCFI extracted essential considerations for setting protection degrees as key words and combined them horizontally and vertically. Consequently, the considerations identified for the protection standard of military facilities include the following six factors: wartime/peacetime mission; omnidirectional threat; stability and resilience of troops; geology and weather; threat detection, alert, reaction and recovery time; military-private combined factor. The design process checklist is shown in Table 2, avoiding excessive design and ensuring the desired performance while considering future diversified battlefield environments and weapon systems. The highest requirements for each item in Table 2 are selected as the final degree of protection and protection level. Table 2 shows an example of determining the degree of protection and protection level for artillery positions.

Table 2. Example of the new design process checklist for artillery fighting position.

Classification	Considerations for the Protection Standard of Military Facilities		Degree of Protection			Protection Level		
	6 Factors	**Detailed Items**	**Direct Hit**	**Contact Explosion**	**Near Misses**	**A**	**B**	**C**
M (Mission)	Wartime/peacetime mission (4 items)	Peacetime mission			•			•
		Wartime mission			•			•
		Importance of mission			•			•
		Camp protection plan			•			•
E (Threat of the enemy)	Omnidirectional threat (4 items)	Present threat (Tactic, arrangement, organization, and activity of enemy)			•			•
		Potential threat			•			•
		Transnational threat			•			•
		Non-military threat			-			-

Table 2. *Cont.*

Classification	Considerations for the Protection Standard of Military Facilities		Degree of Protection			Protection Level		
	6 Factors	Detailed Items	Direct Hit	Contact Explosion	Near Misses	A	B	C
T (Available troops)	Stability and resilience of troops (9 items)	Type of troops			•			•
		Location and capacity of protective facility			-			-
		Possession of explosive weapons			•			•
		Size of troops			•			•
		Smart safety system			-			-
		Possession of reserve facility			•			•
		Loss of lives caused by the destruction of a facility			•			•
		Damage recovery capacity (each unit and higher command)			•			•
		Threat response capacity			•			•
T (Terrain and weather)	Geology and weather (5 items)	Natural terrain			•			•
		Man-maid terrain			•			•
		Avenue of maneuver inside and outside camp			-			-
		Vegetation			•			•
		Weather change			-			-
T (Available time)	Threat detection, alert, reaction, and recovery time (6 items)	Threat detection time			-			-
		Threat alert time			-			-
		Evacuation time			-			-
		Operation time of protective facility			-			-
		Time for neutering threat			•			•
		Damage recovery time			-			-
C (Civil factor)	Military-private combined factor (6 items)	Comprehensive land development plan			•			•
		Military-civil relationship			•			•
		Civil complaint			•			•
		Importance of facility			•			•
		Core technology for establishing protective facilities			-			-
		Civilian protection capacity			-			-
Overall judgment			Close strike of enemy artillery			Protection level C		

• indicates applicable, - indicates not applicable.

4. A Case Study for Artillery Fighting Position Using Finite Element Analysis

4.1. Setting the Protection Degree and Level

When designing a protective structure, it is necessary to consider the dynamic loads of blast waves and impacts of fragments caused by the explosion of a high-energy bomb or a missile. Regarding such dynamic load, the characteristics of a weapon as a means of strike need to be closely examined. A case study for artillery positions in the front-line area was performed by applying the design factors of Table 2. Specifically, the standard type of the existing artillery positions and a new artillery position designed with the new design factors were comparatively evaluated through a FE analysis.

The major threat issue of the new artillery position is not the direct strike of enemy artillery, but rather the protection against blast waves and fragments caused by close explosions. METT-TC of the artillery troops being considered, the protection level was set to Level C, as the fire-and-displace is expected under the enemy's counter-artillery fire. Through the analysis, the protection degrees and levels that were ultimately desired were derived, as shown in Table 2.

4.2. Evaluation of Protective Performance through Numerical Analysis

This study performed a FE analysis to identify the dynamic behavior characteristics of an artillery position under blast waves. It was assumed that 115 kg of TNT, the maximum explosion in enemy's weapon, was exploded 7.6 m away from the artillery fighting position. This is the result of taking into account the accuracy of enemy artillery weapons without using guided weapons during artillery battles.

A numerical analysis mode was developed by using ANSYS AUTODYN®. This program was developed by the Institute for Defense Analysis. It is a very useful tool for solving the ductility issue between fluids and solids through the coupling of Lagrange and Euler Solvers in solid mechanics. For a non-linear dynamic analysis of a structure, a reinforced concrete element was constructed using an explicit FE method. A standard artillery fighting position with a wall length and height of 4500 and, 700 mm, respectively, was selected as the target structure of the analysis. As for the wall thickness, five cases of 300, 350, 400, 450, and 500 mm were considered.

As for the material properties of the concrete wall, as presented in Table 3, the ordinary concrete and reinforcement bar presented tensile strengths of 24 and 400 MPa, respectively. The minimum reinforcement ratio was 0.00306.

Table 3. Material properties used infinite element (FE) models.

Classification	Wall Thickness (mm)	Reinforcement Bar Diameter	Reinforcement Bar Spacing	Number of Reinforcement Bars	Reinforcement Ratio
CASE 1	300	HD16	300	15	0.003161
CASE 2	350	HD16	250	18	0.003070
CASE 3	400	HD19	300	15	0.003119
CASE 4	450	HD19	250	18	0.003213
CASE 5	500	HD22	300	15	0.003216

As illustrated in Figure 3, the detonation point was 7.6 m away from the artillery fighting position. The explosive was TNT with a density of 36.675 kg/m per unit length in the z-direction.

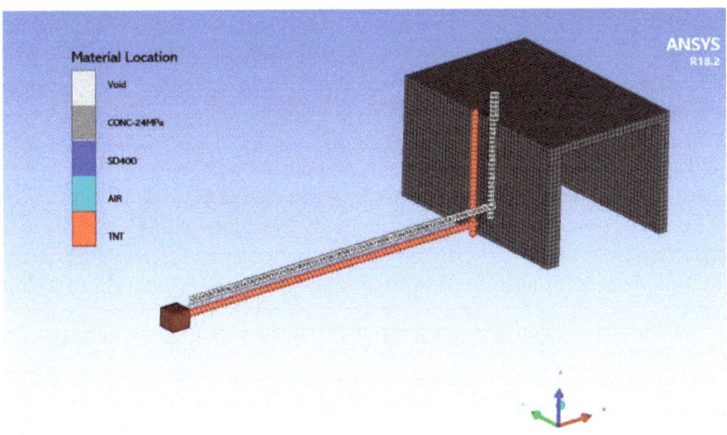

Figure 3. FE model developed.

4.3. Numerical Analysis Result of Protective Performance

Figure 4 shows the impacts of blast waves on the structure over time. As the March front was lower than the height of the structure, the structure was affected by non-uniform pressures

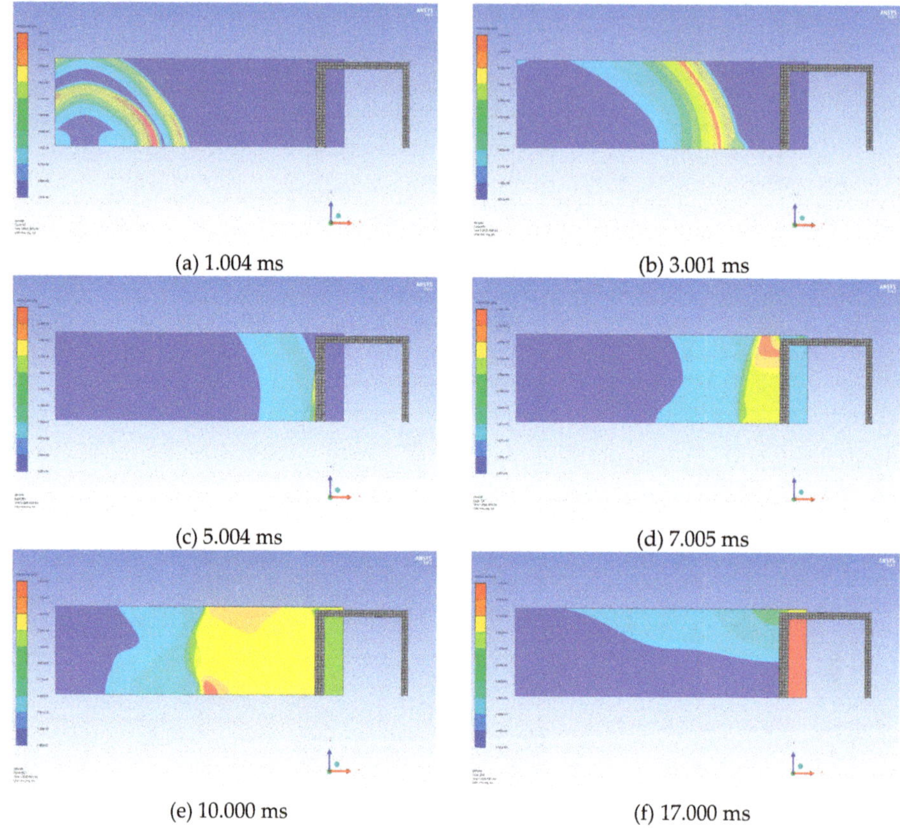

(a) 1.004 ms

(b) 3.001 ms

(c) 5.004 ms

(d) 7.005 ms

(e) 10.000 ms

(f) 17.000 ms

Figure 4. Impact of blast waves on the structure over time.

Figure 5 shows the pressures of blast waves and the displacement of the structure wall over time. Protection degrees and levels desired for the artillery fighting position can be expressed by the maximum displacement and rotation angle of the wall. Table 4 presents protection levels for each case of wall thickness.

As shown in Table 4, the dynamic analysis of the reinforced wall revealed that a sufficient level of protective capacity could be secured, even if the wall thickness was reduced to 300 mm. In other words, if the current design of protective facility reflecting only the elastic displacement of reinforcement structure is replaced by the elasto-plastic design considering the protection levels for each METT + TC factor, as presented in Table 2, protective structures would be more economical and sustainable.

Table 4. Maximum displacement and rotation angle according to wall thickness.

Classification	CASE 1	CASE 2	CASE 3	CASE 4	CASE 5
Maximum Displacement (mm)	18.5800	12.0834	12.08	9.8921	8.3938
Rotation angle (°)	1.9600	0.2564	0.2563	0.2099	0.1781

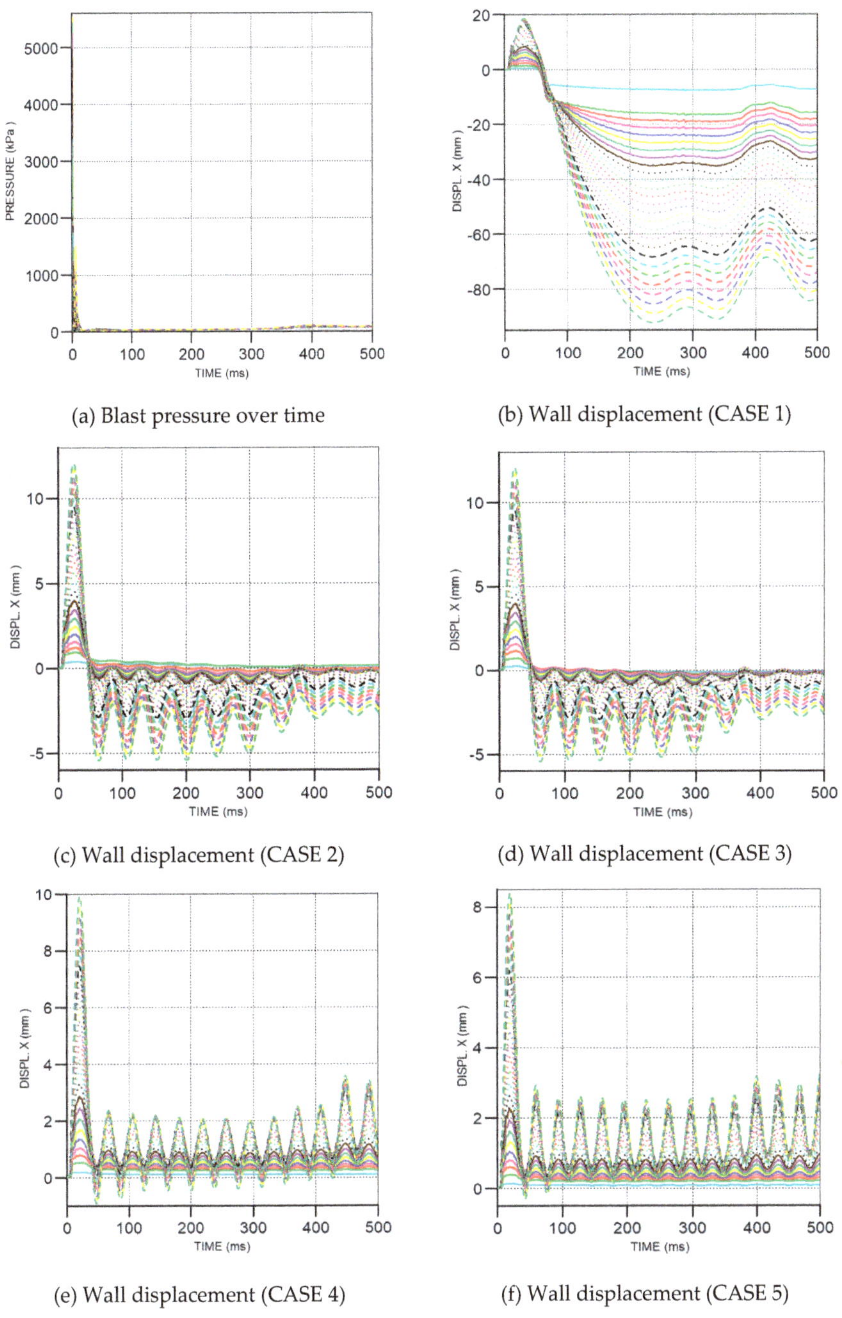

(a) Blast pressure over time

(b) Wall displacement (CASE 1)

(c) Wall displacement (CASE 2)

(d) Wall displacement (CASE 3)

(e) Wall displacement (CASE 4)

(f) Wall displacement (CASE 5)

Figure 5. Blast pressure over time and displacement according to wall height.

4.4. CO_2 Emission Reduction Effects

When using the new design consideration proposed in this paper, the effect of concrete savings should be confirmed. Table 5 shows the calculation results of the CO_2 emissions for concrete saved by the new design procedure as a sustainable approach, when 1000 artillery positions are replaced through the Army transformation plan. When the unit CO_2 emissions of ready-mixed concrete used are 3.152 ton-CO_2/ton [22], the CO_2 emissions from artillery position project Army planed can be reduced by approximately 476,582.4 tons, which is equivalent to 40% of the project. When Korean carbon transaction price of USD 50/ton-CO_2 [23] is applied, the total cost savings of USD 23,829,120 can be calculated. Therefore, if the new design consideration proposed in this study are applied to the entire military protective structure projects, greater cost saving and reduction in CO_2 emissions are expected.

Table 5. Calculation of CO_2 emission reduction effect.

Description	Quantity (ton)	Unit CO_2 Emission (ton-CO_2/ton)	Amount (ton-CO_2)
Existing design procedure (A)	378,000	3.152	1,191,456.0
New design procedure (B)	226,800	3.152	714,873.6
Reduction effect (A–B)	151,200		476,582.4

5. Conclusions

Reducing the amount of concrete used in construction project is very important in terms of sustainability awareness and green planning to reduce carbon and climate change risk globally. The concrete material and reinforcement steel account for about 65% of building greenhouse gas (GHG) emissions, 40% of which is CO_2 emissions from concrete. Therefore, green building planning is very important during construction projects. However, the Korean military's design concept does not take full advantage of the features of reinforced concrete structures, resulting in excessive design. The protection scheme of ROK armed forces consists of four stages. In this scheme, protection degrees are set based on relative protection capabilities against particular weapon systems. Furthermore, the protection degrees established require the protection level A, corresponding to the concept of elastic design. Accordingly, no effective protection using the behavior characteristics of structures for weapon systems is provided. As a result, the degrees of protection currently in use in the ROK armed forces are still grounded in the dated concept of protection focusing on the thickness of heavy-weight structures. This study derived the protective design considerations necessary for future protective facilities to avoid excessive design and to secure a desired level of protection performance. In addition, this study also conducted a Delphi process by organizing a group of experts from the government, military, and private sectors. The result of the Delphi method was combined with the design considerations for protective facilities that were derived by the innovation school of KARCFI and innovation consulting. Thus, sustainable design considerations for protective facilities were obtained.

Using the above considerations, an FE method was performed for the protection performance of the standard artillery position widespread in the frontline area. The protection against close explosion was determined based on METT + TC of the artillery position in each protection degree. A dynamic analysis of a reinforced structure showed that the elasto-plastic design could produce a more sustainable structure.

So far, protective structures have been regarded as heavy-weight structures with thick walls. However, if the new design considerations developed in this study are applied, more economical and sustainable protective facilities can be constructed. In particular, the case study revealed that thousands of artillery positions in the frontline area and DMZ can reduce wall thickness. For instance, if the new design procedure is used to replace 1000 artillery positions through the Army transformation plan, the CO_2 emissions can be reduced by approximately 476,582.4 tons, which is equivalent to as cost of USD 23,829,120. Therefore, if the new design consideration proposed in this study is applied to the entire military protective structure projects, greater cost saving and CO_2 emissions are expected. It confirms that it is possible to provide sustainable protective facilities while satisfying the operational requirements for such artillery positions.

Author Contributions: Conceptualization, Y.P. and K.K.; Methodology, K.K.; Software, Y.P.; Validation, Y.P. and K.K.; Formal Analysis, K.K.; Investigation, K.K.; Resources, Y.P.; Data Curation, K.K.; Writing-Original Draft Preparation, K.K.; Writing-Review & Editing, Y.P.; Visualization, K.K.; Supervision, Y.P.; Project Administration, Y.P.; Funding Acquisition, Y.P. All authors have read and agreed to the published version of the manuscript.

Funding: This research was supported by a grant (18SCIP-B146646-01) from the Korea Agency for Infrastructure Technology Advancement.

Acknowledgments: This work was supported by research fund of the Korea Agency for Infrastructure Technology Advancement. The ROKA Nuclear WMD Protection Research Center at Korea Military Academy is gratefully acknowledged for providing the support that made this study possible.

Conflicts of Interest: The authors declare no conflict of interest. The funders had no role in the design of the study; in the collection, analyses, or interpretation of data; in the writing of the manuscript, or in the decision to publish the results.

References

1. Kim, J.J.; Goodwin, C.W.; Kim, S.K. Communication Turns green construction planning into reality. *J. Green Build.* **2017**, *12*, 168–186. [CrossRef]
2. International Energy Agency and the United Nations Environment Programme. 2018 Global Status Report: Towards a Zero-Emission, Efficient and Resilient Buildings and Construction sector. 2018, p. 9. Available online: https://www.worldgbc.org/sites/default/files/2018%20GlobalABC%20Global%20Status%20Report.pdf (accessed on 3 August 2020).
3. Pacheco-Torgal, F.; Cabeza, L.; Labrincha, J.; De Magalhaes, A. *Eco Efficient Construction and Building Materials: Life Cycle Assessment (LCA), Eco-Labelling and Case Studies*; Elsevier: Cambridge, UK, 2014; pp. 624–630.
4. Clark, D.; Bradley, D. *Information Paper—31: Embodied Carbon of Steel Versus Concrete Buildings*; Cundall Johnston and Partners LLP: Newcastle, UK, 2013; p. 4.
5. Lee, I.J.; Yu, H.; Chan, S.L. Carbon Footprint of Steel-Composite and Reinforced Concrete Buildings, Standing Committee on Concrete Technology Annual Concrete Seminar 2016, Hong Kong, 20 April 2016; Construction Industry Council. Available online: https://www.devb.gov.hk/filemanager/en/content_971/7_Carbon_Footprint_for_Steel_Composite_and_Reinforced_Concrete_Buildings.pdf (accessed on 3 August 2020).
6. Kim, S.K.; Kim, M.H. A Study on the development of the optimization algorithm to minimize the loss of reinforcement bars. *J. Archit. Inst. Korea* **1991**, *7*, 385–390.
7. Lee, D.H.; Lee, S.G.; Kim, S.K. Composite phase-change material mold for cost-effective production of free-from concrete panels. *J. Constr. Eng. Manag.* **2017**, *143*, 1–13. [CrossRef]
8. Kim, S.K.; Hong, W.K.; Joo, J.K. Algorithms for reducing the waste rate of reinforcement bars. *J. Asian Archit Build.* **2004**, *3*, 17–23. [CrossRef]
9. Hwang, J.W.; Park, C.J.; Wang, S.K.; Choi, C.H.; Lee, J.H.; Park, H.W. A Case Study on the Cost Reduction of the Rebar Work through the Bar Loss Minimization. In Proceedings of the KIBIM Annual Conference 2012, Seoul, Korea, 19 May 2012.
10. Department of the Army. ADP 5-0 The Operation Process. 2019. Available online: https://rdl.train.army.mil/catalog-ws/view/100.ATSC/E4166A5D-0FE7-4780-916A-A7E9B227147C-1337689957702/adp5_0.pdf (accessed on 3 July 2020).
11. Ammann and Whitney. *Design of Masonry Structures to Resist the Effects of HE Explosions*; Consulting Engineers: New York, NY, USA; Picatinny Arsenal: Dover, NJ, USA, 1976.
12. Schumacher, R.N. *Air Blast and Structural Response Testing of a Prototype Category III Suppressive Shield*; BRL Memorandum Report No. 2701; U.S. Army Ballistic Research Laboratory: Aberdeen Proving Ground, MD, USA, November 1976.
13. Baker, W.E.; Westine, P.S. *Methods of Predicting Blast Loads Inside and Outside Suppressive Structures*; EM-CR-76026, Report No. 5; Edgewood Arsenal: Aberdeen Proving Ground, MD, USA, 1975.
14. Esparza, E.D. *Estimating External Blast Loads from Suppressive Structures*; Edgewood Arsenal Contract Report EM-CR-76030, Report No. 3; Edgewood Arsenal: Aberdeen Proving Ground, MD, USA, November 1975.
15. Stea, W. *Nonlinear Analysis of Frame Structures to Blast Overpressures*; by Ammann and Whitney; Contractor Report ARLCD-CR-77008; Consulting Engineers: New York, NY, USA; U.S. Army Armament Research and Development Command: Dover, NJ, USA, May 1977.

16. US Army Corps of Engineers, Naval Facilities Engineering Command (Ed.) UFC-3-340-02 (2008) Design of structures to resist the effects of accidental explosions. Air Force Civil Engineer Support Agency, Dept of the Army and Defense Special Weapons Agency: Washington, DC, USA. Available online: https://www.wbdg.org/FFC/DOD/UFC/ARCHIVES/ufc_3_340_02.pdf (accessed on 22 July 2020).

17. Gregory, F.H. *Blast Loading Calculations and Structural Response Analyses of the 1/4-Scale Category I Suppressive Shield*; BRL Report No. 2003; U.S. Army, Ballistic Research Lab.: Aberdeen Proving Ground, MD, USA, 1977.

18. Nelson, K.P. *The Economics of Applying Suppressive Shielding to the M483A1 Improved Conventional Munitions Loading, Assembling and Packing Facility*; Technical Report No. EM-TR-76087; Edgewood Arsenal: Aberdeen Proving Ground, MD, USA, 1977.

19. Thompson, B. *Exploratory and Confirmatory Factor Analysis: Understanding Concepts and Applications*; American Psychological Association: Washington, DC, USA, 2004.

20. Raymond, B. Cattell., The Screen Test for the Number of Factors. *Multivar. Behav. Res.* **2010**, *1*, 245–276.

21. Henry, F.K. The Varimax Considerations for Analytic Rotation in Factor Analysis. *Psychometrika* **1958**, *23*, 187–200.

22. Korea Institute of Construction Technology (KICT). The Environmental Load Unit Composition and Program Development for LCA of Building, The Second Annual Report of the Construction Technology R&D Program. 2004. Available online: http://www.ndsl.kr/ndsl/search/detail/report/reportSearchResultDetail.do?cn=TRKO201000018952 (accessed on 3 August 2020).

23. CDP Worldwide. Carbon Pricing Connect. Available online: https://www.cdp.net/en/climate/carbon-pricing/carbon-pricing-connect (accessed on 3 August 2020).

 *sustainability*

Article

Multidimensional Construction Planning and Agile Organized Project Execution—The 5D-PROMPT Method

David Leicht [1,*], **Daniel Castro-Fresno** [1], **Joaquìn Dìaz** [2] **and Christian Baier** [2]

1   GITECO Research Group, University of Cantabria, 39005 Santander, Spain; castrod@unican.es
2   Department of Information Technology in Construction, University of Applied Sciences, 35390 THM Giessen, Germany; joaquin.diaz@bau.thm.de (J.D.); christian.baier@bau.thm.de (C.B.)
*   Correspondence: david@leicht-info.com

Received: 9 July 2020; Accepted: 3 August 2020; Published: 6 August 2020

Abstract: Although tremendous technological and strategic advances have been developed and implemented in the construction sector in recent years, there is substantial room for improvement in the areas of productivity growth, project performance, and schedule reliability. Thus, the present paper seeks to discover why the currently applied scheduling tools and the latest agile-based project organization approaches have not yet achieved their full potential. A missing interlinkage between the project's design, cost, and time aspects within the project design phase and its sparse utilization throughout project execution were indicated as the driving contributors responsible for the slow progress in development. To fundamentally change this situation, an extensive and coherent project organization solution is proposed. The key process of this solution utilizes a 5D Building Information Model comprising tight concatenations between the individual model objects and the corresponding construction cost and time effort values. The key dates of a waterfall-based construction process simulation, set during the project planning phase, provide particular information to create a structure for agile organized project execution. The implementation of information feedback loops allows target/actual comparisons and contributes to continual improvements in future planning. A comparative case study was conducted with auspicious results on improvements in the overall project performance, and schedule and cost reliability.

Keywords: 5D building information modeling; agile project organization; schedule/cost reliability

1. Introduction

Large-scale construction projects are multi-faceted systems of complex and dynamic processes, which are constantly subjected to a multitude of internal and external influencing factors [1]. Tight time and budget constraints and increasing technical demands create challenging conditions to keep projects within their envisaged timeframes [2]. Insufficient limitations on design changes in later project stages, due to customer requests, increase the risk of postponements and growing time issues. Furthermore, the applied schedules often do not meet project-specific process requirements, frequently run out of time, and are exceeded or totally disregarded. In this way, projects are considerably delayed, costs run out of control, and the failure of the project becomes foreseeable [2–4].

In contrast to other industrial sectors, the construction industry has struggled to achieve high productivity rates over the past years. The contributing factors are manifold and have inimically affected not just national market conditions but the global construction economy as a whole [5–7]. This trend is evidenced by the productivity ratings (gross value added at constant prices) that were continuously recorded between 1995 and 2019 by the statistic department of the Organization for Economic Cooperation and Development (OECD) (comparable EU/US data are provided by the

OECD only between 1995 and 2017). During this time, the average annual productivity growth of the European construction sector was at 0.1%, while the average U.S. value was −0.2%. Compared to the total economy, these values indicate an average deviation of 1.6% in the EU and 2.3% in the US [8].

The annual productivity values of the construction sector in the U.S. increased between 2012 and 2019, highlighting the responsible factors for the error-triggering liabilities in the construction sector. According to the results of many investigations, the tenuous development of the construction industry can be traced back to inefficient working methods. This conclusion is evidenced in Figures 1 and 2, whereupon the efficiency of an economic sector is expressed by the ratio between its aggregated input and output values (based on equivalent and comparable factors (e.g., growth per gross domestic product (GDP), gross value added (GVA), total hours worked, unit labor costs) concerning deviating factors, different activity segments, deflators, and exchange rates) [9].

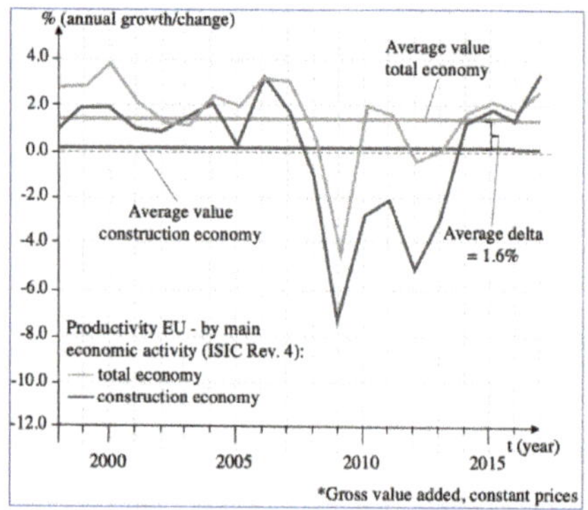

Figure 1. Annual productivity ratings of the EU; total economy vs. construction economy; Figure based on [10].

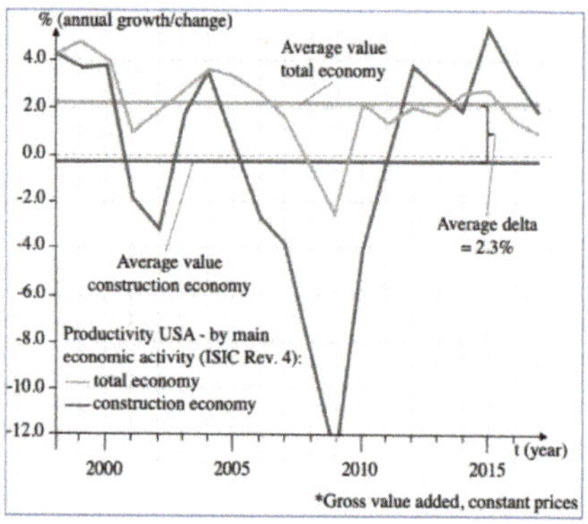

Figure 2. Annual productivity ratings of the US; total economy vs. construction economy; Figure based on [10].

2. Literature Review

An investigation by Kuenzel et al. (2016) indicated that close to 90% of the analyzed construction projects suffered from coordination problems and unsuccessful project management and exceeded project deadlines [11]. In the same year, Oesterreich et al. (2016) revealed organizational issues in construction projects to be a fundamental cause of project failure [2]. Oppong et al. (2017) discovered the insufficient commitment of stakeholders to the project to be an important reason for project failures [1]. Further investigative approaches have shown that planning issues, complications in project organization, and stakeholder disagreements allows projects to exceed their schedules and budgets. Increasing project complexity, constantly changing customer requests, and a wide variety of regulations result in even greater planning and execution efforts [3,12–14]. Moreover, the sophisticated technical requirements and high quantity of project participants greatly affect the deployment efforts in project management and control. Kim et al. discovered a further issue in 2018 that concerns project workflow interruptions caused by sloppy integration of the supply chain to the project execution process. Many planners, contractors, and small and medium sized enterprises (SMEs) source their information, goods, and services via highly fragmented, unstructured supply chains. Moreover, due to the use of mostly impossible just-in-time distribution options, the delivery of goods is rarely in line with the project's progression. Thus, the flow of a project is continually disturbed, which leads to significant project time and budget issues that negatively impact the project's outcome and customer satisfaction [3,4,11].

Sambasivan stated in 2007 that the issue of delays and schedule overruns in construction projects can be understood as a global phenomenon, with conclusive evidence in many studies [15]. A paper by Olawale and Sun (2015) evaluated several international investigations concerning exceeded costs and mismanaged time in construction projects; according to this paper, Hoffman et al. determined in 2007 that 72% of 332 public US-facility projects were delivered late, and 47% exceeded the project timeline by more than 4 months [16,17]. The German Federal Ministry of Construction conducted an analysis between 2000 and 2015 and found exceeded costs and mismanaged timelines in 300 building projects (>10 m EUR). Only 65% achieved their scheduled targets [18]. According to an investigation by Assaf and Al-Hejji (2006), 59% of 76 evaluated projects in Saudi Arabia were considered delayed [14].

However, the examples are not only negative. Salem et al. presented a construction project case study in 2006, where the application of specific agile organization and lean construction approaches (applied lean construction elements: Last Planner System; Increased Visualization; Huddle Meetings; First Run studies; 5S; Fail Save for Quality) brought the project's progression up to three weeks ahead of schedule [19]. Thomas et al. showed as early as 2002 that a significant reduction in project duration of about 30% is achievable through sustainable project management improvements [20]. Hanna et al. (2010) and Hwang et al. (2011) found advantages in thorough pre-planning, leading to improvements in the quality of work execution, increased productivity values, and a reduction in project duration [21,22].

Nevertheless, the main causes for project delays remain under investigation. Doloi (2012), Braimah (2014), and Larsen (2016), in addition to many others, investigated the significant impediments that directly impacted the project's schedule [13,23,24]. The results of these studies indicated weak design elements; poor project planning, site management, and project control; insufficient contractor experience; contract payment problems; equipment availability; weather/environmental conditions; and material supply issues as the primary causes for project delays [13,23,24]. A study by Gebrehiwet et al. (2017) revealed 52 of the most likely reasons for project delays; ineffective project scheduling ranked number two, behind deficient project planning [25].

This investigation shows the international situation of the construction industry and provides information about the general and fundamental problems in construction project planning and execution [7]. Weaknesses in project design and inadequate schedule and cost management appear to be of particular importance regarding the root causes of errors. The inevitable consequences of these differences between planned and actual values lead to unforeseeable and unexpected additional cost and time requirements and thus to an increasing risk of the projects success. In order to further investigate and narrow down the described causes, current project management methods and the most recent solution approaches will be examined in the following.

3. Current Common Project Management and Scheduling Approaches

To understand a project's timeframe, certain project management and scheduling approaches—mostly IT-aided and cross-industry applicable approaches—have been implemented in the construction sector during past decades [15]. The core objective of project scheduling is assigning the start and end dates for individual or cumulative activities and indicating when these activities must be finished to be delivered on time [26]. A valuable method to gather and structure the required project execution activities is given by the Work Breakdown Structure (WBS) method. This method has no time references but provides a general framework for schedule development and enables project management, monitoring, and control tasks [27]. A common and widespread scheduling tool is the *bar chart* or *bar diagram*—also known as a *Gantt chart* or *Gantt diagram*, which graphically represents the connection between planned and actual work performance and whether activities are on schedule, behind schedule, or ahead of schedule [26]. Further common methods include the Critical Path Method (CPM), Line of Balance (LoB), Linear Scheduling Method (LSM), and a Network diagram [28–31].

First and foremost, these tools are based on the *waterfall* principle whose main characteristic is strictly hierarchical embossed organization (priority based) according to the ratio between a chronological task order and appropriate task durations. Each task obtains a clearly defined start- and end-date; dependencies on other tasks can then be determined [31]. Waterfall systems operate according to the push principle, which releases tasks, materials, or information into preassigned procedures or scheduling systems [32]. This method is ideally suited for projects with consistent or repetitive proceedings and recognizable long-term interventions due to regular organization [33]. Due to the predefined structure that is additionally required and/or was previously unconsidered, subsequently added activities could rapidly cause postponements and disturb the flow of a project. Tory et al. stated in 2013 that schedules should be dynamic documents, which can frequently be changed and adjusted in accordance to the project's progression and its various requirements [34]. In 2005, McKay and Wiers (2005) indicated that the amount of dynamic and unpredictable activities and the capacity for compensation should be considered when the scheduling method for a project is determined [31,35]. Thus, a significant disadvantage of the waterfall system is its non-dynamic ability to react quickly to fast-track changing procedures or ad hoc operations triggered by unpredictable events during a project's life cycle. Disregarding these disadvantages, waterfall-based scheduling methods are still widespread in the construction industry [34].

This system is contrasted by the agile methodology, which follows a maximum dynamic mode of operation. Project requirements and tasks are gathered and listed in the initial phase. An iterative process—consisting of task planning, execution, and revision steps—defines the project's organizational structure. Intermediary assessments can also be implemented to revise short-term activities [36]. Sanchez et al. (2001) described agility as a cooperative and synergetic strategy that organizes the processing and delivery of customer-specific high-quality goods and services, even in dynamic or unpredictable project environments. Well-structured organization combines constituent project participants into multi-skilled and cross-functional teams with participating members from both (internal/external) customers and suppliers [37]. The purpose of this method is to streamline project management efforts and to keep flexibility high, even with changes in late project stages [32]. According to Sacks et al. (2010), the basic methodology behind agile management is the lean approach, which was implemented in and adapted to the construction sector to "reduce variation, improve coordination, implement flow, establish pull, and to reduce various forms of waste in construction projects" [38]. The potential deficiencies of agile methods include difficulty in predicting a project's progression and missing the transparency of timescale objectives due to the flexible organization of task execution [7,39]. With the introduction of the Last Planner System (LPS) in 1993/94, the first official agile method was applied to the construction industry [7]. The implementation of master- and phase-schedules have contributed to more organized project execution and have connected production targets with project work structures. Key advantages of the LPS include significant improvements in information exchange and strengthening of the cooperation between project/site managers and foremen, who gather in monthly and weekly meetings to solve upcoming issues before they become critical. Due to its agile characteristics, the LPS improves schedule reliability and is ideally suited for complex, dynamic, and uncertain project conditions [40]. However, a critical aspect of this method is its limited implementation possibilities, as the system was developed mainly for project execution duties and is thus primarily applicable to the execution phase of a project. Further, the insufficient establishment of the lean principle to pursue perfection is a persistent issue, which prevents the continuous improvements and optimization of upcoming projects. Weekly work plans do not provide provisions to conduct any experimentation; thus, the LPS learns from failure rather than from success [38]. Moreover, the knowledge gained through work execution is not stored and organized in databases and cannot be used in further projects [33,34,39]. According to Sacks et al. (2010), the LPS achieves a reduction in variation through the early consideration of upcoming issues but misses the implementation of pull by disregarding important indications (signals) generated by downstream operations. Additionally, the LPS rarely provides a clear evaluation of the actual project status, which may cause imprecise project status indications [38,41]. In order to optimize this kind of project management, the KanBIM method was proposed, which extends the Last Planner system by the use of a 3D Building Information Model, which allows to visualize the construction progress and obstacles in the construction process through Kanban signals and symbols [7,38,42].

The innovation to use virtual 3D CAD/BIM models for the representation of project performance was initially suggested by Songer et al. (2000), who investigated workflow modeling's relationship with virtual 3D modelling to visualize project performance [43]. Later developments produced the 4D method, which connects the virtual 3D model with time-related activity information [44]. Further common approaches have added an additional dimension to offer the benefits of a virtual 3D CAD model that includes the appropriate project cost elements alongside project related time information (4D). This method is commonly referred to as the 5D methodology [45–47]. Figure 3 shows an example of the 5D BIM approach were the allocation of the 3D model objects to the corresponding Bill of Quantity (BoQ) positions as well as schedule activities is conducted manually.

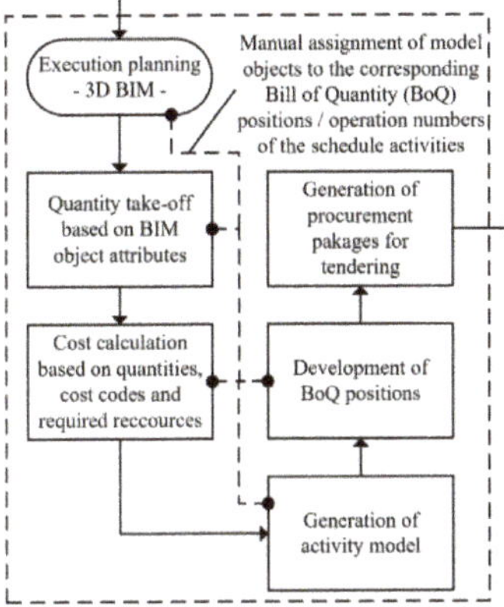

Figure 3. The 5D Building Information Model (5D BIM) approach.

To assign work execution tasks to specific project model parts, Sacks et al. proposed the use of *fine-grained* activity information from a 3D model and the creation of work-packages, which can be split into trade-specific tasks that are manageable by individual workers. These packages are represented within this model by Kanban-symbols or as a group of highlighted objects [38]. Each contractor has to develop an individual trade-specific weekly work plan, which is later synchronized with a general (project-wide) weekly work plan. The *"Kanban card type pull flow control signals and Andon alerts"* display the constraints and workflow interruptions within the 3D model [38]. To avoid interruptions of the execution process, daily on-site inspections and adjustments are conducted by a team of trade-leaders and site managers. The actual project performance statuses are displayed live by the 3D models on various screens at the construction site [38].

Although this methodology improves the reliability of task delivery and reduces variability, it enlarges project management targets drastically, as trade-specific work plans must be developed weekly before trade-related tasks can be negotiated due to synchronization with project-wide weekly work plans. In this way, tasks with lower priority have negative effects on the decision-making process and encourage undesirable discussions. In this scenario, the highly productive and efficient weekly Last Planner meetings threaten to disappear. Moreover, a trade-specific and project-wide evaluation and coordination of constraints could cause unavoidable latencies, which are critical for proper performance status indication and may hinder the flow of the project.

This analysis demonstrates that previously described methods have yielded significant improvements for individual and specific project characteristics. Approaches for repetitive and dynamic process requirements provide helpful possibilities to handle various project execution operations, even with the implementation of fine-grained activity information from 3D models to display the work in progress. Some important factors, however, retain considerable potential for improvements but have been widely disregarded: (a) close cohesion between predesign/design information (3D model objects) and project related time and cost values could be achieved via a tight interdependence between the 3D model objects, the corresponding Bill of Quantity (BoQ) positions (costs), and appropriate execution durations—this is, in the following, referred to as the *5D Building Information Model (5D BIM)*; (b) by using the early division of the 5D BIM into clearly defined project sections (PS), the determination of executing relevant target dates could provide a basic grid, which is necessary to structure the agile-organized project execution; (c) the actual required resources and values, determined during work execution (e.g., the actual required execution durations of specific tasks/actual used resources etc.) could be compared with the planned values. On this basis, continuous improvement strategies could be implemented, thus contributing to sustainable improvements of the planning accuracy of future projects.

4. The *5D-PROMPT* Method

Since the current scientific literature does not provide a coherent solution that combines the advantages of the previously described methods and the unexploited possibilities, this paper presents a comprehensive approach to obtain these goals. This new concept is referred to as the *5D-PROMPT* method. Its main objective is the sustainable reduction of both the deviation and variation between as-planned and as-built construction project targets. The key process consists of:

- Fully applied 5D BIM planning (5D BIM setup, including the interlinkage of 3D model objects, scheduling activities, and corresponding costs);
- Model split—the definition of approximately equal-sized project sections;
- Virtual construction process simulation (waterfall-based);
- Virtual construction process optimization (elimination of collisions/utilization of free resources);
- The establishment of execution-relevant target dates (project start and end date; project section-based delivery dates (milestones); monthly/weekly review/preview meetings);
- The setup of an agile structured project execution planning board;
- Agile work execution organization/agile work execution;
- Determination of the as-built data (quantity-/cost-/time-based effort values);
- The evaluation and comparison of as-built/as-planned data;
- The repatriation of as-built data/update of planning data (companywide);
- 5D BIM-based progress status indication (highlighting of model objects).

A careful selection of these principles is summarized in a multi-crossed hybrid system that operates throughout all project phases in accordance with the individual process requirements.

To provide a broad understanding of the key improvement aspects of the 5D-PROMPT method, and to explain essential enhancements a common and widespread construction design and execution process example should be introduced previously. Its basic structure represents an assumed process of conventional (3D CAD model-based) project planning, tendering, and contracting of subcontractor services, as well as a waterfall-based organization of the work execution. The process is characterized by its appropriate process steps, which are illustrated in Figure 4.

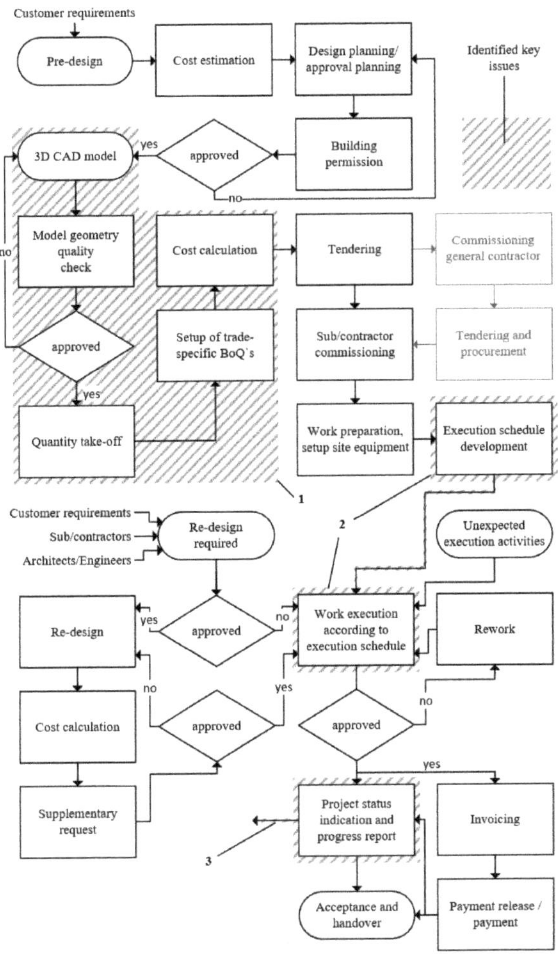

Figure 4. Initial situation: currently common construction design and execution process example.

Potential weak points and missing interconnections that are critical to a coherent and interconnected workflow are represented by the numbers 1–3 and are described as follows: (1) The significant issues include the missing interconnection between the individual 3D CAD model objects/elements, appropriate schedule operations, and the corresponding BoQ positions. In addition, the construction process sequence is not corrected or optimized by simulation. Thus, the project design, project costs, and execution time threaten to drift apart over the course of the project, which will impede project control and are critical to project success; (2) a detailed development of the work execution schedule often takes place immediately before the execution phase starts. Moreover, the utilization of a slightly flexible waterfall-based scheduling method appears inadequate for the numerous unexpected and unpredictable on-site incidents. Further, permanent cooperation between the work execution schedule and on-site operations is required; (3) significant deviations between the planned and used recourse values could be reduced by implementing specific information feedback-loops, which report back crucial as-built values/information in accordance with Deming's Plan-Do-Check-Act cycle [7].

Based on the previously presented workflow, Figure 5 demonstrates the key enhancements and general operating principles of the 5D-PROMPT method. This method consists of five main sections: (1) fully applied 5D BIM planning process; (2) IT-supported connection of the 3D BIM objects, with the associated BoQ positions as well as schedule activities by utilization of linking elements. (3) early determination of project duration and project sections (PS) and definition of target dates for PS deliveries; (5) agile project execution organization according to predefined target dates; and (5) intermediary information feedback loop implementation for project status indication and target/actual comparison by the 5D BIM as the *single source of truth* (*SSOT*).

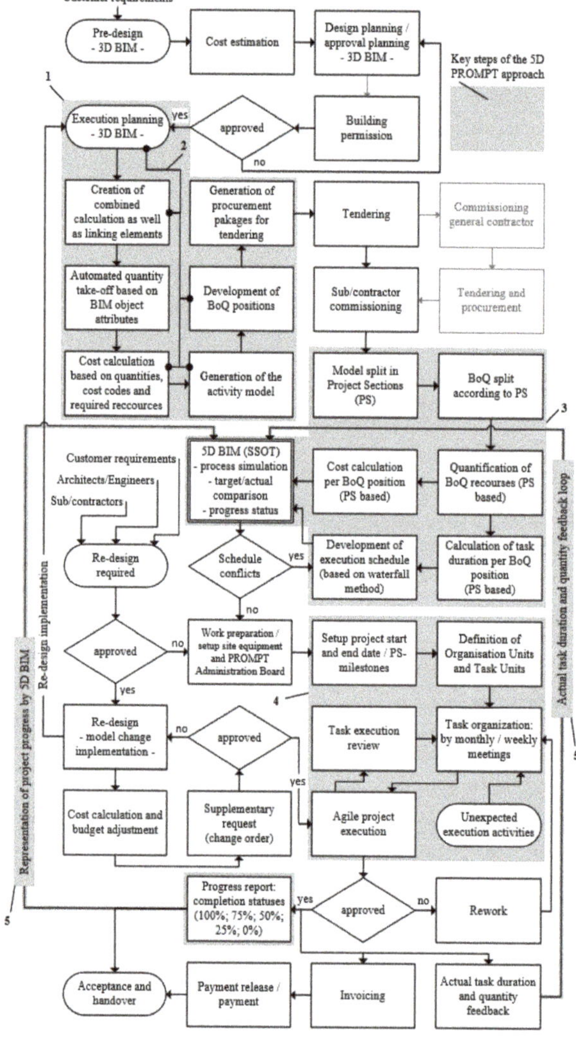

Figure 5. Workflow of the 5D-PROMPT method.

5. Mode of Operation

5.1. 5D BIM Planning Process

To take advantage of the 5D BIM-based planning approach, it is crucial to generate the entire project design using a virtual 3D BIM model (Figure 6(①)). The individual BIM objects are assigned to their corresponding BoQ positions, which contain product characteristics, costs per unit, and quantity information (Figure 6(②,③)). The technical implementation of the 5D BIM approach, which is applied within the 5D-PROMPT method, uses a link position to interconnect the 3D model objects with the associated BoQ positions as well as schedule activities. It is considered as a key element that ensures a close connection between the individual planning elements and contributes to prevent design, project cost and construction time from drifting apart. This innovation was developed by RIB Software SE within the software solution iTWO 4.0. In order to achieve the optimum performance of the 5D-PROMPT method, this methodology was used accordingly. Time-specific effort values, which provide information about the time needed to execute the required construction tasks (activity duration), are extracted from the BoQ position and transferred to a corresponding activity operation in a waterfall-based schedule application, e.g., Gantt Chart (Figure 6(④)). Although the 5D-PROMPT method stipulates agile organized construction work execution, a waterfall-based schedule is developed during the initial project planning phase to determine the project/activity duration and provide a theoretical project execution simulation. In this way, specific start/end dates are assigned to each scheduled activity. To achieve an optimized execution workflow, the methodology of the Critical Path, including the forward pass/backward pass, float calculations, and fast-tracking options, is applied appropriately. In this way, the 5D BIM planning approach enables a virtual construction process simulation, with the concurrent process of project costs and activity durations.

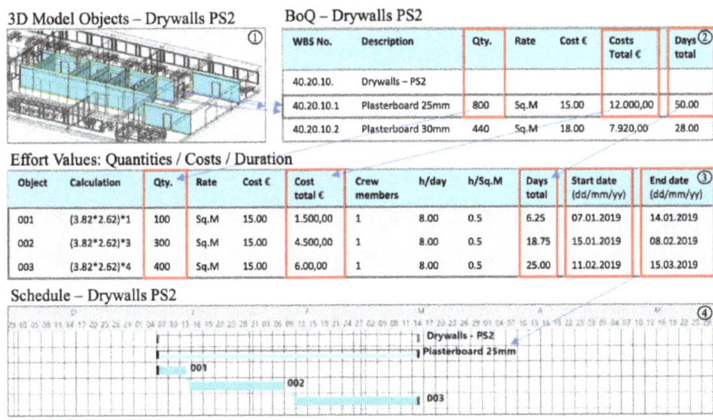

BoQ – Drywalls PS2

WBS No.	Description	Qty.	Rate	Cost €	Costs Total €	Days total
40.20.10.	Drywalls – PS2					
40.20.10.1	Plasterboard 25mm	800	Sq.M	15.00	12.000,00	50.00
40.20.10.2	Plasterboard 30mm	440	Sq.M	18.00	7.920,00	28.00

Effort Values: Quantities / Costs / Duration

Object	Calculation	Qty.	Rate	Cost €	Cost total €	Crew members	h/day	h/Sq.M	Days total	Start date (dd/mm/yy)	End date (dd/mm/yy)
001	(3.82*2.62)*1	100	Sq.M	15.00	1.500,00	1	8.00	0.5	6.25	07.01.2019	14.01.2019
002	(3.82*2.62)*3	300	Sq.M	15.00	4.500,00	1	8.00	0.5	18.75	15.01.2019	08.02.2019
003	(3.82*2.62)*4	400	Sq.M	15.00	6.00,00	1	8.00	0.5	25.00	11.02.2019	15.03.2019

Figure 6. Technical description: 5D BIM planning approach.

5.2. Determination of the Project Duration and Project Sections (PS) and the Definition of Target Dates for PS Deliveries

To form the conditions for an aspired agile organization of construction execution, a basic grid should determine the direction of the work proceedings. For this purpose, the 5D BIM is split horizontally and vertically in *approximately* equal sized Project Sections (PSs). The corresponding BoQ positions, and schedule activities are divided and re-compiled appropriately. After schedule reorganization, the project start/end date (Project Frame—PF) and the start/end dates of each PS can be determined. Any other information provided by the waterfall-based schedule has no further use over the remaining course of the project. Applied effort values within the project planning phase are updated/corrected via actually used (as-built) values.

5.3. Agile Project Execution Organization According to Predefined Target Dates

A core aspect of agile organized work execution is the collaborative competence of the contractors (capacity for teamwork) involved in the execution process. All construction trades should be tendered and contracted/subcontracted at this time of the project; specific project organization requirements must become an integral part of the contractor/subcontractor agreements. To manage/organize project execution onsite, an agile execution organization board (hereafter referred to as the *PROMPT Administration Board*) is formed in close cooperation between the foremen and site/project-manager. The basic approach in this method is similar to the Last Planner/Scrum/Kanban project organization plans or boards; however, the present method differs in its setup and arrangement [39,48]. The PFs and PSs define the general project guidelines and determine the deadlines for PS/total project delivery. Thereafter, fixed time periods are established to manage/review on-site work execution in a monthly and weekly sequence:

- Organization Unit (OU)/Billing Period—monthly cycle
- Task Units (TU)/Report Period—weekly cycle

The form and concept of the PROMPT Administration Board are explained in Table 1.

Table 1. Leading structure for development of the PROMPT Administration Board.

Board Structure	Key Units	Project Management Compendium
Project Start and End Date	Project Frame (PF);	PROMPT Administration Board setup; Set project start-date/end-date for total project duration (derived from waterfall-based project schedule); Aim Project Sections (PSs)
	Organization: administration kick-off meeting	
Project Sections	Project Sections (PSs);	Set Project Section milestones: start-date/end-date (derived from waterfall-based project schedule); Superior task summary; Agile reconciliation of task order per PS (end-to-front organization); Feasibility evaluation—compliance with PF; Commitment to agreements
	Organization: Administration kick-off meeting	
Organizational Units (monthly review of PS compliance)	Organizational Units (OUs)	Set Organization Unit start-date/end-date (duration: 1 month—determination of fixed weekdays is compulsory due to synchronization with weekly Task Unit sequence: 1 month ≈ 4.4 weeks; Set start date: e.g., every last Friday of each month; Set end date/due date for monthly meeting: e.g., every last Thursday of each month; Agile reconciliation of target order per OU (end-to-front organization); Definition of execution scope per Task Unit (TUs are based on model sections: rooms; slabs; objects, etc.); Advanced organization of complex/risky tasks; Feasibility evaluation—compliance with Project Section milestones; Commitment to agreements (post-its)/review of finished Task Units/evaluation of target agreements
	(subordinated organization of project execution tasks);	
	Organization: monthly meeting	
	Billing Periods (period of partial invoicing)	Set Billing Period start-date/due-date: equivalent to OU dates; Invoicing according to finished, reported and approved Task Units

Table 1. *Cont.*

Board Structure	Key Units	Project Management Compendium
Task Units (monthly review of OU compliance)	Task Units (TUs)	Administration task definition—Set Task Unit start-date/end-date (duration: 1 week—determination of fixed weekdays is compulsory due to synchronization with monthly meeting of Organization Unit—in this example: every last Thursday of each month); Set start date: e.g., every Friday of each week; Set end date/due date for weekly meeting: e.g., every Thursday of each week; Agile reconciliation of targets per TU (post-its); Specific organization of complex/risky tasks; Feasibility evaluation—compliance with Organization Unit sequence; Commitment to agreements/review of finished model sections/evaluation of delays and target agreements
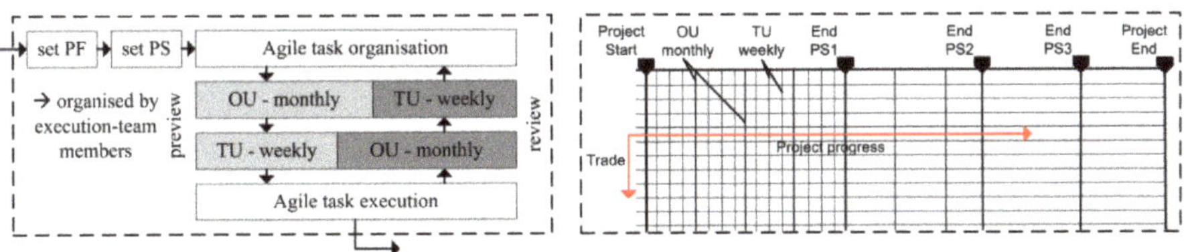	=∑ pre-defined model sections (rooms/walls/slabs, etc.) that could be finished within one week by a trade-specific execution team;	
	Organization: weekly meeting	
	Report Periods (weekly project execution performance report)	Set Billing Period start-date/due-date: equivalent to TU dates; Reports according to finished and approved Task Units—Task Unit finished (100%); Task Unit in progress (50%); Task Unit open (0%)

The task organization is graduated from general to particular. Appropriate descriptions are provided in accordance with the specific levels of organization. The setup and formatting of the organizational structure (PROMPT Administration Board) is done during an administration kick-off meeting conducted by the execution team members (foreman/project manager—contractor side/project manager—client side). The workflow and superstructure are outlined in Figure 7. Since the project start/end dates and milestones for PS delivery are attached to the administration board, the Organizational Units and Task Units can be defined.

Figure 7. Setup and superstructure of the PROMPT Administration Board.

Participation in monthly/weekly meetings is compulsory for each of the execution team members. Individual arrangements and general agreements determined during these meetings must be accomplished within the envisaged time frame.

5.4. Intermediary Information Feedback Loop Implementation for Project Status Indication and Target/Actual Comparison

The construction progress is evaluated and updated during the weekly meetings (Report Period) based on the actual completion status values of the planned (trade-specific) targets. The completion status values are 100%, completed; 50%, partly completed; and 0%, pending. The actual status values are displayed in colored markups (green; yellow; grey) by the 5D BIM. Invoices can be issued in accordance with the Billing Periods (monthly). The basis for payment approval is the actual (trade-specific) competition status of the accumulated monthly activity performance.

To achieve sustainable improvement of the planning accuracy, substantial deviations between the planned and actual execution durations are identified and evaluated by the execution team members and reported back to the project planning department to implement a sustainable correction process for time-based calculation matters. Furthermore, deviations between the planned and applied resources can be determined immediately. Thus, the planning accuracy for future projects could be improved considerably.

6. The Comparative Case Study

An initial implementation and performance test of the 5D-PROMPT method was conducted using an actual practice construction project as a comparative case-study. To pre-classify the workability and expected benefits of the new approach (and to make it comparable to previous methods), one construction project was carried out according to conventional planning and execution methods, as described in the process chart in Figure 4 (hereafter referred to as Project A). During the same time period, an equivalent construction project was carried out according to the 5D-PROMPT method, as described in Figure 5 (hereafter referred to as Project B).

The performance values of both methods were determined using special Key Performance Indicators (KPIs—listed in Table 2) measured during/after the execution phases of both projects and subsequently evaluated by a multi-criteria analysis. The required KPIs were selected based on the findings of the literature study and an analysis of the current state project management approaches.

Table 2. Definition of Key Performance Indicators (criteria); reference units, weighting factors, and determination to target a higher or lower performance value.

Key Performance Indicators—Project A; B	Total	Per trade	UoM	Adjustment Requirements/Remarks	Weightage Factor (ω'_j) b	Category **
Project duration *	x	—	CW	Project comparation criteria—$\Delta < 20\%$	—	—
Gross floor area (GFA) *	x	x	m2	Project comparation criteria—$\Delta < 20\%$	—	—
Gross volume (GV) *	x	x	m3	Project comparation criteria—$\Delta < 20\%$	—	—
Construction costs *	x	x	€	Project comparation criteria—$\Delta < 20\%$	—	—
Average exceeding of construction units	x	—	%	considered: factoring errors by domino eff.	0.25	n-b
Value of cost overruns	x	x	€		0.25	n-b
Value of deadline exceedings	x	x	CW	classification: unprediced/customer wanted	0.15	n-b
Value of supplements/planning changes	x	x	€	classification: unprediced/customer wanted	0.1	n-b
No. of supplements/planning changes	x	x	No.		0.02	n-b

Table 2. *Cont.*

Key Performance Indicators—Project A; B	Total	Per trade	UoM	Adjustment Requirements/Remarks	Weightage Factor (ω'_j) b	Category **
No. construction interruptions	x	x	No.		0.03	n-b
No. of disability complaints	x	x	No.		0.03	n-b
No. of default notifications	x	x	No.		0.03	n-b
No. of construction defects during execution	x	x	No.	classification: self-inflicted/not self-inflicted	0.02	n-b
No. of construction defects after handover	x	x	No.	classification: self-inflicted/not self-inflicted	0.02	n-b
Project commitment of stakeholdersstrongly disagree 1 |2 | 3 | 4 | 5 strongly agree	x	—	Grades	Grade system (*Likert-Scale* [49])	0.05	b
Injuries to workers	x	—	No.		0.05	n-b

* Criteria were used to determine the comparability of Project A and Project B. ** b $\triangleq$ beneficial—higher performance value is desired; n-b $\triangleq$ non-beneficial—lower performance value is desired.

Project A and Project B were carried out by the general contractor company Heinrich Schmid GmbH and Co. KG during the years 2018/19 (Location: South Germany). Twenty-one trades were planned and executed by the project stakeholders over a period of 11 months for each project. As a comparison of the two construction projects would have led to inaccuracies due to a lack of absolute equivalence, a reference value was introduced as a benchmark for data collection. In this study, it is assumed that the process sequence and implementation of Project A are generally known. Therefore, the following section will only describe the procedure for the conduction of Project B. However, to provide a better understanding, the basic process steps of both projects are compared in Table 3.

Table 3. Overall process steps of Project A compared to Project B.

Project Phase+D9:F22	Process Steps-Project A	Process Steps-Project B
Design	2D/3D CAD design 2D/3D-based quantity take-off * BoQ development and cost calculation * Determination of execution activity effort values * Rough estimation of execution durations	3D BIM design 3D BIM-based quantity take-off 3D BIM-based BoQ development linked to model objects 3D BIM-based schedule development linked to model objects 5D Building Information Model split (project sections) 5D BIM-based construction process simulation/optimization
Tendering/Contracting	Software-based tendering and contracting	Intergarted, software-based tendering and contracting
Execution	Construction execution schedule development (waterfall-based) * Construction work execution based on Gantt-chart Work organization and controlling by individual site manager	Agile work organization and controlling - team-based Agile work execution and management 3D model-based execution status representations As-built information feedback loops

* without model linking.

As previously described, the design of Project B was based on a 3D BIM. This model was provided by Contelos GmbH, a virtual-modeling company. The construction software developer RIB Software SE provisioned the software tool iTWO Baseline, which was used to create the required interconnections between the individual model objects, the corresponding BoQ positions, and scheduling activities.

Furthermore, the model split operations, construction process simulation/optimization, the tendering and contracting of sub-contractor services, and 3D model-based execution status representations were conducted/developed by this software. To minimize the risk of method failure, in addition to the 3D BIM, conventional 2D CAD project plans were also created and kept available for execution.

Since the 5D-PROMPT method was introduced to the project stakeholders, continuous team coaching, and individual training were required to acclimate the participating members to run the project. To ensure a solid foundation, all project-related contracts had to comply with the agile project's execution requirements. The modeling company was required to elaborate the 3D BIM to confirm the modeling guidelines of the applied 3D modeling software (Revit) and to make use of a harmonized project-wide BIM attribution. A substantial measure was the awarding and contracting of all execution trades, which had to be completed before work execution started. Next, the *execution organization team* was formed, consisting of foremen, crew leaders, site managers, and the project leader. This team performed the setup of the PROMPT Administration Board using time-regulated "project start and end date as well as milestone date" information, which was determined by the waterfall-based construction execution simulation. Moreover, monthly/weekly time periods for the Organizational Units (OUs) and Task Units (TUs), were added to the board to establish a static structure for the work execution organization.

After work execution started, the team determined, confirmed, and evaluated both upcoming and finished work execution tasks during monthly/weekly meetings. Daily work performance monitoring ensured proper target/actual provisions and provided information about the required execution durations. Actual building statuses were represented by highlighted objects in the 3D BIM. Deviations between the planned and required time and cost values were evaluated and gathered in a database to help make future project planning more precisely predictable and reliable. The core process of the case-study is represented and described in Table 4.

Table 4. Integrated software application and practical 5D-PROMPT workflow implementation.

(a) 3D BIM design and Model Check	Note
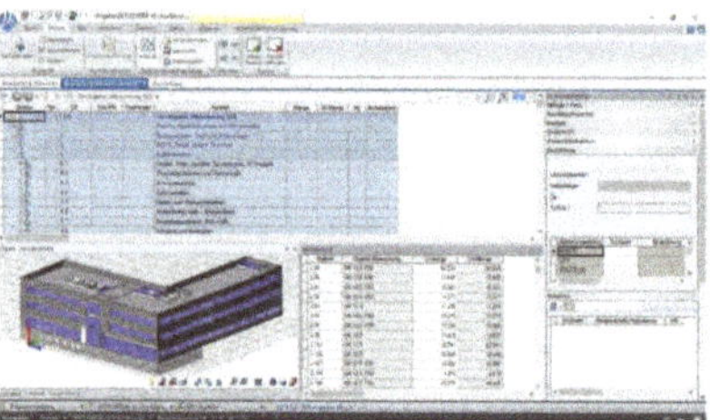	After the 3D BIM was designed using the 3D BIM capable software Revit®, the model was uploaded to iTWO Baseline and checked for errors and comprehensiveness via the integrated BIM Qualifier tool. This picture shows the 3D BIM

(b) BoQ development and connection between BoQ positions and model objects	Note
	The second step was the development of the individual BoQs, including model-based quantity take-off (QTO) and cost calculations based on resource-specific effort values. The BoQ positions were interlinked closely with the individual model objects.

Table 4. *Cont.*

(c) Model split	Note
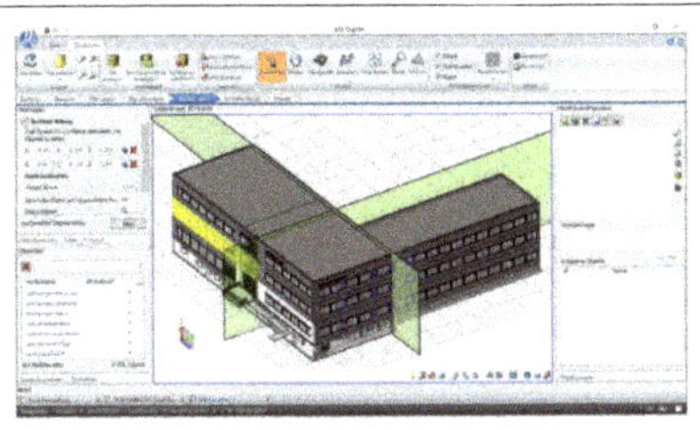	To pre-define the course of the agile organized construction work execution, the model was split into approximately equal sized Project Sections (PSs). This was done using the "Split Model" option.
(d) BoQ split	**Note**
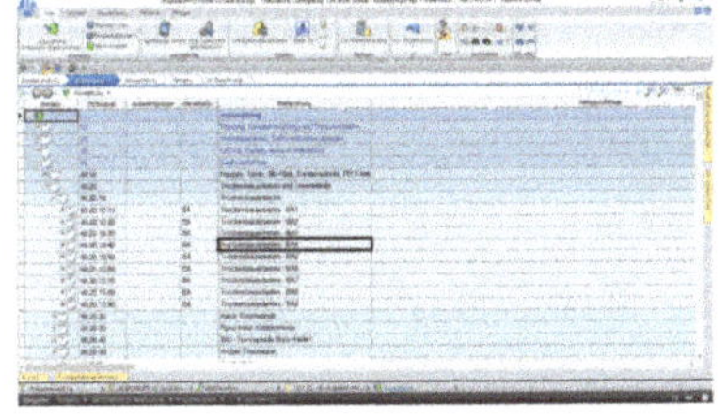	The individual BoQs were split according to the number of Project Sections; the BoQ structure remained unchanged; quantities were distributed in the next step. $\Rightarrow$ 9 Project Sections $\triangleq$ 9 BoQ sections.
(e) Quantity distribution	**Note**
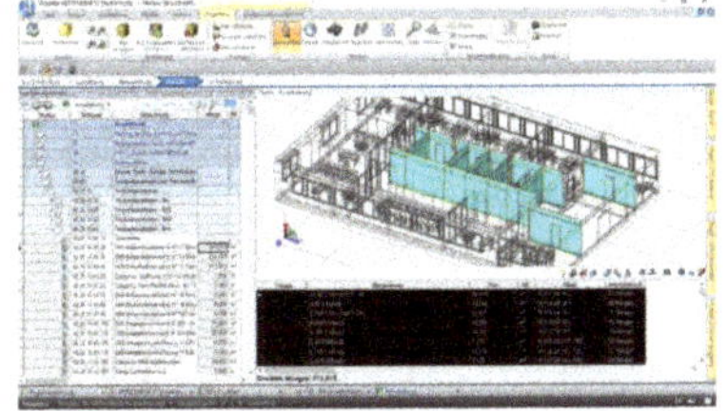	The model object quantities and the assigned BoQ-positions (including resource effort values, corresponding costs, and execution durations) were divided and distributed based on the related Project Sections.
(f) Development of the waterfall-based construction schedule	**Note**
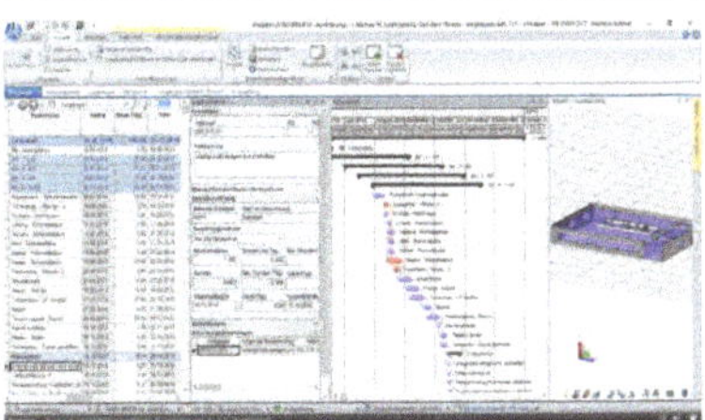	Next, the construction activities and associated execution durations were consecutively and chronologically listed in a waterfall-based project schedule chart. In the following step, the individual scheduling activities were linked to the corresponding model objects and BoQ positions.
(g) Virtual project simulation and process optimization	**Note**
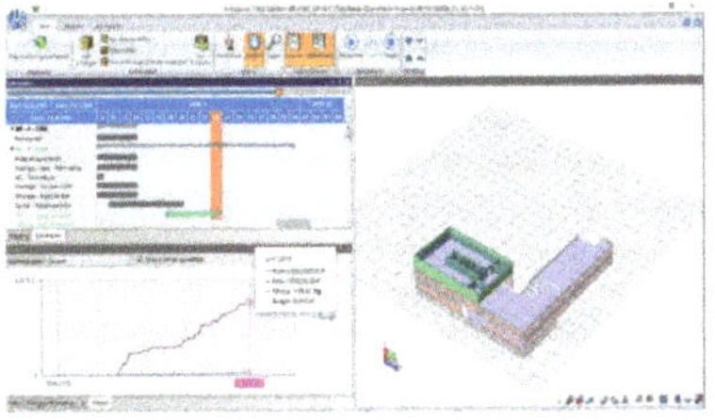	The project execution process was simulated, tested for any collisions/bottlenecks, and optimized accordingly. Moreover, the project start and end dates and milestones for PS delivery were determined.

Table 4. *Cont.*

(h) Agile project planning *	Note
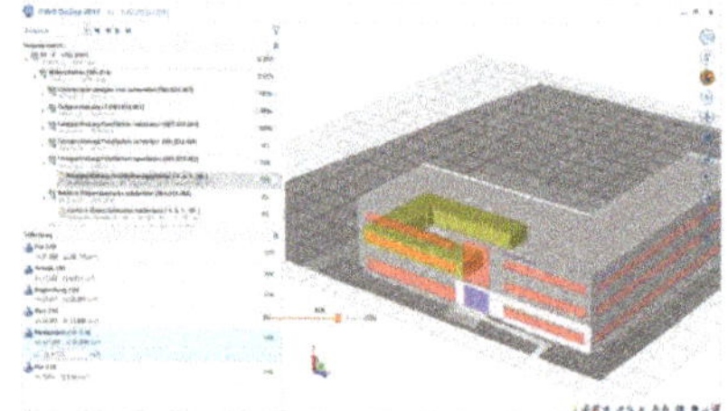	Setup of the PROMPT Administration Board: Transfer of the Project Frame and PS milestones from the waterfall-based project schedule chart. Implementation of Organizational Units and Task Units. Team-based agile work organization and review (→ Table 2) * schematic illustration due to the German project language.

(i) Work in progress measurements	Note
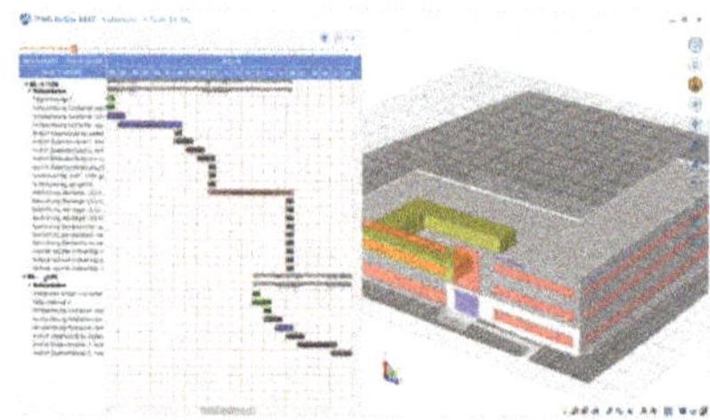	Next, the Task Unit completion statuses (100%; 50%; 0%) were continuously captured onsite, accompanying the agile organized project execution. This was done via the integrated iTwo OnSite® software application. On this basis, trade-specific invoices were generated in accordance with the Organizational Units/Billing Periods.

(j) As-planned/as-built comparison and Feedback Loop implementation	Note
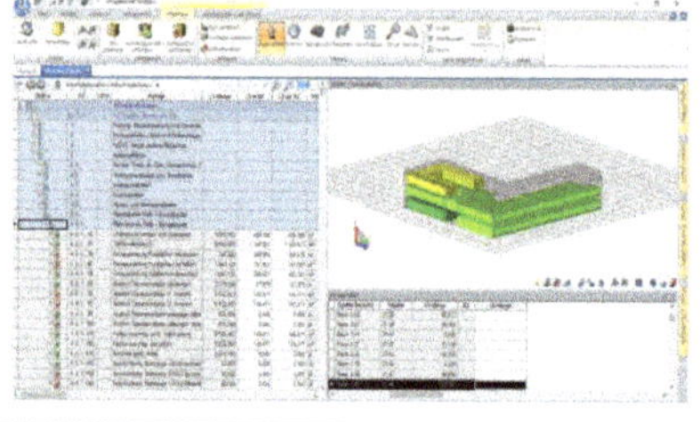	Using the OnSite progress assessment, the actual used quantities/resources and costs were determined. These values were compared to the planned values. Deviations were reported back to the project planning department to be used for quantity/resource effort value updates. In this way, continuous improvement ensures a continual increase in future planning accuracy.

(k) Work in progress representation	Note
	The 5D-PROMPT workflow culminated in the project performance status representation, which was continuously shown by the 5D BIM. The model objects were highlighted by different colors according to the progress status (e.g., 0% grey; 50% yellow; 100% green).

7. Results and Multi Criteria Analysis

To make the two projects comparable, the reference value for the collection of data was set to 1000 m^2 GFA (Gross Floor Area). All measured values were related to this factor. The planned duration of both projects, including planning, tendering, contracting, and execution, was 11 months. To increase the comparability of the two projects, both projects were divided into construction sections of an approximately equal size. These sections were required to optimally allocate the scope of work and to serve as reference points for determination of the project's process status indication.

Project B was completed within the planned construction period. Each construction section was also completed on time, so each predetermined milestone passed on time. Deviations caused by the estimated values determined during the project planning phase were compensated by the agile construction process organization. Moreover, the implementation of the enhanced 5D-BIM approach has contributed to a considerable reduction of the deviations between the as-planned and as-built values. Project A exceeded the planned end date with a delay of about 4 weeks. The project milestones were passed, on average, 3–5 days later than planned. This resulted in a shift of the entire schedule and caused the final deadline to be significantly exceeded. At first glance, Project B matched the budget; however, additional and unexpected extra costs for necessary tablet/computer hardware and servers, including maintenance, hotline, and update services, were required due to the implementation of the new methodology. These costs amounted to a total of EUR 22,825 net per 1000 m^2 per year for Project B. This resulted in an actual cost overrun of 6.04% for this Project. The cost overrun of Project A was 16.2% in total for the reference value of 1000 m^2. This was caused by supplements due to inadequate planning, rework and defect management and the extended construction period (extra costs for provision of site equipment and staff).

To obtain the preliminary assessment of the influence of the new method in terms of its enhancements to project performance, accuracy in project planning, and schedule and cost reliability, a Multi Criteria Analysis was conducted. Based on the two alternatives, Project A (i_A) and Project B (i_B), and the predefined analysis criteria (j_n) previously described by the KPIs (see Table 2), an evaluation matrix was developed. Here, the cell variables represent the project-based performance value (X_{ij}) of each criterion.

To avoid assessment issues and to achieve comparable analysis results, the deviating units of the measurements of the criteria to be compared were unified, and linguistic terms of classification were assigned to a number-based performance value scale. Furthermore, normalization of the analysis matrix was required to obtain only numerical values without any units.

The allocation of each criterion into "beneficial—higher performance value is desired" (e.g., commitment of involved stakeholders) and "non-beneficial—lower performance value is desired" (e.g., exceeding of final project deadline) was next conducted, and the normalized performance value (X'_{ij}) of each cell had to be calculated. For this purpose, the following formulas were applied based on the criteria classification: "beneficial" or "non-beneficial". The minimum and maximum performance values were derived from the lowest and highest performance values of each criterion:

$$\text{It applies} \ \rightarrow \ X'_{ij}(beneficial) = \frac{X_{ij}}{Max\left(X_{ij}\right)} ; \ X'_{ij}(non-beneficial) = \frac{Min\left(X_{ij}\right)}{X_{ij}}.$$

Since the analysis matrix was normalized, a weighting factor (w'_j) was assigned to the normalized performance values (X'_{ij}) of each criterion $(\sum_{j=1}^{n} w'_j = 1)$ (see Table 2) to classify its influence on schedule and cost reliability, accuracy in project planning, and project performance. Each normalized performance value was multiplied by the assigned weighted factor to obtain the weighted normalized analysis matrix.

To calculate the absolute performance scores of Project A and Project B, each weighted normalized performance value (X''_{ij}) of each Project was cumulated. The entire calculation of the performance score per alternative $((i_A)$ and (i_B) (Project A; Project B)) can be described by the following formula:

$$performance \ score_{(i)} = \sum_{j=1}^{n}\left(\frac{X_{ij}}{Max\left(X_{ij}\right)}\right) * w'_j + \sum_{j=1}^{n}\left(\frac{Min\left(X_{ij}\right)}{X_{ij}}\right) * w'_j.$$

After the collection of all measurement data related to the respective Key Performance Indicators and the evaluation of all values, a total performance score of 0.38 was calculated for Project B. For Project A, the score was 0.12. To evaluate the significance of these values, the ranking scale shown in Table 5 was used. This scoring model is generally applied to assess alternatives based on several quantitative and qualitative criteria, objectives, or conditions. It is used to analyze a set of complex alternatives to order the elements of the set according to the analysis preferences based on a multidimensional target system. The order is represented by the performance value of each alternative. The evaluation numbers follow a five-fold scale (in this case, 0.05 to 0.55), where a higher evaluation number stands for a superior evaluation (Table 6) [50–52]. As a result, the project planning and execution of Project B according to the 5D-PROMPT method could generally be rated as "good". In direct comparison, Project A could only be rated as "bad" according to the conventional project planning and execution method. This indicates a considerable improvement of construction planning and execution under the 5D-PROMPT method and suggests an immense enhancement of overall project performance. Further results and differences of both project organization methods are listed and explained in the following section.

Table 5. Evaluation Matrix—Determination of the performance value (X_{ij}). This method applies → $X_{ij} = performance\ value\ of\ the\ i^{th}\ alternative\ over\ the\ j^{th}\ criteria.$

	j_1	j_2	j_3
i_A	$X_{i_A j_1}$	$X_{i_A j_2}$	$X_{i_A j_n}$
i_B	$X_{i_B j_1}$	$X_{i_B j_2}$	$X_{i_B j_n}$

Table 6. Ranking scale [47].

Scale Values	Icon	Definition
0.45–0.55	++	very good
0.35–0.449	+	good
0.25–0.349	0	sufficient
0.15–0.249	-	less sufficient
0.05–0.149	–	bad

The comparison also shows that the number of construction defects during the construction phase of Project B was 82.6% lower than that of Project A. The costs for supplements and changes in planning due to the voice of the customer were also approximately 71.5% lower than those in the comparable project. The commitment of the participants to the newly introduced method of Project B was very high (Grade 5—*strongly agree*, from the Likert scale analysis [49]), but the commitment of the project execution team to the conventional method was also rated "high" (Grade 4—*agree*, from the Likert scale analysis [49]). The number of disability notifications and notices of default was three for Project A, while Project B was completely unaffected by these measures.

There was no absolute stop of construction work in either project. Moreover, an officially decreed construction stop was imposed in neither of the projects, and no workers were injured or killed during the construction process. The costs of implementing the new methodology of Project B were related to the following services: (1) introduction and training of the project organization method, (2) software teaching and training, (3) process consulting, (4) development of project-relevant master data, and (5) operational project support. These services were calculated proportionately according to the reference values of 1000 m2 per year and amounted to EUR 38,500 net in total. These costs include, as described above, the considerable additional costs for extra required hardware equipment and software-related services.

8. Discussion and Conclusions

The initially mentioned productivity growth of the American and European construction sector indicated international structural weaknesses and considerable performance deficiencies over the past twenty years. These preliminary graphs (Figures 1 and 2) represent only a portion of the international situation. Further statistics could be included, but a deeper insight into international conditions was garnered through the literature study. The literature provides a relatively far-reaching overview of the current problems of the construction industry but often refers to different project variants that cannot be compared directly or can only be compared with difficulty. Although some studies included in the literature review of this article show improvements in project organization methods due to advanced innovations, in many cases, these developments fall short of their goals, as they often relate only to the work execution phase and do not sufficiently account for essential criteria and possible improvements during the project design phase. At this point, the 5D-PROMPT method is intended to meet the requirements of transferring essential information from the design phase to the execution phase (e.g., current planning status of the BIM as a single source of truth; project timeframe and key milestones tested and optimized by simulation). In addition, the knowledge gained from the construction phase has to be reintegrated into the construction planning of future projects. The project organization method referred to as the *common* method in this article should only be understood as an example and is only one of countless other possible examples.

The comparative case study was conducted to obtain a preliminary and brief overview about possible improvements in project performance, accuracy in project planning, and scheduling and cost reliability through application of the 5D-PROMPT method. This was not a representative study, as only one sample for each criterion/alternative could be collected due to comparing only two projects. Nevertheless, a range of different Key Performance Indicators (criteria) were measured, which provided the study results with a certain significance. For the comparability of both projects, it was of particular importance that both projects were appropriately similar ($\Delta < 20\%$) and were carried out by the same executing company at the same time. The PROMPT administration board was used as an analog planning board within the case study. To avoid interface problems and a possible loss of information in future investigations, a digital planning board, fully integrated into the overall process, will be required. Automated functions based on a self-learning system should also be developed in future papers.

A main obstacle that emerged during the implementation of the project under the 5D-PROMPT method (Project B) involved defining the Task Unit (TU) content and the corresponding model objects. A general trade-by-trade definition was considered unusable since different trade specifications consist of various completion characteristics in terms of (a) the standardization of target unit measurements, (b) increasing deviations over long project execution periods, and (c) tracking of multiple correlated design changes. These issues were solved by the classification and alignment of trade compliant model sections, which are measurable, de-limitable, and directly extractable from the 5D BIM. This common problem (which is typical for agile organized projects with missing references for project target dates, deadlines, or time limits) was handled by properly determined project start/end dates and Project Section (PS)-related milestones. Thus, the course and frame of the project followed a clear structure and was traced by the participating stakeholders. To solve consequential coordination difficulties, the foremen of adjacent trades took part in the agile organization meetings and contributed to solving forthcoming issues.

The results of the case study indicated a considerable improvement of the objectives pursued. However, since only one project was counted as a sample, this study was relatively limited. Therefore, no statistical evaluation of the results was achievable. However, due to the large number of different criteria examined, it was possible to carry out a multi-criteria analysis, which provided a preliminary impression of the effectiveness of this method. The weighting of the multi-criteria analysis was particularly focused on the criteria that affect exceeding construction time and costs, as well as disruption-free construction processes. One experienced value concerns the oversized storage capacity of the server used for Project B. In future projects, this value this could be reduced considerably, thereby decreasing the direct project costs per 1000 m^2.

Generally, the technical implementation and feasibility of the proposed method was proven to be beneficial, and possible technical improvements have already been derived. To obtain a reliable evaluation of the effects of the entire 5D-PROMPT method, a series of projects must be carried out and examined in accordance with this method in future investigations.

9. Future Perspective

At the present stage, the 5D-PROMPT method indicates promising improvements in project organization and schedule reliability. Moreover, the cooperation of waterfall-, agile-, and lean-based process organization had a positive influence on the project performance. However, to evaluate the real advancements in terms of schedule reliability, project performance, and planning precision improvements, further research and empirical analysis is required. The appropriate project size that best fits the application of the 5D-PROMPT method should also be examined in detail.

Author Contributions: Conceptualization, D.L. and C.B.; Data curation, D.L.; Formal analysis, J.D. and C.B.; Investigation, D.L.; Methodology, D.L. and C.B.; Resources, J.D.; Software, J.D.; Supervision, D.C.-F. and J.D.; Validation, D.C.-F.; Visualization, C.B.; Writing—original draft, D.L. and C.B.; Writing—review & editing, D.C.-F. All authors have read and agreed to the published version of the manuscript.

Funding: This research was supported by Heinrich Schmid GmbH&Co.KG regarding the cooperative conduction of two construction projects used for the case-study; RIB Software SE, who provided the BIM software platform iTWO Baseline and the site management software OnSite, and Contelos GmbH for the provision of the 3D Building Information Model.

Conflicts of Interest: The authors declare no conflict of interest.

References

1. Oppong, G.D.; Chan, A.P.C.; Dansoh, A. A review of stakeholder management performance attributes in construction projects. *Int. J. Proj. Manag.* **2017**, *35*, 1037–1051. [CrossRef]
2. Oesterreich, T.D.; Teuteberg, F. Understanding the implications of digitisation and automation in the context of Industry 4.0: A triangulation approach and elements of a research agenda for the construction industry. *Comput. Ind.* **2016**, *83*, 121–139. [CrossRef]
3. Kim, S.Y.; Nguyen, V.T. A Structural model for the impact of supply chain relationship traits on project performance in construction. *Prod. Plan. Control.* **2018**, *29*, 170–183. [CrossRef]
4. Buvik, M.P.; Tvedt, S.D. The influence of project commitment and team commitment on the relationship between trust and knowledge sharing in project teams. *Proj. Manag. J.* **2017**, *48*, 5–21. [CrossRef]
5. World Economic Forum. *Shaping the Future of Construction—A Breakthrough in Mindset and Technology*; World Economic Forum: Geneva, Switzerland, 2016; pp. 1–64.
6. Dave, B.; Kubler, S.; Framling, K.; Koskela, L. Opportunities for enhanced lean construction management using Internet of Things standards. *Autom. Constr.* **2016**, *61*, 86–97. [CrossRef]
7. Ballard, G. The Last Planner System of Production Control. Ph.D. Thesis, Faculty of Engineering, University of Birmingham, Birmingham, UK, 2000; pp. 1–192.
8. OECD. *Productivity and ULC by Main Economic Activity, Total Economy and Construction, EU vs. US*; Economic Cooperation and Development: Paris, France, 2020.
9. O'Mahony, M.; Van Ark, B. *EU Productivity and Competitiveness: An. Industry Perspective*; Office for Official Publications of the European Communities: Luxembourg, 2003; pp. 1–280.
10. Teicholz, P.; Goodrum, P.M.; Haas, C.T. US Construction labor productivity trends 1970–1998. *J. Constr. Eng. Manag.* **2001**, *127*, 427–429. [CrossRef]
11. Kuenzel, R.; Teizer, J.; Müller, M.; Blickle, A. Intelligent and autonomous environments, machinery, and processes to realize smart road construction projects. *Autom. Constr.* **2016**, *71*, 21–33. [CrossRef]
12. Howell, G.A.; Ballard, G.; Tommelein, I. Construction engineering-reinvigorating the discipline. *J. Constr. Eng. Manag.* **2011**, *137*, 740–744. [CrossRef]
13. Larsen, J.K.; Shen, G.Q.; Lindhard, S.M.; Brunoe, T.D. Factors affecting schedule delay, cost overrun, and quality level in public construction projects. *J. Manag. Eng.* **2016**, *32*, 10. [CrossRef]
14. Assaf, S.A.; Al-Hejji, S. Causes of delay in large construction projects. *Int. J. Proj. Manag.* **2006**, *24*, 349–357. [CrossRef]
15. Sambasivan, M.; Soon, Y.W. Causes and effects of delays in Malaysian construction industry. *Int. J. Proj. Manag.* **2007**, *25*, 517–526. [CrossRef]
16. Olawale, Y.; Sun, M. Construction project control in the UK: Current practice, existing problems and recommendations for future improvement. *Int. J. Proj. Manag.* **2015**, *33*, 623–637. [CrossRef]

17. Hoffman, G.J.; Thal, A.E.; Webb, T.S.; Weir, J. Estimating performance time for construction projects. *J. Manag. Eng.* **2007**, *23*, 193–199. [CrossRef]
18. CIOB. *Managing the Risk of Delayed Completion in the 21st Century*; The Chartered Institute of Building (CIOB): Ascot, UK, 2008; p. 55.
19. BMUB. Construction Projects Often More Expensive and Delayed. Available online: https://www.bundestag.de/presse/hib/201604/418428-418428 (accessed on 3 July 2018).
20. Salem, O.; Solomon, J.; Genaidy, A.; Minkarah, I. Lean construction: From theory to implementation. *J. Manag. Eng.* **2006**, *22*, 168–175. [CrossRef]
21. Hanna, A.S.; Skiffington, M.A. Effect of pre-construction planning effort on sheet metal project performance. *J. Constr. Eng. Manag.* **2010**, *10.1061*, 235–241. [CrossRef]
22. Hwang, B.G.; Ho, J.W. Front-end planning implementation in Singapore: Status, importance and impact. *J. Constr. Eng. Manag.* **2011**, *10.1061*, 567–573. [CrossRef]
23. Doloi, H.; Sawhney, A.; Iyer, K.C.; Rentala, S. Analysing factors affecting delays in Indian construction projects. *Int. J. Proj. Manag.* **2012**, *30*, 479–489. [CrossRef]
24. Braimah, N. Understanding construction delay analysis and the role of pre-construction programming. *J. Manag. Eng.* **2014**, *30*, 10. [CrossRef]
25. Gebrehiwet, T.; Luo, H. Analysis of delay impact on construction project based on RII and correlation coefficient: Empirical study. *Proc. Eng.* **2017**, *196*, 366–374. [CrossRef]
26. Cox, J.F.; Blackstone, J.H.; Spencer, M.S. *APICS Dictionary*; American Production and Inventory Control Society: Falls Church, VA, USA, 1992.
27. Diekmann, J.E.; Thrush, K.B. Project Control. In *Design Engineering*; Construction Industry Institute: Austin, TX, USA, 1986.
28. Tang, Y.J.; Sun, Q.; Liu, R.; Wang, F. Resource leveling based on line of balance and constraint programming. *Comput. Aided Civ. Infrastruct. Eng.* **2018**, *33*, 864–884. [CrossRef]
29. Su, Y.; Lucko, G. Comparison and renaissance of classic line-of-balance and linear schedule concepts for construction industry. *Proc. Eng.* **2015**, *123*, 546–556. [CrossRef]
30. Castro-Lacouture, D.; Süer, G.A.; Gonzalez-Joaqui, J.; Yates, J.K. Construction project scheduling with time, cost, and material restrictions using fuzzy mathematical models and critical path method. *J. Constr. Eng. Manag.* **2009**, *135*, 1096–1104. [CrossRef]
31. Herrmann, J.W. A history of production scheduling. In *Handbook of Production Scheduling*; Herrmann, J.W., Ed.; Springer: New York, NY, USA, 2006; pp. 1–22.
32. VDI. *VDI-Richtlinien 2553—Lean Construction—ICS 91.010.01 (Draft-Version)*; German-Association-of-Engineers—Beuth Verlag GmbH: Berlin, Germany, 2017.
33. Echeverry, D.; Ibbs, C.W.; Kim, S. Sequencing knowledge for construction scheduling. *J. Constr. Eng. Manag. Asce* **1991**, *117*, 118–130. [CrossRef]
34. Tory, M.; Staub-French, S.; Huang, D.; Chang, Y.L.; Swindells, C.; Pottinger, R. Comparative visualization of construction schedules. *Autom. Constr.* **2013**, *29*, 68–82. [CrossRef]
35. McKay, K.N.; Wiers, V.C.S. The human factor in planning and scheduling. In *Handbook of Production Scheduling*; Springer: New York, NY, USA, 2005.
36. Owen, R.; Koskela, L.; Henrich, G.; Codinhoto, R. Is agile project management applicable to construction? In Proceedings of the 14th Annual Conference of the IGLC, Santiago, Chile, 14 July 2006; pp. 51–66.
37. Sanchez, L.M.; Nagi, R. A review of agile manufacturing systems. *Int. J. Prod. Res.* **2001**, *39*, 3561–3600. [CrossRef]
38. Sacks, R.; Radosavljevic, M.; Barak, R. Requirements for BIM based lean production management systems for construction. *Autom. Constr.* **2010**, *19*, 641–655. [CrossRef]
39. Ballard, G.; Howell, G. An update on last planner. In Proceedings of the Annual Conference of the International Group for Lean Construction, Blacksburg, VA, USA, 22–24 July 2003; p. 13.
40. Mossman, A. *Last Planner®: 5 + 1 Crucial & Collaborative Conversations for Predictable Design & Construction Delivery*; The Change Business Ltd.: Stroud, UK, 2015; p. 36.
41. Liker, J.E. *The Toyota Way*; McGraw-Hill: New York, NY, USA, 2003.
42. Radosavljevic, M.; Horner, M. Process planning methodology: Dynamic short-term planning for off-site construction in Slovenia. *Constr. Manag. Econ.* **2007**, *25*, 143–156. [CrossRef]

43. Songer, A.D.; Subramanian, P.S.; Diekmann, J.E. *Integrating Work Flow Modeling and 3D CAD for Visualizing Project Performance*; CIB W78: Reykjavik, Iceland, 2000.

44. Bansal, V.K.; Pal, M. Generating, evaluating, and visualizing construction schedule with geographic information systems. *J. Comput. Civ. Eng.* **2008**, *22*, 233–242. [CrossRef]

45. Fan, S.L.; Wu, C.H.; Hun, C.C. Integration of cost and schedule using BIM. *J. Appl. Sci. Eng.* **2015**, *18*, 223–232.

46. Sánchez-Rivera, O.G.; Galvis-Guerra, J.A.; Porras-Díaz, H.; Ardila-Chacón, Y.D.; Martínez-Martínez, C.A. *"BrIM" 5D Models and Lean Construction for Planning Work Activities in Reinforced Concrete Bridges*; Revista Facultad de Ingeniería: Tunja-Boyacá, Colombia, 2017; pp. 39–50.

47. JOCU. *Guide to the Principles of Comparative Testing*; International Organization of Consumers Unions: The Hague, The Netherlands, 1985.

48. Streule, T.; Miserini, N.; Bartlomé, O.; Klippel, M.; De Soto, B.G. Implementation of Scrum in the Construction Industry. *Proc. Eng.* **2016**, *164*, 269–276. [CrossRef]

49. Westermann, G.; Finger, S. *Kosten-Nutzen-Analyse. Einführung und Fallstudien*; E. Schmidt: Berlin, Germany, 2012.

50. Zangemeister, C. *Nutzwertanalyse in der Systemtechnik—Eine Methodik zur Multidimensionalen Bewertung und Auswahl von Projektalternativen*; Wittemann: Munich, Germany, 1970.

51. Likert, R. A Technique for the Measurement of Attitudes. *Arch. Psychol.* **1932**, *22*, 1–55.

52. Bautsch, M. *Gebrauchstauglichkeit und Gebrauchswert*; Carl Hanser Fachbuchverlag: Munich, Germany; Vienna, Austria, 2014.

 *sustainability*

Article

Special-Length-Priority Algorithm to Minimize Reinforcing Bar-Cutting Waste for Sustainable Construction

Dongho Lee, Seunghyun Son, Doyeong Kim and Sunkuk Kim *

Department of Architectural Engineering, Kyung Hee University, Yongin-si, Gyeonggi-do 17104, Korea; DUL1212@khu.ac.kr (D.L.); seunghyun@khu.ac.kr (S.S.); dream1968@khu.ac.kr (D.K.)
* Correspondence: kimskuk@khu.ac.kr; Tel.: +82-31-201-2922

Received: 30 June 2020; Accepted: 21 July 2020; Published: 23 July 2020

Abstract: Reinforcing bars (rebar), which have the most embodied carbon dioxide (CO_2) per unit weight in built environments, generate a significant amount of cutting waste during the construction phase. Excessive cutting waste not only increases the construction cost but also contributes to a significant amount of CO_2 emissions. The objective of this paper is to propose a special-length-priority cutting waste minimization (CWM) algorithm for rebar, for sustainable construction. In the proposed algorithms, the minimization method by special and stock lengths was applied. The minimization by special length was performed first, and then the combination by stock length was performed for the remaining rebar. As a result of verifying the proposed algorithms through a case application, it was confirmed that the quantity of rebar was reduced by 6.04% compared with the actual quantity used. In the case building, a CO_2 emissions reduction of 406.6 ton-CO_2 and a cost savings of USD 119,306 were confirmed. When the results of this paper are applied in practice, they will be used as a tool for sustainable construction management as well as for construction cost reduction.

Keywords: rebar work; cutting waste; minimization; sustainable construction; CO_2 emission; cutting stock problem

---

1. Introduction

Building construction and operations accounted for 36% of global final energy use and nearly 40% of energy-related carbon dioxide (CO_2) emissions in 2017 [1]. Concrete and reinforcing steel contribute about 65% of building greenhouse gases (GHG), 40% of which are CO_2 emissions generated by concrete [2]. Clark and Bradley described that the mean embodied carbon dioxide (ECO_2) for office buildings is 340 kg-CO_2/m^2, of which the structure accounts for approximately 60% [3]. In their research report, they suggest 95 kg-ECO_2/ton for C25/30 concrete and 872 kg-ECO_2/ton for reinforcing bar (rebar). This suggests that reducing the ECO_2 in the structural frame directly produces a GHG reduction. In addition, in terms of the carbon footprint, efforts to reduce the rebar, which has an ECO_2 of about 9.2 times that of concrete per unit weight [4], are very important.

In general, rebar cutting waste is estimated in the planning stage to be 3–5% [5–8] but more than 5% occurs in the actual construction stage [5,7–13]. This is because there is a lack of optimization technology on construction sites [14]. In order to solve this problem, many studies have been conducted to minimize rebar cutting waste [5–25]. Most studies use a stock length called a standard, or market length to make combinations that minimizes cutting waste [8–16,19–25]. In other words, they combine the rebar indicated in the structural drawings using stock lengths held in the rebar shop or plant in order to minimize cutting waste. If the rebar to be ordered by special length is used in the rebar combinations, cutting waste can be further reduced [5,7,14,17]. Rebar combinations using special lengths and stock lengths can further reduce the cutting waste and CO_2 emissions. However, research on the use of special lengths for cutting waste minimization (CWM) in the construction industry is lacking. The study of Porwal and Hewage introduces the concept of special length combination [7], but constraints for minimization by special length (MSpL) are not clearly described. In addition, several studies have suggested the concept of MSpL but lack a detailed explanation of algorithm operation [7,14,17]. Additionally, from the viewpoint of sustainable construction, the effect of reducing CO_2 emissions through minimization algorithms has not been suggested. Therefore, the objective of this paper is to propose a special-length-priority CWM algorithm for rebar, for sustainable construction.

The study proceeds as shown in Figure 1. First, we describe the originality of this paper and the lessons obtained after reviewing the references on CWM and the cutting stock problem (CSP). Then, we introduce the CWM algorithms, the core content of this paper. We describe in detail the concept of stock and special lengths, the CWM process and algorithms, and MSpL. Next, after applying the proposed algorithms to the case project, we analyze the rebar savings details. In addition, we confirm the CO_2 emission and cost reduction effects associated with the rebar quantity reduction. Finally, we discuss the problems, lessons learned, and opportunities for further studies, and we describe the results of this present study.

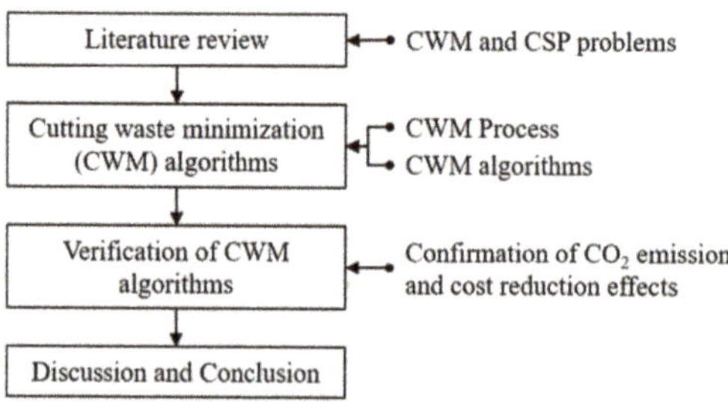

Figure 1. Research process and methodology.

2. Literature Review of CWM Problems

Research on CWM began with a study to solve the CSP. The study of the CSP was first mentioned by Kantorovich in 1939 and first published in *Management Science* in 1960 [26]. The problem consists of determining the best way of cutting a set of large objects into smaller items [27]. In operations research, the CSP is the problem to cut standard-sized pieces of stock material, such as rebar, paper rolls, or sheet metal, into pieces of specified sizes while minimizing the material wasted [28]. Kantorovich provides two examples in his paper to make the CSP easier to understand [26]. Since then, many scholars have conducted research to obtain a solution to the CSP using linear programming [29–42], genetic, or heuristic approaches [43–48].

In the case of rebar CWM, studies using linear programming and/or heuristic algorithms have been conducted [5,7,8,10,12–17,19–25]. In most cases, however, research has been conducted to minimize scrap or cutting waste using stock lengths, and the opportunity to further reduce cutting waste using special lengths has been lost. From the previous studies [5,14], we have confirmed that MSpL reduces cutting waste more than minimization by stock length (MStL). There have been several studies on minimization by special length (MSpL) [5,7,14,17], but various conditions required in practice have not been reflected in algorithms. Furthermore, the application process of MSpL reflecting these conditions was not specifically introduced. In the case of MSpL, variables such as minimum order quantity, rebar lengths for special order, minimum loss rate, and minimum and maximum combination length should be considered in practice. However, in most studies, these conditions were not reflected or sufficiently explained. In the paper of Porwal and Hewage [7], they proposed an algorithm for minimization by market and special lengths using rebar data extracted from building information modeling (BIM). However, detailed descriptions of constraints such as the rebar loss rate and minimum order quantity are not clearly described. In other papers [5,14,17], the MSpL concept was introduced, but detailed application processes were not described in their manuscripts.

In this paper, we propose algorithms that perform minimization by stock length on the rebar that is left after MSpL. However, since many scholars are familiar with MStL, MStL is first introduced, and MSpL is discussed later in the manuscript.

3. Cutting Waste Minimization Algorithms

3.1. Definition of Stock and Special Lengths

Based on the examination of the studies to date, the CWM methods for rebar are largely divided into two types. Minimization by stock length (MStL) [5,7,8,10,12–17,20–25] and MSpL [5,7,14,17]. In these two methods, the target loss ratio and minimum quantity can be added as constraints [5,14,17]. Figure 2 is an example of combinations of stock lengths and special lengths. In the case of cutting pattern 1 in Figure 2a, two reinforcing bars are combined using a stock length of 12 m, and 0.6 m of cutting waste or loss occurs, which corresponds to a loss rate of 5%. In the case of cutting pattern *i*, three reinforcing bars are combined and 0.3 m of cutting waste is generated, which corresponds to a loss rate of 2.5%. If the special length of 11.7 m is used as shown in Figure 2b, in the same cases there is 0.3 m of cutting waste, a 2.6% loss rate, and a loss of zero, respectively. As shown in the examples, when using special lengths, the cutting waste is generally reduced more than with stock lengths.

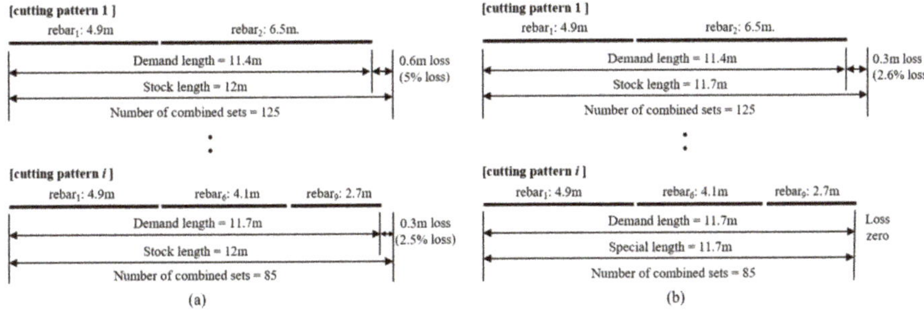

Figure 2. Combination examples of stock and special lengths: (**a**) Combination cases of stock lengths; (**b**) Combination cases of special lengths.

For reference, "special length" means the length determined by the customer's order, not the rebar length sold on the market. For example, stock or market length means the length determined by the producer in regular interval values such as 9, 10, 11, and 12 m, whereas special length includes irregular values such as 8.4, 9.7, and 10.1 m. Although there are differences by country, stock lengths of 7, 8, and up to 12 m are common in many countries, and when ordering rebar with special lengths, conditions for length, minimum quantity, and delivery time must be satisfied. In the case of Korea, orders must be made in 0.1-m intervals with a minimum quantity of 50 tons and a delivery time of two months or more. For example, rebar with a diameter of 25 mm and a length of 8.4 m can be obtained by special order in a quantity of 60 tons and a delivery time of two months.

3.2. Cutting Waste Minimization Process

As mentioned earlier, rebar combination by special length provides an opportunity to reduce cutting waste or trim loss more than by stock length. Therefore, unlike the previous studies, which performed CWM by stock length only, the CWM algorithms proposed in this paper perform an MStL on the rebar that is left after performing an MSpL, as shown in Figure 3.

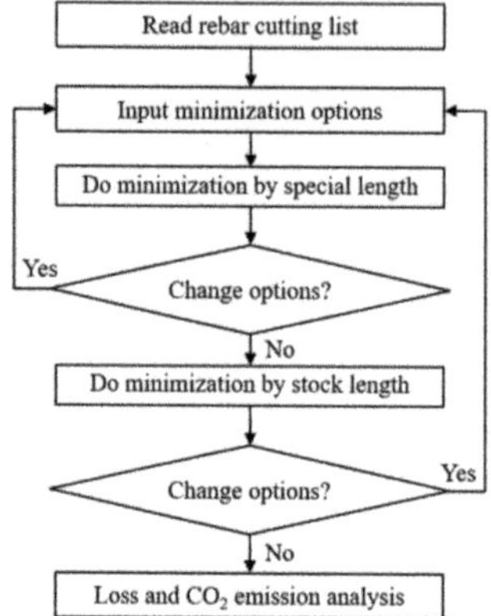

Figure 3. Cutting waste minimization process.

Figure 3 is described briefly as follows: (1) Read the rebar cutting list from the BIM [7] or computerized IPD system [17]; (2) Input options such as minimum and maximum lengths of rebar to be ordered, target loss rate, and minimum rebar quantity to be combined; (3) Execute the MSpL that satisfies the input options; (4) If the desired solution is not derived, decide whether to perform the MSpL again after mitigating options or perform an MStL; (5) If the desired solution cannot be derived from the MStL, decide whether to perform the optimization again by changing options. Otherwise, the process is terminated after analyzing the cutting waste and CO_2 emissions.

3.3. Cutting Waste Minimization Algorithm

In general, CWM is performed for stock lengths using the objective function shown in Equation (1) as introduced in several studies [5,14,17]. This is a mathematical formulation that minimizes the difference between the length of the cutting pattern (l_i) and the stock length (Lst_i) obtained by combining multiple rebars. In this case, the constraints of Equations (2) to (4) must be satisfied. For reference, in Equation (2), l_i corresponds to the demand length in Figure 2a, and r_1, r_2, .., r_n correspond to rebar$_1$, rebar$_2$, rebar$_n$. Equation (3) is not necessary if a single stock length is used, but it must be satisfied if several stock lengths are used. In the case of construction sites, the conditions of Equation (3) are generally valid because rebar of multiple market lengths can be purchased.

$$\text{Minimize } f(X_i) = \sum_{i=1}^{N} (Lst_i n_i - l_i n_i) / Lst_i n_i \tag{1}$$

$$\text{Subject to } l_i \leq Lst_i, \ l_i = r_1 + r_2 + \ldots + r_n \tag{2}$$

$$L_{min} \leq Lst_i \leq L_{max} 0 \tag{3}$$

$$< n_i, \text{ integer, } i = 1, 2, \ldots, N \tag{4}$$

Here,

Lst_i = Stock length of cutting pattern i (m)
l_i = Length of cutting pattern i obtained by combining multiple rebars, demand lengths (m)
n_i = Number of rebar combinations with the same cutting pattern i
r_i = Length of combined rebar (m)
L_{min} = Minimum length of rebar to be ordered (m)
L_{max} = Maximum length of rebar to be ordered (m).

So far, most rebar optimization studies in the construction sector have been conducted for MStL. This is because materials such as rebar, structural steel, pipes, and timber are supplied by the manufacturer in market lengths. If the target loss rate is added as a constraint to the CWM, Equation (5) should be used. In this case, the combination is executed only when the loss rate (ε) caused by the cutting pattern is less than or equal to the target loss rate (ε_t). When MSpL is performed, the loss rate can be further reduced but many algorithms have focused on MStL because of the complexity of the optimization algorithms.

$$\varepsilon = \frac{Lst_i - l_i}{Lst_i} \leq \varepsilon_t \tag{5}$$

Here,

ε_t = Target cutting waste or loss rate (%)
ε = Cutting waste or loss rate (%).

3.4. Minimization by Special Length

The mathematical formulation of CWM by special length is described in Equations (7)–(11), which is similar with previous studies [5,7,14]. Special lengths (Lsp_i) that satisfy constraints such as the target loss (scrap or waste) rate (ε_t) and minimum quantity for special order (Q_{so}) must be searched. In this case, the special length must be within the range of minimum (L_{min}) and maximum (L_{max}) lengths where special orders are possible.

$$\text{Minimize } f(X_i) = \sum_{i=1}^{N} (Lsp_i n_i - l_i n_i)/Lsp_i n_i \tag{6}$$

$$\text{Subject to } l_i \leq Lsp_i, \; l_i = r_1 + r_2 + \ldots + r_n \tag{7}$$

$$L_{min} \leq Lsp_i \leq L_{max}, \tag{8}$$

$$\varepsilon = \frac{Lsp_i - l_i}{Lsp_i} \leq \varepsilon_t, \tag{9}$$

$$Q_{so} \leq Q_{total}, 0 \tag{10}$$

$$< n_i, \text{ integer, } i = 1, 2, \ldots, N \tag{11}$$

Here,

l_i = Length of cutting pattern i obtained by combining multiple rebars, demand lengths (m)
Lsp_i = Special length of cutting pattern i that satisfies the target loss rate (m)
L_{min} = Minimum length of rebar to be ordered (m)
L_{max} = Maximum length of rebar to be ordered (m)
n_i = Number of rebar combinations with the same cutting pattern i
r_i = Length of combined rebar (m)
ε = Cutting waste or loss rate (%)
ε_t = Target cutting waste or loss rate (%)
Q_{total} = Total combined rebar quantity (ton)
Q_{so} = Minimum rebar quantity to be special ordered (ton).

For example, in the case where the loss rate is less than 2%, the length is between 8 and 12 m at intervals of 0.1 m, but the total quantity (Q_{total}) of the same length that will be more than 50 tons is searched. The MSpL that satisfies these conditions proceeds with the process shown in Figure 4. The minimization process in that figure is described in pseudocode, as follows:

(1) After the rebar cutting list is read, in which the number of reinforcing bars by diameter and length is counted, it is sorted in descending order with length and number priority. This is for efficient performance of the quantity–priority combination.

(2) Options such as the maximum (L_{max}) and minimum (L_{min}) lengths of rebar to be ordered, target loss rate (ε_t), and minimum rebar quantity (Q_{so}) to be special ordered are entered. If the target

loss rate is not entered, the combination that satisfies the condition of Q_{so} with a special length priority is executed by default.

(3) The rebar combination (l_i) that satisfies $L_{min} \leq Lsp_i \leq L_{max}$ for rebar of the same diameter is executed in descending order from the maximum length (L_{max}) of rebar to be ordered. If $(Lsp_i - l_i)/Lsp_i \leq \varepsilon_t$ is satisfied, the next combination $(i = i + 1)$ is executed until the end of the list after saving the result of combination, or the combination is performed until the loss rate condition is satisfied. This is because executing the combination in descending order from the maximum length is effective in performing the quantity–priority combination, as described in step (1).

(4) Next, the total quantity of combined rebar is calculated by Equation (12).

$$Q_{total} = w \sum_{i=1}^{N} Lsp_i n_i \tag{12}$$

here, w = unit weight of combined rebar per meter (ton/m).

(5) If $Q_{so} \leq Q_{total}$ is not satisfied, MSpL is repeated while Lsp_i is decreased by 0.1 m until $Lsp_i \leq L_{min}$ is satisfied. If a solution that satisfies the constraints is not found in the process so far, it should be decided whether to perform the minimization again after alleviating the combination conditions. Otherwise, MStL must be subsequently performed.

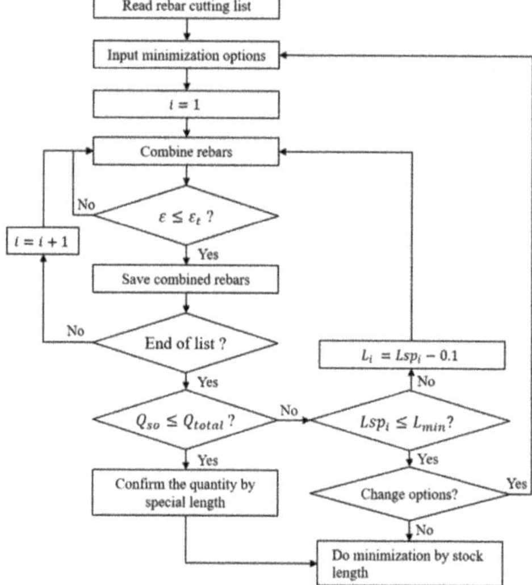

Figure 4. Minimization process by special length.

If $Q_{so} \leq Q_{total}$ is satisfied, the quantity of special length is determined and MStL is executed for the remaining rebar. As all of the rebar is not combined by special length, minimization is performed with stock lengths for the remaining ones.

4. Verification of CWM Algorithms

4.1. Brief Description of the Case Project

The effectiveness of CWM algorithms by stock and special lengths described so far should be verified through case application. To this end, the case project shown in Table 1 was selected in this study. This is a commercial building project constructed in Seoul, Korea, with a total floor area of 66,644 m^2, three basement levels, and 20 floors above ground. The site area of the case project is not large enough for rebar work on site. Moreover, considering the quality, time, and safety, the rebar was supplied from the plant.

Table 1. Description of the case project.

Description	Contents
Location	Seoul, Korea
Site area	8832 m^2
Building area	3970 m^2
Total floor area	66,644 m^2
No. of floors	B3 to F20
Structure	Basement: SRC Superstructure: RC

Reviewing the structure of the case building, the underground structure is steel reinforced concrete (SRC) and the superstructure is reinforced concrete (RC). In addition, the first and second floors are designed as a column-and-beam structure, and as shown in Figure 5, from the 3rd floor to the 20th floor, it is designed as a flat slab structure. That is, the case building includes three types of structures. For effective verification of the proposed CWM algorithms, as shown in Figure 5, the case application is performed on the flat slab structure from the 3rd floor to the 20th floor, which is the largest part of the building.

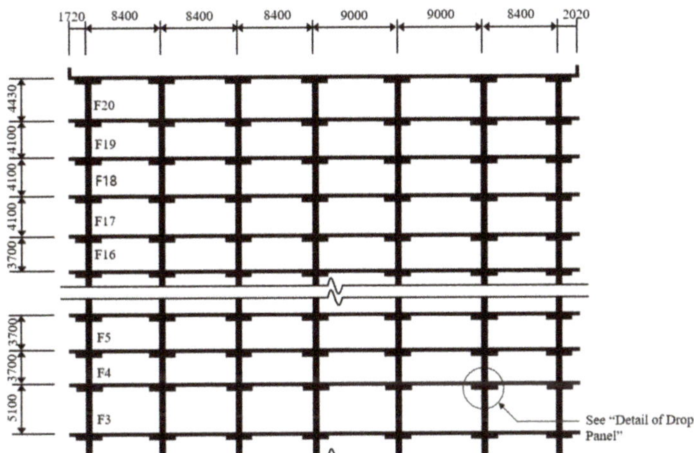

Figure 5. Structural frame of the case building.

The flat slab structure of the case project is composed of columns, slabs, and drop panels. Therefore, as shown in Figures 5 and 6, the top of each column is reinforced by drop panels. Figure 6a is the sectional detail of the drop panel that is most frequently applied to the case structure, and below the drop panel, 11-D16s are reinforced at 300-mm intervals as shown in Figure 6b. In the case of the slab part, D13 is installed at 300-mm intervals in both directions on the upper and lower sides as shown in Figure 6c. Furthermore, at the top of the drop panel, D16 is additionally reinforced at 300-mm intervals over the width of the column strip. As shown in Figure 6a, the slab thickness is 250 mm, and the drop panel thickness is 450 mm (200 mm thicker than the slab). For reference, the cross-section of deformed bars is variously marked in many countries as Y, H, D, etc. In the case of this paper, it is denoted by D (Deformed bar), which is commonly used in Korea.

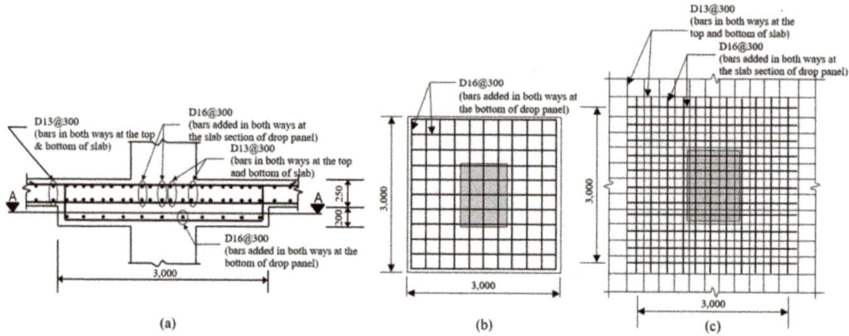

Figure 6. Detail of drop panel reinforcement: (**a**) Sectional detail of drop panel; (**b**) Section A-A; (**c**) Detail at the slab part of drop panel.

For the columns of the case project, as shown in Figures 5 and 7, all of them, including C3, have four sections with different reinforcements, such as 900 × 1200, 800 × 1000, and 600 × 1000, from F3 to F20. This is because the design was optimized according to the change in the load condition of each floor. As shown in Figure 7, the main bars are designed to have 26, 16, and 14 deformed bars with a diameter of 25 mm, gradually decreasing in number. Additionally, the sizes and combinations of tie bars and hoops designed for buckling vary from 5-D10 at F3 to F10, to 2-D10 at F14 to F20, as shown in Figure 7. For reference, the 5-D10 tie bars and hoops consist of five deformed tie bars and one hoop with a diameter of 10 mm. The columns shown in Figures 5 and 7 are connected by mechanical couplers, so there is no splice lapping. Therefore, according to the cross-sectional change, the rebar that is installed continuously in the upper and lower columns is connected by couplers, but the rest of it is anchored to the upper column. The case building has less rebar than the structure size because, in order to reduce the cross-sections of the structural members, super-high-tensile deformed (SHD) bars with a yield strength of 500 MPa were used for D10 and D13, and ultra-high-tensile deformed (UHD) bars with a yield strength of 600 MPa were used for D16, D19, and D25.

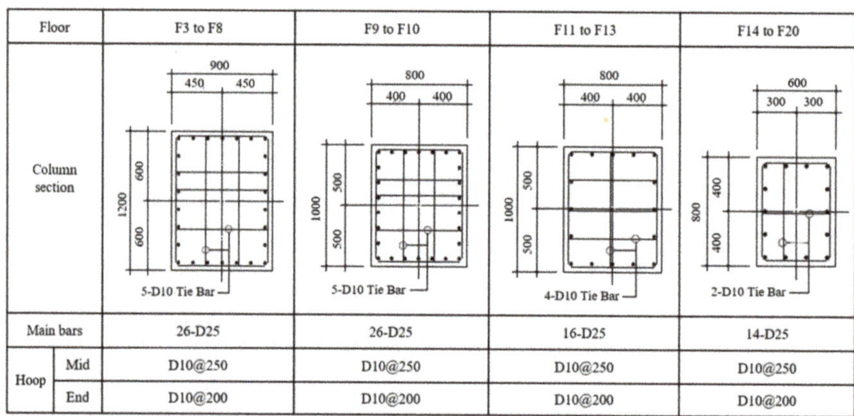

Floor	F3 to F8	F9 to F10	F11 to F13	F14 to F20
Column section				
Main bars	26-D25	26-D25	16-D25	14-D25
Hoop — Mid	D10@250	D10@250	D10@250	D10@250
Hoop — End	D10@200	D10@200	D10@200	D10@200

Figure 7. Rebar details of column C3.

4.2. Application of CWM Algorithms

In this study, for the verification of the proposed algorithm, rebar combinations were performed on structural frames from F3 to F20. The rebar cutting list generated from the bar-bending schedule was used for rebar information. At the case site, various diameters of rebar were used. For example, in the case of the column in Figure 7, 25-mm-diameter rebar was used for the main bar, and 10-mm-diameter rebar was used for the hoop. Tables 2 and 3 show the combination results of the CWM algorithms for the main bars of all of the columns.

Table 2. Combination report for special lengths of 25-mm diameter rebar.

Combination Report for Special Lengths						
Diameter (mm) = 25			Reference files = proj101_bcl.dat			
Combination conditions			Min. length (m) = 6.0	Max. length = 10.0		
			Min. weight (ton) = 50	Max. loss rate (%) = 3.0		
Cutting Pattern	No. of rebars	Combined Length (m)	Order Length (m)	Combined Weight (ton)	Order Weight (ton)	Loss Rate (%)
S1	6121	7400	7400	176.20	176.20	0.00
S2	1196	9200	9200	42.80	42.80	0.00
S3	526	8.530	9200	17.45	18.82	7.85
Sum				236.45	237.82	0.58

Table 3. Combination report for stock lengths of 25-mm-diameter rebar.

Combination Report for Stock Lengths						
Diameter (mm) = 25			Reference files = proj101_bcl.dat			
Combination conditions			Min. length (m) = 6.0	Max. length = 10.0		
			Min. weight (ton) =	Max. loss (%) = 3.0		
Cutting Pattern	No. of Rebars	Combined Length (m)	Stock Length (m)	Combined Weight (ton)	Stock Weight (ton)	Loss Rate (%)
N1	48	8.860	9.000	1.65	1.68	1.58
Sum				1.65	1.68	1.58

Table 2 shows the results of minimization by MSpL according to Equations (6) to (11), and the final loss rate, i.e., cutting waste rate, was calculated to be 0.58%. A detailed description of Table 2 is as follows: (a) Combination is performed on the 25-mm rebar in the bar-cutting list file named "proj101_bcl.dat": (b) The minimum quantity of special length is 50 tons after combination for rebar with a minimum length of 6.0 m and a maximum length of 10.0 m, and the maximum loss rate is not specified as 3.0%; (c) Cutting pattern S1 has the same combined and order lengths of 7.4 m, so the loss rate is zero for the order quantity of 176.2 tons; (d) The combined and order lengths of the cutting pattern S2 are equal to 9.2 m, so the loss rate is zero for the order quantity of 42.8 tons; (e) Finally, in cutting pattern S3, the combined length is 8.53 m, but in order to satisfy the minimum order weight of 50 tons, 18.82 tons must be ordered with a length of 9.2 m, in which case the loss rate is increased to 7.85%. However, as shown in Table 2, the quantity of S3 is relatively small compared with S1 and S2. Therefore, the loss rate of the final MSpL is 0.58%, which corresponds to 1.37 tons of cutting waste.

For reference, in the case structural frame, 25-mm-diameter rebar was used for the columns only. Additionally, there were not many cutting patterns because there were many main rebars of the same length. In other words, as shown in Figure 5, the lengths of the main rebars of all of the columns on the same floor were the same and the length was changed according to the change in floor height. The total number of main rebars in the case frame was 15,734, which were identified in five lengths.

Table 3 shows the results of MStL by Equations (1) to (4), and the final loss rate was calculated as 1.58%. In Table 2, the cutting pattern is combined into one because it is performed on the remaining rebar after combination by special length. For cutting pattern N1, the combined length is 8.86 m and the stock length is 9.0 m. The combined and stock weight are 1.65 and 1.68 tons, respectively, and the loss rate is 1.58%. For reference, in the case of MStL, the combination conditions are the same as for MSpL but the minimum weight is not specified. This is because it is assumed that there is a sufficient quantity in stock.

Comparing the results of Tables 2 and 3, the minimization results by special length have a lower loss rate than by stock length. This is because combination by special length is performed to further reduce the loss rate.

Table 4 shows the results of applying CWM algorithms to all types of rebar used in the case structural frame in Figures 5–7. Five diameters of rebar were used, and a total loss rate of 0.96% was calculated. The total quantity of rebar required for construction is 1.807.45 tons, and the quantity to be supplied in special and stock lengths is 1.824.75 tons. The loss rate is different for each diameter of rebar depending on the design characteristics of the structural member in which each rebar is used. The details are as follows.

Table 4. Combination report by rebar size.

Description	Unit	D10	D13	D16	D19	D25	Sum
Combined weight (C)	ton	335.62	899.93	259.94	73.86	238.10	1.807.45
Supply weight (S)	ton	338.29	906.03	264.01	76.92	239.50	1.824.75
Loss rate (S–C)/S	%	0.80	0.68	1.57	4.14	0.59	0.96

D10, D13, and D16 are mostly rebars that are repeatedly used in various structural members such as slabs, staircase walls, stairs, hoops, and drop panels. In those applications, many rebars of the same length are placed in the same type of structural members. In addition to these characteristics, small-diameter rebar left after cutting for primary use can be used for various other purposes. For example, they can be used as diagonal bars for crack reinforcement, reinforcement of openings, etc. Therefore, the cutting waste rate is lower than that of large-diameter rebar. Moreover, the combined weight was sufficient to cover the various cutting patterns, so MSpL and MStL using the CWM algorithms were performed smoothly.

In the case of D19, MStL was performed because there was no combination that met the minimum weight for a special order (50 tons). As a result, the cutting waste rate increased. Lastly, in the case of D25, most of the rebar was combined with the MSpL algorithm, as described in Tables 2 and 3, so the cutting waste rate was the lowest.

4.3. Comparison of Actual and Optimized Rebar Quantities

In order to verify the effectiveness of CWM algorithms, actual and optimized rebar quantities must be compared. As shown in Table 5, the actual quantity of rebar in the case structural frame is 1.942.05 tons and the quantity optimized by the CWM algorithms is 1.824.75 tons. As a result, 117.30 tons were saved, which is 6.04% of the actual quantity. It can be seen from this table that the reduction rate differs significantly for each diameter of rebar.

Table 5. Comparison of actual and optimized rebar quantities.

Description	Unit	D10	D13	D16	D19	D25	Sum
Actual (A)	ton	377.00	952.43	276.00	87.80	248.82	1.942.05
Optimized (O)	ton	338.29	906.03	264.01	76.92	239.50	1.824.75
Quantity reduction (A–O)	ton	38.71	46.40	11.99	10.88	9.32	117.30
Reduction rate (A–O)/A	%	10.27	4.87	4.34	12.39	3.75	6.04

As described above, small-diameter rebar had more secondary use after cutting. Therefore, it is common that less cutting waste is generated with small-diameter rebar. However, in Table 5, the 10.27% reduction rate for D10 is higher than that of D13 and D16, which means that the loss rate of the actual quantity of rebar is higher than the optimized one. The high quantity reduction rate calculated after optimization by the CWM algorithms means that the loss rate of the actual rebar quantity is high. The reason is presumed to be a problem with the rebar work management. In addition, Table 5 confirms a relatively small loss rate for D13 and D16. This is because there are many rebars with the same lengths repeatedly used in structural members such as slabs, staircase walls, and drop panels.

In the case of D19, it is confirmed that the cutting waste is increased because the quantity required for the work is relatively small and there are not many rebars of the same length placed repeatedly. However, it is confirmed that a cutting waste reduction of 12.39% can be obtained using the CWM algorithms proposed in this study. Lastly, in the case of D25, it is used for the main rebar of the columns, and it is relatively easy to manage the rebar to reduce cutting waste because there are not many changes in length. Therefore, it is confirmed that the quantity reduction of this rebar by optimization is 3.75%, which is smaller than that of the other types of rebar, as shown in Table 5. For reference, when the reduced rebar quantity of 117.3 tons is converted into money, it is about USD 98,976 including material, cutting and bending, and placement costs.

4.4. CO_2 Emission Reduction Effects

When using the CWM algorithms proposed in this study, the contribution to sustainable construction should be confirmed. For this, Table 6 shows the quantitative calculation of the CO_2 emissions for rebar saved by the algorithms. Substituting 3.466 ton-CO_2/ton [49], the unit CO_2 emissions of high-tensile deformed bar published by the Korea Institute of Construction Technology (KICT), show that the CO_2 emissions from the actual rebar work and the optimized result are calculated to be 6.731.15 and 6.324.58 tons, respectively. For reference, the LCI DB (Life Cycle Inventory Database) varies by country, and this study cited the data presented in the research report of the government-funded research institute, KICT. In addition, because the LCI DB for SHD and UHD used in the case project has not been officially provided, the unit CO_2 emission data for high-tensile deformed bar are cited in this paper.

Table 6. Calculation of CO_2 emission reduction effect.

Description	Quantity (ton)	Unit CO_2 Emission (ton-CO_2/ton)	Amount (ton-CO_2)
Actual (A)	1.942.05	3.466	6.731.15
Optimized (O)	1.824.75	3.466	6.324.58
Reduction effect (A–O)	117.30		406.60

As shown in Table 6, when the CWM algorithms are applied, it can be seen that the case project has a CO_2 emission reduction of 406.60 tons which is equivalent to 6.04% of the structure. As mentioned in the introduction, in the case of buildings, the structure accounts for about 65% of building GHGs [3]. Considering this reference, there is a CO_2 emission reduction of 3.93% based on the whole building. From a carbon footprint point of view, the embodied CO_2 per unit weight or volume of rebar is about 9.02 times that of concrete [4]; therefore, the CO_2 emission reduction produced by the CWM algorithms has a great effect on sustainable construction.

It is necessary to confirm the cost reduction effect by converting the CO_2 emission reduction effect to the carbon price. To this end, the cost savings of USD 20,330 can be confirmed when applying the Korean carbon transaction price of USD 50/ton-CO_2 [50] announced by the Carbon Disclosure Project (CDP). When this amount is added to the previously calculated savings of USD 98,976 in construction cost, a total savings of USD 119,306 is confirmed. Similar to LCI DB, the annual price of carbon traded by CDP varies by country. According to CDP data, in the case of Korea, the price was USD 64/ton-CO_2 in 2016 and USD 50/ton-CO_2 in 2017, which is USD 14/ton-CO_2 lower than the previous year.

The proposed algorithms were applied to the 3rd to 20th floors, designed as a flat slab structure, which is part of the case project. The amount of rebar used in the entire structural frame of the case project was found to be 3.444.06 tons. Therefore, if the CWM algorithms proposed in this study are applied to the entire structural frame, greater CO_2 emission and cost reductions are expected.

5. Discussion

In this study, as shown in Figure 3, we proposed an algorithm that performs MStL on the rebar that remains after special length minimization. With this algorithm, cutting waste or trim loss is further reduced because, as illustrated in Figure 2, the special length is combined at 0.1-m intervals, unlike the stock length, which is generally combined at 1-m intervals. This is also confirmed by the results shown in Tables 2 and 3.

Through this study, we confirmed that additional in-depth studies are needed on these two issues. First, it is possible to combine all of the rebar for one project at the same time, but the results may not be practical. For example, rebar placed on the first and 20th floors can be combined. In this case, the inventory management cost may be high because there is a significant time difference between the use of rebar on the 1st floor and the use of the remaining rebar on the 20th floor. Therefore, the combination condition for rebar to be used within a certain time must be added. For example, the condition for performing a combination of rebar to be used within two weeks should be added. So far, most of the papers related to CSP do not consider the time factor. If the required rebar information can be obtained automatically from the BIM [7] or from an integrated project delivery system [17] linked to the schedule, this problem can be easily solved.

Existing CSP-related studies, including this paper, use original rebar information generated after the structural design. In this case, there are different amounts of rebar of various lengths, and numerous combinations are repeated to search for solutions. Additionally, as mentioned above, rebar scattered in various locations are combined. In addition to cutting patterns that are difficult to apply practically, cutting waste rates cannot be reduced below a certain level [5]. However, it was confirmed that near-zero cutting waste could be achieved by realigning the rebar in the drawings created after the structural design in special lengths using heuristic algorithms. It was also confirmed that heuristic algorithms would be more efficient than mathematical algorithms in performing rebar realignment. Therefore, further studies on the rebar alignment algorithm for near-zero cutting waste should be performed for sustainable construction.

During the case study, it was confirmed that significant efforts have been made from the structural design stage to increase the productivity of rebar work and reduce the rebar loss rate. For example, it is common for some Korean companies to use 500- or 600-MPa super- or ultra-tensile bars instead of 400-MPa high-tensile deformed bar, and to use couplers to connect rebar more than 20 mm in diameter. The goal of near-zero cutting waste for sustainable construction is expected to be achieved if heuristic rebar alignment algorithms are applied along with these efforts.

6. Conclusions

Efforts to reduce carbon and climate change risk are being carried out globally and in all industries. In particular, rebar, which has the most ECO_2 per unit weight in built environments, generates a significant amount of cutting waste in the construction phase. Therefore, there is not only the cost of rebar construction but also a considerable amount of CO_2 emissions to be expected. To solve this issue, we proposed rebar CWM algorithms for sustainable construction. The effectiveness of the proposed algorithms was verified through a case project, and the following results were obtained.

First, in the case of the optimization of D25 rebar, the cutting waste rate for special lengths was 0.58%, whereas that for stock lengths was 1.58%. This proved the assumption that combination by special length reduced the loss rate more than combination by stock length.

Furthermore, although the actual quantity of rebar put into the case project was 1942.05 tons, the quantity optimized by the proposed algorithms was 1824.75 tons, which represented a quantity reduction of 117.3 tons. This corresponds to 6.04% of the actual quantity and a savings of USD 98,976 in construction costs.

In addition, CO_2 emissions by the proposed optimization algorithms compared with actual emissions had a reduction of 406.6 ton-CO_2. This corresponds to a CO_2 emission reduction of 3.93% based on the whole building, reflecting that the structure accounts for about 65% of building GHGs [3].

This is a savings of USD 20,330 based on the carbon trade price in Korea, and a total savings of USD 119,306, including a reduction in construction costs. The quantity of rebar used in the entire building of the case project, including the flat slab structure on the 3rd to 20th floors, was found to be 3444.06 tons. If the proposed algorithms had been applied to the entire building, further CO_2 and cost savings would have been expected.

These results confirmed that the proposed CWM algorithms worked as an effective tool for sustainable construction. During this study, it was observed that near-zero cutting waste could be achieved by realigning the rebar in the structural drawings to special lengths. In other words, it was observed that repositioning rebar of a certain length while satisfying the structural design criteria might significantly reduce cutting waste. In order to do this efficiently, heuristic algorithms of a new concept rather than the mathematical algorithms proposed in this study should be developed in the future.

Author Contributions: Conceptualization, S.K.; methodology, S.K.; validation, D.L., S.S. and D.K.; formal analysis, D.L., S.S. and S.K.; investigation, D.K.; data curation, D.K.; writing—original draft preparation, D.L. and S.K.; writing—review and editing, D.L. and S.K.; visualization, S.S.; supervision, S.K.; project administration, S.K.; funding acquisition, S.K. All authors have read and agreed to the published version of the manuscript.

Funding: This work was supported by the National Research Foundation of Korea (NRF) grant funded by the Korean government (MOE) (no. 2017R1D1A1B04033761).

Acknowledgments: The authors thank SK E&C for providing the rebar data of the case project to verify the CWM algorithms.

Conflicts of Interest: The authors declare no conflict of interest. The funders had no role in the design of the study; in the collection, analyses, or interpretation of data; in the writing of the manuscript, or in the decision to publish the results.

References

1. International Energy Agency and the United Nations Environment Programme. 2018 Global Status Report: Towards a Zero-Emission, Efficient and Resilient Buildings and Construction sector. 2018, p. 9. Available online: https://www.worldgbc.org/sites/default/files/2018%20GlobalABC%20Global%20Status%20Report.pdf (accessed on 19 April 2020).

2. Pacheco-Torgal, F.; Cabeza, L.; Labrincha, J.; De Magalhaes, A. *Eco Efficient Construction and Building Materials: Life Cycle Assessment (LCA), Eco-Labelling and Case Studies*; Elsevier: Cambridge, UK, 2014; Volume 1, pp. 624–630.

3. Clark, D.; Bradley, D. *Information Paper—31: Embodied Carbon of Steel Versus Concrete Buildings*; Cundall Johnston & Partners LLP: Newcastle, UK, 2013; p. 4.

4. Lee, I.J.; Yu, H.; Chan, S.L. Carbon Footprint of Steel-Composite and Reinforced Concrete Buildings, Standing Committee on Concrete Technology Annual Concrete Seminar 2016, Hong Kong, 20 April 2016; Construction Industry Council. Available online: https://www.devb.gov.hk/filemanager/en/content_971/7_Carbon_Footprint_for_Steel_Composite_and_Reinforced_Concrete_Buildings.pdf (accessed on 22 April 2020).

5. Kim, S.K.; Kim, M.H. A Study on the development of the optimization algorithm to minimize the loss of reinforcement bars. *J. Archit. Inst. Korea* **1991**, *7*, 385–390.

6. Kim, G.H. A Study on Program of Minimizing the Loss of Re-Bar. Master's Thesis, Korea University, Seoul, Korea, 2002; pp. 8–58.

7. Porwal, A.; Hewage, K.N. Building information modeling-based analysis to minimize waste rate of structural reinforcement. *J. Constr. Eng. Manag.* **2012**, *138*, 943–954. [CrossRef]

8. Poonkodi, N. Development of software for minimization of wastes in rebar in rcc structures by using linear programming. *Int. J. Adv. Res. Trends Eng. Technol. (IJARTET)* **2016**, *3*, 1262–1267.

9. Nadoushani, Z.; Hammad, A.; Akbar Nezhad, A. A Framework for Optimizing Lap Splice Positions within Concrete Elements to Minimize Cutting Waste of Steel Bars. In Proceedings of the 33th International Symposium on Automation and Robotics in Construction (ISARC 2016), Auburn, AL, USA, 21 July 2016. [CrossRef]

10. Shahin, A.A.; Salem, O.M. Using genetic algorithms in solving the one-dimensional cutting stock problem in the construction industry. *Can. J. Civ. Eng.* **2004**, *31*, 321–332. [CrossRef]

11. Chandrasekar, M.K.; Nigussie, T. Rebar Wastage in Building Construction Projects of Hawassa, Ethiopia. *Int. J. Sci. Eng. Res.* **2018**, *9*, 282–287.

12. Nadoushani, Z.S.M.; Hammad, A.W.; Xiao, J.; Akbarnezhad, A. Minimizing cutting wastes of reinforcing steel bars through optimizing lap splicing within reinforced concrete elements. *Constr. Build. Mater.* **2018**, *185*, 600–608. [CrossRef]

13. Zubaidy, S.S.; Dawood, S.Q.; Khalaf, I.D. Optimal utilization of rebar stock for cutting processes in housing project. *Int. Adv. Res. J. Sci.* **2016**, *3*, 189–193. [CrossRef]

14. Kim, S.K.; Hong, W.K.; Joo, J.K. Algorithms for reducing the waste rate of reinforcement bars. *J. Asian Archit Build.* **2004**, *3*, 17–23. [CrossRef]

15. 44/12 Benjaoran, V.; Bhokha, S. Trim loss minimization for construction reinforcement steel with oversupply constraints. *J. Adv. Manag. Sci.* **2013**, *1*, 313–316. [CrossRef]

16. Benjaoran, V.; Bhokha, S. Three-step solutions for cutting stock problem of construction steel bars. *KSCE J. Civ. Eng.* **2014**, *18*, 1239–1247. [CrossRef]

17. Kim, D.; Lim, C.; Liu, Y.; Kim, S. Automatic Estimation System of Building Frames with Integrated Structural Design Information (AutoES). *Iranian J. Sci. Tech. Trans. Civ. Eng.* **2019**, 1–13. [CrossRef]

18. Hwang, J.W.; Park, C.J.; Wang, S.K.; Choi, C.H.; Lee, J.H.; Park, H.W. A Case Study on the Cost Reduction of the Rebar Work through the Bar Loss Minimization. In Proceedings of the KIBIM Annual Conference 2012, Seoul, Korea, 19 May 2012; Volume 2, pp. 67–68.

19. Khalifa, Y.; Salem, O.; Shahin, A. Cutting Stock Waste Reduction using Genetic Algorithms. In Proceedings of the 8th Annual Conference on Genetic and Evolutionary Computation, Seattle, WA, USA; 2006; pp. 1675–1680. [CrossRef]

20. Salem, O.; Shahin, A.; Khalifa, Y. Minimizing cutting wastes of reinforcement steel bars using genetic algorithms and integer programming models. *J. Constr. Eng. Manag.* **2007**, *133*, 982–992. [CrossRef]

21. Gilmore, P.C.; Gomory, R.E. A linear programming approach to the cutting-stock problem. *Oper. Res.* **1961**, *9*, 849–859. [CrossRef]

22. Nanagiri, Y.V.; Singh, R.K. Reduction of wastage of rebar by using BIM and linear programming. *Int. J. Tech.* **2015**, *5*, 329–334. [CrossRef]

23. Zheng, C.; Yi, C.; Lu, M. Integrated optimization of rebar detailing design and installation planning for waste reduction and productivity improvement. *Autom. Constr.* **2019**, *101*, 32–47. [CrossRef]

24. Zheng, C. Multi-Objective Optimization for Reinforcement Detailing Design and Work Planning on a Reinforced Concrete Slab Case. Master's Thesis, Alberta University, Edmonton, Canada, 2018. [CrossRef]

25. Zheng, C.; Lu, M. Optimized reinforcement detailing design for sustainable construction: Slab case study. *Procedia Eng.* **2016**, *145*, 1478–1485. [CrossRef]

26. Kantorovich, L.V. Mathematical methods of organizing and planning production. *Manag. Sci.* **1960**, *6*, 366–422. [CrossRef]

27. Ben Amor, H.; Valério de Carvalho, J. Cutting stock problems. In *Column Generation*; Springier: Boston, MA, USA, 2005; pp. 131–161. [CrossRef]

28. Cutting Stock Problem. Available online: https://en.wikipedia.org/wiki/Cutting_stock_problem (accessed on 20 April 2020).

29. Arbib, C.; Marinelli, F.; Ventura, P. One-dimensional cutting stock with a limited number of open stacks: Bounds and solutions from a new integer linear programming model. *Int. Trans. Oper. Res.* **2016**, *23*, 47–63. [CrossRef]

30. Berberler, M.; Nuriyev, U.; Yildırım, A. A Software for the one-dimensional cutting stock problem. *J. King Saud Univ. Sci.* **2011**, *23*, 69–76. [CrossRef]

31. Belov, G.; Scheithauer, G. Setup and open-stacks minimization in one-dimensional stock cutting. *INFORMS J. Comput.* **2007**, *19*, 27–35. [CrossRef]

32. Dyckhoff, H. A new linear programming approach to the cutting stock problem. *Oper. Res.* **1981**, *29*, 1092–1104. [CrossRef]

33. Feifei, G.; Lin, L.; Jun, P.; Xiazi, Z. Study of One-Dimensional Cutting Stock Problem with Dual-Objective Optimization. In Proceedings of the International Conference on Computer Science and Information Processing (CSIP), Xi'an, Shaanxi, China, 24–26 August 2012. [CrossRef]

34. Fernandez, L.; Fernandez, L.A.; Pola, C. Integer Solutions to Cutting Stock Problems. In Proceedings of the 2nd International Conference on Engineering Optimization, Lisbon, Portugal, 6–9 September 2010.

35. Gilmore, P.C.; Gomory, R.E. A linear programming approach to the cutting-stock problem-part II. *Oper. Res.* **1963**, *11*, 863–888. [CrossRef]
36. Goulimis, C. Optimal solutions for the cutting stock problem. *Eur. J. Oper. Res.* **1990**, *44*, 197–208. [CrossRef]
37. Haessler, R.W.; Sweeney, P.E. Cutting stock problems and solution procedures. *Eur. J. Oper. Res.* **1991**, *54*, 141–150. [CrossRef]
38. Haessler, R.W. Controlling cutting pattern changes in one-dimensional trim problems. *Oper. Res.* **1975**, *23*, 483–493. [CrossRef]
39. Jahromi, M.H.; Tavakkoli, R.; Makui, A.; Shamsi, A. Solving an one-dimensional cutting stock problem by simulated annealing and tabu search. *J. Ind. Eng. Int.* **2012**, *8*, 24. [CrossRef]
40. Lin, P. Optimal Solution of One Dimension Cutting Stock Problem. Master's Thesis, Lehigh University, Bethlehem, PA, USA, 26 April 1994.
41. Roodman, G.M. Near-optimal solutions to one-dimensional cutting stock problems. *Comput. Oper. Res.* **1986**, *13*, 713–719. [CrossRef]
42. Vahrenkamp, R. Random search in the one-dimensional cutting stock problem. *Eur. J. Oper. Res.* **1996**, *95*, 191–200. [CrossRef]
43. Chen, C.; Hart, S.; Tham, W. A simulated annealing heuristic for the one-dimensional cutting stock problem. *Eur. J. Oper. Res.* **1996**, *93*, 522–535. [CrossRef]
44. Cherri, A.C.; Arenales, M.N.; Yanasse, H.H. The one-dimensional cutting stock problem with usable leftover—A heuristic approach. *EUR J. OPER RES.* **2009**, *196*, 897–908. [CrossRef]
45. Cui, Y.; Zhao, X.; Yang, Y.; Yu, P. A heuristic for the one-dimensional cutting stock problem with pattern reduction. *Proc. Inst. Mech. Eng. Part B: J. Eng. Manuf.* **2008**, *222*, 677–685. [CrossRef]
46. Gradišar, M.; Kljajić, M.; Resinovič, G.; Jesenko, J. A sequential heuristic procedure for one-dimensional cutting. *Eur. J. Oper. Res.* **1999**, *114*, 557–568. [CrossRef]
47. Poldi, K.C.; Arenales, M.N. Heuristics for the one-dimensional cutting stock problem with limited multiple stock lengths. *Comput. Oper. Res.* **2009**, *36*, 2074–2081. [CrossRef]
48. Wang, Z.; Rao, G. A multi-stage genetic algorithm for one-dimensional cutting stock problems with one stock material length. *J. Syst. Sci. Inf.* **2005**, *3*, 643–651. [CrossRef]
49. KICT (Korea Institute of Construction Technology). The Environmental Load Unit Composition and Program Development for LCA of Building, The Second Annual Report of the Construction Technology R&D Program. 2004. Available online: http://www.ndsl.kr/ndsl/search/detail/report/reportSearchResultDetail.do?cn=TRKO201000018952 (accessed on 28 May 2020).
50. CDP Worldwide. Carbon Pricing Connect. Available online: https://www.cdp.net/en/climate/carbon-pricing/carbon-pricing-connect (accessed on 29 May 2020).

Article

Analysis of Musculoskeletal Disorders and Muscle Stresses on Construction Workers' Awkward Postures Using Simulation

Shraddha Palikhe [1], Mi Yirong [2], Byoung Yoon Choi [2] and Dong-Eun Lee [2,*]

[1]   Intelligent Construction Automation Center, Kyungpook National University, Daegu 41566, Korea;
    arpsharu@gmail.com
[2]   School of Architecture, Civil, Environment, and Energy Engineering, Kyungpook National University,
    Daegu 41566, Korea; miyirong@knu.ac.kr (M.Y.); jr1381@knu.ac.kr (B.Y.C.)
*   Correspondence: dolee@knu.ac.kr; Tel.: +82-53-950-7540

Received: 15 May 2020; Accepted: 8 July 2020; Published: 15 July 2020

Abstract: The negligence involved in musculoskeletal disorder (MSD) at construction sites results in high rates of muscle injuries. This paper presents findings identified by the MSD for each part of a worker's body, categorizing the awkward postures of each body part, estimating muscle stresses, and establishing the benchmark using anthropometry and hand force data. MSDs and their corresponding frequencies were identified by administering the Nordic Musculoskeletal Questionnaire (NMQ) survey, which solicits responses regarding construction workers' awkward postures. Musculoskeletal stresses were estimated using three-dimensional static strength prediction program (3D SSPP) biomechanical software. The new benchmarks were established for existing preventive measures using the anthropometry and hand force data. Workers suffering from different body muscle pains in awkward postures may be predicted using the compression forces magnitude, strength capability, and body balance. The model was verified by comparing its outputs with the survey analysis results. The study is of value to practitioners because it provided a means to understand the contemporary scenario of MSD and to establish a practical benchmark based on the physical capability of workers. It is relevant to researchers because it digitally predicts MSD and facilitates experimentation with different dimensions, thereby contributing to construction productivity improvement. Test cases validate the prediction method.

Keywords: musculoskeletal disorders; construction workers; muscle stress; standard Nordic questionnaire; awkward posture; simulation

---

1. Introduction

Construction is ranked as the most hazardous operation involving musculoskeletal disorders and injuries. MSDs are caused by sudden exertion or prolonged exposure to physical factors (i.e., high force, repetitive motion, awkward body posture, and vibration) and affects the muscles, nerves, tendons, joints, cartilage, and supporting structures of the upper and lower limbs, neck, and lower back, etc. [1]. MSDs are attributed to handling heavy manual materials, manipulating excessive and repetitive hand tools, performing repetitive screw motions, reinforcing works involving difficult postures, and so on [2]. When the working posture differs from the neutral posture in which the body is aligned and balanced while placing minimal stress on the muscles, tendons, nerves, and bones, the stress on the body parts (i.e., the muscles, tendons, joints, arms, hands, and shoulders) increases, resulting in awkward postures and/or movements of the body parts of the workers, in turn leading to a negative impact on the safety and health of the workers as well as on productivity. The percentage of construction workers exposed to the musculoskeletal hazards in Korea while carrying heavy loads, standing long, and maintaining tiring and painful positions, is about 72%, 83.8%, and 67.9%, respectively [3]. Herein, we identify the factors that either affect ergonomic interventions or reduce MSDs in construction workers (i.e., masons, pavers, and electricians) [4].

Existing studies provide ergonomic analysis methods that employ motion sensing and assessment tools [5] to alleviate MSDs or to implement preventive measures. However, the correlation between anthropometry and the magnitude of hand forces has not been well explored. A new ergonomic analysis method that identifies the correlation between these two will be beneficial to a construction administrator for estimating, say, the compression on the lower back, and for establishing a benchmark of the workload imposed on a worker using BMI and hand forces exerted in diverse working postures (i.e., pushing forward, lifting, stooping, and kneeling). Such estimations may contribute to securing labor safety and health by efficiently identifying competent workers from an ergonomic viewpoint for the given work task. Three dimensions of environment, society, and economy are frequently used to model how sustainability can be incorporated into one's mission, goals, and practices. However, the issues involved in the social dimension of sustainability (i.e., labor relations, diversity, workers benefits/compensation, human rights, the organization of work, etc.) have often been overlooked, resulting in negative impacts (i.e., hazards to workers and creating tension between goals). Many worker issues exist within the concept of sustainability. The proposed posture simulation and benchmarking approach, relevant to the workers' social issues that promotes labor welfare, safety, and health.

The research was conducted in five steps. First, the performance of existing ergonomic modeling and analysis methods for the construction industry was investigated through a literature review to identify new research contributions. Second, a set of Nordic Musculoskeletal Questionnaire surveys was administered to workers from various construction trades to identify the MSD issues of each trade. The ergonomic data, including the MSDs affecting various body parts of workers, were collected from workers engaged in bare-hand manual operations in four Korean construction sites. Third, the new ergonomic model that establishes a benchmark of the workload imposed on construction workers engaged in diverse working postures was implemented in an automated tool by mapping the survey findings into a three-dimensional static strength prediction program (3D SSPP) software. Fourth, the model performance was demonstrated using a set of working postures (i.e., pushing forward, lifting, stooping, and kneeling). The validity and effectiveness of the model were verified by performing a series of case studies, each of which the common awkward body postures were identified and the static strength and compressive forces attributed to each awkward posture were estimated using 3D SSPP. It was confirmed that the model established the benchmark using hand force and body mass index (BMI). Finally, the research contributions and limitations were examined. The material in this paper is organized in the same order. Indeed, the findings will be of help for construction administrators to understand MSD issues experienced by workers employed in a specific operation and will provide clues to identify those tasks that can be semi-automated or fully-automated for better benefits.

2. Current State of Musculoskeletal Disorder Studies

MSD is the highest contributor to global disability, accounting for 16% of all years lived with disability; lower back pain is the single leading cause (Global burden of disease, 2017). In South Korea, the percentage of workers aged 50 years or older was 25% in 2010 and this value is expected to exceed 33% in 2020 [6]. Workers suffering from MSDs include aged construction craftsmen exposed to severe vibrations, construction and mining technicians, and construction finishing workers (61.3%, 47.8%, and 46%, respectively). The prevalence of chronic MSDs and degradation of body parts attributed to aging may lead to decreased physical labor capability. The frequency of back pain, upper extremities, and lower extremities and fatigue are chronically high in construction workers, about 30.7%, 61.3%, 49.2%, and 35.6%, respectively [3]. Existing studies claim that, compared to young workers, aged workers are more likely to suffer from musculoskeletal symptoms.

Meanwhile, existing ergonomics analysis techniques may be classified into self-reporting, manual observation, direct sensing measurement, and vision-based analysis. Self-reporting is a data collection process that involves conducting interviews and web-based questionnaires [7]. Manual observation tools facilitate measuring and/or evaluating MSD via hybridizing body position and movement-tracking tools (e.g., assessment of repetitive task, Ovako working posture analysis system, posture activity tools and handling, rapid upper limb assessment, and rapid entire body assessment (REBA)) [8]. However, they lend themselves neither to a precise posture measurement nor to the recording of delicate movement patterns such as that possible in time-lapse video observation. In direct sensing measurement, various sensor(s) are attached to the body parts of workers; these approaches outperform the two approaches in terms of measurement accuracy. The accuracy may be augmented by hybridizing the measurement method with Microsoft Kinect Cameras for efficient real-time motion analysis [9,10]. Vision-based analysis allows for precise motion tracking along with biomechanical parameter measurements that use devices such as tapes and goniometers, microelectromechanical systems (MEMS), electrodes for electromyography (EMG), and magnetoresistive sensors. Although this method facilitates measuring joint angles, including the angle of the neck, it is cumbersome because construction workers must wear devices while working. Although it outperforms existing methods, methods based on such analysis are still far from the ideal [11].

Existing studies have identified that construction workers suffer from physical fatigue and muscle pain when exposed to excessive energy consumption, resulting in human error, unsafe actions, and productivity loss, etc. In addition, a few studies claim that practical methods that assess MSD risks to all parts of the body are necessary in construction, proposing new technologies that facilitate the identification of preventive measures involved in appropriate body posture [12]. However, they are not yet arrived in a maturity to proper implementation. MSD remains a substantial concern with considerable personal and societal burdens [13]. Indeed, it would be beneficial to enlighten MSD issues to the construction practitioners hybridizing posture simulation and survey methods. It may contribute to identifying human MSD along with a benchmarking approach that supports making preventive MSD tools for construction personnel.

For construction workers, the anthropometric traits and hand forces to which different body parts (i.e., the neck, shoulder, fingers, knee, and wrist) are subjected depend on the task types (i.e., overhead work, ground-floor-level work, and manual material-handling work) [14]. Several musculoskeletal injury prevention measures (e.g., site-specific ergonomics programs, engineering controls, mechanical devices, exercise programs) have been enforced to reduce the burden of manual-lifting hazards. The "best practices" do not cause pain and/or discomfort in the back and wrist, and were identified to increase the productivity [2,15]. These measures encourage the development of initiatives that analyze ergonomic hazards and implement site-specific mitigation strategies and practices. It will be beneficial to reengineer improvement techniques and to upgrade its dynamic condition against work-related MSDs. Existing studies provide methods to identify the body postures of workers and suggest their corresponding preventive measures. However, these studies did not deal with tracking transitory motion changes at an appropriate level of detail or modeling MSDs in a working environment. Therefore, construction safety still incurs considerable personal and socioeconomic burden. A new simulation modeling, analysis, and controlling tool that effectively handles the MSD issues faced by workers involved in a construction operation will be beneficial. A simulation model formulated based on worker survey results may contribute to the construction safety and health by establishing a benchmark for actioning MSD-prevention measures.

3. Materials and Methods

3.1. Research Method

The research method map is shown in Figure 1. Each stage of this research consisted of four "processes" with two "outputs," indicated by numbers (1) to (5). For each "process," a set of criteria (or standard mean) were used to identify the variables and develop a simulation model. First, the variables involved in the MSDs of workers (i.e., anthropometry and hand forces) were identified through comprehensive literature reviews. Second, comprehensive literature reviews were conducted to identify the variables involved in the MSDs of workers (i.e., anthropometry and hand forces). Third, the variables that influenced the MSD symptoms of workers were confirmed by surveying workers actively engaged in construction tasks. Anthropometry and hand force data for awkward postures were obtained from the NMQ survey. The outputs show that three motions (i.e., pushing forward, lifting, and kneeling) among the awkward postures manifested on a specific task deserve special attention. The justification for using these variables for MSDs was confirmed via a survey. Third, the data of these variables were used as the input parameters for simulation using 3D SSPP (Ver. 2017), an easy-to-use model that considers all variables together. This model estimates physical demands by considering input postures in a specified window frame, predicting the changes physical demands as the workers shift from one posture to another, capturing and saving pictures of each awkward posture, creating a digital twin of virtual workers using the photos, duplicating postures, and calculating the lower back compression and body balance. In addition, the anthropometric data, hand load measured for each construction task, and loads obtained from workers' experience, were mapped into the modeled virtual workers. The validity of the survey output was confirmed by comparing it with the simulation output data from a series of simulation experiments. The preventive measures to reduce MSDs were discussed using the obtained static strength for postures. Body balance was assessed by computing the center-of-pressure (COP) and evaluating the location of the COP projected onto the floor while taking into account the limits of the functional stability region using 3D SSPP. Fifth, the benchmark was established according to BMI and corresponding hand forces by changing the magnitude of the hand forces while keeping the body weight constant. In addition, the model was tested under several different sets of variables to estimate lower back compression, percentage of accurate predictions, and body balance. Finally, the contributions and limitations of the model and suggestions for its improvement were discussed.

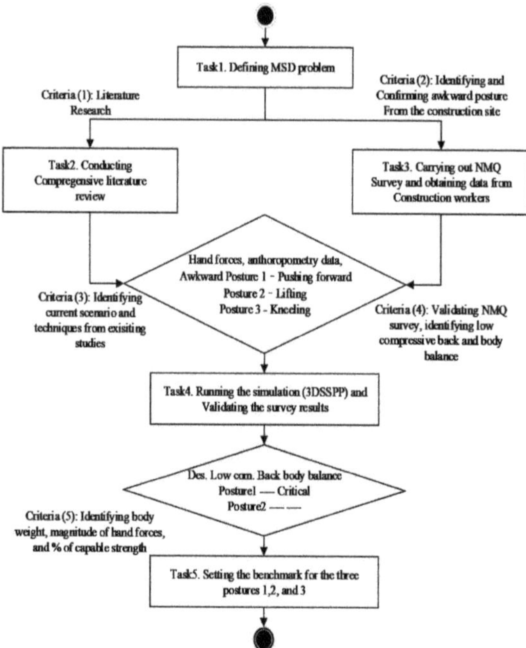

Figure 1. Research method.

3.2. Administering the Nordic Musculoskeletal Questionnaire (NMQ)

The NMQ provides a structured and standardized interview method considering the lower back, neck, and shoulder, studying general complaints from an epidemiological perspective. Its validity and reliability are well accepted in the field of MSD study [16]. The standard questionnaire consists of two parts. One is a general questionnaire of 40 forced-choice items that identify the body parts suffering from musculoskeletal problems; the other is a supplemental questionnaire that considers in depth the problems of the lower back, neck, and shoulder pain [16]. In this study, the NMQ survey was designed and administered to 120 male workers of four high-rise condominium building construction projects in Korea. The participants who were actively engaged in various manual construction operations were identified based on their trade (i.e., carpenters, masons, and ironworkers), task (i.e., ceiling work, material handling, and ground-floor-level helping), and role (i.e., craftsman, journeyman, and helper). The average age of these workers was 48.46 years. Questionnaires were prepared and provided to these workers in envelops. Of the 120 envelops, 28 were returned in either an incomplete or an invalid form. The rate of valid response for the 120 envelopes was 76.66%. In order to obtain accurate data, the objective of the study was clearly explained to the participants before they responded to the survey. After obtaining admissible informed consents from the participants, each criterion, which included a moderate (non-extreme) level of self-reported physical activity, was collected on a daily basis. By adopting the standard NMQ survey process [17], valid anthropometry data along with answers for all questions were obtained. In addition, the extensiveness and precision of the survey were confirmed by the descriptive statistics using the NMQ survey, which included the MSD pain prevalence data for 12 months, frequency of pain over the total working days/weeks, and the distribution of MSD on each body part.

3.3. Biomechanical Assessment Using Three-Dimensional Static Strength Prediction Program (3D SSPP)

The biomechanical assessment software 3D SSPP (Version 8.0) was developed at the University of Michigan, USA, and is well accepted as an effective tool for handling the relationship between various lifting motions and lower back pain [18,19]. It was used to validate the NMQ survey output analysis in this study. The software identifies not only the physical demand attributed to a task (including posture data, force parameters, and male/female anthropometry), but also predicts the static strength requirements for lifting, pushing, and stooping tasks. It provides the percentage of worker strength performance given a designated task, and the spinal compression forces based on the National Institute of Occupational Safety and Health (NIOSH) guidelines.

4. Results

4.1. Descriptive Analysis of the NMQ Survey Outputs

Of the 92 participants (87%) who provided valid questionnaire, responses complained of MSD symptoms attributed to the construction works (Tables 1–3).

Table 1. Height, weight, and body mass index (BMI) of respondents.

Description	Mean
Height (ht.)	174 cm (140–190) cm
Weight (wt.)	75 kg (165 lb)
BMI	24.8 kg/m^2

Table 2. Tenure and working days.

Work Experience	Percentages
<than equal to 10 years	54%
(11–25) years	29%
>25 years	16%
Working Time (Days/Week)	**Percentages**
5	19.5%
6	64.1%
7	16.3%

Table 3. Relation between age groups and MSD pain.

Characteristics	Percentages	Musculoskeletal Disorder	
Age (Mean Age:48.46)		Yes	No
18–30	7%	6%	1%
30–35	30%	27%	3%
50–65	63%	54%	9%
Total	100%	87%	13%

The age distribution of the respondents was as follows: 7% were 18 to 30 years old, 30% were 30 to 50 years old, and 63% were 50 to 65 years old. The worker's age appeared to be a significant contributor to MSD symptoms. The distribution of workers according to their work experience was as follows: 54% with less than 10 years' tenure, 29% with 11 to 25 years' tenure, and 16% with more than 25 years' tenure. The average body weight and BMI of the population was 75 kg and 24.8, respectively, which indicates that the construction workers had normal BMI. The longest duration for MSD pain was more than a month for the lower back, followed by the neck and shoulder during the past 12 months (Table 4).

Table 4. Duration of MSD pain in the past 12 months.

Description	Half Day (Nos)	Within 1 Week	Within 1 Month	More than 1 Month	Daily
Lower back	20	10	10	32	2
Neck pain	17	25	8	19	2
Shoulder pain	17	9	8	20	1

4.2. Reconfirming Awkward Postures Contributing to MSDs

Existing research portends that pushing, lifting, and kneeling are major awkward postures contributing to MSDs. The NMQ questionnaire survey found that nearly 43%, 38%, 16%, and 16% of the studied population suffered from pain in the shoulder, lower back, neck, and knee, respectively (Table 5). The main awkward postures (e.g., lifting, pulling, kneeling posture) obtained by these surveys on construction workers' muscle pain were chosen for biomechanical simulation to validate the survey results. Since the amount of pain for upper back, hip, and wrist were nominal, these postures were not included in the simulation. In addition, the main motions contributing to each MSD were identified. Shoulder pain was the outstanding MSD complaint during daily working hours. It was mainly attributed to the bending and/or twisting of the body. Working in a bent or twisted body posture for long hours daily may increase this MSD symptom significantly. It was found that leg squatting while performing tasks on the ground or floor was an awkward motion that involved the knee acutely. In addition, the most common awkward postures at construction job sites were pushing forward (posture 1), lifting (posture 2), and kneeling (posture 3).

Table 5. MSD profile among respondents.

MSDs	Number	Percentage (%)	Motions
Neck pain	15	16	Groundwork
Shoulder pain	40	43	Lifting, pushing, pulling
Arm pain	15	16	Pushing
Wrist pain	13	14	Pulling
Upper back pain	5	5	Push/pull
Lower back pain	35	38	Lifting, pushing, pulling
Hip pain	4	4	–
Knee pain	15	16	Kneeling
Ankle/foot pain	10	11	–

4.3. Simulation Modeling and Analysis Using 3DSSPP

Anthropometry data were obtained from the survey at the aforementioned four Korean construction sites and provided the posture details and input parameters for workers (including average height (174 cm), weight (165 lb), and left- and right-hand forces). These data are listed in Table 6. They were used as input data for the simulation model.

Table 6. Anthropometry data of each posture.

Posture Detail	Gender	Average Height (cm)	Average Weight (lb)	Hand Forces Left (lb)	Right (lb)
Pushing forward	Male	174	165	20	20
Lifting	Male	174	165	25	25
Kneeling	Male	174	165	10	10

4.3.1. Analyzing Motion of Pushing Forward

The pushing forward motion shown in Figure 2a did not cause a severe risk of injury to the lower back because it demanded a low lumbar disc compression force (L4/L5) that was less than the NIOSH back compression action limit of 770 lb (3400 N). The compression force on a disc of the spine was recommended by the NIOSH. The safety level for disc compression force during lifting objects in manual material handling should be less than 3400 N (Waters et al. 1993). While pushing forward against a force of 295 lb (1338 N), the worker did not bend his torso. Therefore, high flexion of the back was not needed to move forward an object of weight up to 9 kg (20 lb) in the simulation experiment, as shown in Figure 2b. The heavier the object that the worker pushes and/or the greater bender the extent to which the worker bends his/her torso, the higher the compression force. The simulation output analysis confirmed that only 35% and 62% of the surveyed workers could perform the posture of the wrist joint and that of the knee joint, respectively, and manifested in the pushing forward motion. Further, the other joints fell within the critical zone, indicating the influence of the pushing forward motion on body balance (see Table 7).

Table 7. Simulation output analysis of pushing forward.

S.N.	Posture Type	Description	Body Weight	Hand Forces (H.F.)	Knee	Shoulder	Hip	Wrist	Body Balance	Low Back Compression
1	Pushing forward	Pushing 20 lb load	75 kg	20 lb	62%	99%	98%	35%	Critical	295 lb (<770 lb)

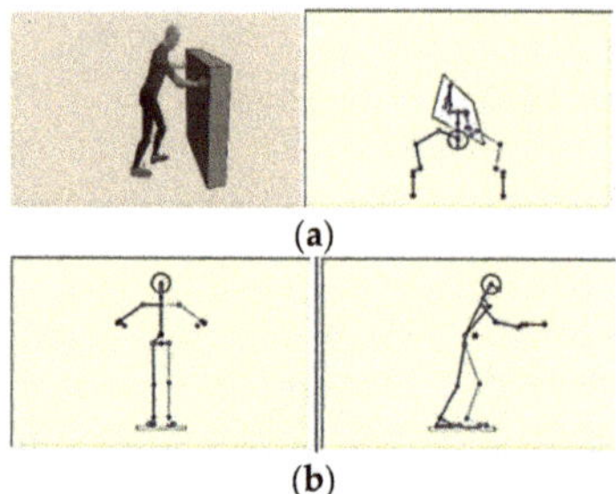

(a)

(b)

Figure 2. Pushing forward motion (**a**), and limb angles in pushing forward motion (**b**) obtained by the three-dimensional static strength prediction program (3D SSPP).

The change in location of the center of gravity of the worker's body dictated the functional stability region obtained while releasing the pushing forward posture and was projected on the body balance graph by obtaining 30 window frames within a second, as shown in Figure 3. The virtual manikin retained static balance when the value of hand forces was decreased. The manikin could bend further forward if its center of gravity was located further backward from its base support. Thus, it may be beneficial either to hire a stronger worker or to decrease the hand force according to the BMI of the workers in order to avoid falling accidents.

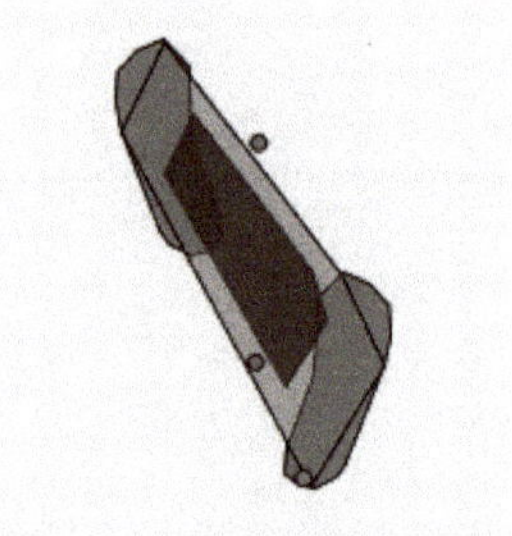

Figure 3. Center of gravity of body in pushing forward motion.

4.3.2. Analyzing Lifting Motion

Four different postures of lifting a 25 lb box may cause a severe injury to the low back and were thus modeled in Figure 4a–d. The compression force (L4/L5), i.e., 3821 N (859 lb), exceeded the NIOSH back compression action limit of 3400 N. Since the worker bent his torso, these postures required high flexion to move an object (25 lb weight in the simulation experiment) forward. It was confirmed that the other joints fall within the critical yellow zone.

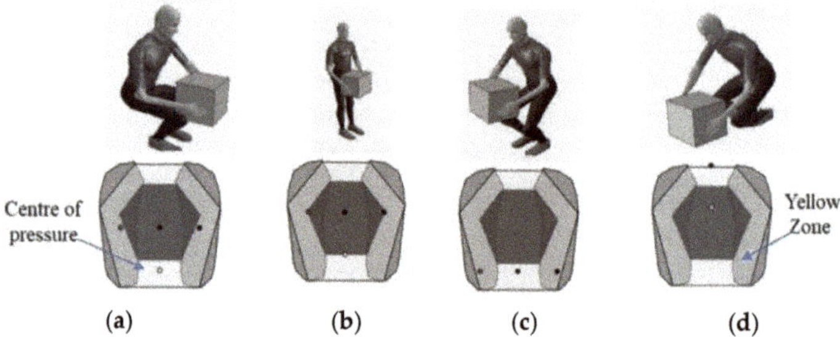

Figure 4. Body balance in lifting postures based on center-of-pressure (COP)—acceptable (**a**), acceptable (**b**), critical (**c**), and unacceptable (**d**).

These postures may cause severe low back injuries since the compression force (L4/L5) exceeds the NIOSH back compression action limit of 3400 N (Table 8).

Table 8. Simulation output analysis of lifting.

S.N.	Posture Type	Description	Body Weight	Hand Forces (H.F.)	Knee	Shoulder	Hip	Wrist	Body Balance	Low Back Compression
1		Carrying 25 lb box	75 kg	25 lb	87%	88%	92%	77%	A	824 lb (>770 lb)
2		Standing and carrying 25 lb box	75 kg	25 lb	100%	70%	82%	80%	A	859 lb (>770 lb)
3		Going to put the 25 lb box on the floor	75 kg	25 lb	100%	87%	99%	85%	C	343 lb (<770 lb)
4		Put the 25 lb box onto the floor	75 kg	25 lb	98%	94%	90%	78%	U	723 lb (<770 lb)

Note: A = acceptable, C = critical, U = unacceptable.

The maximum and the minimum compressive forces exerted while performing the lifting motion were 859 lb (3821 N) and 343 lb (1525 N), respectively. Since the worker did not bend his/her torso, these postures did not require high flexion to move forward an object weighing 12 kg (25 lb) used in the simulation experiment. The compression force increased if the worker bent his/her torso to push a heavier object. Only 77%, 87%, and 70% of the population could perform the corresponding postures of the wrist, knee, and shoulder joints, respectively. A posture may have static balance, fall within the yellow zone, or tend to cause a fall. The change in the location of the center of gravity while releasing the lifting posture was projected on the body balance graph, as shown in Figure 5. Body balance was categorized as acceptable, critical, or unacceptable by 3D SSPP when the COP was within, on the boundary, or outside the functional stability region, respectively, as depicted in Figure 4a–d. The virtual manikin retained static balance when the value of hand force was decreased from 25 lb to 15 lb The further backward the center of gravity of the manikin was located from its base support, the farther the manikin bent. Thus, it is beneficial to either hire a stronger worker or decrease the hand force according to the worker's BMI to avoid a dropping accident.

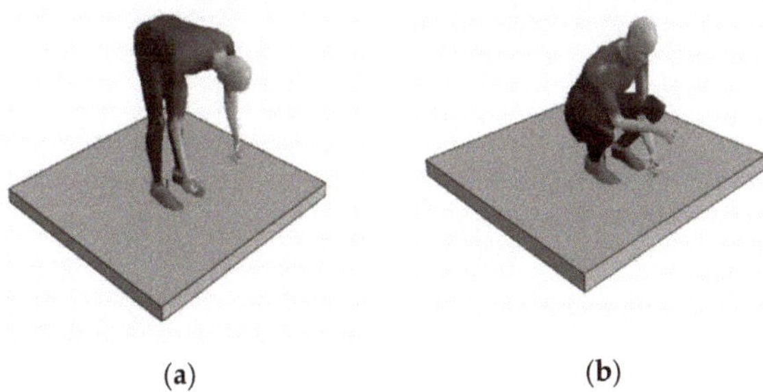

(a)   (b)

Figure 5. Reinforcing postures for (**a**) 25 fps and (**b**) 20 fps.

4.3.3. Analyzing Kneeling Posture

Two different postures reinforcing rebar for 25 frames per second and 20 frames per second were modeled, as shown in Figure 5a,b, respectively. The compression force (L4/L5), i.e., 742.6 lb (3303 N), was within the NIOSH back compression action limit of 3400 N (see Table 9), resulting in a margin of a lower back injury. Since the worker must bend his torso, these postures require high flexion to move an object weighing 9 kg (20 lb) forward in the simulation experiment. Only 74%, 72%, and 84% of the population could perform the holding posture of the knee, ankle, and torso joints, respectively. Further, only 70% and 52% of the population could perform the reinforcing posture of the wrist and knee joints, respectively. It was confirmed that the other joints engaged in the holding posture were unacceptable, but those engaged in the reinforcing posture were acceptable.

Table 9. Simulation output analysis of kneeling.

S.N.	Posture Type	Description	Body Weight	Hand Forces (H.F.)	Knee	Shoulder	Hip	Wrist	Body Balance	Low Back Compression
1		Holding position (before kneeling position)	75 kg	10 lb	74%	100%	81%	99%	U	742.6 lb (<770 lb)
2		Reinforcing position	75 kg	10 lb	100%	70%	82%	80%	A	56.9 lb (<770 lb)

Note: U = unacceptable, A = acceptable.

Two different postures involving stooping (bending at the waist) with a hand tool for 25 frames per second and squatting down to reinforce rebars for 20 frames per second were modeled, as shown in Figure 6a,b. The center of gravity of the body while performing these postures was located away from the support, leading to a tendency to fall. The center of gravity of the virtual manikin remained in the base and the manikin maintained static balance. Indeed, either decreasing the hand force or maintain a constant hand force will be a good preventive measure to avoid a falling accident for a given BMI.

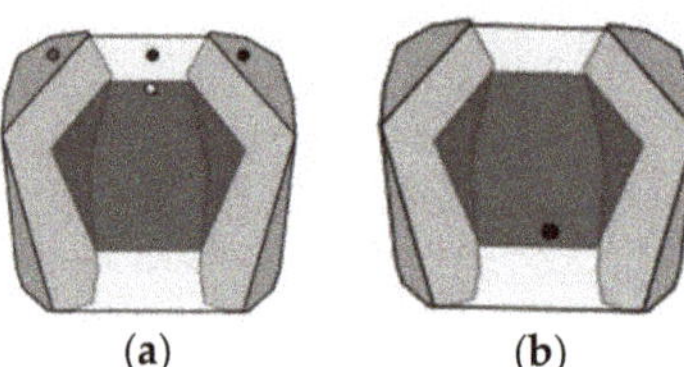

(a)   (b)

Figure 6. Stooping postures with hand tool for (**a**) 25 fps and (**b**) reinforcing rebar for 20 fps.

4.4. Tradeoff between BMI and Magnitude of Force

While decreasing hand force when pushing forward, lifting, and kneeling, the low back compression, body balance, and the percentage of strength capability were obtained. These values are listed in Table 10. The body balance in the lifting posture was critically unacceptable, but it became stable as the hand force decreases. The lower back compression decreased from 942 lb to 752 lb, which was less than the standard level (770 lb), as the hand force decreased. The percentage of strength capability increased remarkably to more than 90% for all the body joints, including the knee, shoulder, wrist, and hip, as seen from the data in Table 10. The benchmark provided admissible evidence that a Korean worker with an average weight of 75 kg can carry 16 lb, 19 lb, and 16 lb of loads when performing tasks involving pushing forward, lifting, and kneeling, respectively. When the hand force was greater than these loads applied to the manikin (i.e., the virtual worker) weighing 75 kg, the body tended to be unbalanced in those postures.

Table 10. Tradeoff between BMI and the magnitude of forces in different postures.

S.N.	Posture Type	Description	Body Weight	Hand Forces (H.F.)	Knee	Shoulder	Hip	Wrist	Body Balance	Low Back Compression
1	Pushing forward	P-1	75 kg	20 lb	62%	99%	98%	35%	C	295 lb
	Pushing forward (after reducing H.F.)	P-1	75 kg	16 lb	98%	99%	96%	63%	A	309 lb
2	Lifting	P-2	75 kg	25 lb	99%	96%	86%	91%	C and U	942 lb
	Lifting (after reducing H.F.)	P-2	75 kg	19 lb	99%	90%	88%	91%	A	752 lb
3	Kneeling	P-3	75 kg	20 lb	74%	100%	81%	98%	U	742.6 lb
	Kneeling (after reducing H.F.)	P-3	75 kg	16 lb	90%	100%	96%	99%	A	700 lb

Note: P-1 = posture 1, P-2 = posture 2, P-3 = posture 3, C = critical, A = acceptable, U = unacceptable.

5. Discussion

The method combining the NMQ survey, biomechanical analysis, and benchmark approach facilitates quantitative MSD control over the muscle stress of construction workers. It encourages informed decision making on recruiting appropriate workers considering their physical merits (i.e., muscle strength) for a specific job function in a construction operation. It fills gaps the computational method handling different parts MSD that the existing methods had not adequately described to access the health risks of construction workers by doing simulation on construction worker postures. It may replicate specific task, finding construction workers' MSD issues attributed to using semi-automatic or fully automatic tools. The biomechanical analysis outputs involved in the unacceptable and awkward postures (Tables 5–7) that impose high risk provide a tool to field employment managers to identify the tradeoff between BMI and the magnitude of the hand forces to execute preventive measures (i.e., exercise programs and engineering controls) [20]. Few studies provide an insight into reducing work-related musculoskeletal injuries given the existing preventive measures. This lack of research may be attributed to the fact that analysis of work tasks at a job site is a complex task because of various factors (e.g., organizational, human, task factors). The new hybrid method allows an elaborate analysis of work postures associated with construction tasks by considering job-specific risks attributed to process, motion, and posture. Note that the method identifies potential MSDs associated with the awkward postures of a worker performing a job function while controlling other job-site variables.

The limitations of the method are related to the biomechanical issues as follows: First, it is desirable that sophisticated postures are considered jointly by accommodating 3D motion analysis in a future version of the method. A worker's muscle strength may not be determined by the biomechanical simulation model alone. However, the model may provide a control tool for MSD safety and health of workers by validating the physical demands (e.g., lower back compressive strength, percentage of strength capability, and body balance) obtained by expert group surveys of the construction community. Second, the momentary and transitory issues involved in motions have been intensified among the workers involved in construction. A controlled experiment on construction workers is not feasible because it is not easy to have many workers perform identical motions at a construction job site. In addition, their motions are momentary, transitory, and involve multiple tasks at a time. It will be commendable to perform controlled experiments in a simulated construction job site to generalize the outputs obtained by the method. Third, it will be beneficial to track each motion activated by a participant performing a specific task. It may encourage the elaborative evaluation of MSD. For instance, biomechanical human simulation may effectively predict the relation between muscle strength for a construction task and the workplace dimension by using identified awkward postures. Fourth, extensive controlled experiments with different exercise protocols (i.e., the type of working layouts, frequency of postures, intensity of motion, and duration of posture) may contribute to identifying unknown variables that influence the relationship between two variables and to secure the validity of the method and its corresponding data.

6. Conclusions

The main contribution of this study is that the hybrid method lends itself to scientific fact-finding. A set of benchmarks was established using the model by manipulating the BMI and hand forces of the workers. The method provides a means to not only understand the contemporary scenario of MSD in construction workers but also to establish a practical benchmark based on the physical capability of workers that is helpful to construction managers during recruitment. It confirms that 87% of respondents suffering from MSD had three common awkward postures. Further, the simulation output analysis provided admissible evidence that the muscle stress involved in lower back compression exceeds the tolerance. The body of a worker suffering from back pain may be unstable while performing a work task. The awkward postures in which the body balance is proportional to the loads aggravate the situation. Indeed, decreasing the hand forces makes the posture static, thereby reducing the MSD. It will be beneficial to incorporate these findings into computer-based predictions to secure the effectiveness and validity of biomechanical human simulations. The current version of the developed method handles static postures, not dynamic movements. It is desirable to extend the new method to assess real-time work processes to identify the dynamics in real practices in the future study. The new method promotes academic division in the multi-paradigm computing approach and may contribute to the advancement of the construction workers' health assessment when monitoring the next generation.

Author Contributions: Conceptualization, S.P. and D.-E.L.; Data curation, M.Y. and B.Y.C.; Formal analysis, S.P., M.Y., B.Y.C. and D.-E.L.; Funding acquisition, D.-E.L.; Investigation, S.P. and M.Y.; Methodology, S.P.; Project administration, D.-E.L.; Resources, S.P.; Software, S.P. and B.Y.C.; Supervision, D.-E.L.; Validation, D.-E.L.; Visualization, S.P. and B.Y.C. Writing—original draft, S.P.; Writing—review & editing, D.-E.L. All authors have read and agreed to the published version of the manuscript.

Funding: This work was supported by the National Research Foundation of Korea (NRF) grant funded by the Korea government (MSIT) (NRF-2018R1A5A1025137) and (NRF-2019R1I1A1A01062006).

Conflicts of Interest: The authors declare no conflict of interest.

References

1. OSHA. Ergonomics: Prevention of Musculoskeletal Disorders in the Workplace. Available online: https://www.osha.gov/SLTC/ergonomics/index.html (accessed on 13 July 2020).
2. Choi, S.D.; Yuan, L.; Borchardt, J.G. Critical Analyses of Work-related Musculoskeletal Disorders and Practical Solutions in Construction. *Proc. Hum. Factors Ergon. Soc. Annu. Meet.* **2014**, *58*, 1633–1637. [CrossRef]
3. Park, J.; Kim, S.G.; Park, J.-S.; Han, B.; Kim, K.B.; Kim, Y. Hazards and health problems in occupations dominated by aged workers in South Korea. *Ann. Occup. Environ. Med.* **2017**, *29*, 27. [CrossRef] [PubMed]
4. Salas, E.A.; Vi, P.; Reider, V.L.; Moore, A.E. Factors affecting the risk of developing lower back musculoskeletal disorders (MSDs) in experienced and inexperienced rodworkers. *Appl. Ergon.* **2016**, *52*, 62–68. [CrossRef]
5. Chen, J.; Qiu, J.; Ahn, C. Construction worker's awkward posture recognition through supervised motion tensor decomposition. *Autom. Constr.* **2017**, *77*, 67–81. [CrossRef]
6. Kim, H.; Choi, J. The relation between obesity and cancer of gastrointestinal tract in Korea: The data from Statistic Korea between 2001 and 2016. *Ann. Oncol.* **2019**, *30*, ix55. [CrossRef]
7. Golabchi, A.; Han, S.; Seo, J.; Han, S.; Lee, S.; Al-Hussein, M. An Automated Biomechanical Simulation Approach to Ergonomic Job Analysis for Workplace Design. *J. Constr. Eng. Manag.* **2015**, *141*, 04015020. [CrossRef]
8. Lee, W.; Seto, E.; Lin, K.-Y.; Migliaccio, G. An evaluation of wearable sensors and their placements for analyzing construction worker's trunk posture in laboratory conditions. *Appl. Ergon.* **2017**, *65*, 424–436. [CrossRef] [PubMed]
9. Alwasel, A.; Elrayes, K.; Abdel-Rahman, E.M.; Haas, C. Sensing Construction Work-Related Musculoskeletal Disorders (WMSDs). In Proceedings of the 28th International Symposium on Automation and Robotics in Construction (ISARC 2011), Seoul, Korea, 29 June—2 July 2011; International Association for Automation and Robotics in Construction (IAARC): Seoul, Korea, 2011; pp. 164–169.

10. Chang, J.D.; Bennis, F.; Ma, L. Muscle Fatigue Analysis Using OpenSim. In *Digital Human Modeling. Applications in Health, Safety, Ergonomics, and Risk Management: Ergonomics and Design; Proceedings of the International Conference on Digital Human Modeling, Vancouver, BC, Canada, 9–14 July 2017*; Duffy, V., Ed.; Springer International Publishing: Cham, Switzerland, 2017; Volume 10286, pp. 95–106.

11. Valero, E.; Sivanathan, A.; Bosché, F.; Abdel-Wahab, M. Musculoskeletal disorders in construction: A review and a novel system for activity tracking with body area network. *Appl. Ergon.* **2016**, *54*, 120–130. [CrossRef] [PubMed]

12. Kong, L.; Li, H.; Yu, Y.; Luo, H.; Skitmore, M. Quantifying the physical intensity of construction workers, a mechanical energy approach. *Adv. Eng. Inform.* **2018**, *38*, 404–419. [CrossRef]

13. Wells, R. Why have we not solved the MSD problem? *Work* **2009**, *34*, 117–121. [CrossRef] [PubMed]

14. Choi, S.D. Investigation of Ergonomic Issues in the Wisconsin Construction Industry. 2016. Available online: https://www.researchgate.net/publication/252554958 (accessed on 14 July 2020).

15. Burgess-Limerick, R. Participatory ergonomics: Evidence and implementation lessons. *Appl. Ergon.* **2017**, *68*, 289–293. [CrossRef] [PubMed]

16. Kuorinka, I.; Jönsson, B.; Kilbom, A.; Vinterberg, H.; Biering-Sorensen, F.; Andersson, G.; Jørgensen, K. Standardised Nordic questionnaires for the analysis of musculoskeletal symptoms. *Appl. Ergon.* **1987**, *18*, 233–237. [CrossRef]

17. Dickinson, C.; Campion, K.; Foster, A.; Newman, S.; O'Rourke, A.; Thomas, P. Questionnaire development: An examination of the Nordic Musculoskeletal questionnaire. *Appl. Ergon.* **1992**, *23*, 197–201. [CrossRef]

18. Chaffin, D.B. Development of computerized human static strength simulation model for job design. *Hum. Factors Ergon. Manuf.* **1997**, *7*, 305–322. [CrossRef]

19. University of Michigan. Centre for Ergonomics: 3D Static Strength Prediction Program Version 7.0.4. 2017. Available online: https://c4e.engin.umich.edu/tools-services/3dsspp-software/ (accessed on 7 March 2017).

20. Holmström, E.; Ahlborg, B. Morning warming-up exercise—Effects on musculoskeletal fitness in construction workers. *Appl. Ergon.* **2005**, *36*, 513–519. [CrossRef] [PubMed]

 *sustainability*

Article

Evaluation of Space Service Quality for Facilitating Efficient Operations in a Mass Rapid Transit Station

I-Chen Wu * and Yi-Chun Lin

Department of Civil Engineering, National Kaohsiung University of Science and Technology,
Kaohsiung 80778, Taiwan; x791220@gmail.com
* Correspondence: kwu@nkust.edu.tw; Tel.: +886-7381-4526 (ext. 15238)

Received: 29 May 2020; Accepted: 29 June 2020; Published: 30 June 2020

Abstract: In an urban public transport system, mass rapid transit (MRT) stations play an important role in the concentration and deconcentration of passengers. Spatial conflicts and unclear routes may lead to crowding in MRT stations and reduce their operational efficiency. For this reason, this study proposes a space service quality evaluation method based on agent-based simulation by employing spatial information from building information modeling (BIM) systems as boundary constraints. Moreover, passengers and trains are simulated as interacting agents with complex behaviors in a limited space. This method comprehensively assesses congestion, noise, and air quality to determine service quality in different spaces. Moreover, the results are visualized in different ways for decision making about space planning. Finally, this research demonstrates and verifies the functions of the proposed system with an actual MRT station. Such simulation results can be used as a reference for management personnel to adjust space/route plans to increase passenger satisfaction environment quality, and operational efficiency in the operation stage of an MRT station. The evaluation method establishes valid and reliable measures of service performance and passenger satisfaction as well as other performance outcomes.

Keywords: building information modeling; agent-based simulation; space service quality; efficient operation

---

1. Introduction

Mass rapid transit (MRT) stations play an important role in hosting and distributing passengers through an urban transport system. However, station space is a limited resource, and passengers move through or temporarily halt in this limited space. Many studies have examined how to effectively configure and use space. For example, Bahrehmand et al. [1] present an interactive layout solver that can assist designers in layout planning by recommending personalized space arrangements based on architectural guidelines and user preferences. Guo and Li [2] present a method for the automatic generation of a spatial architectural layout from a user-specified architectural program. But the quality of an open public space may be significantly negatively associated with psychological distress. Therefore, emphasis must be placed on space planning and service quality in building such spaces [3]. The quality of space planning for MRT stations will affect passengers' evaluations of the space service quality. Poor space planning may, to a large extent, negatively impact passengers' perceptions of the space service quality. In addition to the space in the infrastructure, time and user experience are additional factors to consider in space and route planning to improve the overall levels of service quality and passenger satisfaction. To enhance space planning and the degree of user satisfaction with station services, Li et al. [4] employed a scientific method to assess a building's space performance while emphasizing its influence on environmental quality and passenger satisfaction. They also developed an evaluation tool to continuously monitor the overall sustainable performance in the operation stage. Wang et al. [5] used a questionnaire survey to understand the overall level of satisfaction with the interior environment of a flight terminal, and the outcomes can be used to assist in the future design and planning of airports and their operations. Tomé et al. [6] stated that buildings are complex dynamic systems composed of sub-systems and components in continuous interaction with human behavior. Therefore, information needs to be obtained from records to understand passenger concentrations and levels of space usage. The above research studies belong to the post-occupancy evaluation (POE) method. The question which we must consider next is the cost and operational impacts of reconstruction caused by future use problems due to poor design. Hayek et al. [7] and te Brömmelstroet et al. [8] both argued that planning integration should occur in the early stages of design to avoid any imperfections or conflict problems in the public transport system. However, planning is often based on the instinct and experience of the decision-makers, even if they lack the ability to interpret modeling results. Therefore, it is a challenge to provide models and evaluation results that are easy to understand so that they can assist planning personnel in the decision-making process to achieve a reasonable balance between planning design and evaluation. Evaluation methods can be divided into two types: non-parametric and parametric evaluation methods. Data envelopment analysis (DEA) is a non-parametric method in operations research and economics for the estimation of production frontiers [9,10].

Another method is the parametric evaluation method, as Indraprastha and Shinozaki [11] present in the computational model to analyze and assess the quality of architectural space by using visual distance combined with viewing angle to obtain the spatial quality. Understanding and evaluating space quality at the design stage can assist in making modifications at the pre-construction stage. Zawidzki et al. [12] propose a framework wherein the architectural functional layout is optimized for the following objectives: functionality (defined by users), insolation (calculated according to geographical conditions), outside view attractiveness (assessed on-site) and external noise (measured on-site). Although we can use mathematics or optimization methods to evaluate the quality of the space design, these research approaches ignore the influence of human interaction and grouping in a confined space. The agent-based modeling (ABM) [13,14] technique is also widely adopted to simulate real social conditions and human psychological reactions to determine the problems that may arise. A multi-agent system consists of multiple independent agents interacting with each other. They can result in different sorts of complex and interesting behaviors. It is a method to model real-life situations. Research has supported the validity of ABM in modeling human behaviors. Lee and Malkawi [15] utilized ABM to predict passenger behaviors, demonstrating that passenger behaviors impact both comfort and energy management activities. Osman [16] applied ABM to predict infrastructure asset management activities to study the effects of the social-psychological behaviors of users on how the users spend time on infrastructure services. Langevin et al. [17] developed and validated ABM for occupant behaviors using data from a one-year-long field study in a medium-sized, air-conditioned office building. Building information modeling (BIM) [18,19] is another popular technique in the construction industry, where it has been applied in the design and planning and the operation and maintenance phases. BIM can provide a visual modeling environment to assist in space planning, thus reducing discrepancies between design and actual construction outcomes. BIM models consist of comprehensive engineering attributes and spatial information. In view of these merits, in this study, BIM and ABM are applied to the evaluation of space service quality for providing dynamic visualization of the interactions between passengers and space. In simulating the actual conditions, the effects of spatial topology and human perception factor on the service quality of a space are considered. The results can not only assist planning personnel in adjusting current spatial designs but also provide feedback for the planning of station space and routes and in the analysis of alternative options at the design stage. This method can reduce labor and other financial costs associated with making changes in the design to improve passenger satisfaction levels.

2. Evaluation Method for Space Service Quality

Human beings have long endeavored to create indoor environments in which they can feel comfortable. Durmisevic and Sariyildiz [20] pointed out key features that influence underground space design, including accessibility and nearest surroundings, orientation and way finding, spatial proportion, contact with the outside world, natural and artificial lighting, materials and colors, noise level, and air quality, among others. In addition to hardware equipment and environmental factors, other features are predominantly related to subjective feelings. In determining the impact of an indoor space on the comfort of the human body, the most common discrimination item is the indoor environment quality (IEQ), which is a benchmark for residential quality performance. It includes four items: thermal comfort, air quality, noise level, and lighting level [21]. Some research scholars have shown that indoor environmental quality factors can affect occupant satisfaction [22,23]. Among them, the thermal comfort and lighting level can be improved by adjusting the hardware equipment to improve the quality of space services. However, noise and air quality are more difficult to improve. The main reason is that different measurements result from the interactions and states of crowds of people, so they are difficult to quantify. Therefore, this study considers the impact of human interaction and grouping on space to propose a novel method for evaluating space service quality for facilitating the efficient operations of an MRT Station. This method employs BIM technology combined with agent-based simulation to simulate the behavioral patterns of passengers at MRT stations. Through dynamic modeling of possible scenarios, an examination of spaces crowded by large volumes of passengers and the associated noise and air quality issues is performed to understand service quality at MRT stations. Figure 1 illustrates a flowchart of the evaluation method for space service quality. A major feature of this method is the reuse of the BIM models, which reduces the time and cost associated with the data preparation required for simulation. The BIM model contains geometric shapes, spatial locations, and boundaries, which are important simulation constraints. Subsequently, this research establishes agent-based models based on BIM and the passenger and train movement conditions. The actual conditions are simulated by setting the relevant influencing factors and behavioral patterns. This system can simulate the status of space usage, usage level, air quality, noise, etc. Moreover, the 2D and 3D visualization and statistical charts are used to present the simulation results. Finally, the simulation results of the space service quality measurement can be exported into Excel spreadsheets to assist planning personnel in evaluation and decision making in the design and operation stage during the building's life cycle. In this way, the potential impact of future activities on service quality can be predicted and managed in advance.

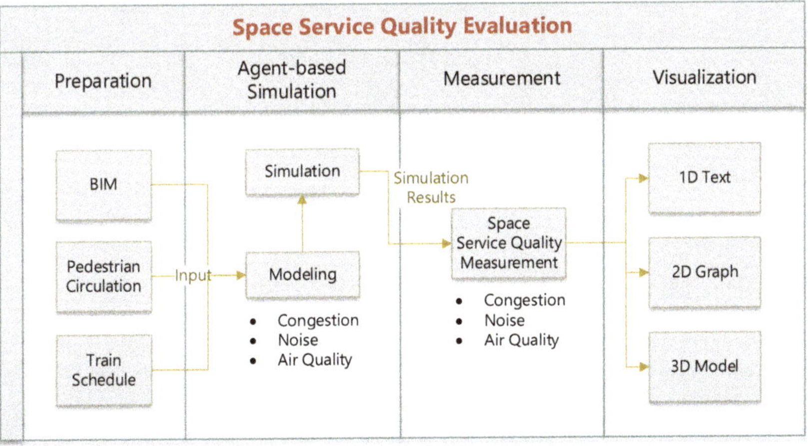

Figure 1. Flowchart of the evaluation method for space service quality.

3. Data Preparation

For space and pedestrian circulation planning, the dynamic simulation of virtual display methods needs to involve the timing between moving and stationary objects, the space configuration, and user participation characteristics. This research uses AnyLogic simulation software for agent-based simulation. Although it can support the 3D object formats VRML (Virtual Reality Modeling Language) and X3D (Extensible 3D), the VRML format has too many restrictions, and no updated version beyond VRML 97 exists. Therefore, this study employs X3D as the model conversion object. Since X3D is developed in XML format, it can be verified or modified using XML-related editing tools. Having extensible characteristics, it is a highly readable format due to the interaction between cross-platforms and is one of the current unified exchange formats for 3D data. In addition, to improve the sustainability value of the conversion program, the BIM model information exchange format called the Industry Foundation Classes (IFC), a data format released by BuildingSMART, is used as the standard format for converting all BIM models in the process. It can be used by different modeling software, such as Bentley AECOsim, Autodesk Revit, ArchiCAD, and Tekla, and this conversion mechanism can be used to export files in the X3D 3.0 version format for subsequent analysis and simulation. Thus, the consistent format conversion of the data model in the data preparation stage is a problem that must be solved. This study develops a BIM model data capture and format conversion tool, as shown in Figure 2. The user can directly retrieve the floor plan and compartment data of the BIM model, convert it into the X3D virtual reality file format, and directly import it into AnyLogic for conversion to the active space boundary condition of the system. This study reuses the BIM model established in the design planning stage to ensure the accuracy of the simulation and to avoid the labor cost of rebuilding the model. Moreover, the data on pedestrian circulation were collected from historical data of the operation stage. In addition to traffic at the exits and entrances of the station, passengers board and alight from MRT trains at the platforms. Thus, this research collected the train schedules to obtain train capacities at different times and consider the overall pedestrian circulation.

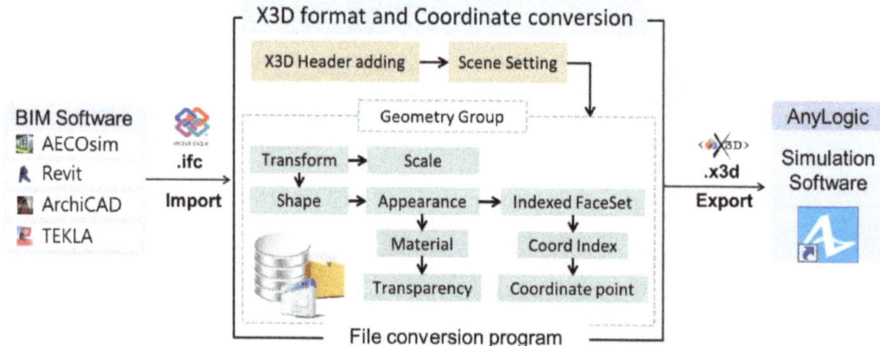

Figure 2. Concept of X3D format and coordinate conversion.

4. Agent-Based Simulation

Space quality, similar to service quality, is subject to user perceptions. Crowded spaces or spaces with noise and/or bad air quality directly negatively affect people's perception of the service quality. Agent-based modeling can deal with continual temporal and spatial states of events. Agents can make decisions on space boundaries, destinations, entrances, exits, and route disturbances as well as identify potential problems. A multi-agent system consists of multiple independent agents interacting with each other. A multi-agent system simulation can be applied to society, biological bodies, mechanical processes, human beings, or any movable object. The social force model proposed by Helbing et al. [24] can be used to promote or influence agents' physical and psychological states, generating distance between the agents and resulting in socio-psychological, physical, and reaction forces. This model is widely applied for the simulation of behavioral patterns of agents. Therefore, in addition to using BIM to understand the walking behaviors of passengers in an MRT station space, this study uses an agent-based method to simulate congestion, noise, and air quality according to the number of passengers, determine the extent of their effect, and evaluate service quality. The results can serve as a reference and a basis for decision making in the planning process.

4.1. Modeling for Congestion

MRT stations serve male and female passengers of different ages. The passenger's speed, grouping, behavior, etc. produce different interactions within the space. Factors such as the location of entrances and exits and the placement of equipment affect the passengers' circulation within the space. They also have different behavioral impacts on other agents, which are reflected in the results of subsequent decisions. Understanding the relationship between passengers and MRT station spaces allows reductions of relative obstacles and increases in circulation speed. Therefore, this study uses the BIM technology to capture the boundary conditions of the MRT station model, integrates the agent's virtual role to simulate the flow of people, and reflects the behavioral state and judgment logic of the passengers in different environments in different spaces. Through the establishment of influencing factors, the simulation results can be presented in dynamic 2D and 3D visualization without static assumptions. It is hoped that based on the specific situation analysis and the actual situation, by the setting of parameters and simulation of the agent, whether the existing space can cope with varying crowd sizes can be analyzed. In addition, through relevant settings, the possible behavior results of various agents in different spaces, the flow of people and the state of congestion are evaluated, and data results are provided to improve the impact of service quality within the space.

Pedestrian agents are intelligent in the state of social force model agent simulation. Passengers are simulated through continuous calculation and judgment for each step they take; this study also adds the calculation parameters in the space to the agent calculation equation, making the agent more reliable in the simulated state. The parameters include simulated walking targets, walking velocity, walking distance, walking speed, passenger influence range, obstacle avoidance, and other factors affecting pedestrian agents between spaces and obstacles. In addition, because the agent system needs to first generate a category during the construction, this category generates the agent character objects according to the conditions based on the parameters and state settings defined in the study. This ensures that the agent character objects are independent of each other. Different state behaviors are additionally set in the category, resulting in different behavioral rules for different pedestrian agents. In this study, pedestrian agents are distinguished by age as the object parameters of adults, the elderly, and children, and they are set in groups or partnered so that the pedestrian agents not only walk independently but also may be in the group movement state. However, in situations such as queuing, waiting, ascending and descending stairs, and boarding, the states that produce a separate pass or use conditions need to be changed. This study presents the above-mentioned conditions based on the basic conditions of pedestrian agents in the simple behavioral state, and they are presented in flow charts, as shown in Figure 3.

At the platform level of the MRT station, a track area caters to the trains' demand outside the passenger use area. Therefore, simulation is performed while considering the train entry and exit statuses and the passengers' boarding behaviors in the same space. If the simulation is not performed simultaneously, it will not be able to meet the changes in trains' spatial demands for different passengers. However, trains and passengers are objects of different agent types and have different characteristics. Therefore, it is necessary to establish the various agent types for different agents. Moreover, because the train travels on a track, it is divided into two service states—inbound and outbound—so there are no roaming and collision problems. The number of train cars is set to 8 according to the number of platform doors. The length of each car is 23.5 m, and the train's arrival and travel times are controlled by parameters in the simulation. The basic traveling speed of the train is 20 m/s, and the train is set to have an acceleration/deceleration state when entering and leaving the station. The transition mode of the cyclic state is shown in Figure 4. In this study, the train agent is connected to the process in the initial state of the train, and as shown in the flowchart, the position of the inbound and outbound track, as well as the length and running speed of the car, are set. Before entering the model, we establish that no train is on the track, and we set the speed of entry and departure as well as the stops for passengers. In addition, we set the time for the boarding of passengers in a delayed state for simulation of both passengers and cars, and then we set a fixed cyclic state after the train leaves and change the cycle time according to demand. This is used as a train agent simulation process.

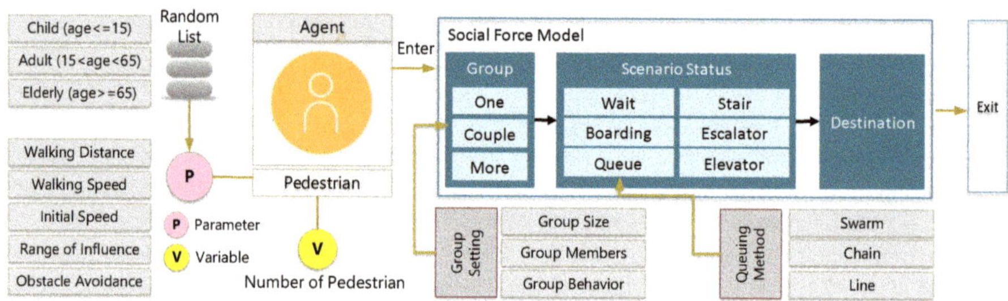

Figure 3. Pedestrian agent basic settings and behavior flow chart.

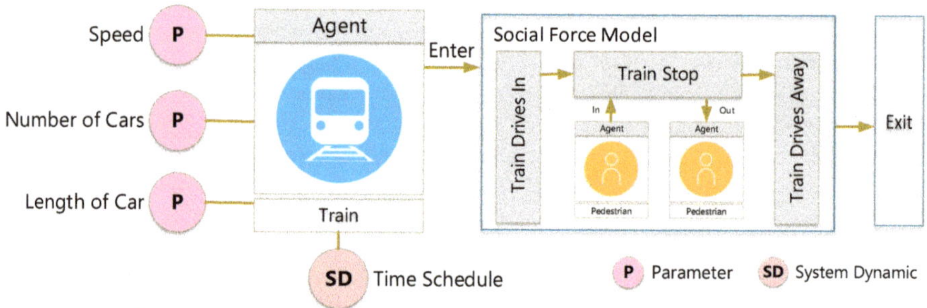

Figure 4. Train agent basic settings and state transition diagram.

Simulation of the congestion conditions requires knowing the number of people entering the model, the number of users of each space, train arrival times, the number of passengers brought in by each train, the hourly passenger volume, the area of the space, etc. The walking routes of agents are recorded and used in the simulation to derive the results.

4.2. Modeling for Noise

Balaras et al. [25] studied the indoor environment quality of Greek airports in 2003. The study showed that noise is a major problem, with a dissatisfaction rate of 78%. This study reflects that noise is one of the main factors affecting the quality of space service. Sound is a perception of human hearing, and noise pollution in the space will cause discomfort to people. Passengers walking and talking in public environments produce basic sounds, which can have a superimposing effect in the space. From an acoustic point of view, the human ear can detect sound due to rapid pressure changes in the air transmitting the sound. Therefore, the noise in this study is calculated in terms of the Sound Pressure Level (SPL) in decibels (dB) [26,27]. It is defined as the common logarithm of the ratio of the effective value of the measured sound pressure $p(e)$ and the reference sound pressure $p(ref)$, multiplied by 20, as given by Equation (1). The general value of the reference sound pressure $p(ref)$ in the air is 2×10^{-5} Pa.

$$SPL = 20log_{10}\left(\frac{p(e)}{p(ref)}\right)$$ (1)

In the MRT space, passengers will create other basic sounds, such as speaking, phone calls, or footsteps, which affect the environment. In this study, the SPL is added to the pedestrian agent's self-behavior with a random parameter number such that the passenger gains a decibel value of sound when entering the space. At the same time, to evaluate the total noise amount in each space, the total SPL generated by the cluster is calculated according to Equation (2).

$$SPL_{(toatl)} = 10log_{10}\sum_{i=1}^{n} 10^{\frac{SPLi}{10}}$$ (2)

This study constructs a noise model based on the above description. Passengers talking to each other, footsteps, and phone sounds are added to pedestrian agent behaviors as variables, and each agent randomly generates only one type of sound. Pedestrian agents walk in different spaces according to specific behaviors and routes. The number of people and different sound states in each space is shown in Figure 5. About parameter settings for affecting the space, the total SPL obtained is considered as the basis for evaluating the decibel levels of the sound generated by each passenger, and other sounds increasing the decibel level result are considered. Then the impact score due to noise in each space is calculated to facilitate follow-up space service quality result measurement. The parameter settings required for the simulation are shown in Figure 6.

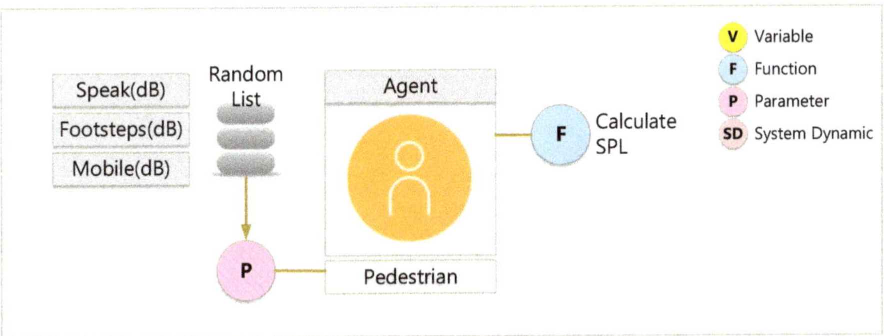

Figure 5. Related settings and methods of Pedestrian agent voice influence.

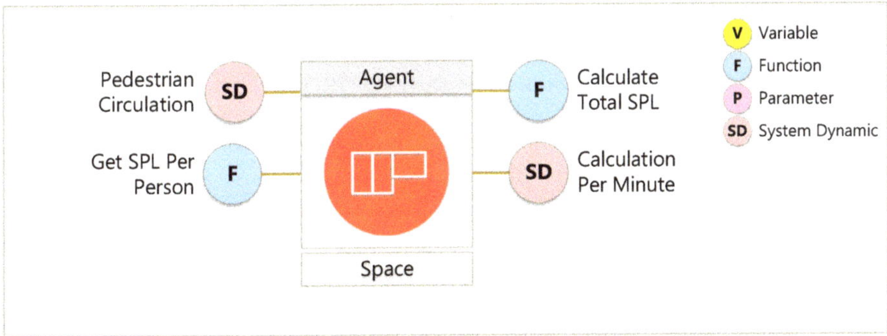

Figure 6. Relevant settings and methods of Space agent noise impact.

4.3. Modeling for Air Quality

Air quality is one of the main factors affecting the space environment. A concentration of carbon dioxide of 1000 ppm or higher in an indoor environment will cause dizziness and tiredness in people and affect their work mood. If the content of carbon dioxide is too high, it will harm the human body, causing hypoxia, numbness in the hands and feet, and loss of consciousness, or even difficulty in breathing, coma, and possibly suffocation. Therefore, this study considers air quality for space service quality and uses the concentration of carbon dioxide as the main simulation subject. To calculate the carbon dioxide equivalent produced by each passenger every minute during the simulation, the amount of air inhaled in each breath, the number of breaths, the amount of ventilation per minute per person, and the space area are set as variables in this study. Based on the simulation time for the method, the carbon dioxide content exhaled per minute per person can be calculated as shown in Equation (3).

$$C_p = (N_{breath} \times V_{breath}) \times R_{CO_2} \tag{3}$$

where

C_p: The concentration of carbon dioxide produced per person per minute
N_{breath}: Number of breaths per minute
V_{breath}: Volume of each breath
R_{CO_2}: The proportion of carbon dioxide in the air

Since the space has been set to the agent type, the spatial parameter can be set in the pedestrian agent through a variable reference for the calculation. Owing to the movement of air and passengers, there is no fixed result, and it is necessary to focus on the causal feedback relationship between the overall simulation process and a large number of variables. To understand the mutual influence of the movement of people on the carbon dioxide level in each space, from the perspective of system dynamics, the passengers in the simulation process are considered to have a pure level initially, which can be accumulated or reduced as time goes. During the simulation, through the interactive relationship between carbon dioxide level and passenger behavior, the feedback of the information obtained from the interaction results in the change of the carbon dioxide volume and the behavior of the impact rate.

In the planning process of the carbon dioxide model, we must first understand the setting parameters of the carbon dioxide air exchange required by the pedestrian simulation, as shown in Figure 7. This allows determining the amount of carbon dioxide generated by each passenger in the space. Next, the passengers are randomly generated in the space, and the carbon dioxide emissions are continuously calculated. The emissions are then fed back to the space agent to calculate the overall carbon dioxide concentration. In addition to the carbon dioxide produced by passengers, each space has the original carbon dioxide value generated by environmental equipment, this must also be included in the calculation. Furthermore, considering the poor ventilation environment of an MRT station, most of the air conditioning systems use forced ventilation to improve space ventilation efficiency. Therefore, this study also takes into account the ventilation rate to obtain the total value of carbon dioxide concentration accumulated in the space, as shown in Figure 8. The space will have reduced air quality due to the increase in the number of passengers. Therefore, the number of passengers in each space is obtained through simulation, and the current carbon dioxide concentration in each space is calculated with the carbon dioxide equation given by Equation (4).

$$C_{space} = \left(\sum_{0}^{n} C_p + C_o - R_v \times T \right) \tag{4}$$

where

C_{space}: Carbon dioxide concentration in the space
n: Number of persons in the space C_{CO_2}: The amount of carbon dioxide exhaled by each person
C_o: Original CO_2 content in the space (ppm)

R_v: Ventilation Rate (ppm/minute)

T: Time (minute)

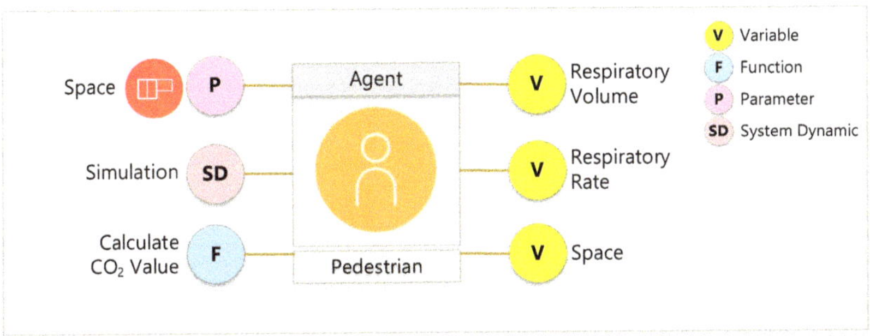

Figure 7. Basic settings for CO_2 emissions by Pedestrian agents.

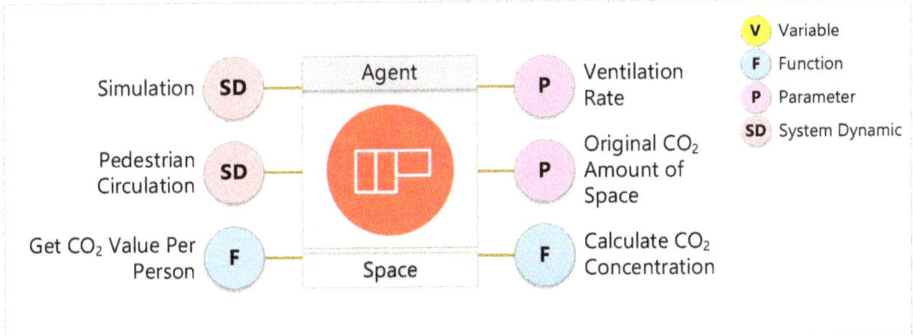

Figure 8. Relevant parameter settings of total CO_2 produced by Space agents during the simulation.

4.4. Space Service Quality Measurement

In addition to congestion, noise and air quality are important factors influencing the evaluation of space service quality. Overcrowding will lead to greater noise and air pollution. These three factors are simulated separately in this research, and the results are combined to derive the final score for overall space quality.

Table 1 indicates the influence score (Qc) for congestion conditions [28]. In the color schema, blue corresponds to a score of 1, indicating a sparse density of less than 0.31 persons/m^2, the non-congested condition of 0.32–0.72 persons/m^2 is represented by green, corresponding to a score of 2. A score of 3 denotes a normal condition of 0.72–1.08 persons/m^2, shown in yellow. Orange corresponds to a score of 4, representing a slightly congested condition with 1.09–2.5 person/m^2, and red, with the highest score of 5, means a highly congested condition with a distribution of greater than 2.5 persons/m^2.

Table 1. Score table for congestion conditions.

Color					
Density (persons/m^2)	Sparse (< 0.31)	Non-congested (0.32–0.72)	Normal (0.72–1.08)	Slightly congested (1.09–2.5)	Highly congested (> 2.5)
Influence score (Qc)	1	2	3	4	5

When passengers enter the space, the system will calculate the sound of one person, and the result will be used to analyze the effect of the entry of agents on the noise level. This system refers to a WHO research report [29] in defining the influence scores (Qn) for the noise levels and effects as shown in Table 2. The noise of less than 40 dB is scored 0, while the noise of more than 120 dB is scored 5. The effect of noise on space service quality is measured as such.

Table 2. Noise levels and effects.

Noise Level (dB)	< 40	40–60	60–80	80–100	100–120	> 120
Effects	Suitable for sleep	Affecting study	Disturbing conversation	Low work efficiency	Hearing damage	Permanent hearing loss
Influence score (Qn)	0	1	2	3	4	5

The system based on the ASHRAE standard [30] defines five levels corresponding to different colors and scores, as shown in Table 3. "Good" is scored 1 point and represented by green, "unhealthy" is scored 3 points and represented by red, "hazardous" is scored 5 and represented by brown. A higher score implies lower quality.

Table 3. Color scheme and influence score for air quality.

Concentration (ppm)	Air Quality Description	Influence Score (Qa)	Color
> 700	The CO_2 level at which people can stay in the room	1	Good
> 1000	The CO_2 level in normal situations	2	Moderate
> 1500	Air pollution	3	Bad
> 2500	Headache, drowsiness, difficulty in concentrating	4	Unhealthful
> 5000	Hypoxia, brain damage, or even death	5	Hazardous

This method of space service quality measurement is illustrated in Figure 9. The highest score for the overall space service quality is 15. A higher score indicates poorer space service quality. The scores are provided to relevant parties for modification and adjustment to achieve high-quality planning of space service.

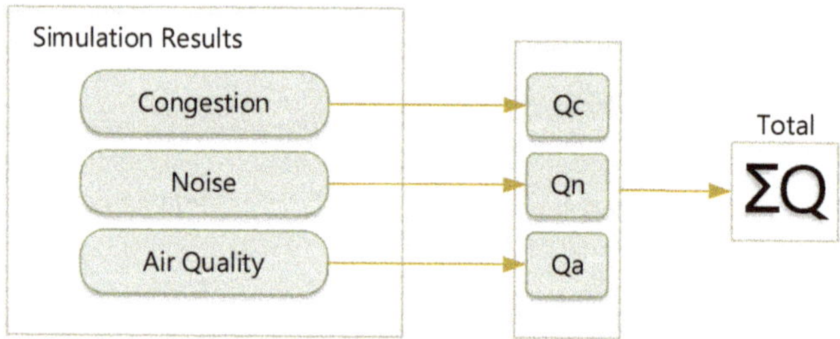

Figure 9. Measurement of space service quality.

5. Demonstration

In this study, 3800 people are imported into the simulation system to represent the peak traffic time of the Daan Park metro station, and the possible scenarios are set. For example, considering passengers entering and exiting the station at entrances, cashing out, purchasing tickets, checking tickets, boarding the trains, and even moving from location to location allows for more realism in the simulation, thus enabling potential problems and difficulties to be evaluated and observed. This allows management and decision-makers to produce more accurate judgments and analysis before the actual implementation. Before setting the congestion state, the space configuration and planning must be completed, the pedestrian agent process and logic settings must be completed, and the corresponding train agent and boarding behavior agent must be set to understand the state of congestion in the simulation. The space configuration planning status, such as the platform level in this study, is divided into pedestrian agent needs and train agent needs, in which track area, waiting area, and other area configurations and planning are completed, as shown in Figures 10 and 11.

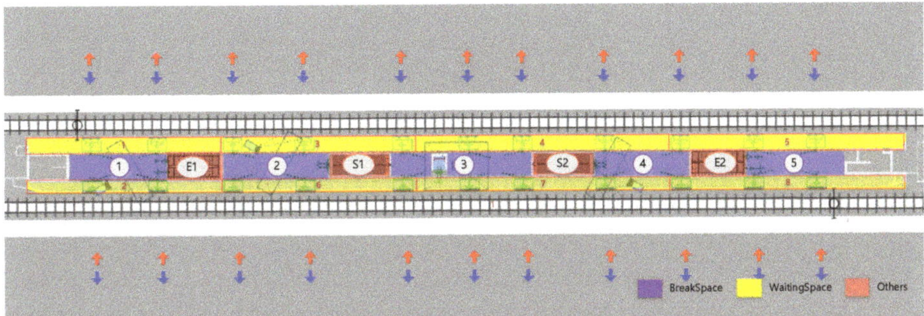

Figure 10. Planned platform space configuration of the Daan Park Station.

	BreakSpace1	BreakSpace2	BreakSpace3	BreakSpace4	BreakSpace5	Stair1	Stair2	Escalator1	Escalator2
Area(M^2)	70.6	50.68	66.125	48.4	63.67	30.2	33.2	23.3	33.2
Volume(M^3)	190.62	136.836	178.5375	130.68	171.909	81.54	89.64	62.91	89.64
	WaitingSpace1	WaitingSpace2	WaitingSpace3	WaitingSpace4	WaitingSpace5	WaitingSpace6	WaitingSpace7	WaitingSpace8	
Area(M^2)	293	282.875	293	293	293	282.875	282.875	282.875	
Volume(M^3)	791.1	763.7625	791.1	791.1	791.1	763.7625	763.7625	763.7625	

Figure 11. The area and volume of the space configuration in the Daan Park Station.

This study presents all the data in the main editor of the software system after the space, passengers, trains, and congestion density are set. One of the simulation results is shown in Figure 12. This system simulates different floor spaces separately. The simulation results of the Hallway indicate that the ATM location, ticket machine location, entrance and exit locations, and changes in pedestrian circulation greatly affect the degree of space usage. In addition, the circulation chaos caused by the device locations increases the level of crowding in the space. Moreover, the sizes of the entrances and exits are a factor in congestion. If the equipment locations were set according to the circulation requirements, the practical function and quality of the space of this MRT station would be greatly improved.

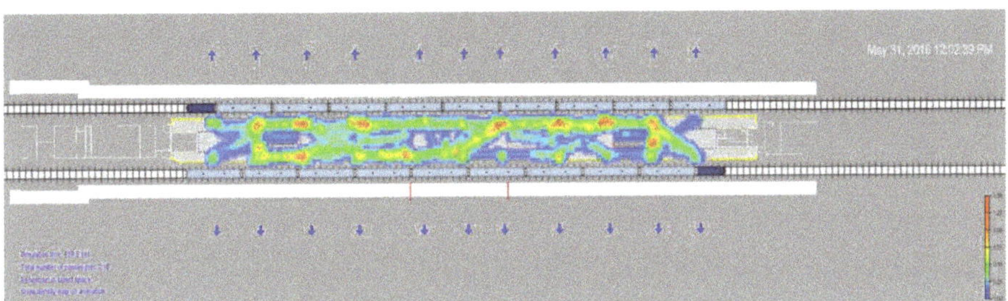

Figure 12. Simulated density distribution of space congestion on the platform.

In terms of noise, this study adds sound factors to the pedestrian agent's self-behavior, so the passengers themselves have sound parameters. There are different volume levels according to different parameters of the sound, and the range of influence of the sound will vary according to the decibel level. During the simulation, you can clearly see the decibel presentation status issued by each person, and the influence range of the agent's own noise will be visualized in the simulation, as shown in Figure 13. In reality, the passengers have a multiplying relationship with the sounds in the space. Therefore, in this study, the total number of people in the space is simulated, and then the total sound pressure value is calculated for all decibel values; the simulation results of each space are obtained through Equation (2) and are shown in Figure 14. The results clearly indicate the locations and distributions of places where the noise gathers.

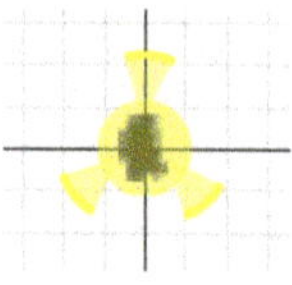

Figure 13. Extent of the sound volume generated by the agents.

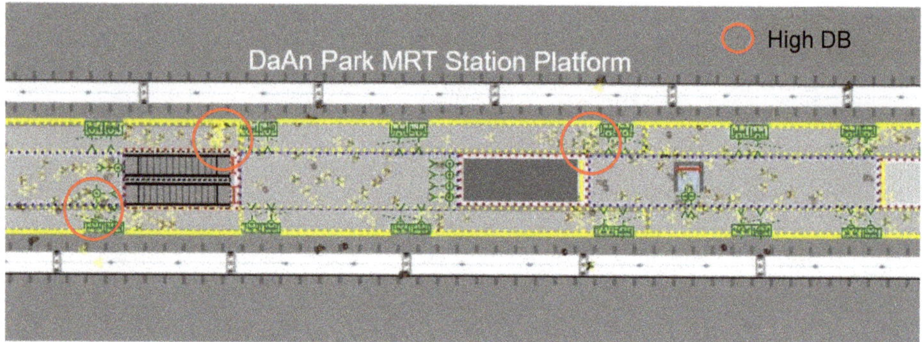

Figure 14. Visual representation of noise agents.

The system also calculates the decibel level of each space after the simulation is performed and presents the calculation results in the form of a bar chart, as shown in Figure 15. These results provide managers with an understanding of how the decibel levels change in the spaces in the simulation.

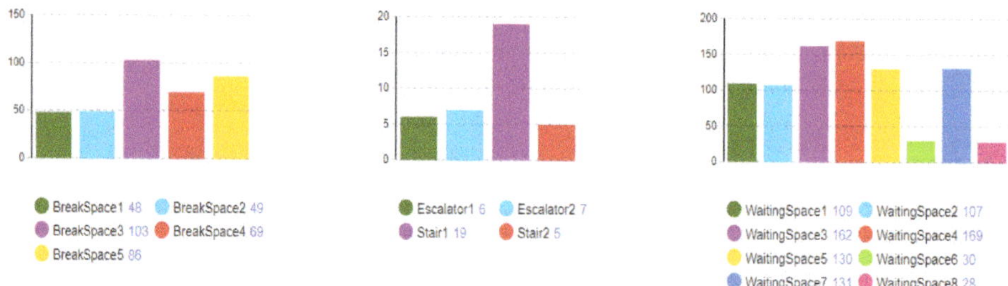

Figure 15. Changes in decibel levels in spaces in the simulation.

To simulate the space air service quality, the space is set as an agent type, and the carbon dioxide concentration is used. According to the above-mentioned parameters and settings required for the simulation of the congestion state and the setting for the carbon dioxide in the air, including carbon dioxide concentration, space area, number of people, etc., the air quality-related parameters are connected through the space environment agent. The number of passengers in each space during the simulation is obtained, and the current carbon dioxide concentration in each space is calculated according to Equation (4). During the simulation, the user can mouse-click any space to select it and obtain the current number of users in the space and the current carbon dioxide concentration. The actual simulation results are shown in Figure 16.

In the system in this study, the impact of an agent's carbon dioxide emissions caused when the agent enters each space during the simulation process is presented through a line graph, as shown in Figure 17. The system calculates the data changes every 15 s for planners to understand the current status of the space visually; the data vary for different time periods and simulation times. Users can understand the changes in carbon dioxide concentrations from the data recorded in this graph and then return to the model to understand the relationship between the state of the space and the change in air quality. This enables planning personnel to change the design as well as the circulation needs.

Figure 16. Carbon dioxide concentration and passenger simulation results in the space.

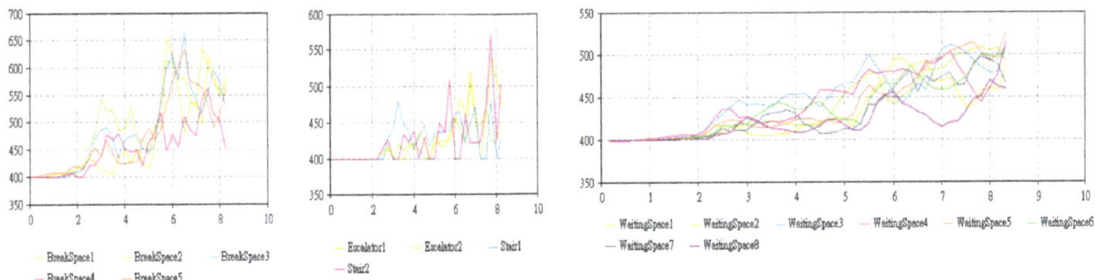

Figure 17. Changes in carbon dioxide concentration in the simulation.

The highest overall space service quality score is 15, based on the congestion state score of Table 1, the noise and decibel impact state evaluation of Table 2, and the carbon dioxide concentration evaluation score of Table 3. The three evaluation scores are summed for the overall score, with higher scores indicating lower space service quality. This study uses statistical bar graphs for the scores obtained for each space, as shown in Figure 18. Then it provides relevant units for modification and adjustments to achieve high-quality space service planning.

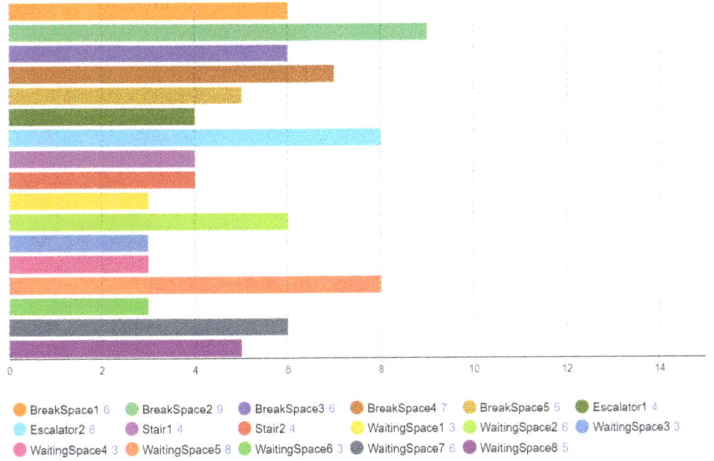

Figure 18. Space service quality measurement results.

The system can also export the relevant information service quality measurement results of each space to an Excel spreadsheet, which can be used by subsequent personnel in related fields for decision-making, as shown in Figure 19.

	BreakSpace1	BreakSpace2	BreakSpace3	BreakSpace4	BreakSpace5	Stair1	Stair2	Escalator1	Escalator2	WaitingSpace1	WaitingSpace2	WaitingSpace3	WaitingSpace4	WaitingSpace5	WaitingSpace6	WaitingSpace7	WaitingSpace8
Density	0.2975	0.1381	0.2117	0.2066	0.1885	0.1987	0.0904	0.1288	0.0602	0.0375	0.0424	0.0887	0.0590	0.0663	0.0804	0.0530	0.0389
Noise	125	43	86	62	75	39	20	20	10	68	73	163	173	124	150	90	61
CO2	695.3513	522.2488	752.5142	744.0779	713.7848	641.5472	509.7763	556.5231	473.1364	412.1662	413.9555	430.4824	434.1457	423.1559	430.3977571	417.7498463	412.6906819
Qc	1	1	1	1	1	1	1	1	1	1	1	1	1	1	1	1	1
Qn	5	2	3	2	2	0	0	0	0	2	2	5	5	5	5	3	2
Qa	2	1	2	2	2	1	1	1	1	1	1	1	1	1	1	1	1
Space Service Quality (Q)	8	4	4	5	5	2	2	2	2	4	4	7	7	7	7	5	4

Figure 19. Excel report of space service quality measurement.

When space planners or decision-makers receive the information, they can accordingly adjust or change spaces with low service quality. For example, BreakSpace2, which has a space service quality score of 9, originally has an area is 50.68 m^2. If its area is increased to 75 m^2, the original settings of passenger flow, noise, and carbon dioxide settings will still affect the parameter values for simulation after the modified model is imported. In larger spaces, more people can be accommodated. This implies that the noise and carbon dioxide concentration will be relatively increased, but the overall service quality score after the simulation will be significantly reduced to 4 due to changes in space conditions affecting pedestrians' circulation, which in turn affects the adjacent space quality score.

6. Conclusions

To study the current space usage, this research used Daan Park Station as a case study to simulate streams of people entering and exiting the station from trains or from the outside. It also proposed combining building information modeling technology and an agent-based model to simulate the interaction of agents in the space. A study of the published literature revealed that, in addition to space planning and route interruptions, factors that can lead to a low quality of space service include noise and air quality. Therefore, this research set these factors as agents, including passengers, space, noise, and air quality. The results on space service quality were presented in 2D and 3D visualizations. Possible scenarios were visualized to provide solutions for the space design of an MRT station and route planning. Different colors were used to show and distinguish the space usage so as to provide decision-makers with a better understanding of the actual space usage and service quality at MRT stations through visual presentation. Simple equations were used to combine simulation results for the derivation of the space service quality score.

In the present study, the three influencing factors were simulated comprehensively. We expect to integrate various relevant factors and provide various infographics and dashboards in the future to present results that bear a better resemblance to reality. Good visualization results will be used as a bridge to facilitate communication with other relevant parties so that planning personnel can make space adjustments and other modifications. We would also like to provide these results as feedback for the space designs of MRT stations and routes and the analysis of alternative options with the aim of reducing the labor and costs associated with design variations.

Author Contributions: The work described in this article is the collaborative development of all authors. I.-C.W. devised the project, the main conceptual ideas, and proof outline. Y.-C.L. carried out the implementation. I.-C.W. took the lead in writing the manuscript. Both authors provided critical feedback and helped shape the research, analysis, and manuscript. All authors have read and agreed to the published version of the manuscript.

Funding: This research was funded by Taiwan Ministry of Science and Technology, grant number 103-2221-E-151-021-MY3.

Acknowledgments: The authors are grateful for the support of the CeIT Laboratory, BIM Research Center, and University.

Conflicts of Interest: The authors declare no conflict of interest.

References

1. Bahrehmand, A.; Batard, T.; Marques, R.; Evans, A.; Blat, J. Optimizing layout using spatial quality metrics and user preferences. *Graph. Model.* **2017**, *93*, 25–38. [CrossRef]
2. Guo, Z.; Li, B. Evolutionary approach for spatial architecture layout design enhanced by an agent-based topology finding system. *Front. Arch. Res.* **2017**, *6*, 53–62. [CrossRef]

3. Francis, J.; Wood, L.; Knuiman, M.; Giles-Corti, B. Quality or quantity? Exploring the relationship between Public Open Space attributes and mental health in Perth, Western Australia. *Soc. Sci. Med.* **2012**, *74*, 1570–1577. [CrossRef]

4. Li, J.; Song, Y.; Lv, S.; Wang, Q. Impact evaluation of the indoor environmental performance of animate spaces in buildings. *Build. Environ.* **2015**, *94*, 353–370. [CrossRef]

5. Wang, Z.; Zhao, H.; Lin, B.; Zhu, Y.; Ouyang, Q.; Yu, J. Investigation of indoor environment quality of Chinese large-hub airport terminal buildings through longitudinal field measurement and subjective survey. *Build. Environ.* **2015**, *94*, 593–605. [CrossRef]

6. Tomé, A.; Kuipers, M.; Pinheiro, T.; Nunes, M.; Heitor, T. Space–use analysis through computer vision. *Autom. Constr.* **2015**, *57*, 80–97. [CrossRef]

7. Hayek, U.W.; Efthymiou, D.; Farooq, B.; Von Wirth, T.; Teich, M.; Neuenschwander, N.; Grêt-Regamey, A. Quality of urban patterns: Spatially explicit evidence for multiple scales. *Landsc. Urban Plan.* **2015**, *142*, 47–62. [CrossRef]

8. Brömmelstroet, M.T.; Bertolini, L. Developing land use and transport PSS: Meaningful information through a dialogue between modelers and planners. *Transp. Policy* **2008**, *15*, 251–259. [CrossRef]

9. Russo, F.; Rindone, C. Container maritime transport on an international scale: Data envelopment analysis for transhipment port. *WIT Trans. Ecolog. Environ.* **2011**, *501*, 831–843.

10. Musolino, G.; Rindone, C.; Vitetta, A. Evaluation in Transport Planning: A Comparison between Data Envelopment Analysis and Multi Criteria Decision Making Methods. In Proceedings of the European Simulation and Modelling Conference, Lisbon, Portugal, 25–27 October 2017.

11. Indraprastha, A.; Shinozaki, M. Computational models for measuring spatial quality of interior design in virtual environment. *Build. Environ.* **2012**, *49*, 67–85. [CrossRef]

12. Zawidzki, M.; Szklarski, J. Multi-objective optimization of the floor plan of a single story family house considering position and orientation. *Adv. Eng. Softw.* **2020**, *141*. [CrossRef]

13. Abar, S.; Theodoropoulos, G.K.; Lemarinier, P.; O'Hare, G.M. Agent Based Modelling and Simulation tools: A review of the state-of-art software. *Comput. Sci. Rev.* **2017**, *24*, 13–33. [CrossRef]

14. Macal, C.M. Everything you need to know about agent-based modelling and simulation. *J. Simul.* **2016**, *10*, 144–156. [CrossRef]

15. Lee, Y.S.; Malkawi, A.M. Simulating multiple occupant behaviors in buildings: An agent-based modeling approach. *Energy Build.* **2014**, *69*, 407–416. [CrossRef]

16. Osman, H. Agent-based simulation of urban infrastructure asset management activities. *Autom. Constr.* **2012**, *28*, 45–57. [CrossRef]

17. Langevin, J.; Wen, J.; Gurian, P.L. Simulating the human-building interaction: Development and validation of an agent-based model of office occupant behaviors. *Build. Environ.* **2015**, *88*, 27–45. [CrossRef]

18. Bradley, A.; Li, H.; Lark, R.; Dunn, S. BIM for infrastructure: An overall review and constructor perspective. *Autom. Constr.* **2016**, *71*, 139–152. [CrossRef]

19. Santos, R.; Costa, A.A.; Silvestre, J.D.; Pyl, L. Informetric analysis and review of literature on the role of BIM in sustainable construction. *Autom. Constr.* **2019**, *103*, 221–234. [CrossRef]

20. Durmisevic, S.; Sariyildiz, S. A systematic quality assessment of underground spaces—Public transport stations. *Cities* **2001**, *18*, 13–23. [CrossRef]

21. Lee, M.C.; Mui, K.W.; Wong, L.T.; Chan, W.Y.; Lee, E.W.M.; Cheung, C.T. Student learning performance and indoor environmental quality (IEQ) in air-conditioned university teaching rooms. *Build. Environ.* **2012**, *49*, 238–244. [CrossRef]

22. Geng, Y.; Yu, J.; Lin, B.; Wang, Z.; Huang, Y. Impact of individual IEQ factors on passengers' overall satisfaction in Chinese airport terminals. *Build. Environ.* **2017**, *112*, 241–249. [CrossRef]

23. Tang, H.; Ding, Y.; Singer, B.C. Interactions and comprehensive effect of indoor environmental quality factors on occupant satisfaction. *Build. Environ.* **2020**, *167*. [CrossRef]

24. Helbing, D.; Farkas, I.; Vicsek, T. Simulating dynamical features of escape panic. *Nature* **2000**, *407*, 487–490. [CrossRef] [PubMed]

25. Balaras, C.A.; Dascalaki, E.; Gaglia, A.; Droutsa, K. Energy conservation potential, HVAC installations and operational issues in Hellenic airports. *Energy Build.* **2003**, *35*, 1105–1120. [CrossRef]

26. Liu, C.; Chen, J.; Zhang, Y.-B.; Zhang, X.-Z.; Li, J.-Z. A method of measuring the powertrain noise for the indoor prediction of pass-by noise. *Appl. Acoust.* **2019**, *156*, 289–296. [CrossRef]

27. Liu, X.; Li, L.; Chen, G.-D.; Salvi, R. How low must you go? Effects of low-level noise on cochlear neural response. *Hear. Res.* **2020**, *392*. [CrossRef]
28. Fruin, J. *Pedestrian Planning and Design*; Metropolitan Association of Urban Designers and Environmental Planners: New York, NY, USA, 1971.
29. Goelzer, B.; Hansen, C.H.; Sehrndt, G. *Occupational Exposure to Noise: Evaluation, Prevention and Control*; World Health Organization: Geneva, Switzerland, 2001.
30. ASHRAE. *Ventilation for Acceptable Indoor Air Quality*; American Society of Heating, Refrigerating and Air-Conditioning Engineers Inc.: Atlanta, GA, USA, 2016.

Article

Natural Hazard Influence Model of Maintenance and Repair Cost for Sustainable Accommodation Facilities

Sang-Guk Yum [1], Ji-Myong Kim [2] and Kiyoung Son [3],*

[1] Department of Civil Engineering and Engineering Mechanics, Columbia University, New York, NY 10027, USA; sy2509@columbia.edu

[2] Department of Architectural Engineering, Mokpo National University, Mokpo 58554, Korea; jimy@mokpo.ac.kr

[3] School of Architectural Engineering, Ulsan University, Ulsan 44610, Korea

* Correspondence: sky9852111@ulsan.ac.kr

Received: 28 May 2020; Accepted: 17 June 2020; Published: 18 June 2020

Abstract: To optimally maintain buildings and other built infrastructure, the costs of managing them during their entire existence—that is, lifecycle costs—must be taken into account. However, due to technological improvements, developers now build more high-rise and high-performance buildings, meaning that new approaches to estimating lifecycle costs are needed. Meanwhile, an accelerating process of industrialization around the world means that global warming is also accelerating, and the damage caused by natural disasters due to climate change is increasing. However, the costs of losses related to such hazards are rarely incorporated into lifecycle-cost estimation techniques. Accordingly, this study explored the relationship between, on the one hand, some known parameters of natural disasters, such as earthquakes, high winds, and/or flooding, and on the other hand, the data on exceptional maintenance costs, represented by gross loss costs, generated by a large international hotel chain from 2007 to 2017. The regression model used revealed a correlation between heavy rain and insurance-claim payouts. This and other results can usefully inform safety and design guidelines for policymakers, both in disaster management and real estate, as well as in insurance companies

Keywords: natural disaster; risk management; accommodations; operations and maintenance; lifecycle cost; disaster management

---

1. Introduction

As the sizes and heights of buildings continue to increase, new approaches for estimating and managing their lifecycle costs have become necessary [1,2]. Construction's impacts on development, society, the environment, and the economy should all be considered as fundamental to considerations of long-term building sustainability. Accordingly, an increasing number of studies are being conducted on buildings' social impacts, including numbers of fatalities during disasters; environmental ones such as CO_2 emissions during deconstruction/demolition; and economic ones such as natural-disaster-related repair costs [3–6]. According to the Intergovernmental Panel on Climate Change [7], average global temperatures have been rising, making natural disasters such as typhoons both more frequent and less predictable. It is therefore very important to assess the maintenance and repair costs that have been associated with such natural disasters in the past as a means of anticipating such costs going forward. Due not only to the increased likelihood of various types of damage related to global warming but also to public demand for greater urban-system resilience, effective estimation of such future costs should comprehensively incorporate those factors that may require structural repair or complete replacement [8]. For this study, hotel facilities were chosen because the hotel business is perhaps uniquely vulnerable to the negative consequences of both poorly maintained facilities and natural disasters [9]. Yet, despite the profound impact that the cosmetic appearance of a building can have on hotel revenues, and despite the long lifespans and considerable age of many hotel buildings, their natural-disaster management tends to be passive rather than proactive, unsystematic, and poorly funded relative to their overall budgets [10,11].

1.1. Research Background

Building maintenance costs are increasing due to the greater frequency of natural disasters and the generally greater heights of new commercial buildings [8,12]. To address this challenge, effective management should take into account the specific features of every building, along with a comprehensive range of factors that might cause that building's functionality to deteriorate. In recent decades, many property managers have applied asset-management techniques to more efficiently deal with the maintenance costs incurred in the operating stage, which account for the highest proportion of any building's total lifecycle costs [13], the other stages being planning, feasibility studies, basic design, execution design, construction, and demolition. However, asset management can be difficult for many entities to implement since, as well as being building-specific, it requires information compatibility across all stages of the building's lifecycle, massive quantities of maintenance materials, and long-term investment. A considerable body of asset-management research is devoted to mitigating these drawbacks, but such studies rarely consider the relationship between natural disasters and maintenance costs. The present paper addresses this gap in the literature, using 11 years of data on an international hotel chain's insurance-claim payouts.

1.2. Research Objective

A hotel chain was chosen as this study's research case because an insurance company was willing to provide the researchers with gross loss data on that chain's claim payouts. First, this paper contains a review of the prior research on asset management as it relates to building lifecycles. Second, it features hotel property-loss data on the 2007–2017 period, including the gross loss, loss factor, and date of loss, to explore the relationships between natural-hazard factors and operation and management costs. Finally, this paper includes a regression analysis of the data collected to identify the correlations among maintenance costs, damage, and the incidence and intensity of earthquakes, high winds, and flooding.

2. Literature Review

Facility management comprises professional methodologies aimed at ensuring the functionality of properties and built environments (International Facility Management Association, 1992, 2015). Its techniques, which include lifecycle-cost estimation, address safety and durability, as well as economic considerations [14]. Lee and Jung's [15] comparison of facility-management practices in the United States, Canada, Australia, and South Korea suggests that, although all four countries focused on the operation and management stage, which occupies, on average, 85% of the building lifecycle [13], only Australia emphasized a lifecycle-cost approach to managing costs. Specifically, Lee and Jung [15] conducted a high-volume review of the existing literature on applied facility management and categorized this discipline's functions into 19 types, covering property, service, space, communication, energy, environment, equipment, moves, quality, security, costs, documents, human resources, materials, outsourcing, regulations, schedules, technology, and general management.

Yu et al. [16] proposed a methodology for developing facility-management functions and their computerization. Foster [17] emphasized the importance of operation and management costs, especially energy costs, which account for 25% of all operation costs, but which many U.S. federal buildings were found to neglect. The same author also advocated the establishment of a strategy to reduce energy costs in the operation and management stage. Williams et al. [18] investigated the potential usage of building information modeling (BIM) in facility management using a survey and interviews to explore the gaps between real-world applications and common perceptions. They found that, although there was still a need to improve and educate facility-management professionals about real-world utilization of BIM, this approach to facility management should be adopted due to its usefulness not only for information exchange but also for collaboration among construction stakeholders.

Lifecycle costs have also been recognized as an important basis for the improvement of structural resilience, and in that context, American Society for Testing and Materials (ASTM) [19,20] developed a standard for the quantification and specification of costs at each stage of a building's lifecycle to produce more accurate lifecycle-cost estimates. Several researchers have also developed methodologies for improving lifecycle-cost assessments, with some focusing on costs during the early design process of buildings and other types of infrastructure. For example, matrix-based frameworks for choosing cost-efficient materials were investigated by Pettang et al. [21], whose estimates of projects' construction costs included labor, materials, and operation and management costs, in an effort to support decision making by construction stakeholders in a range of material scenario. Later, Günaydın and Doğan [22] proposed a novel cost-estimation approach based on artificial neural networks (ANNs) but, again, focused on the early stage of building construction. Another approach to creating an accurate construction-cost estimation model, developed by Kim et al. [23], was built around three different methods—statistical analysis (regression modeling), ANNs, and case-based reasoning—and established that it could effectively manage construction projects' costs in their early stages. However, their approach did not consider long-term operation and management costs despite the fact that they account for 85% of lifecycle costs [7].

The effects of aging on buildings were investigated by Rahman et al. [24], who proposed a decision framework for simultaneously evaluating various criteria, including resilience, energy, cost-effectiveness, durability, and environment. They concluded that multiple aspects of building performance should be considered during the operation and management stage, and the materials should be altered accordingly. Another perspective on lifecycle costs focused on energy consumption has arisen amid the development of advanced building materials and technologies with the potential to make buildings more energy-efficient, which in turn, would likely reduce costs and lessen environmental impacts [12]. For example, Hasan [25] used lifecycle-cost assessment to optimize wall thickness for purposes of insulation; Kneifel [26] investigated the effects of energy-efficient design on commercial buildings' lifecycle costs, energy consumption, and carbon emissions; Morrissey and Horne [27] studied the interrelationships of new buildings' thermal properties, initial construction costs, and whole-life energy costs; and Gluch and Baumann [28] applied the concept of lifecycle costs to a proposed framework for eco-friendly decision making.

In addition to investigating tools and techniques for the effective management for high-performance buildings, such as increasing their energy efficiency as discussed above, a comprehensive lifecycle assessment still needs to take into account the repair costs arising from natural hazards if overall asset management is to be effective. Prior research has utilized building characteristics such as height, area, and price as variables for the extent of damage to properties [29,30]. According to those studies, building height had a clear statistical relationship with financial losses caused by natural hazards, notably windstorms.

As well as the relationships between building features such as height and losses from natural hazards, some researchers have emphasized the importance to lifecycle costs of damage by such hazards, despite the inherent randomness with which such events strike, both by building type and geographically. As noted by Chang and Shinozuka [31] in connection with the 1994 Northridge (USA) and 1995 Kobe (Japan) earthquakes, it is tempting to neglect the potential for damage by natural disasters when estimating the lifecycle costs of infrastructure systems due to these many uncertainties. However, through a case study of pipeline systems, they demonstrated the value of extending traditional lifecycle-cost assessment to include potential repair costs and related user costs arising from earthquake damage. Similarly, Wei et al. [32] argued that the potential cost of damage from earthquakes should be added to lifecycle costs when evaluating long-term building performance. Nevertheless, it remains very difficult to estimate property losses caused by natural hazards, as both their frequency and intensity are inherently random and uncertain [12].

Due to global warming, unexpected natural disasters have been increasing in frequency, driving up lifecycle costs during buildings' operation and management stage. In addition, high-performance and high-rise buildings—which are increasingly prevalent due to accelerating urbanization and population growth—have special vulnerabilities to natural hazards, as shown in previous studies on this aspect of lifecycle cost [12,30]. The present paper tackles these problems directly, by proposing a lifecycle-cost assessment method that covers not only expected costs such as routine repairs, but also the exceptional ones associated with natural hazards across buildings' entire lifecycles.

3. Research Methods

3.1. Case-Study Approach and Research Process

In this study, we investigated the relationship between natural disasters and the operation and management costs of a hotel chain that is currently one of the largest of its kind in the world, comprising more than two dozen brands and over 5000 properties around the globe. Despite their geographic dispersal, these properties are similar in terms of construction quality, construction type, and exterior design, among other characteristics. Therefore, their guidelines and methods for estimating lifecycle costs, including operation and management costs, should also be similar. This research relied on the data on 725 incidents of gross losses from natural hazards that this hotel chain incurred from early 2007 to late 2017. The most prevalent type of damage was water-related—a category including floods, overflow, and water-supply facility failures—which comprised 44% of all damage by the number of reported incidents. The second most prevalent was wind-related, including but not limited to hurricanes and typhoons, which comprised 17% of all damage. Other natural disaster-related damage, including but not limited to earthquakes, hail, and wildfires together made up an additional 1%. Prominent among the non-disaster-related incidents that accounted for the remaining 38% of all damage included HVAC failures (13%), fires (6%), and extreme cold (2%).

First, the characteristics of the particular natural disasters that affected the hotel chain's properties were categorized as independent variables. Second, the gross loss data were categorized according to the natural hazards that caused them. In this step, claim-payout amounts served as the dependent variable as a proxy for operation and management costs, while the causes of damage were utilized as the independent variables. Third, a regression analysis was conducted to establish the relationships of the independent and dependent variables. Those variables are described below, along with this study's data-collection procedures and statistical analysis methods.

3.2. Dependent Variable

Losses from individual events that caused damage during the period of interest ranged from less than $10,000 to $57,445,698. The smallest single payout was $37.

3.3. Independent Variables

Although lifecycle-cost assessment can be utilized to design buildings to cope with natural hazards, the expected costs of natural-hazard damage related specifically to building loading are often minimized, which could cause problems [33–35].

According to Harvard's Joint Center for Housing Studies [36], repair costs related to all types of natural disasters made up 8.2% of all improvement expenditures by homeowners in the United States in 2013. At USA $15.8 billion, these hazard repairs were also among the most costly of the 54 categories of expenditure in the same study.

Ayyub [37] emphasized that, among all types of natural disasters, earthquakes were the most severe from the point of view of damage to buildings and infrastructure systems while also having substantial effects on society, the economy, and the environment. Wei et al. [32] noted that many researchers have sought to reduce structures' seismic response, but relatively few have focused on the costs of earthquake-related damage over the course of a building's lifecycle. The intensity of earthquakes is usually represented as peak ground acceleration (PGA) [12].

Hurricanes, for their part, can also be very damaging to buildings and infrastructure systems. Their severity can be characterized according to their maximum wind speed radius, forward-motion speed, and sustained maximum wind speed [38–40]. Among these, sustained maximum wind speed and maximum wind speed radius are the main factors utilized to estimate hurricane damage [38,40,41].

However, some research has emphasized the importance of rainfall in the accurate evaluation of the extent of hurricane damage [42,43]. Recently, damage from flooding has also been on the rise, in part due to the effects of urbanization, including the reshaping of river systems [44,45]. Brody et al. [42] highlighted the importance of effective flood control, given that water systems can easily overflow when heavy rain occurs, magnifying flood damage. Therefore, altitude and distance from such systems are useful indicators of water-related hazards that were adopted for this study.

3.4. Data Collection and Statistical Analysis

In this study, the wind was measured by wind speed, and earthquakes by PGA, while flooding was measured as a combination of precipitation, the distance from water systems, and the difference in altitude from the nearest water system (Table 1). Data on the first three of these (i.e., wind speed, earthquakes, and flooding) independent variables were provided by the National Oceanic and Atmospheric Administration (NOAA), while the latter two were computed using Google Maps. Data on the dependent variable were provided by the insurers of the hotel chain that participated in this research. To establish correlations between the dependent and independent variables, the ordinary least squares regression method was used.

Table 1. Data types and sources.

Variable	Explanation	Measure	Data Source
Wind speed	Sustained maximum wind speed over 10 min	m/s	National Oceanic and Atmospheric Administration (NOAA)
Peak ground acceleration (PGA)	Value of PGA	g	
Precipitation	Total amount of rainfall	mm/day	
Distance from water systems	Linear distance from nearest river, lake, or coastline	m	Google Maps
Altitude difference compared to water systems	Altitude difference compared to the nearest river, lake, or coastline	m	

4. Results

4.1. Descriptive Statistics

Table 2 presents the descriptive statistics of the variables, with N standing for the number of data points (i.e., insurance-claim payouts corresponding to at least one of the types of natural disaster considered in this study).

Table 2. Descriptive statistics of variables.

Category	N	Mean	SD
Dependent Variable			
Maintenance and repair costs (Gross loss) (USD)	725	271,567.5	2,213,233.9
Independent Variables			
Wind speed (m/s)	725	36.6	5.1
PGA (g)	725	0.9	1.0
Precipitation (mm/day)	725	228.4	75.0
Distance from water system (m)	725	55,819.6	741,193.8
Altitude difference compared to water system (m)	725	368.5	3424.3

4.2. Multiple Regression Analysis

Normality test of the dependent variable was conducted to verify if the variable followed normal distribution or not before multiple linear regression analysis. The result showed that the significant level of 0.000 was smaller than 0.05, which means that the dependent variable did not follow a normal distribution. Therefore, the dependent variable was transformed to natural log as follows;

$$\text{Transformed gross loss} = \text{Log (Gross loss (\$))} \tag{1}$$

As seen in Table 3, the result of the normality test with the transformed gross loss showed that the significant value of 0.232 was greater than 0.000. It was proved that the dependent variable was normally distributed. The histogram and Q-Q plot of Figure 1 confirm that the gross loss followed a normal distribution.

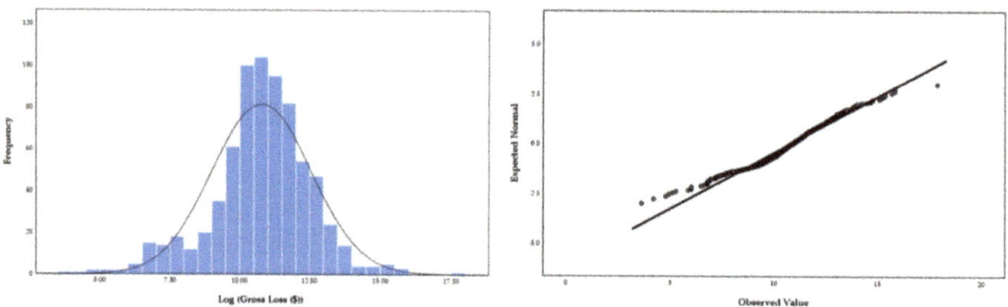

Figure 1. Histogram and Q-Q plot, transformed gross loss.

Table 3. Normality-test results, transformed gross loss.

	Statistic	df	Sig.
Log Gross loss	0.972	725	0.232

The results of our multiple regression analysis for the gross loss connected with natural disasters are shown in Figures 2 and 3 and Table 4. The histogram and *P-P* plot in Figure 2 indicate that the residual of the regression model was normally distributed. The scatter residual plot of the regression model in Figure 3, meanwhile, shows that the variable of the residual was constant, confirming homoscedasticity. In addition, the significance level of 0.000 in Table 4, being smaller than 0.05, indicates that the regression model was statistically significant. It confirms that the relation of the dependent variable to the independent variables was linear. The regression model's R^2 value was 0.342, meaning that it can explain 34.2% of the variation in the dependent variable. The p values indicated that precipitation and the distance from water systems were correlated with the dependent variable, but that the other three independent variables were not. The variance inflation factor (VIF) values of this study's variables ranged from 1.002 to 1.114, which means there was no significant multicollinearity among them.

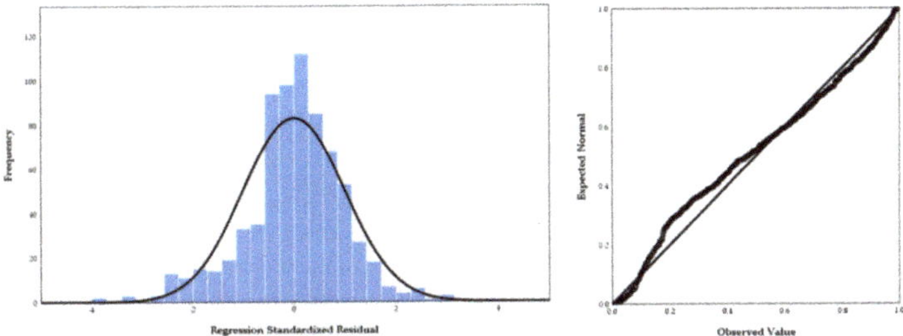

Figure 2. Histogram and *P-P* residual plot, regression model.

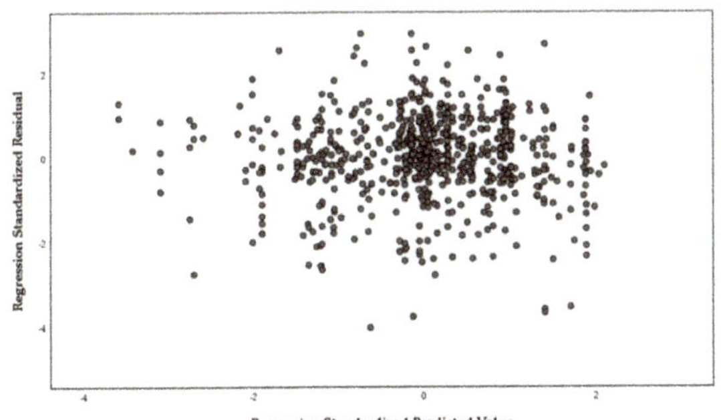

Figure 3. Scatter residual plot, regression model.

Table 4. Regression analysis: final results. VIF: variance inflation factor.

Variables	Coef.	Std. Error	Beta Coef.	$p > |z|$	VIF
Wind speed (m/s)	−0.001	0.001	−0.033	0.394	1.087
PGA (g)	−0.106	0.001	−0.060	0.107	1.034
Precipitation (mm/day)	0.059	0.013	0.168	0.000 *	1.084
Distance from water system (m)	1.761×10^{-7}	0.000	0.074	0.045 *	1.002
Altitude difference compared to water system (m)	1.268×10^{-5}	0.000	0.024	0.527	1.114
Number of observations	725				
F	5.350				
Sig.	0.000				
Adjusted R^2	0.342				

Note. * denotes *p*-value which was smaller than 0.05.

A beta coefficient (standard coefficient) was utilized to compare the independent variables, with the highest absolute value being recognized as having the strongest effect on the dependent variable. Table 4 shows that precipitation (0.168) and the distance from water systems (0.074) had higher beta-coefficient values than the other independent variables did.

5. Discussion

The research method proposed in this paper offers an opportunity to incorporate loss costs arising from natural hazards into the lifecycle cost, specifically by relating operation and management costs to prior natural disasters. Total worldwide gross property damage caused by high winds, flooding, and/or earthquakes during such events cost the insurer of the hotel chain that participated in this study around U.S. $200 million in the 2007–2017 period. The correlation between this gross loss and the full set of chosen variables was confirmed as significant.

Among this set, however, the significance of this correlation was accounted for by just two loss factors, precipitation and distance from water systems, both of which had p values < 0.05 (0.000 for precipitation and 0.045 for distance from water systems). The regression's adjusted R^2 value (0.342) indicated that 34.2% of the variation in the dependent variable, gross loss, can be explained by these two loss factors, while the other 65.8% of the variation in the dependent variable was due to loss factors that were not covered by this research. Thus, through statistical analysis, we discovered that the adopted natural-hazard variables had an important relation to the hotel chain's gross loss.

The findings above reinforce those of previous studies [34,35], which suggest that heavy rainfall and built environments—construction activities or flood-control facilities—are the significant factors in losses arising from natural hazards. Precipitation and distance from water systems are commonly related to flooding damage and, in this case, indicated that heavy rain is likely to cause considerable damage to the hotel chain's properties. An unexpectedly high volume of rain can seep into existing cracks in buildings, leading to severe damage to their interiors, including furniture, partition walls, and other internal structural components. Usually, hotels' basements are used to store essential equipment, but when heavy rain occurs to the point that water systems overflow, basements are very susceptible to inundation.

The identified correlation between two types of natural hazards and gross loss is potentially useful to professionals and policymakers concerned with hotel operations and catastrophe management, as this finding goes some way in addressing the absence of disaster losses in operation and management costs in traditional lifecycle-cost estimation. The present study's findings should also enable insurance companies to modify their business models and/or premium prices based on natural-hazard loss factors and estimates of maximum loss, risk exposure, the probabilities of certain events occurring, and so forth. Likewise, construction companies building hotel facilities may wish to reassess their designs, materials, building features, and safety guidelines from the point of view of vulnerability to precipitation and distance from water systems. In short, the present study confirms that facility management will be greatly enhanced if due consideration is given to natural disasters as important factors in lifecycle costs, especially when—as is the case here—actual gross loss costs can be used in lifecycle-cost estimation.

6. Conclusions

The purpose of this study was to explore the relationship between costs arising from natural hazards, both in the broad context of lifecycle-cost estimation and the narrower one of operation and management. It demonstrated the value of incorporating the most damaging types of frequent natural disasters as lifecycle-cost variables through quantitative analysis of the actual gross losses suffered. Even though this research was limited to properties belonging to a single hotel chain, the global nature of that chain increases the likelihood that its findings may be generalizable to other such chains and other types of property portfolios.

Nevertheless, future research should incorporate more resources related to building features such as area, height, material, and price, as well as type-of-damage data (e.g., at a minimum, whether the damage is structural vs. non-structural), as part of the ongoing quest for optimally effective means of managing lifecycle costs. Additionally, to broaden the concept of buildings' long-term sustainability, environmental risks such as proximity to land areas with mountain slopes altered by human activity should be taken into account, since such changes can increase the chance of avalanches [44,46]. Additional loss factors such as the radius and forward-movement speeds of hurricanes, areas of basin, and vegetation types should also be considered, especially in light of the present work's relatively low adjusted R^2.

A balanced consideration of social, economic, and environmental impacts is necessary if a complete picture of buildings' long-term sustainability, and thus their lifecycle cost, is to be achieved. Accordingly, future research should give due consideration to energy consumption, CO_2 emissions, and other environmentally relevant factors during construction and end-of-life demolition, as well as costs such as water, lighting, garbage disposal, and mechanical, electrical, and plumbing (MEP) systems during the operation and management phase. It should also be noted that the present research did not account for variations in the climate or local economies of the locations of the hotel's different properties, which may mean that its approach cannot be generalized to all locations. Thus, future research should give greater consideration to geographic variation in climate, local economies, and construction/repair costs to ensure that the proposed approach to lifecycle-cost estimation can be applied accurately in a full range of global contexts. In such future projects, artificial neural networks (ANNs) would be a useful tool for investigating complex non-linear relationships among research variables, for identifying independent variables, and for optimizing the process through training and testing phases in such future research; this would be valuable not only for initial-stage cost estimation but also during other phases of construction and other aspects of construction-project management.

Researchers could also use BIM in such research by including natural disasters as n-D, followed by 4D modeling that includes construction scheduling in the 3D model. The insurance industry is already using catastrophe-modeling techniques to predict damage from natural hazards, estimate maximum losses, and adjust premium prices. Similarly, facility-management companies looking to manage their properties more effectively by reducing unexpected costs could use the results of the present research as a basis for hazard mapping or hazard-prediction modeling at regional and national levels, combining n-D modeling or catastrophe modeling as mentioned above, because such models can estimate the value of unexpected potential loss from natural disasters. Moreover, fragility or vulnerability curves including building information such as building history, number of floors, locations, and building codes corresponding to wind speed and/or distance from water systems could be usefully included in future research on risk assessment for hotel properties.

Author Contributions: Conceptualization, S.-G.Y.; Data curation, S.-G.Y. and J.-M.K.; Funding acquisition, J.-M.K.; Investigation, S.-G.Y.; Methodology, S.-G.Y.; Project administration, K.S.; Software, J.-M.K.; Supervision, K.S.; Validation, S.-G.Y. and J.-M.K.; Resources, J.-M.K.; Writing—original draft, J.-M.K.; Writing—review & editing, S.-G.Y. and K.S. All authors have read and agreed to the published version of the manuscript.

Funding: This research was funded by the Basic Science Research Program through the National Research Foundation of Korea (NRF) funded by the Ministry of Education (NRF-2019R1F1A1058800).

Conflicts of Interest: The authors declare no conflict of interest.

References

1. Kim, S.; Jin, R.; Hyun, C.; Cho, C. Development of operation & maintenance cost estimating model for facility management of buildings. *J. Archit. Inst. Korea* **2013**, *29*, 11–21.
2. Sung, M.; Kim, K.; Yu, J. Development of BIM-based information management model for efficient building maintenance. *Proc. Korean Inst. Build. Constr.* **2011**, *11*, 137–140.
3. FEMA (Federal Emergency Management Agency). *HAZUS-MH Estimated Annualized Earthquake Losses for the United States*; FEMA: Washington, DC, USA, 2008.
4. Rein, A.; Corotis, R.B. An overview approach to seismic awareness for a 'quiescent' region. *Nat. Hazard. J.* **2013**, *67*, 335–363. [CrossRef]
5. Remo, J.W.; Pinter, N. HAZUS-MH earthquake modeling in the central USA. *Nat. Hazard. J.* **2012**, *63*, 1055–1081. [CrossRef]
6. Tantala, M.W.; Nordenson, G.J.; Deodatis, G.; Jacob, K. Earthquake loss estimation for the New York City metropolitan region. *Soil Dyn. Earthq. Eng. J.* **2008**, *28*, 812–835. [CrossRef]
7. Intergovernmental Panel on Climate Change (IPCC). *Contribution of Working Group I to the Fourth Assessment Report of the Intergovernmental Panel on Climate Change*; Cambridge University Press: Cambridge, UK, 2007.
8. Sissine, F. *Energy Independence and Security Act of 2007: A Summary of Major Provisions*; Library of Congress Washington DC Congressional Research Service: Washington, DC, USA, 2007.

9. Brown, N.A.; Orchiston, C.; Rovins, J.E.; Feldmann-Jensen, S.; Johnston, D. An integrative framework for investigating disaster resilience within the hotel sector. *J. Hosp. Tour. Manag.* **2018**, *36*, 67–75. [CrossRef]

10. Brown, N.A.; Rovinsa, J.E.; Feldmann-Jensenb, S.; Orchistonc, C.; Johnston, D. Exploring disaster resilience within the hotel sector: A systematic review of literature. *Int. J. Disaster Risk Reduct.* **2017**, *22*, 362–370. [CrossRef]

11. Brown, N.A.; Rovinsa, J.E.; Feldmann-Jensenb, S.; Orchistonc, C.; Johnston, D. Measuring disaster resilience within the hotel sector: An exploratory survey of Wellington and Hawke's Bay, New Zealand hotel staff and managers. *Int. J. Disaster Risk Reduct.* **2019**, *33*, 108–121. [CrossRef]

12. Noshadravan, A.; Miller, T.R.; Gregory, J.G. A lifecycle cost analysis of residential buildings including natural hazard risk. *J. Constr. Eng. Manag.* **2017**, *143*, 04017017-1-10. [CrossRef]

13. Teicholz, E. Bridging the AEC/FM technology gap. *J. IFMA Facil. Manag.* **2004**, *2*, 1–8.

14. Chae, M.; Lee, G.; Kim, J.; Cho, M. Analysis of domestic and international infrastructure asset management practices and improvement strategy. *Korean J. Constr. Eng. Manag.* **2009**, *10*, 55–64.

15. Lee, K.-J.; Jung, Y.-S. Assessment of facility management functions for life-cycle information sharing. *Korean J. Constr. Eng. Manag.* **2016**, *17*, 40–52. [CrossRef]

16. Yu, K.; Froese, T.; Grobler, F. A development framework for data models for computer-integrated facilities management. *J. Autom. Constr.* **2000**, *9*, 145–167. [CrossRef]

17. Foster, B. BIM for facility management: Design for maintenance strategy. *J. Build. Inf. Model.* **2011**, *2011*, 18–19.

18. Williams, R.; Shayesteh, H.; Marjanovic-Halburd, L. Utilizing building information modeling for facilities management. *Int. J. Facil. Manag.* **2014**, *5*, 1–19.

19. American Society for Testing and Materials (ASTM). *Standard Practice for Measuring Payback for Investments in Buildings and Building Systems*; ASTM E1121-12; ASTM: West Conshohocken, PA, USA, 2012.

20. American Society for Testing and Materials (ASTM). *Standard Practice for Measuring Life-Cycle Costs of Buildings and Building Systems*; ASTM E917-13; ASTM: West Conshohocken, PA, USA, 2013.

21. Pettang, C.; Mbumbia, L.; Foudjet, A. Estimating building materials cost in urban housing construction projects, based on matrix calculation: The case of Cameroon. *J. Constr. Build. Mater.* **1997**, *11*, 47–55. [CrossRef]

22. Günaydın, H.M.; Doğan, S.Z. A neural network approach for early cost estimation of structural systems of buildings. *Int. J. Proj. Manag.* **2004**, *22*, 595–602. [CrossRef]

23. Kim, G.-H.; An, S.-H.; Kang, K.-I. Comparison of construction cost estimating models based on regression analysis, neural networks, and case-based reasoning. *J. Build. Environ.* **2004**, *39*, 1235–1242. [CrossRef]

24. Rahman, S.; Perera, S.; Odeyinka, H.; Bi, Y. A Conceptual Knowledge-Based Cost Model for Optimizing the Selection of Materials and Technology for Building Design. In Proceedings of the 24th Annual Association of Research in Construction Management, Cardiff, UK, 1–3 September 2008.

25. Hasan, A. Optimizing insulation thickness for buildings using life cycle cost. *J. Appl. Energy* **1999**, *63*, 115–124. [CrossRef]

26. Kneifel, J. Life-cycle carbon and cost analysis of energy efficiency measures in new commercial buildings. *J. Energy Build.* **2010**, *42*, 333–340. [CrossRef]

27. Morrissey, J.; Horne, R. Life cycle cost implications of energy efficiency measures in new residential buildings. *J. Energy Build.* **2011**, *43*, 915–924. [CrossRef]

28. Gluch, P.; Baumann, H. The life cycle costing (LCC) approach: A conceptual discussion of its usefulness for environmental decision-making. *J. Build. Environ.* **2004**, *39*, 571–580. [CrossRef]

29. De Silva, D.G.; Kruse, J.B.; Wang, Y. Spatial dependencies in wind-related housing damage. *Nat. Hazard. J.* **2008**, *47*, 317–330. [CrossRef]

30. Khanduri, A.; Morrow, G. Vulnerability of buildings to windstorms and insurance loss estimation. *J. Wind Eng. Ind. Aerodyn.* **2003**, *91*, 455–467. [CrossRef]

31. Chang, S.E.; Shinozuka, M. Life-cycle cost analysis with natural hazard risk. *J. Infrastruct. Syst.* **1996**, *2*, 118–126. [CrossRef]

32. Wei, H.-H.; Shohet, I.M.; Skibniewski, M.J.; Shapira, S.; Yao, X. Assessing the lifecycle sustainability costs and benefits of seismic mitigation designs for buildings. *J. Archit. Eng.* **2016**, *22*, 04015011. [CrossRef]

33. Liu, M.; Burns, S.A.; Wen, Y. Optimal seismic design of steel frame buildings based on life cycle cost considerations. *J. Earthq. Eng. Struct. Dyn.* **2003**, *32*, 1313–1332. [CrossRef]

34. Wen, Y.K.; Kang, Y.J. Minimum building life-cycle cost design criteria. II: Applications. *J. Struct. Eng.* **2001**, *127*, 338–346. [CrossRef]
35. Wen, Y.K.; Kang, Y.J. Minimum building life-cycle cost design criteria. I: Methodology. *J. Struct. Eng* **2001**, *127*, 330–337. [CrossRef]
36. JCHS (Joint Center for Housing Studies of Harvard University). *Improving America's Housing 2015: Emerging Trends in the Remodeling Market*; Harvard University: Cambridge, MA, USA, 2015.
37. Ayyub, B.M. *Risk Analysis in Engineering and Economics*; Chapman & Hall/CRC: Boca Raton, FL, USA, 2003.
38. Burton, C.G. Social vulnerability and hurricane impact modeling. *J. Nat. Hazards Rev.* **2010**, *11*, 58–68. [CrossRef]
39. Dunion, J.P.; Landsea, C.W.; Houston, S.H.; Powell, M.D. A reanalysis of the surface winds for Hurricane Donna of 1960. *J. Mon. Weather Rev.* **2003**, *131*, 1992–2011. [CrossRef]
40. Vickery, P.J.; Skerlj, P.F.; Lin, J.; Twisdale, L.A., Jr.; Young, M.A.; Lavelle, F.M. HAZUSMH hurricane model methodology. II: Damage and loss estimation. *J. Nat. Hazards Rev.* **2006**, *7*, 94–103. [CrossRef]
41. Watson, C., Jr.; Johnson, M.E. Hurricane loss estimation models: Opportunities for improving the state of the art. *J. Bull. Am. Meteorol. Soc.* **2004**, *85*, 1713–1726. [CrossRef]
42. Brody, S.D.; Zahran, S.; Highfield, W.E.; Grover, H.; Vedlitz, A. Identifying the impact of the built environment on flood damage in Texas. *J. Disasters* **2008**, *32*, 1–18. [CrossRef]
43. Choi, O.; Fisher, A. The impacts of socioeconomic development and climate change on severe weather catastrophe losses: Mid-Atlantic Region (MAR) and the US. *J. Clim. Chang.* **2003**, *58*, 149–170. [CrossRef]
44. Cui, B.; Wang, C.; Tao, W.; You, Z. River channel network design for drought and flood control: A case study of Xiaoqinghe River basin, Jinan City, China. *J. Environ. Manag.* **2009**, *90*, 3675–3686. [CrossRef]
45. Zhai, G.; Fukuzono, T.; Ikeda, S. Multi-attribute evaluation of flood management in Japan: A choice experiment approach. *Water Environ. J.* **2007**, *21*, 265–274. [CrossRef]
46. Dai, F.; Lee, C.; Ngai, Y.Y. Landslide risk assessment and management: An overview. *J. Eng. Geol.* **2002**, *64*, 65–87. [CrossRef]

Article

System Dynamics Model for the Improvement Planning of School Building Conditions

Suhyun Kang [1], Sangyong Kim [1], Seungho Kim [2] and Dongeun Lee [3,*]

[1] School of Architecture, Yeungnam University, 280 Daehak-ro, Gyeongsan-si, Gyeongbuk 38541, Korea; yp043422@ynu.ac.kr (S.K.); sangyong@yu.ac.kr (S.K.)

[2] Department of Architecture, Yeungnam University College, 170 Hyeonchung-ro, Nam-gu, Daegu 42415, Korea; kimseungho@ync.ac.kr

[3] School of Architecture & Civil and Architectural Engineering, Kyungpook National University, 80 Daehak-ro, Buk-gu, Daegu 41566, Korea

* Correspondence: dolee@knu.ac.kr; Tel.: +82-53-950-7540

Received: 5 April 2020; Accepted: 13 May 2020; Published: 21 May 2020

Abstract: As the number of aged infrastructures increases every year, a systematic and effective asset management strategy is required. One of the most common analysis methods for preparing an asset management strategy is life cycle cost analysis (LCCA). Most LCCA-related studies have focused on traffic and energy; however, few studies have focused on school buildings. Therefore, an approach should be developed to increase the investment efficiency for the performance improvement of school buildings. Planning and securing budgets for the performance improvement of school building is a complex task that involves various factors, such as current conditions, deterioration behavior and maintenance effect. Therefore, this study proposes a system dynamics (SD) model for the performance improvement of school buildings by using the SD method. In this study, an SD model is used to support efficient decision-making through policy effect analysis, from a macro-perspective, for the performance improvement of school buildings.

Keywords: school buildings; system dynamics; deterioration; rehabilitation; lifecycle cost analysis; budget allocation

---

1. Introduction

Recently, due to the rapid increase in deteriorated social infrastructures, the significance of long-term planning for sustainability and performance improvement has been noted. In the United States, the facility deterioration problem was noted in the 1980s, and in 2017, the condition grade of infrastructure was confirmed to be "D+" on average. In particular, according to the '2017 infrastructure report card', which was published by the American Society of Civil Engineers (ASCE), the required restoration cost is approximately KRW 1,120 trillion. In Japan, 63% of roads and bridges, 62% of river management facilities, and 58% of harbors and wharves in 2033 will have passed over 50 years of age after construction. Therefore, in major advanced countries, the rapid deterioration of social infrastructures has been caused by the lack of appropriate measures and investments, despite the increase in deteriorated social infrastructures [1]. Furthermore, because of climate change and the frequent occurrence of natural disasters (i.e., earthquakes and storms) worldwide, many human lives are lost during disasters such as the collapse of deteriorated bridges [2]. Therefore, the importance of life extension and performance improvement of existing deteriorated social infrastructures is stressed, to ensure the safety of people from these disasters and catastrophes.

Systematic and effective asset management strategies are required to solve these problems. One of the most common analysis methods for preparing an asset management strategy is life cycle cost analysis (LCCA). Most LCCA-related studies have focused on traffic [3], pavement [4] and energy [5,6], however, few studies have focused on school buildings accommodating a population of over 50 million people on a daily basis.

In recent years, many studies have discussed the performance improvement of school buildings; however, the governments of various countries are experiencing difficulties in planning and financing, because of the lack of comprehensive data regarding school buildings [1]. Until recently, the maintenance of most school buildings was conducted using a breakdown maintenance method, instead of a preventive maintenance method. This method has led to the rapid deterioration of school buildings because the appropriate maintenance period was missed. Currently, the governments of advanced countries are hurriedly allocating excessive budgets; however, because executing these budgets within a financial year is an impossible task, the budget is customarily carried over to the following year, every year. This phenomenon is seemingly a result of short-term and emergency response, instead of investments based on mid-/long-term planning for the performance improvement of school buildings. Therefore, an approach should be developed to increase the investment efficiency for the performance improvement of school buildings. Planning and securing budgets for the performance improvement of social infrastructures is a complex task that involves various factors, such as current conditions, deterioration behavior, and maintenance effect [7]. Based on these complexities, it will be advantageous to predict the policy effect by using a simulation, to ensure that the policy-makers can plan the changes in policy direction in advance.

In this study, a system dynamics (SD) model is proposed to support efficient decision-making through policy effect analysis, from a macro-perspective for the performance improvement of school buildings. The SD model performs LCCA simulation based on performance improvement scenarios, to predict the deterioration pattern of school buildings and respond to it. Based on the simulation results, this study evaluates the long-term effects of rehabilitation policy on the performance grades of school buildings. Moreover, this study identifies an effective policy scenario that can achieve performance improvement.

2. Literature Review

Common methods of analyzing the complexity of asset management of social infrastructure include agent-based simulation (ABS) and SD. ABS is a micro-simulation method that can model interactions between agents; this method is used in various fields related to social infrastructures [8–12]. Echaveguren, Chamorro, and De Solminihac [13] modeled the interaction among agents (state, private and public) related to road infrastructure management systems, and analyzed the effects of the decisions made by agents regarding maintenance plans. Mallory, Crapper, and Holm [14] developed agent-based models (ABM) for fecal sludge (FS) recycling and proved the efficacy of the model by using case studies. Zechman [15] developed ABM for a water distribution system, and analyzed the interaction of systems. However, the ABS has a limitation in terms of modeling strategies, because the simulation results can differ based on small changes in the interaction method. Moreover, the level of detailed factors is high.

On the other hand, SD is a macro-simulation method that can decipher all the behaviors of complex systems [16]. In general, SD is used for modeling problems, such as performance measurement related to a social system, and estimating the effects of strategies and alternatives, as well as those of various social policies [17]. SD describes the interrelationships between factors causing changes in complex system growth estimates and patterns of change. SD emphasizes the causal relationships and feedback among individual components in a system [18]. Therefore, all causal relationships are recognized as circular relationships, without distinguishing between independent and dependent variables. This method focuses on the types of dynamic trends in changes among variables based on the flow of time, instead of obtaining the accurate value of the variable. Furthermore, SD is helpful when decision-makers examine the behavior of complex systems and evaluate the long-term policy effects [16]. Therefore, many studies have applied the SD modeling approach to determine the asset management strategy of social infrastructures in various fields.

Rehan [19] applied the SD modeling approach to develop asset management strategies for water and wastewater systems, and demonstrated the advantages of the SD model for modeling the interactions between physical, social, and financial systems. Mohammadifardi [20] also verified the applicability and efficacy of the SD model for wastewater collection (WWC). Hong, Frangmin, and Rongbei [21] developed an SD model related to highway maintenance issues, and proved that the SD method is effective during decision-making for a long-term plan by using case simulations and analyses. Soetjipto, Adi, and Anwar [22] developed a bridge deterioration model, and used it to simulate the possibility of bridge failure and detect that components that cause bridge failures. Furthermore, they used the SD model to analyze environmental pollution and energy problems, such as CO2 emission in the transport industry [23,24]. Sing, Love, and Liu [25] proved that adopting the SD modeling approach is useful for dealing with the long-term rehabilitation policy of existing building stock, related to the sustainability of a city. Wang and Yuan [26] used an SD simulation to determine an optimal measure for effective risk management in infrastructure projects.

Therefore, various studies have shown that the SD modeling approach is an effective tool for exploring asset management strategies and the policy effects of social infrastructures. This approach is also used in various types of social infrastructures. However, studies are yet to use the SD modeling approach to investigate the performance improvement of school buildings. Therefore, this study proposes a model for the performance improvement of school buildings via the SD method.

3. Research Methodology

The overall study procedure is shown in Figure 1. A literature review is conducted to determine the conventional modeling methods of asset management, and to find a suitable model for this study. A decision-making model is then developed for performance improvement of the school building. The SD method is applied as a modeling method, and it is developed by considering the correlations among the deterioration, rehabilitation and finance models. The SD model used in this study is developed using the following sequence: (1) define the problem, (2) create a causal loop diagram (CLD), (3) create a stock and flow diagram (SFD), and (4) verify the model. For the completed SD model, the effectiveness of the decision-making model is proven by using case studies including data from school buildings. Moreover, suggestions for long-term planning and financing are provided for future performance improvement of school buildings, based on the test results of various policy scenarios.

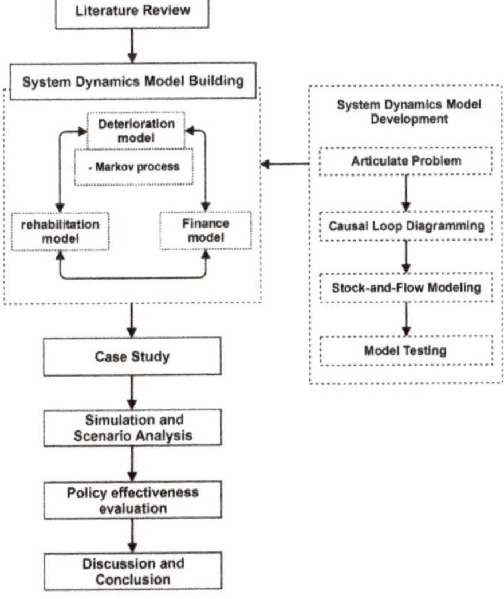

Figure 1. Research procedure.

4. Causal Loop Diagram of School Building Rehabilitation Management

During the first stage of SD model development, the causal relationships between key variables are determined to define the problem and compose the CLD for the school building rehabilitation system. A dynamic hypothesis is developed to explain the dynamic behavior of key variables in the structure. The overall system must be understood to establish a dynamic hypothesis, thereby emphasizing the need for conducting a literature review, expert group discussion, and survey. This study derived key variables and a dynamic hypothesis to understand the system by conducting a literature review. As a result of the relevant literature review, the SD model proposed in this study considers three major functions (asset deterioration, rehabilitation action, and total repair cost) for the macro-analysis of the rehabilitation system. Based on the literature review [7,19,27], nine variables, composing the three major functions, were derived. Figure 2 presents the CLD showing the causal relationships between the nine variables.

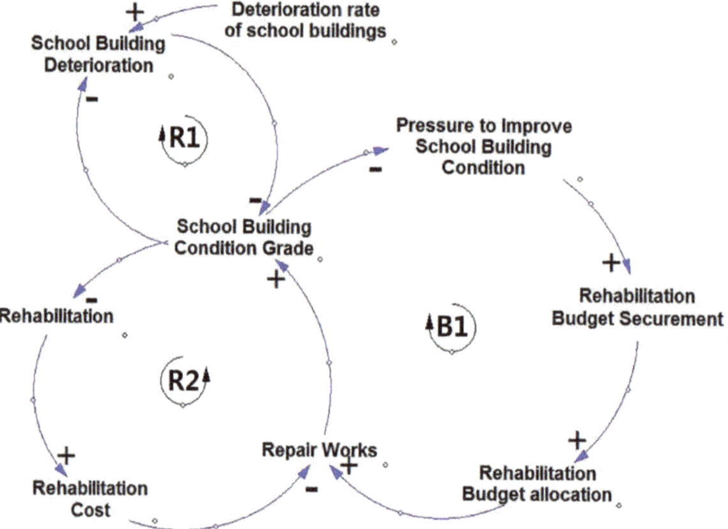

Figure 2. Causal loop diagram of the school building rehabilitation network management.

The CLD shown in Figure 2 consists of arrows, "+" or "−" signs, and feedback loops. A causal relationship between variables is expressed using the "+" or "−" signs through an arrow. These signs indicate the relationship between variables. The "+" link indicates that two variables (var1 and var2) are changing in a similar direction in the model. In other words, if the independent variable increases, the dependent variable also increases [Equation (1)].

$$\frac{\Delta \text{Var2}}{\Delta \text{Var1}} > 0 \tag{1}$$

The "−" link indicates that two variables (var1 and var2) are changing in different directions in the model. In other words, if the independent variable increases, the dependent variable decreases [Equation (2)].

$$\frac{\Delta \text{Var2}}{\Delta \text{Var1}} < 0 \tag{2}$$

The arrows of Figure 2 form the feedback loop, thereby indicating the characteristics of the loop. There are two types of loop, based on the characteristics of the loop: (1) reinforcing loop, and (2) balancing loop. The CLD presented in this study consists of two reinforcing loops (R1 and R2) and one balancing loop (B1). Each feedback loop shows the dynamic behaviors of deterioration, rehabilitation, and rehabilitation finance (expenses and budget) for school buildings.

4.1. Feedback Loop in School Building Deterioration

The deterioration loop (R1) shows the representative physical deterioration process of social infrastructure. The two variables ("school building condition grade" and "school building deterioration") that form this loop are connected by the "−" link, thereby indicating that the "school building condition grade" negatively affects the "school building deterioration", and the "school building condition grade" is affected by the "school building deterioration". If the "school building condition grade" decreases (e.g., in the scale of A–E, whereby A is the optimal condition and E is the poor condition), the deterioration increases. If the deterioration increases, the "school building condition grade" decreases. Furthermore, the "deterioration rate of school building" is connected with the "school building deterioration" by the "+" link. Therefore, if the "deterioration rate of school building" increases, the "school building deterioration" increases. A combination of these links produces a reinforcing loop (R1), as depicted in Figure 2. A reinforcing loop includes the feature of reinforcing to the extreme of a certain side (thus causing an index growth/decrease behavior). Therefore, the deterioration loop (R1) establishes a cycle, wherein the condition deterioration of the school building accelerates as time elapses. For this dynamic behavior, a similar process has also been reported in many asset management studies and related references [7,19].

4.2. Feedback Loop in Rehabilitation

The rehabilitation loop (R2) shows the rehabilitation process of the school building. Because the deterioration loop (R1) causes the exponential deterioration of the school building, the "school building condition grade" decreases and the "rehabilitation action" increases. Therefore, the relationship between the two variables is connected with a "−" link. In the real world, monetary payments are required to perform maintenance and repair tasks during rehabilitation. Therefore, "rehabilitation cost" has a positive relationship ("+" link) with "rehabilitation action". If the "rehabilitation action" increases, the "rehabilitation cost" also increases. On the other hand, "rehabilitation cost" and "repair works" have a negative relationship ("−") link. This is because repair works can be performed only when sufficient rehabilitation budgets are supplied. Therefore, the rehabilitation loop (R2) shows the rehabilitation process of the school building, and the decrease of "repair works" indirectly indicates the decrease of "school building condition grade".

4.3. Feedback Loop in Rehabilitation Finance

The finance loop (B1) shows a budgeting process. If the "school building condition grade" decreases, users' condition improvement demands (those by students, teachers, staff, and local residents), and the managers of school facilities is increase. If the need for the school building's condition improvement is noted, the government can secure a budget for rehabilitation. The secured budget is appropriately allocated, based on the policies and plans. According to the final budget, the maintenance and repair tasks are performed, thereby improving the condition grade of the school building. This combination of links generates a balancing loop (B1), as described in Figure 2, and the finance loop (B1) mitigates the condition grade decline of the school building by the deterioration loop (R1) and the rehabilitation loop (R2).

Therefore, the "rehabilitation cost" of the feedback loop (R2), and the "rehabilitation budget allocation", directly affect the maintenance and repair tasks. Therefore, they affect the "school building condition grade".

5. Stock and Flow Modeling for System Dynamics Simulation

After understanding the overall feedback loop through the CLD, it should be converted into an SFD to perform computer simulations. System dynamics is a diagram-based programming language, and the following diagrams constitute an SFD in the Vensim software: Stock, Flow, Valve, and Cloud (Figure 3a).

(a) Component of stock and flow diagram

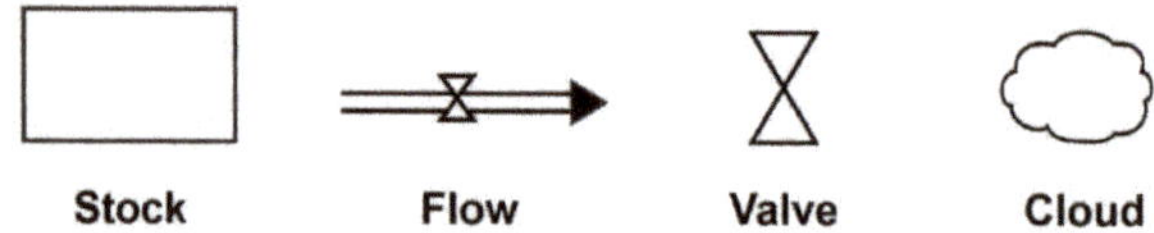

(b) stock and flow diagram

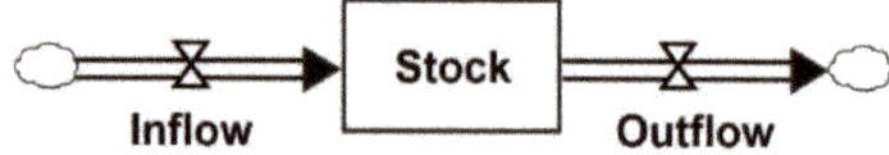

Figure 3. Stock and flow diagram.

Stock is a variable that accumulates or integrates the state of systems based on time. Flow is a variable that changes the value of the stock variable, and consists of inflow and outflow. Valve is a variable that controls the amount of inflow and outflow, and shows a boundary point of entry and exit of cloud. The relationship of stock and flow can be expressed using Equation (3), thus showing the value of the stock variable, based on the simulation time [16]. In this equation, t_0 is the initial time, t is the current time, and stock (t_0) is the initial value of stock. Inflow and outflow refer to flow coming into, and going out from, the stock, for an arbitrary duration (s) between the initial time (t_0) and the current time (t). Equation (4) determines changes in the rate of stock, based on time [16].

$$\text{Stock}(t) = \int_{t_0}^{t} [\text{Inflow}(s) - \text{Outflow}(s)]ds + \text{Stock}(t_0) \tag{3}$$

$$\frac{d(\text{Stock})}{dt} = \text{Inflow}(t) - \text{Outflow}(t) \tag{4}$$

The relationship of stock and flow can be expressed according to the aforementioned, as shown in Figure 3b.

5.1. School Building Deterioration Sector

The school building deterioration model in this study is developed with the goal of simulating the overall deterioration pattern. Most assets are managed based on the school building condition; deterioration models using data regarding the condition have been presented using various methods. Based on the results of the literature review, deterioration models are primarily classified into three categories: deterministic, stochastic, and artificial intelligence [28,29] (Figure 4).

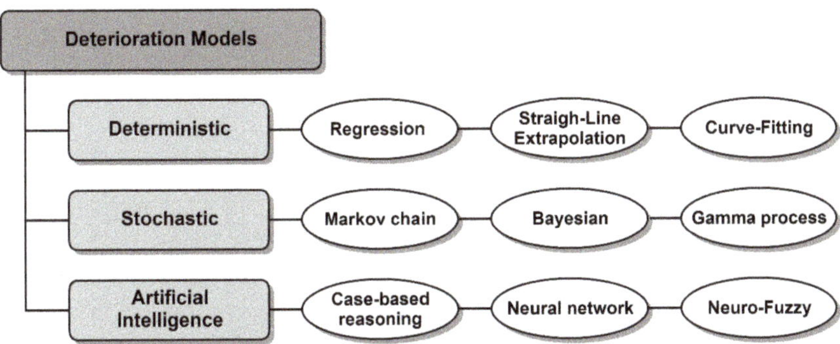

Figure 4. Classification of deterioration models.

Among the categories of deterioration models, Markov chain is a stochastic method for predicting the future condition state of assets in a social infrastructure management system. This method is also most frequently used [30–33]. Therefore, this study attempts to predict the deterioration pattern of school buildings by using the Markov chain.

The Markov chain indicates a case wherein the probability of reaching a specific state for a stochastic variable depends only on the state of the preceding time point. This study classifies the physical condition of school buildings using grades A–E, according to the condition evaluation criteria provided by the Ministry of Education in South Korea (Table 1).

Table 1. Physical condition grade of school buildings.

Grade	Grade Point	Description
A	5	Exceptional: Fit for the future
B	4	Good: Adequate for now
C	3	Mediocre: Requires attention
D	4	Poor: At risk
E	1	Critical: Unfit for the future

The deterioration model is developed based on the assumption that a school building deteriorates to the next condition state only from a specific condition state (e.g., from condition A to B, and B to C). The five stock variables (A–E) shown in Figure 5 indicate the number of school buildings for each condition grade. A transition probability variable is derived, based on case study data, that serves as an auxiliary variable of each flow variable. Moreover, to induce a pattern that is similar to the actual deterioration behavior of assets, the stock variable has a feedback relationship that affects the flow variable. This variable can be expressed as shown in Equation (5) (X indicates the condition grade, and X-1 refers to the condition grade that is one step lower than that of the condition X).

$$\text{Deterioration } X \text{ to } X - 1 = X * \text{Transition Probability } X \text{ to } X - 1 \tag{5}$$

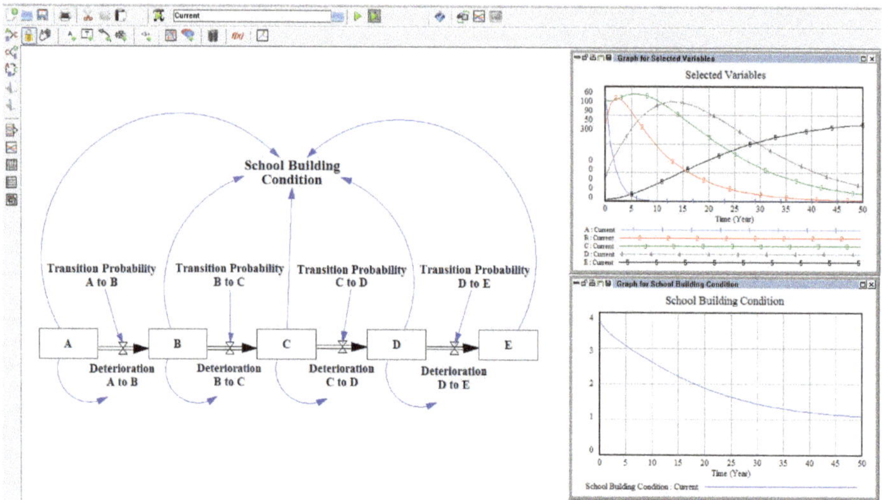

Figure 5. System dynamics (SD) model of deterioration and simulation graph.

Moreover, to identify the condition grade of overall school buildings based on time, Equation (6) was applied, based on the grade score shown in Table 1.

$$\text{School building condition} = (A*5 + B*4 + C*3 + D*2 + E*1) \,/\, \text{Total number of school building} \tag{6}$$

To test the completed deterioration model, data regarding the safety inspection and condition assessment of school buildings for the winters of 2014–2018, from the Education Office in Daegu metropolitan city, was used in this study. The condition grades of 214 school buildings in total showed: grade A = 55, grade B = 67, grade C = 79, grade D = 10, and grade E = 3 buildings. The transition probability was set as: A to B = 0.45, B to C = 0.1, C to D = 0.09, and D to E = 0.15 (the transition probability of school buildings that are applied in the case studies are described in detail in Section 5.1). The result of testing the deterioration model using the case study data is shown in the graph on the right side of Figure 5.

The result of testing the deterioration model using the case study data is shown in the graph on the right side of Figure 5. As time elapses, the number of school buildings with the condition grades A, B, C, and D decreases, and the number of school buildings with the condition grade E increases. The curve illustrating the comprehensive condition of school buildings based on these dynamic changes has an initial value of 3.75, which is close to the grade B. However, after 50 years, the value deteriorates to 1.09, thus the school building condition grade deteriorates to grade E. Therefore, this study verified the validity of the deterioration SD model as a tool that predicts deterioration patterns, using the number of assets by grade.

5.2. Rehabilitation Sector

The rehabilitation model shows the rehabilitation action based on the condition grades of school buildings. The model proposed in this study assumes that schools categorized under the three grades, C, D, and E, which do not indicate a good condition, will be repaired to ensure the school building is categorized under the best grade A. Based on this assumption, the rehabilitation action was integrated with the deterioration models shown in Figure 6.

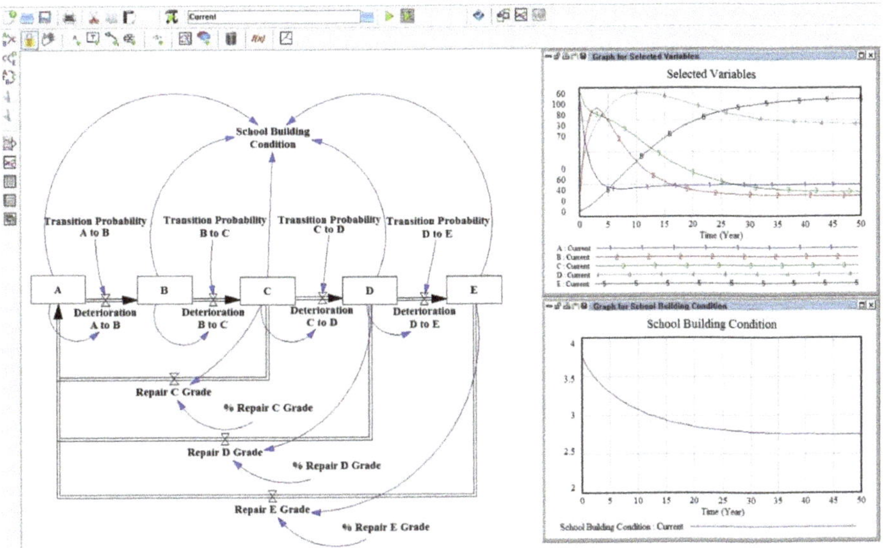

Figure 6. SD model of rehabilitation and simulation graph.

The dynamic flow of flow variables (e.g., Repair C Grade), that expresses the rehabilitation action in Figure 6, is pointed toward an improved condition state (grade A) from a specific condition state (grades C, D, or E). The value of the rehabilitation flow variable is determined based on an auxiliary variable (e.g., % Repair C Grade). This variable shows the proportion of repairing from condition grade X to grade A in the number of school buildings of a specific condition grade. The flow variable showing the rehabilitation action of the SD model is calculated by Equation (7).

$$\text{Repair X Grade } = \text{ X} * \text{\% Repair X Grade} \tag{7}$$

Equation (7) determines the number of buildings of each condition X (grades C, D, or E). Stock A—the number of school buildings that secured grade A—increases through the rehabilitation action. This is expressed using Equation (8)

$$\text{Stock A}(t) \;=\; \int_{t_0}^{t} [\text{Repair C Grade}(s) + \text{ Repair D Grade}(s) + \text{Repair E Grade}(s) \\ -\text{Deterioration A to B}(s)] ds + \text{Stock A}(t_0) \tag{8}$$

Equation (8) indicates the inflow into Stock A, which refers to the number of school buildings that have been improved from the grades C, D, and E, during an arbitrary time period between the initial time t_0 and the current time t. Equation (8) also indicates the outflow to Stock B caused by deterioration as time elapses.

The auxiliary variables—% Repair X Grade—were set to 5% to test the model that included rehabilitation action. For a case concerning the performing of rehabilitation actions to the school buildings with grades C, D, and E every year, the result of simulating the condition grade changes over 50 years is shown in a graph on the right side of Figure 6. A comparison of the number of school buildings in each grade, described on the right sides of Figures 5 and 6, indicates that the graph curves of grades A, B, C, and D in Figure 5 decrease rapidly, whereas the grades A, B, C, and D in Figure 6 maintain specific levels for approximately 25 years.

Table 2 shows the simulation results, on a yearly basis, for the condition grade values of all school buildings for 50 years, targeting the deterioration (Det.) and rehabilitation (Reh.) models. Based on the 50 year period, the condition grade value in the deterioration model was 1.09, which was critical.

However, the value improved to 2.75 in the rehabilitation model, thus indicating a poor condition. The results of repairing 5% of the school buildings in grades C, D, and E to grade A, respectively, every year are shown in Table 2.

Table 2. Comparison of simulation results of school building physical condition grade between deterioration model and rehabilitation model.

Model	Simulation Time (Year)						State
	0	5	10	15	25	50	
Det.	3.75	3.07	2.60	2.21	1.63	1.09	critical
Reh.	3.75	3.29	3.07	2.93	2.80	2.75	poor

Moreover, Table 3 compares the differences in the deterioration model (Det.) and rehabilitation model (Reh.), based on their respective grades for the same simulation results. Although the initial value (0 year) was identical, the number of school buildings for each grade indicated a significant difference between the two models as time elapsed. Based on the 50 year period, the Det. model showed that most school buildings deteriorated to grade E, whereas the number of school buildings was evenly distributed in the Reh. model. Therefore, it is proven that the rehabilitation SD model in Figure 6 can quantitatively analyze the effects of school buildings' deterioration and rehabilitation action on the increase or decrease of the physical conditions of all school buildings.

Table 3. Comparison of simulation results between deterioration model and rehabilitation model.

		0 Year		5 Year		10 Year		15 Year		25 Year		50 Year	
Grade	State	Det.	Reh.	Det.	Reh.	Det.	Reh.	Det.	Reh.	Det.	Reh.	Det.	Reh.
A	Exceptional	55	55	2.76	12.9	0.13	13.47	0.007	14.21	0	14.67	0	14.75
B	Good	67	67	77.76	87.1	47.83	75.46	28.3	70.7	9.88	66.99	0.7	66.42
C	Mediocre	79	79	84.22	69.5	79.26	63.39	65.5	57.28	37.07	50.43	5.7	47.54
D	Poor	10	10	31.45	26.6	41.69	28.90	42.9	27.71	31.79	24.05	6.6	21.48
E	Critical	3	3	17.79	17.7	45.06	32.75	77.1	44.72	135.2	57.04	200	63.78

5.3. Finance Sector

One of the crucial tasks in a maintenance and rehabilitation (M&R) plan is efficiently distributing a limited fund to achieve an optimal outcome. This section aims to propose an SD model that has added a cost model to the deterioration and rehabilitation models for efficient budget allocation.

Figure 7 is an integrated SFD, whereby the cost model is included in Figure 6. To fulfill the rehabilitation action of a deteriorated school building, maintenance costs are required, which are provided from a limited budget. The "Available Rehabilitation Policy Budget" variable refers to the total budget that can be used for the rehabilitation action in the model. The "Allocated Budget to Repair X Grade" variable refers to a budget allocated from the limited total budget to repair the school buildings belonging to the respective grades X (C, D, and E) [Equation (9)].

$$\text{Allocated Budget to Repair X Grade} = \text{Available Rehabilitation Policy Budget} * \text{"\% Budget to Repair X Grade"} \tag{9}$$

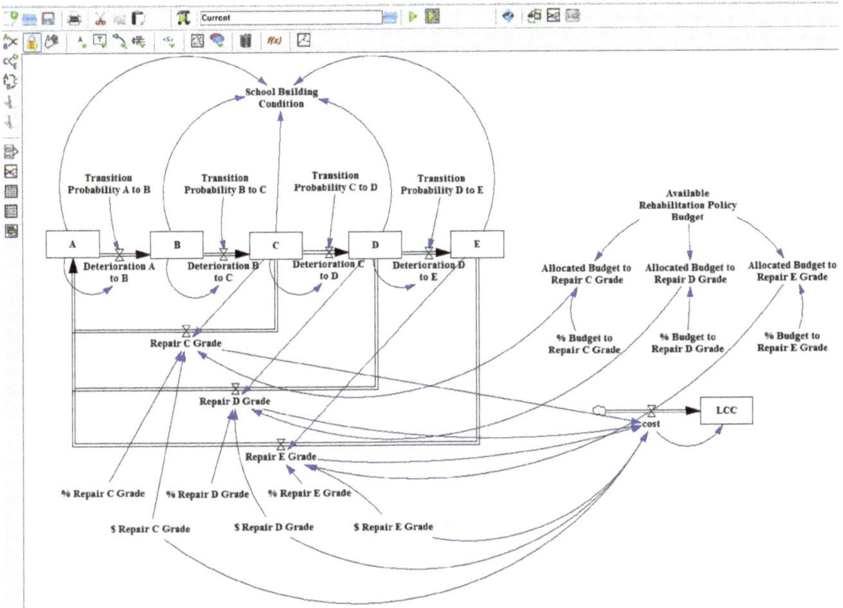

Figure 7. SD model for rehabilitation cost and budgeting analysis.

The value of this variable is determined by a variable "% Budget to Repair X Grade", which shows the percentage of budget allocated to each grade X. Moreover, unlike the rehabilitation model, the cost model shows that the number of school buildings restored is limited according to the allocated budget. This is determined through Equation (10).

$$\text{Repair X Grad} = \text{IF THEN ELSE}(X * \%\ \text{Repair X Grade} * \$\ \text{Repair X Grade} \\ < \text{Allocated Budget to Repair X Grade, } X * \%\ \text{Repair X Grade,} \\ X * \%\ \text{Repair X Grade} * 0) \tag{10}$$

Equation (10) used the IF THEN ELSE({cond}, {ontrue}, {onfalse}) function, which is a built-in function of Vensim, to derive different values, based on the condition. The variable "$ Repair X Grade" shows the repair cost required to rehabilitate the school buildings of each grade. If the cost for repairing from a condition grade X (C, D, or E) to the grade A (X * "% Repair X Grade" * "$ Repair X Grade") is less than the limited budget ("Allocated Budget to Repair X Grade"), the rehabilitation action is conducted; otherwise, it is stopped. Moreover, the life cycle cost by time t shows an equation similar to Equation (11).

$$\text{LCC}(t) = \int_{t_0}^{t} [\$\ \text{Repair C Grade} * \text{Repair C Grade}(s) + \$\ \text{Repair D Grade} \\ * \text{Repair D Grade}(s) + \$\ \text{Repair E Grade} * \text{Repair E Grade}(s)] ds \tag{11}$$

The completed integrated SD model is used as a model for determining the life cycle cost analysis and future outcome prediction, to improve the performance of school buildings via use of a case study. This study performs a scenario analysis, to investigate the effects of budget allocation by grade on the total outcome and cost, based on simulations considering various values for the "% Budget to Repair X Grade" variable.

6. Application of the Developed SD Model

This section performs the budget allocation scenario analysis, to predict the deterioration behavior of school buildings and improve performance by simulating the SD model proposed in Section 5. This study used the safety inspection and condition assessment data provided by the Ministry of Education in South Korea, in investigating 214 school buildings in a metropolitan city, Daegu. Based this data, this study acquired data regarding the 5 year (2014–2018) condition assessment and maintenance cost of school buildings, categorized by condition grade. At present, the Ministry of Education in South Korea designates only grades D and E, among the five condition grades A–E, as disaster-prone buildings, and conducts performance improvement primarily for these buildings. However, the SD simulation sets the rehabilitation scenarios by considering buildings up to the grade C for preventive maintenance. Finally, the simulation analysis is performed by applying the deterioration rate variable (transition probability matrix, TPM), derived by using the Markov chain stochastic process, to the integrated SD model (Figure 7).

Markov Approach

Markov chain is a discrete time stochastic process, and the conditional probability of a specific future event changes according to only the current condition; it is unrelated to past conditions [34]. Because five condition states exist in the case study data, the transition probability from one condition state to another is expressed in a 5×5 matrix, and the simplified transition probability matrix (TPM) is shown in Equation (12).

$$\text{TPM} = \begin{bmatrix} 0.88 & 0.12 & 0 & 0 & 0 \\ 0 & 0.96 & 0.04 & 0 & 0 \\ 0 & 0 & 0.91 & 0.09 & 0 \\ 0 & 0 & 0 & 0.86 & 0.14 \\ 0 & 0 & 0 & 0 & 1 \end{bmatrix} \tag{12}$$

Each element (P_{ij}) of TPM shows a probability (P) of transitioning from a state "i" to another state "j". For example, '0.88' indicates the probability of transitioning from state A to state A (the probability of a school building remaining in state A), and '0.12' refers to the probability of transitioning from state A to state B. It is assumed that the condition state of school buildings shift from one condition state to the next condition state only. Suppose the initial condition state's value is CS0, and the distribution of the condition state by year is n; then, Equation (13) can be derived.

$$\text{CS}_n = \text{CS}_0 \times \text{TPM}^n \tag{13}$$

Equation (13) shows that a future state (CS0) can be estimated when the TPM and the initial state (CSn) are known.

7. SD Model Simulation Results of Scenario Analysis

The developed SD model (Figure 7) analyzes the effect of budget allocation strategy scenarios to determine cost-effective rehabilitation actions for school buildings. Table 4 shows the budget allocation proportions of grades C, D, and E, based on the average annual educational environment improvement budget provided for Daegu city. The results of simulating 10 scenarios using the Vensim software are shown in Figure 8.

Table 4. Budget allocation scenarios.

Scenario	Budget to Repair C Grade	Budget to Repair D Grade	Budget to Repair E Grade
S1	0 %	0 %	100%
S2	0%	100%	0%
S3	100%	0%	0%
S4	33%	33%	34%
S5	0%	50%	50%
S6	50%	50%	0%
S7	50%	0%	50%
S8	25%	25%	50%
S9	25%	50%	25%
S10	50%	25%	25%

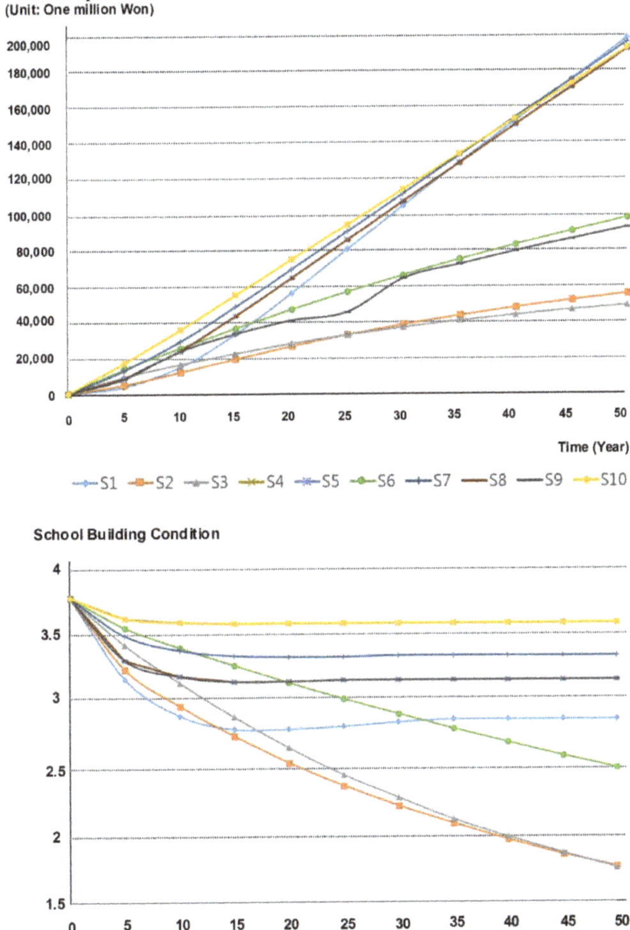

Figure 8. SD model simulation analysis results for budget allocation scenarios.

Figure 8 shows the scenario analysis results for the condition grades (based on Table 1) and Total life cycle cost (TLCC) of all school buildings. S2 and S3 can reduce the TLCC in the long term. However, because the condition grade of school buildings declines gradually to 1.75 (grade E: Critical), they can be perceived as the worst-case scenarios. TLCC of approximately KRW 20 billion is expected for S1, S4, S5, S7, S8, and S10, and TLCC of approximately KRW 9 billion is expected for S6 and S9. Therefore, S9, S10, and S4 are picked as scenarios with good performance improvement effects relative to the cost. From these results, it can be noted that the appropriate budgets were allocated to school buildings in the condition grade C. In the case of S10, wherein the condition grade was the highest, 50% of the total budget was allocated to the condition grade C and the remaining 50% was equally allocated to the condition grades D and E. This result shows that, when repairing is performed primarily on buildings in condition grade C, the rehabilitation cost required is less than that of buildings in grades D and E, and in the long term, a preventive maintenance effect can be obtained. Therefore, by using the 10 different scenario analyses, it is ascertained that budget allocation based on condition grade has a crucial impact on the total school building performance.

8. Discussion and Conclusions

This study proposed an integrated SD model for the rehabilitation policy analysis of school buildings. To validate the SD model, 10 rehabilitation budget allocation scenarios were analyzed, based on the simulations. The results show that the integrated SD model can support strategic decision-making, by identifying the school building condition grades and TLCC behavior for each scenario in the long-term perspective. According to the scenario analysis, the rehabilitation action of preventive maintenance that primarily repairs the buildings in condition grade C showed the best performance improvement effect relative to the cost.

The Ministry of Education in South Korea currently performs post-event maintenance management, to repair buildings when performance deterioration occurs (grades D and E). However, the preventive maintenance method should be adopted to reduce the deterioration speed of school buildings. The costs calculated based on the SD simulation can be used for the long-term planning of rehabilitation action, by estimating the cost that will be injected into repairing the deteriorated school buildings for 50 years in the future. However, the proposed SD model has several limitations. The available case study data for this study was insufficient, and increasingly accurate deterioration modeling will be possible if it is supplemented with an optimal method for estimating accurate TPM with limited data. Moreover, the budget of the Ministry of Education in South Korea, which is the subject of the case study, is in a situation wherein continuous investment for the performance improvement of school buildings is difficult because of other educational policies, such as free school meals and the New University for Regional Innovation (NURI) project. Therefore, if the proposed SD model is expanded to consider the effects of other educational policies, the crucial performance improvement budget can be estimated in the long-term perspective.

Author Contributions: Conceptualization, S.K. (Suhyun Kang), S.K. (Sangyong Kim), S.K. (Seungho Kim), and D.L.; data curation, D.L.; formal analysis and investigation, S.K. (Suhyun Kang), S.K. (Sangyong Kim); methodology, S.K. (Suhyun Kang), S.K. (Sangyong Kim), S.K. (Seungho Kim); resources, D.L.; software, S.K. (Suhyun Kang) and S.K. (Seungho Kim); supervision, S.K. (Sangyong Kim) and D.L.; validation, S.K. (Suhyun Kang), S.K. (Sangyong Kim), S.K. (Seungho Kim), and D.L.; visualization, S.K. (Suhyun Kang); writing—original draft, S.K. (Suhyun Kang), S.K. (Seungho Kim); writing—review and editing, S.K. (Sangyong Kim) and D.L. All authors have read and agreed to the published version of the manuscript.

Funding: This work was supported by the National Research Foundation of Korea (NRF) grant funded by the Korea government (MSIT) (No. NRF-2018R1A5A1025137).

Conflicts of Interest: The authors declare no conflict of interest.

References

1. ASCE. *2017 Infrastructure Report Card*; ASCE: Reston, VA, USA, 2017.

2.	Morris, J.C.; Little, R.G. Symposium Issue: Climate Change and Infrastructure—The Coming Challenge. *Public Works Manag. Policy* **2019**, *24*, 3–5. [CrossRef]

3.	Shen, Y.; Goodall, J.L.; Chase, S.B. Condition State–Based Civil Infrastructure Deterioration Model on a Structure System Level. *J. Infrastruct. Syst.* **2018**, *25*, 04018042. [CrossRef]

4.	Lu, Q.; Mannering, F.L.; Xin, C. *A Life Cycle Assessment Framework for Pavement Maintenance and Rehabilitation Technologies: An Integrated Life Cycle Assessment (LCA)–Life Cycle Cost Analysis (LCCA) Framework for Pavement Maintenance and Rehabilitation*; University of South Florida: Tampa, FL, USA, 2018.

5.	Thomas, A.; Mantha, B.R.; Menassa, C.C. Pipelines 2016: Out of Sight, Out of Mind, Not Out of Risk. A framework to evaluate the life cycle costs and environmental impacts of water pipelines. In Proceedings of the Pipelines 2016, Kansas City, MI, USA, 17–20 July 2016.

6.	Lee, H.; Shin, H.; Rasheed, U.; Kong, M. Establishment of an inventory for the Life Cycle Cost (LCC) analysis of a water supply system. *Water* **2017**, *9*, 592.

7.	Rashedi, R.; Hegazy, T. Holistic analysis of infrastructure deterioration and rehabilitation using system dynamics. *J. Infrastruct. Syst.* **2016**, *22*, 04015016. [CrossRef]

8.	Osman, H. Agent-based simulation of urban infrastructure asset management activities. *Automat. Constr.* **2012**, *28*, 45–57. [CrossRef]

9.	Mostafavi, A.; Abraham, D.; DeLaurentis, D.; Sinfield, J.; Kandil, A.; Queiroz, C. Agent-based simulation model for assessment of financing scenarios in highway transportation infrastructure systems. *J. Comput. Civ. Eng.* **2015**, *30*, 04015012. [CrossRef]

10.	Rasoulkhani, K.; Logasa, B.; Reyes, M.P.; Mostafavi, A. Agent-based modeling framework for simulation of complex adaptive mechanisms underlying household water conservation technology adoption. In Proceedings of the 2017 Winter Simulation Conference, Las Vegas, NV, USA, 3–6 December 2017.

11.	Rasoulkhani, K.; Logasa, B.; Presa Reyes, M.; Mostafavi, A. Understanding fundamental phenomena affecting the water conservation technology adoption of residential consumers using agent-based modeling. *Water* **2018**, *10*, 993. [CrossRef]

12.	Guo, Y.; Hawkes, A. Asset stranding in natural gas export facilities: An agent-based simulation. *Energy Policy* **2019**, *132*, 132–155. [CrossRef]

13.	Echaveguren, T.; Chamorro, A.; De Solminihac, H. Concepts for modeling road asset management systems using agent-based simulation. *Rev. Ing. Constr.* **2017**, *32*, 47–56. [CrossRef]

14.	Mallory, A.; Crapper, M.; Holm, R.H. Agent-Based Modelling for Simulation-Based Design of Sustainable Faecal Sludge Management Systems. *Int. J. Environ. Res. Public Health* **2019**, *16*, 1125. [CrossRef]

15.	Zechman, E.M. Agent-based modeling to simulate contamination events and evaluate threat management strategies in water distribution systems. *Risk Anal.* **2011**, *31*, 758–772. [CrossRef] [PubMed]

16.	Sterman, J.D. *Business Dynamics: Systems Thinking and Modeling for a Complex World*; Irwin/McGraw-Hill: Boston, MA, USA, 2000.

17.	Lyneis, J.M.; Cooper, K.G.; Els, S.A. Strategic management of complex projects: A case study using system dynamics. *Syst. Dyn. Rev.* **2001**, *17*, 237–260. [CrossRef]

18.	Forrester, J.W. Currently available from Pegasus Communications. In *Industrial Dynamics*; MIT Press: Waltham, MA, USA, 1961.

19.	Rehan, R.; Knight, M.A.; Haas, C.T.; Unger, A.J. Application of system dynamics for developing financially self-sustaining management policies for water and wastewater systems. *Water Res.* **2011**, *45*, 4737–4750. [CrossRef] [PubMed]

20.	Mohammadifardi, H.; Knight, M.A.; Unger, A.A. Sustainability Assessment of Asset Management Decisions for Wastewater Infrastructure Systems—Implementation of a System Dynamics Model. *Systems* **2019**, *7*, 34. [CrossRef]

21.	Hong, Z.; Fangmin, R.; Rongbei, Z. Simulation Research for Highway Maintenance Management System Based on System Dynamics. *J. Syst. Simul.* **2016**, *3*, 23.

22.	Soetjipto, J.W.; Adi, T.J.W.; Anwar, N. System Dynamics Approach for Bridge Deterioration Monitoring System. *Int. J. Eng. Technol. Innov.* **2016**, *6*, 264–273.

23.	Feng, Y.Y.; Chen, S.Q.; Zhang, L.X. System dynamics modeling for urban energy consumption and CO2 emissions: A case study of Beijing, China. *Ecol. Model* **2013**, *252*, 44–52. [CrossRef]

24. Liu, X.; Ma, S.; Tian, J.; Jia, N.; Li, G. A system dynamics approach to scenario analysis for urban passenger transport energy consumption and CO2 emissions: A case study of Beijing. *Energy Policy* **2015**, *85*, 253–270. [CrossRef]

25. Sing, M.C.; Love, P.E.; Liu, H.J. Rehabilitation of existing building stock: A system dynamics model to support policy development. *Cities* **2019**, *87*, 142–152. [CrossRef]

26. Wang, J.; Yuan, H. System dynamics approach for investigating the risk effects on schedule delay in infrastructure projects. *J. Manag. Eng.* **2016**, *33*, 04016029. [CrossRef]

27. Elbehairy, H.; Elbeltagi, E.; Hegazy, T.; Soudki, K. Comparison of two evolutionary algorithms for optimization of bridge deck repairs. *Comput. Aided Civil Infrastruct. Eng.* **2006**, *21*, 561–572. [CrossRef]

28. Morcous, G.; Rivard, H.; Hanna, A.M. Modeling Bridge Deterioration Using Case-Based Reasoning. *J. Infrastruct. Syst.* **2002**, *8*, 86–95. [CrossRef]

29. Elkhoury, N.; Hitihamillage, L.; Moridpour, S.; Robert, D. Degradation Prediction of Rail Tracks: A Review of the Existing Literature. *Open Transp. J.* **2018**, *12*, 1. [CrossRef]

30. Wellalage, N.K.W.; Zhang, T.; Dwight, R. Calibrating Markov chain–based deterioration models for predicting future conditions of railway bridge elements. *J. Bridge Eng.* **2014**, *20*, 04014060. [CrossRef]

31. Wellalage, N.W.; Zhang, T.; Dwight, R.; El-Akruti, K. Bridge deterioration modeling by Markov Chain Monte Carlo (MCMC) simulation method. In *Engineering Asset Management-Systems, Professional Practices and Certification*; Springer: Berlin, Germany, 2015; pp. 545–556.

32. Mohseni, H.; Setunge, S.; Zhang, G.; Wakefield, R. Markov Process for Deterioration Modeling and Asset Management of Community Buildings. *J. Constr. Eng. Manag.* **2017**, *143*, 6. [CrossRef]

33. Shen, L.; Soliman, M.; Ahmed, S.; Waite, C. Life-Cycle Cost Analysis of Reinforced Concrete Bridge Decks with Conventional and Corrosion Resistant Reinforcement. *MATEC Web Conf.* **2019**, *271*, 01009. [CrossRef]

34. Ross, S.M. *Introduction to Probability Models*, 11th ed.; Academic Press: San Diego, CA, USA, 2014.

Article

Sustainable Application of Hybrid Point Cloud and BIM Method for Tracking Construction Progress

Seungho Kim [1], Sangyong Kim [2] and Dong-Eun Lee [3,*]

[1] Department of Architecture, Yeungnam University College, Nam-gu, Daegu 42415, Korea; kimseungho@ync.ac.kr
[2] School of Architecture, Yeungnam University, Gyeongsan-si, Gyeongbuk 38541, Korea; sangyong@yu.ac.kr
[3] School of Architecture & Civil and Architectural Engineering, Kyungpook National University, Buk-gu, Daegu 41566, Korea
* Correspondence: dolee@knu.ac.kr; Tel.: +82-53-950-7540

Received: 18 April 2020; Accepted: 14 May 2020; Published: 18 May 2020

Abstract: Compared to the past, the complexity of construction-project progress has increased as the size of structures has become larger and taller. This has resulted in many unexpected problems with an increasing frequency of occurrence, such as various uncertainties and risk factors. Recently, research was conducted to solve the problem via integration with data-collection automation tools of construction-project-progress measurement. Most of the methods used spatial sensing technology. Thus, this study performed a review of the representative technologies applied to construction-project-progress data collection and identified the unique characteristics of each technology. The basic principle of the progress proposed in this study is its execution through the point cloud and the attributes of BIM, which were studied in five stages: (1) Acquisition of construction completion data using a point cloud, (2) production of a completed 3D model, (3) interworking of an as-planned BIM model and as-built model, (4) construction progress tracking via overlap of two 3D models, and (5) verification by comparison with actual data. This has confirmed that the technical limitations of the construction progress tracking through the point cloud do not exist, and that a fairly high degree of progress data which contains efficiency and accuracy can be collected.

Keywords: building information modeling; drone; LIDAR; point cloud; progress tracking

---

1. Introduction

Data related to the progress of construction projects are very useful both to determine whether timelines are being kept and to assess the quality of the work done, and these data are essential to improving the productivity of construction management [1,2]. However, the progress of construction projects is currently tracked in various ways, such as scheduling, utilizing construction methods, expenditure management, and resource/quality management, and it is difficult to accurately track and record all of those activities [3].

The information required to measure the progress of a construction project can be classified in two categories. The first one is information related to the plan and design and can be acquired at the end of the design phase. The second one is information related to the current construction progress. The latter type cannot be easily collected, and continuously changes. Unfortunately, for most construction project sites, data acquisition depends on the manual recording of information on paper, and the use of photos and documents causes many constraints in time and space. Automation is considered to be the most economical solution to these data acquisition-related problems [4,5].

The goal of this study is to improve the efficiency and accuracy with which progress data is acquired, as this is important to the overall management of the progress of each construction project. To achieve this goal, the study considers recent trends with some construction projects and proposes an alternative process for solving the problems related to data collection on project sites. This study conducts verification procedures on various buildings to confirm the validity of the proposed measures as well as to identify methods of post-processing the acquired data. The contents of each of the performance stages presented in this study are shown in Figure 1.

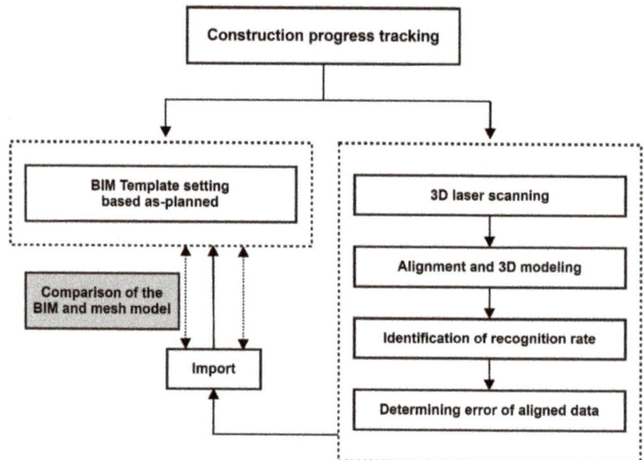

Figure 1. Construction progress tracking procedure.

2. Existing Studies on Automated Progress Data Acquisition

Conventional processes employed to acquire data related to the progress of construction projects are inefficient, both in terms of time and cost, and this has led to many studies in the field of automation technologies [6–9]. Various mobile IT devices were initially proposed as a way to automate data acquisition because they can transmit information via the Internet. Initial representative studies involved the development of various automated methods to perform field data acquisition using data acquisition technologies (DATs) such as radio frequency identification (RFID), global positioning systems (GPSs), bar codes, time-lapse cameras, and ultra-wideband (UWB) [5,10–13]. The above-mentioned studies generally indicated that the introduction of mobile-based IT could enhance the efficiency with which data are collected from project sites in real time. Nevertheless, in practice, the proposed methods had technical limitations and they were therefore not commonly applied to construction projects. Typical problems include the cost of purchasing equipment and software for construction projects and the cost of upgrading hardware for maintenance. Furthermore, it has been confirmed that the technical limitations have not yet deviated from the conceptual stage in terms of automation and that the usefulness of the information collected has been poorer than information collected through other techniques [14].

Photogrammetry is a technique that involves the development of a point cloud model from digital photos in order to acquire data about the progress of a construction project. El-Omari and Moselhi [15] is one of the representative studies on photogrammetry, where the amount of work done for a certain time was estimated based on images captured over the corresponding time interval. However, as progress data need to be stored over time, the memory space reserved for data storage should be secured.

The video-based measurement collects progress data by filming the construction project site. This method is effective as it is possible to extract sequential video frame data [16]. Studies on construction projects that have utilized the video-based measurement to acquire data have focused on civil engineering projects such as roads and bridges. The damage detection and safety assessment of facilities and the detection of mobile equipment have been the main areas of focus [17–19]. In particular, video-based measurements are affected by many factors, including temperature changes of objects, focus, the data-capture range, and camera resolution.

In 3D laser scanning, laser lights are emitted onto an object, and the distance to the object is calculated using the return time of travel of the light. This method is widely used in the engineering field [16]. Representative studies in this area have examined monitoring methods for the process and interference of mechanical, electrical, and plumbing (MEP) by comparing two 3D models, or by utilizing other methods to create 3D models using actual progress data acquired by LIDAR [20,21]. However, as data acquisition using LIDAR requires the emission of laser lights, if an object has a high reflectance, the efficiency decreases [22]. Besides, the high cost and limited applicability of LIDAR in complex indoor environments are obstacles to its popularity.

Augmented reality (AR) is a combination of various technologies, where virtual images from a computer are added to a real environment [23]. BIM is the representative software used for AR, and it is also applicable to visualization, simulation, information modeling, and safety testing [24,25]. The advantages of AR are that the construction progress and potential defects can be easily determined during the decision-making process, and if necessary, corrections can be made. While AR has been adopted by a large number of studies, there are still many problems related to user convenience, noise, and data interference filtering. Accordingly, practical methods of solving those problems need to be developed. Table 1 shows the characteristics of the data acquisition technologies and is based on elements that should be considered for technical use in measuring the progress in a construction project.

Table 1. Comparison of data acquisition technologies.

	Mobile IT	Photogrammetry Videogrammetry	LIDAR	Augmented Relity
Cost	Medium	Low	High	High
Automation level	Medium	Medium	High	High
Educational necessity	Low	Low	High	Medium
Portability	Medium	Medium	High	High
Potentiality	Low	Low	Medium	High

In recent years, studies have been conducted to verify the progress by comparing as-built 3D models collected through LIDAR with those produced during the design phase [20,26]. Representative studies in this area have examined the visualization of process rate monitoring through a 4D simulation model conducted in combination with modeling based on laser scanning.

Han and Golparvar-Fard [27] developed a progressive model via laser scanning and studied the construction progress through the BIM. Patraucean et al. [28] also conducted research on the modeling method for the progressive status of a project through the BIM. Meanwhile, Adan et al. [29] focused on the recognition of objects. After segmenting the point clouds corresponding to the walls of a building, a set of candidate objects was detected independently in the color and geometric spaces, and an original consensus procedure integrated both results to infer recognition. In addition, the recognized object was positioned and inserted in the as-is semantically rich 3D model, or BIM model. Wang et al. [30] developed a technique to automatically estimate the dimensions of precast concrete bridge deck panels and create as-built building information modeling (BIM) models to store the real dimensions of the panels. Bueno et al. [31] presented a novel automatic coarse registration method that is an adaptation of as-is 3D point clouds with 3D BIM models. Rebolj et al. [32] proposed methodology including three parameters (minimum local density, minimum local accuracy, and level of scatter) to measure the quality of point cloud data for construction progress tracking. While a recent study investigated the relationship between the quality of point cloud data and the successful identification of building elements, research is still lacking that can identify the required point cloud data quality for each specific application.

Therefore, Wang et al. [33] suggested three main future research directions within the scan-to-BIM framework. First, the information requirements for different BIM applications should be identified, and the quantitative relationships between the modelling accuracy or point cloud quality and the reliability of as-is BIM for its intended use should be investigated. Second, scan planning techniques should be further studied for cases in which an as-designed BIM does not exist and for UAV-mounted laser scanning. Third, as-is BIM reconstruction techniques should be improved with regard to accuracy, applicability, and level of automation. Puri and Turkan [34] mentioned that future work should focus on multiple larger construction projects that contain elements with complex geometrical shapes.

3. Point Cloud-Based Progress Data Acquisition

3.1. LIDAR-Based Point Cloud Data

Image scanning is a technique that involves optically reading images and converting them into data, information and objects, and a LIDAR device is a device that supports image scanning [35]. LIDAR emits a laser beam to an object at specific intervals, and expresses the shape of the object in a set of 3D coordinates by using the direction of the reflected laser and the distance measured [36].

Points that are obtained in this way have 3D X, Y, and Z coordinates, including geo information, and each constituent point is formed at a point where the laser of the LIDAR is reflected from the object. Accordingly, although no geometric information of the object is given, the surface coordinates of the object are included, from which the length, height, and other similar attributes of the object can be acquired (Figure 2). Consequently, a point cloud that includes the geoinformation of an object can provide high-resolution data without distortion using a 3D mesh model. More information can be acquired by modeling points that are obtained from each scanning task into a shape.

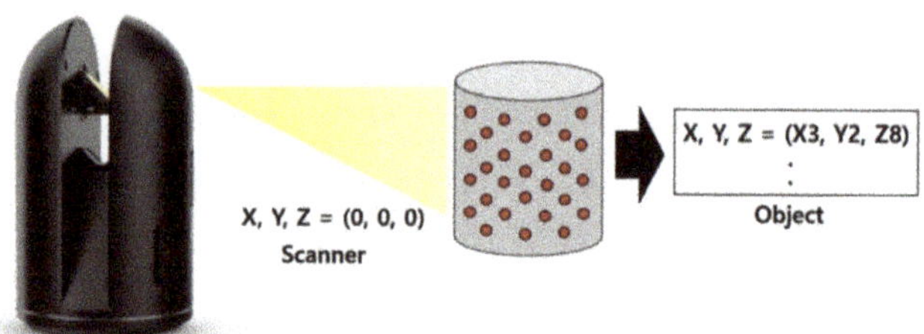

Figure 2. Geometric information acquired by LIDAR.

In each scanning iteration, LIDAR enables only the object seen in the straight line from it to be scanned. If there is another object between the LIDAR and the object, no scanning data are acquired. For this reason, in the case where a laser beam does not arrive at a point from the measurement point, information about the point cannot be determined. In addition, as shown in Figure 3, LIDAR radially emits the source of light and thus generates a shadow area. In other words, even if a projection plane is created vertical to the scanning direction of LIDAR, there may be an overlap such as the one shown in the dotted line inside the circle. To prevent such a phenomenon from occurring behind the object to be measured, all of the whole information pertaining to the appearance of the object needs to be scanned. This means that an object should be scanned at least two times.

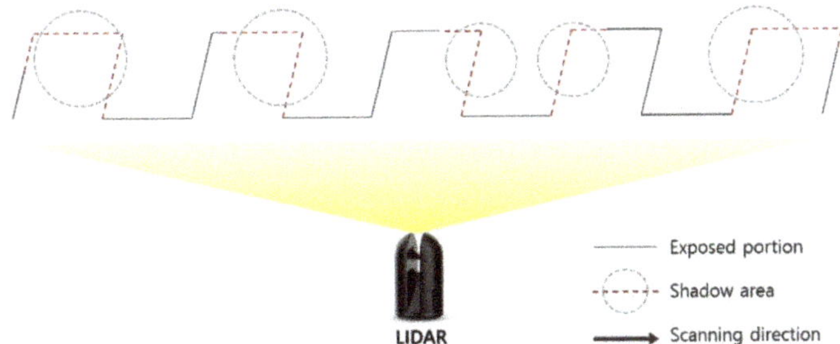

Figure 3. Example of shadow area due to LIDAR scanning.

LIDAR is classified mainly as contact LIDAR and noncontact LIDAR. Contact scanning is a measurement method that attaches a contact sensor called a Touch Probe to an object. The coordinate measuring machine (CMM) is a representative device. Nevertheless, because the sensor directly touches the surface of the object, the object may be easily deformed, thus making the measurement either impossible or time consuming for materials that are likely to become deformed [35].

The first principle of noncontact scanning is that 3D coordinates are formed by timing the return of a laser beam emitted to and reflected from an object surface on the basis of the time-of-flight (TOF) measurement [37]. As this method does not require the sensor to contact the surface of the object, wide areas can be measured at much faster speeds [38]. The TOF measurement installs a measurement device on an axis of rotation and rotates it by a certain angle for horizontal scanning. Meanwhile, for vertical scanning, the laser reflection mirror inside the measurement device is moved by a certain angle. The second principle of noncontact scanning is laser-based triangulation, as illustrated in Figure 4. The reflected light from a target object on which a line-shaped laser beam is irradiated, is measured at a specific cell of a charge-coupled device (CCD) or a complementary metal-oxide semiconductor (CMOS). In other words, this method restores points obtained by scanning a target object into a 3D plane or figure.

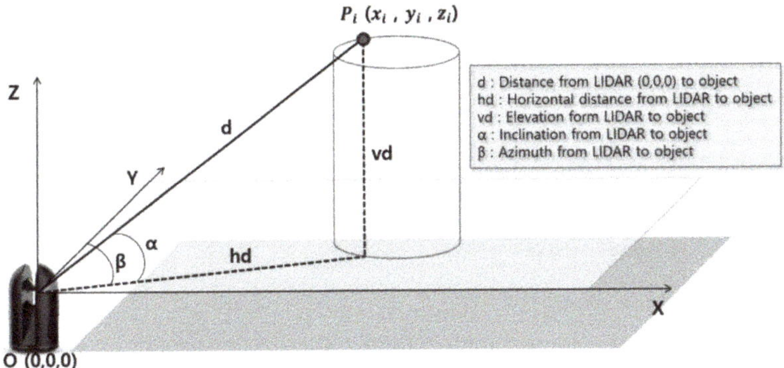

Figure 4. Laser measurement by triangulation.

The distance between the laser oscillator and the optic sensor is specified, and the oscillation angle is also given. Thus, in a triangle consisting of the laser oscillator, the optic sensor, and the target object, the lengths of two sides can be obtained from the remaining side and two angles. A larger number of points can be measured within a given time period when compared with the TOF method. However, rotation is needed to scan the whole area. Other methods that are used to acquire 3D shapes of objects are the shape from shades (SFS) and the structured light system (SLS). SFS restores the 3D shape of an object by illuminating it with light, and then measuring the intensity of the reflected light source. SLS identifies the outer shape of an object by projecting a light source with a regular pattern on a target object and using the shape of pattern reflected.

The point cloud data of each scene, which are obtained by scanning a target object, need to be combined into a single coordinate system to measure the object dimensions, and to analyze points with nonuniform curvatures, and modeling of shapes. The alignment target is the criterion for the alignment process. Generally, the data of a single scanned scene consist of numerous points, and several hundreds of millions or billions of points are given after the alignment process. Accordingly, it takes a long time to accurately align data. However, depending on users' demands, the scanning or alignment time is preferred relative to the alignment accuracy. The scanning or alignment time may vary according to the alignment method employed. Table 2 presents the characteristics of alignment methods for point cloud data.

Table 2. Alignment methods for point clouds and their characteristics.

Alignment Method	Scan Time	Alignment Time	Alignment Accuracy
Cloud to Cloud	High	Medium-Low	High
Target to Target	Low	High	High
Auto Registration	High	Low	Medium
Visual Registration	High	Medium	High-Medium

Cloud-to-cloud alignment does not require any specific target but utilizes a particular point in a point cloud. After selecting the model space of two stations to be aligned, a particular point is picked up in the same place, and individual points are selected in a multi-pick mode and are aligned. Here, a station is a scanning point, that is, a point raised by a laser scanner. When selecting the feature points of scanned scenes, which are obtained at each station, they need to be maximally magnified so that an identical point can be selected and picked, and a fixed point with a nonreflecting material should be selected. In addition, accurate picking is required because the alignment quality and error rate are affected. When stations are aligned, at least three pairs of identical points are needed between each scan, and three points need to constitute as large a triangle as possible to minimize the alignment error. In the case where three or more stations are to be aligned, this alignment process should be repeated.

The target-to-target alignment utilizes targets to align two scanned scenes and to combine them as a single scene Targets are installed beforehand on the plane or bottom, wall, and edge of a target object, and the central points or edges of targets are used for alignment. Targets are to be firmly installed. In the case where a shadow area is included in a scene captured by an installed LIDAR, the object needs to be scanned in a different direction so that at least three common points can be recognized between two scan datasets. In this way, accurate alignment can be possible. If a target is installed on the ground that may be inclined or uneven, care is required because the alignment software program may not recognize the target. Besides, as the size of a target varies according to the scanning points, the alignment program may not recognize the target. Accordingly, if the target object is far from the scanner, the target needs to be larger.

Visual alignment is a manual alignment performed by a user who imports two scanning stations to be aligned into the same space. With this method, after two scanning stations are aligned in the X-axis and Y-axis from the user's perspective, they are also aligned by being moved on the Z-axis and rotated. Visual alignment is most effective for the same or similar features and is also easy for beginners to master.

Cloud-to-cloud alignment and target-to-target alignment, which identify the coordinates of each point and are basically manual operations, are representative methods for the geometric modeling of scanning data. However, if the scanned object is complex, a lot of time and alignment works are required. In such a case, a specific reverse engineering program is usually implemented to automatically extract and align the parts desired by the user. Nevertheless, the use of automatic extraction by employing reverse engineering software has a limitation in terms of the reverse engineering shape, and shapes are often wrongly recognized, which results in inaccurate data alignment. For this reason, the user needs to confirm the result of the automatic alignment using the reverse engineering software and needs to manually remove the wrongly extracted parts. In other words, manual modeling is necessary.

3.2. Drone-Based Point Cloud Data

Drone-based photogrammetry can acquire data of large buildings and terrains. As this method is applicable to large areas, it is recognized as an alternative or supplementary approach for conventional measuring devices [39]. With this advantage, drone-based photogrammetry has been used to measure tasks performed in diverse fields such as building construction, civil work, cultural property management, disaster prevention, and agriculture [40–42]. However, this method that employs a drone produces different outputs depending on the weather and brightness of the photo. Besides, it is difficult to obtain close-up images, and a large relative error tends to occur according to the skill of workers and the performance of equipment. In recent times, there have been numerous studies that aim to improve such disadvantages of drone-based photogrammetry in several ways. The majority of those studies focus on verifying the accuracy of data and enhancing it up to a suitable level. In particular, a marker is used for point matching in order to reduce the error range of drone-based scanning [37]. As shown in Figure 5, drone-based photogrammetry can extract point clouds by implementing various software programs such as Pix4D and Context Capture (Bentley), and it can also capture hardly accessible sites at high altitudes. Thus, this method is being more widely used for data acquisition while monitoring, managing, and inspecting facilities.

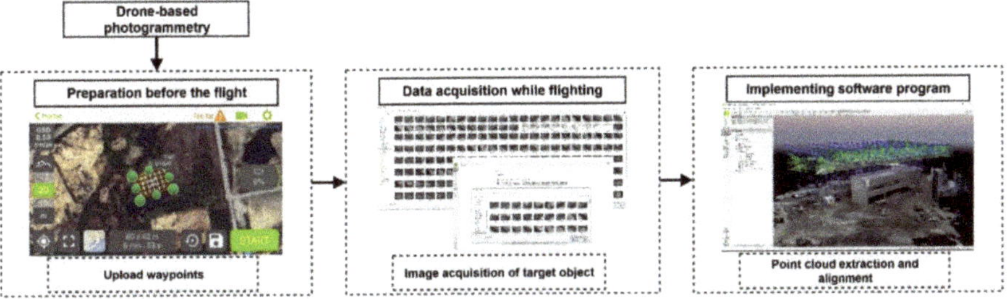

Figure 5. Geometric information obtained by a drone.

4. Verification of Accuracy of Point Cloud Data

4.1. Selection of Target Object and Identification of Recognition Rate

This study acquired point cloud data and verified the accuracy of data obtained using LIDAR, which can scan both the exterior and interior of buildings and using a drone that could capture an area inaccessible to managers. This study also examined a method of acquiring available data using post-processing, and finally determined the accuracy and error of the data acquisition according to building shapes.

In this study, three buildings were selected to acquire point cloud data, which were obtained from the framework of those buildings, that is, from columns, girders, beams, and slabs. Building A consisted of two stories and a rooftop. The framework of this building included 12 columns, 20 girders, 23 beams, and 17 slabs. Building B also comprised two stories and a rooftop. The framework of this building included 25 columns, 36 girders, 40 beams, and 62 slabs. Building C consisted of five stories and a rooftop. For Building C, after point cloud data were acquired by using a drone, ultimate data were obtained by image matching. The base data employed for accuracy verification were acquired by comparing the data that were generated by aligning point clouds with measurements. Table 3 presents details of Buildings A, B, and C, where point cloud data were collected for accuracy verification.

Table 3. Target buildings used to acquire point cloud data.

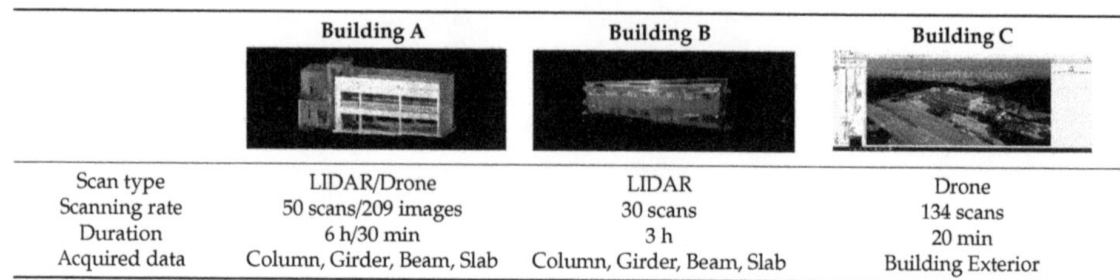

	Building A	**Building B**	**Building C**
Scan type	LIDAR/Drone	LIDAR	Drone
Scanning rate	50 scans/209 images	30 scans	134 scans
Duration	6 h/30 min	3 h	20 min
Acquired data	Column, Girder, Beam, Slab	Column, Girder, Beam, Slab	Building Exterior

This study adopted the visual alignment where data were visually aligned and rotated in the X, Y, and Z axes of the same space. After 3D point cloud data of Building A were completely aligned, the recognition rate of data acquisition was determined for the members of the framework, which include columns, girders, beams, and slabs. In the case of Building A, when the scanning was conducted, the framework had already been completed, but the finishing work had not yet started. For this reason, data for the member of the framework could be easily acquired, and the scanning conditions were similar to those of real construction sites. Fifty rounds of scanning were carried out, and the total duration was 6 h. To prevent incomplete alignment, the scanning interval was set with an overlap of at least 50–60%. Thus, the recognition rate of the members was 100%, and the cloud data could be reliably acquired using LIDAR.

Building B was not under construction but had already been completed. However, as special finishing work had not yet been conducted, the members of the framework could be determined and were thus under similar conditions to those of real construction sites. Thirty rounds of scanning were carried out, and the total duration was 3 h. The LIDAR scanning of this building was performed by setting the acquisition density to "medium." To prevent incomplete alignment, the scanning interval was set with an overlap of at least 50–60%. While the acquisition density of LIDAR scanning was set to medium, the recognition rate of the members was 100%. Thus, the point cloud data acquisition using LIDAR was found to be reliable.

Building A was selected to verify the accuracy of object recognition. However, this building had a rooftop that could not be scanned using a terrestrial LIDAR. Therefore, a drone was needed to obtain aerial photos, from which data of the overall external building shape could be aligned. This study utilized a rotary wing drone for an experiment to acquire point cloud data. The aerial shots obtained using the drone for data acquisition should be as accurate as possible to minimize alignment errors. However, there are limitations with realizing data acquisition using a drone. These limitations pertain to battery, safety, and GPS technology, making it almost impossible to acquire the high-quality data required by the user. In the case where the need is for accurate engineering, the quality of the scanned data needs to be examined using an appropriate criterion before the data are applied. For Buildings A and C, 209 and 134 photos were, respectively, obtained by operating the drone. Then, point cloud alignment was carried out using those photos.

4.2. Determining Error of Aligned Data

The error of the point cloud data was determined by comparing the measurement data of Buildings A and B and the LIDAR-based alignment models of the scanned data. For the measurement data, the real distances between each building were measured using a measuring device. For alignment model data, the distance between point clouds was measured by implementing a software program. The error was measured by comparing the dimensions of the external width, the distance between columns, and the height of the column, which corresponded to the width, length, and height of the building, respectively. Table 4 presents the errors obtained by comparing measurements and LIDAR scanning results for Buildings A and B. It shows that in the case of Building A, the average error values were 0.011 m, 0.012 m, and 0.019 m in the external width, the distance between columns, and the column height, respectively. Meanwhile, in the case of Building B, the average error values were 0.012 m, 0.011 m, and 0.012 m in the external width, the distance between columns, and the column height, respectively.

Table 4. Errors of LIDAR-based point cloud data.

		Measured Distance (m)	LIDAR Scan (m)	Average Error (m)
Building A	External width	29.28, 6.62, 10.45	29.289, 6.608, 10.465	0.011
	Distance between columns	7.53, 6.67, 1.48	7.547, 6.673, 1.495	0.012
	Column height	2.76	2.779	0.019
	External width	35.81, 18.49, 8.08	35.829, 18.491, 8.098	0.012
Building B	Distance between columns	3.09, 3.09, 3.02	3.115, 3.097, 3.021	0.011
	Column height	6.25, 6.71	6.261, 6.724	0.012

According to the BIM guide for 3D imaging, which is published by the General Service Administration (GSA) of the USA, the error range needs to be a maximum of 51 mm for urban design projects and a maximum of 13 mm for architectural designs. Otherwise, the practical accuracy cannot be maintained. In this study, the errors for each item, which were identified by performing comparative measurements, were between 11 mm and 19 mm. This result was remarkably close to 13 mm, which is recommended by GSA for the application of point cloud data to architectural designs. The distance between the two end points of a target member in the scanned data was measured by mouse picking. As this method implies an unavoidable error, the above errors indicate that very accurate data were acquired by this study. Consequently, based on the cases of Buildings A and B, the LIDAR-based measurement and alignment of this study is shown to be accurate.

Errors that were present in the point cloud data obtained using a drone were compared in the same way as the LIDAR-based error verification. The target was Building A. As the drone could capture only the external building shape, the measurement data of external members were compared with the drone-based alignment model of scanned data. Table 5 presents the errors between the measurements and the drone-based alignment data for Building A. It shows that the average errors for the width, length, and height of the building were 0.378 m, 0.358 m (distance between columns), and 0.072 m (column height), respectively. These values were far below 13 mm, which is recommended by GSA for the application of point cloud data to architectural designs. Such a large gap is attributable to the following intrinsic characteristics of drones. First, because each drone captures a target while flying, it is difficult to acquire accurate data. Second, images obtained by a drone need to be converted to point clouds and then imported into a software program that can measure distance. These steps result in significant gaps. Accordingly, this study used the drone-based point cloud data only for the parts for which data could not be acquired using LIDAR.

Table 5. Errors of drone-based point cloud data.

		Measured Distance (m)	LIDAR Scan (m)	Average Error (m)
Building A	External width	29.28, 6.62, 10.45	28.173, 6.638, 10.46	0.378
	Distance between columns	7.25, 2.85, 5.46	6.651, 2.839, 4.994	0.358
	Column height	2.76	2.688	0.072

5. 3D Modeling of Point Cloud Data

5.1. Creation of 3D Model of Point Cloud Date for Target Object

Upon verification of the accuracy of point cloud data, which had been acquired using drone- and LIDAR-based scanning, respectively, it was shown that the data obtained by LIDAR scanning had a higher accuracy than those acquired by the drone. However, the progress data acquisition is likely to include inaccessible areas such as rooftops, and there may be a risk factor when acquiring data. In such a case, the application of LIDAR to data acquisition may be restricted, which will result in uncertain parts in the alignment of the whole point clouds. In this regard, by combining the datasets that have been acquired by a drone and a LIDAR, respectively, the loss of data can be prevented, thus improving the accuracy of progress data for construction sites. As shown in Figure 6, to improve the accuracy of progress data, this study combined two types of point cloud data. The mixing process can be summarized as follows.

1. As the two types of point cloud data acquired by LIDAR and a drone have different file formats, their file attributes need to be unified prior to mixing those data. In this study, the data acquired by a drone had the p4d format and were thus converted to xyz coordinates, which indicated GPS coordinates, in order to be combined with the LIDAR-based data.
2. As the drone-based data thus converted to xyz coordinates had fixed coordinates, automatic alignment was possible without the need for any additional alignment tasks. Accordingly, when the files were imported into the program for LIDAR, the alignment was automatically completed.
3. As the drone-based data were acquired by an aerial shot, they included not only the target object but also the surrounding area. Therefore, the noise was removed to obtain only the necessary part.
4. From the drone-based data, which had been completely imported, only the part available for mixed data was selected, and the remaining parts were removed.
5. The final mixed data were completed by conducting the cloud-to-cloud alignment between the selected drone-based data and the LIDAR-based data.

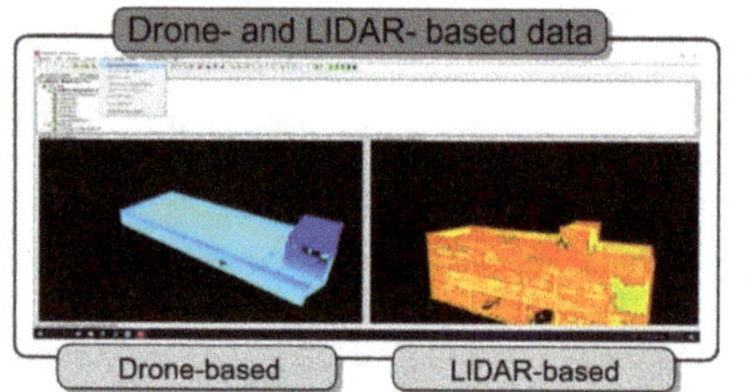

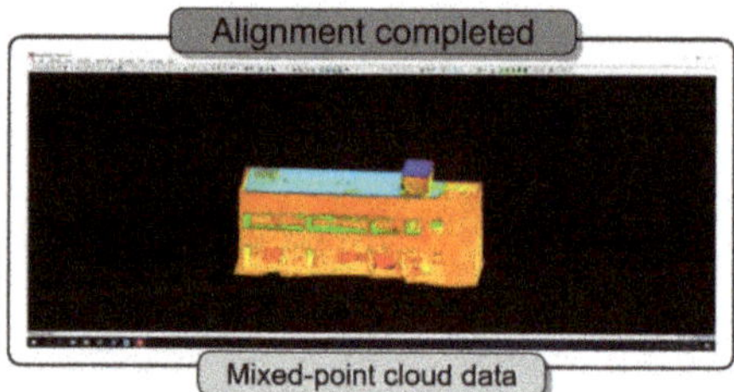

Figure 6. Combination of drone-based data and LIDAR-based data.

5.2. 3D Polygon Mesh Modeling

Delanay triangulation (DT) and the Voronoi diagram (VD) are basic concept for 3D polygon mesh modeling. DT is a division in which points on a plane are connected in triangles to divide the space such that the minimum value of the cabinet is maximal, and the outer source of any triangle does not include anything other than the three vertices of the triangle. In other words, of various triangulations, the division in which each triangle is as close as possible to the regular triangle. Meanwhile, a VD is a division of a plane into polygons that contain each of these points when there are points on the plane. When there are points on the plane, the two adjacent generation points should be connected to the line, and a vertical equal division of this line should be drawn. In this way, a vertical isomeric line is drawn, creating a polygon with a vertical isomeric line that divides the sides into polygons. DT and VD are in a dual relationship, and if one is known, the other can immediately be obtained.

Figure 7a shows VD and the DT of the same set of points. The VD is created by sequentially linking the center of the circumcircle of DTs with generating points as a common vertex, and by linking points between adjacent VD areas, DT can be generated for these points. For 3D stereoscopic modeling from 3D point clouds, the use of DT allows for polygon mesh to be obtained from a collection of points on the surface. The triangulation in 3D is called tetrahedralization or tetrahedrization [43]. A tetrahedralization is the partitioning of the input domain into a collection of tetrahedra that meet at only shared faces (vertices, edges, or triangles). Polygons are typically ideal for accurately representing the results of measurements, providing an optimal surface description. However, the results of tetrahedralization are much more complicated than those of a 2D triangulation. Therefore, this study was performed by utilizing the Commercial Modeling Software Package. The Leica Cyclone platform was used for 3D point cloud data visualization and processing, and the Leica 3D Reshaper platform was used for polygon mesh model generation Figure 7b,c.

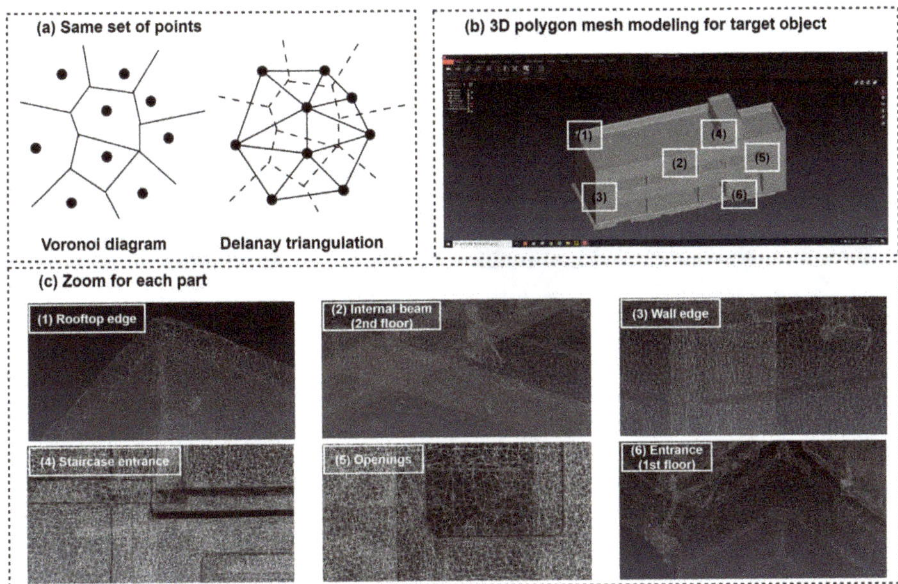

Figure 7. Combination of drone-based data and LIDAR-based data.

Mixed-point cloud data can be configured into a 3D model using a modeling process. This process generates a polygon from the outline of the point cloud. After the polygon model of each member is generated, an ultimate 3D model is completed by an editing process. However, this modeling method cannot reflect the details of the acquired data. Construction projects usually include the installation of molds and the casting of concrete, which may cause some errors or bent surfaces that were not originally planned. Manual 3D modeling sets the surfaces of each member and allocates heights in the form of a straight line. Accordingly, detailed errors such as a small slope or bends on a target surface cannot be modeled. Nevertheless, such errors can be detected compared with the actual plan. Figure 8 illustrates a representative process of 3D modeling for a completely aligned point cloud.

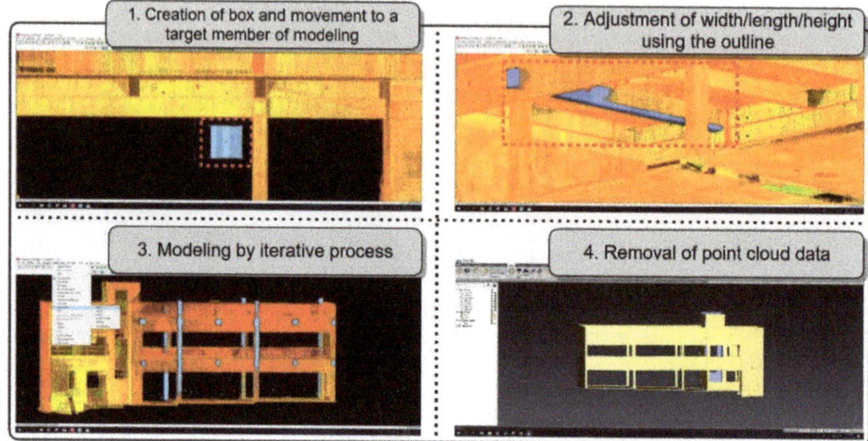

Figure 8. Three-dimensional modeling process for point cloud.

5.3. Determination of Errors in the Created 3D Model

The acquisition of accurate data is the most essential part of the reverse engineering using the progress data acquired from a construction site. The acquisition of accurate data during the proposed process in this study is based on the 3D shapes of buildings. Accordingly, error identification is necessary for the 3D shape of a target building. This study also determined the shape of the created 3D model. The volume of the 3D model was compared with the actual data. The amount of concrete poured for the construction was used as the actual data. Table 6 presents the locations, date, and volume (m^3) of concrete poured in Building A. Concrete was poured 6 times, and the total volume was 522 m^3.

Table 6. Details of concrete pouring in building A.

	Location						
	Subslab Concrete	**Foundation**	**PIT**	**1 F**	**2 F**	**Protective Concrete and Rooftop**	**Total**
Date	D + 0	D + 8	D + 21	D + 47	D + 68	D + 131	-
Volume (m^3)	36	18	174	134	126	34	522

In the case of Building A, the initial data acquired by LIDAR were limited to the above-ground part, and back filling parts, such as sub-slab concrete and foundation, were excluded. In other words, the 3D model was generated by acquiring the point cloud data for the above-ground part. Accordingly, the volume data of the poured concrete were compared over 468 m^3, which included PIT, 1F, 2F, protective concrete, and the rooftop.

As with other commercial software programs, a software program that automatically creates a 3D model after a point cloud is imported enables the length, area, and volume of each object to be identified. The volume of the 3D model of Building A was measured to be 479 m^3. When the actual data were compared with the volume data of the 3D model, the difference was 11 m^3. This result corresponded to a difference of less than 3% compared with the actual data. Thus, the 3D model data showed relatively little error compared with the volume based on the actual data, demonstrating high accuracy.

5.4. Visualization of Construction Progress

In order to track the progress of a project, the current status should be compared with the planned status. This study examined an overlap-based method of comparing the BIM model, which provides the as-planned data of a project, and the point cloud-based 3D model, which shows as-built data. For the comparison, the point cloud-based 3D model needs to be imported into the BIM model. However, in the case where two types of 3D models were used, they were implemented on different software bases. File conversion is required for the importing of data. Figure 9 shows the process involved in comparing two models.

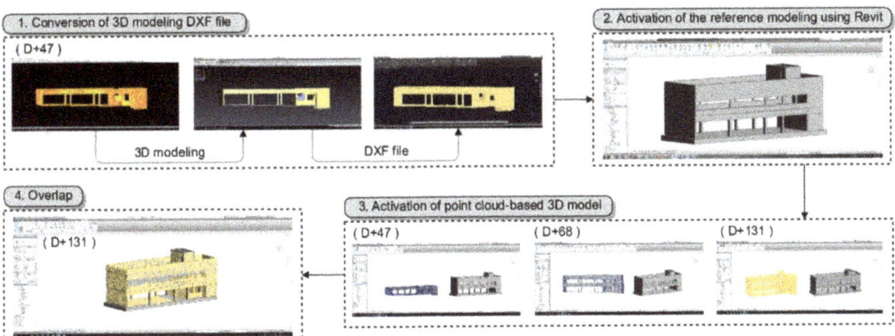

Figure 9. Comparison of the BIM model and the point cloud-based model.

6. Conclusions

This study proposed methods that can be used to track the progress of construction projects, and each of the proposed methods was verified. With respect to for data acquisition, the drone- and LIDAR-based point cloud data acquisition methods were examined, and the accuracy of data was verified with respect to their application to actual construction projects. LIDAR-based point cloud data had error rates of roughly 11–19 mm, indicating a high accuracy level. However, the drone-based data showed a considerably low accuracy level. Because the progress data are based on the 3D shapes of buildings, errors in the 3D shapes were also examined. In the case of Building A, the 3D model based on point cloud data had a difference of 11 m^3 when compared to the actual data. This value represented a difference of less than 3% compared with the actual data, thereby demonstrating a low error rate.

In order to track the progress of a project, the current status should be compared with the planned status. The proposed overlapping method used in this study for the BIM model and the point cloud-based 3D model enabled the actual progress to be visualized and compared to the corresponding plan. Therefore, it is expected to permit project managers to more easily track project progress and identify precise statues when progress has not proceeded as planned. This provides the advantage of progress management being carried out through the establishment of future construction plans and the review of schedules. It is also believed that various reports and related data using visualized three-dimensional models will help project participants and stakeholders greatly. All additional accumulated data could also be used as the basis for the maintenance phase after the end of the project or for similar projects in the future.

Based on results obtained, the data acquisition method proposed in this study appears to be very efficient and can enable project managers to assess the progress and comprehensively manage projects. In particular, as decisions can be made quickly based on rapid information delivery, the incidence of workers' errors and the accompanying need for reconstruction can thus be prevented, leading to reductions in time and cost overruns. However, this study showed that errors and omissions in the alignment of point cloud data caused the poor-quality data alignment. The representative causes were the reflexivity of laser emitted on the surfaces of objects, the distance, and the atmospheric environment. The path of laser was also problematic. If it is possible to omit a specific section or to utilize an independent one that does not need to be aligned with other sections, the problem may be trivial. However, if the section is an essential one that interfaces with different sections, the problem should be resolved.

Author Contributions: In this paper, S.K. (Seungho Kim) collected the data and wrote the paper. S.K. (Sangyong Kim) analyzed the data and conceived the methodology. D.-E.L. developed the ideas and designed the research framework. All authors have read and agreed to the published version of the manuscript.

Funding: This work was supported by the National Research Foundation of Korea (NRF) grant funded by the Korean government (MSIT) (No. NRF-2018R1A5A1025137).

Conflicts of Interest: The authors declare no conflict of interest.

References

1. Akinci, B.; Boukamp, F.; Gordon, C.; Huber, D.; Lyons, C.; Park, K. A formalism for utilization of sensor systems and integrated project models for active construction quality control. *Autom. Constr.* **2006**, *15*, 124–138. [CrossRef]

2. Tang, P.; Huber, D.; Akinci, B.; Lipman, R.; Lytle, A. Automatic reconstruction of as-built building information models from laser-scanned point clouds: A review of related techniques. *Autom. Constr.* **2010**, *19*, 829–843. [CrossRef]

3. Nahangi, M.; Haas, C. Automated 3D compliance checking in pipe spool fabrication. *Adv. Eng. Inform.* **2014**, *28*, 360–369. [CrossRef]

4. Cheng, T.; Venugopal, M.; Teizer, J.; Vela, P. Automated trajectory and path planning analysis based on ultra wideband data. *J. Comput. Civ. Eng.* **2011**, *26*, 151–160. [CrossRef]

5. Woo, S.; Jeong, S.; Mok, E.; Xia, L.; Choi, C.; Pyeon, M.; Heo, J. Application of wifi-based indoor positioning system for labor tracking at construction sites: A case study in Guangzhou mtr. *Autom. Constr.* **2011**, *20*, 3–13. [CrossRef]

6. El-Omari, S.; Moselhi, O. Integrating automated data acquisition technologies for progress reporting of construction projects. *Autom. Constr.* **2011**, *20*, 699–705. [CrossRef]

7. Golparvar-Fard, M.; Peña-Mora, F.; Savarese, S. Automated progress monitoring using unordered daily construction photographs and IFC-based building information models. *J. Comput. Civ. Eng.* **2015**, *29*, 04014025. [CrossRef]

8. Kim, C.; Son, H.; Kim, C. Automated construction progress measurement using a 4D building information model and 3D data. *Autom. Constr.* **2013**, *31*, 75–82. [CrossRef]

9. Kim, S.; Kim, S.; Tang, L.; Kim, G. Efficient Management of Construction Process Using RFID+PMIS System: A Case Study. *Appl. Math. Inf. Sci.* **2013**, *7*, 19–26. [CrossRef]

10. Navon, R.; Berkovich, O. Development and on-site evaluation of an automated materials management and control model. *J. Constr. Eng. Manag.* **2005**, *131*, 1328–1336. [CrossRef]

11. Cheng, T.; Venugopal, M.; Teizer, J.; Vela, P. Performance evaluation of ultra wideband technology for construction resource location tracking in harsh environments. *Automat. Constr.* **2011**, *20*, 1173–1184. [CrossRef]

12. Giretti, A.; Carbonari, A.; Naticchia, B.; DeGrassi, M. Design and first development of an automated real-time safety management system for construction sites. *J. Civ. Eng. Manag.* **2009**, *15*, 325–336. [CrossRef]

13. Golparvar-Fard, M.; Peña-Mora, F.; Arboleda, C.; Lee, S. Visualization of construction progress monitoring with 4D simulation model overlaid on time-lapsed photographs. *J. Comput. Civ. Eng.* **2009**, *23*, 391–404. [CrossRef]

14. Wang, Q.; Kim, M.K. Applications of 3D point cloud data in the construction industry: A fifteen-year review from 2004 to 2018. *Adv. Eng. Inf.* **2019**, *39*, 306–319. [CrossRef]

15. El-Omari, S.; Moselhi, O. Data acquisition from construction sites for tracking purposes. *Eng. Constr. Archit. Manag.* **2009**, *16*, 490–503. [CrossRef]

16. Omar, T.; Nehdi, M. Data acquisition technologies for construction progress tracking. *Autom. Constr.* **2016**, *70*, 143–155. [CrossRef]

17. Bosché, F.; Ahmed, M.; Turkan, Y.; Haas, C.; Haas, R. Tracking the built status of MEP works: Assessing the value of a Scan-vs-BIM system. *J. Comput. Civ. Eng.* **2014**, *28*, 05014004. [CrossRef]

18. Koch, C.; Jog, G.; Brilakis, I. Automated pothole distress assessment using asphalt pavement video data. *J. Comput. Civ. Eng.* **2013**, *27*, 370–378. [CrossRef]

19. Zhu, Z.; Brilakis, I. Machine vision-based concrete surface quality assessment. *J. Constr. Eng. Manag.* **2010**, *136*, 210–218. [CrossRef]

20. Bosché, F.; Ahmed, M.; Turkan, Y.; Haas, C.; Haas, R. The value of integrating Scan-to-BIM and Scan-vs-BIM techniques for construction monitoring using laser scanning and BIM: The case of cylindrical MEP components. *Autom. Constr.* **2015**, *49*, 201–213. [CrossRef]

21. Turkan, Y.; Bosche, F.; Haas, C.; Haas, R. Automated progress tracking using 4D schedule and 3D sensing technologies. *Autom. Constr.* **2012**, *22*, 414–421. [CrossRef]

22. Dai, F.; Rashidi, A.; Brilakis, I.; Vela, P. Comparison of image-based and time-of-flight-based technologies for three-dimensional reconstruction of infrastructure. *J. Constr. Eng. Manag.* **2013**, *139*, 69–79. [CrossRef]

23. Wang, X.; Truijens, M.; Hou, L.; Wang, Y.; Zhou, Y. Integrating Augmented Reality with Building Information Modeling: Onsite construction process controlling for liquefied natural gas industry. *Autom. Constr.* **2014**, *40*, 96–105. [CrossRef]

24. Rankohi, S.; Waugh, L. Review and analysis of augmented reality literature for construction industry. *Vis. Eng.* **2013**, *1*, 9. [CrossRef]

25. Shirazi, A.; Behzadan, A.H. Design and assessment of a mobile augmented reality-based information delivery tool for construction and civil engineering curriculum. *J. Prof. Issues Eng. Ed. Pr.* **2015**, *141*, 04014012. [CrossRef]

26. Liu, Z.; Lu, Y.; Peh, L.C. A Review and Scientometric Analysis of Global Building Information Modeling (BIM) Research in the Architecture, Engineering and Construction (AEC) Industry. *Buildings* **2019**, *9*, 210. [CrossRef]

27. Han, K.; Golparvar-Fard, M. Appearance-based material classification for monitoring of operation-level construction progress using 4D BIM and site photologs. *Autom. Constr.* **2015**, *53*, 44–57. [CrossRef]

28. Patraucean, V.; Armeni, I.; Nahangi, M.; Yeung, J.; Brilakis, I.; Haas, C. State of research in automatic as-built modelling. *Adv. Eng. Inform.* **2015**, *29*, 162–171. [CrossRef]

29. Adan, A.; Quintana, B.; Prieto, S.A.; Bosche, F. Scan-to-BIM for 'secondary' building components. *Adv. Eng. Inf.* **2018**, *37*, 119–138. [CrossRef]

30. Wang, Q.; Sohn, H.; Cheng, J.C. Automatic as-built BIM creation of precast concrete bridge deck panels using laser scan data. *J. Comput. Civ. Eng.* **2018**, *32*, 04018011. [CrossRef]

31. Bueno, M.; Bosche, F.; Gonzalez-Jorge, H.; Martinez-Sanchez, J.; Arias, P. 4-Plane congruent sets for automatic registration of as-is 3D point clouds with 3D BIM models. *Autom. Constr.* **2018**, *89*, 120–134. [CrossRef]

32. Rebolj, D.; Pučko, Z.; Babič, N.C.; Bizjak, M.; Mongus, D. Point cloud quality requirements for Scan-vs-BIM based automated construction progress monitoring. *Autom. Constr.* **2017**, *84*, 323–334. [CrossRef]

33. Wang, Q.; Guo, J.; Kim, M.K. An application oriented scan-to-BIM framework. *Remote Sens.* **2019**, *11*, 365. [CrossRef]

34. Puri, N.; Turkan, Y. Bridge construction progress monitoring using lidar and 4D design models. *Autom. Constr.* **2020**, *109*, 102961. [CrossRef]

35. Kang, T. *3D Scanning Vision Reverse Engineering*; CIR Publishing: Seoul, Korea, 2017.

36. Kwon, S. Object Recognition and Modeling Technology Using Laser Scanning and BIM for Construction Industry. *AIK* **2009**, *53*, 31–38.

37. Choi, G. *A Study on the Comparison and Utilization of 3D Point Cloud Data for Building Objects Using Laser Scanning and Photogrammetry*; Sungkyunkwan University: Seoul, Korea, 2017.

38. Tonon, F.; Kottenstette, J.T. Laser and photogrammetric methods for rock face characterization. In Proceedings of the 41st US Rock Mechanics Symposium, Golden, CO, USA, 17–18 June 2006.

39. Siebert, S.; Teizer, J. Mobile 3D mapping for surveying earthwork projects using an Unmanned Aerial Vehicle (UAV) system. *Autom. Constr.* **2014**, *41*, 1–14. [CrossRef]

40. McLeod, T.; Samson, C.; Labrie, M.; Shehata, K.; Mah, J.; Lai, P.; Elder, J.H. Using video acquired from an unmanned aerial vehicle (UAV) to measure fracture orientation in an open-pit mine. *Geomatica* **2013**, *67*, 173–180. [CrossRef]

41. Park, M.H.; Kim, S.G.; Choi, S.Y. The Study about Building Method of Geospatial Informations at Construction Sites by Unmanned Aircraft System (UAS). *Korean Assoc. Cadastre Inf.* **2013**, *15*, 145–156.

42. Zarco-Tejada, P.J.; Diaz-Varela, R.; Angileri, V.; Loudjani, P. Tree height quantification using very high resolution imagery acquired from an unmanned aerial vehicle (UAV) and automatic 3D photo-reconstruction methods. *Eur. J. Agron.* **2014**, *55*, 89–99. [CrossRef]

43. Remondino, F. From point cloud to surface: The modeling and visualization problem. *Int. Arch. Photogramm. Remote Sens. Spat. Inf. Sci.* **2003**, *34*, W10.

 *sustainability*

Article

A Simple and Sustainable Prediction Method of Liquefaction-Induced Settlement at Pohang Using an Artificial Neural Network

Sung-Sik Park [1], Peter D. Ogunjinmi [1], Seung-Wook Woo [1] and Dong-Eun Lee [2,*]

[1] Department of Civil Engineering, Kyungpook National University, 80 Daehakro, Bukgu, Daegu 41566, Korea; sungpark@knu.ac.kr (S.-S.P.); ogunjinmipeter@gmail.com (P.D.O.); geowsw@knu.ac.kr (S.-W.W.)

[2] Department of Architectural Engineering, Kyungpook National University, 80 Daehakro, Bukgu, Daegu 41566, Korea

* Correspondence: dolee@knu.ac.kr; Tel.: +82-53-950-7540

Received: 8 April 2020; Accepted: 9 May 2020; Published: 13 May 2020

Abstract: Conventionally, liquefaction-induced settlements have been predicted through numerical or analytical methods. In this study, a machine learning approach for predicting the liquefaction-induced settlement at Pohang was investigated. In particular, we examined the potential of an artificial neural network (ANN) algorithm to predict the earthquake-induced settlement at Pohang on the basis of standard penetration test (SPT) data. The performance of two ANN models for settlement prediction was studied and compared in terms of the R^2 correlation. Model 1 (input parameters: unit weight, corrected SPT blow count, and cyclic stress ratio (CSR)) showed higher prediction accuracy than model 2 (input parameters: depth of the soil layer, corrected SPT blow count, and the CSR), and the difference in the R^2 correlation between the models was about 0.12. Subsequently, an optimal ANN model was used to develop a simple predictive model equation, which was implemented using a matrix formulation. Finally, the liquefaction-induced settlement chart based on the predictive model equation was proposed, and the applicability of the chart was verified by comparing it with the interferometric synthetic aperture radar (InSAR) image.

Keywords: settlement; artificial neural network; liquefaction

1. Introduction

The Pohang earthquake (M_W = 5.4) that struck the Heunghae Basin, around Pohang City, on 15 November 2017, had a damaging effect, leading to liquefaction and lateral spreading. Since the event, several attempts have been made to study the post-earthquake damage [1–5]. However, little attention has been paid to the settlement resulting from the liquefaction. This study tried to predict the liquefaction-induced settlement of Pohang by applying a machine learning algorithm to a standard penetration test (SPT) data and proposes a liquefaction settlement chart based on the results. Before constructing a structure on the ground, design is performed based on the ground investigation results. In addition, many sites, including Pohang, have a lot of SPT data. The SPT is a common method to get ground investigation data.

Assessing liquefaction-induced settlements is a major challenge in geotechnical earthquake engineering since a variety of phenomena such as re-sedimentation or reconsolidation (volumetric strain) of the liquefied soil, ground loss due to venting of liquefied soil (i.e., sand boils or ejecta), lateral spreading under zero volume change, soil-structure interaction ratcheting, and bearing capacity failure are associated with them [6]. For numerical analysis, earthquake-induced liquefaction in the free-field can be interpreted as a 1D phenomenon occurring along a vertical soil column in which earthquake-induced cyclic shear and compressive forces increase the pore pressure and thereby cause a reduction in the transient stiffness and strength of the soil. After liquefaction, reconsolidation occurs in the soil owing to the dissipation of the excess pore pressure (Δu) by means of water flow, and it results in the vertical settlement of the ground surface [7].

Tang et al. [8] classified the significant parameters controlling seismic soil liquefaction into seismic parameters, site conditions, and soil parameters. Out of 22 influence factors, they identified 12 as being significant, and they were the magnitude, epicentral distance, duration, fines content, particle size, grain composition, relative density, drainage condition, degree of consolidation, thickness of the sand layer, depth of the sand layer, and groundwater table. Over the years, researchers have considered some of these significant influence factors for predicting earthquake-induced liquefaction and its effects through machine learning techniques [9,10].

Therefore, simple artificial neural network (ANN) models were adopted to predict liquefaction-induced settlement on the basis of SPT database from the Korea Geotechnical Information DB system [11] and the Pohang earthquake. In the following sections, the research methodology and findings are presented.

2. Motivation and Study Objective

Liquefaction-induced settlement is often calculated by considering numerous parameters and following several complex analytical and numerical procedures. However, obtaining such parameters in the field may not be practicable in most cases, as some of the required data may not be available. Hence, there is a need for an alternative simple settlement prediction procedure that requires a few parameters that are readily obtained from a field observations database. Therefore, the objective of this study is to fill this gap by presenting a tool to predict liquefaction-induced settlement that may occur when an earthquake occurs in the field using SPT data obtained in the past.

3. Methodology

The database used in this study was collected from the Korea Geotechnical Information DB system [11] and the UBCSAND constitutive effective stress model [12]. Through a 1D column analysis, the UBCSAND model estimates the shear-induced deformation from SPT data and earthquake information. SPT data were obtained for five different borehole sites near the epicenter of the earthquake at Pohang. The summary statistics of the data set are presented in Table 1 and the details of the database are in Table A1.

Table 1. Summary statistics of the data set.

	Input Parameters				Output
	Depth (m)	**Unit Weight (kN/m^3)**	$N_{1(60)}$	**CSR**	**Settlement (mm)**
Count	100	100	100	100	100
Mean	10.500	18.960	13.620	0.314	0.898
Std.	5.795	1.869	8.722	0.045	0.874
Min.	1	16	0	0.21	0
25%	5.75	18	7	0.29	0.3
50%	10.5	20	11	0.32	0.6
75%	15.25	21	25	0.34	1.4
Max.	20	21	25	0.39	3.4

The data set comprised 100 data points (20 data for each borehole) along with the corresponding settlement values. The locations of the boreholes considered in the study are shown in Figure 1.

Figure 1. Locations of boreholes in Pohang City considered in this study.

3.1. Data Division and Preprocessing

The settlement prediction process comprises training and testing. Seventy percent of the entire data set was used for training, and the remaining 30% was used for testing. The data were preprocessed before training the algorithm, to ensure quick convergence and minimize the generalization error. This involved scaling the input variables to the range −1 to +1 by using Equation (1).

$$X_n = \left(\frac{(b-a)}{(B-A)} \times X_{unscaled} \right) + \left(a - \left[A \times \frac{(b-a)}{(B-A)} \right] \right) \tag{1}$$

where *A* and *B* are the minimum and maximum values of the unscaled data set, respectively, and *a* and *b* are the minimum and maximum values of the scaled data set, respectively.

3.2. Overview of the Artificial Neural Network Model

3.2.1. Basic Concept of ANN

Artificial neural networks (ANNs) are complex mathematical models inspired by biological neurons, and they emulate biological neural networks. They are widely used for nonlinear system modeling and system identification [13]. A typical ANN consists of an input layer, one or more hidden layers, and an output layer. The numbers of layers and neurons in each layer depend on the complexity of the problem under consideration.

3.2.2. Mathematical Representation of ANN Architecture

A neural network in its simplest form can be used to model the relationship between data points *x* and the corresponding real-valued targets *y*. Mathematically, if our inputs (*x*) comprise *n* features, we can choose weights (*w*) and bias (*b*) such that our prediction (*y′*) is given by Equation (2).

$$y' = w_1 \cdot x_1 + \cdots + w_n \cdot x_n + b \tag{2}$$

For easy computation, all the features can be collected into a vector **x** and all weights into a vector **w** to express our model compactly using the dot product notation—Equation (3).

$$y' = w^{\tau}x + b \tag{3}$$

ANNs can learn by example (supervised learning). In ANNs, a set of input variables are multiplied by adjustable connection weights to produce the output. When input data are fed to an ANN, the ANN adjusts through a feed-forward back-propagation technique to determine the rules governing the relationship between the concerned variables. Figure 2 shows a graphical depiction of a typical feedforward ANN architecture. A neural network is trained using error back-propagation.

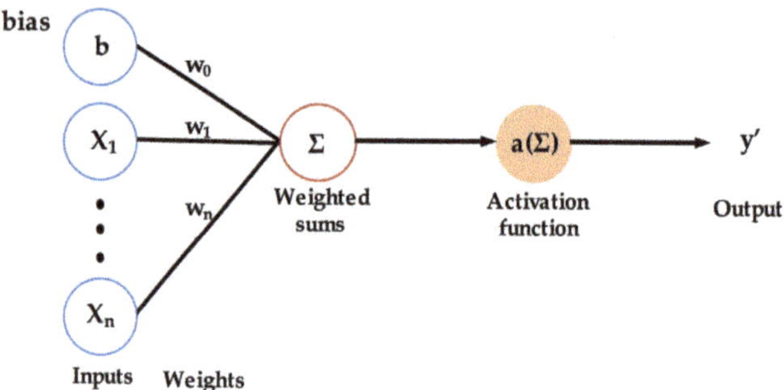

Figure 2. Feedforward neural network architecture.

Two ANN models were considered in this study, and they are shown in Figure 3. Both models had three input variables. The input variables of model 1 were unit weight (γ), corrected SPT blow count ($N_{1(60)}$), and cyclic stress ratio (CSR), while those of model 2 were depth of the soil layer (d), $N_{1(60)}$, and CSR.

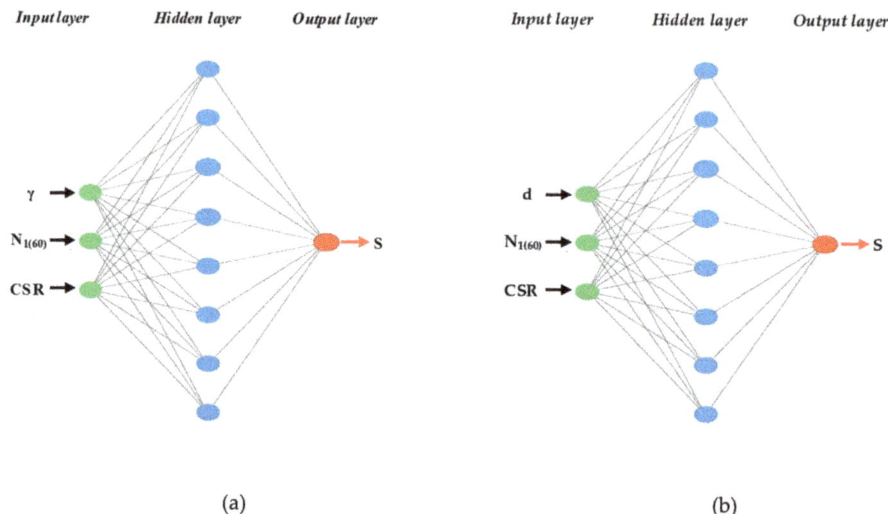

Figure 3. Architecture of the artificial neural network (ANN); (**a**) model 1 and (**b**) model 2.

The choice of input parameters was based on domain knowledge. They were chosen by considering how the seismic and soil properties influence liquefaction-induced settlement. The soil properties considered were γ, $N_{1(60)}$, and d, while the CSR represented the seismic property. The CSR quantifies the demand imposed on the critical soil layer as a result of the seismic ground motion.

4. Results and Discussion

Table 2 summarizes the performance statistics of the two ANN models used for settlement prediction. For the test data set, models 1 and 2 had R^2 (coefficient of determination) values of 0.8601 and 0.7352 and MAE (mean of absolute errors) values of 0.1941 and 0.3136, respectively.

Table 2. Performance statistics of the ANN models.

Model	Input Parameters	Output	No. of Epochs	R^2	MAE
1	γ, $N_{1(60)}$, CSR		60	0.8601	0.1941
2	d, $N_{1(60)}$, CSR	S	60	0.7352	0.3136

After the models were trained, the root mean square error (RMSE) and loss were plotted to check the models' performance for the training and test data sets, as shown in Figures 4 and 5. The x-axis represents the number of epochs (i.e., the number of times the model ran through the entire training/test data set and updated the weights).

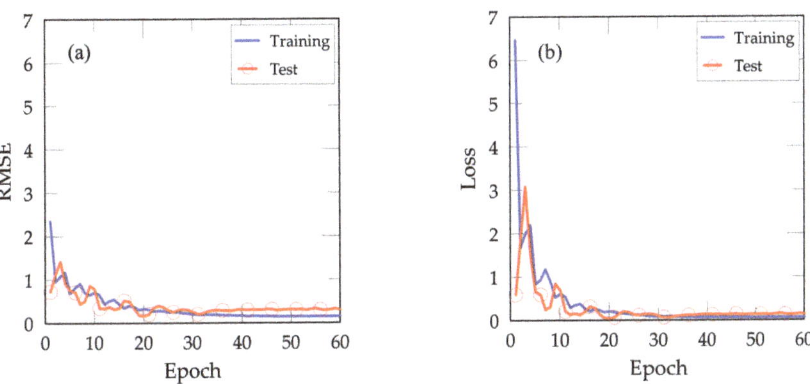

Figure 4. Plot of the (**a**) root mean square error (RMSE) and (**b**) loss for ANN model 1.

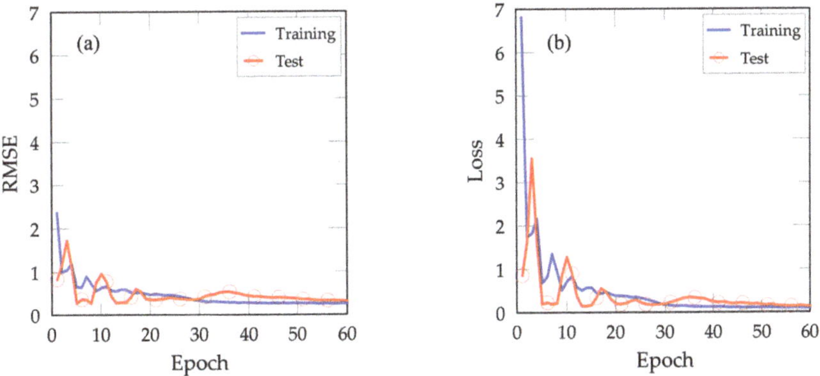

Figure 5. Plot of the (**a**) RMSE and (**b**) loss for ANN model 2.

Figures 6 and 7 show the performance of the ANN models in terms of R^2 for the test data set.

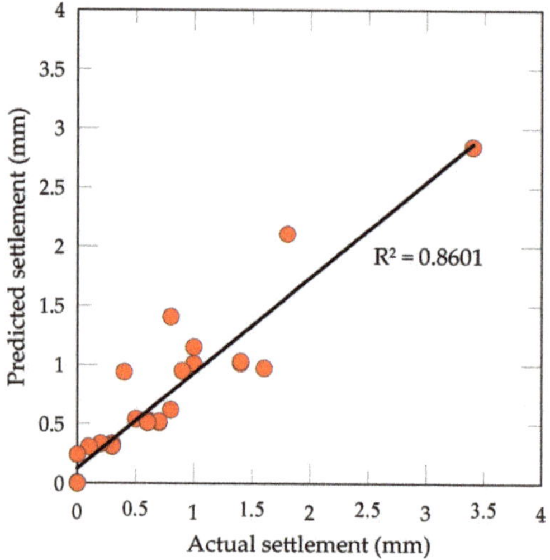

Figure 6. Scatter plot showing the performance of ANN model 1 for the test data set.

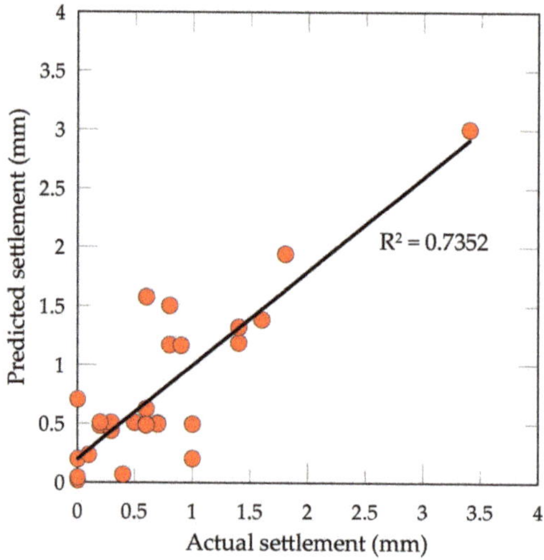

Figure 7. Scatter plot showing the performance of ANN model 2 for the test data set.

A comparison of models 1 and 2 in terms of the prediction accuracy shows that the prediction accuracy of the former is higher. The difference in the R^2 correlation between the two models is about 0.12.

From the results shown in Figures 5 and 6, it can be concluded that there exists a strong correlation between the model predictions and the actual settlement in both cases considered.

In this study, ANN models composed of two or more hidden layers were considered, and it was found that the difference in accuracy between models with two or more hidden layers and the model with the single hidden layer was not significant. Therefore, an ANN model using one layer was used.

4.1. ANN-Based Numerical Equation

A simple equation was developed to predict the liquefaction-induced settlement. The optimal ANN model structure used for the purpose is shown in Figure 8, and its associated weights with biases are presented in Table 3.

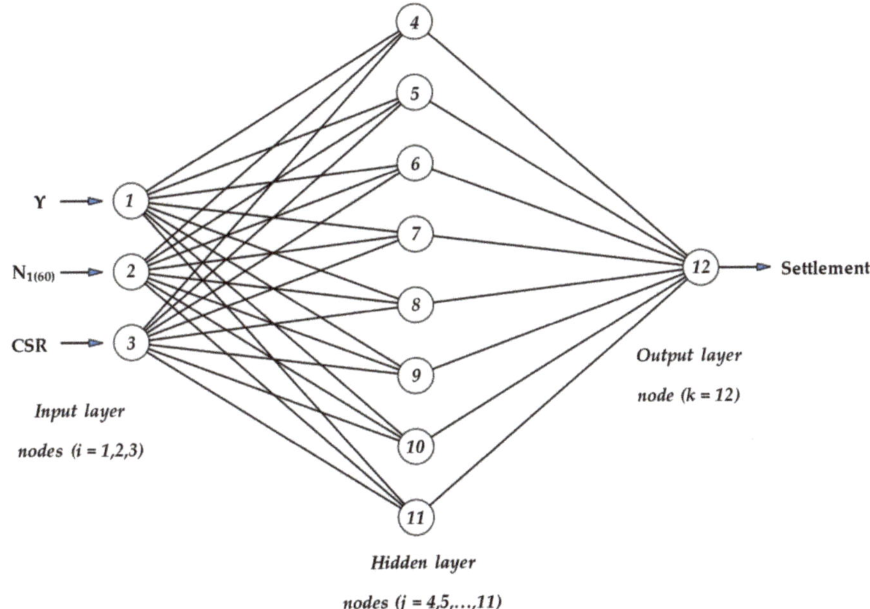

Figure 8. Structure of the optimal ANN model.

Table 3. Weight matrix and bias vector for the ANN model.

W$_1$		−2.231		2.729		−2.500			**B$_1$**	−9.631
		−8.874		−3.629		−15.703				−6.179
		−6.271		−5.433		−1.570				−5.334
		−1.100		5.470		−3.295				−8.614
		5.617		7.774		1.701				−7.317
		−1.866		−4.224		−9.756				−0.926
		−2.116		7.453		−1.157				−7.489
		0.314		−1.285		−4.980				−6.722
W$_2$	0.579	−1.853	−1.058	−1.591	1.006	1.964	−0.852	−0.320	**B$_2$**	1.006

Note: Matrices W_1 (8×3), B_1 (8×1), W_2 (1×8) and B_2 (1×1) were used in Equation (2)

The optimal-ANN-model-based numerical equation for settlement prediction can be expressed as Equation (4).

$$f_{sig} = \sigma(z) = \frac{1}{1 + e^{-z}} \tag{4}$$

where T_{12} is the output variable, namely, the predicted settlement value (S), B_k is the bias value at the output layer, W_{kj} is the connection weight between the jth node in the hidden layer and the kth node in the output layer, B_j is the bias value of the jth hidden node, W_{ji} is the connection weight between the ith input node and the jth hidden node, X_i is the ith input variable, and f_{sig} is the sigmoid transfer function given by Equation (5).

$$S = T_{12} = B_k + \sum_{j=4}^{11} \left\{ W_{kj} \times f_{sig}\left[B_j + \sum_{i=1}^{3}(W_{ji}X_i) \right] \right\} \tag{5}$$

For the simplification of the calculation process, the weights and biases were arranged in a matrix form.

4.2. Example of Settlement Calculation Using the ANN Model

For $\gamma = 18$ kN/m^3, $N_{1(60)} = 13$, and CSR $= 0.34$, the input vector **X** is

$$X = \begin{bmatrix} 18 \\ 13 \\ 0.34 \end{bmatrix}$$

The normalized input vector (Xn) is calculated from Equation (1) by using the A and B values in Table 1:

$$X_n = \begin{bmatrix} -0.200 \\ 0.040 \\ 0.444 \end{bmatrix}$$

Note: $(a, b) = (-1, 1)$

The settlement (S) is calculated using the normalized input vector as follows:

$$W1 \times Xn + B1 = \begin{bmatrix} -2.231 & 2.729 & -2.500 \\ -8.874 & -3.629 & -15.703 \\ -6.271 & -5.433 & -1.570 \\ -1.100 & 5.470 & -3.295 \\ 5.617 & 7.774 & 1.701 \\ -1.866 & -4.224 & -9.756 \\ -2.116 & 7.453 & -1.157 \\ 0.314 & -1.285 & -4.980 \end{bmatrix} \times \begin{bmatrix} -0.200 \\ 0.040 \\ 0.444 \end{bmatrix} + \begin{bmatrix} -9.631 \\ -6.179 \\ -5.334 \\ -8.614 \\ -7.317 \\ -0.926 \\ -7.489 \\ -6.722 \end{bmatrix} = \begin{bmatrix} -10.187 \\ -11.529 \\ -4.995 \\ -9.640 \\ -7.374 \\ -5.058 \\ -7.282 \\ -9.049 \end{bmatrix}$$

$$f_{sig}(W_1 \times X_n + B_1) = \begin{bmatrix} 3.76E-05 \\ 9.84E-06 \\ 6.73E-03 \\ 6.51E-05 \\ 6.27E-04 \\ 6.32E-03 \\ 6.87E-04 \\ 1.17E-04 \end{bmatrix}$$

$$S = [0.579 \ -1.853 \ -1.058 \ -1.591 \ -0.423 \ 1.964 \ -0.852 \ -0.320]$$

$$\times \begin{bmatrix} 3.76 \times 10^{-5} \\ 9.84 \times 10^{-6} \\ 6.73 \times 10^{-3} \\ 6.51 \times 10^{-5} \\ 6.27 \times 10^{-4} \\ 6.32 \times 10^{-3} \\ 6.87 \times 10^{-4} \\ 1.17 \times 10^{-4} \end{bmatrix} + [1.006]$$

$$S = [1.010]$$

The actual value of the settlement was 1 mm, and the value predicted using the ANN model was 1.010 mm.

4.3. Sensitivity Analysis

Sensitivity analysis was performed to determine the effect of the input parameters on the settlement prediction. The measure of variable importance was obtained using the permutation importance approach for random forests, described by Breiman [14]. This approach involves measuring the drop in the ANN model performance when a feature is unavailable.

As shown in Figures 9 and 10, the unit weight had the strongest influence on the settlement prediction in the case of ANN model 1, while the depth of the soil layer had the strongest influence on the predicted settlement in the case of model 2. In both cases, $N_{1(60)}$ had a stronger influence than the CSR.

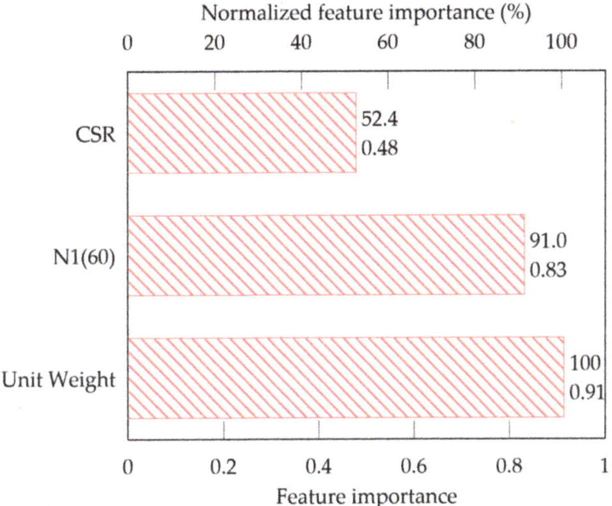

Figure 9. Relative importance of the input parameters of ANN model 1.

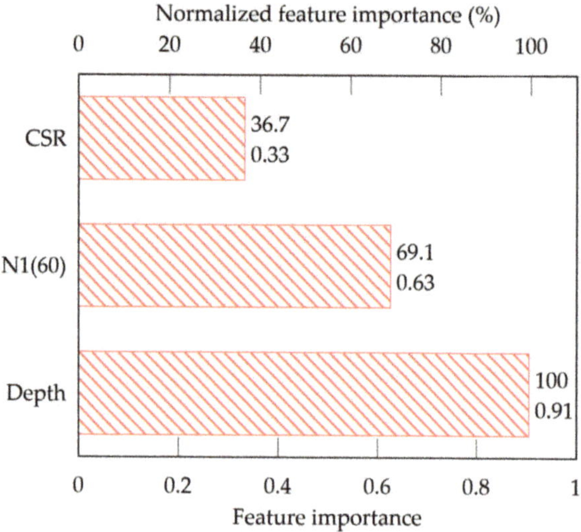

Figure 10. Relative importance of the input parameters of ANN model 2.

4.4. Parametric Study and Extrapolation beyond the Training Data

A parametric study was conducted to verify the validity and robustness of the optimal ANN model, and it involved generating a synthetic data set within the range of the training data set to test the model. For a given unit weight of soil, the settlement was determined based on the unit thickness of each layer. As shown in Figure 11a, the amount of predictive settlement generally increased with increasing a CSR and decreased with an increase in $N_{1(60)}$. However, it is necessary to expand the range of $N_{1(60)}$ and CSR obtained through the parametric study due to some field data being beyond the range. Therefore, this study proposed a simple settlement chart based on a parametric study as shown in Figure 11b.

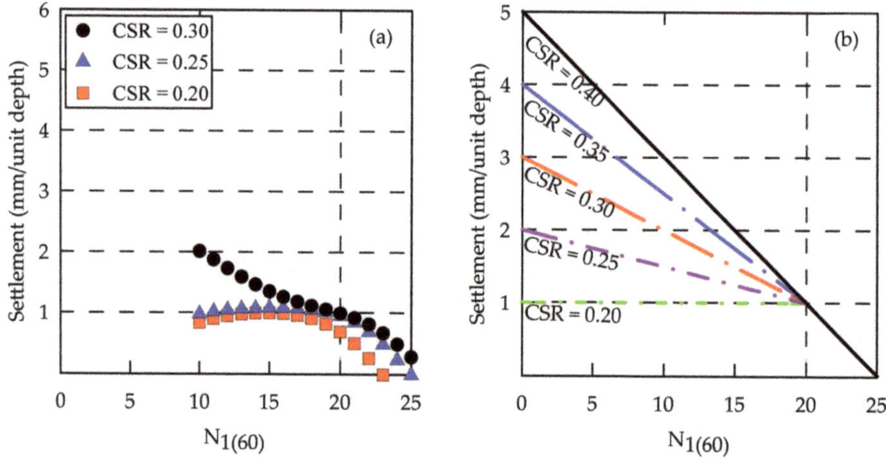

Figure 11. Variation of settlement with $(N_1)_{60}$ and CSR for $\gamma = 18$ kN/m^3. (**a**) Settlement relationship between $(N_1)_{60}$ and CSR; (**b**) settlement chart based on the ANN method.

4.5. Application of Settlement Chart Based on the ANN Method

The proposed settlement chart from the optimal ANN model was assessed using the SPT data obtained from three additional boreholes at the Pohang site. The locations of the boreholes and the measured settlement obtained from interferometric synthetic aperture radar (InSAR) imaging are shown in Figure 12.

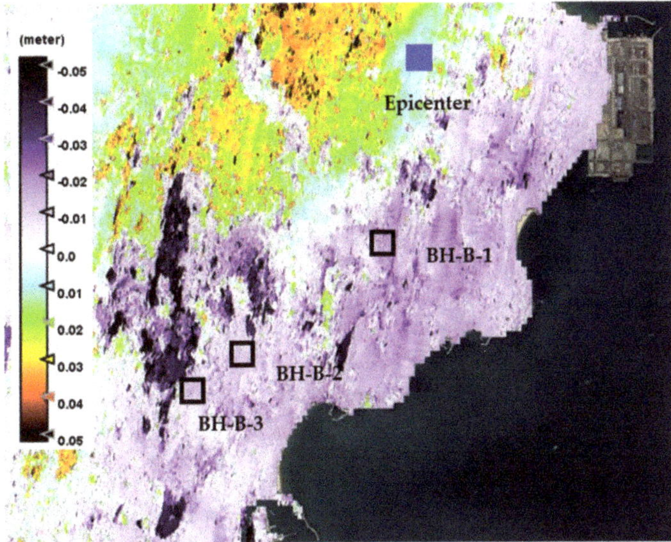

Figure 12. A settlement map from interferometric synthetic aperture radar (InSAR) and a location of extra boreholes (BHs).

The InSAR procedure was recommended by the Remote Sensing Lab at Kangwon National University, Korea [15]. Following the procedure, the settlement was analyzed by the Pohang satellite images between November 4 and 16, 2017, from Google Earth. Such Google Earth images were used to generate the settlement map in Figure 12 by using a freely distributed SentiNel Application Platform (SNAP) program by the European Space Agency [16]. With an average unit weight of 18 kN/m^3, $N_{1(60)}$ values were converted from the SPT blow count (N_{SPT}) of boreholes [17]. The CSR can be calculated from Equation (6) [18].

$$CSR = (\tau_{av}/\sigma'_{vo}) = 0.65(a_{max}/g)(\sigma_{vo}/\sigma'_{vo})\gamma_d \tag{6}$$

where a_{max} = peak acceleration at the ground surface from the earthquake (this study used the Pohang Earthquake, 0.2712 g); g = acceleration of gravity; σ_{vo} and σ'_{vo} are total and effective vertical overburden stresses, respectively; and γ_d = stress reduction coefficient.

The calculated total settlement for the additional boreholes, 1, 2, and 3, using the optimal ANN model are 17.14, 19.77, and 13.88 mm, respectively, as shown in Table 4. It can be observed that these settlement values are close to those measured by the InSAR imaging. Unlike the numerical analysis approach, the proposed chart between $(N_1)_{60}$-CSR-Settlement from the optimal ANN model has been proven to estimate settlement values with minimal input parameters. For an earthquake with similar impact and magnitude, this simple ANN model can be deployed as a handy tool to obtain liquefaction-induced settlement in the field.

Table 4. Predicted settlement due to Pohang earthquake using the proposed settlement chart.

Depth	BH-B-1				BH-B-2				BH-B-3			
	N_{spt}	$(N_1)_{60}$	CSR	S(mm)	N_{spt}	$(N_1)_{60}$	CSR	S(mm)	N_{spt}	$(N_1)_{60}$	CSR	S(mm)
0	3	4	0.39	4.04	4	7	0.39	3.47	4	6	0.39	3.66
1	4	5	0.39	3.85	5	8	0.39	3.28	14	23	0.39	0.4
2	5	8	0.39	3.28	6	10	0.39	2.9	7	11	0.39	2.71
3	7	11	0.39	2.71	7	12	0.39	2.52	7	11	0.39	2.71
4	9	14	0.38	2.08	9	13	0.38	2.26	8	12	0.38	2.44
5	13	19	0.38	1.18	10	13	0.38	2.26	13	18	0.38	1.36
6	21	28	0.38	0	11	14	0.38	2.08	17	22	0.38	0.6
7	29	36	0.37	0	17	20	0.37	1	21	25	0.37	0
8	37	43	0.37	0	42	48	0.37	0	24	27	0.37	0
9	44	49	0.36	0	36	39	0.36	0	28	31	0.36	0
10	50	53	0.35	0	50	53	0.35	0	47	50	0.35	0
11									50	51	0.34	0
	Total Settlement			17.14	Total Settlement			19.77	Total Settlement			13.88

5. Conclusions

In this study, the potential of an ANN to predict the liquefaction-induced settlement at Pohang was examined. Two ANN models were trained using a back-propagation algorithm. Both models had three input variables. The input variables of model 1 were unit weight, corrected SPT blow count ($N_{1(60)}$), and CSR, while those of model 2 were depth of the soil layer, $N_{1(60)}$, and CSR. The output of the models was the settlement (S). After the training and testing of the models, it was evident that model 1 had higher prediction accuracy, and the difference in the R^2 correlation between the two models was about 0.12. Subsequently, the weights and biases of an optimal ANN model were used to develop a simple predictive model equation, which was implemented using a matrix formulation.

Sensitivity analysis performed using the permutation importance algorithm indicated that the corrected SPT blow count had a stronger influence than the CSR on the predicted settlement. Furthermore, a parametric study showed that for a given unit weight of soil, the settlement decreased with an increase in $N_{1(60)}$.

Finally, the simplified relationship between $(N_1)_{60}$-CSR-Settlement was proposed using the optimal ANN model, and the cumulative settlement was predicted by applying the proposed relationship to additional boreholes and compared with the InSAR results. The cumulative settlement had a similar range as the InSAR displacement map. Thus, the simplified relationship of this study can be deployed as a handy tool to obtain liquefaction-induced settlement in the field.

Author Contributions: Conceptualization, S.-S.P. and P.D.O.; methodology, S.-S.P. and P.D.O.; software, P.D.O. and S.-W.W.; validation, S.-S.P., P.D.O., and S.-W.W.; formal analysis, P.D.O., S.-S.P., and D.-E.L.; investigation, P.D.O. and S.-S.P.; resources, S.-S.P. and P.D.O.; data curation, S.-S.P. and P.D.O.; writing—original draft preparation, S.-S.P. and P.D.O.; writing—review and editing, S.-S.P. and P.D.O.; visualization, P.D.O. and S.-W.W.; supervision, S.-S.P.; project administration, S.-S.P. and D.-E.L.; funding acquisition, S.-S.P. All authors have read and agreed to the published version of the manuscript.

Funding: This work was supported by a National Research Foundation of Korea (NRF) grant funded by the Korean government (MSIT) (No. NRF-2018R1A5A1025137).

Conflicts of Interest: The authors declare no conflicts of interest.

Appendix A

Table A1. Details of the liquefaction-induced settlement database.

Borehole	Depth (m)	Unit Weight (kN/m^3)	$N_{1(60)}$	CSR	Settlement (mm)
BH-A-1	1	20	11	0.33	0.50
	2	20	11	0.31	0.50
	3	20	14	0.29	0.80
	4	20	16	0.28	1.40
	5	20	5	0.27	3.30
	6	20	10	0.26	3.40
	7	20	5	0.27	3.40
	8	20	6	0.29	2.50
	9	20	9	0.3	1.60
	10	20	9	0.31	1.00
	11	18	25	0.31	0.40
	12	18	25	0.31	0.30
	13	18	25	0.32	0.20
	14	18	25	0.32	0.30
	15	18	25	0.32	0.30
	16	18	25	0.32	0.30
	17	18	25	0.32	0.30
	18	18	25	0.32	0.20
	19	18	25	0.31	0.30
	20	18	25	0.31	0.30
BH-A-2	1	20	15	0.35	0.40
	2	20	17	0.32	0.80
	3	20	17	0.3	1.60
	4	20	7	0.29	3.10
	5	20	6	0.27	2.80
	6	21	13	0.31	1.40
	7	21	18	0.34	0.80
	8	21	13	0.36	0.90
	9	21	11	0.37	1.10
	10	21	13	0.36	0.80
	11	16	2	0.36	0.00
	12	16	1	0.37	0.00
	13	16	1	0.38	0.00
	14	16	1	0.39	0.00
	15	16	1	0.39	0.00
	16	16	1	0.39	0.00
	17	16	1	0.39	0.00
	18	16	1	0.39	0.00
	19	16	1	0.38	0.00
	20	16	1	0.37	0.00

Table A1. *Cont.*

Borehole	Depth (m)	Unit Weight (kN/m^3)	$N_{1(60)}$	CSR	Settlement (mm)
BH-A-3	1	18	6	0.24	0.60
	2	18	8	0.28	1.40
	3	18	10	0.3	2.00
	4	18	10	0.29	2.30
	5	18	11	0.28	2.00
	6	18	10	0.3	1.80
	7	18	11	0.32	1.40
	8	18	11	0.33	1.30
	9	18	12	0.34	1.20
	10	18	13	0.34	1.00
	11	21	25	0.34	0.70
	12	21	25	0.33	0.60
	13	21	25	0.33	0.60
	14	21	25	0.33	0.50
	15	21	25	0.32	0.50
	16	21	25	0.32	0.40
	17	21	25	0.31	0.50
	18	21	25	0.31	0.40
	19	21	25	0.3	0.40
	20	21	25	0.3	0.50
BH-A-4	1	20	5	0.23	1.10
	2	20	7	0.27	1.90
	3	20	18	0.27	1.60
	4	20	9	0.27	2.80
	5	20	6	0.26	2.80
	6	20	11	0.31	1.60
	7	20	9	0.34	1.40
	8	21	25	0.36	0.60
	9	21	25	0.38	0.60
	10	21	25	0.38	0.60
	11	21	25	0.38	0.60
	12	21	25	0.37	0.50
	13	16	7	0.22	0.00
	14	16	1	0.21	0.00
	15	16	0	0.21	0.00
	16	16	2	0.22	0.00
	17	16	3	0.22	0.00
	18	16	3	0.22	0.00
	19	16	3	0.22	0.00
	20	16	3	0.22	0.00

Table A1. *Cont.*

Borehole	Depth (m)	Unit Weight (kN/m^3)	$N_{1(60)}$	CSR	Settlement (mm)
	1	20	11	0.32	0.50
	2	20	10	0.31	1.60
	3	20	9	0.29	2.60
	4	20	11	0.3	1.90
	5	20	11	0.32	1.50
	6	20	10	0.33	1.40
	7	20	15	0.34	0.90
	8	20	15	0.35	0.80
	9	21	25	0.34	0.60
	10	21	25	0.34	0.60
BH-A-5	11	21	25	0.34	0.60
	12	21	25	0.33	0.60
	13	21	25	0.33	0.70
	14	21	25	0.33	0.70
	15	18	15	0.33	1.10
	16	18	11	0.33	1.40
	17	18	12	0.32	1.00
	18	16	14	0.32	0.10
	19	16	10	0.32	0.00
	20	16	7	0.32	0.00

References

1. Choi, J.H.; Ko, K.; Gihm, Y.S.; Cho, C.S.; Lee, H.; Song, S.G.; Bang, E.S.; Lee, H.J.; Bae, H.K.; Kim, S.W.; et al. Surface Deformations and Rupture Processes Associated with the 2017 Mw 5.4 Pohang, Korea, Earthquake. *Bull. Seismol. Soc. Am.* **2019**, *109*, 756–769. [CrossRef]
2. Gihm, Y.S.; Kim, S.W.; Ko, K.; Choi, J.-H.; Bae, H.; Hong, P.S.; Lee, Y.; Lee, H.; Jin, K.; Choi, S.; et al. Paleo seismological implications of liquefaction-induced structures caused by the 2017 Pohang Earthquake. *Geosci. J.* **2018**, *22*, 871–880. [CrossRef]
3. Kang, S.; Kim, B.; Bae, S.; Lee, H.; Kim, M. Earthquake-Induced Ground Deformations in the Low-Seismicity Region: A Case of the 2017 M5.4 Pohang, South Korea, Earthquake. *Earthq. Spectra* **2019**, *35*, 1235–1260. [CrossRef]
4. Kim, H.-S.; Sun, C.-G.; Cho, H.-I. Geospatial Assessment of the Post-Earthquake Hazard of the 2017 Pohang Earthquake Considering Seismic Site Effects. *ISPRS Int. J. Geo-Inf.* **2018**, *7*, 375. [CrossRef]
5. Naik, S.P.; Kim, Y.-S.; Kim, T.; Su-Ho, J. Geological and Structural Control on Localized Ground Effects within the Heunghae Basin during the Pohang Earthquake (Mw 5.4, 15th November 2017), South Korea. *Geosciences* **2019**, *9*, 173. [CrossRef]
6. National Academies of Sciences, Engineering, and Medicine. *State of the Art and Practice in the Assessment of Earthquake-Induced Soil Liquefaction and Its Consequences*; The National Academies Press: Washington, DC, USA, 2016.
7. Da Fonseca, A.V.; Millen, M.; Gómez-Martinez, F.; Romão, X.; Quintero, J. State of the Art Review of Numerical Modelling Strategies to Simulate Liquefaction-Induced Structural Damage and of Uncertain/Random Factors on the Behaviour of Liquefiable Soils. Deliverable D3.1, Version 1.0. 2017. Available online: http://www.liquefact.eu/disseminations/deliverables/ (accessed on 11 January 2020).
8. Tang, X.-W.; Hu, J.-L.; Qiu, J.-N. Identifying significant influence factors of seismic soil liquefaction and analyzing their structural relationship. *KSCE J. Civ. Eng.* **2016**, *20*, 2655–2663. [CrossRef]
9. Lee, C.Y.; Chern, S.-G. Application of a Support Vector Machine for Liquefaction Assessment. *J. Mar. Sci. Technol.* **2013**, *21*, 318–324. [CrossRef]

10. Xue, X.; Liu, E. Seismic liquefaction potential assessed by neural networks. *Environ. Earth Sci.* **2017**, *76*, 192. [CrossRef]
11. Integrated DB Center of National Geotechnical Information, SPT Database Available at 542. 2015. Available online: http://www.geoinfo.or.kr (accessed on 15 May 2015).
12. Park, S.S. Liquefaction evaluation of reclaimed sites using an effective stress analysis and an equivalent linear analysis. *KSCE J. Civ. Eng.* **2008**, *28*, 83–94.
13. Lai, J.; Qiu, J.; Feng, Z.; Chen, J.; Fan, H. Prediction of Soil Deformation in Tunnelling Using Artificial Neural Networks. *Comput. Intell. Neurosci.* **2016**, *2016*, 6708183. [CrossRef]
14. Breiman, L. Random Forests. *Mach. Learn.* **2001**, *45*, 5–32. [CrossRef]
15. Lee, H.Y. Pohang Earthquake (15 November 2017 Mw 5.4) observed by Sentinel-1A/B SAR Interferometry. Available online: http://sar.kangwon.ac.kr/pohang.htm (accessed on 11 December 2017).
16. Foumelis, M.; Blasco, J.M.D.; Desnos, Y.L.; Engdahl, M.; Fernández, D.; Veci, L.; Lu, J.; Wong, C. ESA SNAP-StaMPS Integrated Processing for Sentinel-1 Persistent Scatterer Interferometry. In Proceedings of the IGARSS 2018—2018 IEEE International Geoscience and Remote Sensing Symposium, Valencia, Spain, 22–27 July 2018; pp. 1364–1367. [CrossRef]
17. Youd, T.L.; Idriss, I.M. Liquefaction resistance of soils: Summary report from the 1996 NCEER and 1998 NCEER/NSF workshops on evaluation of liquefaction resistance of soils. *J. Geotech. Geoenvironmental Eng.* **2001**, *127*, 297–313. [CrossRef]
18. Seed, H.B.; Idriss, I.M. Simplified procedure for evaluating soil liquefaction potential. *J. Soil Mech. Found. Div.* **1971**, *97*, 1249–1273.

 *sustainability*

Article

Analysis of Deformation Characteristics of Foundation-Pit Excavation and Circular Wall

Xuhe Gao [1,*], Wei-ping Tian [1] and Zhipei Zhang [2]

[1] Key Laboratory of Highway Engineering in Special Region, Ministry of Education, Chang'an University, Xi'an 710064, China; fz02@gl.chd.edu.cn

[2] College of Geology and Environment, Xi'an University of Science and Technology, Xi'an 710054, China; zhipeizhangt@126.com

* Correspondence: 2017021008@chd.edu.cn

Received: 7 March 2020; Accepted: 10 April 2020; Published: 14 April 2020

Abstract: The surrounding ground settlement and displacement control of an underground diaphragm wall during the excavation of a foundation pit are the main challenges for engineering safety. These factors are also an obstacle to the controllable and sustainable development of foundation-pit projects. In this study, monitoring data were analyzed to identify the deformation law and other characteristics of the support structure. A three-dimensional numerical simulation of the foundation-pit excavation process was performed in Midas/GTS NX. To overcome the theoretical shortcomings of parameter selection for finite-element simulation, a key data self-verification method was used. Results showed that the settlement of the surface surrounding the circular underground continuous wall was mainly affected by the depth of the foundation-pit excavation. In addition, wall deformation for each working condition showed linearity with clear staged characteristics. In particular, the deformation curve had obvious inflection points, most of which were located deeper than 2/3 of the overall excavation depth. The characteristics of the cantilever pile were not obvious in Working Conditions 3–9, but the distribution of the wall body offset in a D-shaped curve was evident. Deviation between the monitoring value of the maximal wall offset and the simulated value was only 4.31 %. The appropriate physical and mechanical parameters for key data self-verification were proposed. The concept of the circular-wall offset inflection point is proposed to determine the distribution of inflection-point positions and offset curves. The method provides new opportunities for the safety control and sustainable research of foundation-pit excavations.

Keywords: circular foundation pit; construction monitoring; numerical simulation; underground continuous wall

---

1. Introduction

Underground continuous walls have been widely applied as foundation-pit supports due to their high stability, rigidity, and impermeability, in addition to their predictable deformation characteristics. However, for circular anchor foundation pits with underground continuous walls as the predominant retaining structure, monitoring and predicting wall displacement and surface settlement around the foundation pit remain challenges. As such, these factors need to promptly and consistently be monitored, and monitoring data should be accordingly analyzed. Challenges have also inspired scholars to explore new research methods, promoting the application of computer technology in the construction of foundation pits.

Studies on this topic have been conducted. Bolton and Powriet [1] carried out various laboratory tests to study the deformation characteristics of an underground continuous wall under different soil conditions and foundation-pit parameters. They also calculated the deformation and failure conditions of the foundation pit. However, they did not discuss the validity of the used parameters in the calculation. Pohetal. [2] collated the monitoring data of two foundation pits, and used real-world data to calculate the bending moment generated by the underground continuous wall. Results showed that the bending moment of the underground continuous wall was largely generated due to the cracking of the wall, and that the lateral displacement of the wall was not affected by this factor. However, the study did not provide any description of the monitoring-data collection, nor did it further demonstrate the factors and characteristics of the lateral displacement of the wall. Bose and Som [3] created a more accurate finite-element-analysis program based on the Cambridge model to address deficiencies of the existing model, which is mainly used for calculations and analysis of the internal supports of a foundation pit. However, that study also lacked a demonstration of the validity of the model parameters. Whittle et al. [4] innovatively integrated two-dimensional seepage into the deep-foundation-pit calculation model, and examined soil stress in the deep-foundation-pit engineering of a postal building in Boston on the basis of the finite-element method. However, the study did not discuss the displacement and surrounding settlement of the support structure during excavation of the foundation pit. To better analyze foundation-pit support systems, Kishnani and Borja [5] conducted detailed analysis of the soil structure and seepage into the foundation pit, and analyzed the impact of these two factors on the support system. They determined that the seepage affected earth pressure behind the wall and caused the surrounding ground to settle. The effect of seepage on wall displacement, however, was not discussed. After summarizing multiple theories and practical experiences, Alejano et al. [6] conducted a related investigation on the factors affecting the displacement of typical structural types (filling and excavation). Soil traits were regarded as ideally elastoplastic, and it was noted that the displacement of the soil was not the only factor; the physical properties of the soil and the wall, as well as the location of the erected supports also contributed. That study did not involve ground settlement around the foundation pit, and the quantitative analysis of wall displacement was insufficient. Faheem et al. [7] focused on the poor stability of foundation pits in areas with soft soil from a two- and a three-dimensional perspective. Their study particularly focused on the stability of the bottom of the pit, and presented a detailed simulation using the finite-element method. However, there was no analysis of ground settlement around the foundation pit and the deformation of the supporting structure, and the validity of the parameters in the simulation process was not verified.

Liu and Ding [8] used the finite-element method to study the stiffness coefficient of the Goodman unit, which was determined to affect surface settlement outside the foundation pit and the displacement of the underground continuous wall. That study also failed to verify the validity of the finite-element calculation parameters. Chen et al. [9] investigated the deep foundation pit of a steel plant in Shanghai on the basis of collected monitoring data during foundation-pit construction. They analyzed the deformation and internal structural forces of the circular underground continuous wall supporting the foundation pit that was subjected to the pressure of confined groundwater. The study focused on analysis of existing monitoring data, and did not use finite-element analysis to further demonstrate the deformation characteristics of the supporting structure. Xu et al. [10] collected monitoring data from foundation pits in Shanghai that used underground continuous wall supports, and calculated the deformation law of the underground continuous wall to study the influence of various factors on these laws. Their study focused on regional data collected by statistical analysis, and had limited applicability to early warnings on surrounding surface settlement and wall displacement in special geological environments. Wang and Hu [11] studied the double-layer elliptical supporting structure in the foundation pit of the China Petroleum Building, and aimed to reduce the number of layers supported by the internal structure during the excavation of the foundation pit. The structure was analyzed by force-deformation calculations. It was concluded that a T- or I-shaped underground continuous wall could be used instead of the elliptical wall shape, which could reduce the number of required internal supports. That study lacked monitoring data or finite-element simulation to validate the results, and there was no analysis of surface settlement and wall displacement around the foundation pit. Hu et al. [12] used the foundation pit of a subway station as a research subject, and monitored the variation of the horizontal displacement of the underground continuous wall at the excavation depth during the construction of the foundation pit. A three-dimensional finite-element model was established to simulate the foundation-pit excavation of the subway station, and the calculated deformation characteristics were compared with the monitoring results. Results showed that the difference between the simulated maximal horizontal displacement of the underground continuous wall and the measured value was small, and that the trend in displacement was comparable. However, the study also failed to verify the validity of the finite-element calculation parameters, and did not analyze ground settlement around the foundation pit. Zheng et al. [13] used finite-difference software FLAC3D to numerically simulate the horizontal deformation and surface settlement of a foundation-pit-excavation support structure and compared it with the measured values. Results showed that the maximal horizontal displacement of the underground continuous wall appeared at the top of the wall, and the horizontal displacement curve exhibited a "half-cup" composite shape with multiple inflection points. The settlement curve of the ground surface beyond the wall was an asymmetrical groove-type curve. Similarly, that study verified the validity of the finite-element calculation parameters.

In summary, the existing literature has conducted a large number of theoretical calculations and finite-element analysis of underground continuous walls (including self-programming and commercial software). However, there is almost no argument concerning the method of obtaining parameters. This shows that the method of parameter selection needs further study. If only research results are pursued, and access to key parameters is ignored, such research is questioned by other disciplines, and the sustainability of that work is also threatened. In this study, the anchored circular underground continuous wall of the Humen Second Bridge West foundation-pit project was monitored and simulated. Monitoring data were analyzed to identify the deformation law and other characteristics of the support structure. Three-dimensional numerical simulation of the foundation-pit excavation was conducted in Midas/GTS NX. To overcome the theoretical shortcomings of parameter selection for finite-element simulation, the key data self-verification method was used, and a layer-by-layer algorithm was employed to determine more accurate simulation parameters. The deviation rate was used to quantify the difference between simulated results and measured values. The appropriate physical and mechanical parameters for key data self-verification were proposed and utilized to compensate for the shortcomings of the on-site monitoring data. The concept of the "circular-wall offset inflection point" was proposed to determine the distribution of inflection-point positions and offset curves. The method provides new opportunities for the safety control and sustainable research of foundation-pit excavations.

2. Materials and Methods

2.1. Project Overview

The rock and soil layers in the foundation pit were silt, muddy soil, fine sand, medium sand, coarse sand, strong weathered mudstone, middle weathered mudstone, and microweathered mudstone (Figures 1 and 2). According to these geological conditions and the design requirements of the anchor body, the underground continuous wall adopted a circular structure with an outer diameter of 82.0 m and a wall thickness of 1.5 m. The elevation of the top surface of the pit was 1.00 m, and the elevation of the bottom of the pit was −35.00 to −43.00 m. The bottom of the pit was embedded in mud, siltstone, and moderately weathered mudstone strata. The underground continuous wall was divided into two sections (Sections 1 and 2). Section 1 was three-milled, with a side groove length of 2.8 m, a middle slot length of 1.47 m, and a slot length of 7.07 m; Section 2 had a slot length of 2.8 m. The length of the Sections 1 and 2 groove sections was 0.25 m on the axis of the ground wall, and Sections 1 and 2 had 27 slots. Thus, the trough section was divided into 54 sections (Figure 2). The designed maximal trough depth was 46.0 m. On both sides of the underground continuous wall, a 50 cm diameter cement-powder spray was used to create a pile to reinforce the silt soil with a spacing of 40 cm and a reinforcement depth of 15.0 m. After construction of the underground continuous wall was completed, the bottom of the wall was grouted.

After construction of the underground continuous wall had been completed, the soil was excavated by the reverse method, and the lining of the pit was layered and constructed. The construction period of each layer was controlled by the excavation of the soil. The excavation depth of the soil was 27 m, and the lining and soil-stratification height were controlled within 3 m. The lining of the pit was constructed from top to bottom. The top and bottom plates were 6 m thick with a concrete-filled core in the middle.

Figure 1. Cross-sectional view of geological section along the bridge.

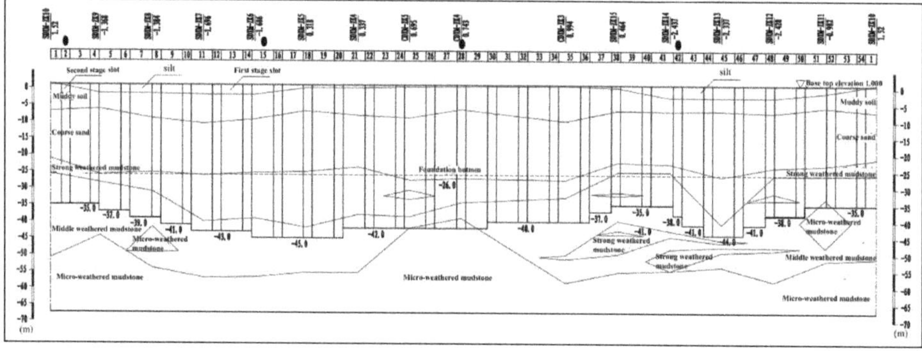

Figure 2. Expanded view of slots.

2.2. Surface-Deformation Monitoring around Underground Continuous Wall

Because of the need for surface-settlement monitoring during construction, a group of sensors were arranged to the east, south, west, north, southeast, northeast, southwest, and northwest of the foundation pit. Typical settlement monitoring started from the outside of the foundation pit with 10 monitoring points arranged at equal intervals of 5 m numbered D1-i to D8-i (with i = 1–10). Due to on-site construction-monitoring points that were actually available, only the first five points of valid data were obtained. A total of eight settlement-monitoring sections and 80 surface-settlement monitoring points were set. If the points encountered obstacles, they could be moved in parallel, as shown in Figure 3.

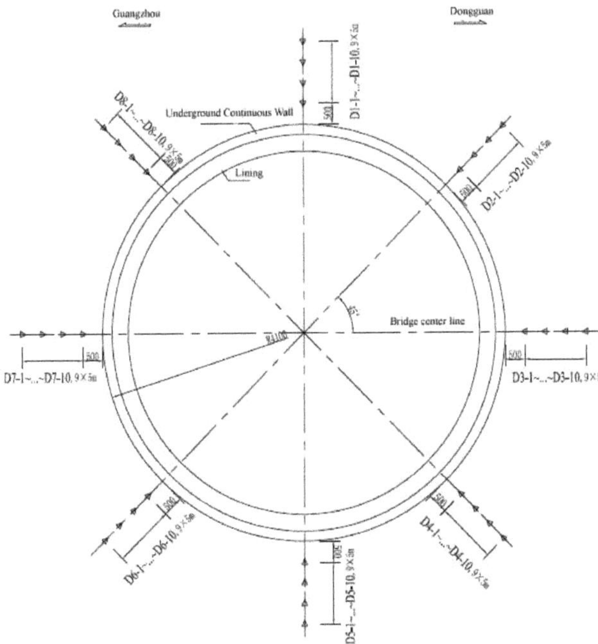

Figure 3. Layout of surface-settlement monitoring sites.

2.3. Deep-Lateral-Deformation Monitoring of Underground Continuous Wall

Deep-lateral-deformation monitoring of the underground continuous wall is a key component of monitoring and measuring the deformation of the foundation-pit support, which can directly reflect the safety and stability of the foundation pit and its supporting structures (Figure 4). To ensure the effective functioning of the inclined pipe fitting under the effects of high-pressure concrete, the inclined measuring holes were repeatedly arranged according to the spare hole position in the groove section where the four inclined measuring pipe parts of P1, P3, P5, and P7 were located (P1′, P3′, P5′, P7′). There were a total of 12 inclinometer tubes.

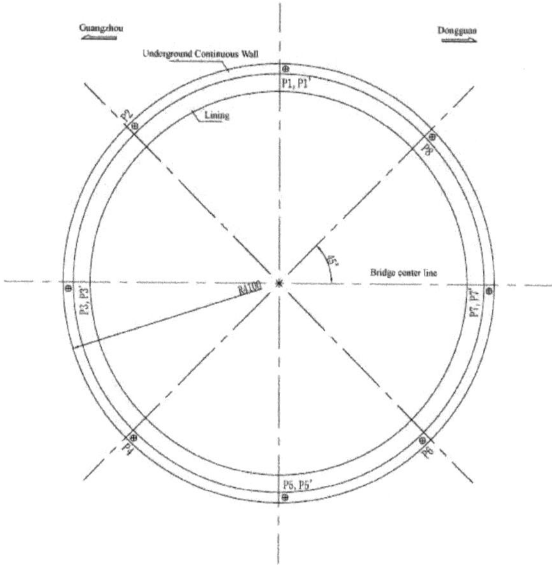

Figure 4. Deep-deformation monitoring site layout for underground continuous wall.

2.4. Monitoring-Data Analysis

The underground continuous wall was divided into 54 slot segments for analysis, as shown in Figure 2. To facilitate the statistical data processing, surrounding-settlement and wall-offset data corresponding to slot segments 2, 15, 28, and 42 were selected. In working-condition simulations, these four slot segments were defined to correspond to the calculation results of the four diagonal directions of the model.

2.5. Mohr–Coulomb Strength Criterion

The Mohr–Coulomb strength criterion states that shear failure is the most fundamental cause of soil failure. The shear strength of any point in the soil is only related to normal stress σ_n on the plane, such that

$$\tau_f = f(\sigma_n). \tag{1}$$

This function is a curve in τ_f-σ co-ordinates, known as the molar-intensity line. The Moore envelope can be approximated as a linear relationship, known as the Coulomb equation, as follows:

$$\tau_f = c + \sigma_n \tan\phi, \tag{2}$$

where τ_f is the shear strength at any point in the soil, and σ_n is the normal stress on the calculated plane.

Equation (3) is the stress condition at any point in the soil under the limit equilibrium state (stress is positive with compression). This equation is known as the Mohr–Coulomb strength criterion. The radius of the stress Mohr circle is

$$R = \left(\frac{c}{\tan\phi} + \frac{\sigma_{11} + \sigma_{22}}{2}\right)\sin\phi = c\cos\phi + \frac{\sigma_{11} + \sigma_{22}}{2}\sin\phi, \tag{3}$$

where σ_{11} and σ_{22} are the maximal and minimal principal stress when the plane-soil mass under goes shear failure, respectively.

When the Mohr envelope is tangential to the most stressed Mohr circle in the material, the soil undergoes shear failure. In other words, the magnitude of σ_{22} has no effect on shear strength. The Mohr–Coulomb strength criterion is an irregular hexagonal cone in the principal stress space. The projection of the hexagonal cone on the π plane is an irregular hexagon.

The Mohr–Coulomb criterion is widely used, as the constitutive model can accurately reflect the unequal tensile and compressive characteristics of geotechnical materials. However, numerical calculations for this model are prone to nonconvergence due to the discontinuous corners of the hexagonal cone.

2.6. Establishing Model of Foundation-Pit Excavation

Since the classical yield criterion ignores the frictional component of soil shear strength, such criteria can be used for the undrained analysis of saturated soils, such that $\varphi = 0$. The Mohr–Coulomb criterion surpasses classical criteria and considers the frictional component of the soil, which is more suitable for most scientific research and engineering practice. It is also more widely used in numerical simulation. Finite-element software Midas/GTS NX was used for numerical simulation analysis on the basis of the Mohr–Coulomb constitutive model.

The excavation project described in this study included a two-part supporting structure consisting of the underground continuous wall and the lining. The lower end of the underground continuous wall was embedded in the middle weathered-rock layer, and the embedded depth range was 10–20 m. In numerical simulation, it is necessary to simplify the foundation-pit excavation support model and the construction steps to ensure computational capacity and accuracy. The underground continuous wall retaining structure was constructed before the foundation pit was excavated. The excavation method selected a single layer of flat excavation and added a layer lining after the excavation of each layer was completed. This process continued until all construction steps were performed.

The soil layers are described in Section 2.1. Each soil layer was distinguished by a natural planar interface. According to the construction conditions and the topography of the project, the top surface of the calculation model was selected as the ground and defined as a free surface. The four lateral sides of the design model were also defined to limit horizontal displacement. The bottom plane of the pit was defined to limit vertical displacement. The initial self-weight stress field was the main model load condition. The design calculation model used the Mohr–Coulomb elastoplastic strength criterion. In addition, the river levee was approximately 50 m away from the foundation pit. In the numerical-calculation process, this levee was considered according to the most unfavorable situation for the excavation project.

The size of the design-calculation model was carefully selected to be 300 m long, 300 m wide, and 100 m deep. Errors in slot sections at segmentation were caused by errors on the construction site. The channel sections neighboring certain modelling errors were collected by overlap. Thus, the thickness of the simplified underground continuous wall was calculated as 1.3 m. The model was divided into various sections (Figures 5 and 6).

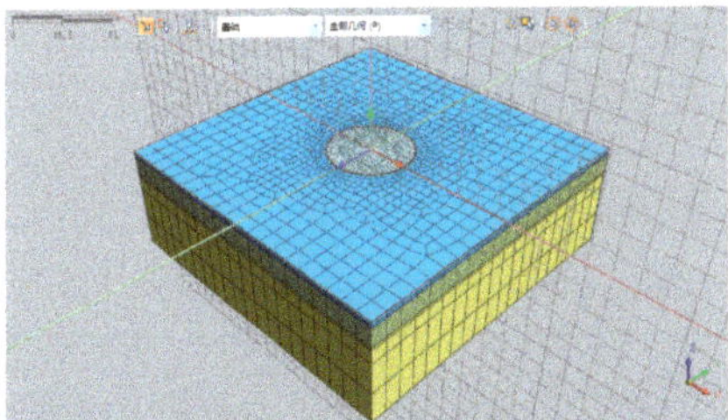

Figure 5. Pit-model grid diagram.

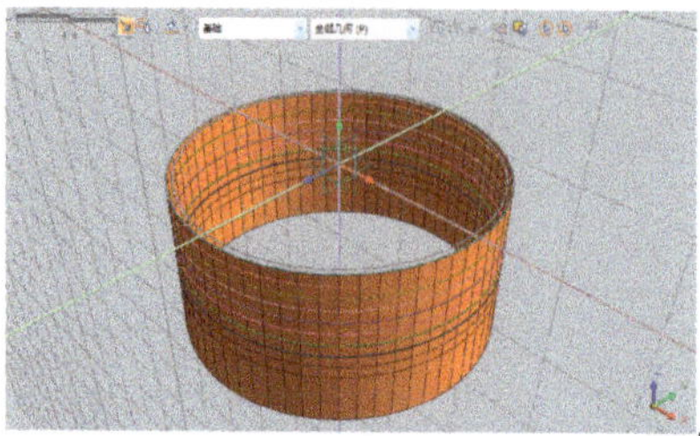

Figure 6. Support-structure grid diagram.

The model had a total of 15,840 units and 17,680 nodes. The first layer in the model was a silt layer with a thickness of 2 m; the second layer was a silty-clay layer with a thickness of 5 m; the third, fourth, and fifth layers of silt, and the medium and coarse-sand layers had a thickness of 6 m; the sixth, seventh, and eighth layers were strongly weathered mudstone, moderately weathered rock, and slightly weathered rock layers, with thicknesses of 15, 30, and 30 m, respectively.

The thickness of the underground continuous wall was calculated as 1.3 m. The thickness of the inner lining was 1.5 m in the range of 0–6 m depth, and thickness was 2 m below 6 m depth.

2.6.1. Selection of Physical and Mechanical Parameters

In the finite-element model, the parameters of the concrete material were assigned according to the defined specifications. The mechanical parameters of the rock layer were determined by geotechnical testing and the key data self-validation method. The required physical and mechanical parameters to calculate the constitutive equations in the model are shown in Tables 1 and 2.

Table 1. Soil parameters and indices.

Soil Layer	Elastic Modulus (kN/ m^2)	Poisson Ratio	Angle of Internal Friction (°)	Cohesive Forces (kN/m^2)	Unit Weight (kN/m^3)
Silt	3000	0.30	3	5	15.40
Muddy soil	50,000	0.27	5	8	16.50
Fine sand	80,000	0.23	18	0	19.00
Medium sand	120,000	0.24	25	0	19.50
Coarse sand	200,000	0.22	28	0	18.80
Strong weathered mudstone	500,000	0.19	20	50	19.99
Middle weathered mudstone	1,000,000	0.17	30	450	20.50
Microweathered mudstone	1,400,000	0.15	35	600	20.70

Table 2. Structural and mechanical parameters.

Structure	Elastic Modulus (kN/ m^2)	Unit Weight (kN/ m^3)	Poisson Ratio
Underground Continuous Wall	3.0×10^7	25	0.2
Lining	3.0×10^7	25	0.2

Note: Elastic modulus: ratio of stress to corresponding strain when ideal material has small deformation. Poisson ratio: ratio of absolute value of transverse normal strain to axial normal strain when material is under uniaxial tension or compression. Angle of internal friction: friction characteristics caused by mutual movement of particles and gluing. Cohesive forces: mutual attraction between adjacent parts within same substance. Unit weight: gravity characteristic of an object due to its gravitation in the natural state.

2.6.2. Calculation Process for Excavation-Pit Model

According to the support and excavation process for the circular-underground-continuous-wall foundation pit, pit simulations were calculated and analyzed for nine working conditions. Specifically, steps shown in Figure 7 were performed.

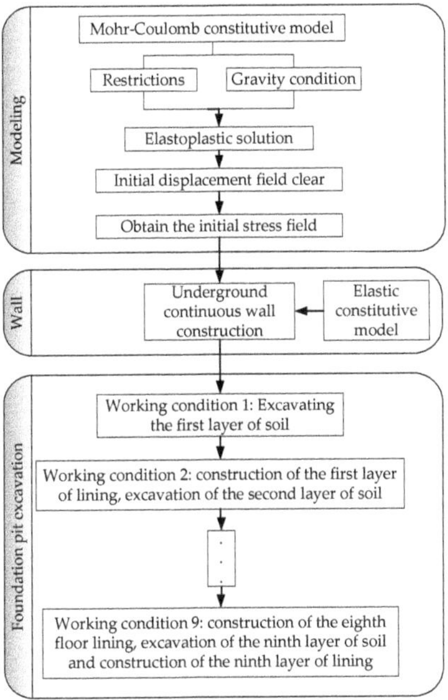

Figure 7. Modeling and calculation workflow.

2.6.3. Key Data Self-validation and Divisional-Condition Calculations

Stability analysis and the quantitative calculation of the supporting structure of existing foundation-pit engineering are mainly controlled by several key geotechnical parameters. The determination of parameters has always been a matter of debate in this field. Current practices are ① obtained by geotechnical tests, ② based on statistical data obtained from a large number of similar strata, and ③ empirical data. Because obtained parameters by geotechnical tests are different from the actual project, they need to be corrected. The method of statistical data is only applicable to ordinary strata and requires a lot of construction accumulation. Empirical data are easy to use, but are obviously less scientific. In addition, the three existing parameter-acquisition methods have a fatal disadvantage for engineering special geological environments: the parameter-selection method is not universal, and it is less sustainable.

Therefore, for traditional theoretical calculations and finite-element analysis, obtaining a method that could self-verify key data on the basis of project-site-monitoring data is critical to the sustainable development of foundation-pit and geotechnical engineering.

This paper proposes a key data self-validation theory. More specifically, we propose the selection of physical and mechanical parameters for numerical simulation that should be as reasonable as possible. However, due to many potential sources of uncertainty in these values, including theoretical defects that simplify soil and rock into ideal homogeneous materials, acquisition processing, and data conversion, when the parameter-selection basis was not sufficiently convincing, the key data obtained by monitoring were used to verify the results obtained by the simulation. When the deviation rate of the data obtained by the simulation was within a reasonable error range, the physical and mechanical parameters selected for the calculation model were deemed reasonable. Following this, large-scale data calculations were performed.

This method requires trial calculation. During research, parameters obtained from the literature and background data were used for trial calculations, and we calculated the deviation rate multiple times. Finally, the maximal simulated offsets of the walls under the second and third working conditions were 1.31 and 2.25 mm, respectively. The maximal monitored offsets of the wall for the second and third working conditions were 1.58 and 2.57 mm, respectively. That is, the difference between the simulated and monitored values was calculated. The absolute value of the difference divided by the monitoring value was used to quantify the credibility of the simulated value. It was further verified that the parameters used in the simulation were feasible. The deviation rate was calculated as follows:

$$\text{Deviation rate} = (\text{analog value} - \text{monitored value})/\text{monitored value}. \tag{4}$$

On the basis of this equation, the deviation rate of the wall was −10.39% for the second working condition and −14.22% for the third working condition. Thus, the obtained data from the simulation demonstrated a limited deviation, and the preliminary verification data were valid.

After having determined the appropriate parameters, the input parameters were calculated to obtain the force-deformation characteristics of other conditions. The calculation results were confirmed by result monitoring. Another benefit of this method is that it could expand the scope of the simulation calculations to compensate for the lack of on-site monitoring data.

3. Results

3.1. Surface-Settlement-Monitoring Analysis

The maximal settlement value of the monitoring points was 9.9 mm. For excavation Working Conditions 1–3, surface settlement at each monitoring point increased linearly. For Working Conditions 4–6, the monitoring points generated relatively stable settlement. For Working Conditions 7–9, the settlement at each monitoring point increased linearly. The growth rate in Working Conditions 7–9 was greater than in Working Conditions 1–4 (Figure 8).

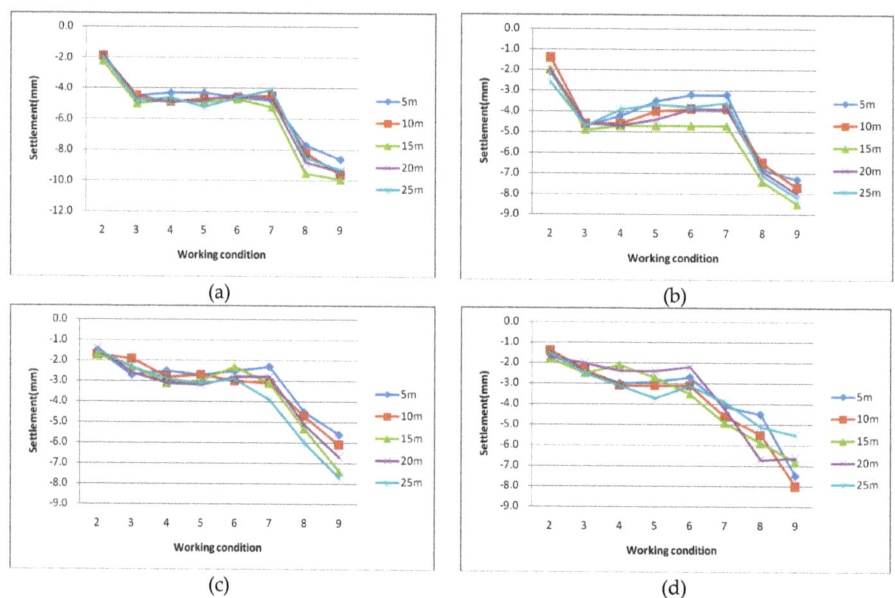

Figure 8. Settlement at outer edge of Slot Sections (**a**) 2, (**b**) 15, (**c**) 28, and (**d**) 42.

3.2. Wall-Body-Migration Analysis

Analysis of data presented in Figure 9 yielded the following results. First, the wall deformation of each working condition was linear at an excavation depth of 27 m, and the deformation curve had segmental characteristics. The displacement of the wall body had an inflection point at a certain depth; that is, there was a peak in the displacement curve of the wall. This point gradually moved deeper with increasing depth of excavation and was generally located near the maximal excavation depth. This differed from the deformation characteristics of a cantilever pile (the lower part of the pile is fixed, and the upper part is subject to lateral thrust) because performance of the circular underground continuous wall arose from its own annular restraining force. We termed this point for each working condition the "round-underground-continuous-wall-deformation inflection point". Additionally, for Working Conditions 2 and 3, at some stage of excavation, the bottom of the wall body deviated away from the direction of the foundation pit. This was similar to the deformation characteristics of a cantilever pile. Working Conditions 4–9 did not exhibit a reverse offset at the bottom of the wall, and the final forward offset gradually increased with excavation depth. Finally, the wall-offset curve exhibited D-type distribution (Figure 9), and the maximal offset appeared at approximately 2/3 of the excavation depth. The maximal value of the inflection point was Working Condition 9, which had an offset of 6.1 mm.

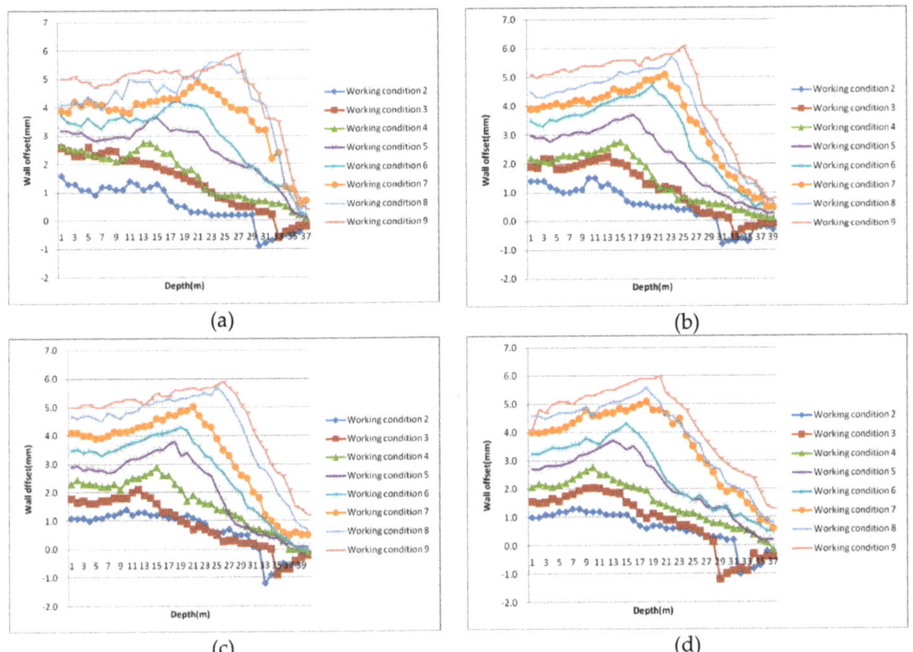

Figure 9. Offset around wall of Slots (a) 2, (b) 15, (c) 28, and (d) 42.

3.3. Settlement Analysis around Foundation Pit

There were only a few buildings and communities around the foundation pit. Thus, the construction machinery and the soil load near the foundation pit were the main factors for settlement. Settlement around the foundation pit is shown in Figure 10 for excavation Working Conditions 2–9.

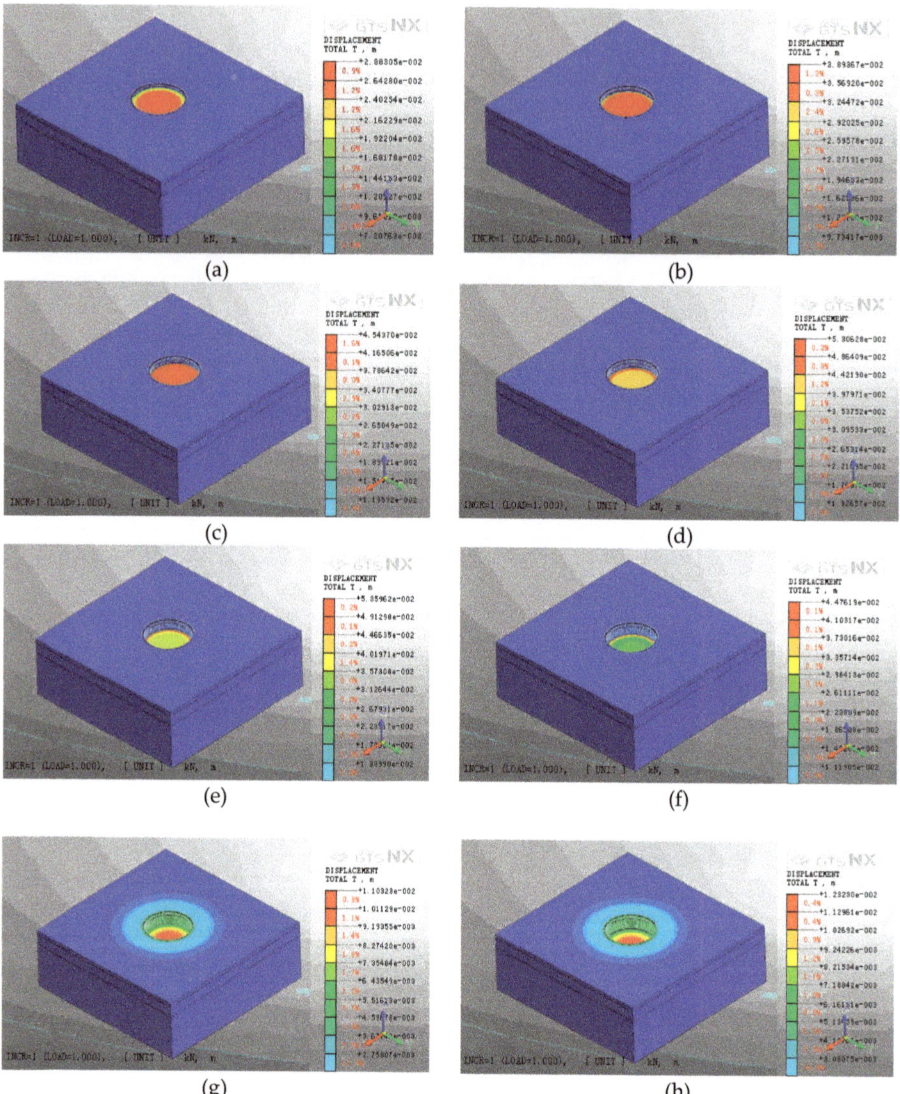

Figure 10. Settlement cloud around foundation pit for Cases (**a**) 2, (**b**) 3, (**c**) 4, (**d**) 5, (**e**) 6, (**f**) 7, (**g**) 8, and (**h**) 9.

Surface settlement of the outer edge of Slots 2, 15, 28, and 42 was also investigated. Due to limitations of the grid and the calculation of the model, settlement analysis was performed for 4, 8, 12, 17, 22, 27, 37, 47, 57, and 70 m depth (Figure 11).

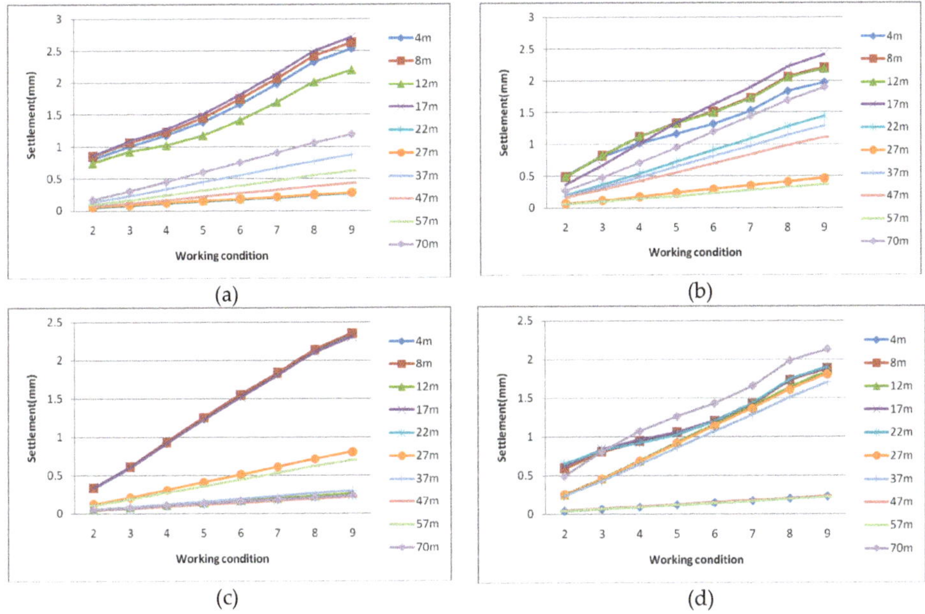

Figure 11. Settlement of outer edge of Slot Sections (**a**) 2, (**b**) 15, (**c**) 28, and (**d**) 42.

Figure 11 shows that surface settlement was linear and increased with the excavation depth of the foundation pit. Surface settlement within a range of about 27 m outside the foundation pit rapidly increased with the increase of excavation depth. The amount of ground settlement beyond the surface of the foundation pit, about 50 m, was slightly affected by the excavation depth of the foundation pit. Maximal surface settlement was located near the edge of the foundation pit, with a maximal value of 2.715 mm.

3.4. Displacement of Underground Continuous Wall

During the excavation of the foundation pit, the underground continuous wall was affected by soil stress and became offset. The wall deviation of the foundation pit for each working condition of the excavation is shown in Figure 12.

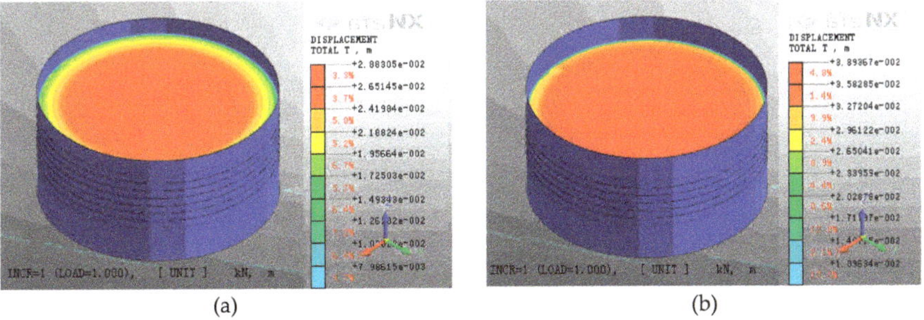

Figure 12. *Cont.*

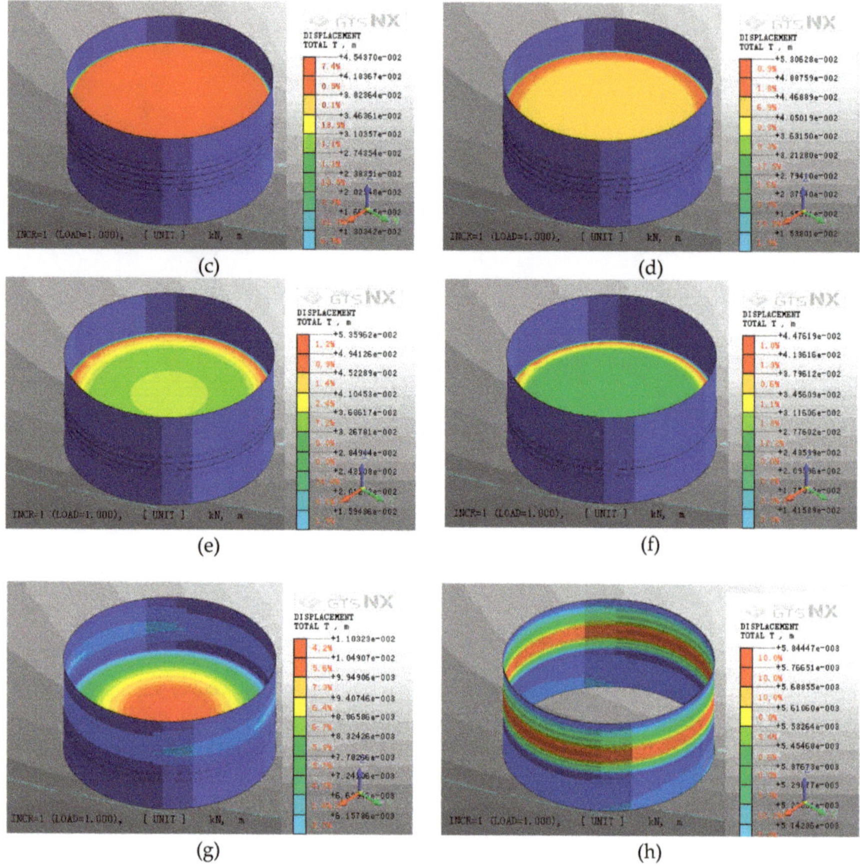

Figure 12. Underground diaphragm wall deviation for Cases (**a**) 2, (**b**) 3, (**c**) 4, (**d**) 5, (**e**) 6, (**f**) 7, (**g**) 8, and (**h**) 9.

The displacement model of the underground continuous wall at Slots 2, 15, 28, and 42 was selected for data processing. Due to limitations of the grid and the operation of the model, analysis of the displacement was performed for the 3, 6, 9, 12, 15, 18, 21, 24, 27, 30, and 40 m positions (Figure 13).

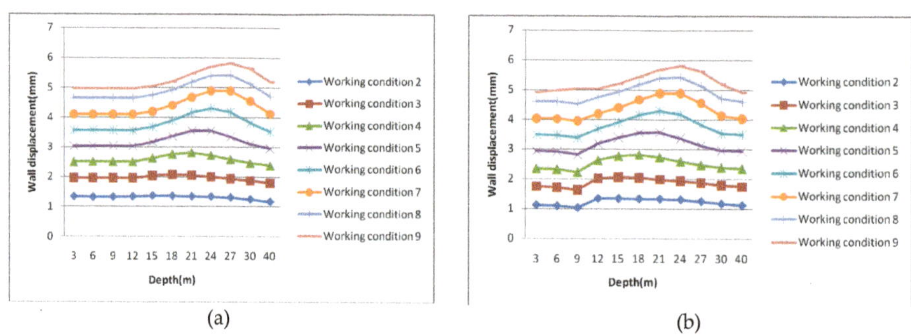

Figure 13. *Cont.*

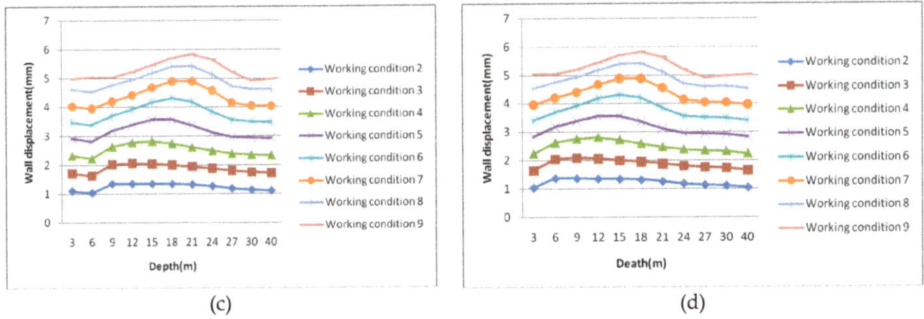

Figure 13. Wall offset of Slots (**a**) 2, (**b**) 15, (**c**) 28, and (**d**) 42.

These results revealed the following. First, the wall-body offset linearly increased with the depth of the excavation. Furthermore, the wall offset of each working condition showed a peak, after which the wall-body offset began to decrease. As excavation depth increased, the maximal offset of the wall shifted deeper. Second, there was no reverse offset calculation; the maximal offset of the wall was concentrated at a depth of approximately 2/3 of the total excavation depth. Third, as the depth of the excavation increased, the wall-offset curve showed D-shaped distribution. The simulated maximal offset was 5.837 mm.

Existing analysis of the deformation of the supporting wall of underground-continuous-wall foundation pits and the surrounding surface settlement mostly focuses on simple theoretical calculations [1,2,9–11] or finite-element analysis [3,4,7,8,12,13] that lack(s) validation of the used parameters. In this study, monitoring data were analyzed to identify the deformation law and other characteristics of the support structure. Three-dimensional numerical simulation of the foundation-pit excavation was conducted in Midas/GTS NX. In the process of realizing the analysis of ground settlement and wall offset around the circular underground continuous wall during construction, this paper demonstrated a key data self-verification method based on monitoring data, a breakthrough in difficulties in the selection of construction-safety calculation and finite-element-analysis parameters of foundation-pit engineering. It provides a new way of parameter selection for the sustainability study of foundation-pit and geotechnical engineering. In addition, we obtained the characteristics of surface settlement and wall offset around the circular underground continuous wall. The inflection point of the displacement of the circular underground continuous wall was proposed. These results are of great significance for the construction guidance of special-shaped underground continuous walls, providing an important reference for the continuous promotion of circular underground continuous walls.

4. Discussion

Monitoring and simulation results and analysis were as follows. The settlement of the surface surrounding the circular underground continuous wall was mainly affected by the depth of the foundation-pit excavation. As excavation progressed, both monitoring and simulation data showed good linearity. Monitored maximal settlement showed that the simulated value was a conservative calculation.

In addition, the deformation of the wall for each working condition showed linearity with clear staged characteristics. In particular, the deformation curve had obvious inflection points, most of which were located deeper than 2/3 of the overall excavation depth. The characteristics of the cantilever pile were not obvious in Working Conditions 3–9, but the distribution of the wall body offset in a D-shaped curve was evident. Deviation between the monitoring value of the maximal wall offset and the simulated value was only 4.31%; thus, monitoring and simulation data were in good agreement. Furthermore, force-deformation characteristics were different from those of the cantilever pile. The monitored value showed more convergence at the bottom of the wall, while analog values were not obvious. Preliminary analysis suggests that this was because monitoring data showed increased rock-embedded rock mass at the bottom of the wall compared to the simulated data.

5. Conclusions

This study drew three main conclusions. First, it was determined that the surface settlement of a circular underground continuous wall is mainly controlled by the depth of foundation-pit excavation. Both monitoring and simulation data demonstrated increased linearity as excavation progressed. Appropriate physical and mechanical parameters for key data self-verification were proposed and utilized to compensate for the shortcomings of on-site monitoring data, and the extent of surface settlement caused by construction excavation was determined. Second, analysis, monitoring, and simulation results showed that the deformation of the circular underground continuous wall had unique constraint characteristics. The wall offset of each working condition showed a peak, after which the wall-body offset began to decrease. On this basis, the concept of a round-underground-continuous-wall deformation inflection point was proposed. Finally, we determined that the deformation pattern of the circular underground continuous wall showed distinct linearity, the deformation curve had an inflection point, and most of the inflection points were located below 2/3 of the excavation depth. In addition, wall-offset distribution showed an evident D-shaped curve.

The key data self-verification method proposed in this paper can be used as a method to check the validity of simulation parameters, and subsequent research can extend this method to other computing systems. This method is expected to build a bridge between monitoring data and simulation results. The concept of a round-underground-continuous-wall deformation inflection point, proposed in the paper, needs to further be applied to the quantitative relationship between the displacement of the inflection point and excavation depth.

Author Contributions: X.G.; conceptualization, methodology, software, validation, data analysis, investigation, resources, data curation, writing—original-draft preparation, review and editing, data visualization, and project supervision. W.-p.T.; conceptualization, validation, and funding acquisition. Z.Z.; validation and project administration. All authors have read and agreed to the published version of the manuscript.

Funding: This research was funded by the Western Transportation Construction Science and Technology Project (2006-318-000-07), the China Communications Construction Co., Ltd (CCCC) Technology Research and Development Project (2011-ZJKJ-01), the National Natural Science Foundation of China (51708043), and the Fundamental Research Funds for the Central Universities, CHD (300102219106).

Acknowledgments: We would like to thank Editage (www.editage.com) for the English language editing.

Conflicts of Interest: The authors declare no conflicts of interest.

References

1. Bolton, M.D.; Powriet, W. Behaviour of diaphragm walls in clay prior to collapse. *Géotechnique* **1988**, *38*, 167–189. [CrossRef]
2. Poh, T.Y.; Wong, I.H.; Chandrasekaran, B. Performance of Two Propped Diaphragm Walls in Stiff Residual Soils. *J. Perform. Constr. Fac.* **1997**, *11*, 190–199. [CrossRef]
3. Bose, S.K.; Som, N.N. Parametricstudy of a braced cut by finite element method. *Comput. Geotech.* **1998**, *22*, 91–107. [CrossRef]
4. Whittle, A.J.; Hashash, Y.M.A.; Whitman, R.V. Analysis of Deep Excavation in Boston. *J. Geotech. Eng.* **2001**, *119*, 69–89. [CrossRef]
5. Kishnani, S.S.; Borja, R.I. Seepage and Soil-structure Interface Elects in Braced Excavations. *J. Geotech. Eng.* **2003**, *119*, 912–929. [CrossRef]
6. Alejano, L.R.; Pons, B.; Bastante, F.G.; Alonso, E.; Stockhausen, H.W. Slope geometry design as a means for controlling rockfalls in quarries. *Int. J. Rock Mech. Min. Sci.* **2007**, *44*, 903–921. [CrossRef]
7. Faheem, H.; Cai, F.; Ugai, K.; Hagiwara, T. Two-dimensional base stability of excavations in soft soils using FEM. *Comput. Geotech.* **2003**, *11*, 141–163. [CrossRef]
8. Liu, X.Z.; Ding, W.Q. Analysis of the influence of initial stiffness coefficient of contact element on deep foundation pit deformation in soft soil. *Chin. J. Rock Mech. Eng.* **2001**, *20*, 118–122.

9.   Chen, Y.F.; Zhang, Q.H.; Zhou, J.; Wang, Y. Construction and monitoring of ultra-deep circular foundation pits by semi-reverse construction method. *Chin. J. Geotech. Eng.* **2006**, *28*, 1865–1869.

10.   Xu, Z.H.; Wang, J.H.; Wang, W.D. Deformation behavior of underground continuous wall in deep foundation pit engineering in Shanghai area. *China Civ. Eng. J.* **2008**, *41*, 81–86.

11.   Wang, P.; Hu, L. Design and Emergency Treatment of Underground Continuous Wall of China Petroleum Building. *Constr. Tech.* **2009**, *38*, 107–109.

12.   Hu, A.F.; Zhang, G.J.; Wang, J.C.; Song, H. Deformation monitoring and numerical simulation of foundation pit retaining structure for subway transfer station. *Chin. J. Geotech. Eng.* **2012**, *34*, 77–81.

13.   Zheng, J.M.; Xie, J.Q.; Yang, P.; Zhang, Z. Numerical simulation of influence of horizontal deformation of deep excavation and support structure on settlement of ground. *Mod. Tunn. Technol.* **2013**, *50*, 102–108.

 *sustainability*

Review

Cutting Waste Minimization of Rebar for Sustainable Structural Work: A Systematic Literature Review

Keehoon Kwon, Doyeong Kim and Sunkuk Kim *

Department of Architectural Engineering, Kyung Hee University, Yongin-si 17104, Korea;
charade0820@naver.com (K.K.); dream1968@khu.ac.kr (D.K.)
* Correspondence: kimskuk@khu.ac.kr; Tel.: +82-31-201-2922

Abstract: Rebar, the core resource of reinforced concrete structures, generates more carbon dioxide per unit weight than any other construction resource. Therefore, reducing rebar cutting wastes greatly contributes to the reduction of greenhouse gas (GHG). Over the past decades, many studies have been conducted to minimize cutting wastes, and various optimization algorithms have been proposed. However, the reality is that about 3 to 5% of cutting wastes are still generated. In this paper, the trends in the research on cutting waste minimization (CWM) of rebar for sustainable work are reviewed in a systematic method with meta-analysis. So far, the literature related to cutting waste minimization or optimization of rebar published has been identified, screened, and selected for eligibility by Preferred Reporting Items for Systematic Reviews and Meta-Analyses, and the final 52 records have been included in quantitative and qualitative syntheses. Review by meta-analysis was conducted on selected literatures, and the results were discussed. The findings identified after reviewing the literature are: (1) many studies have performed optimization for the market length, making it difficult to realize near-zero cutting wastes; (2) to achieve near-zero cutting wastes, rebars must be matched to a specific length by partially adjusting the lap splice position (LSP); (3) CWM is not a one-dimensional problem but an n-dimensional cutting stock problem when considering several rebar combination conditions; and (4) CWM should be dealt with in terms of sustainable value chain management in terms of GHG contributions.

Keywords: rebar cutting waste; minimization; optimization; structural work; systematic literature review

Citation: Kwon, K.; Kim, D.; Kim, S. Cutting Waste Minimization of Rebar for Sustainable Structural Work: A Systematic Literature Review. *Sustainability* **2021**, *13*, 5929. https://doi.org/10.3390/su13115929

Academic Editor: Nicholas Chileshe

Received: 14 April 2021
Accepted: 21 May 2021
Published: 24 May 2021

Publisher's Note: MDPI stays neutral with regard to jurisdictional claims in published maps and institutional affiliations.

1. Introduction

Reinforced concrete (RC) structures, such as buildings and infrastructure, use enormous amounts of concrete and rebar during the construction phase. In 2012, global concrete and concrete constituent consumption reached about 10 billion m^3 [1], and the amount is rapidly increasing every year due to increased demand for RC structures along with global economic development. Rebar, the core resource of RC structures, generates more CO_2 per unit weight than any other construction resource. For example, C25/30 concrete generates embodied CO_2 (ECO_2) of 95 kg-ECO_2/t, but reinforcement bar (rebar) generates ECO_2 of 872 kg-ECO_2/t, which is equivalent to about 9.2 times of the concrete [2]. Therefore, reducing the cutting waste of rebars greatly contributes to the reduction of GHG [3]. Over the past few decades, numerous studies have been conducted on minimizing cutting wastes, and various optimization algorithms have been proposed. However, in reality, cutting wastes are still generated in the process of cutting and bending of rebars, which are at least 3% to 5% [3–7], and as much as 5% [4,6–11] to 8% [12], compared to the volume shown in the structural drawings.

Estimating how much rebar cutting wastes contribute to global GHG is a very difficult task, but to confirm the need for sustainable structural work, the authors follow a three-step estimation process after surveying literature and actual data: (1) analyzing the concrete and rebar ratio after surveying actual project data for concrete and rebar in Korea; (2) estimating the global annual use of concrete and rebar, and CO_2 emissions by rebar using global

459

concrete consumption in 2012, world GDP growth rate [13], and analyzed concrete and rebar ratio; and (3) estimating the global annual rebar cutting wastes and the resulting CO_2 emissions, applying the relatively conservative waste rates of 3 to 5% rates identified in the literature mentioned above.

Although the construction environment varies from country to country, in case of high-rise residential buildings in Korea, the analysis of 30 projects, as shown in Table 1, showed that the rebar quantity compared to concrete was found to be about 0.070 ton/m^3, and commercial buildings have long-span heavily loaded attributes compared to residential buildings. The analysis of 12 projects showed a result of about 0.119 ton/m^3. The average of these amounts is calculated at about 0.077 ton/m^3. If this average value is applied to 10.058 billion m^3 [1], as of 2012 as shown in Table 2, about 778.9 million ton of rebar usage is calculated. Applying the world GDP growth rate as shown in Table 2, rebar usage increases every year, which is estimated to be about 947 million tons in 2019.

Table 1. Estimation of rebar quantity compared to concrete in reinforced concrete structures.

Description	No. of Projects	Concrete (m^3)	Rebar (Ton)	Rebar/Concrete (Ton/m^3)
Residential buildings	30	4,680,573	327,489	0.070
Commercial buildings	12	835,514	99,698	0.119
Sum	42	5,516,087	427,187	0.077

Source: authors' research results.

Table 2. Estimated global annual use of concrete and rebar, and CO_2 emissions of rebar.

Year	World GDP Growth Rate (%)	Concrete (Billion m^3)	Rebar (Ton)	CO_2 Emission (Ton·CO_2)
2012	2.52	10.058	778,930,801	266,082,762
2013	2.66	10.326	799,650,360	273,160,563
2014	2.85	10.620	822,440,395	280,945,639
2015	2.88	10.926	846,126,679	289,036,873
2016	2.59	11.209	868,041,360	296,522,929
2017	3.26	11.574	896,339,508	306,189,576
2018	3.10	11.933	924,126,033	315,681,453
2019	2.48	12.229	947,044,359	323,510,353

Source: authors' research results.

For reference, it is impossible to investigate all RC structures around the world to estimate global rebar usage by year. Therefore, despite some errors, it is meaningful to have applied some data of high-rise residential and commercial buildings in Korea. Later, when the data of investigation into various RC structures are added, the range of error will gradually decrease. The application of the world GDP growth in 2012 was also estimated in the same context, as shown in Table 2, because data on global concrete and concrete constituent consumption by year could not be obtained.

Using an estimated global annual use of rebar, if about 0.3416 ton·CO_2/ton [14], which is the unit value of rebar CO_2 in Korea, is applied, the generation of about 323.5 million tons of CO_2 in 2019 is estimated, starting with 266.1 million ton·CO_2 in 2012. For reference, the unit value of CO_2 is different according to industrial structure by country, so it is not possible to obtain a unified value. Therefore, in this study, the calculation was performed based on data analyzed in Korea.

If a rebar cutting waste rate of about 3 to 5% is applied based on this value, about 23.368 to 38.947 million tons of wastes are generated as of 2012, as shown in Table 3, and the amount keeps increasing every year to reach about 28.411 to 47.352 million tons in 2019. When calculating the corresponding CO_2 emission, the amount increases annually from about 7.982 to 13.304 million ton·CO_2 in 2012 to reach about 9.705 to 16.176 million ton·CO_2 in 2019, as shown in Table 3. If the near-zero cutting waste of rebars is realized, the effect of CO_2 emission reduction of up to 16.176 million tons can be achieved, and the corresponding

GHG is reduced. For reference, since the rebars placed into structures vary in length, diameter, and number, it is impossible to combine them without cutting wastes, called zero cutting wastes, by the length of rebars supplied by the steel mill. However, by combining rebars with special lengths supplied by the steel mill, cutting wastes can be reduced to close to zero, which is called near-zero cutting wastes.

Table 3. Estimation of global annual rebar cutting wastes and CO_2 emissions.

Year	Rebar (Ton)	Cutting Wastes of Rebar (Ton)		CO2 Emissions from Cutting Wastes (Ton·CO_2)	
		3%	5%	3%	5%
2012	778,930,801	23,367,924	38,946,540	7,982,483	13,304,138
2013	799,650,360	23,989,511	39,982,518	8,194,817	13,658,028
2014	822,440,395	24,673,212	41,122,020	8,428,369	14,047,282
2015	846,126,679	25,383,800	42,306,334	8,671,106	14,451,844
2016	868,041,360	26,041,241	43,402,068	8,895,688	14,826,146
2017	896,339,508	26,890,185	44,816,975	9,185,687	15,309,479
2018	924,126,033	27,723,781	46,206,302	9,470,444	15,784,073
2019	947,044,359	28,411,331	47,352,218	9,705,311	16,175,518

Source: authors' research results.

As shown in Table 2, demand for buildings and infrastructure increases in line with global economy growth and corresponding demand for RC structures increases every year. The increase in RC structures leads to demand chains that increase demand for concrete and rebars, as shown in Table 2, resulting in an annual increase in rebars cutting waste and CO_2 emissions such as Table 3. In particular, it is expected in the future to be more concentrated in developing countries where the population is concentrated [15,16]. The increase in global cutting waste of rebars not only causes unnecessary cost losses but also problems of generating large amounts of CO_2 in the production, transportation, and processing phases. Therefore, research to realize near-zero cutting waste is critical to implement sustainable rebar work.

So far, many studies have been conducted to optimize the use of rebars or to reduce the cutting waste. However, the near-zero cutting waste has not yet been realized. However, the near-zero cutting waste has not yet been implemented. The study on cutting stock problem (CSP), which is considered to be the beginning of cutting waste minimization (CWM), was first mentioned by Kantorovich in 1939 and was first published in Management Science in 1960 [17]. Therefore, CSP-related literature has been searched for since 1960 in this study. The literature on the optimization of rebar cutting waste was basically targeted from 1990 to 2020, because the CSP-related research in rebar work started in earnest from 1991. In this paper, we performed a search and review of studies related to optimization or cutting waste minimization of rebars that have been conducted so far and identified the status and problems of existing studies. We then proposed the direction of future research to implement near-zero cutting waste and identify its potential.

2. Data Sources and Methodology

2.1. Data Sources

There are literature databases of various fields around the world, but for the search of articles related to minimal cutting waste of rebars, Scopus, Science Direct, Web of Sciences (WoS), Taylor and Francis Online, Springer Link, American Society of Civil Engineers (ASCE) Library, and Willey Online Library were used. Some dissertations or literature published in internationally uncertified journals were searched for using Google or Google Scholar databases.

2.2. Systematic Literature Review

SLR is an exact and reproducible method for identification, evaluation, and interpretation of predefined fields of study [18]. Kitchenham and Charters defined "a systematic literature review is a means of identifying, evaluating, and interpreting all available research relevant to a particular research question, topic area, or phenomenon of interest. Individual studies contributing to a systematic review are called primary studies; a systematic review is a form of secondary study" [19]. Since similar SLR methodologies have been proposed by several scholars [20,21], MDPI publisher based in Basel, Switzerland recommends that the authors follow Preferred Reporting Items for Systematic Reviews and Meta-Analyses (PRISMA) [22], checklist, and flow diagram for reporting systematic reviews and meta-analyses.

In construction field, numerous literature review articles have been published, not based on SRL [23–28], before 2010. The reason for this is that the perception of SLR in the construction field was not high. Since 2010, with the exception of some articles [29–34], most review articles have been written based on SLR [18,35–49]. After 2018, many review articles have been written according to PRISMA [18,46–49], and this study also follows the PRISMA statement for systematic reviews and meta-analysis.

2.3. Methodology

The previously mentioned literature database was sequentially searched using Boolean operator "AND" by keywords, such as rebar, rebar work, rebar optimization, and rebar cutting waste. As a result, Google Scholar found about 79,100 search results for literature related to rebar work, about 16,000 cases of rebar optimization, about 14,000 cases of rebar cutting waste, and about 4410 cases of rebar cutting waste optimization, as shown in Table 4. Google Scholar has confirmed that it includes various reports such as books, content, and dissertations along with academic papers in most databases, as shown in Table 4. In addition to construction, literature of almost all fields, such as medicine and chemistry, is searched by keywords. Additionally, it is confirmed that the search works even if there is a rebar or work in the name. Therefore, searching and reviewing all relevant literature in Google Scholar is an inefficient approach. Since the minimum cutting waste of rebars to be dealt with in this review article is a very specific topic, most of literature is searched in databases such as ScienceDirect, WoS, and ASCE Library. However, Google Scholar was used to search for books, magazines, and documents such as dissertations, which are not well searched for in databases such as WoS and ASCE, and when original text could be downloaded from these databases.

Table 4. Search by keyword in literature database (as of 1 December 2020).

Literature Database	Literature Keywords			
	Rebar Work	Rebar Optimization	Rebar Cutting Waste	Rebar Cutting Waste Optimization
Google Scholar	79,100	16,000	14,000	4410
ScienceDirect	9041	3191	409	211
Springer Link	4292	672	384	88
ASCE Library	3896	962	233	73
Willey Online Library	2796	776	265	116
Taylor and Francis Online	2002	628	176	60
Scopus	892	163	13	9
Web of Science	572	89	10	6

As shown in Table 4, a search for literature was performed on Google Scholar, ScienceDirect, Springer Link, and ASCE Library. It has been confirmed through the literature search process that the number of literatures searched for varies depending on the characteristics of each database. For example, topics such as rebar cutting waste correspond to construction engineering; therefore, literature is frequently searched for in databases of engineering fields. In particular, the ASCE Library is a database dedicated to the construction field; hence, many literatures related to this review paper have been searched for. When searching the literature with a keyword of rebar work, which includes all rebar-related work, many articles are retrieved as shown in Table 4. However, many literatures include corrosion of rebar, rebar tying tool, rebar cutting and bending machine, and rebar work schedule, and are not related to CWM. When search range is narrowed to rebar optimization and rebar cutting waste, the number of records is reduced dramatically. For reference, rebar optimization is the general word of rebar minimization, and rebar cutting waste literally means the waste remaining after cutting the rebar. Finally, in most databases, searching with rebar cutting waste optimization that has the same definition as CWM results in fewer records. In the case of Scopus and WoS, it is reduced to 9 and 6 records, respectively, but all records are valid. In other databases, many records are identified as RC design optimization.

Based on literature searched for on 1 December 2020, 1811 records were finally identified, as shown in Figure 1, excluding duplicated literature or literature not related to the subject of this study. Among them, duplicated 384 records were screened and 638 and 386 records that were not relevant to the rebar cutting work and rebar cutting optimization were excluded, respectively, to finally select 403 full-text articles. Then, 351 literatures related to design optimization of the RC components or frames were excluded. The reason is that design optimization of RC corresponds to pre-processing research of rebars optimization, while CWM of rebars covered in this study corresponds to post-processing research. As a result of reviewing some of the literature [50–81] related to design optimizations of rebars, it was confirmed that they are related to design optimization of RC components such as slab [50–57], beam [58–61], column [62–65], foundation [64], and wall [66–68], and design optimization of RC frames such as bridges [69–71] and building [72]. In addition, many studies related to design optimization have been well-organized in the review article [37].

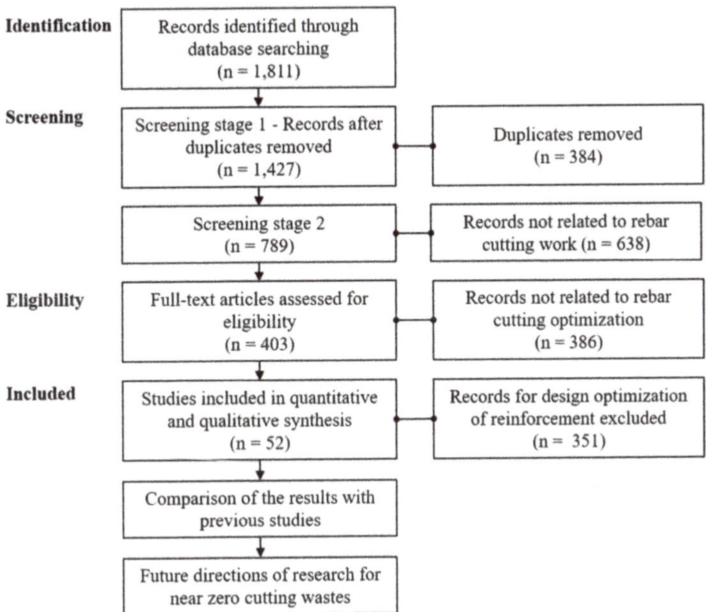

Figure 1. Flow diagram of the literature review and the analysis process. Source: authors' research results.

2.4. Descriptive Analysis

Studies in the field of construction project management vary widely, including time, cost, quality, and safety. In the Project Management Body of Knowledge, there are 13 knowledge areas [82], and there is countless management research connected with engineering technology, and post-processing research such as minimum cutting wastes of rebars is a very narrow and special topic. Therefore, it is confirmed that there are not many articles directly dealing with this topic. As shown in Table 5, 37 articles were published in the journal [3,4,6,8–12,63,83–110] and 11 articles were published in peer reviewed conference publication [7,55,62,71,111–117]. The rest have three dissertations [5,118,119] and one book chapter [120].

Table 5. Number of literatures by publication type.

Publication Type	Number of Literatures	Percent
Journal	37	71.2%
Proceedings	11	21.1%
Dissertation	3	5.8%
Book chapter	1	1.9%
Total	52	100.0%

Source: authors' research results.

When examining papers published in internationally certified SCI or SCIE journals, as shown in Table 6, the biggest number of papers, seven, were published in *Journal of Construction Engineering and Management (JCEM)* [6,84,87,95,97,101,102]. As of 2019, *Journal Impact Factor (JIF)* of JCEM is not as high as 2.347, but it is one of the most popular ASCE journals. In addition, two papers were published in *Automation in Construction* [63,94,105], and *Journal of Computing in Civil Engineering* [86,103], and one was published each in the remaining journals. It is notable that each paper was also published in high *JIF* journals, such as the *International Journal of Engineering Science* [85], *Computer-Aided Civil and Infrastructure Engineering* [110], and the *Journal of Advanced Research* [100]. It is assumed as such because the problem of minimizing the rebar cutting waste is important and difficult. *Construction Management and Economics* is not an SCI journal classified by JCR but was included in the list because it is internationally popular [99].

Table 6. List of the most popular journals.

Journal Title	Papers	Published Year	JIF 2019
Journal of Construction Engineering and Management	7	1993, 1994, 1996, 2000 (2 papers), 2007, and 2012	2.347
Automation in Construction	3	1995, 2019, and 2021	5.669
Journal of Computing in Civil Engineering	2	1995, 2013	2.979
International Journal of Engineering Science	1	2016	9.219
Computer-Aided Civil and Infrastructure Engineering	1	2014	8.552
Journal of Advanced Research	1	2016	6.992
Construction and Building Materials	1	2018	4.419
International Journal of Computer-Integrated Manufacturing	1	1998	2.861
Sustainability	1	2020	2.576
KSCE Journal of Civil Engineering	1	2014	1.515
Canadian Journal of Civil Engineering	1	2004	0.985
Construction Management and Economics	1	2014	-
Others	16	-	-
Sum	37		

Source: authors' research results.

Table 7 shows 37 articles that are classified by countries based on lead authors. According to Table 7, Korea has the largest number of publications, which is 16 articles, followed by Canada with 5, Israel with 4, and Turkey with 3; 5 countries, including Bangladesh, published 2 papers each. Eight countries, including Albania, published 1 paper each. In Korea, the number of rebars per unit area of RC structure has more than doubled due to the strengthening of seismic design standards in 1988 [121], the strengthening of noise standards between floors in 2000 [122], and the rapid increase in the number of high-rise buildings over 20 stories. Therefore, by continuously conducting studies on rebars design optimization and CWM of RC structures since 1999, the reduction in the use amount of rebar was confirmed.

Table 7. Number of literatures by country.

Country	Number of Articles	Remarks
Korea	16	
USA	6	
Canada	5	
Israel	4	
Turkey	3	
Bangladesh, India, UK, Taiwan, and Thailand	2	Five countries presented 2 papers each
Albania, Australia, Ethiopia, Germany, Iraq, Malaysia, Spain, and Ukraine	1	Eight countries presented 1 paper each
Total	37	

Source: authors' research results.

Figure 2 shows the number of articles published by year. One or two articles were published every year until 2004, but after 2012, the number of articles increased until 2016 with the development of various techniques, including computer-aided design (CAD) and building information models (BIM). It is confirmed that the number of articles dropped sharply to one in 2017 and increased again. In the past, cutting wastes were approached from an economic perspective; however, recently, research has been conducted from a sustainable construction perspective.

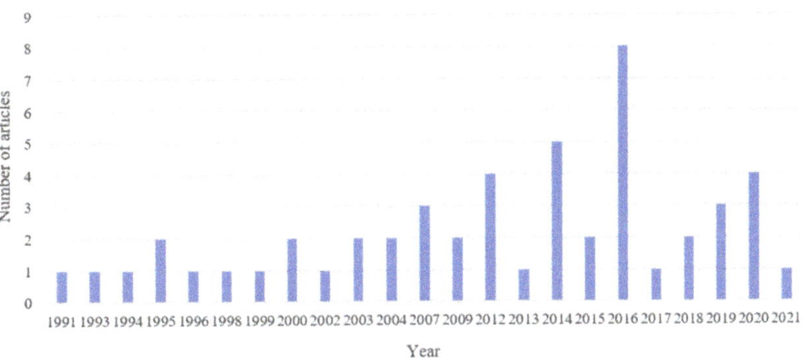

Figure 2. Number of published articles by year (Source: research results).

3. Review Results and Discussion

3.1. Selection of the Papers

As shown in Figure 1, the number of literatures corresponding to rebar cutting optimization through the identification, screening, and eligibility process was a total of 403, and 52 literatures related to RC design optimization were selected, excluding 351 literatures that fall under the category of pre-processing research of rebar optimization. They address the problems of rebars cutting waste, corresponding to post-processing research of rebars optimization after RC design. These literatures will be reviewed by factors such as application of optimization techniques and graphic solutions for CWM, range of rebar work process, consideration of lap space position, reflection of length in a special order, consideration of bending margin, and consideration of schedule. The review of selected literature will not only measure characteristics and trends of the research for CWM of rebars but also present a direction of future research. In addition to the selected literature, there is rebar modeling [123], using BIM solutions for optimized rebar work, and software that creates rebar details using the information generated after the structural design [124]. However, these were excluded in this paper as they are articles written for the promotion of commercial software and are not described mainly on CWM.

3.2. Identification of Cutting Waste Minimization-Related Factors

One-dimensional CSP has been studied not only in rebars but also in all areas of cutting linear stock material such as pipe and timber. Since the publication of Kantorovich's article [17], many articles have been published in various fields related to CSP [8,125–148]. In the case of rebar, research has been vigorously conducted after 1991 with the development of computer science, since it was first introduced as an example of CSP by Kantorovich [3]. This is mainly because the need for CSP in the construction field to build a single building on site was not highlighted much, unlike general manufacturing, which mass-produces large quantities of the same or similar products in factories. Moreover, it was not easy to develop algorithms to deal with varying variables, such as length, diameter, number of required, and point of use of rebar, which are subject to CSP, and algorithms to consider variables, such as bending margin, various market length, and special length.

In this study, factors that influence the analysis of attributes of the literature should be identified for quantitative and quantitative analysis of the final selected literature. The following is a summary of the variables identified during screening and eligibility assessment of full-text articles related to rebar optimization along with the authors' research experience.

- Applied optimization techniques: integer programming (IP), linear programming (LP), genetic algorithm (GA), simulated annealing (SA), binary search algorithm (BSA), heuristic algorithm (HA), and harmony search (HS) algorithm.
- Rebar work process: preparation of a drawing, quantity take-off, rebar production such as cutting and bending, and in-site rebar placement.

In addition, literature can be reviewed by lap splice position (LSP), use of special length (SpL) or stock length (StL) rebars, and schedule.

3.3. Results of Quantitative and Qualitative Review

3.3.1. Description by Optimization Techniques

Table 8 summarizes the optimization techniques adopted by the literature selected for rebar's CWM. Afzal et al. [37] introduces the definitions, advantages, disadvantages, and application cases of various algorithms in the study of RC structural design optimization. However, the problem of rebar cutting waste is relatively limited in the scope of study compared to RC design, so the literature is summarized by seven optimization techniques, as shown in Table 8. The advantages and disadvantages associated with rebar CWM are described by optimization techniques, and the classification of literature that adopted these techniques is presented in Table 8.

Table 8. Summary of the adopted optimization techniques.

Optimization Techniques	Advantage	Disadvantage	References
Linear programming (LP)	Flexibility to be paired up with other approximation to improve convergence	Slower in finding special-length-priority or waste-rate-priority solutions under multiple search conditions	[3–5,7,9,12,63,86,93,96,109,114]
Integer programming (IP)	Rapid generation of solutions under limited search conditions	Difficulty in search of solutions under complex conditions or in search of float number solutions	[7,12,63,100,101,104–106]
Genetic algorithm (GA)	Simplicity in programming, proof in finding the global optimum, applicable to diverse problem domains, computing performance, and diversity of solutions	Time consuming for formulating a CSP problem under complex combination conditions	[8,11,71,83,101,113,118]
Binary search algorithm (BSA)	Quick search for rebars of a specific length to be used in combination	Long CPU run-time for global search as the increase of rebar combination conditions	[10,116]
Simulated annealing (SA)	Use for combinatorial optimization problems in a discrete search space and simplicity in implementation	Large computing time and cost if boundary conditions are not provided	[6]
Heuristic algorithm (HA)	Low computing cost to obtain near-zero cutting waste solution	Large computing time and cost to obtain an optimized solution for all conditions	[62]
Harmony search (HS)	Easy to build and fast convergence for the optimal solution in a reasonable amount of computational time	Randomness, instability, and uncertainty of search direction	[117]

Source: authors' research results.

In the case of LP, it has an advantage in terms of flexibility to be paired up with other approximations to improve convergence, but there is a disadvantage in terms of being slower in finding special-length-priority or waste-rate-priority solutions under multiple search conditions. In studies of rebar's CWM, 12 articles were selected as optimization techniques [3–5,7,9,12,63,86,93,96,109,114]. The reason for this is that CSP or CWM problems have been adopted most often in modeling and have become more common.

Integral programming (IP) has the advantage of quickly generating solutions under limited search conditions, while many search conditions or requiring a float number solution are challenging. In rebar's CWM study, the second largest number of research articles is adopted in eight research articles [7,12,63,100,101,104–106], despite the fact that it takes considerable time to formulate the problem [101]. IP is one of the long-standing optimization techniques used as one-dimensional cutting stock problems like LP and is said to be the most common in modeling CWM problems.

GA has advantages such as simplicity in programming, proof in finding the global optimum, applicability to diverse problem domains, computational performance, and diversity of solutions, but it also has disadvantages such as being time consuming for formulating a CSP problem under complex combination conditions. As a result of reviewing the literature, seven articles have started to adopt GA since 2004 [8,11,71,83,101,113,118], and most of them have adopted LPs and IPs before. Salem, Shahin, and Khalifa [101] conducted a study comparing CWM using GA and IP models and then verified that GA further reduces the cutting waste of the rebar through a case study. Computational time of GA models is practical for everyday use and, in some cases, the GA model was able to lump the wastes in bigger lengths, thus achieving more savings.

Binary search algorithm (BSA) has an advantage in terms of providing quick-iterated local search for rebars of a specific length to be used in combination but has a disadvantage of long CPU run-time for global search as the increase of rebar combination conditions. BSA has the advantage of quickly performing iterated local search, so the CWM problem should be divided according to the supply schedule of rebars. In this case, there is a problem that the CWM effect is not greater than global search. BSA has been confirmed to have been adopted by two articles, as shown in Table 8 [10,116].

SA has advantages such as use for combinatorial optimization problems in a discrete search space and simplicity in implementation, while it has disadvantages of large computing time and cost if not providing boundary conditions. Porwal and Hewage asked "which conditions are more suitable?" when LP, IP, GA, BSA, sequential heuristic procedure, and SA are applied to a combination of rebar cutting patterns [6], and Porwal and Hewage proposed integration with BIM and combination with special-ordered length, available market lengths, and SpL of rebars by SA. In addition, the case study suggested that SA models are successful in complex combinatorial optimization programs through controlled randomization.

Heuristic algorithms are algorithms that solve problems in a more efficient way than conventional methods at the expense of optimality, completeness, and accuracy to obtain rapid solutions. Heuristic algorithms are expensive for accurate calculations and are frequently used if approximate solutions are sufficient. Bekdas and Nigdeli [62] optimized RC columns using a metaheuristic algorithm, called a bat algorithm.

HS is a metaheuristic search algorithm that tries to mimic the improvisation process of musicians in finding a pleasing harmony [149,150]. Although HS algorithm has better global optimization capability, its disadvantages are randomness and instability. Nonetheless, search direction of the algorithm is uncertain [150], and HS requires higher number of iterations [37]. HS was applied to optimize costs of materials, including concrete and rebars, by implementing design variables such as width and height of RC column, including diameters and number of reinforcements, and loading condition variables are implemented as harmony vectors [117].

HA is divided into local search-based metaheuristic algorithms, such as SA, and global search-based metaheuristic algorithms, such as GA and HS, to find better solutions. Therefore, HA or SA is more efficient if the target of rebar CWM is a local-search-based problem, and GA or HS is more effective if the target is a global-search-based problem.

Reference numbers written in Table 8 indicate that the corresponding optimization techniques are used in combination, for example, references [7,12,63], used a combination of IP and LP, and reference [101], performed rebar CWM using IP and GA.

3.3.2. Description by Rebar Work Process

Rebar work process consists of structural design, drawing work, quantity take-off (QTO), rebar production, and rebar placement. The literature related to structural design optimization of the RC component or frame has been sufficiently reviewed in other articles, and this paper reviews the literature that performed post-processing CWM from drawing creation to rebar placement. Table 9 shows literature review by rebar work process. Strategies to reduce rebar cutting waste are effective only when implemented from the drawing work stage. Accordingly, the top 20 literature, as shown in Table 9 references, suggested reducing cutting wastes in conjunction with drawings, and some literature included a mathematical algorithm that automatically writes rebar drawings using CWM algorithm [6,84,91,105,108,111].

Table 9. Summary by rebar work process.

Work Stage	Contents	References
Drawing work	Using the information provided in the drawing, rebar CWM-related tasks are performed. Alternatively, a rebar drawing is created by applying CWM algorithm.	[3,6,55,84,86–89,91,95,97,98,103,105,108,110,111,114,119,120]
Quantity take-off	Rebar CWM algorithm is connected to the QTO task and progress.	[3,5,6,85–87,90,91,98,103,107,108,111,113,115,116]
Rebar production	After completing the bar bending schedule, the work is performed to combine cutting patterns using CWM algorithm.	[6,8,87,88,91,92,94–96,99,107,111,112,116]
Rebar placement	For CWM, the cutting wastes are reduced by adjusting the lap splice position or length of the rebars in the range of satisfying the structural code.	[84,85,89,97,98,102,103,105,110,112,118,120]

Unlike ordinary materials, quite many variables should be considered in the case of the exact QTO work of rebars. It is a complex task that should reflect the size of stock material, concrete cover, and lap splice length, as well as variables not shown in the drawing, such as bending margin. Thus, studies have been conducted to develop algorithms to automate QTO in several studies [3,6,90,91,107,108,115], where variables such as the length and number of rebars applied could be directly utilized for rebar CWM algorithm. Thus, the second largest number of articles, as shown in Table 8, for reducing rebar cutting waste at the QTO stage is 16.

CWM algorithm has been widely applied to rebar production stages, including cutting and bending [6,8,87,88,91,92,94–96,99,107,111,112,116]. This is because the bar-bending schedule is prepared first before supplying rebars to the site, the cutting list is prepared, and the bar combination is performed by cutting patterns using the list. Several studies have indicated that the cutting wastes of rebars start from the purchase order stage [6,90,91,104]. This is because ordering by market length without analysis of optimal cutting patterns is a major factor in increasing cutting waste.

As for the study on reducing cutting waste in rebar replacement, 12 articles were published, as shown in Table 9 references. In the rebar placement stage, studies are divided into two—one is to prevent loss or waste caused by a mismatch in field installation after cutting, bending, and fabrication of rebars [84,89,97,105,112,118], and the other is to perform a detailing design and installation planning as an optimization method considering the productivity of the rebar placement [2,10].

The reference numbers written multiple times in Table 9 indicate that the corresponding article was performed on several stages of rebar works. For example, references [98] and [103] were performed for drawing work, QTO, and rebars placement stages, and references [84,89,97,105,110,120] were performed for drawing work and rebars placement stages.

3.3.3. Description by Other Factors

Because the location, size, and structural performance of RC components such as column, beam, foundation, slab, and wall are different, the length of the rebar generated after structural design is very diverse. Since rebar has the characteristic of being repeatedly installed, if the rebar is determined to have a certain length after structural design, the cutting patterns appear simple and the effect of CWM is also significant. LSP should be partially adjusted to satisfy the structural design codes to arrange the length of rebar in RC components constant. Several studies, as shown in Table 10, have revealed that the effect of CWM is significant when LSP was adjusted [6,7,12,86,105,108,112,119].

Table 10. Summary by other factors.

Factor	Contents	References
LSP	Adjusting lap splice position to satisfy structural code to make the length of rebars used for cutting patterns constant	[6,7,12,86,105,108,112,119]
StL	Performing CWM on rebar in stock or market length	[3,4,6,9–11,86,90,104,105]
SpL	Performing CWM on rebars with special ordered length	[3,4,6,10,90]

As shown in Table 10, 10 rebar CWM-related articles are described for rebars of stock length (StL) or market length. This is because the CSP study on the optimization of the material of one-dimensional stock length stored in the factory is the beginning of rebar CWM. In particular, in the case of factory production, materials sold at a fixed length in the market are stored, and the combination of cutting pattern optimization is performed for mass production. However, in the case of a construction project, various lengths and number of rebars must be combined, so it is difficult to reduce cutting wastes using stock length. Therefore, rebar combination is needed by SpL [3,4,6,10,90].

The use of SpL can further reduce cutting wastes, in contrast to the use of StL [3,6,10]. However, the minimum quantity and pre-order time must be satisfied to order rebars with SpL. The length of rebars should be adjusted so that it is combined with SpL. Eventually, additional algorithms should be developed to adjust LSPs easily and quickly. The references in Table 10 are written several times, because the corresponding articles considered multiple factors for CWM. For example, Porwal and Hewage [6] incorporated StL and SpL as well as LSP into the study for CWM.

3.4. Discussion

If near-zero cutting waste of rebar, one of the most ECO_2 generating resources in construction materials, is realized, environment-friendly sustainable construction is implemented and the waste of high-cost resources is prevented. The results of SLR analysis on studies that attempted to reduce rebar cutting wastes showed that there were relatively many design optimizations of RC literature corresponding to pre-processing research, while the rebar CWM-related literature corresponding to post-processing research was 52 articles. The results of systematic critical review on CWM of rebar are summarized as follows:

- Applied optimization techniques: LP, IP, and GA were the most frequently adopted 27 articles, 84.4% of the total, and the remaining BSA, SA, HA, and HS were selected in 5 articles. Initially, optimization algorithms were adopted based on LP and IP, but recently, the adoption of HS and GA has been confirmed to increase. This is because HS and GA, which are operated based on expertise, can perform the task of realigning reinforcing bars with special lengths more easily and faster than mathematical algorithms. HS and GA can realign rebars that are repeatedly installed with special lengths more efficiently than IP, LP, and BSA, while satisfying structural requirements.
- Rebar work process: CWM studies have been conducted in many literatures linking four stages of work processes such as drawing work, QTO, rebar production, and rebar placement. The reason is that rebar CWM is linked to all four stages from drawing work to rebar placement. In addition, since the importance of information is determined according to the order of the rebar work process, it was confirmed that 20 papers, 16 papers, 14 papers, and 12 papers were associated with each work stage. In consideration of the characteristics that the initial information affects the information generated later, it was confirmed that performing CWM in the drawing work stage is most effective. If CWM is performed in the drawing work stage, the results are sequentially reflected in the QTO, rebar production, and rebar placement stages to minimize cutting wastes.
- Other factors: partially adjusted LSPs, while satisfying structural design codes, expect the related research to increase [12], due to the large effectiveness of CWM.

Although there have been many CWM studies on StL so far, it has been confirmed that the research focusing on SpL will be expanded in the future.

In general, rebars are sold as linear rods in the market. Therefore, most studies on minimum cutting waste have focused on algorithms to solve one-dimensional cutting stock problems because contractors purchase, cut, and bend them. With the development of software and hardware solutions for engineering programs, techniques such as GA, BIM, AR, VR, and integrated project delivery have been added. It is assumed that there is a fundamental problem that the cutting waste rate is not yet reduced to near-zero despite those studies.

In this study, we have confirmed that cutting wastes can be significantly reduced in the following two cases. First, the use of coiled rebars can reduce cutting wastes to near-zero. The coiled rebar, which has been used since the 1990s, automatically performs rebar cutting and bending by machine. Initially, coiled rebars with a diameter of 10 mm to 16 mm were processed, but recently, coiled rebars with a diameter of 50 mm are automatically cut and bent by machine [151]. Global near-zero cutting waste can be achieved if machines are available that automatically perform straightening, cutting, and bending after coiled rebars are produced and supplied in all countries. However, not many countries have an industrial structure that satisfies such a supply chain. Except for some countries in Europe and North America, most countries do not yet supply coiled rebars. Therefore, CWM research should be continued until the rebar supply chain is globally established.

Second, if mechanical rebar couplers are used, steel quantities are reduced compared to lap spaces and ECO_2 is also reduced proportionally. This study summarizes the comparisons between lap splices and mechanical couplers, as shown in Table 11. Lapping length and weight are different from high-tensile deformed bars 10 mm (D10) to 32 mm (D32) in diameter. The weight of couplers of each diameter is different, but the overall effect of reducing ECO_2 is significant when couplers are used. In the small case of D10, a weight of 0.166 kg/EA and an ECO_2 difference of 0.145 kg-ECO_2/EA are generated. In the large case of D32, a weight of 14.237 kg/EA and an ECO_2 difference of 12.415 kg-ECO_2/EA are generated, respectively. In particular, with mechanical rebar couplers, ECO_2 reduction effect can be achieved from at least 84.7% to up to 96.4% compared to splice lap.

Table 11. ECO_2 comparison between splice lap and coupler.

Size	Building Element	Unit Weight (KG/M)	Splice Length (M/EA)	Splice Weight (KG/EA)	Coupler Weight (KG/EA)	Weight Difference (KG/EA)	ECO_2 Difference (kg-ECO_2/EA)	Reduction Rate (%)
D10	Wall	0.560	0.350	0.196	0.030	0.166	0.145	84.7
D13	Wall	0.995	0.450	0.448	0.042	0.406	0.354	90.6
D16	Wall	1.560	0.660	1.030	0.060	0.970	0.845	94.2
D19	Wall	2.250	0.730	1.643	0.109	1.534	1.337	93.4
D22	Column	3.040	1.450	4.408	0.160	4.248	3.704	96.4
D25	Column	3.980	1.650	6.567	0.260	6.307	5.500	96.0
D29	Column	5.040	2.150	10.836	0.390	10.446	9.109	96.4
D32	Column	6.230	2.370	14.765	0.528	14.237	12.415	96.4

Source: authors' research results.

However, when comparing the cost of rebar splice and mechanical couplers, couplers cost is higher up to D25, as shown in Table 12, but couplers are cheaper in rebars of above D29. For reference, splice cost is calculated by multiplying rebar cost per ton and splice weight, and rebar cost includes material, shop drawing work, cutting and bending, and installation cost. RC structures use various diameters of rebars. As shown in Table 11, mechanical couplers for all sizes of rebar are more advantageous for ECO_2 reduction than splice laps. However, as shown in Table 12, the cost of rebar couplers from D10 to D25 is more expensive than lap splice. However, the analysis results in Tables 11 and 12 may be different in each country because the types of mechanical couplers are diverse in shape and the rebar work cost is different. In Korea, the use of a mechanical coupler for rebars with D25 mm or larger in diameter has little cost reduction effect, but it has been confirmed that the ECO_2 reduction effect is remarkable. In the U.S., despite the fact that mechanical butt splices provide a variety of benefits, it is realized that the cost is still higher than lap splices [152]. If carbon tax is applied, the cost benefit is generated as much as the ECO_2 reduction in Table 11, and related research should be added.

Table 12. Cost comparison between splice lap and coupler.

Size	Rebar Cost (USD/Ton)	Splice Cost (USD/EA)	Coupler Cost (USD/EA)	Difference (USD/EA)
D10	930.36	0.18	4.46	−4.28
D13	921.43	0.41	4.91	−4.50
D16	934.82	0.96	5.36	−4.39
D19	934.82	1.54	5.80	−4.27
D22	934.82	4.12	6.25	−2.13
D25	934.82	6.14	6.70	−0.56
D29	934.82	10.13	7.14	2.99
D32	934.82	13.80	7.59	6.21

Exchange rate: 1120 Won/USD as of 25 February 2021, Bank of Korea. (Source: authors' research results).

4. Conclusions

During this review research, several facts have been identified in addition to findings identified from meta-analysis, as described in the Section 3. Various CWM algorithms have been developed and have contributed to reducing cutting wastes. However, it was confirmed that many algorithms have two principal problems to be applied in the field. First, although some algorithms can theoretically implement CWM, it is difficult to apply in practice when various site conditions are reflected. Second, some algorithms can reduce cutting wastes of some or major RC components, such as columns, beams, and slabs, but cannot reduce the cutting wastes of the entire structure of a project to near-zero. A description of identified findings after literature review is as follows:

1. Although cutting wastes can be reduced using SpLs of rebars rather than market lengths, many studies conducted optimization on market lengths to minimize cutting wastes of some rebars, but near-zero cutting waste of the entire construction project was difficult to realize. This phenomenon has been clearly identified on rebars with a diameter of above D19.

2. To achieve near-zero cutting waste by SpL, research should be conducted (a) by determining an SpL that meets the minimum order quantity conditions during RC structure design, or (b) by finding a SpL that meets the minimum order quantity conditions after structural design. In both cases, partial adjustment of LSPs requires a specific length of rebars, and it has been confirmed to be more efficient to apply it during RC structure design.

3. Considering the conditions such as different use schedules of combined rebar, combinations by special length, and minimum quantity for special order, CWM is not a one-dimensional problem but an n-dimensional CSP. Therefore, it is difficult to realize near-zero cutting waste with algorithms for existing one-dimensional cutting stock problems.

4. It should be dealt with from a sustainable value chain management perspective beyond supply chain management. In particular, if research has been developed to (1) optimization by cutting or combination pattern, (2) optimization of rebar information generated after structural design results are drawn, and (3) optimization of the amount of rebars in the structural design stage, in the future, structural design and construction-integrated management should be developed. Structural design should be used to combine special lengths rather than market lengths or stock lengths. This requires GA-based near-zero cutting waste algorithms for developing and integrating them into the RC design process.

Author Contributions: Conceptualization, K.K. and S.K.; methodology, S.K.; validation, K.K., D.K., and S.K.; formal analysis, K.K. and S.K.; investigation, K.K. and D.K.; resources, K.K. and D.K.; data curation, K.K. and D.K.; writing—original draft preparation, K.K. and S.K.; writing—review and editing, K.K. and S.K.; supervision, S.K.; project administration, S.K.; funding acquisition, S.K. All authors have read and agreed to the published version of the manuscript.

Funding: This work was supported by the National Research Foundation of Korea (NRF) grant funded by the Korea government (MOE) (No. 2017R1D1A1B04033761).

Institutional Review Board Statement: Not applicable.

Informed Consent Statement: Informed consent was obtained from all subjects involved in the study.

Data Availability Statement: Data sharing is not applicable to this article.

Conflicts of Interest: The authors declare no conflict of interest.

Abbreviations

AR	Augmented Reality
ASCE	American Society of Civil Engineers
BIM	Building Information Model
BSA	Binary Search Algorithm
C25/30	Compressive Strength of Concrete in N/mm2 when Tested with Cylinder/Cube
CAD	Computer-Aided Design
CAM	Computer-Aided Manufacture
CSP	Cutting Stock Problem
CWM	Cutting Waste Minimization
D10, D32	High-Tensile Deformed Bar in Diameter. D10: 10 mm, D32: 32 mm
EA	Each
ECO_2	Embodied Carbon Dioxide
GA	Genetic Algorithm
GDP	Gross Domestic Product
GHG	Greenhouse Gas
HA	Heuristic Algorithm
HS	Harmony Search
IFC	Industry Foundation Classes
IP	Integer Programming
JCEM	Journal of Construction Engineering and Management
JCR	Journal Citation Reports
LP	Linear Programming
LSP	Lap Splice Position
PRISMA	Preferred Reporting Items for Systematic Reviews and Meta-Analyses
QTO	Quantity Take-Off
RC	Reinforced Concrete
SA	Simulated Annealing
SCI/SCIE	Science Citation Index/Science Citation Index Expanded
SLR	Systematic Literature Review
SpL	Special Length
StL	Stock Length
VR	Virtual Reality
WoS	Web of Sciences

References

1. Miller, S.A.; Horvath, A.; Monteiro, P.J.M. Readily implementable techniques can cut annual CO2 emissions from the production of concrete by over 20%. *Environ. Res. Lett.* **2016**, *11*, 074029. Available online: https://iopscience.iop.org/article/10.1088/1748-9326/11/7/074029/meta (accessed on 8 December 2020).
2. Lee, I.J.; Yu, H.; Chan, S.L. Carbon Footprint of Steel-Composite and Reinforced Concrete Buildings, Standing Committee on Concrete Technology Annual Concrete Seminar 2016, Hong Kong, 2016, Construction Industry Council. Available online: https://www.devb.gov.hk/filemanager/en/content_971/7_Carbon_Footprint_for_Steel_Composite_and_Reinforced_Concrete_Buildings.pdf (accessed on 26 December 2020).
3. Lee, D.; Son, S.; Kim, D.; Kim, S. Special-Length-Priority Algorithm to Minimize Reinforcing Bar-Cutting Waste for Sustainable Construction. *Sustainability* **2020**, *12*, 5950. [CrossRef]
4. Kim, S.K.; Kim, M.H. A Study on the development of the optimization algorithm to minimize the loss of reinforcement bars. *J. Archit. Inst. Korea* **1991**, *7*, 385–390. Available online: http://www.dbpia.co.kr/journal/articleDetail?nodeId=NODE00358879 (accessed on 26 December 2020).
5. Kim, G.H. A Study on Program of Minimizing the Loss of Re-Bar. Master's Thesis, Korea University,, Seoul, Korea, 2002; pp. 8–58.
6. Porwal, A.; Hewage, K.N. Building information modeling-based analysis to minimize waste rate of structural reinforcement. *J. Constr. Eng. Manag.* **2012**, *138*, 943–954. [CrossRef]
7. Nadoushani, Z.; Hammad, A.; Akbar Nezhad, A. A Framework for Optimizing Lap Splice Positions within Concrete Elements to Minimize Cutting Waste of Steel Bars. In Proceedings of the 33th International Symposium on Automation and Robotics in Construction (ISARC 2016), Auburn, AL, USA, 18–21 July 2016. [CrossRef]
8. Shahin, A.A.; Salem, O.M. Using genetic algorithms in solving the one-dimensional cutting stock problem in the construction industry. *Can. J. Civ. Eng.* **2004**, *31*, 321–332. [CrossRef]
9. Zubaidy, S.S.; Dawood, S.Q.; Khalaf, I.D. Optimal utilization of rebar stock for cutting processes in housing project. *Int. J. Adv. Res. Sci. Eng. Technol.* **2016**, *3*, 189–193. [CrossRef]
10. Kim, S.K.; Hong, W.K.; Joo, J.K. Algorithms for reducing the waste rate of reinforcement bars. *J. Asian Arch. Build.* **2004**, *3*, 17–23. [CrossRef]
11. Benjaoran, V.; Bhokha, S. Trim loss minimization for construction reinforcement steel with oversupply constraints. *J. Adv. Manag. Sci.* **2013**, *1*, 313–316. Available online: http://www.joams.com/uploadfile/2013/1024/20131024100240137.pdf (accessed on 26 December 2020). [CrossRef]
12. Nadoushani, Z.S.M.; Hammad, A.W.A.; Xiao, J.; Akbarnezhad, A. Minimizing cutting wastes of reinforcing steel bars through optimizing lap splicing within reinforced concrete elements. *Constr. Build. Mater.* **2018**, *185*, 600–608. [CrossRef]
13. Macrotrends LLC. World GDP Growth Rate 1961–2020. Available online: https://www.macrotrends.net/countries/WLD/world/gdp-growth-rate (accessed on 27 December 2020).
14. Choi, J.; Lee, D.; Kwon, G.; Kim, S. A Study on energy consumption and CO2 emission of rebar. *KIEAE J. Constr. Ind. Counc.* **2010**, *10*, 101–109. Available online: http://www.auric.or.kr/User/Rdoc/DocRdoc.aspx?returnVal=RD_R&dn=242412#.X_UgKtgzaUk (accessed on 6 January 2021). (In Korean)
15. Trading Economics. GDP Annual Growth Rate—Forecast 2020–2022. Available online: https://tradingeconomics.com/forecast/gdp-annual-growth-rate (accessed on 9 January 2021).
16. Schneeweiss, Z. The Data Show a Bleak Outlook for Global Economic Growth—Developing Countries Show Signs of Emerging from the Pandemic More Quickly Than Advanced Economies. *Market Magazine, Bloomberg*. Available online: https://www.bloomberg.com/news/articles/2020-10-09/global-economic-data-paint-a-bleak-picture-for-the-future (accessed on 9 January 2021).
17. Kantorovich, L.V. Mathematical methods of organizing and planning production. *Manag. Sci.* **1960**, *6*, 366–422. [CrossRef]
18. Wawak, S.; Ljevo, Ž.; Vukomanović, M. Understanding the Key Quality Factors in Construction Projects—A Systematic Literature Review. *Sustainability* **2020**, *12*, 10376. [CrossRef]
19. Kitchenham, B.; Charters, S. *Guidelines for Performing Systematic Literature Reviews in Software Engineering*; Version 2.3; EBSE Technical Report EBSE-2007-01; School of Computer Science and Mathematics, Keele University: Keele, UK; University of Durham: Durham, UK, 2007.
20. Denyer, D.; Tranfield, D. Producing a Systematic Review. In *The Sage Handbook of Organizational Research Methods*; Sage Publications: Thousand Oaks, CA, USA, 2009; pp. 671–689.
21. Petersen, K.; Feldt, R.; Mujtaba, S.; Mattsson, M. Systematic mapping studies in software engineering. In Proceedings of the 12th International Conference on Evaluation and Assessment in Software Engineering (EASE), Bari, Italy, 26–27 June 2008; Volume 8, pp. 68–77.
22. Moher, D.; Liberati, A.; Tetzlaff, J.; Altman, D.G. Preferred Reporting Items for Systematic Reviews and Meta-Analyses: The PRISMA Statement, The PRISMA Group. *PLoS Med.* **2009**, *6*, e1000097. [CrossRef]
23. Carnwell, R.; Daly, W. Strategies for the Construction of a Critical Review of the Literature. *Nurse Educ. Pract.* **2001**, *1*, 57–63. [CrossRef]
24. Isilay, T.; Attila, D. Competitiveness in Construction—A literature Review of Research in Construction Management Journals. In Proceedings of the 24th Annual ARCOM of the Conference, Cardiff, UK, 1–3 September 2008; Association of Researchers in Construction Management: Cardiff, UK, 2009; pp. 799–812.

25. Ortiz, O.; Castells, F.; Sonnemann, G. Sustainability in the Construction Industry: A Review of Recent Developments Based on LCA. *Constr. Build. Mater.* **2009**, *23*, 28–39. [CrossRef]

26. Yang, H.; Yeung, J.F.; Chan, A.P.; Chiang, Y.H.; Chan, D.W. A Critical Review of Performance Measurement in Construction. *J. Facilities Manag.* **2010**, *8*, 269–284. [CrossRef]

27. De Muynck, W.; De Belie, N.; Verstraete, W. Microbial Carbonate Precipitation in Construction Materials: A review. *Ecol. Eng.* **2010**, *36*, 118–136. [CrossRef]

28. Bygballe, L.E.; Jahre, M.; Swärd, A. Partnering Relationships in Construction: A literature Review. *J. Purch. Supply Manag.* **2010**, *16*, 239–253. [CrossRef]

29. Zhou, W.; Whyte, J.; Sacks, R. Construction Safety and Digital Design: A Review. *Autom. Constr.* **2012**, *22*, 102–111. [CrossRef]

30. Zuo, J.; Zillante, G.; Wilson, L.; Davidson, K.; Pullen, S. Sustainability Policy of Construction Contractors: A Review. *Renew. Sustain. Energy Rev.* **2012**, *16*, 3910–3916. [CrossRef]

31. Evins, R. A Review of Computational Optimisation Methods Applied to Sustainable Building Design. *Renew. Sustain. Energy Rev.* **2013**, *22*, 230–245. [CrossRef]

32. Wang, Z.; Lim, B.T.; Kamardeen, I. Change Management Research in Construction: A Critical Review. In Proceedings of the 19th CIB World Building Congress, Brisbane, Australia, 5–9 May 2013; Queensland University of Technology: Brisbane, Australia, 2013; pp. 1–14.

33. Buyle, M.; Braet, J.; Audenaert, A. Life Cycle Assessment in the Construction Sector: A review. *Renew. Sustain. Energy Rev.* **2013**, *26*, 379–388. [CrossRef]

34. Taroun, A. Towards a Better Modelling and Assessment of Construction Risk: Insights from A literature Review. *Int. J. Proj. Manag.* **2014**, *32*, 101–115. [CrossRef]

35. Tay, Y.W.D.; Panda, B.; Paul, S.C.; Noor Mohamed, N.A.; Tan, M.J.; Leong, K.F. 3D Printing Trends in Building and Construction Industry: A Review. *Virtual Phys. Prototyp.* **2017**, *12*, 261–276. [CrossRef]

36. Hanafi, A.G.; Nawi, M.N.M. Critical Success Factors for Competitiveness of Construction Companies: A Critical Review. *AIP Conf. Proc.* **2016**, *1761*, 020042. [CrossRef]

37. Afzal, M.; Liu, Y.; Cheng, J.C.; Gan, V.J. Reinforced Concrete Structural Design Optimization: A Critical Review. *J. Clean. Prod.* **2020**, *260*, 120623. [CrossRef]

38. Wuni, I.Y.; Shen, G.Q. Barriers to the adoption of modular integrated construction: Systematic review and meta-analysis, integrated conceptual framework, and strategies. *J. Clean. Prod.* **2020**, *249*, 119347. [CrossRef]

39. Abdirad, H.; Dossick, C.S. BIM curriculum design in architecture, engineering, and construction education: A systematic review. *J. Inf. Technol. Constr.* **2016**, *21*, 250–271.

40. Abd Jamil, A.H.; Fathi, M.S. Contractual challenges for BIM-based construction projects: A systematic review. *Built Environ. Proj. Asset Manag.* **2018**, *8*, 372. [CrossRef]

41. Marcher, C.; Giusti, A.; Matt, D.T. Decision Support in Building Construction: A Systematic Review of Methods and Application Areas. *Buildings* **2020**, *10*, 170. [CrossRef]

42. Babalola, O.; Ibem, E.O.; Ezema, I.C. Implementation of lean practices in the construction industry: A systematic review. *Build. Environ.* **2019**, *148*, 34–43. [CrossRef]

43. Saieg, P.; Sotelino, E.D.; Nascimento, D.; Caiado, R.G.G. Interactions of building information modeling, lean and sustainability on the architectural, engineering and construction industry: A systematic review. *J. Clean. Prod.* **2018**, *174*, 788–806. [CrossRef]

44. Alencar, L.; Alencar, M.; Lima, L.; Trindade, E.; Silva, L. Sustainability in the Construction Industry: A Systematic Review of the Literature. *J. Clean. Prod.* **2020**, *29*, 125730. [CrossRef]

45. Schwartz, Y.; Raslan, R.; Mumovic, D. The life cycle carbon footprint of refurbished and new buildings—A systematic review of case studies. *Renew. Sustain. Energy Rev.* **2018**, *81*, 231–241. [CrossRef]

46. Charef, R.; Alaka, H.; Emmitt, S. Beyond the third dimension of BIM: A systematic review of literature and assessment of professional views. *Renew. Sustain. Energy Rev.* **2018**, *19*, 242–257. [CrossRef]

47. Martínez-Aires, M.D.; Lopez-Alonso, M.; Martinez-Rojas, M. Building information modeling and safety management: A systematic review. *Saf. Sci.* **2018**, *101*, 11–18. [CrossRef]

48. Gharbia, M.; Chang-Richards, A.; Lu, Y.; Zhong, R.Y.; Li, H. Robotic technologies for on-site building construction: A systematic review. *J. Build. Eng.* **2020**, *32*, 101584. [CrossRef]

49. Araújo, A.G.; Carneiro, A.M.P.; Palha, R.P. Sustainable construction management: A systematic review of the literature with meta-analysis. *J. Clean. Prod.* **2020**, *256*, 120350. [CrossRef]

50. Fiala, C.; Hájek, P. Environmentally based optimisation of RC slabs with lightening fillers. In Proceedings of the International Conference "Advanced Engineering Design (AED) 2006", Prague, Czech Republic, 11–14 June 2006; p. 6.

51. Borkowski, A. Optimization of slab reinforcement by linear programming. *Comput. Methods Appl. Mech. Eng.* **1977**, *12*, 1–17. [CrossRef]

52. Ahmadkhanlou, F.; Adeli, H. Optimum cost design of reinforced concrete slabs using neural dynamics model. *Eng. Appl. Artif. Intell.* **2005**, *18*, 65–72. [CrossRef]

53. Ahmed, M.; Datta, T. Minimum volume of curtailed reinforcement in rectangular slabs. *Civ. Eng. Syst.* **1984**, *1*, 151–159. [CrossRef]

54. Sahab, M.G.; Ashour, A.F.; Toporov, V.V. Cost optimisation of reinforced concrete flat slab buildings. *Eng. Struct.* **2005**, *27*, 313–322. [CrossRef]

55. Bavafa, M.; Kiviniemi, A. Optimised strategy by utilising BIM and set-based design: Reinforced concrete slabs. In Proceedings of the CIB W78 2012: 29th International Conference, Beirut, Lebanon, 17–19 October 2012; Corpus ID: 56437315. Available online: https://livrepository.liverpool.ac.uk/2007737/ (accessed on 12 February 2021).

56. Hájek, P. Integrated environmental design and optimization of concrete floor structures for buildings. In Proceedings of the 2005 World Sustainable Building Conference, Tokyo, Japan, 27–29 September 2005.

57. Eleftheriadis, S.; Duffour, P.; Stephenson, B.; Mumovic, D. Automated specification of steel reinforcement to support the optimisation of RC floors. *Autom. Construct.* **2018**, *96*, 366–377. [CrossRef]

58. Coello Coello, C.A.; Christiansen, A.D.; Hernandez, F.S. A simple genetic algorithm for the design of reinforced concrete beams. *Eng. Comput.* **1997**, *13*, 185–196. [CrossRef]

59. Leps, M.; Sejnoha, M. New approach to optimization of reinforced concrete beams. *Comput. Struct.* **2003**, *81*, 1957–1966. [CrossRef]

60. Govindaraj, V.; Ramasamy, J.V. Optimum detailed design of reinforced concrete continuous beams using genetic algorithms. *Comput. Struct.* **2005**, *84*, 34–48. [CrossRef]

61. Ferreira, C.; Barros, M.; Barros, A. Optimal design of reinforced concrete T-sections in bending. *Eng. Struct.* **2003**, *25*, 951–964. [CrossRef]

62. Bekdas, G.; Nigdeli, S.M. Bat algorithm for optimization of reinforced concrete columns. *Proc. Appl. Math. Mech.* **2016**, *16*, 681–682. [CrossRef]

63. Khondoker, M.T.H. Automated reinforcement trim waste optimization in RC frame structures using building information modeling and mixed-integer linear programming. *Autom. Constr.* **2021**, *124*, 103599. [CrossRef]

64. Bouassida, M.; Carter, J.P. Optimization of design of column-reinforced foundations. *Int. J. Geomech.* **2014**, *14*, 04014031. [CrossRef]

65. Rafiq, M.Y.; Southcombe, C. Genetic algorithms in optimal design and detailing of reinforced concrete columns supported by a declarative approach for capacity checking. *Comput. Struct.* **1998**, *69*, 443–457. [CrossRef]

66. Al-Mosawi, S.; Saka, M. Optimum design of single core shear walls. *Comput. Struct.* **1999**, *71*, 143–162. [CrossRef]

67. Camp, C.V.; Akin, A. Design of retaining walls using big bang-big crunch optimization. *J. Struct. Eng.* **2012**, *138*, 438–448. [CrossRef]

68. Yepes, V.; Alcalá, J.; Perea, C.; Gonzalez-Vidosa, F. A parametric study of optimum earth retaining walls by simulated annealing. *Eng. Struct.* **2008**, *30*, 821–830. [CrossRef]

69. Cohn, M.Z.; Lounis, Z. Optimal design of structural concrete bridge systems. *J. Struct. Eng.* **1994**, *120*, 2653–2674. [CrossRef]

70. Perea, C.; Alcala, J.; Yepes, V.; Gonzalez-Vidosa, F.; Hospitaler, A. Design of reinforced concrete bridge frames by heuristic optimization. *Adv. Eng. Softw.* **2008**, *39*, 676–688. [CrossRef]

71. Perea, C.; Baitsch, M.; Gonzalez-Vidosa, F.; Hartmann, D. Optimization of reinforced concrete frame bridges by parallel genetic and memetic algorithms. In Proceedings of the Third International Conference on Structural Engineering, Mechanics and Computation, Cape Town, South Africa, 10–12 September 2007; pp. 1790–1795.

72. Paya, I.; Yepes, V.; Gonzalez-Vidosa, F.; Hospitaler, A. Multiobjective optimization of concrete building frames by simulated annealing. *Comput. Aided Civ. Infrastruct. Eng.* **2008**, *23*, 596–610. [CrossRef]

73. Akin, A.; Saka, M.P. Harmony search algorithm based optimum detailed design of reinforced concrete plane frames subject to ACI 318-05 provisions. *Comput. Struct.* **2015**, *147*, 79–95. [CrossRef]

74. Amir, O. A topology optimization procedure for reinforced concrete structures. *Comput. Struct.* **2013**, *114*, 46–58. [CrossRef]

75. Balling, R.J.; Yao, X. Optimization of reinforced concrete frames. *J. Struct. Eng.* **1997**, *123*, 193–202. [CrossRef]

76. Bekdas, G.; Nigdeli, S.M. Modified harmony search for optimization of reinforced concrete frames. In *International Conference on Harmony Search Algorithm*; Springer: Singapore, 2017; pp. 213–221. [CrossRef]

77. Bond, D. The optimum design of concrete structures. *Eng. Optim.* **1974**, *1*, 17–28. [CrossRef]

78. Leite, J.P.B.; Topping, B.H.V. Improved genetic operators for structural optimization. *Adv. Eng. Softw.* **1998**, *29*, 529–562. [CrossRef]

79. Rajeev, S.; Krisnamoorthy, C.S. Genetic algorithm-based methodology for design optimization of reinforced concrete frames. *Comput. Aided Civ. Infrastruct. Eng.* **1998**, *13*, 63–74. [CrossRef]

80. Lee, C.; Ahn, J. Flexural design reinforced concrete frames by genetic algorithm. *J. Struct. Eng* **2003**, *129*, 762–774. [CrossRef]

81. Koumousis, V.K.; Arsenis, S.J. Genetic algorithms in optimal design of reinforced concrete members. *Comput. Aided Civ. Infrastruct. Eng.* **1998**, *13*, 43–52. [CrossRef]

82. PMI. *Construction Extension to the PMBOK_ Guide*, 3rd ed.; Project Management Institute, Inc.: Newtown Square, PA, USA, 2016.

83. Benjaoran, V.; Bhokha, S. Three-step solutions for cutting stock problem of construction steel bars. *KSCE J. Civ. Eng.* **2014**, *18*, 1239–1247. [CrossRef]

84. Bernold, L.E.; Salim, M. Placement-oriented design and delivery of concrete reinforcement. *J. Constr. Eng. Manag.* **1993**, *119*, 323–335. [CrossRef]

85. Chandrasekar, M.K.; Nigussie, T. Rebar Wastage in Building Construction Projects of Hawassa, Ethiopia. *Int. J. Sci. Eng. Res.* **2018**, *9*, 282–287. Available online: https://www.ijser.org/researchpaper/Rebar-wastage-in-building-construction-projects-of-Hawassa-Ethiopia.pdf (accessed on 30 December 2020).

86. Chen, Y.; Yang, T. Lapping pattern, stock length, and shop drawing of beam reinforcements of an RC building. *J. Comput. Civ. Eng.* **2013**, *29*, 04014028. [CrossRef]

87. Dunston, P.S.; Bernold, L.E. Adaptive control for safe and quality rebar fabrication. *J. Constr. Eng. Manag.* **2000**, *126*, 122–129. [CrossRef]

88. Ham, C.S.; Park, J.P.; Park, J.K.; Chung, J.Y.; Kim, I.H. A Study on the Development of Shop Drawings and Bar-bending Schedule Automation System for Apartment Housing. *J. Archit. Inst. Korea* **1999**, *15*, 111–119.
89. Jung, H.O.; Cho, H.H.; Park, U.Y. A Study on the Improvement of Erection Bar Detailing in Domestic Building Construction. *J. Korea Inst. Build. Constr.* **2009**, *9*, 39–46. [CrossRef]
90. Kim, D.; Lim, C.; Liu, Y.; Kim, S. Automatic Estimation System of Building Frames with Integrated Structural Design Information (AutoES). *Iran. J. Sci. Tech. Trans. Civ. Eng.* **2020**, *44*, 1145–1157. [CrossRef]
91. Kim, S.K.; Kim, C.K. Integrated automation of structural design and rebar work in RC structures. *J. Archit. Inst. Korea* **1994**, *10*, 113–121. Available online: http://www.dbpia.co.kr/journal/articleDetail?nodeId=NODE00359422 (accessed on 26 December 2020). (In Korean)
92. Mishra, S.P.; Parbat, D.K.; Modak, J.P. Field data-based mathematical simulation of manual rebar cutting. *J. Constr. Dev. Ctries.* **2014**, *19*, 111. Available online: http://web.usm.my/jcdc/vol19_1_2014/JCDC%2019 (accessed on 28 December 2020).
93. Nanagiri, Y.V.; Singh, R.K. Reduction of wastage of rebar by using BIM and linear programming. *Int. J. Tech.* **2015**, *5*, 329–334. [CrossRef]
94. Navon, R.; Rubinovitz, Y.; Coffler, M. Development of a fully automated rebar-manufacturing machine. *Autom. Constr.* **1995**, *4*, 239–253. [CrossRef]
95. Navon, R.; Rubinovitz, Y.; Coffler, M. Fully automated rebar CAD/CAM system: Economic evaluation and field implementation. *J. Constr. Eng. Manag.* **1996**, *122*, 101–108. [CrossRef]
96. Navon, R.; Rubinovitz, Y.A.; Coffler, M. Reinforcement-bar manufacture: From design to optimized production. *Int. J. Comput. Integr. Manuf.* **1998**, *11*, 326–333. [CrossRef]
97. Navon, R.; Shapira, A.; Shechori, Y. Automated rebar constructability diagnosis. *J. Constr. Eng. Manag.* **2000**, *126*, 389–397. [CrossRef]
98. Park, U.Y. BIM-Based Simulator for Rebar Placement. *J. Korea Inst. Build. Constr.* **2012**, *12*, 98–107. [CrossRef]
99. Polat, G.; Arditi, D.; Ballard, G.; Mungen, U. Economics of on-site vs. off-site fabrication of rebar. *Constr. Manag. Econ.* **2006**, *24*, 1185–1198. [CrossRef]
100. Poonkodi, N. Development of software for minimization of wastes in rebar in RCC structures by using linear programming. *Int. J. Adv. Res. Trends Eng. Technol.* **2016**, *3*, 1262–1267. Available online: https://www.researchgate.net/publication/316031463_Development_of_software_for_minimization_of_wastes_in_rebar_in_rcc_structures_by_using_linear_programming (accessed on 30 December 2020).
101. Salem, O.; Shahin, A.; Khalifa, Y. Minimizing cutting wastes of reinforcement steel bars using genetic algorithms and integer programming models. *J. Constr. Eng. Manag.* **2007**, *133*, 982–992. [CrossRef]
102. Salim, M.; Bernold, L.E. Effects of design-integrated process planning on productivity in rebar placement. *J. Constr. Eng. Manag.* **1994**, *120*, 720–738. [CrossRef]
103. Salim, M.; Bernold, L.E. Design-integrated process planner for rebar placement. *J. Eng.* **1995**, *9*, 157–167. [CrossRef]
104. Stainton, R.S. The Cutting Stock Problem for the Stockholder of Steel Reinforcement Bars. *Oper. Res. Q.* **1977**, *28*, 139–149. [CrossRef]
105. Zheng, C.; Yi, C.; Lu, M. Integrated optimization of rebar detailing design and installation planning for waste reduction and productivity improvement. *Autom. Constr.* **2019**, *101*, 32–47. [CrossRef]
106. Zheng, C.; Lu, M. Optimized reinforcement detailing design for sustainable construction: Slab case study. *Procedia Eng.* **2016**, *145*, 1478–1485. [CrossRef]
107. Kim, T.H.; Hong, C.G.; Kim, S. Quantity take-off algorithm for the reinforced concrete frame. *Korea J. Constr. Eng. Manag.* **2003**, *4*, 114–121. Available online: https://www.kicem.or.kr/html/sub08_01.jsp?tbnm=r&organCode2=kicem01&yearmonth=200303 (accessed on 30 December 2020).
108. Lee, S.; Kim, S.; Lee, G.; Kim, S.; Joo, J. Automatic Algorithms of Rebar Quantity Take-Off of Green Frame by Composite Precast Concrete Members. *Korea J. Constr. Eng. Manag.* **2012**, *13*, 118–128. (In Korean) [CrossRef]
109. Choi, H.; Lee, S.E.; Kim, C.K. Parametric Design Process for Structural Quantity Optimization of Spatial Building Structures. *J. Comput. Struct. Eng. Inst. Korea* **2017**, *30*, 103–110. [CrossRef]
110. Cho, Y.S.; Lee, S.I.; Bae, J.S. Reinforcement placement in a concrete slab object using structural building information modeling. *Comput. Aided Civ. Infrastruct. Eng.* **2014**, *29*, 47–59. [CrossRef]
111. Park, H.Y.; Lee, S.H.; Kang, T.K.; Lee, Y.S. Developing an Automatic System for Quantity Taking-off Cut and Bent Rebar and Making a Placing Drawing. In Proceeding of Conference in Architectural Institute of Korea, Cheongju, Korea, 26–27 October 2007; pp. 358–363. (In Korean)
112. Bendaj, B. Investigating the accuracy of fabricated rebar and rebar's placement in beams, and their impact in weight and cost of a building. In Proceedings of the 3rd International Balkans Conference on Challenges of Civil Engineering, 3-BCCCE, Tirana, Albania, 19–21 May 2016; Available online: https://core.ac.uk/reader/152490180 (accessed on 28 December 2020).
113. Chang, S.; Shiu, R.S.; Wu, I.C. Applying an A-Star Search Algorithm for Generating the Minimized Material Scheme for the Rebar Quantity Takeoff. In Proceedings of the 36th International Symposium on Automation and Robotics in Construction (ISARC 2019), Banff, AB, Canada, 21–24 May; 2019; 36, pp. 812–817. [CrossRef]

114. Hwang, J.W.; Park, C.J.; Wang, S.K.; Choi, C.H.; Lee, J.H.; Park, H.W. A Case Study on the Cost Reduction of the Rebar Work through the Bar Loss Minimization. In Proceedings of the KIBIM Annual Conference 2012, Seoul, Korea, 19 May 2012; Volume 2, pp. 67–68. (In Korean)

115. Hong, C.G.; Kim, T.H.; Kim, S.; Han, C.H. A study on the quantity estimation algorithm of reinforced concrete frame. In Proceedings of the 3rd Conference on Construction Engineering and Management, Korea Institute of Construction Engineering and Management, Seoul, Korea, 26 October 2002; p. 358. (In Korean). Available online: http://www.auric.or.kr/User/Rdoc/DocRdoc.aspx?returnVal=RD_R&dn=157040#.X-wJstgzaUk (accessed on 30 December 2020). (In Korean)

116. Matviyishyn, Y.; Janiak, T. Minimization of steel waste during manufacture of reinforced concrete. In *AIP Conference Proceedings*; AIP Publishing LLC: Melville, NY, USA, 2019; Volume 2077, p. 020040. [CrossRef]

117. Bekdas, G.; Nigdeli, S.M. The Optimization of Slender Reinforced Concrete Columns. *Proc. Appl. Math. Mech.* **2014**, *14*, 183–184. [CrossRef]

118. Zheng, C. Multi-Objective Optimization for Reinforcement Detailing Design and Work Planning on a Reinforced Concrete Slab Case. Master's Thesis, Alberta University, Edmonton, AB, Canada, 2018. [CrossRef]

119. Hasan, G.N.; Das, C.; Sumon, S.M.F.H. Splice Length of Reinforcing Bars Calculated in Different Design Codes. Bachelor's Thesis, Ahsanullah University of Science and Technology, Dhaka, Bangladesh, December 2015.

120. Mellenthin Filardo, M.; Walther, T.; Maddineni, S.; Bargstädt, H.J. Installing Reinforcement Rebars Using Virtual Reality and 4D Visualization. In *Proceedings of the 18th International Conference on Computing in Civil and Building Engineering*; Springer: Berlin/Heidelberg, Germany, 2021; pp. 1200–1216. [CrossRef]

121. Lee, D.G. Establishment direction of seismic structure design standards. *Rev. Archit. Build. Sci.* **1988**, *32*, 26–29. Available online: http://www.auric.or.kr/dordocs/cart_rdoc2_.asp?db=CMAG&dn=59623 (accessed on 26 February 2021). (In Korean)

122. Jo, S.J.; Kim, T.H.; Haan, C.H. Investigation of the Standards for Dispute Resolution of the Floor Noises of Apartments. In Proceedings of the Architectural Institute of Korea-RA, Jeju, Korea, 2–3 December 2010; pp. 488–493. Available online: http://www.auric.or.kr/dordocs/cart_rdoc2_.asp?db=RD_R&dn=245401 (accessed on 26 February 2021). (In Korean)

123. Kochummen, J.; Colcombe, J. Revit and RebarCAD 3D: The BIM Solution for Rebar Modeling to Production Drawing. *Autodesk University*. Available online: https://www.autodesk.com/autodesk-university/class/Revit-and-RebarCAD-3D-BIM-Solution-Rebar-Modeling-Production-Drawing-2018#downloads (accessed on 13 January 2021).

124. Tekla. Making Rebar Detailing an Ingredient of your Construction Project Overall Success. Available online: https://www.tekla.com/resources/blogs/making-rebar-detailing-an-ingredient-of-your-construction-project-overall-success (accessed on 13 January 2021).

125. Arbib, C.; Marinelli, F.; Ventura, P. One-dimensional cutting stock with a limited number of open stacks: Bounds and solutions from a new integer linear programming model. *Int. Trans. Oper. Res.* **2016**, *23*, 47–63. [CrossRef]

126. Ben Amor, H.; Valério de Carvalho, J. Cutting stock problems. In *Column Generation*; Springier: Boston, MA, USA, 2005; pp. 131–161. [CrossRef]

127. Belov, G.; Scheithauer, G. Setup and open-stacks minimization in one-dimensional stock cutting. *Inf. J. Comput.* **2007**, *19*, 27–35. [CrossRef]

128. Berberler, M.; Nuriyev, U.; Yildırım, A. A Software for the one-dimensional cutting stock problem. *J. King Saud Univ. Sci.* **2011**, *23*, 69–76. [CrossRef]

129. Chen, C.; Hart, S.; Tham, W. A simulated annealing heuristic for the one-dimensional cutting stock problem. *Eur. J. Oper. Res.* **1996**, *93*, 522–535. [CrossRef]

130. Cherri, A.C.; Arenales, M.N.; Yanasse, H.H. The one-dimensional cutting stock problem with usable leftover—A heuristic approach. *Eur. J. Oper. Res.* **2009**, *196*, 897–908. [CrossRef]

131. Cui, Y.; Zhao, X.; Yang, Y.; Yu, P. A heuristic for the one-dimensional cutting stock problem with pattern reduction. *Proc. Inst. Mech. Eng. Part. B J. Eng. Manuf.* **2008**, *222*, 677–685. [CrossRef]

132. Dyckhoff, H. A new linear programming approach to the cutting stock problem. *Oper. Res.* **1981**, *29*, 1092–1104. [CrossRef]

133. Gradišar, M.; Kljajić, M.; Resinovič, G.; Jesenko, J. A sequential heuristic procedure for one-dimensional cutting. *Eur. J. Oper. Res.* **1999**, *114*, 557–568. [CrossRef]

134. Feifei, G.; Lin, L.; Jun, P.; Xiazi, Z. Study of One-Dimensional Cutting Stock Problem with Dual-Objective Optimization. In Proceedings of the International Conference on Computer Science and Information Processing (CSIP), Xi'an, China, 24–26 August 2012. [CrossRef]

135. Fernandez, L.; Fernandez, L.A.; Pola, C. Integer Solutions to Cutting Stock Problems. In Proceedings of the 2nd International Conference on Engineering Optimization, Lisbon, Portugal, 6–9 September 2010; Available online: https://scholar.google.com/scholar?q=related:Mk60IvrPB0kJ:scholar.google.com/&scioq=&hl=ko&as_sdt=0,5 (accessed on 30 December 2020).

136. Gilmore, P.C.; Gomory, R.E. A linear programming approach to the cutting-stock problem. *Oper. Res.* **1961**, *9*, 849–859. [CrossRef]

137. Gilmore, P.C.; Gomory, R.E. A linear programming approach to the cutting-stock problem-part II. *Oper. Res.* **1963**, *11*, 863–888. [CrossRef]

138. Goulimis, C. Optimal solutions for the cutting stock problem. *Eur. J. Oper. Res.* **1990**, *44*, 197–208. [CrossRef]

139. Haessler, R.W.; Sweeney, P.E. Cutting stock problems and solution procedures. *Eur. J. Oper. Res.* **1991**, *54*, 141–150. [CrossRef]

140. Haessler, R.W. Controlling cutting pattern changes in one-dimensional trim problems. *Oper. Res.* **1975**, *23*, 483–493. [CrossRef]

141. Jahromi, M.H.; Tavakkoli, R.; Makui, A.; Shamsi, A. Solving an one-dimensional cutting stock problem by simulated annealing and tabu search. *J. Ind. Eng. Int.* **2012**, *8*, 24. [CrossRef]
142. Khalifa, Y.; Salem, O.; Shahin, A. Cutting Stock Waste Reduction using Genetic Algorithms. In Proceedings of the 8th Annual Conference on Genetic and Evolutionary Computation, Seattle, WA, USA, 8–12 July 2006; pp. 1675–1680. [CrossRef]
143. Lin, P. Optimal Solution of One Dimension Cutting Stock Problem. Master's Thesis, Lehigh University, Bethlehem, PA, USA, 1994. Available online: https://preserve.lehigh.edu/cgi/viewcontent.cgi?referer=1&article=1286&context=etd (accessed on 30 December 2020).
144. Poldi, K.C.; Arenales, M.N. Heuristics for the one-dimensional cutting stock problem with limited multiple stock lengths. *Comput. Oper. Res.* **2009**, *36*, 2074–2081. [CrossRef]
145. Roodman, G.M. Near-optimal solutions to one-dimensional cutting stock problems. *Comput. Oper. Res.* **1986**, *13*, 713–719. [CrossRef]
146. Suliman, S. Pattern generating procedure for the cutting stock problem. *Int. J. Prod. Econ.* **2001**, *74*, 293–301. [CrossRef]
147. Vahrenkamp, R. Random search in the one-dimensional cutting stock problem. *Eur. J. Oper. Res.* **1996**, *95*, 191–200. [CrossRef]
148. Reinertsen, H.; Vossen, T.W.M. The one-dimensional cutting stock problem with due dates. *Eur. J. Oper. Res.* **2010**, *201*, 701–711. [CrossRef]
149. Askarzadeh, A.; Rashedi, E. Harmony Search Algorithm: Basic Concepts and Engineering Applications. In *Intelligent Systems: Concepts, Methodologies, Tools, and Applications*; Khosrow-Pour, M., Ed.; IGI Global: Hershey, PA, USA, 2018; pp. 1–30. [CrossRef]
150. Tian, Z.; Chao Zhang, C. An Improved Harmony Search Algorithm and Its Application in Function Optimization. *J. Inf. Process. Syst.* **2018**, *14*, 1237–1253. [CrossRef]
151. BFT International. New Generation of Rotor Straightening Machines with Hyperbolic Profiled Rollers Processing Diameters up to 36 mm Rebar and 50 mm Wire. Available online: https://www.bft-international.com/en/artikel/bft_New_generation_of_rotor_straightening_machines_with_hyperbolic_profiled_2551059.html (accessed on 27 December 2020).
152. Hurd, M.K. Mechanical vs. Lap Splicing. *Concrete Construction Magazine*. Available online: https://www.concreteconstruction.net/how-to/construction/mechanical-vs-lap-splicing_o (accessed on 14 January 2021).

Technical Note

Inter-Floor Noise Monitoring System for Multi-Dwelling Houses Using Smartphones

Suhyun Kang [1], Seungho Kim [2], Dongeun Lee [3] and Sangyong Kim [1,*]

[1] School of Architecture, Yeungnam University, 280 Daehak-ro, Gyeongsan-si, Gyeongbuk 38541, Korea; yp043422@ynu.ac.kr

[2] Department of Architecture, Yeungnam University College, 170 Hyeonchung-ro, Nam-gu, Daegu 42415, Korea; kimseungho@ync.ac.kr

[3] School of Architecture & Civil and Architectural Engineering, Kyungpook National University, 80 Daehak-ro, Buk-gu, Daegu 41566, Korea; dolee@knu.ac.kr

* Correspondence: sangyong@yu.ac.kr; Tel.: +82-53-810-2425

Received: 12 May 2020; Accepted: 19 June 2020; Published: 22 June 2020

Abstract: The noise between the floors in apartment buildings is becoming a social problem, and the number of disputes related to it are increasing every year. However, laypersons will find it difficult to use the sound level meters because they are expensive, delicate, bulky, etc. Therefore, this study proposes a system to monitor the noise between the floors, that will measure the sound and estimate the location of the noise using the sensors and applications in smartphones. To evaluate how this system can be used effectively within an apartment building, a case study has been performed to verify its validity. The result shows that the mean absolute error (MAE) between the actual noise generating position and the estimated noise source location was measured at 2.8 m, with a minimum error of 1.2 m and a maximum error of 4.3 m. This means that smartphones, in the future, can be used as low-cost monitoring and evaluation devices to measure the noise between the floors in apartment buildings.

Keywords: inter-floor noise; multi-dwelling houses; smartphone application; real-time monitoring system

---

1. Introduction

1.1. Background

Population concentration due to urbanization has led to housing shortages, and many cities opted for the construction of multi-dwelling houses, which can be supplied in large quantities at a relatively low cost, as a solution [1]. In multi-dwelling houses, however, the residents are easily exposed to the noises of neighbors, as the walls and slabs are shared with other households. The continuous exposure to external noises of the residents of multi-dwelling houses may cause physical and mental health problems, such as high blood pressure, annoyance, and sleep disorders [2–4]. As such, inter-floor noise has also caused discord amongst neighbors, including an elevated number of disputes, assaults, and even arson [5–7].

To address disputes related to inter-floor noise, it is essential to secure objective noise data. Sound level meters are generally used to obtain objective noise data. It is difficult, however, for non-experts to use sound level meters, because they are expensive, delicate, and bulky [8]. The recent technical development of smartphones has opened up a possibility where they can be used as substitutes for sound level meters [9–11].

Smartphones are powerful mini-computers with various sensors (e.g., microphones, accelerometers, gyroscopes, and GPS) and are owned by the majority of the population. They can be used as low-cost noise monitoring tools with available broadband internet access [12].

A number of studies have been conducted lately to examine the accuracy of smartphone noise measurement applications (apps). Murphy and King [11] tested the accuracy of several noise measurement apps on two platforms (Android and iOS) using 100 smartphones. The test results showed that one of the apps was very accurate in measuring the noise levels with errors less than ±1 dB from the actual sound levels in the reference value range. The conducted study indicated that noise measurement apps have a potential to be used as sound level meters in the future. Zamora et al. [13] proposed environmental noise-sensing units using smartphones. According to these experimental results, if the smartphone application is well tuned, it is possible to measure noise levels with an accuracy degree comparable to professional devices for the entire dynamic range typically supported by microphones embedded in smartphones. Garg et al. [8] proposed an averaging method for accurately calibrating the noise acquired through a smartphone microphone. This method achieves an accuracy of 0.7 dB.

Smartphones also provide an inexpensive and flexible infrastructure for the measurement of overall environmental noise (e.g., noise and air pollution) in cities. Various related studies have shown that smartphone apps are useful for environmental monitoring evaluation [14–17]. Although the aforementioned studies verified the accuracy of smartphone noise measurement apps and their potential as environmental monitoring tools, studies on the possibility of using smartphones to address the inter-floor noise problem are not sufficient.

The problem to be solved in relation to inter-floor noise is to identify the noise types and locations of those noise sources [18]. This is important, since some disputes have resulted from misunderstanding of the noise sources by listeners [18]. Most studies on inter-floor noises, however, are focused on noise measurement [3,19], noise reduction measures [20,21], and annoyance measurement [22,23]. If smartphones can identify objectively and reliably the noise source locations and noise types in real time, they can contribute to dispute mediation.

1.2. Motivation and Objective

Inter-floor noise is transmitted to neighboring households in multi-dwelling houses, and unpleasant sounds disturb other house residents. In South Korea, where most people live in multi-dwelling houses, 88% of the apartment residents are under stress due to inter-floor noise [24]. In South Korea, most apartments have been constructed in the wall column structure style since the 1980s, due to reasons of constructability, economic efficiency, and a reduction in the construction period. In apartments with the wall column structure, all four apartment sides are made of concrete, with a large vibration transfer coefficient. Thus, the airborne sound that is generated on the upper floor and the vibration that is generated at the bottom of the upper floor are easily transferred to the lower floor [25].

In particular, the wall column structure apartments built before 2005 in Korea generally used a concrete slab thickness ranging from 135 mm to 150 mm, but in recent years, with the emergence of frequent inter-floor noise problems, a new regulation was established to standardize the slab thickness to be at least 210 mm [3]. Despite the legal regulations on the slab thickness, the number of complaints related to inter-floor noise has increased from 8795 in 2012 to 28,231 in 2018 (Figure 1).

This phenomenon appears to have occurred because there was no solution for noise mitigation for the existing apartments built before 2005, when the regulations on the slab thickness were enacted. The regulations can be applied only to the newly built apartments because improved construction methods, such as reinforced thicknesses of the walls and floor slabs and application of floating floors, have not been made available for the existing apartments. However, there has been an increase in the number of complaints related to inter-floor noise in new apartments built under new regulations. The study conducted by Park, Lee and Lee [3] verified that the slab thickness did not have any effect in lowering the indoor noise level.

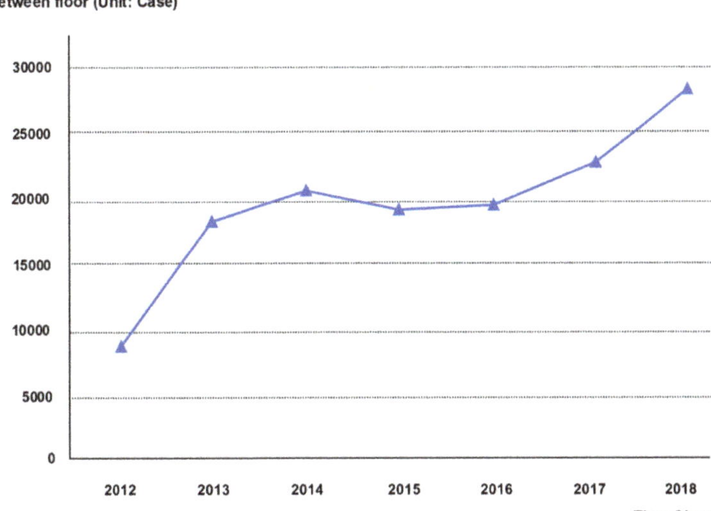

Figure 1. Trend of inter-floor noise complaints (Korea Environment Corporation).

The increase in the inter-floor noise complaints has led to conflicts and disputes among neighbors [26]. Emotional reactions to noise problems even led to a number of retaliatory crimes between neighbors, such as arson and murder [27]. As the conflicts caused by inter-floor noise expanded to a social problem, the South Korean government established a 'center for inter-floor noise mitigation between neighbors' in 2012, to oversee the disputes related to inter-floor noise. The center, however, has no legal rights and on-site investigation for objective noise measurement and shows some limitations in solving the inter-floor noise problem, due to a lack of manpower. The inter-floor noise problem is still unsolved, and thus, more effective measures are required to resolve the occurring disputes.

As noise is judged from a subjective perspective due to its environmental nature, conflicts due to a difference in opinions cannot be avoided. To resolve such conflicts, it is necessary to prove the fact that a noise level higher than the inter-floor noise criterion occurred, state its duration, and the degree of damage caused. Therefore, this study proposes an inter-floor noise monitoring system for measuring the inter-floor noise and estimating the noise time and location, by utilizing sensors and mobile applications of widely available smartphones. The proposed system enables recording various data related to inter-floor noise, and it is expected to be used as an important tool for resolving disputes related to inter-floor noise in the future.

2. Research Method

In this study, a system to monitor inter-floor noise using smartphones is proposed. To verify the validity of the system, apartment B, completed in 1996 and located in Gyeongsan City, Gyeongsangbuk-do, South Korea, was selected as a case study site. For inter-floor noise monitoring, an inter-floor noise monitoring application was developed using sensors built into smartphones. To this end, the functions of such sensors were identified and used to achieve the target functions for the inter-floor noise monitoring system.

Table 1 shows the smartphone sensors and their functions, that were used in this study in order to implement the developed application. The microphone was used to obtain the sound pressure level (SPL). The accelerometer and gyroscope were used to measure the vibration acceleration level (VAL) created by a heavy impact on part of a building. Moreover, GPS was used to locate the smartphone and to measure the timing of the occurring noise. Wi-Fi was used to transfer the obtained inter-floor noise information to a server.

Table 1. Smartphone sensor features and their utilization for the application.

Sensor Type	Description	Application
Microphone	Detects sound signals and converts them into an electrical signal	Sound detecting and record/Sound Pressure Level (SPL) measurement
Accelerometer	Measures the acceleration force in m/s^2 on all three physical axes (x, y, z)	Distinguish between air-borne sound and floor impact sound/Vibration Acceleration Level (VAL) measurement
Gyroscope	Measures a device's rate of rotation in rad/s on all three physical axes (x, y, z)	
GPS	Positioning and provides time information	Identify the location of the noise measuring device and the time of noise occurrence
Wi-Fi	Wireless networking	Noise Data Transmission

The developed inter-floor noise monitoring application requires a certain level of sound as a baseline for determining inter-floor noise. In this study, the legal criteria existing for the case study site (i.e., for South Korea) were applied. Inter-floor noise is largely divided into floor impact noise (e.g., running and walking sounds), which is generated when the energy is applied directly to the floor, and airborne sound (e.g., conversation and musical instrument sounds). Therefore, when a floor impact occurs, inter-floor noise must be determined by measuring the SPL of the lower floor and the vibration acceleration level generated by construction components (e.g., ceilings, walls, and windows). Table 2 shows the criteria for each type of inter-floor noise, as specified by the Ministry of Environment and the Ministry of Land, Infrastructure and Transport of South Korea.

Table 2. Criteria of noises between floors (Korea Ministry of Government Legislation).

Classification		Standard Value (Unit: dB)	
		Day Period	Night Period
Floor impact sound	A minute equivalent sound level (LAeq 1 min)	43	38
	The highest sound level (LAmax)	57	52
Air-borne sound	A five minute equivalent sound level (LAeq 5 min)	45	40

In the case of floor impact noises, inter-floor noise is determined when 'LAeq 1 min' exceeds 43 dB in the daytime and 38 dB at night, or when 'LAmax' exceeds 57 dB in the daytime and 52 dB at night. LAeq 1 min corresponds to the average value of noise measured for one minute, using a sound level meter. LAmax denotes noise with the highest dB value among the noises generated during the measurement period. In the case of airborne sounds, inter-floor noise is determined when 'LAeq 5 min' exceeds 45 dB in the daytime and 40 dB at night. The length of airborne noise detection was extended to five minutes, to reflect the long-lasting characteristics of television noise or musical instrument sounds. Therefore, in this study, inter-floor noise was determined by applying the above-mentioned criteria to the smartphone application.

3. Construction of the Monitoring System for Measurement of Inter-Floor Noise and Estimation of Noise Source Location

3.1. System Design

Figure 2 shows the configuration of the proposed monitoring system for the measurement of inter-floor noise levels and the estimation of noise source locations. In general, the system contains four steps. In the first step (the inter-floor noise sensing step), noise and vibration data are obtained from the place where data acquisition is required. Data is collected using the microphone, gyroscope, and accelerometer embedded in a smartphone. The decibel value and vibration velocity (i.e., noise data) are acquired every second, and the surrounding noise is recorded every minute. The acquired noise and vibration data are then transferred to a web server through Wi-Fi wireless communication in the second step (the inter-floor noise data transfer step). In this case, the transferred data consist of the ID and location of the measuring device, noise acquisition time, decibel level (dB) values, and vibration velocity (m/s^2). The web server stores the transferred data in a database in real time.

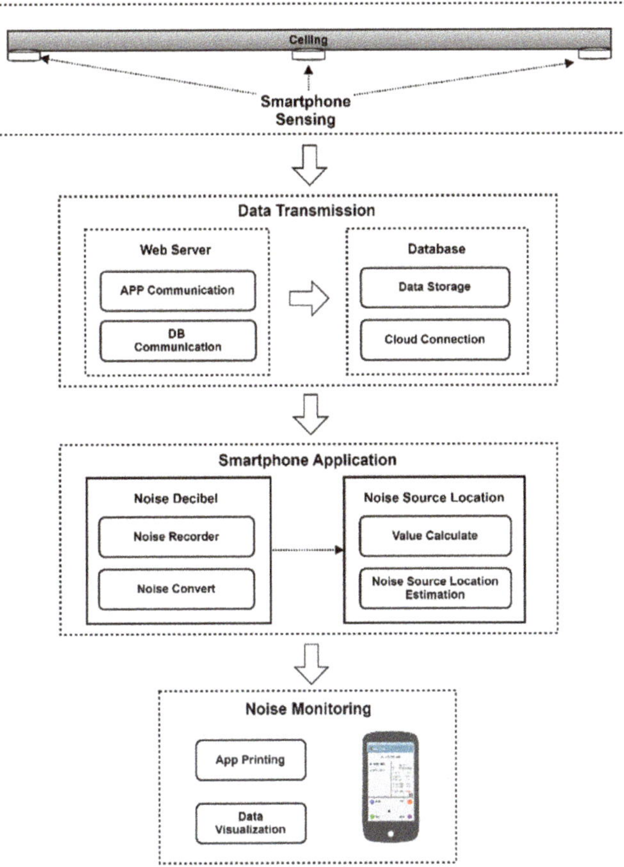

Figure 2. System Architecture.

Figure 3 shows a schema of tables that are stored in a database. The database consists of a number of tables, such as the NoiseHistory, DeviceList, and RecordList. Each table contains noise data, information on noise measuring devices, and recorded files. In the NoiseHistory table, the ID of the device that transferred the data, acquisition time, decibel values, and vibration velocity are stored. When the decibel value is higher than the threshold, "1" is recorded in the noise field. In this instance, noise is determined using the criteria displayed in Table 1. Information on the ID and location of each device is stored in the DeviceList table. Information on the files recorded by each device is stored in the RecordList table. In the third step, the developed application estimates the location of the noise source, based on the records stored in the database. The application stores the noise data values in real time, converts them into decibel values, and determines the noise location using the estimation algorithm. In the final step, the acquired inter-floor noise information is visualized on the user's smartphone screen.

NoiseHistory, Major key : ID + date

Field name	Data Type	Explanation
ID	int	Measurement device ID
date	datetime	Data acquisition time (yyyyMMddHHmmss)
dB	float	Decibel
Speed	float	Vibration acceleration(m/s^2)
noise	int	In regards of noise (Y=1, N=0)

DeviceList, Major key : ID

Field name	Data Type	Explanation
ID	int	Measurement device ID
loc_x	int	x-axes with device installed
loc_y	int	y-axes with device installed

RecordList, Major key : ID

Field name	Data Type	Explanation
ID	int	Measurement device ID
date	datetime	Recording time(yyyyMMddHHmm)
path	varchar	File storage path

Figure 3. Database Schema.

Figure 4 shows the application execution screen. The information that can be found in the application includes the timing of occurring noise, the noise measurements at that time, the estimated noise location and the noise type. The location at which the noise occurred is displayed on the floor plan of the measurement site and is located at the bottom of the application. The noise type (e.g., floor impact or airborne noise) can be determined using the recorded vibration values. It is determined as floor impact noise if there is vibration information when the noise occurred, or as airborne noise if there is no vibration information available.

3.2. Noise Source Location Estimation Method Used in This Study

Previous studies on sound source location estimation have been conducted using specialized equipment, such as microphone arrays. Those studies were also arranged for limited experimental environments [28,29]. The proposed system, however, uses only smartphones, thereby providing a method for many people to easily estimate noise source locations. In this study, an attempt was made to estimate noise source locations using differences in the sound intensity. For this method, hardware configuration and operation are very simple, even though it is difficult to calculate the exact distance to the sound source. The purpose of this study is not in finding the exact location of noise, but rather in estimating the approximate noise source occurrence area.

Due to the nature of sound, a lower decibel value is measured as the distance increases. Based on this phenomenon, a method of estimating noise sources using the proportions of the decibel values measured through four smartphones is described. As shown in Figure 5a, it is assumed that noise measurement devices ($T = \{T_1, T_2, T_3, T_4\}$) are placed in the form of a grid in two-dimensional coordinates. Each device has a decibel value (dB) and coordinate information (x, y). In this study, among the noise measuring devices (T), three devices (S_1, S_2, S_3) are arbitrarily selected according to the decibel level to locate the noise source. As shown by Equation (1), among the devices (T), the device with the largest decibel value (dB) is designated as S_1.

$$S_1 = \text{Max.db}(T) \tag{1}$$

For example, when a noise or vibration takes place, assuming that the highest decibel value was observed in T_1 among the devices (T), the T_1 device is set as S_1. Subsequently, as shown by Equation (2), the device (T) located on the horizontal line of S_1 is selected as S_2.

$$S_2 = \begin{cases} if\ S_1{\cdot}y = T_2{\cdot}y\ then\ T_2 \\ else\ T_3 \end{cases} \tag{2}$$

Here, S_2 is a device which has the same y-coordinate value as, but a different x-coordinate value to, S_1. Lastly, as expressed by Equation (3), the device having the largest decibel value among the devices other than the devices designated as S_1 and S_2 is selected as S_3.

$$S_3 = \{t{\cdot}y = S_1{\cdot}y \wedge t{\cdot}x \neq S_1{\cdot}x \mid t \in T\} \tag{3}$$

When it is assumed that $T_1{\cdot}db = 80$, $T_2{\cdot}db = 40$, and $T_3{\cdot}db = 60$, the placement of S_1, S_2, and S_3 can be expressed as shown in Figure 5b. In this case, the approximate values of X and Y that serve as the estimated location coordinates of the noise source are obtained using Equations (4) and (5).

$$X = \begin{cases} if\ S_1{\cdot}x > S_2{\cdot}x\ then\ \frac{S_1{\cdot}db}{S_1{\cdot}db+S_2{\cdot}db}{\cdot}width \\ else\ \frac{S_2{\cdot}db}{S_1{\cdot}db+S_2{\cdot}db}{\cdot}width \end{cases} \tag{4}$$

$$Y = \begin{cases} if\ S_1{\cdot}y > S_3{\cdot}y\ then\ \frac{S_1{\cdot}db}{S_1{\cdot}db+S_3{\cdot}db}{\cdot}height \\ else\ \frac{S_3{\cdot}db}{S_1{\cdot}db+S_3{\cdot}db}{\cdot}height \end{cases} \tag{5}$$

Width means the distance between S_1 and S_2, and height is calculated as the distance between S_1 and S_3. Figure 5 shows the estimated noise source locations using Equations (4) and (5).

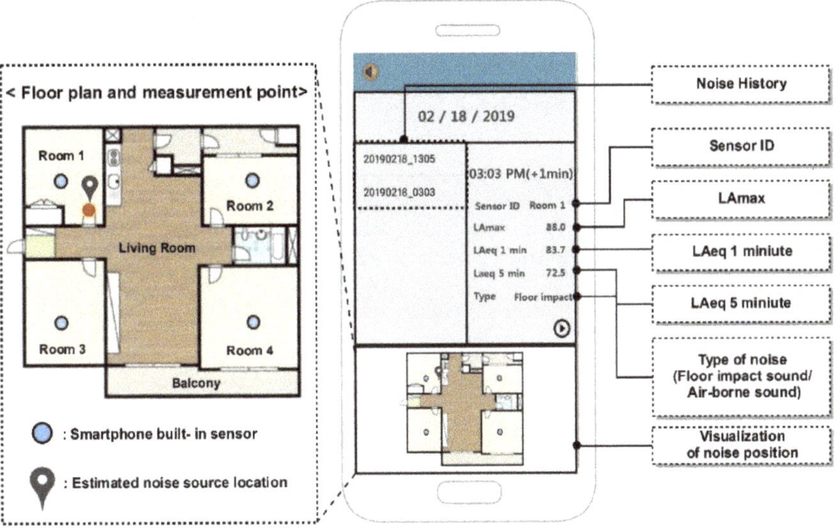

Figure 4. Application execution screen.

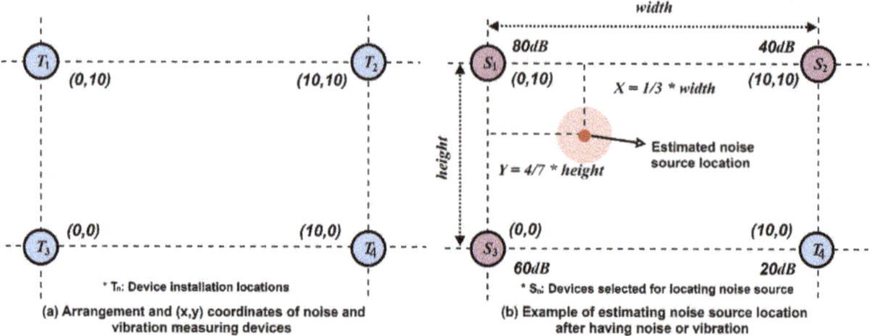

Figure 5. Methods for estimating the noise source location.

4. Experiment and Performance Evaluation

4.1. Experiment Overview

In this study, inter-floor noise data were acquired using four smartphones to estimate the noise source locations, and one smartphone was used to display the inter-floor noise data in real time for the user. Thus, a total of five smartphones were used in the experiment.

4.1.1. Software/Hardware Configuration

Table 3 shows the software components used in the experiment. In this study, JSP programming language was used based on Apache Tomcat (a web application server—WAS) in a Windows 10 Pro operating system for system development. Moreover, the database was managed by linking Apache Tomcat with MySQL. Android 5.0 APIs was used as an operating system to control smartphones.

Table 3. Software component.

Component	Explanation
Computer OS	Windows 10 pro
Web Application Server (WAS)	Apache Tomcat
Programming Language	JavaServer Pages(JSP)
Database management system (DBMS)	MySQL
Mobile OS	Android 5.0 APIs

Table 4 shows the hardware components used in the experiment. As the noise source locations were estimated using the differences in the sound intensity acquired from four measuring devices, only one smartphone model was used for the same conditions. Hardware was easily obtained, and devices with the sensors required for system implementation were selected.

Table 4. Hardware component.

Classification	Component	Specification
Built	Dimensions, Weight	146.8 × 75.3 × 8.9 mm, 163 g
	Display Size	13.3 cm (5.25 inches)
Platform	OS	Android
	AP	4 Core, 1.2 Ghz
	CPU	Quad-core 1.2 GHz Cortex-A7
	GPU	Adreno 305/400 MHz
	RAM	1.5 GB/LPDDR2 SDRAM
Communications	Network	4 G LTE
	WIFI	802.11 b/g/n/ac, dual band
	GPS	○
Sensor	Microphone	○
	Accelerometer	○
	Gyroscope	○

4.1.2. Experimental Environment and Method

To evaluate the performance and applicability of the proposed system to measure inter-floor noise and track the noise source locations, the experiment was performed in an apartment that serves as a representative for the residential type of multi-dwelling houses. Table 5 shows the overview of the experiment site.

Table 5. Profile of the experiment place.

Type (P'yong)	45 (148 m^2)	
	Lenght	11,500
Dimension [mm]	Width	12,100
	Ceiling height	2200
	Room 1	10.51
	Room 2	13.06
Area of measurement room [m^2]	Room 3	13.85
	Room 4	5.62
	Livingroom	28.36
Thickness of slab [mm]	180	
Measurement point	4 point	

The floor of the experiment site consisted of a reinforced concrete slab (180 mm), insulating materials (20 mm), lightweight concrete (40 mm), cement mortar (40 mm), and floor finishing materials (Figure 6). To collect noise and vibration data, smartphones were installed on the ceiling of each room (Figure 7). The exact installation locations can be found on the floor plan (Figure 4). The smartphone located at the bottom left corner was then designated as the origin, and the scales were marked at 24.2 cm intervals in the horizontal direction and at 23 cm intervals in the vertical direction.

As for the noise generation type, real impact sources (e.g., human footsteps and dropped objects) were used rather than standard impact sources (i.e., impact balls), to create an environment similar to real inter-floor noise in the experiment. At certain points over the ceiling, random noises were generated for over 20 s at a time (i.e., impacts of >70 dB, human voices, musical instrument sounds).

The experiment was repeated 100 times, whilst the noise occurrence locations were randomly changed, and the actual noise occurrence locations were then compared to the estimated locations displayed in the application.

4.2. Experimental Evaluation Method and Results

To evaluate the performance of the system, the errors between the actual noise occurrence locations and the estimated noise source locations were obtained using the mean absolute error (*MAE*). *MAE* was calculated using Equation (6).

$$\text{MAE} = \frac{1}{n} \sum_{i=1}^{n} distance(rpoint_i, ePoint_i) \tag{6}$$

where *rpoint$_i$* is the epicenter of the *i*-th actual noise and *ePoint$_i$* is the estimated location of the *i*-th noise. Figure 8 shows the *distance* function to obtain the absolute error between the actual noise epicenter and the estimated location.

Table 6 shows the experiment results. The calculated mean absolute error (MAE) was 2.8 m, while the minimum and maximum errors were 1.2 and 4.3 m, respectively.

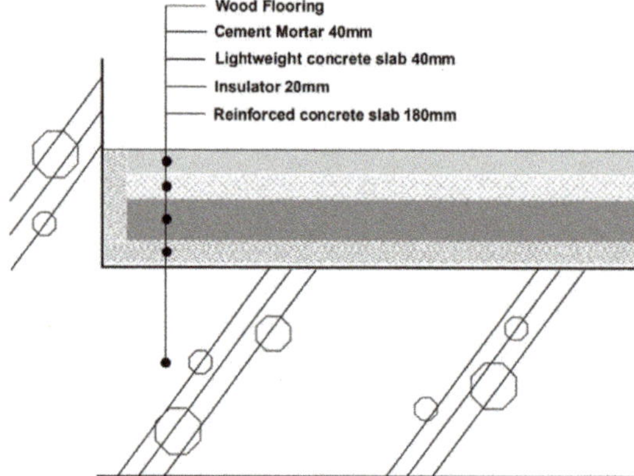

Figure 6. Cross section of the slab.

Figure 7. A smartphone installed on the ceiling of the multi-dwelling house.

	function : $distance(Point1, Point2)$
1	$df\_x = Point1.x - Point2.x$ // The difference in x-cordinates
2	$df\_y = Point1.y - Point2.y$ // The difference in y-cordinates
3	$distance = root(df\_x * df\_x + df\_y * df\_y)$ //Distance between two points using Pythagoras
4	$return\ distance$

Figure 8. Distance function.

Exact noise source locations could not be identified with the calculated values, but they were sufficient to distinguish among noise occurrence areas (Room 1, Room 2, Room 3, or Room 4) of the study site. Therefore, the proposed system performed the following four target functions using the smartphone sensors and the developed application: (1) it displayed the degree of inter-floor noise (dB) and recorded its values in the application by using the smartphone microphone devices; (2) it detected vibration using accelerometers and gyroscopes and classified the types of inter-floor noise (e.g., floor impact noise, airborne noise); (3) it estimated the noise source locations using the differences in the sound intensity and visualized the locations on the apartment floor plan; and (4) it provided reports of inter-floor noise on an hourly, daily, and monthly basis. Such reports are generated based on the information stored in the database, so that the recorded data can be accessed if a dispute occurs.

Table 6. Measured differences (m) between the actual and estimated noise sources.

	$rpoint_i$ (x, y)	$ePoint_i$ (x, y)	Distance $(rpoint_i, ePoint_i)$ (m)
1	(24.2, 23.0)	(25.3, 24.4)	1.8
2	(1640.3, 5.1)	(1638.6, 4.4)	1.9
4	(765.6, 121.0)	(767.5, 123.0)	2.8
7	(538.5, 98.9)	(532.4, 96.1)	4.3
10	(364.1, 483.6)	(36.0, 482.7)	1.2
14	(1161.6, 1031.0)	(1164.4, 1031.6)	2.9
18	(219.5, 643.5)	(217.8, 641.0)	3.1
20	(721.8, 469.3)	(723.5, 472.7)	3.8
24	(836.2, 563.7)	(838.9, 565.5)	3.3
35	(1638.6, 4.4)	(1641.3, 7.5)	3.8
42	(121.3, 689.9)	(122.2, 690.2)	1.4
50	(766.7, 123.2)	(765.6, 121.3)	2.2
65	(25.6, 25.2)	(24.2, 23.4)	2.3
70	(839.5, 657.8)	(836.2, 655.8)	3.9
75	(387.0, 583.5)	(383.5, 581.4)	4.0
80	(217.8, 641.88)	(220.8, 648.2)	4.1
85	(741.4, 1021.4)	(740.0, 1019.1)	2.7
90	(482.2, 65.6)	(485.0, 67.5)	3.4
95	(38.8, 139.9)	(38.7, 193.1)	1.3
Mean			2.8 m

5. Conclusions

This study proposed a system capable of monitoring inter-floor noise in real time, using smartphone sensors and a developed application. The designed noise monitoring system makes it possible to record the timing of the noise and its type (i.e., floor impact noise or airborne noise), acquire the exact noise values (e.g., LAeq 1 min, LAmax, and LAeq 5 min), estimate the location where the noise took place, and keep record of noise files by using the smartphone application.

An experiment was performed to evaluate the performance of the system and its applicability to multi-dwelling houses. The experiment results showed that the mean absolute error (MAE) was 2.8 m, and the minimum and maximum errors constituted 1.2 and 4.3 m, respectively. Although the exact locations of the noise sources could not be identified with these values, it was possible to establish the noise occurrence areas by a room on the apartment floor plan. Therefore, it is concluded here that the tested system can easily acquire objective noise data without any help of agencies specializing in inter-floor noise measurements. It is also expected that this system can reduce unnecessary misunderstandings among neighboring residents, by estimating the types and locations of inter-floor noise. Accordingly, in the case of having an inter-floor noise dispute, the inter-floor noise data stored in the database can be accessed through the application.

While a recent increase in the number of discarded smartphones has caused problems such as the waste of resources and pollution of soil by heavy metals, recycling the discarded smartphones using the results of this study is expected to contribute to solving social problems. However, given that the proposed system does not have any noise-data filtering feature, there is a possibility of violating the privacy of others. Therefore, a number of criteria is yet to be met for the future application usage: (1) a method of estimating exact noise locations using smartphone sensors has be developed, (2) a calibration method to measure the accuracy of sound should be administered, and (3) the privacy of neighbors and personal data collection should be sufficiently protected.

Author Contributions: Conceptualization, S.K. (Suhyun Kang), S.K. (Sangyong Kim), S.K. (Seungho Kim), and D.L.; data curation, D.L.; formal analysis and investigation, S.K. (Suhyun Kang), S.K. (Sangyong Kim); methodology, S.K. (Suhyun Kang), S.K. (Sangyong Kim), S.K. (Seungho Kim); resources, D.L.; software, S.K. (Suhyun Kang) and S.K. (Seungho Kim); supervision, S.K. (Sangyong Kim) and D.L.; validation, S.K. (Suhyun Kang), S.K. (Sangyong Kim), S.K. (Seungho Kim), and D.L.; visualization, S.K. (Suhyun Kang); writing—original draft, S.K. (Suhyun Kang), S.K. (Seungho Kim); writing—review and editing, S.K. (Sangyong Kim) and D.L. All authors have read and agreed to the published version of the manuscript.

Funding: This work was supported by the National Research Foundation of Korea (NRF) grant funded by the Korea government (MSIT) (No. NRF-2018R1A5A1025137).

Conflicts of Interest: The authors declare no conflict of interest.

References

1. Yang, H.J.; Choi, M.J. Apartment complexes in the Korean housing market: What are the benefits of agglomeration. *J. Hous. Built Environ.* **2019**, *34*, 987–1004. [CrossRef]
2. Basner, M.; Babisch, W.; Davis, A.; Brink, M.; Clark, C.; Janssen, S.; Stansfeld, S. Auditory and non-auditory effects of noise on health. *Lancet* **2014**, *383*, 1325–1332. [CrossRef]
3. Park, S.H.; Lee, P.J.; Lee, B.K. Levels and sources of neighbour noise in heavyweight residential buildings in Korea. *Appl. Acoust.* **2017**, *120*, 148–157. [CrossRef]
4. Jensen, H.A.; Rasmussen, B.; Ekholm, O. Neighbour and traffic noise annoyance: A nationwide study of associated mental health and perceived stress. *Eur. J. Public Health* **2018**, *28*, 1050–1055. [CrossRef]
5. Lee, C.W. A study on dispute of apartment noises between floors. *J. Aggreg. Build. Law* **2015**, *15*, 45–64.
6. Shin, J.; Song, H.; Shin, Y. Analysis on the characteristic of living noise in residential buildings. *J. Korea Inst. Build. Constr.* **2015**, *15*, 123–131. [CrossRef]
7. Bin, C.K. Urban noise management in Singapore. In Proceedings of the Institute of Noise Control Engineering, Reston, VA, USA, 21–24 August 2016; Volume 253, pp. 1744–1749.
8. Garg, S.; Lim, K.M.; Lee, H.P. An averaging method for accurately calibrating smartphone microphones for environmental noise measurement. *Appl. Acoust.* **2019**, *143*, 222–228. [CrossRef]
9. Kanjo, E. Noisespy: A real-time mobile phone platform for urban noise monitoring and mapping. *Mob. Netw. Appl.* **2010**, *15*, 562–574. [CrossRef]
10. D'Hondt, E.; Stevens, M.; Jacobs, A. Participatory noise mapping works! An evaluation of participatory sensing as an alternative to standard techniques for environmental monitoring. *Pervasive Mob. Comput.* **2013**, *9*, 681–694. [CrossRef]
11. Murphy, E.; King, E.A. Testing the accuracy of smartphones and sound level meter applications for measuring environmental noise. *Appl. Acoust.* **2016**, *106*, 16–22. [CrossRef]
12. Aumond, P.; Lavandier, C.; Ribeiro, C.; Boix, E.G.; Kambona, K.; D'Hondt, E.; Delaitre, P. A study of the accuracy of mobile technology for measuring urban noise pollution in large scale participatory sensing campaigns. *Appl. Acoust.* **2017**, *117*, 219–226. [CrossRef]
13. Zamora, W.; Calafate, C.T.; Cano, J.C.; Manzoni, P. Accurate ambient noise assessment using smartphones. *Sensors* **2017**, *17*, 917. [CrossRef] [PubMed]
14. Milošević, M.; Shrove, M.T.; Jovanov, E. Applications of smartphones for ubiquitous health monitoring and wellbeing management. *J. Inf. Technol. Appl.* **2011**, *1*, 7–15. [CrossRef]
15. Aram, S.; Troiano, A.; Pasero, E. Environment sensing using smartphone. In Proceedings of the 2012 IEEE Sensors Applications Symposium, Berscia, Italy, 7–9 February 2012; pp. 1–4.
16. Murphy, E.; King, E.A. Smartphone-based noise mapping: Integrating sound level meter app data into the strategic noise mapping process. *Sci. Total Environ.* **2016**, *562*, 852–859. [CrossRef] [PubMed]
17. Sagawe, A.; Funk, B.; Niemeyer, P. Modeling the intention to use carbon footprint apps. In *Information Technology in Environmental Engineering*; Springer: Cham, Germany, 2016; pp. 139–150.
18. Choi, H.; Lee, S.; Yang, H.; Seong, W. Classification of noise between floors in a building using pre-trained deep convolutional neural networks. In Proceedings of the 2018 16th International Workshop on Acoustic Signal Enhancement, Tokyo, Japan, 17–20 September 2018; pp. 535–539.
19. Jeon, J.Y.; Ryu, J.K.; Jeong, J.H.; Tachibana, H. Review of the impact ball in evaluating floor impact sound. *Acta Acust. United Acust.* **2006**, *92*, 777–786.
20. Kim, H.R.; Lee, B.S.; Kim, S.M. Using a shear connector to stiffen a floor structure to reduce the heavy-weight floor-impact noise. In Proceedings of the Institute of Noise Control Engineering, Reston, VA, USA, 21–24 August 2016; Volume 253, pp. 3228–3233.
21. Hwang, N.; Chun, Y.; KIM, S. Evaluation of floor impact sound reduction effect of the buffer-type floor structure according to the shape of shear connect. In Proceedings of the Institute of Noise Control Engineering, Reston, VA, USA, 18 December 2018; Volume 258, pp. 6068–6075.

22. Park, S.H.; Lee, P.J.; Yang, K.S.; Kim, K.W. Relationships between non-acoustic factors and subjective reactions to floor impact noise in apartment buildings. *J. Acoust. Soc. Am.* **2016**, *139*, 1158–1167. [CrossRef]
23. Park, S.H.; Lee, P.J. Effects of indoor and outdoor noise on residents' annoyance and blood pressure. In Proceedings of the Euronoise 2018 Conference, Heraklon, Greece, 27–31 May 2018.
24. Anti-Corruption & Civil Rights Commission. Available online: http://www.acrc.go.kr/acrc/board.do?command=searchDetailTotal&method=searchDetailViewInc&menuId=05050102&boardNum=34367 (accessed on 3 December 2013).
25. Quirt, J.D. Controlling air-borne and structure-borne sound in buildings. *Noise News Int.* **2011**, *19*, 37–47. [CrossRef]
26. Korea Environment Corporation. *Floor Noise Management Centre*; Korea Environment Corporation: Incheon, Korea, 2018.
27. Yoo, S.Y.; Jeon, J.Y. Investigation of the effects of different types of interlayers on floor impact sound insulation in box-frame reinforced concrete structures. *Build Environ.* **2014**, *76*, 105–112. [CrossRef]
28. Zietlow, T.; Hussein, H.; Kowerko, D. Acoustic source localization in home environments-the effect of microphone array geometry. In Proceedings of the 28th Conference on Electronic Speech Signal, Saarbrücken, Germany, 15–17 March 2017; pp. 219–226.
29. Lamelas, F.; Swaminathan, S. Laboratory demonstration of acoustic source localization in two dimensions. *Am. J. Phys.* **2019**, *87*, 24–27. [CrossRef]

 *sustainability*

Technical Note

Development and Application of Precast Concrete Double Wall System to Improve Productivity of Retaining Wall Construction

Seungho Kim [1], Dong-Eun Lee [2], Yonggu Kim [3] and Sangyong Kim [3],*

[1] Department of Architecture, Yeungnam University College, Nam-gu, Daegu 42415, Korea;
    kimseungho@ync.ac.kr
[2] School of Architecture & Civil and Architectural Engineering, Kyungpook National Universit, Buk-gu,
    Daegu 41566, Korea; dolee@knu.ac.kr
[3] School of Architecture, Yeungnam University, Gyeongsan-si, Gyeongbuk 38541, Korea; kyg3355@ynu.ac.kr
*   Correspondence: sangyong@yu.ac.kr; Tel.: +82-53-810-2425

Received: 31 March 2020; Accepted: 10 April 2020; Published: 23 April 2020

Abstract: The construction of most apartment underground parking lots utilizes reinforced concrete (RC) structures composed mainly of rebar work and formwork. RC structures lower construction efficiency and significantly delay the construction because they require a large number of temporary materials and wooden formwork. In this study, a precast concrete double wall (PCDW) system was developed to address the existing problems of RC structures and to improve the productivity of retaining wall construction. PCDW is a precast concrete (PC) wall in which two thin concrete panels are connected parallel to each other with truss-shaped reinforcement between them. PCDW can contribute to securing integrity, reducing the delay in construction, and improving quality. An overall process for the member design and construction stage of the PCDW system was proposed, and its improvement effects were examined regarding various aspects in comparison to the RC method.

Keywords: reinforced concrete; precast concrete double wall; retaining wall; lateral pressure; lateral bending

---

1. Introduction

Recent construction projects have actively used various improved methods to shorten the construction period and improve efficiency. However, the construction of most apartment underground parking lots utilizes reinforced concrete (RC) structures mainly composed of rebar work and formwork [1]. The construction of these parking lots affects the entire construction period of a project. Their construction must be completed early because underground parking lots are used as rebar workplaces for the construction of ground parts, and as storage yards for building finishing materials. However, RC structures have low construction efficiency and, most significantly, delay construction because they require temporary materials in large quantities and wooden formwork [2,3]. Therefore, there has been a growing need for measures to improve underground parking lot structure systems capable of addressing these problems. Employment of precast concrete (PC) method has been gradually increasing for this purpose [4,5]. The PC method enables efficient construction management, such as shortening the construction period and saving labor cost, because high-quality standardized members are produced in factories and assembled at sites [6,7]. It has also become advanced and has been widely used since its development in the mid-1800s owing to its higher constructability and productivity than the RC method [8–10]. However, for the construction of most retaining walls, the PC method is replaced with a combination of PC and RC processes resulting in frequent defects due to the occurrence of various cracks at the joints [11]. Furthermore, studies have been conducted on various methods, including joining methods and performance verification, to be applied to special members involving difficult construction such as apartment framework, balconies, stairs, railings, and underground parking lots [12–16].

Ji and Choi [17] researched a method of manufacturing an integrated wall by installing a link beam on the inside and outside walls of a PC and applied the method to common and reservoir walls. Furthermore, Park [18] proposed a method of forming a wall by fastening a PC panel and a panel with anchor bolts, while Oh et al. [19] conducted a study to confirm the advantages of the corresponding wall in the area of air shortening. In addition, Yang et al. [20] produced a double-synthetic precast wall with a double T-shaped PC panel facing each other to secure the economy and safety of the basement wall construction and then conducted experiments on the bending and shear behavior of the specimen. The method of pouring topping concrete after PC installation was applied also to slabs and columns. For slabs, double tee slabs, hollow slabs, and half PC slabs were identified [21]. In the case of columns, the hollowed precast concrete (HPC) column was produced by centrifugally molding a hollow PC part in the factory and pouring concrete into the field [22]. It was confirmed that the difference in performance between the existing RC structure and the HPC column was applicable to the seismic structure system. In addition, Roh and Hashlamon [23] and Kim and Kang [24] presented a development for piers and bridge columns through pouring concrete in the hollow precast and further conducted a study to analyze the seismic performance. In the case of a typical PC method, stress discontinuity due to inter-component disconnection is formed at the joint; thus, it is not easy to achieve the same performance as that of the RC structure. Furthermore, such a method may fail if the external wall support is insufficient, and there is risk of a safety accident. Therefore, the composite method of combining PC and topping concrete is increasing [17].

This study intends to present an application method for the precast concrete double wall (PCDW) system that is more suitable for retaining wall construction than the existing method. PCDW refers to a PC wall in which two thin concrete panels are connected parallel to each other with a truss-shaped reinforcement between them. As PCDW is connected to adjacent panels by pouring concrete between the panels, the completed wall achieves integrity. Furthermore, shortening of the construction period, quality management, and waste reduction can also be expected.

In this study, important factors in the processes of the design, production, installation, and completion of the PCDW system are examined to propose measures to activate the method. Further, an overall process for the member design and construction stage of the PCDW system was proposed, and its improvement effects were examined by applying it to actual construction sites. During the member design stage, the main examination items were analyzed considering the mechanical behavior of the joints, and appropriate member connection and joining methods were derived. Therefore, measures of securing the integrity of the joints of each PCDW member with vertical, corner, horizontal, and foundation concrete were presented. Furthermore, a pull-out test of headed bar was conducted in this study to evaluate the connection performance of the vertical and horizontal joints of PCDW. PCDW should resist the lateral pressure of concrete during the pouring process and curing period. Hence, the PCDW member design was examined based on the criteria suggested by the South Korean Building Code (KBC2009) and the Structural Design Standards and Commentary for Precast Concrete Prefabricated Buildings (1992). During the PCDW construction stage, an overall construction process, from the installation process to the pouring of concrete into the PCDW void, was established and verified through a case study. The benefits of the PCDW system were then examined based on various aspects via a comparison with the reinforced concrete (RC) method, which was applied to the construction of most apartment underground parking lots.

2. Development of Precast Concrete Double Wall System

2.1. Securing the Integrity of PCDW Joints

Retaining wall construction through the PCDW system requires appropriate geometry and reinforcement of the joints. Examination of the retaining wall construction cases that used RC structures showed that the retaining wall thickness was in the range of 400–600 mm. In addition, the vertical and horizontal rebars of walls were reinforced with wall-rebar ratio in the range of 0.002–0.007 to resist external forces such as earth pressure. In some cases, the upper and lower parts of walls required shear reinforcement. This study aims to propose the geometry, details, and reinforcement method of panels for the retaining walls of a structure based on the commonly used 400 mm wall thickness. In a PCDW system, two thin concrete panels are connected parallel to each other. Therefore, to secure the integrity of the panels, lattice bars were fabricated and placed at the center of these panels as shown in Figure 1.

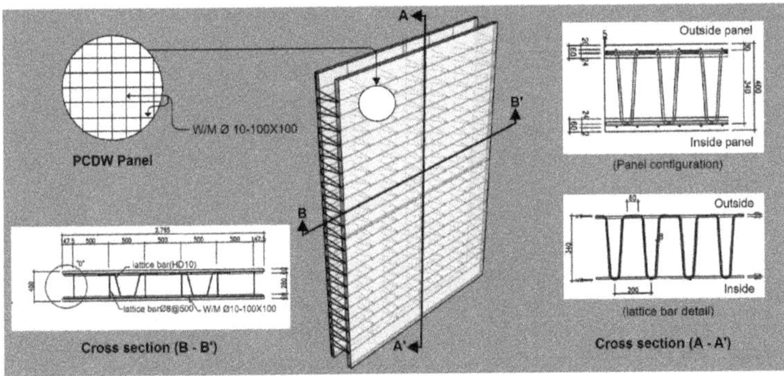

Figure 1. Panel configuration and lattice bar details of the precast concrete double wall (PCDW) system.

2.2. PCDW Joint Configuration

The PCDW system requires panel-to-panel joints with vertical joints to connect the left and right panels, horizontal joints to connect the upper and lower panels, and wall-foundation joints to connect the panels and foundation concrete. The joints require appropriate reinforcement to achieve integrated behavior against the stress and deformation caused by out-of-plane loading applied to the walls on both sides.

When the PCDW system is applied to the basement, it is necessary for the vertical joints to secure resistance performance against the bending moment through separate resistance mechanisms for safety against loads such as earth pressure. To address this problem, connection using standard hook (180° hook type) rebars, headed bars, or wire welding can be used. For the horizontal joints between the upper and lower walls composed of PCDW panels, sufficient resistance performance is required against the bending moment and shear force that may occur at the joints under vertical forces such as earth and hydraulic pressures. However, as the vertical wire welding applied to panels is discontinuous, separate resources are required at the joints to resist the bending moment. Connection using standard hook rebars or headed bars or lap splice using straight rebars can be used for the purpose. The joints between PCDW and the foundation can be constructed with concrete after assembling the dowel bars in the cast-in-place concrete foundation plate or PC foundation plate to be placed in the void of the PCDW. As the retaining walls of a structure are subjected to large wall end moment and shear force due to loads such as earth pressure, the joints between PCDW and the foundation can sufficiently resist such stress. In this instance, the dowel bars can provide the tensile force due to the bending moment, and the required shear strength can be obtained by the concrete filling the void of PCDW and the vertically arranged lattice bars. Figure 2 shows the lattice bar types and joining methods available for each joint.

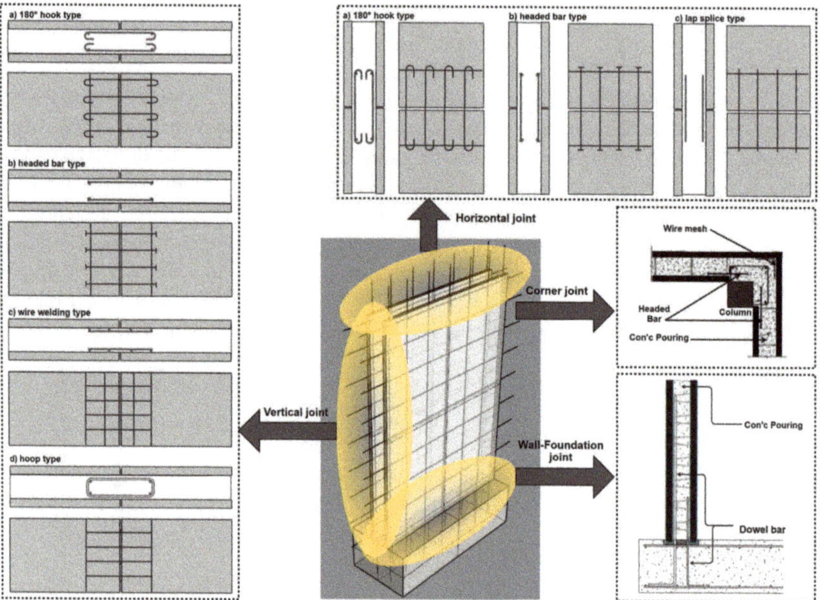

Figure 2. Joining methods for each PCDW joint.

2.3. Headed Bar Performance Evaluation

Although the headed bars used at PCDW joints may vary in size and geometry, appropriate guidelines are not sufficient in South Korea. Therefore, analysis is required for specific geometry. Hence, a pull-out test was conducted in this study to evaluate the connection performances of the vertical and horizontal joints of PCDW. In the pull-out test, the tensile strength and anchorage capacity of the ten test pieces of the developed headed bar were evaluated by burying them in concrete and applying pull-out loads (Figure 3).

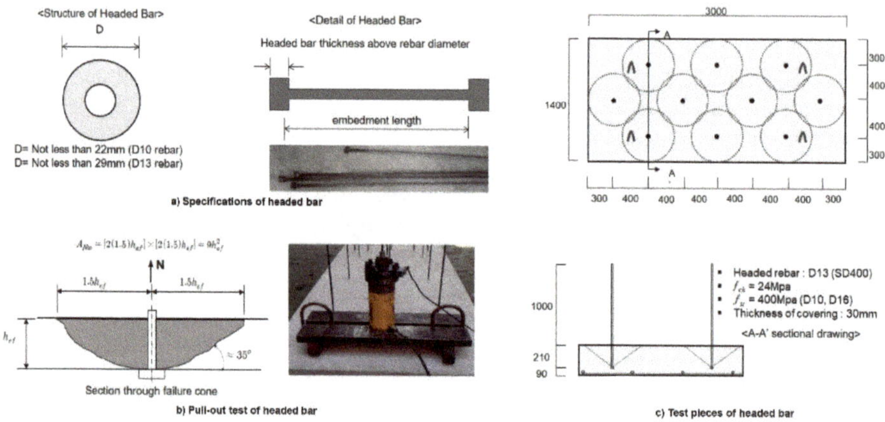

Figure 3. Headed bar conduct analysis.

The specified strength of the concrete used for the test pieces was 24 MPa, and the size of the test pieces was Ø 100 × 200 mm. Tests on the compressive strength of concrete were conducted on the 7th, 14th, and 28th days after the fabrication of the test pieces. The strength of concrete was determined by averaging the values obtained from three test pieces. For the fabrication of the headed bar, screw threads were machined at the end of the D13 (Deformed bar, Yield Strength = 400 MPa, Unit weight = 0.995 kg/m) deformed bar and a head was attached.

The results of the pull-out test on the headed bar showed that the ten test pieces did not exhibit any cracks or fractures in concrete during the pull-out test, and most of them failed at the position of the strain measuring gauge attached in the middle of the deformed bar (Table 1). Figure 4 shows the load-strain relationship of the pull-out test. The average of the maximum loads was 73.8 kN, which was higher than the yield strength. It was confirmed that failure occurred in a plastic deformation state that exceeded the yield strain. This indicates that the developed headed bar is suitable for securing the yield strength of rebars. However, machining the screw heads reduces the cross-sectional area of the deformed bar of the headed bar by approximately 10%. Hence, it is necessary to set 90% of the cross-sectional area of the deformed bar as the effective cross-sectional area for the headed bar that is to be used as a joint reinforcement.

Table 1. Results of the pull-out test on the headed bar.

Basic Data of Specimen	f_{ck} (MPa) 44.7	f_y (MPa) 516	h_{ef} (mm) 210	D (mm) 30	A (mm) 127	Additional Information
T_y	$A_s \times f_y = 127 \times 516 = 65.5$				(kN)	Yield strength of the headed bar
N_{sa}	$A_s \times f_u = 127 \times 640 = 81.3$				(kN)	Rupture Strength of the headed bar
N_{cb}	$\frac{A_{Nc}}{A_{Nco}}\varphi_{cd}\varphi_c\varphi_{cp}N_b = 254.3$				(kN)	Concrete Cone Breakout
specimen-1	73.3				(kN)	yield and fracture
specimen-2	73.6				(kN)	yield and fracture
specimen-3	77.6				(kN)	yield and fracture
specimen-4	68.2				(kN)	yield and fracture
specimen-5	76.1				(kN)	yield and fracture
specimen-6	74.9				(kN)	yield and fracture
specimen-7	73.6				(kN)	yield and fracture
specimen-8	71.4				(kN)	yield and fracture
specimen-9	70.7				(kN)	yield and fracture
specimen-10	78.7				(kN)	yield and fracture
Overall average	73.8				(kN)	
Standard deviation	3.2				(kN)	
Coefficient of variation	4.3				(%)	

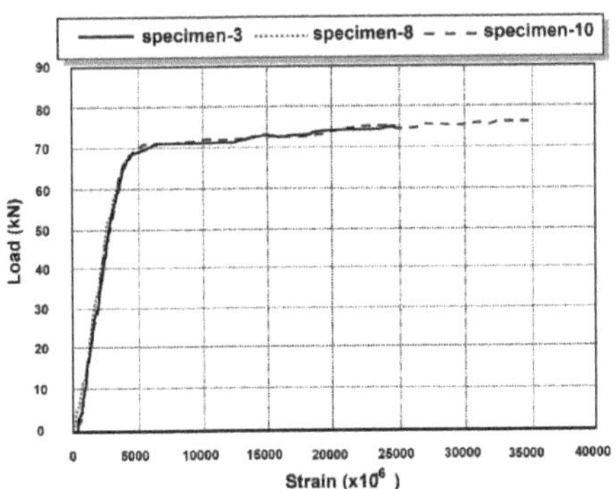

Figure 4. Load-strain relationship of the headed bar.

3. PCDW Design through the Examination of Lateral Pressure and Bending

3.1. PCDW Member Design

For PCDW, cast-in-place concrete poured into the space between PC panels. Therefore, PCDW should resist the lateral pressure of concrete during the pouring process and the curing period. The lateral pressure is determined by the unit weight, pouring height, pouring speed, and temperature. Detailed examination of pouring plans and partition height calculation is required before the concrete pouring. In this study, the PCDW member design was examined based on the criteria suggested by the South Korean Building Code (KBC2009) and the Structural Design Standards and Commentary for Precast Concrete Prefabricated Buildings (1992). Figure 5 shows the PCDW member design conditions.

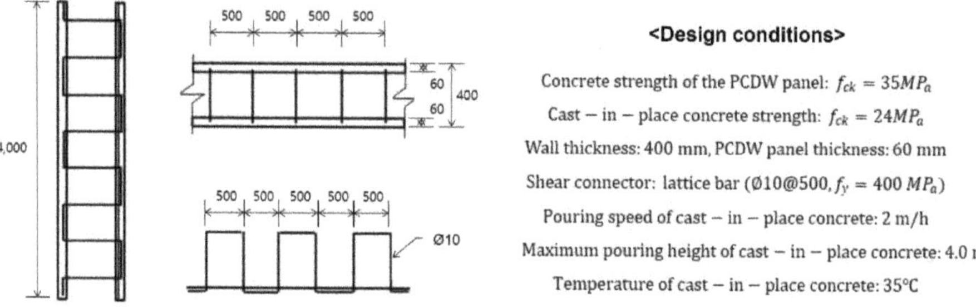

Figure 5. PCDW member design conditions.

3.2. Examination of Lpressure and Bending

Equations (1)–(3) is the lateral pressure calculation formula for concrete poured by general internal vibro-compaction for which the concrete slump is ≤175 mm and the depth is ≤1.2 m. The equation can be used for walls when the pouring speed is <2.1 m/h and the pouring height is <4.2 m. In the equation, "p" is the horizontal pressure (Kn/m^2), "R" is the pouring speed (m/h), and "T" is the concrete temperature in the formwork (°C). "C_w" is the unit weight factor with a value of 1 corresponding to the unit weight ranging from 22.5 to 24 N/m^3, which was used based on the South Korean Building Code (KBC2009). "C_c" is the chemical additive factor with a value of 1 corresponding to the type 1, 2, and 3 cement of KS L 5201 that uses no retarder.

$$\text{p} = C_w C_c 7.2 + \frac{790R}{T+18} \tag{1}$$

$$C_w = C_c = 1.0 \tag{2}$$

$$\text{p} = 7.2 + \frac{790 \times 2}{35+18} = 37.1 \text{ kN/m}^2 \tag{3}$$

The flexural strength of PCDW was calculated using a panel thickness of 60 mm and a lattice bar spacing of 500 mm. Equations (4)–(7) shows the results of the working load moment (M), bending stress (σ), allowable tensile stress under crack width limitation (f_t), and flexural reinforcement (M_u). Equations (8)–(12) shows the results when the inside of the PCDW panel was reinforced with wire welding (Ø 8 × 150 × 150 (f_y = 400 MPa)).

$$M = \frac{pL^2}{8} = \frac{37.1(0.5)^2}{8} = 1.159 \text{ km} \tag{4}$$

$$\sigma = \frac{M}{Z} = \frac{M}{bh^2/6} = \frac{6(1.159)}{1(0.06)^2} = \frac{1931.7 \text{ kN}}{\text{m}^2} = 1.9 \text{ MPa} \tag{5}$$

$$f_t = 0.63 \sqrt{f_{ck}} = 0.63 \sqrt{35} = 3.73 \text{ MPa} > 1.4 \text{ MPa} - o.k. \tag{6}$$

$$M_u = 1.2 \times 1.159 = 1.39 \text{ kNm} \tag{7}$$

$$A_s = 333 \text{ mm}^2/\text{m} \tag{8}$$

$$a = \frac{A_s f_y}{0.85 f_{ck} b} = \frac{333 \times 400}{0.85 \times 35 \times 1000} = 4.5 \text{ mm} \tag{9}$$

$$d = \frac{60}{2} = 30 \text{ mm} \tag{10}$$

$$\varnothing M_n = \varnothing A_s f_y \left(d - \frac{a}{2}\right) = 0.85 \times 333 \times 400 \times \left(30 - \frac{4.5}{2}\right) \times 10^{-6} \tag{11}$$

$$= 3.14 \text{ kNm} > 1.39 \text{ kNm} - o.k. \tag{12}$$

Equations (13)–(16) shows the shear performance based on the PCDW lateral pressure examination results. "V_u" is the ultimate shear force in the cross section, and "$\varnothing$" is the strength reduction factor.

$$V = \frac{pL}{2} = \frac{37.1(0.5)}{2} = 9.275 \text{ kN} \tag{13}$$

$$V_u = 1.2 \times 9.275 = 11.13 \text{ kN} \tag{14}$$

$$\varnothing V_n = \varnothing \left(\frac{1}{6}\right)\sqrt{f_{ck}} b_w d = 0.75\left(\frac{1}{6}\right)\sqrt{35}(1000)(30)\left(10^{-3}\right) \tag{15}$$

$$= 22.19 \text{ kN} > 11.13 \text{ kN} - o.k. \tag{16}$$

Equations (17)–(19) shows the shear connector examination results, and the safety factor (n) according to the tensile force ($\varnothing T_n$) and working load (T_u) of the shear connector (lattice bar $\varnothing10$ at 500, f_y = 400 MPa). Equations (20) and (21) shows the deflection examination (δ) results for the lateral pressure of PCDW.

$$\varnothing T_n = \varnothing A_s f_y = 0.85 \times 71 \times 400 = 24.1 \text{ kN} \tag{17}$$

$$T_u = 1.2 \times p \times L \times @Tie - bar = 1.2 \times 37.1 \times 0.5 \times 0.5 = 11.13 \text{ kN} \tag{18}$$

$$n = 24.1/11.13 = 2.16 > 2 - o.k. \tag{19}$$

$$\delta = \frac{5pL^4}{384BI} = \frac{5PL^4}{384\left(6500\sqrt[3]{f_{ck}}\right)\left(\frac{bh^3}{12}\right)} = \frac{5(37.1)(500)^4}{384\left(8500\sqrt[3]{35}\right)(1000)\dfrac{(60)^3}{12}} \tag{20}$$

$$= 0.06 \text{ mm} < \frac{L}{360}(= 1.39 \text{ mm}) - o.k. \tag{21}$$

4. Field Application of the PCDW System

4.1. PCDW Construction Sequence

The case study site for this study was a new apartment construction site, which included six buildings (two basement stories and 25 ground stories). The PCDW system was applied to the retaining walls of underground parking lots, and a total of 100 units were used. Figure 6 shows the layout of the site and the installation plan by section and the PCDW construction sequence.

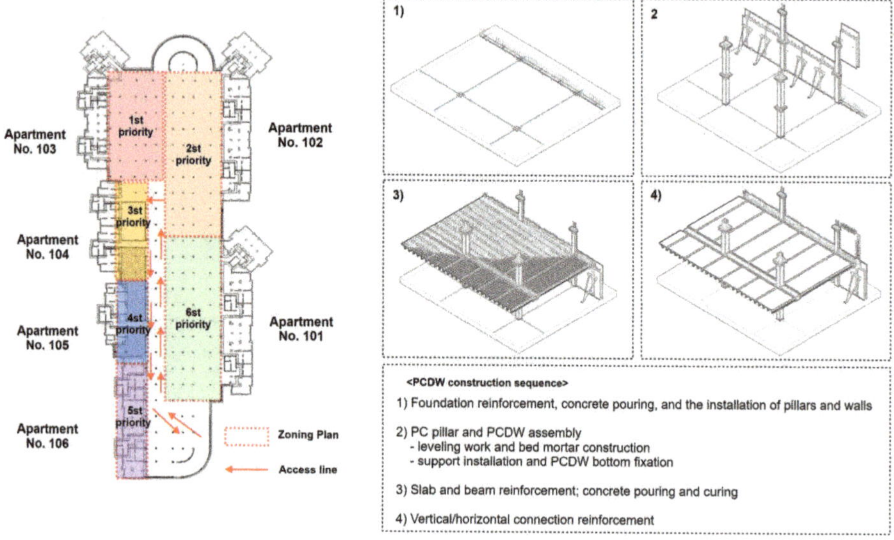

Figure 6. PCDW construction plan and flowchart.

Figure 7 shows the main construction process of PCDW, and its contents are as follows:

1. Before the installation of PCDW, foundation rebars and the anchorage rebars of PCDW are placed and the recess metal lath for pad mortar pouring are installed at the top for accurate connection between PCDW and the foundation. In this case, the cover thickness of the upper part of the foundation must be approximately 50 mm.

2. Two liner shims are installed on the floor per PCDW system. After examination of the liner shim level, pad mortar is applied in two rows and PCDW is installed on top of them.

3. After the assembly of PCDW, its vertical state is examined using an inclinometer. Two or more prop supports are firmly installed to prevent any gaps or misalignment.

4. After inspection of the assembly state, the reinforced state, and the installation of the other parts, concrete is poured in the PCDW void. Before concrete pouring, the inside is cleaned to remove foreign substances, and water is sprayed to keep the inside wet. In addition, compaction is performed using a rod-type vibrator or a form vibrator to prevent poor-compacted concrete, and then PCDW is assembled and prop supports are installed. After the assembly of the PCDW system, the assembly accuracy is inspected. Table 2 shows the inspection methods and the judgment criterion.

Table 2. Assembly accuracy inspection criterion for the PCDW system.

Category		Test Method	Frequency	Judgment Criterion
PCDW system	Installation position	The difference from the reference line marked on the floor is measured using a steel ruler	After assembly	±5 mm or less
	Inclination	Measured using a plumb or a slope scale		
	Ceiling height	Measured using a level		

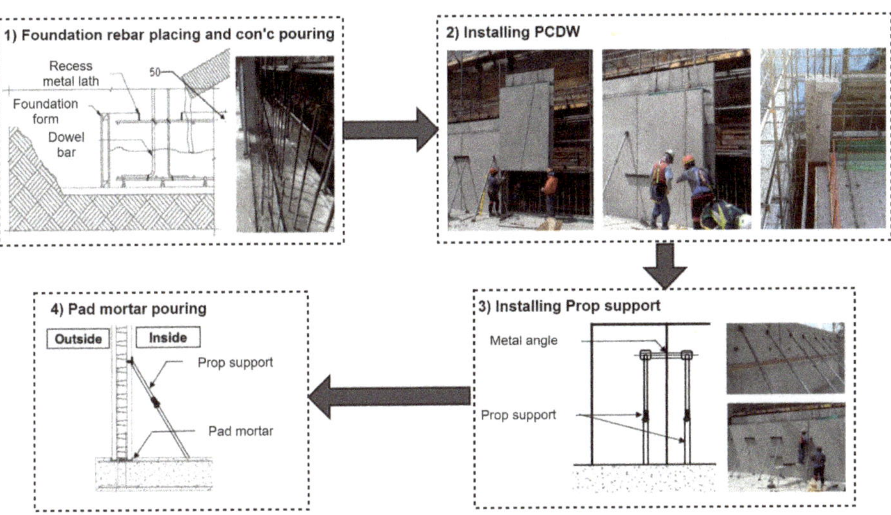

Figure 7. Main construction process of PCDW.

4.2. Analysis of the Effect of PCDW System Application

In this study, the actual effects of the application of the PCDW system, which improved the existing PC method, were examined on the basis of various aspects via a comparison with the RC method. Table 3 shows the effects of the PCDW system that were verified through the case study.

Table 3. Comparison between the RC and PCDW methods.

Category	RC	PCDW	Remark
Construction cost	100%	99%	1% reduction
Construction/ safety	- Formwork for concrete pouring requires a considerable amount of time - Work safety must be examined for pouring	- Site work can be simplified without formwork - Construction safety can be secured without external scaffold and temporary facilities	
Construction period	100%	60%	40% reduction
Quality	- Quality significantly varies depending on the type and condition of formwork	- Factory production ensures excellent quality	
Others	- No lifting equipment required - Easy connection to the bottom wall rebars - Labor-intensive structure, lack of skilled workers - Highly difficult formwork	- Eco-friendly because of on-site waste reduction - Member size limited by the transport and lifting conditions - Constructible regardless of the climate - Increased durability due to steam curing	

The actual cost comparison was calculated based on the material and labor costs. The material cost of the RC method consisted of the concrete, form rebar, grinding, and plastering works. Based on this, labor costs were calculated according to the number of workers required to perform each task. The material cost of the PCDW method consisted of the PC panels and concrete poured into the PCDW void, and the labor cost was calculated according to the number of workers required for each work performance. The PCDW method was able to reduce the cost by omitting formwork and rebar work compared with the RC method, but the PC panel cost was added. As a result, the cost difference between the two methods was approximately 1%. In addition, both methods required lifting equipment, but no additional cost was required, as T/C had already been installed at the site.

The primary benefit of the PCDW system compared to the existing RC method is the shortening of the construction period. As the PCDW system is 100% produced in factories for on-site installation, it does not need formwork for concrete pouring, which requires a considerable amount of time as in the case of the RC method. It can also reduce the framing construction period by approximately 40%. Figure 8 compares the progress schedule of the RC method with that of the PCDW system to examine the construction period of apartment retaining walls (pillar + beam + wall). The progress schedules show the number of days required for each process for a 30.8 m × 4 m (one floor with 4 spans) floor size, and it was calculated based on one formwork team (seven persons).

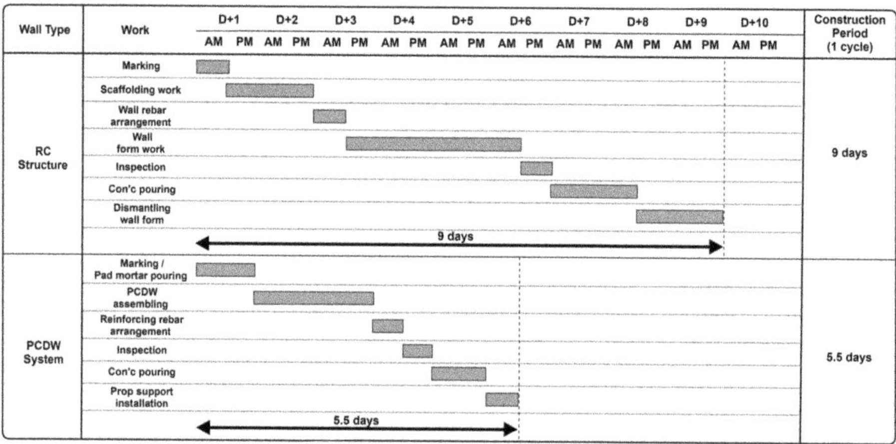

Figure 8. Comparison of basement framing construction periods.

5. Conclusions

This study proposed an overall process for applying the precast concrete double wall (PCDW) system, which addressed the drawbacks of the precast concrete (PC) method, to actual construction sites. Particularly, measures to secure the integrity of the joints of each PCDW member with vertical, corner, horizontal, and foundation concrete were presented. Member design was performed considering concrete lateral pressure, and pouring plans and partition height calculation were examined in detail. In addition, the benefits of the PCDW system were examined based on various aspects via a comparison with the reinforced concrete (RC) method, which has been applied to the construction of most apartment underground parking lots.

The currently applied PC method has the disadvantages that the PC and RC processes are mixed, workability is poor, and construction management is cumbersome because only the inner columns, beams, and slabs are applied, except for the retaining walls. Therefore, the introduction of the PCDW system is expected to simplify construction management and improve the constructability because PC can be used for the entire framework of apartment underground parking lots. In addition, it is expected to enable active improvements to the construction site situation, in which the lack of skilled workers, such as form and reinforcement workers, worsens the situation. However, as the application of the PCDW system was limited to the external walls of apartment underground parking lots in this study, additional case studies are required for its application to entire buildings. Furthermore, the examination of economic efficiency presents limitations because only the construction cost of basement framing construction was identified. Although the shortening of the construction period reduces the total construction cost due to the reduction of indirect cost, construction cost analysis is required considering other elements in addition to the cost of basement framing construction analyzed in this study.

Author Contributions: In this paper, S.K. (Seungho Kim) designed the research framework and wrote the paper. D.-E.L. and Y.K. collected and analyzed the data. S.K. (Sangyong Kim) conceived the methodology and developed the ideas. All authors have read and agreed to the published version of the manuscript.

Funding: This work was supported by the National Research Foundation of Korea (NRF) grant funded by the Korea government (MSIT) (No. NRF-2018R1A5A1025137).

Conflicts of Interest: The authors declare no conflict of interest.

References

1. Kim, S.Y.; Lee, B.S.; Park, K.Y.; Lee, S.B.; Yoon, Y.H. Foreign parking structures and practical using method of PC system for the domestic underground parking structures. *Mag. Korea Concr. Inst.* **2008**, *20*, 34–40. [CrossRef]

2. Hwang, J.H. Development of Integrated Management Process for Precast Concrete Construction Method based on BIM. Master's Thesis, School of Urban Science, University of Seoul, Seoul, Korea, 2019.

3. Qin, Y.; Shu, G.P.; Zhou, G.G.; Han, J.H. Compressive behavior of double skin composite wall with different plate thicknesses. *J. Constr. Steel Res.* **2017**, *157*, 297–313. [CrossRef]

4. Chai, S.; Guo, T.; Chen, Z.; Jun, Y. Monitoring and simulation of long-term performance of precast concrete segmental box girders with dry joints. *J. Bridge Eng.* **2019**, *24*, 04019043. [CrossRef]

5. Park, J.H.; Choi, J.W.; Jang, Y.J.; Park, S.K.; Hong, S.N. An experimental and analytical study on the deflection behavior of precast concrete beams with joints. *Appl. Sci.* **2017**, *7*, 1198. [CrossRef]

6. Chen, S.; Feng, K.; Lu, W. A Simulation-Based optimisation for contractors in precast concrete projects. In *10th Nordic Conference on Construction Economics and Organization*; Emerald Publishing Limited: Bingley, UK, 2019; Volume 2, pp. 137–145. [CrossRef]

7. Zhai, X.; Reed, R.; Mills, A. Factors impeding the offsite production of housing construction in China: An investigation of current practice. *Constr. Manag. Econ.* **2014**, *32*, 40–52. [CrossRef]

8. Ahmad, S.; Soetanto, R.; Goodier, C.I. Lean approach in precast concrete component production. *Built Environ. Proj. Asset Manag.* **2019**, *9*, 457–470. [CrossRef]

9. Kang, T.S. Precast concrete construction of SAMPYO engineering & construction LTD, R&D institute. *Mag. Korea Concr. Inst.* **2019**, *25*, 46–49.

10. Kim, S.Y.; Yoon, Y.H.; Park, K.Y.; Lee, B.S.; PC Council. A Research for Practical Using Method of PC Structural System for the Underground Parking Garage in an Apartment Housing Site. Korea National Housing Corporation Housing & Urban Research Institute. 2006. Available online: https://dl.nanet.go.kr/SearchDetailView.do?cn=MONO1200827476 (accessed on 9 April 2020).

11. Kim, H.D.; Lee, S.S.; Park, K.S.; Bae, K.W. A study on plant certification program for precast concrete products. *KSMI* **2014**, *18*, 131–138. [CrossRef]

12. Augusto, T.; Mounir, K.; Melo, A.M. A cost optimization-based design of precast concrete floors using genetic algorithms. *Autom. Constr.* **2012**, *22*, 348–356. [CrossRef]

13. Castilho, V.C.; Lima, M.C.V. Comparative costs of the production, transport and assembly stages of prestressed precast slabs using genetic algorithms. *IJOCE* **2012**, *2*, 407–422.

14. Ko, C.H. Material transshipment for precast fabrication. *J. Constr. Eng. Manag.* **2013**, *19*, 335–347. [CrossRef]

15. Sacks, R.; Eastman, C.M.; Lee, G. Parametric 3D modeling in building construction with examples from precast concrete. *Autom. Constr.* **2004**, *13*, 291–312. [CrossRef]

16. Suh, J.I.; Park, H.G.; Hwang, H.J.; Im, J.H.; Kim, Y.N. Development of PC double wall for staircase construction. *J. Korea Inst. Build. Constr.* **2014**, *14*, 571–581. [CrossRef]

17. Ji, K.H.; Choi, B.J. Improvement of Underground Wall Design and Construction Safety Using Mega Double Wall Construction Method. *J. Korean Soc. Hazard Mitig.* **2019**, *19*, 1–12. [CrossRef]

18. Park, K.M. A Study on the Production and Construction of Precast Concrete Walls. Master's Thesis, Kyonggi University, Suwon, Korea, 2017.

19. Oh, S.Y.; Hong, S.Y.; Park, K.M. A study on the site work of precast concrete double composite wall. *Proc. Korea Concr. Inst.* **2017**, *29*, 407–408.

20. Yang, H.M.; Han, S.J.; Lee, S.H.; Choi, S.H.; Chung, J.H.; Kim, K.S. Out of plane behavior of double composite PC walls. *Proc. Korea Concr. Inst.* **2018**, *30*, 149–150.

21. Kim, J.S.; Jung, J.W.; Lee, K.H.; Ahn, J.M. Recent domestic precast concrete slab technologies. *Mag. Korea Concr. Inst.* **2004**, *16*, 16–19. [CrossRef]

22. Seo, S.Y.; Yoon, S.J.; Lee, W.J. Evaluation of structural performance the hollow PC column joint subjected to cyclic lateral load. *J. Korea Concr. Inst.* **2008**, *20*, 335–343. [CrossRef]

23. Roh, H.S.; Hashlamon, I.H. Hysteretic model and seismic response of partial precast concrete piers with cast-in-place for base and outside of hollow cross section. *Proc. Korea Concr. Inst.* **2016**, *28*, 123–124. Available online: http://www.riss.kr/link?id=A101902151 (accessed on 9 April 2020).

24. Kim, T.H.; Kang, H.T. Seismic performance assessment of hollow circular reinforced concrete bridge columns with confinement steel. *J. Earthq. Eng. Soc. Korea* **2012**, *16*, 13–25. [CrossRef]

MDPI
St. Alban-Anlage 66
4052 Basel
Switzerland
Tel. +41 61 683 77 34
Fax +41 61 302 89 18
www.mdpi.com

Sustainability Editorial Office
E-mail: sustainability@mdpi.com
www.mdpi.com/journal/sustainability